ELECTRICAL, ELECTRONICS, AND DIGITAL HARDWARE ESSENTIALS FOR SCIENTISTS AND ENGINEERS

ELECTRICAL, ELECTRONICS, AND DIGITAL HARDWARE ESSENTIALS FOR SCIENTISTS AND ENGINEERS

Ed Lipiansky

Cisco Systems, Inc.
San Jose, California, USA

IEEE

IEEE PRESS

IEEE PRESS SERIES ON MICROELECTRONIC SYSTEMS

WILEY

A JOHN WILEY & SONS, INC., PUBLICATION

Cover Designer: Denise Lipiansky

Library of Congress Cataloging-in-Publication Data:

Lipiansky, Ed.
Electrical, electronics, and digital hardware essentials for scientists and engineers / Ed Lipiansky.
p. cm.
ISBN 978-1-118-30499-0 (hardback)
1. Electronic circuits. 2. Electronic apparatus and appliances. I. Title.
TK7876.L484 2012
621.3–dc23

2012012321

Printed in the United States of America

10 9 8 7 6 5 4 3 2 1

To my lovely wife Ruty

CONTENTS

Preface **xvii**

About the Author **xix**

1 From the Bottom Up: Voltages, Currents, and Electrical Components **1**

1.1 An Introduction to Electric Charges and Atoms / 1
1.2 Electric DC Voltage and Current Sources / 3
1.2.1 Electric Current and Voltage / 3
1.2.2 DC Voltage and Current Sources / 5
1.2.3 Sources Internal Resistance / 9
1.3 Electric Components: Resistors, Inductors, and Capacitors / 12
1.3.1 Resistors / 12
1.3.2 Resistors in Series and in Parallel / 19
1.3.3 Resistivity: A Physical Interpretation / 23
1.3.4 Resistance of Conductors / 25
1.4 Ohm's Law, Power Delivered and Power Consumed / 25
1.4.1 Voltage Source Internal Resistance / 30
1.5 Capacitors / 33
1.5.1 Physical Interpretation of a Parallel-Plate Capacitor Capacitance / 35
1.5.2 Capacitor Voltage Current Relationship / 36
1.5.3 Capacitors in Series / 37
1.5.4 Capacitors in Parallel / 39
1.5.5 Energy Stored in a Capacitor / 40
1.5.6 Real Capacitor Parameters and Capacitor Types / 41
1.6 Inductors / 44
1.6.1 Magnetism / 45
1.6.2 Magnetic Field around a Coil / 49
1.6.3 Magnetic Materials and Permeability / 52
1.6.4 Electromagnetic Induction and Inductor Current–Voltage Relationship / 53

1.6.5 Inductors in Series / 58
1.6.6 Inductors in Parallel / 60
1.6.7 Mutual Inductance / 62
1.6.8 Energy Stored by an Inductor / 69
1.6.9 Inductor Nonlinearity / 70
1.6.10 Inductor Component Selection / 71
1.7 Kirchhoff's Voltage Law (KVL) and Kirchhoff's Current Law (KCL) / 73
1.8 Summary / 87
Further Reading / 87
Problems / 88

2 Alternating Current Circuits **98**

2.1 AC Voltage and Current Sources, Root Mean Square Values (RMS), and Power / 98
2.1.1 Ideal and Real AC Voltage Sources / 99
2.1.2 Ideal and Real AC Current Sources / 106
2.2 Sinusoidal Steady State: Time and Frequency Domains / 111
2.2.1 Resistor under Sinusoidal Steady State / 112
2.2.2 Inductor under Sinusoidal Steady State / 112
2.2.3 Capacitor under Sinusoidal Steady State / 113
2.2.4 Brief Complex Number Theory Facts / 115
2.3 Time Domain Equations: Frequency Domain Impedance and Phasors / 123
2.3.1 Phasors / 123
2.3.2 The Impedance Concept / 124
2.3.3 Purely Resistive Impedance / 126
2.3.4 Inductive Impedance: Inductive Reactance / 127
2.3.5 Purely Capacitive Impedance: Capacitive Reactance / 131
2.3.6 R, L, and C Impedances Combinations / 133
2.4 Power in AC Circuits / 136
2.4.1 AC Instantaneous Power Drawn by a Resistor / 137
2.4.2 AC Instantaneous Power Drawn by a Capacitor / 137
2.4.3 AC Instantaneous Power Drawn by an Inductor / 139
2.5 Dependent Voltage and Current Sources / 145
2.5.1 Voltage-Controlled Voltage Source (VCVS) / 145
2.5.2 Current-Controlled Voltage Source (CCVS) / 146
2.5.3 Voltage-Controlled Current Source (VCCS) / 147
2.5.4 Current-Controlled Current Source (CCCS) / 148
2.6 Summary of Key Points / 149
Further Reading / 149
Problems / 150

3 Circuit Theorems and Methods of Circuit Analysis 155

3.1 Introduction / 155
3.2 The Superposition Method / 156
3.2.1 Circuits Superposition / 160
3.3 The Thévenin Method / 165
3.3.1 Application of the Thévenin Method / 167
3.4 Norton's Method / 172
3.4.1 Source Transformations / 174
3.4.2 Finding the Norton Equivalent Circuit Directly from the Given Circuit / 175
3.5 The Mesh Method of Analysis / 179
3.5.1 Establishing Mesh Equations. Circuits with Voltage Sources / 180
3.5.2 Establishing Mesh Equations by Inspection of the Circuit / 186
3.5.3 Establishing Mesh Equations When There Are also Current Sources / 189
3.5.4 Establishing Mesh Equations When There Are also Dependent Sources / 196
3.6 The Nodal Method of Analysis / 199
3.6.1 Establishing Nodal Equations: Circuits with Independent Current Sources / 199
3.6.2 Establishing Nodal Equations by Inspection: Circuits with Current Sources / 201
3.6.3 Establishing Nodal Equations When There Are also Voltage Sources / 205
3.6.4 Establishing Nodal Equations When There Are Dependent Sources / 207
3.7 Which One Is the Best Method? / 210
3.7.1 Superposition Theorem Highlights / 210
3.7.2 Thévenin Theorem Highlights / 211
3.7.3 Norton's Theorem Highlights / 211
3.7.4 Source Transformations Highlights / 212
3.7.5 Mesh Method of Analysis Highlights / 212
3.7.6 Nodal Method of Analysis Highlights / 212
3.8 Using all the Methods / 213
3.8.1 Solving Using Superposition / 213
3.8.2 Example 3.21: Solving the Circuit of Figure 3.36 by Thévenin / 216
3.8.3 Example 3.22: Solving the Circuit of Figure 3.36 by Norton / 219
3.8.4 Example 3.23: Solving the Circuit of Figure 3.36 Using Source Transformations / 221

3.8.5 Example 3.24: Solving the Circuit of Figure 3.36 Using the Mesh Method / 223
3.8.6 Example 3.25: Solving the Circuit of Figure 3.36 Using the Nodal Method / 224
3.9 Summary and Conclusions / 225
Further Reading / 225
Problems / 226

4 First- and Second-Order Circuits under Sinusoidal and Step Excitations **233**

4.1 Introduction / 233
4.2 The First-Order *RC* Low-Pass Filter (LPF) / 235
4.2.1 Frequency Domain Analysis / 235
4.2.2 Brief Introduction to Gain and the Decibel (dB) / 236
4.2.3 *RC* LPF Magnitude and Phase Bode Plots / 238
4.2.4 *RC* LPF Drawing a Bode Plot Using Just the Asymptotes / 241
4.2.5 Interpretation of the *RC* LPF Bode Plots in the Time Domain / 244
4.2.6 Why Do We Call This Circuit a LPF? / 245
4.2.7 Time Domain Analysis of the *RC* LPF / 245
4.2.8 First-order *RC* LPF under Pulse and Square-Wave Excitation / 248
4.2.9 The *RC* LPF as an Integrator / 251
4.3 The First-Order *RC* High-Pass Filter (HPF) / 252
4.3.1 *RC* HPF Frequency Domain Analysis / 253
4.3.2 Drawing an *RC* HPF Bode Plot Using Just the Asymptotes / 254
4.3.3 Interpretation of the *RC* HPF Bode Plots in the Time Domain / 257
4.3.4 Why Do We Call This Circuit an HPF? / 258
4.3.5 Time Domain Analysis of the *RC* HPF / 258
4.3.6 First-Order *RC* LPF under Pulse and Square-Wave Excitation / 260
4.3.7 The *RC* HPF as a Differentiator / 263
4.4 Second-Order Circuits / 265
4.5 Series RLC Second-Order Circuit / 266
4.6 Second-Order Circuit in Sinusoidal Steady State: Bode Plots / 275
4.7 Drawing the Second-Order Bode Plots Using Asymptotic Approximations / 278
4.8 Summary / 279
Further Reading / 279
Problems / 280

5 The Operational Amplifier as a Circuit Element 287

5.1 Introduction to the Operational Amplifier / 287
5.2 Ideal and Real Op Amps / 288
5.3 Brief Definition of Linear Amplifiers / 290
5.4 Linear Applications of Op Amps / 294
5.4.1 The Inverting Amplifier / 294
5.4.2 The Noninverting Amplifier / 306
5.4.3 The Buffer or Noninverting Amplifier of Unity Gain / 309
5.4.4 The Inverting Adder / 314
5.4.5 The Difference Amplifier / 316
5.4.6 The Inverting Integrator / 321
5.4.7 The Inverting Differentiator / 322
5.4.8 A Practical Integrator and Differentiator Circuit / 326
5.5 Op Amps Nonlinear Applications / 331
5.5.1 The Open-Loop Comparator / 332
5.5.2 Positive and Negative Voltage-Level Detectors Using Comparators / 332
5.5.3 Comparator with Positive Feedback (Hysteresis) / 336
5.6 Operational Amplifiers Nonidealities / 341
5.7 Op Amp Selection Criteria / 343
5.8 Summary / 347
Further Reading / 348
Problems / 348
Appendix to Chapter 5 / 353

6 Electronic Devices: Diodes, BJTs, and MOSFETs 354

6.1 Introduction to Electronic Devices / 354
6.2 The Ideal Diode / 355
6.2.1 The Half-Wave Rectifier / 357
6.2.2 The Full-Wave Bridge Rectifier / 360
6.2.3 The Real Silicon Diode *I-V* Characteristics: Forward-Bias, Reverse-Bias, and Breakdown Regions / 363
6.2.4 Two More Realistic Diode Models / 367
6.2.5 Photodiode / 369
6.2.6 Light Emitting Diode (LED) / 369
6.2.7 Schottky-Barrier Diode / 371
6.2.8 Another Diode Application: Limiting and Clamping Diodes / 371
6.2.9 Diode Selection / 372
6.3 Bipolar Junction Transistors (BJT) / 374
6.3.1 Basic Concepts on Intrinsic, *n*-type and *p*-type Silicon Materials / 374
6.3.2 The BJT as a Circuit Element / 376

6.3.3 Bipolar Transistor *I-V* Characteristics / 377
6.3.4 Biasing Techniques of Bipolar Transistors / 382
6.3.5 Very Simple Biasing / 385
6.3.6 Resistor Divider Biasing / 387
6.3.7 Emitter Degeneration Resistor Biasing / 391
6.3.8 Self-Biased Staged / 394
6.3.9 Biasing Techniques of PNP Bipolar Transistors / 396
6.3.10 Small Signal Model and Single-Stage Bipolar Amplifier Configurations / 397
6.3.11 Common Emitter (C_E) Configuration / 399
6.3.12 Common Emitter (C_E) Configuration with Emitter Degeneration / 404
6.3.13 Common-Base (CB) Configuration / 407
6.3.14 The Common-Collector (CC) Configuration / 415
6.4 Metal Oxide Field Effect Transistor (MOSFET) / 420
6.4.1 MOSFET *I-V* Characteristics / 424
6.4.2 MOSFET Small Signal Model / 427
6.4.3 MOSFET Biasing Techniques / 428
6.4.4 Common Source (CS) Configuration / 434
6.4.5 Common Source (CS) Configuration with Degeneration / 436
6.4.6 Common Gate (CG) Configuration / 437
6.4.7 Common Drain (CD) Configuration or Source Follower / 439
6.4.8 Other MOSFETs: Enhancement Mode *p*-Channel and Depletion Mode (*n*-Channel and *p*-Channel) / 439
6.5 Summary / 443
Further Reading / 446
Problems / 446

7 Combinational Circuits **456**

7.1 Introduction to Digital Circuits / 456
7.2 Binary Numbers: a Quick Introduction / 456
7.3 Boolean Algebra / 460
7.3.1 *AND* Logic Operation 460
7.3.2 *OR* Logic Operation (Also Called *Inclusive OR, or XNOR*) / 461
7.3.3 NOT Logic Operation or Inversion—*NAND* and *NOR* / *462*
7.3.4 Exclusive *OR* Logic Operation or *XOR* / 463
7.3.5 DeMorgan's Laws, Rules, and Theorems / 464
7.3.6 Other Boolean Algebra Postulates and Theorems / 465
7.3.7 The Duality Principle / 466
7.3.8 Venn Diagrams / 467

7.4 Minterms: Standard or Canonical Sum of Products (SOP) Form / 467
7.5 Maxterms: Standard or Canonical Product of Sums (POS) Form / 472
7.6 Karnaugh Maps and Design Examples / 473
7.6.1 Two-Variable Karnaugh Maps / 474
7.6.2 Three-Variable Karnaugh Maps / 479
7.6.3 Four-Variable Karnaugh Maps / 484
7.6.4 Five-Variable Karnaugh Maps / 487
7.7 Product of Sums Simplifications / 490
7.8 Don't Care Conditions / 491
7.9 Logic Gates: Electrical and Timing Characteristics / 495
7.9.1 Gates Key Electrical Characteristics / 497
7.9.2 Gates Key Timing Characteristics / 499
7.10 Summary / 500
Further Reading / 500
Problems / 500

8 Digital Design Building Blocks and More Advanced Combinational Circuits 503

8.1 Combinational Circuits with More than One Output / 503
8.2 Decoders and Encoders / 510
8.2.1 Making Larger Decoders with Smaller Ones / 515
8.2.2 Encoders / 517
8.3 Multiplexers and Demultiplexers (MUXes and DEMUXes) / 519
8.3.1 Multiplexers / 521
8.3.2 Building Larger Multiplexers / 522
8.3.3 De-Multiplexers / 526
8.4 Signed and Unsigned Binary Numbers / 527
8.4.1 One's Complement Representation of Binary Numbers: Addition / 527
8.4.2 Two's Complement Representation of Binary Numbers: Addition / 531
8.4.3 Other Numbering Systems / 533
8.5 Arithmetic Circuits: Half-Adders (HA) and Full-Adders (FA) / 533
8.5.1 Building Larger Adders with Full-Adders / 536
8.5.2 Notes about Full-Adder Timing / 540
8.5.3 Subtracting with a *4*-bit Adder Using *1's* Complement Representation / 540
8.5.4 Subtracting with a *4*-bit Adder Using *2's* Complement Representation / 542

8.6 Carry Look Ahead (CLA) or Fast Carry Generation / 543
8.7 Some Short-Hand Notation for Large Logic Blocks / 546
8.8 Summary / 547
Further Reading / 548
Problems 548

9 Sequential Logic and State Machines **550**

9.1 Introduction / 550
9.2 Latches and Flip-Flops (FF) / 552
9.2.1 *NAND*-Implemented $\bar{R}/\bar{S}$ Latch / 557
9.2.2 *SR-Latch* with Enable / 559
9.2.3 Master/Slave *SR*-Flip-Flop / 561
9.2.4 Master/Slave *JK* Flip-Flop / 565
9.2.5 Master/Slave *T* and *D* Type Flip-Flops / 568
9.3 Timing Characteristics of Sequential Elements / 571
9.3.1 Timing of Flip-Flops with Additional Set and Reset Control Inputs / 571
9.4 Simple State Machines / 574
9.4.1 *SR* Flip-Flop Excitation Table / 578
9.4.2 *T* Flip-Flop Excitation Table / 580
9.4.3 *D* Flip-Flop Excitation Table / 581
9.5 Synchronous State Machines General Considerations / 592
9.5.1 Synchronous State Machine Design Guidelines / 592
9.5.2 Timing Considerations: Long and Short Path Analyses / 595
9.6 Summary / 599
Further Reading / 600
Problems / 600

10 A Simple CPU Design **603**

10.1 Our Simple CPU Instruction Set / 603
10.2 Instruction Set Details: Register Transfer Language (RTL) / 605
10.3 Building a Simple CPU: A Bottom-Up Approach / 607
10.3.1 The Registers / 607
10.3.2 The Memory Access Path or Memory Interface / 610
10.3.3 The Arithmetic and Logic Unit (ALU) / 611
10.3.4 The Program Counter (PC) / 611
10.4 Data Path Architecture: Putting the Logic Blocks Together / 615
10.4.1 Data Path: *LDA* Instruction Fetch, Decode and Execution RTL / 615
10.4.2 All Other Instructions: Fetch, Decode and Execution: RTL / 618

10.5 The Simple CPU Controller / 620
10.5.1 State Assignments and Controller Implementation / 622
10.6 CPU Timing Requirements / 626
10.7 Other System Pieces: Clock, Reset and Power Decoupling / 628
10.7.1 Clock / 628
10.7.2 Reset / 628
10.7.3 Power Decoupling / 631
10.8 Summary / 633
Further Reading / 633
Problems / 633

Index **637**

PREFACE

For several years I taught an introductory analog and digital essentials course for the University of California Extensions at Berkeley and Santa Cruz. Teaching there motivated me to put together, under one cover, a textbook that contains fundamentals of electrical, electronics, analog, and digital circuits. That is the reason for the word "essentials" in the title. There are not that many books in the market that try to accomplish this task in about 600 pages.

The book is divided into 10 chapters. It is useful for surveys of electrical and electronics courses, for college students as well as practicing scientists and engineers; it is also useful for introductory circuit courses at the undergraduate level. The book provides many examples from beginning to end. Within the examples, specific components part numbers were avoided to prevent this book from becoming obsolete. The book can be used by students who have some to no previous knowledge of the material, and for graduate-level and working professionals' circuit courses. The prerequisites for using this book are freshman-level calculus and algebra. Nevertheless, the level of math needed is quite light. The book is a gentle introduction to electrical and electronic circuit analysis with many examples.

Physical concepts are emphasized not only with text but also with specially prepared figures that should help the first-time readers study the material.

This book emphasizes problem solving, using different circuit analysis methodologies. These techniques allow readers to understand when one method is more appropriate than another. Ultimately, it is the student who is responsible for adopting the methods that make the most sense. No one thinks exactly in the same way. An example is differentiation and integration. For some people, differentiation is simpler than integration; for others, is the other way around.

Chapter 1 covers the three basic circuit elements: resistors, inductors, and capacitors. Additionally, *ideal* and *real* independent DC current and voltage sources are addressed. Chapter 2 emphasizes AC circuits, as they are applied to the three basic circuit elements. Their *time-domain* and *frequency-domain* behavior is seen throughout examples. A brief refresher on *operations with complex numbers* is embedded in this chapter and not in an appendix for reasons of reading continuity. The concept of power drawn by a circuit and its different types are addressed. The chapter ends with the coverage of dependent voltage and current sources. Chapter 3 addresses methods to solve circuits; it should be studied with the greatest attention and as many problems with different circuit analysis methods as possible should be solved. From a practical point of view, this is a core chapter to master.

Chapter 4 describes with plenty of detail the behavior of first-order and second-order circuits in the time and frequency domains. Many textbooks do not put as much emphasis on first-order circuits, because they are considered too simple. It has been my experience with students that first-order high-pass filters are particularly more difficult to understand than first-order low-pass filters. Chapter 5 is dedicated to *operational amplifiers*. Even though op amps consist of to-be-covered electronic components, it is useful to have the reader think with some high level of abstraction. Under some conditions op amps are seen as functional blocks and not as circuits with transistors and resistors. Linear and nonlinear applications with op amps are covered with many examples.

Chapter 6 covers electronic devices. Much information on devices is provided. One can say that entire books have been written just on the electronic components addressed by this chapter. The textbook takes a systematic approach to study the circuits using diodes and transistors, hardly dwelling on device physics. Chapter 7 begins with digital logic. *Combinational* (and not "combinatorial," as it is sometimes mistakenly called) logic circuits do not have any memory. Logic operations or Boolean algebra is presented, and logic simplification methods such as the Karnaugh map method are illustrated. Chapter 8 deals with more advanced combinational circuits such as multiplexers, decoders, and some arithmetic circuits. A method to produce a very fast arithmetic sum of two operands is covered. Chapter 9 is about state machine design or sequential logic. Sequential logic has memory, unlike combinational logic, and it is the core subject when designing logic circuits that perform useful and complete functions. Chapter 10 describes piece by piece the construction of a simple CPU. The CPU basic functional blocks, such as its instruction set, the data path architecture, its memory interface, and the control logic, are described step by step. Some insights into capacitor power decoupling and reliable reset circuits are also presented. The problems at the end of this chapter provide tremendous insight into the CPU functionality. This chapter can be thought as a very light introduction to a computer architecture course.

Writing this book has been a very rewarding experience for me. This book should be very useful to college students and those professionals who need an essential analog and digital source.

I want to thank my wife Ruty and daughter Denise for their infinite patience and support while I was preparing the manuscript.

Eduardo (Ed) M. Lipiansky

ABOUT THE AUTHOR

Eduardo (Ed) Lipiansky received his undergraduate degree in electrical engineering from the National University of La Plata, Argentina (UNLP). He performed graduate studies at the University of California, Berkeley, obtaining a master of science degree in electrical engineering. Mr. Lipiansky has 25 years of industry experience, having worked at Varian Associates, Tandem Computer, Sun Microsystems, Cisco Systems, and Google.

He is author or co-author of six patents; four have been issued by the U.S. Patent Office, and two more have been submitted. Mr. Lipiansky's key interests are maintenance and diagnostics subsystems for servers, networking line cards, analog and digital electronic design for medical instrumentation and computers, and power engineering. Ed wrote *Embedded Systems Hardware for Software Engineers* (2011), which deals with more advanced hardware concepts and can be used as an add-on to the present textbook. For about 20 years, he taught a variety of courses such as circuit analysis, digital design, operational amplifiers, and microprocessor interfacing techniques at the University of California Berkeley and Santa Cruz Extensions. Mr. Lipiansky lives with his family in the San Francisco Bay area in northern California.

1

FROM THE BOTTOM UP: VOLTAGES, CURRENTS, AND ELECTRICAL COMPONENTS

1.1 AN INTRODUCTION TO ELECTRIC CHARGES AND ATOMS

The ancient Greek philosophers knew that when amber was rubbed against wool, it would attract lightweight particles of other materials like small pieces of paper or lint. Also, little pieces of paper get attracted to a plastic comb when the weather is dry. These experiments reveal that electric charge exists. If we rub one end of a glass rod with silk, charges will move toward that end of the rod. Rubbing a second glass rod in the same fashion and placing it close to the rubbed end of the first glass rod will exhibit a repelling force between the rods. However, when a plastic rod is rubbed with fur and it is placed near the rubbed glass rod, the plastic and the glass rods will attract each other. These simple experiments prove the existence of two different types of charge. Benjamin Franklin* called one of them positive and the other one negative. Most charge in an everyday object appears to be nonexistent because there is an equal amount of positive and negative charge. The word *electron* is derived from the Greek word "elektron," which means amber. From the above experiments the following can be asserted:

> *Charges of the same sign repel each other, while charges of opposite signs attract each other.*

* Benjamin Franklin: American scientist, writer, and politician (1706–1790).

Electrical, Electronics, and Digital Hardware Essentials for Scientists and Engineers, First Edition.
Ed Lipiansky.

All matter is made of the basic elements, those elements listed in the periodic table of chemical elements. As of 2006, there are 117 elements of which 94 are found naturally on the Earth. The remaining elements are synthesized in particle accelerators. Loosely speaking, all matter is made of some combination of atoms, where an atom is the basic unit of matter. An atom contains a nucleus surrounded by a cloud of electrons. The nucleus consists of positively charged protons and electrically neutral neutrons. Neutrons have no electrical charge, but their mass is about 1800 times the mass of electrons. The electronic cloud around the nucleus is negatively charged, and an atom with an equal number of protons and electrons is said to be neutral. Protons have a positive charge and a mass about 1800 times larger than the mass of electrons. Different element atoms are different from each other because of the different numbers and arrangements of the atom's basic particles: electrons, neutrons, and protons. Traditionally in elementary physics and chemistry, the atom was compared to our planetary system. The nucleus is in the center of the atom, like the sun is the center of our system. The electrons are like the planets, orbiting around the sun. Electrons occupy different layers or shells that are at different distances away from nucleus. The outermost shell is referred to as the valence shell. The valence shell electrons determine the electrical characteristics of an atom.

Table 1.1 presents the elementary charge, which has a positive sign for a proton and a negative sign for an electron. Values for the mass of the electron, proton, and neutron are also tabulated.

From an electrical point of view, there are four main types of materials: conductors, nonconductors or insulators, semiconductors, and superconductors. The fourth type of material, the superconductor, is beyond the scope of this book.

Conductors are materials through which charge can move quite freely, such as copper or gold. Insulators are materials through which charge cannot move freely such as plastic or rubber. Semiconductors are materials that have an intermediate behavior between that of conductors and insulators. More on semiconductors will be covered in Chapter 6.

Table 1.1 Some atomic constants

	Abbreviation	Mass Value	Units	Relative Mass to the Electron Mass (m_e)	Charge in C (coulombs)
Elementary charge	e				1.602×10^{-19}
Electron	m_e	9.109×10^{-31}	kg	1	-1.602×10^{-19}
Proton	m_p	1.673×10^{-27}	kg	1800 (approx.)	$+1.602 \times 10^{-19}$
Neutron	m_n	1.675×10^{-27}	kg	1800 (approx.)	0

1.2 ELECTRIC DC VOLTAGE AND CURRENT SOURCES

Two types of independent sources are available, voltage and current sources. A source is said to be independent when either its nominal voltage or current is constant and does not depend on any other voltage or current present in a circuit. In a later section, we will cover the concept of dependent sources. The ideal voltage source produces a constant voltage across its terminals, regardless of the current that is being drawn from it by a load. Conversely, an ideal current source produces a constant current to a load connected across its terminals regardless of the voltage that is developed across the load. Let us now address the concepts of electric current and voltage.

1.2.1 Electric Current and Voltage

A net flow of electric charges through a circuit establishes an electric current. Note that conductors in isolation, such as a piece of copper not connected to anything else, contain free electrons or conduction electrons that randomly move. Such electrons do not constitute an electric current since in any cross section of the copper wire, the net amount of charge moved through the wire is zero. The emphasis here is on the word "net"; the net flow of charge constitutes an electric current. Current is defined as

$$i(t) = \frac{dq}{dt}, \tag{1.1}$$

where $i(t)$ represents electric current as a function of time and dq/dt is the net variation of charge with respect to time. Traditionally, electric current was referred to as current intensity. In most places, the term "current" is used, which is a short form of current intensity. The letter i denotes current, while dq differential of charge over dt differential of time refers to the net passage of charge during a time interval through a cross section of the conductor. On the other hand, a voltage can be interpreted as the "pressure" that needs to be asserted in a circuit in order to cause electric current to flow.

Throughout the book, we will assume that a conductor or a wire is ideal and will have zero resistance to the flow of current, unless it is stated otherwise. The unit of resistance is the ohm (Ω). Electric components that have greater than $0\ \Omega$ resistance are called resistors. The current that flows through a resistor times the resistance value equals the voltage drop that is produced across such resistor. Conventional current in a resistor flows from higher voltages or potentials to lower voltages or potentials.

Figure 1.1 depicts a resistor, a current flowing through it, and the voltage with its polarities that is produced across the resistor. The current through the resistor times its resistance value equals the voltage obtained across the resistor terminals. Mathematically,

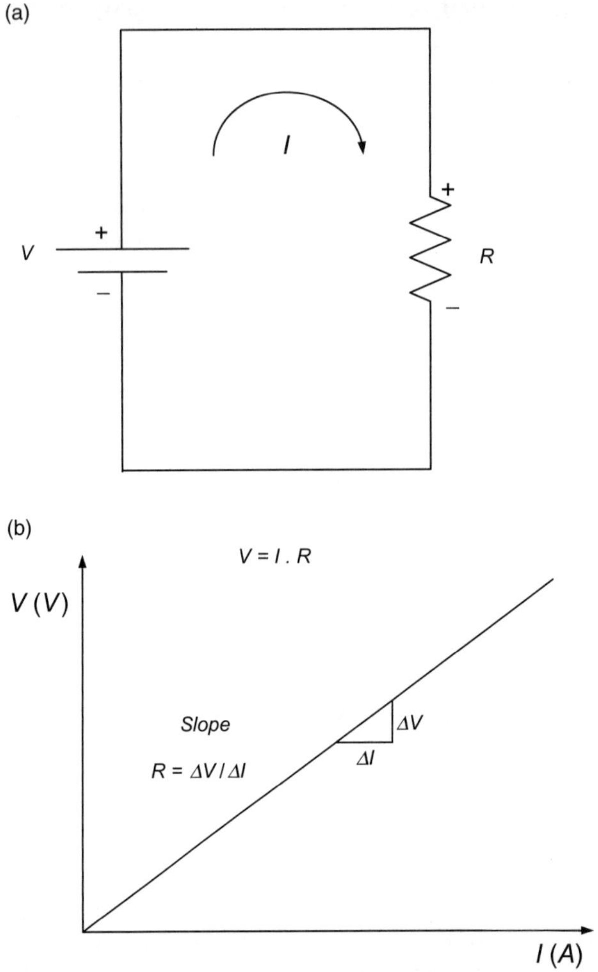

Figure 1.1 Ohm's Law: (a) DC voltage source powering a resistor; (b) linear variation of resistor voltage versus current.

$$V = IR. \tag{1.2}$$

Equation (1.2) states the voltage across a resistor is proportional to the current flowing through it. The constant of proportionality is the resistor value R. Equation (1.2) is *Ohm's Law*. In Figure 1.1a, a resistor powered by a DC source is shown; Figure 1.1b depicts the variation of resistor voltage versus current variation. The slope of the line $V = I\,R$ is the resistance value. Ohm's law in Equation (1.2) denotes a linear variation of the voltage across the resistor versus the current flowing through it.

Example 1.1 Given a 10-V DC voltage source that is connected across a 0.1 Ω resistor, calculate for the current that will flow through the resistor.

Solution to Example 1.1

From Equation (1.2),

$$V = IR.$$

And since

$$I = V/R,$$

$$I = 10\text{ V}/0.1\,\Omega = 100\text{ A}.$$

1.2.2 DC Voltage and Current Sources

We all have some familiarity with electricity and electronics. We have seen flashlights, batteries, battery chargers, lightbulbs, portable electronic devices, and electrical and electronic appliances such as toasters and microwave ovens.

Flashlight batteries, toy batteries, and automobile batteries are all examples of DC voltage sources. DC stands for direct current, and what this means is that the current polarity that the source supplies does not change; that is, the current always flows in the same direction through the load.

An idealization of the DC voltage source is that its DC voltage is always constant with respect to time and independent of the amount of current that it may supply. In practical devices such as batteries, that voltage is "somewhat" constant, and it varies based on factors such as temperature, environmental factors, mechanical vibration, age of the battery, and use of the battery. However, unless we state otherwise, the first-order approximation of a battery is that of a constant or DC voltage source.

Current sources are as also idealized like DC voltage sources. An everyday example of a current source is a battery charger. A battery charger provides a constant current to recharge a battery with rechargeable chemistry. Note that not all batteries are rechargeable. No attempt should be made to recharge batteries that are not of the rechargeable kind, since this causes a hazard to the user. Another example of a current source is that of a transistor hooked up to operate as a current source.

A DC voltage source may not always be a chemical battery. It may, for example, be built with electronic components that behave largely like a DC source. An example of this is a DC power supply (see Figure 1.2).

When a DC voltage source is not being used, it must be stored in an open-circuit condition (refer to Figure 1.3a). That means nothing is connected to the positive and the negative electrical terminals. Upon connecting an element

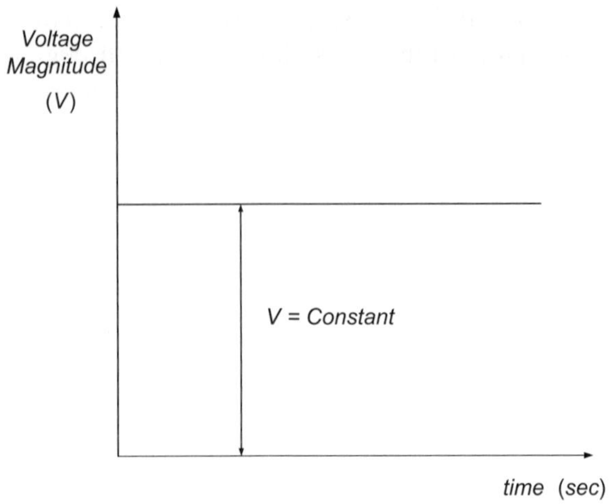

Figure 1.2 Mathematical representation of a DC voltage source as a function of time.

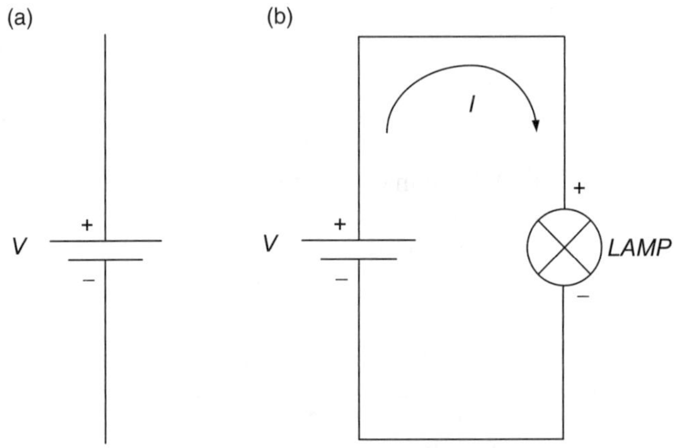

Figure 1.3 DC voltage source in (a) open-circuit condition and (b) loaded with a lightbulb.

such as a lightbulb across the voltage source terminals, a current flows through the circuit that was just established. Figure 1.3b shows a DC voltage source, which in this case is actually a battery connected with wires to a lightbulb.

The battery exerts "pressure" into the circuit by displacing charges. The net flow of charge with respect to time is called an electric current. Physically, an electric current consists of a net flow of electrons. That is, the electronic current leaves the negative terminal of the source, goes through the lightbulb, and returns back into the positive terminal of the source. However, the traditional interpretation is that current flows from the positive terminal of the source through the lightbulb and back into the negative terminal of the source. Throughout this book, the traditional or conventional current flow will be used. This is what most of the electrical engineering literature assumes.

The lightbulb depicted in Figure 1.3 is in effect a resistor. The voltage applied and the resistance of the lightbulb determines the current that will be present in the closed circuit. Resistance is the opposition that a resistor presents to the net flow of current. In other words, the DC voltage source voltage is basically constant, regardless of the amount of current that is being drawn form the source. Naturally, this is an idealization of what a DC voltage source is, or what we would like it to be. Real voltage sources do not behave that way; their output voltage is quite constant as long as the current flowing through the circuit is considerably less than what the total current pumping capability of the source is. More details on this topic will be provided when the internal resistance of a source is addressed, later in this chapter.

DC current sources, on the other hand, produce a constant current when a lightbulb or a resistive element establishes a closed loop circuit and the voltage across it will depend strictly on the resistive value placed across the current source and the current value. Just like with the DC voltage source, the DC current source is an idealization. Real current sources can provide a constant current as long as the voltage across the resistor does not produce an excessive voltage. Figure 1.4 depicts a constant DC current as a function of time.

Figure 1.5a depicts a DC current source in a standby condition, that is, with its terminals short-circuited to each other. A current source should not be left open-circuited because the voltage that gets developed across its terminals would grow without bound. A real or physical current source would self-destruct or become severely damaged if its terminals were left in an open-circuit condition. Figure 1.5b depicts a DC current source with a resistor connected across its terminals. In Figure 1.5a,b, both states of the current source are benign states or normal states. In both cases, the current supplied by the current source is identical.

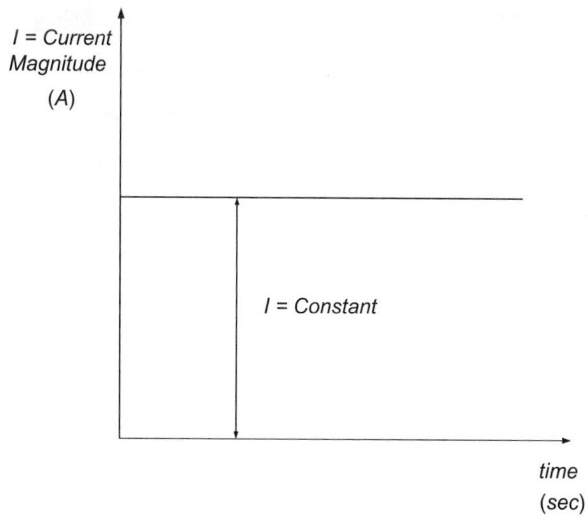

Figure 1.4 Mathematical representation of a DC current source as a function of time.

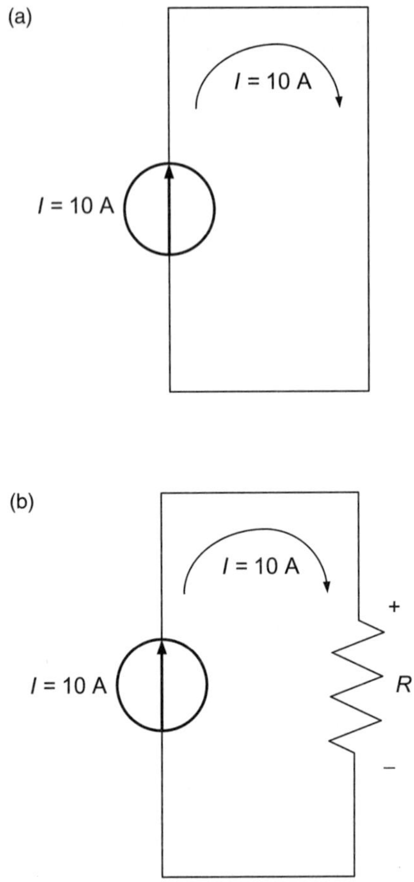

Figure 1.5 DC current source (a) in short-circuit condition and (b) loaded with a resistor.

The ideal current source with a resistive element in a closed circuit (Fig. 1.5) provides a constant current, and the voltage across the terminals of the current source depends on the value of the resistor across the current source times the current supplied by the source. Changes of the resistor values across the current source will produce proportional changes of the voltage across the current source. Note that the resistor (or load) across a current source produces higher voltages as the load resistor increases in value, because the current remains constant. For the case that the load resistor is very large, the voltage across the current source will be very large. When current sources are in open-circuit condition, the voltage across its terminal grows without bound. Real current sources would self-destruct quickly under an open-circuit condition. The current source must always be short-circuited when not in use (refer to Figure 1.5a). The voltage across a shorted current source is zero because the wire across the current source has zero resistance. However, the current supplied by a shorted current source is its nominal value of current. For example, a 10 A current source must be shorted when not supplying any current to a load. This current has a value of 10 A. When the current source is loaded by a resistor (Fig. 1.5b), the 10 A times the value of the resistor determines the voltage across the resistor. Note that since the current supplied by a current source remains constant as the resistor becomes larger in value, the voltage becomes larger as well.

In the extreme case if the value of the resistor is an open circuit, or infinitely many ohms, the voltage developed across the ideal current source is infinite.

In mathematical terms, an open-circuit current source produces an indetermination. The voltage source on the other hand must never be short-circuited because if we did, the current that the voltage source would supply to the short circuit is infinitely large, that is, also an indetermination (unbounded current). A real or physical short-circuited voltage source would also self-destruct rather quickly. Figure 1.6 depicts an open-circuit voltage source and a short-circuited current source. These are the benign or idle states for the voltage and the current sources. Figure 1.7 depicts ill-defined or undesirable states for a voltage and current source, respectively. A voltage source cannot survive short-circuited conditions, like a current source cannot survive an open-circuited condition. If they did, their reliability would be severely affected after that. Finally, let us be 100% clear about it: Figure 1.7 depicts circuits that fall under the "do-not-do-this-at-home-or-at-work" category.

1.2.3 Sources Internal Resistance

No real voltage source has a capacity of generating an infinite current upon being short-circuited. Similarly, no real current source can produce an infinite voltage across its terminals if left open-circuited. Real sources have their

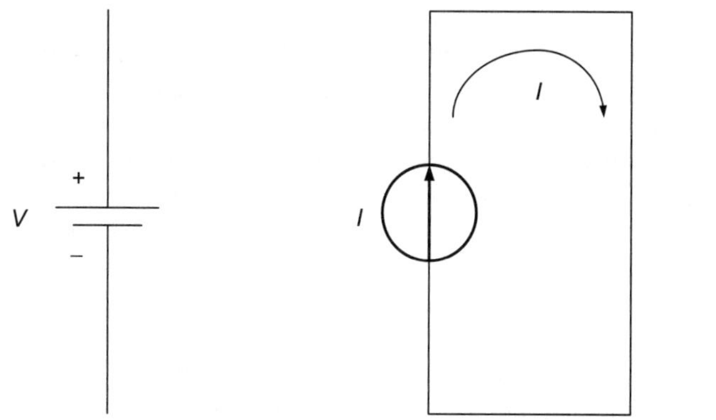

(a) *Open-Circuited Voltage Source* (b) *Short-Circuited Current Source*

Figure 1.6 Sources benign states: (a) for a voltage source; (b) for a current source.

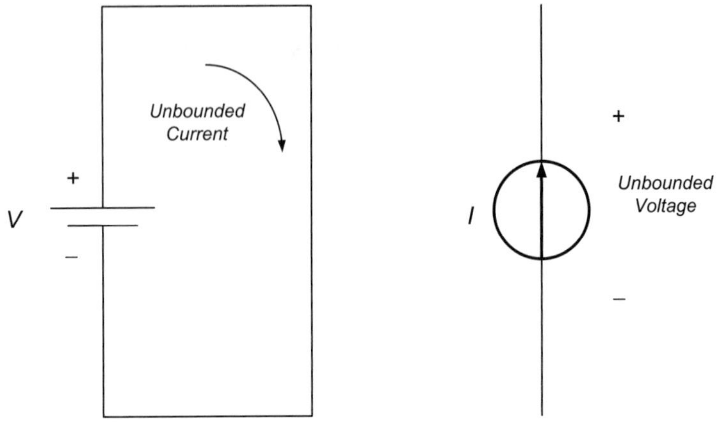

(a) *Short-Circuited Voltage Source* (b) *Open-Circuited Current Source*

Figure 1.7 Ill-defined (hazardous) states: (a) for a voltage source; (b) for a current source.

physical limitations. To account for these limitations, a voltage source is modeled with a current-limiting resistor in series, and a current source is modeled with a voltage-limiting resistor in parallel with the current source. Figure 1.8a,b depicts models of physical voltage and current sources with their respective series and parallel resistors. The series resistance in the voltage source represents the current limitation characteristic that a voltage source has. Note that if the voltage source of Figure 1.8a is 10 V and has an internal resistance of 1 Ω, the total short-circuit current capability of this source is its open-circuit voltage V divided by its internal resistance $r_{internal}$:

$$I_{short\text{-}circuit} = V_{constant\_source\_open\text{-}circuit\_voltage}/r_{internal}. \tag{1.3}$$

In particular for the example stated above, this short-circuit current is 10 V/1 Ω = 1 A. The internal resistance distinguishes a real voltage source from an ideal voltage source, which is assumed to have an infinite capability of generating current. Its internal resistance limits the current that can be drawn from a real or physical voltage source. This limitation is stated by Equation (1.3). When referring to Figure 1.8a, it is important to say that the internal resistance is an integral part of the real DC source, and the real DC source is modeled by an ideal DC source in series with the source internal resistance in series.

Similarly for the DC current source (refer to Figure 1.8b), the paralleled resistor with the current source represents or models the finite voltage-generating capability that an open-circuited current source has. We can write the current voltage relationship for the real current source modeled in Figure 1.8b as follows:

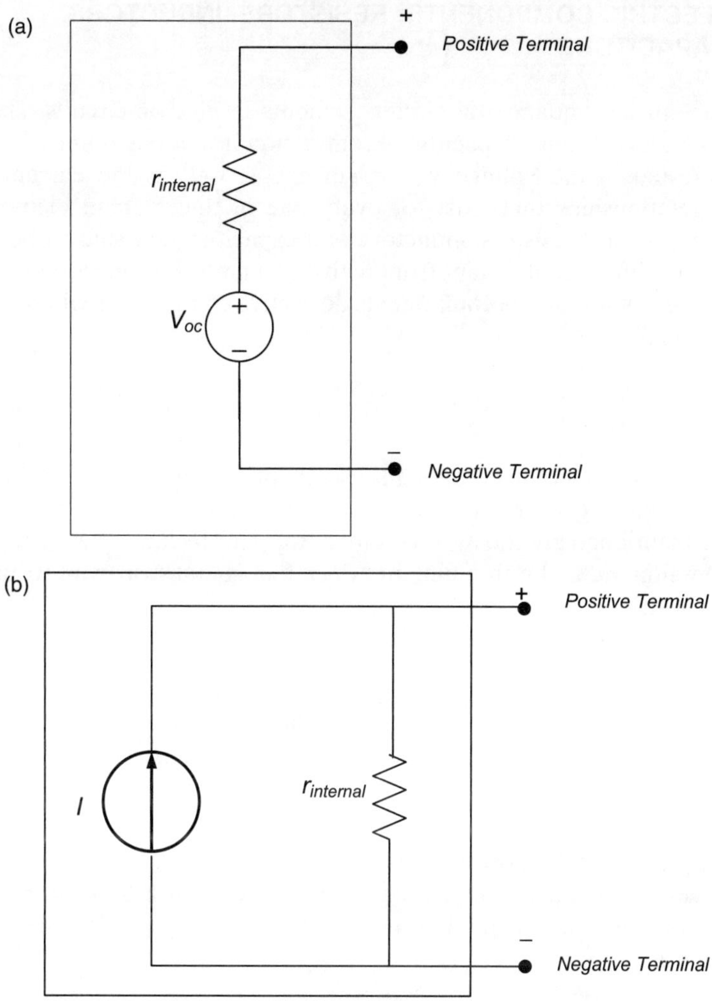

Figure 1.8 Modeling of real sources: (a) voltage source; (b) current source.

$$I_{constant\text{-}source\text{-}current} = V_{current\ source}/r_{internal} \tag{1.4}$$

Note that Equations (1.3) and (1.4) are governed by Ohm's law.

So, if we have a 10 A current source with its internal resistance of 10 Ω, using Equation (1.4), we determine that the maximum output voltage that this current source produces is 10 A·10 Ω = 100 V. Note that the open-circuit voltage of a real current source cannot exceed the limits imposed by Equation (1.4). The open-circuited ideal current source would produce a very large voltage across its terminals.

1.3 ELECTRIC COMPONENTS: RESISTORS, INDUCTORS, AND CAPACITORS

There are three fundamental circuit elements in electric circuits. These are resistors, inductors, and capacitors. From a circuit analysis point of view, we are interested in the voltage versus current as well as the current versus voltage relationships that exist for every one of these circuit elements. All three components, resistors, inductors, and capacitors, are said to be passive elements to differentiate them from active elements. Passive components do not have *gain*, while active components do. Active components will be covered in Chapters 5 and 6.

1.3.1 Resistors

The resistor is an electric component usually made with some of the resistive materials such as carbon, metal film, or paste. Other materials are used, but the ones mentioned are the most common. Resistor technologies vary, and the most common are metal thin film and thick film, metal strip, wire wound, foil, and composition.

A resistor opposes the flow of electric current. As the current flows through the resistor, a voltage is developed across such resistor. The voltage drop across the resistor follows Ohm's law, which states that the voltage across the resistor equals the current flowing through it, times the resistance value. Figure 1.9 shows the circuit symbol that represents a resistor. Figure 1.9 depicts a positive current and a positive voltage on the resistor (a), a negative current and a negative voltage (b), a positive current and a negative voltage (c), and a negative current and a positive voltage (d).

The current voltage and voltage current relationship on a resistor is given by Ohm's law, Equation (1.2), which we repeat for the reader's convenience:

$$V = IR = f(I),$$

where $f(I)$ is a linear function of I equal to $I\ R$.

And conversely,

$$I = V/R \tag{1.5}$$

$$I = g(V), \tag{1.6}$$

where $g(V)$ is a linear function of V and equals V/R. The term $1/R$, the inverse of the resistance, is called the conductance G. Using G in Equation (1.6), it becomes

$$I = GV. \tag{1.7}$$

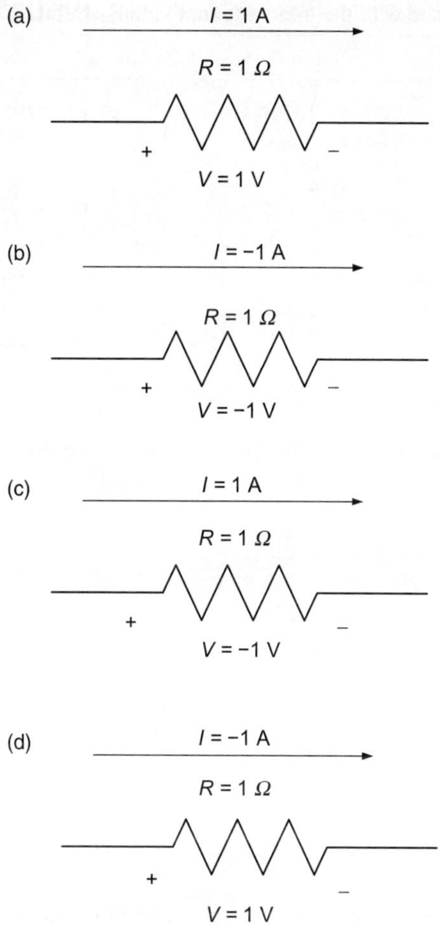

Figure 1.9 Circuit symbol of a resistor showing current and voltage polarities.

In Equation (1.5), V is the voltage across the resistor in volts, I is the current flowing through the resistor in amperes, and R is the resistor value in ohms. In Equation (1.7), I is in amperes, G in siemens (or mho), and V in volts.

The unit of resistance is the Ω, and R represents resistance. The inverse of the resistance ($1/R$) in DC circuit analysis is called conductance (G), and it is measured in siemens (S), where $1\ \mathrm{S} = \Omega^{-1}$, formerly also referred to as a mho (ohm spelled backward). Common values of resistors range from very small fractions of an ohm up to several megaohms ($1\ \mathrm{M}\Omega = 10^6\ \Omega$). This is a good point to introduce the most common prefixes used in the SI system of units.

Table 1.2 shows the internationally accepted power of 10 factors, their names, and their symbols. The most commonly used prefixes in electrical engineering and computer sciences range from 10^{24} (yotta) down to 10^{-24} (yocto). Within such range, some prefixes are more common than others.

Table 1.2 Prefixes used with the International System of Units (SI)

Factor	Name	Symbol	Factor	Name	Symbol
10^{24}	yotta	Y	10^{-1}	deci	d
10^{21}	zetta	Z	10^{-2}	centi	c
10^{18}	exa	E	10^{-3}	milli	m
10^{15}	peta	P	10^{-6}	micro	μ
10^{12}	tera	T	10^{-9}	nano	n
10^{9}	giga	G	10^{-12}	pico	p
10^{6}	mega	M	10^{-15}	fempto	f
10^{3}	kilo	k	10^{-18}	atto	a
10^{2}	hecto	h	10^{-21}	zepto	z
10^{1}	deka	da	10^{-24}	yocto	y

When selecting a resistor from a data sheet, there are key parameters that cannot be ignored to make a good component selection. Such resistor parameters are

(a) Resistance value in Ω
(b) Resistor power rating in W
(c) Resistor tolerance in ±%
(d) Temperature coefficient of resistance in ±ppm/°C, which stands for parts per million per Celsius degree or 10^{-6}/°C.

Other defined parameters that characterize resistors may vary a little bit from manufacturer to manufacturer. For a more complete list of precision resistor parameters, access the first reference of the Further Reading section at the end of the chapter.

Resistance values range from a small fraction of an ohm all the way up to a few MΩ (1 M = 10^6 Ω). The electronic industry standardized the resistor values based on the E-series according to Table 1.3. For example, for 1% tolerance resistors, 96 values per decade are chosen in an equally spaced basis. The resistor values follow the following geometric progression:

$$N = 10^{n-1/k}. \tag{1.8}$$

In Equation (1.8), N is the nominal resistance value at position n, and k is 96 for the E96 series. For example, for series E96, using Equation (1.8) for n = 1, 2, 3 . . . , the progression of resistance values becomes

$$1.00, 1.02, 1.05\ldots \tag{1.9}$$

The resistor values of (1.9) have been rounded.

In Equation (1.8), k refers to a decade of resistors values such as 1 Ω, 10 Ω, 100 Ω, and so on, while "n" is the series number. For example, for the E96

Table 1.3 Common Values of Resistor Tolerances

Series	Tolerance	Comments
E3	50%	3 steps per decade. No longer used.
E6	20%	6 steps per decade. No longer used.
E12	10%	12 steps per decade. No longer used.
E24	5%	24 steps per decade. Not commonly used.
E48	2%	48 steps per decade. Commonly used.
E96	1%	96 steps per decade. Commonly used.
E192	0.5%, 0.25%, 0.1% or higher	192 steps per decade. Commonly used in high accuracy and precision designs.

series, there are 96 (i.e., the reason for Equation (1.8) is to obtain the same number of steps within each decade of resistor values). Ultimately, the goal is to limit the number of resistor values or the inventory that manufacturers and distributors would have to have to handle otherwise.

For example for the E48 series of standardized resistor values, using Equation (1.8), the first three values are 100, 105, and 110 Ω. E48 is the ±2% tolerance series. Note that each of the three values mentioned plus and minus their 2% tolerance are

$$98\,\Omega < 100\,\Omega < 102\,\Omega. \tag{1.10}$$

$$102.9\,\Omega < 105\,\Omega < 107.1\,\Omega. \tag{1.11}$$

$$107.8\,\Omega < 110\,\Omega < 112.2\,\Omega. \tag{1.12}$$

It is easy to see that none of the values including their tolerance overlap. The same applies to all values for all ohmic decades of all other series. The tolerance of a resistor is the deviation in percent that the actual value of a resistor can deviate from its nominal value at room temperature. Room temperature for some manufacturers is defined as 20°C, for others is 25°C. However, when the resistor is in use, the temperature surrounding the resistor will cause its nominal value to increase or decrease. Resistor manufacturers specify the maximum/minimum resistance variation of their components for a certain temperature span.

Example 1.2 Assume that a 1 MΩ resistor has a ±100 ppm/°C temperature coefficient within an operating temperature range of –25°C to +125°C. Further assume that the resistor is exactly 1 MΩ at 20°C (i.e., assume the resistor has zero tolerance ±0%):

(a) What will the resistor value range be for the above temperature range?
(b) What will the resistor value range be if in addition to the temperature range, a ±1% tolerance is assumed?

Solution to Example 1.2a

When the resistor operates at +125°C,

$$\begin{aligned} &1\,\text{M}\Omega \times [+100\ \text{ppm/°C}] \times (125°\text{C} - 20°\text{C}) = \pm 1\,\text{M}\Omega \times 100 \times 10^{-6} \times 105 \\ &\quad = \pm 10{,}500\ \Omega = \pm 1.05\%\ \text{of the resistor nominal value } 125°\text{C}. \end{aligned} \tag{1.13}$$

When the resistor operates at –25°C,

$$\begin{aligned} &1\,\text{M}\Omega \times [-100\ \text{ppm/°C}] \times (-25°\text{C} - 20°\text{C}) = \pm 1\,\text{M}\Omega \times 100 \times 10^{-6} \times 45 \\ &\quad = \pm 4500\ \Omega = \pm 0.45\%\ \text{of the resistor nominal value} - 25°\text{C}. \end{aligned} \tag{1.14}$$

From Equations (1.13) and (1.14), we conclude that for a –25°C to +125°C temperature range, the resistor varies from –0.45% up to 1.05% from its nominal value.

Solution to Example 1.2b

In Example 1.2a, the resistance variation was just due to the resistor temperature coefficient over the operating temperature range. If in addition the resistor nominal value will be its value ±tolerance (%). That means that if the resistor has a ±2% tolerance, its value can vary between –2% (or 980,000 Ω) to +2% (or 1,020,000 Ω); a total of 1.05% due to its temperature coefficient operating in the –25°C to +125°C temperature range. Thus, at the high end the resistor value can be +1% due to tolerance and +1.05% operating at +125°C; that is a total of 2.05%. At the low end, the resistor value can be –1% due to tolerance and –0.45% operating at –25°C, that is, a total of –1.45% from its nominal value.

Resistor Tolerances: The most commonly used resistor tolerances for electrical and electronic applications are ±1% and ±2%. Many years ago, ±5%, ±10%, and ±20% were commonly used. When dealing with high precision analog electronics, ±0.1% tolerance is available. When higher precision is required, metal foil resistors and ultrahigh precision metal film resistors are available. Metal foil resistor tolerances of up to ±0.005% tolerance are available, and special metal film resistors of up to ±0.01% are also available.

Example 1.3 Resistor Value, Tolerance, and Power Rating Selection
So let us assume that we need to select a resistor R to establish a constant current of 100 mA. The resistor will have 12 V DC ±120 mV applied across its terminals. Assume that the DC source and resistor R will always operate at 20°C and no temperature changes will occur. Determine (1) a reasonable resistor value, (2) its tolerance, and (3) its power rating that can keep the load current at 100 mA ±2% under all voltage variations and resistor variations due to its tolerance.

Solution to Example 1.3

If the voltage (V) were exactly 12 V without any variations and the resistor were exactly 120 Ω, the current would be, by virtue of Equation (1.2),

$$I = 12\ \text{V}/120\ \Omega = 100\ \text{mA}. \tag{1.15}$$

The statement that the resistor will always operate at 20°C is equivalent to saying that its temperature coefficient is zero or that there are no resistor variations due to temperature changes.

Let us start adding the real requirements to the problem. We are told that the voltage can vary ±120 mV or ±1% from its nominal value of V = 12 V. Thus,

$$11.88\ \text{V} < \text{V} < 12.12\ \text{V}. \tag{1.16}$$

If we assume that we have a perfect resistor of 120 Ω with a 0% tolerance, then

$$11.88\ \text{V}/120\ \Omega < 12\ \text{V}/120\ \Omega < 12.12\ \text{V}/120\ \Omega, \tag{1.17}$$

which leads to

$$99\ \text{mA}\Omega < \text{I} < 101\ \text{mA}. \tag{1.18}$$

From Equation (1.16) we can state that having a ±120 mV voltage variation and a perfect resistor of 120 Ω, the current *I* will be bounded between 99 mA and 101 mA, or 100 mA ±1%.

Now let us introduce the concept that the resistor *R* is not perfect, and let us assume that it has a tolerance of ±1%.

Then the resistor value will range from

$$118.80\ \Omega < R < 121.20\ \Omega. \tag{1.19}$$

Using the resistor *R* range obtained in Equation (1.19) and combining it with all possible variation of the voltage V, it yields

$$V_{\text{max}}/R_{\text{max}} \tag{1.20}$$

$$V_{\text{min}}/R_{\text{min}} \tag{1.21}$$

$$V_{\text{max}}/R_{\text{min}} \tag{1.22}$$

$$V_{\text{min}}/R_{\text{max}} \tag{1.23}$$

where

$$V_{\text{max}} = V + (1\%\ \text{of}\ V) = 12\ \text{V} + 0.12\ \text{V} = 12.12\ \text{V}. \tag{1.24}$$

$$V_{min} = V - (1\% \text{ of } V) = 12\text{ V} - 0.12\text{ V} = 11.88\text{ V}. \tag{1.25}$$

$$R_{max} = R + (1\% \text{ of } R) = (120\ \Omega + 1.2\ \Omega) = 121.2\ \Omega. \tag{1.26}$$

$$R_{min} = R - (1\% \text{ of } R) = (120\ \Omega - 1.2\ \Omega) = 118.8\ \Omega. \tag{1.27}$$

Using Equations (1.24) through (1.27) in Equations (1.20) through (1.23), we obtain

$$V_{max}/R_{max} = 12.12\text{ V}/121.2\ \Omega = 100.0\text{ mA}. \tag{1.28}$$

$$V_{min}/R_{min} = 11.88\text{ V}/118.8\ \Omega = 100.0\text{ mA}. \tag{1.29}$$

$$V_{max}/R_{min} = 12.12\text{ V}/118.8\ \Omega = 102.0\text{ mA}. \tag{1.30}$$

$$V_{min}/R_{max} = 11.88\text{ V}/121.2\ \Omega = 98.0\text{ mA}. \tag{1.31}$$

Equations (1.28) through (1.31) provide all the possible extreme variations of current I. And from Equations (1.22) and (1.21), it can be seen that current I varies approximately ±2% from its nominal value of 100 mA.

Finally, the power dissipated by resistor R will be V^2/R. To account for voltage and resistor variations, we need to calculate

$$P_1 = V_{max}^2/R_{min} = 1.236\text{ W (maximum)}. \tag{1.32}$$

$$P_2 = V_{min}^2/R_{max} = 1.164\text{ W (minimum)}. \tag{1.33}$$

$$P_3 = V_{max}^2/R_{max} = 1.212\text{ W}. \tag{1.34}$$

$$P_4 = V_{min}^2/R_{min} = 1.188\text{ W}. \tag{1.35}$$

By inspection of Equations (1.32) through (1.35), Equation (1.32) shows the maximum dissipated power, and Equation (1.33) shows the smallest dissipated power.

Commonly available resistor power ratings are 0.063 W, 0.1 W, 0.125 W, 0.2 W, 0.25 W, 0.5 W, 1 W, 2 W, and 5 W for most electronic and some electrical applications. For special high-power electrical applications, the power ratings go well beyond 5 W, such as 10 W, 20 W, 50 W, 75 W, 100 W, 500 W, 1 kW, and above.

For our example, the most logical choice is to select a 2-W power-rated resistor.

Answers to Example 1.3

(a) $R = 120\ \Omega$, (b) ±1%, (c) 2 W.

R can be 120 Ω ±1%, 2 W. When the voltage varies from 12 V ±1%, current I remains within ±2% of its nominal value of 100 mA. Note that choosing a 1 W resistor is not an option since the minimum power dissipation exceeds 1 W; 2 W resistor is the next given resistor power rating that provides a headroom of 0.764 W or 76.4% more with respect to 1 W.

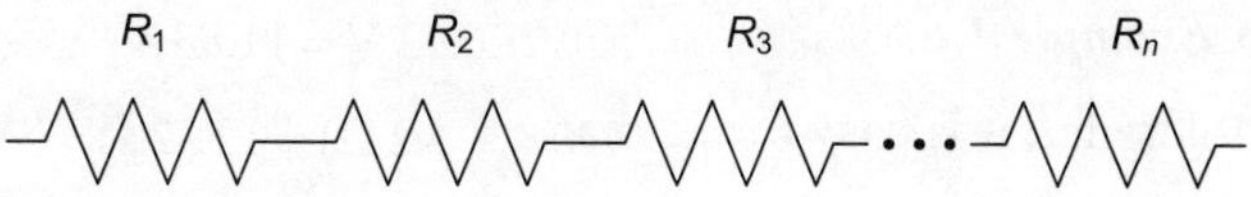

Figure 1.10 Resistors in series.

1.3.2 Resistors in Series and in Parallel

Two or more resistors are connected in series when the same current flows through all of them. Referring to Figure 1.10, resistors R_1, R_2, and R_3 are in series. Why? Because if a positive terminal of a voltage source is applied to the free end of one resistor in the series, and the negative terminal of the source is applied to the free terminal of the last resistor in the series, the current flowing through such circuit is identical for all resistors.

Generalizing the above concept, "n" resistors in series are equivalent to the sum of all n resistors. n is an integer and the total number of resistors in series.

$$R_1 + R_2 + R_3 + \ldots + R_n = \sum_i^n R_i. \tag{1.36}$$

Two resistors in series are equivalent to the sum of each of the resistors.

Example 1.4 Resistors in Series
Given a 1 kΩ, a 3 kΩ, and a 100 Ω resistor in series, find the series equivalent resistance.

Solution to Example 1.4

1000 Ω + 3000 Ω + 100 Ω = 4100 Ω = 4.1 kΩ.

Example 1.5 Resistors of Significantly Different Values
Given 1 MΩ and 1 kΩ resistors, find their series equivalent resistance.

Solution to Example 1.5

Not different from the previous example, the solution is 1,000,000 Ω + 1000 Ω = 1.001 MΩ.

Example 1.6 Let us assume that both resistors of the previous example have ±1% tolerance. What is the series equivalent resistance of both resistors with an error of approximately ±1%?

Solution to Example 1.6

With ±1% tolerance, it is easy to see that

$$\pm 1\% \text{ of } 1\ \text{M}\Omega = \pm 10{,}000\ \Omega \tag{1.37}$$

and

$$\pm 1\% \text{ of } 1\ \text{k}\Omega = \pm 10\ \Omega. \tag{1.38}$$

Since the value of the 1 kΩ resistor is much smaller than 1% of the value of the series equivalent resistor, found in the earlier part of this example to be 1.001 MΩ, the entire value of the small resistor can be neglected. The approximate answer is 1 MΩ ±10,000 kΩ, which is within a ±1% error. Note that a 1% error of 1 MΩ from Equation (1.37) is 10,000 Ω.

Resistors in parallel are those resistors that are connected such that the voltage across all of them is the same. Figure 1.11, depicts "n" resistors in

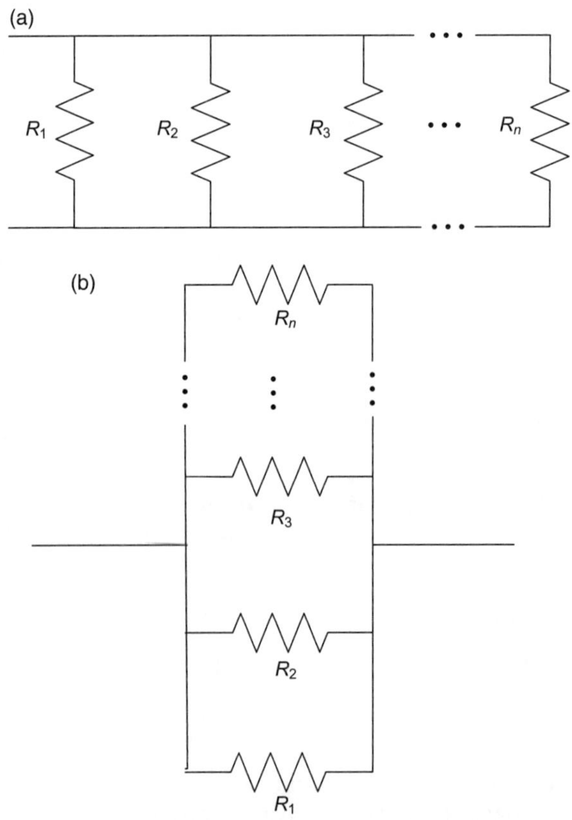

Figure 1.11 Resistors in parallel.

parallel. Note that part (a) and part (b) of the figure represent the exact same circuit.

Given two resistors R_1 and R_2 in parallel, the total parallel equivalent resistance ($R_{parallel\text{-}equiv}$) is

$$R_{parallel\text{-}equiv} = \frac{product-of-both-resistor-values}{sum-of-both-resistor-values}. \tag{1.39}$$

Example 1.7 Given two resistors in parallel, $R_1 = 3\ \Omega$ and $R_2 = 6\ \Omega$, find the total equivalent resistance.

Solution to Example 1.7

Applying Equation (1.39),

$$R_{parallel\text{-}equiv} = 3 \times 6/(3+6) = 2\ \Omega. \tag{1.40}$$

Equation (1.39) can be arithmetically expressed as follows:

$$\frac{1}{R_{parallel\text{-}equiv}} = \frac{1}{R_1} + \frac{1}{R_2} \tag{1.41}$$

$$R_{parallel\text{-}equiv} = \frac{R_1 \cdot R_2}{R_1 + R_2}, \tag{1.42}$$

where $R_{parallel\text{-}equiv}$ refers to the parallel equivalent resistor of R_1 and R_2. Note that Equations (1.41) and (1.42) are equivalent.

Generalizing from Equation (1.41), the parallel equivalent resistance of n (where *n* is an integer) that represents the number of resistors equals

$$1/R_{parallel\text{-}equiv} = 1/R_1 + 1/R_2 + 1/R_3 + \ldots + 1/R_n. \tag{1.43}$$

Upon covering Kirchhoff's laws in the next section we will justify the computations to find series and parallel equivalent resistors.

Example 1.8 Given three resistors in parallel, where $R_1 = 3\ \Omega$, $R_2 = 6\ \Omega$, and $R_3 = 2\ \Omega$, calculate the parallel equivalent resistor.

Solution to Example 1.8

Using Equation (1.43), we obtain

$$\frac{1}{R_{parallel\text{-}equiv}} = \frac{1}{R_1} + \frac{1}{R_2} + \frac{1}{R_3}, \tag{1.44}$$

and using the corresponding values for R_1, R_2, and R_3, we get that

$$1/R_{parallel\text{-}equiv} = 1/3 + 1/6 + 1/2, \tag{1.45}$$

from where

$$R_{equiv} = 1\,\Omega. \tag{1.46}$$

Example 1.9a Given 10 resistors in parallel of equal value, find the parallel equivalent resistor of the group of 10.

Solution to Example 1.9a

Using Equation (1.43) for "$n = 10$" resistors in parallel, we find that

$$1/R_{parallel\text{-}equiv} = 1/R + 1/R + 1/R + \ldots + 1/R, \tag{1.47}$$

where Equation (1.43) has 10 equal terms because all 10 resistors have the same value.

From Equation (1.43), we obtain

$$1/R_{parallel\text{-}equiv} = 10/R \tag{1.48}$$

or

$$R_{parallel\text{-}equiv} = R/10. \tag{1.49}$$

Example 1.9b Given two resistors in parallel where one is 1 kΩ and the other one is 1 Ω, find the total equivalent resistance.

Solution to Example 1.9b

Using Equation (1.43) one more time, we obtain

$$1/R_{parallel\text{-}equiv} = 1/1 + 1/1000, \tag{1.50}$$

from where we obtain that

$$R_{parallel\text{-}equiv} = 1000/1001 = 0.999001\,\Omega. \tag{1.51}$$

Corollary from Example 1.9

> *The parallel of one resistor with another one that is several orders of magnitude larger than the first one is approximately equal to the smaller resistor value.*

Example 1.10 Given three resistors, where $R_1 = 1\ \Omega$, $R_2 = 27\ \Omega$, and $R_3 = 500\ \Omega$, calculate the parallel equivalent resistance of the three resistors.

Solution to Example 1.10

Using Equation (1.43) from above,

$$1/R_{parallel\text{-}equiv} = 1/R_1 + 1/R_2 + 1/R_3$$

$$1/R_{equiv} = 1/1 + 1/27 + 1/500 = 1/0.9624$$

$$R_{equiv} = 0.9624\ \Omega$$

Note that the parallel equivalent resistor of 0.9624 Ω is smaller than the smallest given resistor, which is 1 Ω.

Corollary from Example 1.10

The reader should be convinced that

> *Given* n *resistors,* $R_1, R_2, \ldots, R_n$ *where* $R_1 < R_2 < \ldots < R_n$*, the total parallel equivalent resistor is always smaller than* R_1*. In other words, a number of resistors in parallel has a parallel equivalent which is numerically smaller than the smallest resistor value.*

1.3.3 Resistivity: A Physical Interpretation

Figure 1.12 shows a conductor made of a conductive material. The most common conductive materials are metals; copper is the most abundant and the least expensive metal to mass-produce electrical wire. Although in many practical applications it is reasonable to assume that the total resistance of a conductor is close to 0 Ω, actually, it is not 0. So for some applications, the true resistance of a conductor needs to be taken into account.

The resistance of a conductor is given by

$$R = \rho L / A, \tag{1.52}$$

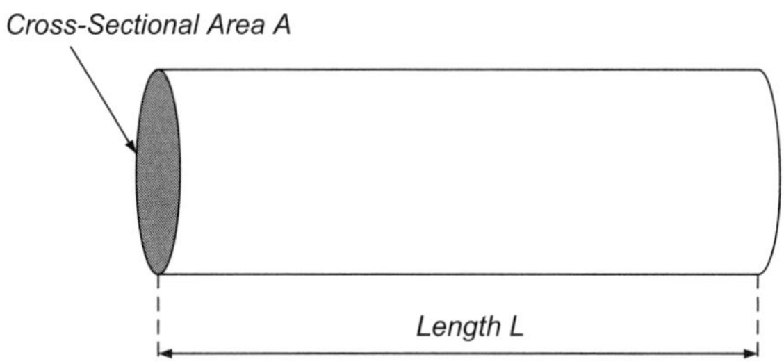

Figure 1.12 Conductor of length *L* and cross-section *A*.

Table 1.4 Resistivity of some materials at room temperature (20°C) [2]

Material	Resistivity ρ in (Ω·m) at 20°C	Conductor or Insulator?
Silver	1.62×10^{-8}	Conductor
Copper	1.69×10^{-8}	Conductor
Gold	2.35×10^{-8}	Conductor
Aluminum	2.75×10^{-8}	Conductor
Magnanin[a]	4.82×10^{-8}	Conductor
Tungsten	5.25×10^{-8}	Conductor
Iron	9.68×10^{-8}	Conductor
Platinum	10.6×10^{-8}	Conductor
Glass	10^{10} to 10^{14}	Insulator
Fused quartz	$\sim 10^{16}$	Insulator

[a] Magnanin is an alloy with a very small temperature coefficient of resistivity.

where ρ (lowercase Greek letter rho) is the material resistivity in ohm meter (Ωm), L is the length of the conductor in meters (m), and A is the cross-sectional area of the conductor in meters squared (m^2).

Figure 1.12 depicts a conductor of length L and cross-section A. Resistivity ρ is an electric characteristic of the material used, and it varies with temperature. Most commonly, resistivity is specified at room temperature of 20°C.

Table 1.4 lists some of the most common conductor and insulator materials and their resistivity.

Note that the range in resistivity between a conductor and an insulator such as glass minimally ranges from 10^{-8} to 10^{10}; this is 19 orders of magnitude!

Example 1.11 Assume that we have a DC voltage source that can produce 100 A of current at a constant 12 V. What voltage level will be present across the resistive load, without neglecting the voltage drop across the copper wires? Assume that you are using 10 mm^2 cross-section copper wires and that the one-way wire length (from source to load) is 1 m and the ambient temperature is 20°C. For illustration purposes of this problem, the reader is strongly referred to the circuit diagram of Figure 1.1 at the beginning of this chapter.

Solution to Example 1.11

If the resistance of the wires was 0 Ω, the load would see exactly 12 V at 100 A. We need to take into account the wire resistance, then:

From Equation (1.52), $R = \rho\, L/A$.

Since one-way length of the wire is 1 m, the round-trip length of the wire is 2 m. From Table 1.4, copper resistivity is 1.69×10^{-8} Ωm at a room temperature of 20°C. A cross section of 10 mm^2 equals 0.0001 m^2. Plugging all the values into Equation (1.52) yields

$$R_{wire} = (1.69\times10^{-8}\ \Omega\text{m})\times 2\ \text{m}/10^{-4}\ \text{m}^2 = 338\ \mu\Omega. \tag{1.53}$$

Since the current flowing through the wires is 100 A, the voltage drop across the load wires is

$$V_{drop\text{-}across\text{-}wires} = IR_{\text{wire}} = 100\times 338\times 10^{-6}\ \text{A}\cdot\Omega = 33.8\ \text{mV}. \tag{1.54}$$

Answer to Example 1.11

The voltage that the load resistor will see is 12 V – 0.0338 V = 11.9662 V.

1.3.4 Resistance of Conductors

It is interesting to observe that for a given conductor material, for example, copper, the resistance of the conductor equals to its resistivity, which depends on the material, times the length of the conductor (L), and it is inversely proportional to the conductor cross-sectional area (A) according to Equation (1.52).

So if the length of a conductor is doubled, all other factors remaining equal, the resistance of such conductor doubles. If the thickness (i.e., cross section) of a conductor doubles, while all other factors remain equal, the resistance of the conductor becomes half of the original resistance.

Resistivity ρ, is a temperature-dependent parameter, and it is a characteristic of the material. A good empirical approximation of how resistivity varies with temperature is

$$\rho - \rho_0 = \rho_0\alpha(T - T_0), \tag{1.55}$$

where α is called the temperature coefficient of resistivity, ρ_0 is the resistivity at the reference temperature, usually 20°C (or 293K [kelvin degrees]), T_0 is the reference temperature (20°C in our case), and ρ and T are respectively the resistivity and the temperature of the conductor at the temperature of interest, or at the unknown temperature. Equation (1.55) is linear and remains linear for most engineering problems over a wide temperature range around 20°C. Table 1.5 lists the temperature coefficient of resistivity α for some metals.

1.4 OHM'S LAW, POWER DELIVERED AND POWER CONSUMED

A voltage source happens to behave very much like a constant pressure water pump. The voltage source pushes the current through the electric circuit very much like a water pump pushes a volume of water through the closed-loop hydraulic case as depicted in Figure 1.13a, which shows an electrical circuit with a DC voltage source, a conductor or wire and a resistor, and Figure 1.13b

Table 1.5 Temperature coefficients of resistivity for some metals [2]

Metal	(α) Temperature Coefficient of Resistivity [K^{-1}]
Silver	4.1×10^{-3}
Copper	4.3×10^{-3}
Gold	4.0×10^{-3}
Aluminum	4.4×10^{-3}
Manganin[a]	0.002×10^{-3}
Tungsten	4.5×10^{-3}
Iron	6.5×10^{-3}
Platinum	3.9×10^{-3}

[a] An alloy with an extremely low value of α.

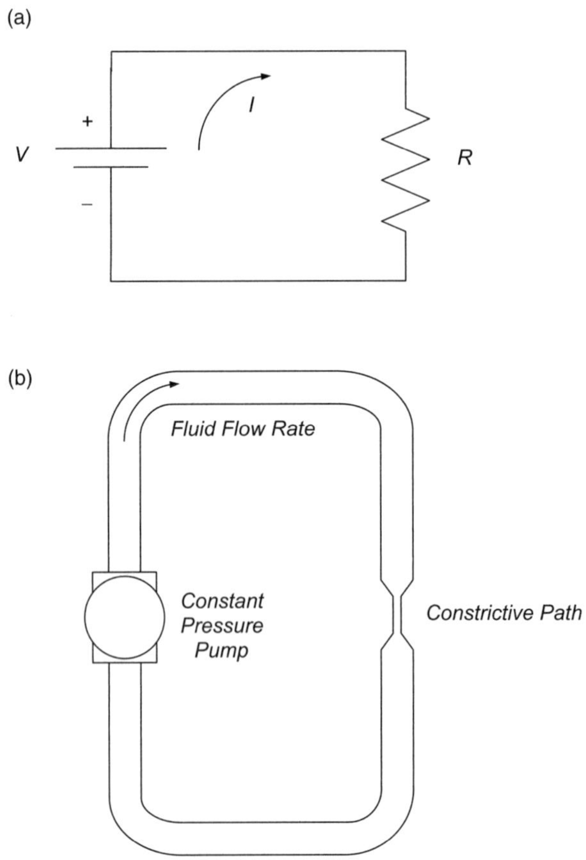

Figure 1.13 (a) An electric circuit and (b) its hydraulic analogy.

which depicts the hydraulic analogy of the electric circuit. The constant pressure pump is the analog to the electric voltage source. The pipe is analogous to the wiring, and the pipe with flow restriction is analogous to the resistor.

In Figure 1.13a, we can state that the current that flows through resistor R is proportional to the voltage applied across its terminals. Equation (1.2) is repeated here for the reader's convenience.

Ohm's Law

$$V = IR. \tag{1.56}$$

In Equation (1.56), V is the voltage across the resistor, I is the current flowing through resistor R, and R is the resistor value. Note from Figure 1.13a that V is the same as the voltage of the DC source applied.

The summary and a side-by-side comparison between the circuit elements and those of the hydraulic analogy in Figure 1.13 are presented in Table 1.6.

Rearranging terms of Equation (1.56), it yields

$$I = V/R \tag{1.57}$$

and

$$R = V/I. \tag{1.58}$$

SI units in Table 1.1 refer to the international system of units (in French *Le Système International d'Unités*) that was established worldwide by the General Conference of Weights and Measures in 1960. Note that Ohm's law can be expressed in any of the three forms given by Equations (1.56), (1.57), or (1.58), where in all three equations, I is the current flowing through resistor R, and V is the voltage across resistor R terminals.

Now back to Joule's law, a resistor will consume or dissipate in the form of heat an amount of power given in watts (W) in the SI system. This power

Table 1.6 Electrical and hydraulic analogies

Electric Circuit		Hydraulic Analogy	
Element	SI Units	Element	SI Units
DC voltage source	V (volts)	Constant pressure pump	Kg/m^2
Current	A (coulomb/s)	Volume flow rate	Liter/s
Resistor	Ω (ohms)	Flow constrictive pipe	There is no unit of resistance for a water flow constrictive pipe.

equals the product of the voltage across the resistor times the current flowing through it or

$$P = VI. \tag{1.59}$$

Plugging Ohm's law Equation 1.56 into power Equation 1.59, one can see that the power consumed by a resistor can also be expressed according to Equation (1.60):

$$P = I^2V. \tag{1.60}$$

Finally, using Ohm's law Equation (1.57) into power Equation (1.59) yields

$$P = V^2/R. \tag{1.61}$$

In terms of units, note that

$$[\text{watts}] = [\text{volts}]\cdot[\text{amperes}] = [\text{amperes}]^2\cdot[\text{ohms}] = [\text{volts}]^2/[\text{ohms}], \tag{1.62}$$

and using the appropriate SI abbreviations for each unit in Equation (1.62) becomes

$$[\text{W}] = [\text{V}][\text{A}] = [\text{A}]^2[\Omega] = [\text{V}]^2/[\Omega]. \tag{1.63}$$

It is important to emphasize from Equations (1.60) and (1.61) that the power dissipated or consumed by a resistor increases with the square of the current flowing through it, and the power also increases with the square of the voltage applied directly across such resistor.

Example 1.12 Given the circuit of Figure 1.14 where the DC voltage source V = 10 V, the load resistor 0.1 Ω, find the current in the circuit.

Solution to Example 1.12

The current I in the circuit is calculated using Ohm's law from Equation (1.57), and it becomes

$$I = V/R = 10\ \text{V}/0.1\ \Omega = 100\ \text{A}. \tag{1.64}$$

Answer to Example 1.12

100 A

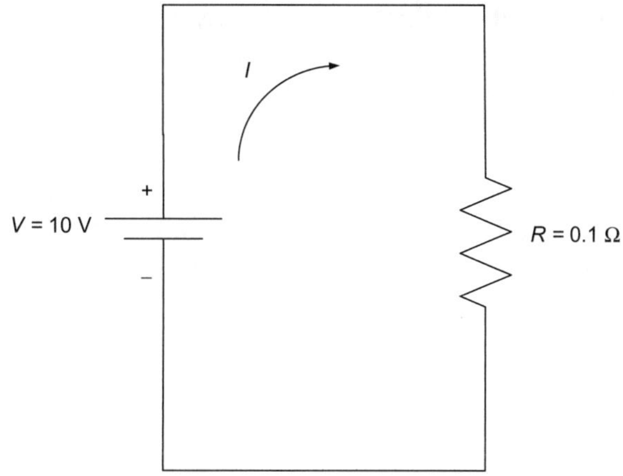

Figure 1.14 Circuit for Example 1.12 with a DC voltage source and a load resistor.

Example 1.13 Using the circuit of Figure 1.14, assume that the source is still 10 V, but the value of the resistor is now just 1 mΩ (or 0.001 Ω). Recalculate the current developed in the circuit.

Solution to Example 1.13

From Equation (2.3), the current is

$$I = V/R = 10\ \text{V}/0.001\ \Omega = 10{,}000\ \text{A}. \tag{1.65}$$

Answer to Example 1.13

10,000 A (ten thousand amperes!)

The purpose of Examples 1.12 and 1.13 is twofold. First, note that regardless of the value of the load resistance, which is 0.1 Ω in Example 1.12 and 0.001 Ω in Example 1.13, the voltage across the resistor is constant and equal to 10 V. Second, note that the ideal voltage source violates Ohm's law under the extreme case, that is, when the load resistor value is 0 Ω. The voltage source produces the same voltage at its output terminals regardless of the current being drawn from it. However, when the load is 0 Ω, meaning when the terminals of the voltage source are short-circuited, the current cannot be determined, because the resistance across its terminals is 0 Ω, and the voltage of the source is a finite number (10 V in our example); thus, 10 V divided by 0 is an undetermined quantity. In theory, the current approaches an infinitely large value. In actuality, the source will attempt to deliver a very large current, but it will only deliver a maximum number of amperes only for a short period of time. If this short-circuit condition is indefinitely applied, then the most likely outcome is at least one or more of the listed events: a damaged voltage source, burnt wires, smoke, even fire, and a serious hazard to people in the neighborhood. Do not try this at home or at any other place.

From Example 1.12 we find that the voltage source has to provide 100 A and maintain its 10 V across its terminals. From Example 1.13, the source has to provide 10,000 A and maintain its 10 V across its terminals. Clearly, the ideal model of a voltage source cannot hold up for extremely large currents because generating 10,000 A is an almost an unreal amount of current, too high for most standards.

The association of a resistor in series with the ideal voltage source adds a dose of reality to the modeling of a voltage source. This added resistor is referred to as the internal resistance of the voltage source.

1.4.1 Voltage Source Internal Resistance

No real battery or DC voltage source can generate an infinite amount of current for any length of time. Even if the current is not infinite, no battery can supply a fixed amount of current at a constant voltage indefinitely. These concepts seem pretty familiar because most of us probably had a flashlight or a car battery replaced, even if the battery is of the rechargeable type like a lead-acid car battery. Real batteries, unlike ideal DC voltage sources, have a finite lifetime. The ability of a fully charged and good battery to supply a given amount of energy, which means supplying a current at a voltage for a finite amount of time, depends on the battery construction, battery type, materials used, size, weight, discharge rate, temperature, resistance, time, and age of the battery. Clearly, real batteries neither have the capability of generating an infinite current nor that of generating a constant current indefinitely. The first-order approximation that we need to introduce into the ideal voltage source model is a non-zero Ohm internal resistance. In the real world, such resistance is a function of all of the factors mentioned such as temperature and discharge rate or usage of the battery. But for most practical purposes, it is reasonable enough to assume that the internal resistance of the battery is not zero, and under normal battery operating conditions, such internal resistance remains fairly constant. Figure 1.15 represents a circuit model of a real battery.

The real battery is depicted within the dotted box, and it is composed of two basic components, an ideal voltage source in series with a current limiting resistance or what we refer to as the battery internal resistance. Note that there is no access to the internal node, where the positive terminal of the battery connects to the internal resistance.

Note that the model in Figure 1.15 is a mathematical representation of the battery and is just a first-order approximation. The model represents the finite amount of current capability of the battery. Why is it only a first-order approximation? Because the model that includes the internal resistance does not account for the increase of its internal resistance as the battery looses

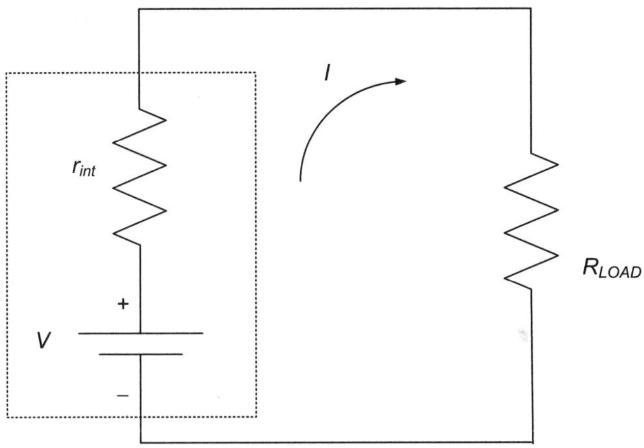

Figure 1.15 First-order approximation model of a real battery, showing its non-zero internal resistance.

current-generating capability due to usage, temperature, and any other factors over time.

As the battery ages or becomes discharged, its capacity of generating current diminishes and that can be modeled as the internal resistance increasing in value with respect to time. Note that as time goes by, the ampere-hour rating of the battery diminishes; the battery is being discharged. Figure 1.16 depicts an alkaline battery discharge characteristics. Ampere-hours versus power delivered by the battery are plotted for different battery output voltages.

Example 1.14 For example, let us assume we get a battery that is rated to provide 12 V in open-circuit mode, that is, no load. Let us further assume that we model such battery to have an internal resistance of 1 Ω. What does this mean?

Solution to Example 1.14

The internal resistance of the battery is in a way a figure of merit of the battery. It expresses what can conceivably be the absolute largest current that the battery can supply if its terminals are short-circuited. The battery short-circuit current is given by

$$\textit{Short-Circuit Current } I_{shc} = \textit{Open-Circuit Voltage } V_{oc} / \textit{Internal Resistance } r_{int}. \tag{1.66}$$

$$I_{shc} = V_{oc} / r_{int}. \tag{1.67}$$

$$I_{shc} = 12\ \text{V}/1\ \Omega = 12\ \text{A}. \tag{1.68}$$

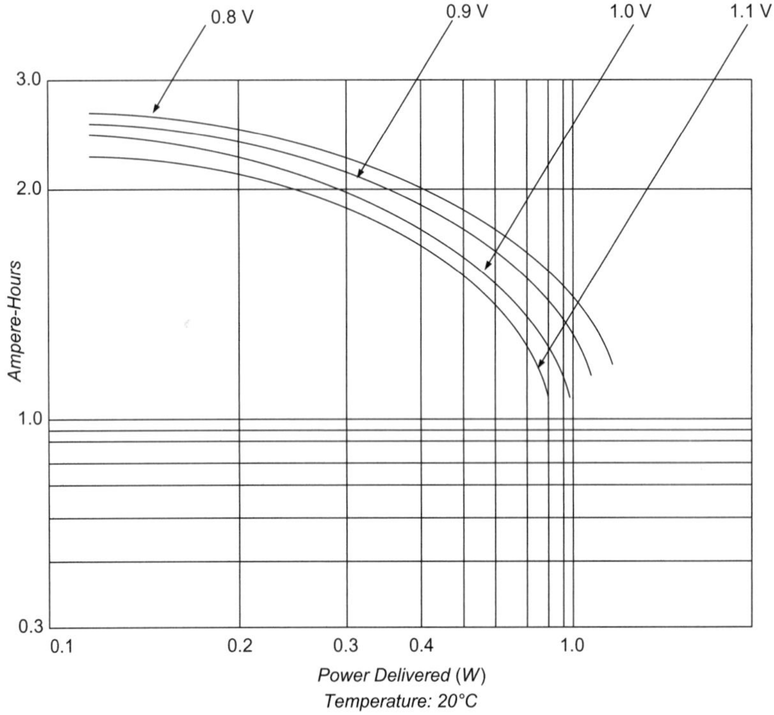

Figure 1.16 Alkaline battery discharge characteristic curves.

In practical terms, real batteries will not be able to supply their short-circuit current for very long (maybe for just a few seconds if that long). However, the internal resistance inclusion within the ideal voltage source model provides a touch of realism when modeling the battery. Note that modeling the battery with an ideal voltage source means that the short-circuit current that the source can supply is infinitely large. Inclusion of an internal resistance limits the current to a finite number. As the battery supplies power to a load, it becomes discharged; which is equivalent to the battery internal resistance to grow in ohmic value as time progresses.

Example 1.15 Assume that we have five different batteries all of which have a 10 V open-circuit voltage. However, each one of the batteries has a different internal resistance:

(a) 0.1 Ω, (b) 1 Ω, (c) 2 Ω, (d) 5 Ω, and (e) 10 Ω.
(b) What is the short-circuit current for each one of the batteries?

Solution to Example 1.15

Since the short-circuit current of a battery is its open-circuit voltage divided by its internal resistance from Equation (1.67), it yields that

(a) $I_{shc} = 10\ \text{V}/0.1\ \Omega = 100\ \text{A}$
(b) $I_{shc} = 10\ \text{V}/1\ \Omega = 10\ \text{A}$
(c) $I_{shc} = 10\ \text{V}/2\ \Omega = 5\ \text{A}$
(d) $I_{shc} = 10\ \text{V}/5\ \Omega = 2\ \text{A}$
(e) $I_{shc} = 10\ \text{V}/10\ \Omega = 1\ \text{A}$.

It is important to emphasize from Example 1.15 that the larger is the numerical value of the battery's internal resistance, the smaller is its short-circuit current. Shortly we will see that a battery with higher internal resistance also has less capacity of generating a voltage closer to its open-circuit voltage when a load is connected across the battery terminals. This example will be addressed again when we cover Kirchhoff's laws.

1.5 CAPACITORS

The most basic capacitor consists of two metallic or conducting plates in parallel, separated by a dielectric. A dielectric is an insulator, and it can be air, mica, polystyrene, transformer oil, glass, porcelain, or many others. Figure 1.17 depicts a parallel-plate capacitor with its two terminals. Each plate is connected to a capacitor terminal.

When we apply a DC voltage E across the capacitor plates, electrical charges (q) become accumulated on both plates of the capacitor. The positive side of the DC source will accumulate positive charges, and the negative side of the source will accumulate negative charges. Figure 1.18 depicts the parallel-plate capacitor, energized by a battery V. The voltage difference that exists across the capacitor is identical to the voltage V produced by the battery after all transients are over.

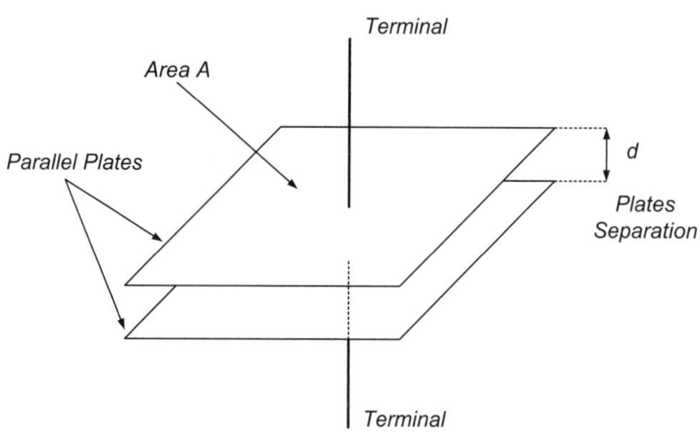

Figure 1.17 A parallel-plate capacitor: the area of the plates is A, and the separation between the plates is d; the dielectric used is air.

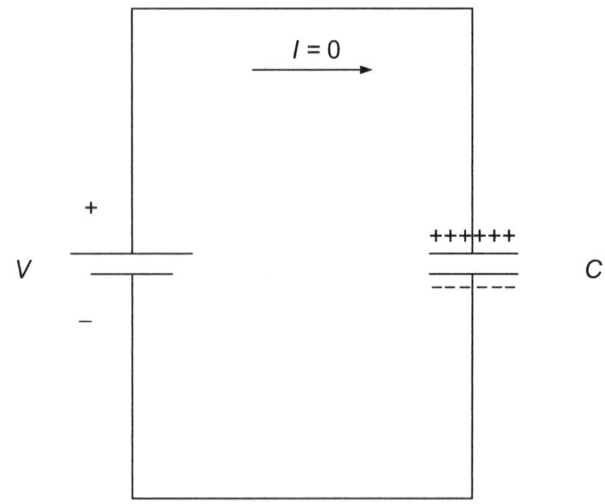

Figure 1.18 Capacitor with battery applied across its terminals.

It can be experimentally proven that the charge accumulated in the capacitor is proportional to the voltage applied across the capacitor. The constant of proportionality is referred to as *C*, the capacitance of the capacitor. The unit of capacitance is the farad (F), which practically speaking is a very large unit. More common values of everyday use capacitors are in μF, nF, or pF.

In mathematical terms,

$$q = CV, \tag{1.69}$$

where q is electric charge, C is capacitance, and V is the voltage impressed across the capacitor terminals. From units perspective from Equation (1.69), it can be seen that

$$[\text{coulombs}] = [\text{farads}][\text{volts}] \quad \text{or} \quad [\text{C}] = [\text{F}][\text{V}]. \tag{1.70}$$

The voltage V applied across the capacitor plates creates an electric field set up by the battery. Once the capacitor is charged, it takes some finite, but nonzero amount of time for the capacitor charging to occur; the plate connected to the negative of the battery accumulates negative charges or electrons, while the plate connected to the positive of the battery lacks negative charges. Should the wires and the battery be quickly removed, the charges on the plates of the capacitor will remain for as long as the capacitor is not discharged. Figure 1.19 shows two ways to discharge the capacitor after removing the battery across its terminals. Discharging the capacitor by short-circuiting its terminals may be a very hazardous operation, in particular when the capacitance value is large, and the voltage stored in the form of charge inside the capacitor is large. Discharging the capacitor, with an appropriately sized resistor, is safer. The discharge is more gradual and the resistor has to be sized to

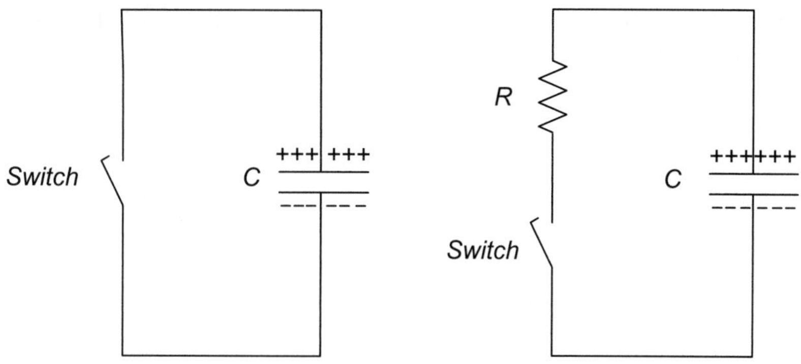

(a) *Unsafe Discharge: Not Current Limited*

(b) *Safer Discharge: Current Limited*

Figure 1.19 Discharging of a charged capacitor: (a) short-circuiting its terminals; (b) placing a resistor across its terminals.

handle the capacitor charge that will be dissipated by the resistor in the form of heat.

1.5.1 Physical Interpretation of a Parallel-Plate Capacitor Capacitance

It can be experimentally determined that the capacitance exhibited by the parallel-plate capacitor with air as dielectric is proportional to the dielectric constant of free space, to the area of the parallel plates, and inversely proportional to the separation d of the plates; see Equation (1.72). The dielectric constant of air is quite close to that of free space (1.0006 ε_o). And the dielectric constant of free space ε_o is a physical constant determined to be

$$\varepsilon_o = 8.85 \times 10^{-12} \text{ F/m} = 8.85 \text{ pF/m}. \tag{1.71}$$

The parallel-plate, air-dielectric, capacitor capacitance equals

$$C_{air} = \varepsilon_o \frac{A}{d}, \tag{1.72}$$

where C_{air} is capacitance in farads with air dielectric, A is the plate area in m^2, and d is plate's separation in meters. If instead of air or free space in between the capacitor plates we introduce a dielectric, the capacitance of the new capacitor with the dielectric different from air will be k times bigger than the capacitance of the original capacitor of capacitance given by Equation (1.74). What is the value of k? k is the relative dielectric constant of the used dielectric material.

For example, if the dielectric is FR4 material, frequently used to fabricate printed circuit boards, the relative dielectric constant is approximately

$$k = \varepsilon_{FR4}/\varepsilon_o = 4.5. \tag{1.73}$$

k is the relative dielectric constant of the material, FR4, in our example; ε_{FR4} is the dielectric absolute dielectric constant. k is always greater than one for dielectrics other than air and vacuum.

Combining Equation (1.72) with Equation (1.73), we obtain the capacitance of a parallel-plate capacitor with a dielectric in between its plates:

$$C = k\varepsilon_o \frac{A}{d} = kC_{air}, \tag{1.74}$$

where k is the relative dielectric constant of the material used, the other parameters are the same ones described before. Looking at Equations (1.72) and (1.74), it is interesting to observe that with a dielectric other than air (or vacuum), a capacitor of the same structure (e.g., parallel-plate capacitor with area A and plate separation d) has a capacitance k times higher than the same capacitor with air as its dielectric.

Table 1.7 shows the dielectric constants for some materials relative to free space dielectric constant.

1.5.2 Capacitor Voltage Current Relationship

From Equation (1.69), charge in a capacitor is proportional to the voltage applied across its terminals.

Thus,

$$q = CV. \tag{1.75}$$

Table 1.7 Table of dielectric constants of some materials measured at room temperature (20°C) [2]

Material	Relative Dielectric Constant k
Vacuum (free space)	1
Air (at 1 atmosphere of pressure)	1.0006
Polystyrene	2.6
Paper	3.4
Porcelain	6.5
Silicon	12
Germanium	16
Standard FR4 Epoxy Glass[a]	4.5
Cyanate Ester[a]	3.8
Teflon[a]	2.2

[a] Dielectric material commonly used to fabricate printed circuit boards (PCBs).

Differentiating the above equation with respect to time yields

$$\frac{dq}{dt} = \frac{d}{dt}(CV), \tag{1.76}$$

and since the capacitance is a constant parameter, then

$$i(t) = \frac{dq}{dt} = C\frac{dV}{dt}, \tag{1.77}$$

where $i(t)$ is the electric current flowing through the capacitor, dq/dt is the definition of current, C is the capacitance value, and dV/dt is the variation of the voltage across the capacitor with respect to time.

Equation (1.77) is of utmost important equation and describes the voltage and current behavior on a capacitor. It is an experimentally determined expression and is valid for all current and voltage waveforms on a capacitor.

Example 1.16 Given the voltage across a capacitor is a sinusoidal function of time,

$$v(t) = V_{peak}\sin(2\pi ft - \varphi), \tag{1.78}$$

where:

- Frequency (f) is the inverse of the sinusoidal waveform period T.
- Phase angle (φ) is the angle with respect to the origin of the time axis that the waveform is shifted from. For a left-shifted sine waveform, the phase angle is positive; for a right-shifted waveform, the phase angle is negative.
- Amplitude (V_{peak}) is also called the *peak* value of the waveform.

Figure 1.20 depicts a sinusoidal voltage waveform and all of its parameters.

1.5.3 Capacitors in Series

Capacitors are said to be in series when they are in a circuit where the same current flows through all of them. Figure 1.21 depicts n capacitors in series, where n is an integer.

A DC voltage source is connected across the three capacitors in series. The positive terminal of the source is connected to the left-most terminal of the the left hand side capacitor; the negative terminal of the source is connected to the right-most terminal of right hand side capacitor.

Since the current is the same flowing through all three capacitors, they all have identical charge (q). The sum of all the voltage differences across each

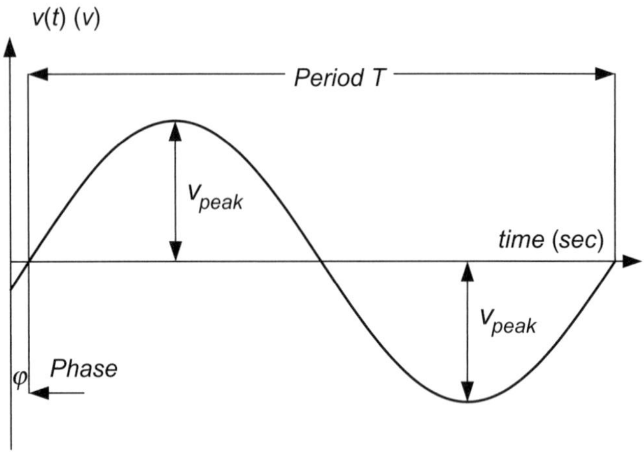

Figure 1.20 Sinusoidal voltage waveform as a function of time.

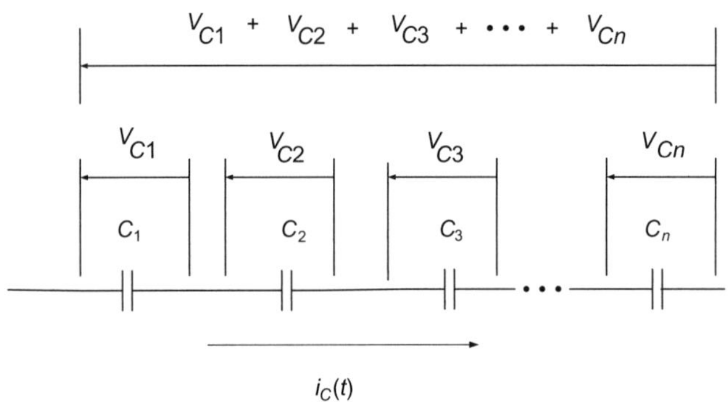

Figure 1.21 Capacitors in series.

capacitor equals to the DC source voltage whose positive terminal is placed across the first series capacitor free terminal and its negative terminal connected to the last capacitor free terminal:

$$V_{BATT} = \mathrm{V}_1 + \mathrm{V}_2 + \mathrm{V}_3 + \ldots + \mathrm{V}_\mathrm{n}. \tag{1.79}$$

We are now interested in finding the series equivalent value of all n capacitors connected in series. For each capacitor we have that

$$V_1 = q/C_1; V_2 = q/C_2; V_2 = q/C_3; \ldots; \quad ; V_n = q/C_n, \tag{1.80}$$

since all have the same amount of charge q. Plugging Equation (1.80) into Equation (1.79) we get

$$\begin{aligned} V_{BATT} &= V_1 + V_2 + V_3 \ldots + V_n \\ &= q/C_1 + q/C_2 + q/C_3 + \ldots + q/C_n \\ &= q(1/C_1 + 1/C_2 + 1/C_3 + \ldots + 1/C_n). \end{aligned} \tag{1.81}$$

From above we know that the series equivalent capacitance $C_{series\text{-}equiv}$ has the total voltage V_{BATT} across it and charge q within it. Thus, from Equation (1.81) we have

$$1/C_{series\text{-}equiv} = (1/C_1 + 1/C_2 + 1/C_3 + \ldots + 1/C_n) \tag{1.82}$$

or

$$\frac{1}{C_{series\text{-}equiv}} = \sum_{i=1}^{n} \frac{1}{C_i}. \tag{1.83}$$

Example 1.17 Capacitors in series: Given three capacitors in series, $C_1 = 2\ \mu F$, $C_2 = 3\ \mu F$, and $C_3 = 6\ \mu F$, find the series equivalent capacitance.

Solution to Example 1.17

Applying Equation (2.83) we obtain that

$$1/C_{series\text{-}equiv} = 1/2 + 1/3 + 1/6, \text{ which leads to } C_{series\text{-}equiv} = 1\ \mu\text{F}.$$

1.5.4 Capacitors in Parallel

Capacitors are said to be in parallel when they are in a circuit where the same voltage is applied to their terminals. Figure 1.22 depicts n capacitors in parallel.

A voltage source is connected across the n capacitors in parallel. All n capacitors have the exact same voltage applied across their terminals.

For each capacitor we have that

$$V = q_1/C_1 = q_2/C_2 = q_3/C_3 = \ldots = q_n/C_n. \tag{1.84}$$

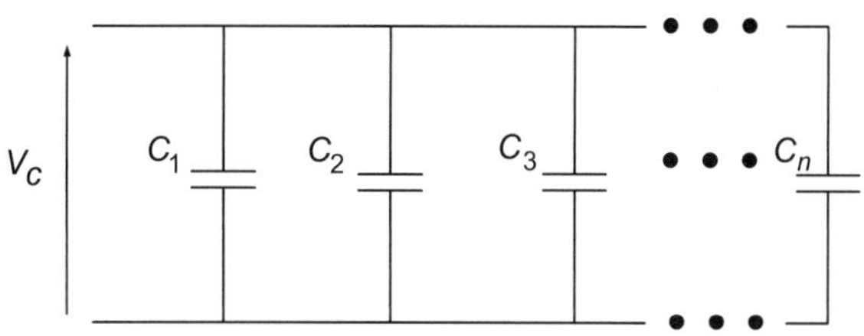

Figure 1.22 Capacitors in parallel.

Since all have the same voltage across, a parallel equivalent capacitance $C_{parallel\text{-}equiv}$ such that its charge equals to the sum of the charges of each one of the capacitors in parallel, exists. Thus,

$$q_1 + q_2 + q_3 + \ldots + q_n = q_{parallel\text{-}equiv} = C_{parallell\text{-}equiv} V, \tag{1.85}$$

and since

$$q_1 = C_1 V;\, q_2 = C_2 V;\, q_3 = C_3 V;\ldots \text{and } q_n = C_n V. \tag{1.86}$$

Plugging Equation (1.86) into Equation (1.85),

$$C_1 V + C_2 V + C_3 V + \ldots + C_n V = C_{parallell\text{-}equiv} V, \tag{1.87}$$

which leads to

$$C_{parallell\text{-}equiv} = C_1 + C_2 + C_3 + \ldots + C_n. \tag{1.88}$$

Or in a more compact form,

$$C_{parallell\text{-}equiv} = \sum_{i=1}^{n} C_i. \tag{1.89}$$

Example 1.18 Given a 10 nF, 0.1 μF, and 140 pF capacitor, calculate the total parallel equivalent capacitance of all three capacitors in parallel.

Converting all capacitances to pF, we obtain
10 nF = 10,000 pF and 0.1 μF = 100,000 pF.
Applying Equation (1.89),

$$C_{parallell\_equiv} = 10{,}000 + 100{,}000 + 140 = 110{,}140 \text{ pF}.$$

1.5.5 Energy Stored in a Capacitor

The electrical energy is

$$\begin{aligned} Energy = &(capacitor\ average\ voltage) \times \\ &(average\ current\ through\ the\ capacitor) \times time. \end{aligned} \tag{1.90}$$

From Equation (1.75),

$$C = qV \tag{1.91}$$

and

$$q = It = I_{average}t, \tag{1.92}$$

where I is the average current through the capacitor, $I_{average}$.

From Equations (1.91) and (1.92) we obtain the average value of current through the capacitor:

$$I_{average} = CV/t. \tag{1.93}$$

When a capacitor is charged with a constant current, the voltage across it grows linearly with time from zero to V volts, so that the average voltage across the capacitor is

$$V_{average} = \frac{1}{2}V. \tag{1.94}$$

Plugging Equations (1.93) and (1.94) into Equation (1.90), the energy stored in the capacitor is

$$Energy_{capacitor} = \frac{1}{2}CV^2, \tag{1.95a}$$

where C is the capacitance value in farads, V is the voltage across the capacitor in volts, and the energy is in joules.

Example: Energy Stored by a Capacitor

A 100 μF capacitor has been connected to a 400-V DC source for a very long time. What amount of energy will the capacitor be holding upon removal of the 400-V source?

From Equation (1.95a),

$$Energy = \frac{1}{2}CV^2$$

Thus,

$$Energy = \frac{1}{2} \times 100 \times 10^{-6} \times 400^2 = 8\text{ J}.$$

1.5.6 Real Capacitor Parameters and Capacitor Types

Real capacitors have the following parameters that characterize them:

C, their capacitance expressed in farads, this is the most important parameter of a capacitor; like the resistance is to a resistor.

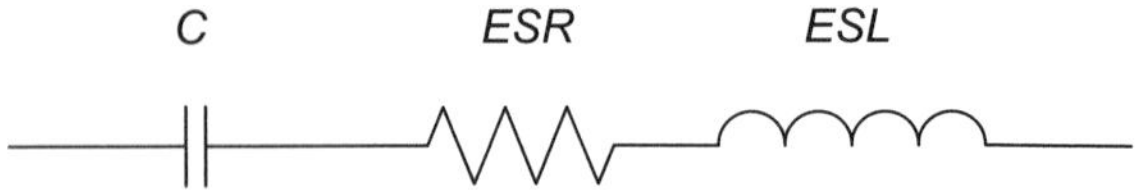

ESR = Equivalent Series Resistor

ESL = Equivalent Series Inductance

Figure 1.23 Simple model of a real capacitor.

The maximum constant voltage that can be applied to a capacitor dielectric is what capacitor manufacturers call the capacitor voltage rating. Depending on the type of capacitor dielectric the voltage can be from a few volts to hundreds of volts.

Some dielectric materials used are, for example, ceramics, mica, or polystyrene.

A real or physical capacitor does not behave entirely like a capacitor; it also has a small value of resistance in series referred to as the equivalent series resistance (ESR). This ESR is due to the fact that the dielectric is not a perfect insulator and inevitably has some resistance. Capacitors also have some unavoidable stray inductance because they have leads or terminals that have a small non-zero inductance. Figure 1.23 depicts a possible model used to describe the imperfections of a real capacitor.

The ESR accounts for the non-zero resistance of the dielectric material while the equivalent series inductance (ESL) mostly accounts for the inductive parasitic effects that the capacitor leads have. Of course other electrical capacitor models are possible. Capacitor manufacturers sometimes indicate the capacitor model they use for the data sheet specifications. Some capacitor manufacturers specify a *dissipation factor DF in %*. DF is defined as

$$DF = \tan\delta(\text{loss angle}) = \text{ESR}/X_C = 2\pi fC \times \text{ESR}. \qquad (1.95b)$$

In Equation (1.95b), X_C is the absolute value of the capacitive reactance, which equals $1/2\pi fC$, where f is the frequency of the sinusoidal signal applied to the capacitor. Chapter 2 will cover capacitive reactance in more detail.

Table 1.8 lists some of the most popular types of capacitors, their dielectric, and some of their electrical characteristics.

Capacitor Component Selection To select capacitors, we need to start by finding out their key purpose in the application. Let us assume for simplicity that we look at only three basic kinds of applications: bulk decoupling, high frequency decoupling, and precise timing control. Bulk decoupling requires large capacitance values; generally, the exact value is not as important as the fact of obtaining large amounts of capacitance. Moreover, these capacitors need to operate with frequencies that are typically well under 100 kHz.

Table 1.8 Some capacitor types

Capacitor Type	Capacitance Range	Operating Voltage Range (V)	Temp. Range (°C)	Tolerance (%)	Typical Uses	Polarized (Yes/No)
Aluminum electrolytic	From low μF to ~68,000 μF	~16 to ~450	–40 to ~85	–10 to 150	Bulk decoupling	Yes
Tantalum electrolytic	From ~2.7 pF to ~1500 μF	~2 to ~125	–55 to ~105	–70 to ~20	Bulk decoupling	Yes
Poscap	4.7 μF to ~1500 μF	~2 to ~25	–55 to ~125	±20	Bulk decoupling, low ESR	Yes
Oscon	~10 μF to ~2700 μF	~2 to ~25	–55 to ~105	±20	Bulk decoupling, low ESR	Yes
Multilayer chip ceramic (MLCC) and ceramics	~1 pF to ~1 μF	10 to 700	–55 to 125, –55 to 150	±1, ±2, ±10, ±20	High frequency decoupling	No
Mylar	1 nF to 0.47 μF	50 to 630	–20 to 75	±2 to ±20	Timing	No
Polyester	1 nF to 10 μF				Timing	No
Polycarbonate	1 nF to 1 μF; 1 μF to 50 μF;	~30 to ~400	–55 to 125	±1, ±2, ±5, ±10, ±20	Timing	No
Polystyrene		various	up to 85	±1, ±2.5, ±5, ±10, ±20	Timing	No
Glass case, hermetically sealed with glass dielectric					Ultrareliable, ultrastable, and resistant to nuclear radiation. Typically used in space missions.	No
Oil impregnated paper	0.1 μF to 20 μF	100 to 10,000			Older technology for industrial controls	No

Practically speaking, one *farad* is an extremely large unit of capacitance. Large capacitance values are in the several thousands of microfarads (μF) and up.

The capacitors' ESR is usually not as important as it is to obtain a huge amount of capacitance for the bulk capacitance application. The operating voltage at which the capacitor will be subjected to in the application should, under no circumstances, be more than about 50% of the capacitor-rated voltage for good reliability. For example, if a number of capacitors will be connected in parallel across a 5-V source, the capacitors must be rated at a voltage of at least 10 V or more. The higher the voltage rating of the capacitor, the bigger the capacitor usually is, so if volume or real estate on the printed circuit board is a prime factor, a very large voltage-rated capacitor may not be suitable for such application. On the other hand, the closer the capacitor voltage rating is to the voltage that the capacitor will be withstanding (5 V in our example), the shorter will be the long-term reliability of the capacitor. As a good rule of thumb, people pick at least a rated voltage that is twice the maximum voltage at which the capacitor will be subjected to. Naturally, the higher is such voltage, the harder it is to obtain a more generous voltage margin. As another example, in electronic circuits, a very common voltage used is 3.3 V. Capacitance values of a few pF are easily obtained with voltage ratings of 50 V or higher. Microfarad-valued capacitors are easily obtained in 10 V, 16 V, and some higher voltage ratings. But when the capacitance values are many hundreds or even thousands of microfarads, the size of capacitor becomes very large with a high voltage rating such as 100 V and above.

High-frequency decoupling capacitors do not necessarily require as large a capacitance value, but they require good behavior of the capacitors at higher frequencies. High frequencies for decoupling capacitor mean frequencies of many thousands of kilohertz and above. The capacitors need to preferably be more stable in value and their ESR needs to be much smaller than those of capacitors used for bulk decoupling. Finally, in timing applications, capacitors are used as analog timing elements. It is required for the capacitance to be as accurate in value as possible; their capacitance value should also be stable under temperature variations, aging of the component, and changing operating conditions. And finally, it is also required that the ESR of the capacitor be small. This produces capacitor smaller losses and also plays a role in the capacitor accuracy characteristics.

1.6 INDUCTORS

An inductor mainly consists of a copper wire wound around either a nonmagnetic core (air-core inductor) or a magnetic core. The core is the physical medium that holds the copper wire in place in relatively large inductors. Most importantly, the iron-core of an inductor allows one to obtain a much higher inductance per unit volume. For example, assume two inductors of the same dimensions, same wire gauge, same number of turns, and same volume. One has an air-core and the other one has an iron core. The iron-core inductor inductance value will be thousands of times the inductance of the same inductor built with an air-core. This directly depends on the relative permeability (μ_R) of the iron-core ferromagnetic material. We will cover more on magnetism and permeability of materials later in this chapter. Small air-core inductors generally do not need any media to hold the copper turns since they usually have a handful of turns. A thin layer of enamel electrically isolates the wound wire. The purpose of the insulating enamel is to keep multiple coil turns and multiple layers of turns from short-circuiting each other. Inductors used in power applications typically require higher inductance values than those inductors used in radio frequency (RF) applications. In RF, usually the inductor is used for the purpose of building a high-frequency tank circuit or tuned circuit and high frequencies (usually many megahertz) drive down the size of the inductor. On the other hand, on power applications, the high current requirements and the relatively low switching frequencies of the switching power supplies (in the order of hundreds of kilohertz) drive the size of the inductors up.

1.6.1 Magnetism

There are some minerals that have magnetic properties. Figure 1.24 shows lines of force created by the magnetic field produced by a naturally magnetized bar. The two ends of the magnetic bar are referred to as the north and the south poles. Note that opposite poles produce attracting lines of force (Fig. 1.24a), and like poles produce opposing lines of force (Fig. 1.24b).

When an electric current flows through a conductor, such current produces a magnetic field in the surroundings of the conductor. Figure 1.25 depicts a current-carrying conductor that penetrates a piece of cardboard in a perpendicular direction to the cardboard surface. Iron filings that are arbitrarily placed on the cardboard and in proximity to the current-carrying conductor will become rearranged in the shape of concentric rings around the conductor.

What is the direction of the concentric magnetic force field lines around the conductor? Let us refer one more time to Figure 1.25. When the current flows from top to bottom, the concentric magnetic field lines will be in a clockwise direction looking at the concentric rings from the top of the cardboard. This is also referred to as the right-hand rule. Referring now to Figure 1.26, a current is flowing from below the surface of this sheet of paper toward the reader. Concentric lines of magnetic force are formed in the surroundings of the current on all the planes, which are perpendicular to the current flow. Assume that you embrace with your right hand the current-carrying conductor with the right-hand side thumb pointing upward or toward the viewer. The right thumb represents the current direction, and the other four fingers embracing the current-carrying conductor represent the direction of the magnetic field.

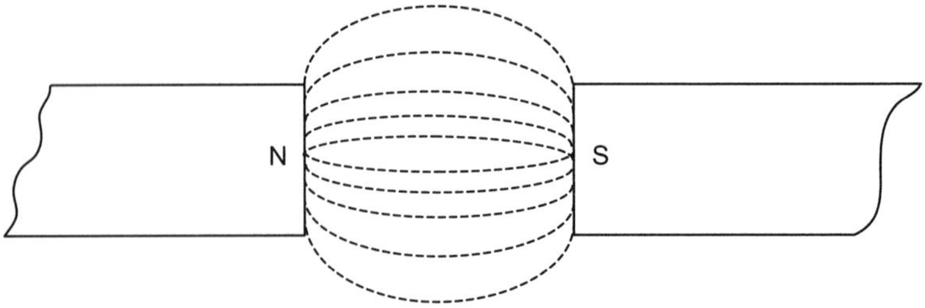

(a) *Magnetic Lines of Force Produced by Unlike Poles*

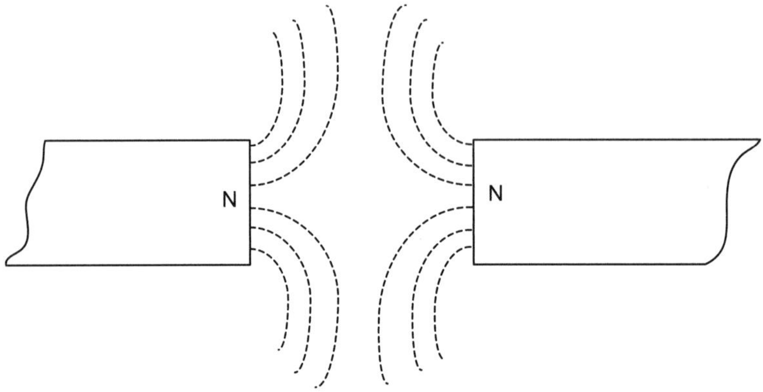

(b) *Magnetic Lines of Force Produced by Like Poles*

Figure 1.24 (a) Magnetic bar with opposite polarity poles produces attracting lines of force; (b) magnetic bar with same polarity poles produces repelling lines of force.

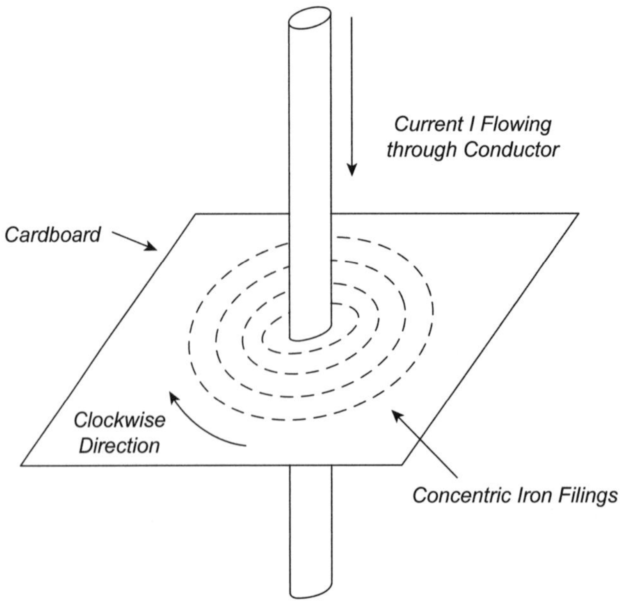

Figure 1.25 Magnetic field force produced around a current-carrying conductor.

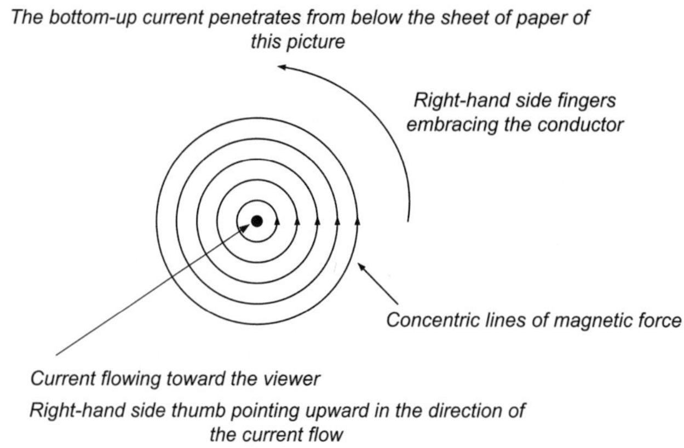

Figure 1.26 Right-hand rule.

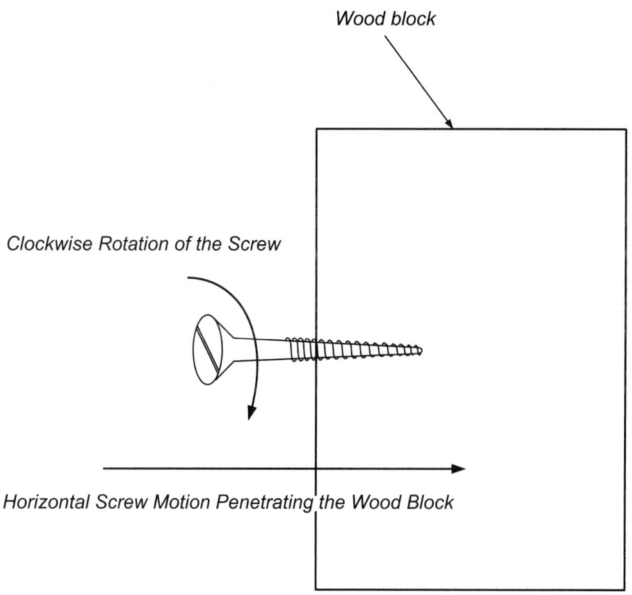

Figure 1.27 Right-hand screw rule.

The right-hand rule is analogous to the rotational and longitudinal motions of a screw into a piece of wood. Referring to Figure 1.27, when an observer looks at the head of the screw, and the screw is rotated in the clockwise direction, the screw length will penetrate the piece of wood. The analogy here is that the longitudinal displacement of the screw is analogous to the current flowing through a conductor, and the clockwise rotation of the screw is analogous to the direction of the magnetic field created around the current-carrying conductor.

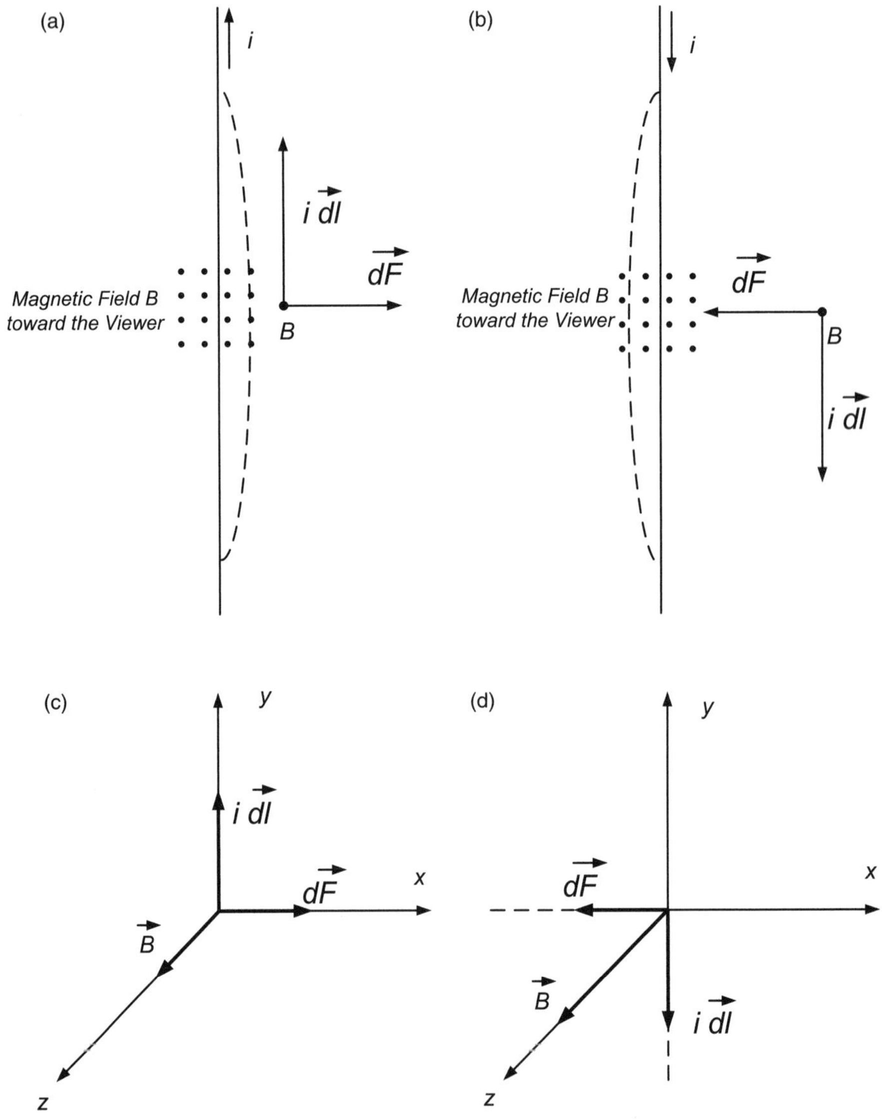

Figure 1.28 Magnetic forces produced by a current-carrying conductor: (a) current *i* flows from the bottom up; (b) current *i* flows from the top to the bottom; (c) isometric view of vectors *dF*, *idl*, *B* that corresponds to part (a); (d) isometric view of vectors *dF*, *idl*, *B* that corresponds to part (b).

Figure 1.28a shows a current "*i*" carrying conductor, the current flows from the bottom up, through a perpendicularly applied magnetic field ***B***. Magnetic field ***B*** is perpendicular to the surface of this sheet, coming toward the viewer. Small black circles represent the tips of the vector field arrows. A force on the conductor is exerted in the direction shown by ***dF***. Mathematically, the force equals the vector or cross product of ***idl*** with magnetic field ***B***. Figure 1.28a is associated with the Biot–Savart equation: ***dF*** = ***idl*** × ***B***, and it is graphically represented in Figure 1.28c. Note that the three vectors, ***dF***, ***idl***, and ***B*** are shown in three perpendicular planes in three dimensions. For Figure 1.28a, its corresponding vectors are shown in Figure 1.28c: ***dF*** along the x-axis; ***idl*** along the y-axis, and ***B*** along the z-axis. The operator × stands for cross product or vectorial product, a mathematical operation among vectors.

Similarly, Figure 1.28b shows a conductor with the current flowing from top to bottom; the magnetic field ***B*** is perpendicular to the sheet of paper on this book. Figure 1.28d is associated with Figure 1.28b which graphically expresses the vector or cross product of ***idl*** and magnetic field ***B*** to generate force ***dF***.

Biot–Savart's law is fundamental in the understanding of electric motors and it is expressed by

Biot–Savart Law

$$\boldsymbol{dF} = \boldsymbol{idl} \times \boldsymbol{B}, \tag{1.96}$$

where ***dF*** is a differential of force, ***idl*** is current times differential of length, and ***B*** is the magnetic field. The × sign is the mathematical symbol for the cross product, also referred to as the vector product.

The units of force (***dF***) are newtons, units of ***idl*** in amperes-meters, and ***B*** in webers (Wb). Biot–Savart's Law can be expressed in its equivalent form, substituting ***idl*** with

q v, where q is the charge of the particle and v its velocity. This yields

Lorentz Law

$$\boldsymbol{dF} = \boldsymbol{qv} \times \boldsymbol{B}. \tag{1.97}$$

In Equation (1.97) ***dF***, ***v***, and ***B*** are vectors. Finally, it is important to emphasize that the vector or cross products graphically shown in Figure 1.28c,d follow the convention adopted by the right-hand rule, discussed earlier.

Looking one more time at Figure 1.28a,b, the dotted line drawn conductor is the deflection that the wire will experience due to the forces produced by the given current and magnetic field directions.

1.6.2 Magnetic Field around a Coil

An alternating current (AC) flowing through an inductor or coil (Fig. 1.29) produces a magnetic flux from the contribution of each one of its turns. An AC voltage or current is one whose polarity alternates between a positive and a negative direction at some constant frequency. Deeper coverage of AC circuits is addressed in Chapter 2.

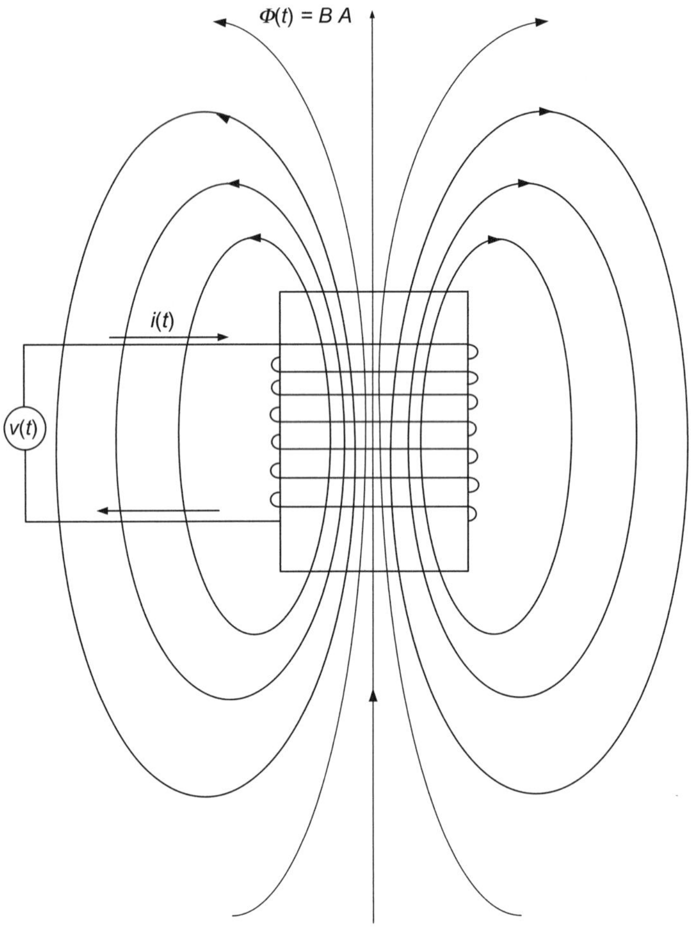

Figure 1.29 Iron-core inductor with AC excitation.

The larger the number of turns, the larger will be the magnetic flux. The magnetization effect from each one of the coil turns is additive, as if each turn was a small magnet. The contribution of all turns produces a magnetic field (B), also called the magnetic induction field B or flux density B. Flux density for a magnetic field perpendicular to a cross-sectional area is

$$B = \frac{\phi}{A}, \tag{1.98}$$

where B is the magnetic field or flux density in teslas (T) or weber/m^2 (Wb/m^2), ϕ is magnetic flux in webers, and A in m^2 is the cross-sectional area through which the lines of flux travel through.

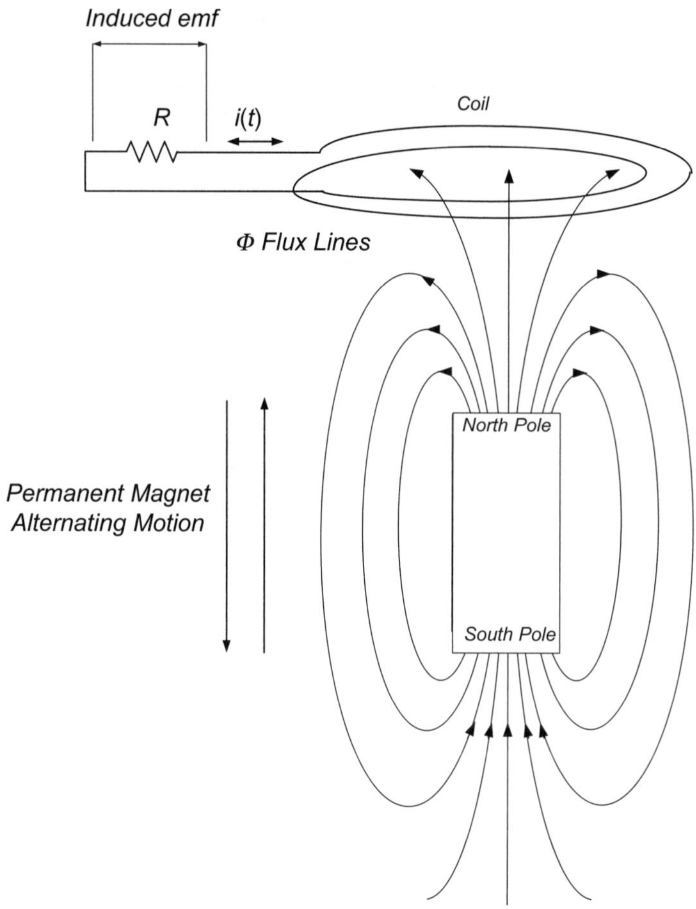

Figure 1.30 Magnetic induction produced by movement of permanent magnet bar.

Figure 1.30 depicts a natural permanent magnet bar in proximity to an air-core coil. A back and forth motion of the permanent magnet causes the flux density to vary with respect to time. This time, varying flux induces an *electromotive force* (*emf*) on the upper coil of Figure 1.30. An important difference between an AC-induced magnetic field and the magnetic field of a permanent magnet is that the permanent magnet field is constant with respect to time when the bar is stationary, and in such case, we can call it a DC field. The AC-induced field B is alternating with respect to time (Fig. 1.29). The AC current produced by the induced *emf* (Fig. 1.30) is produced by a permanent magnet because its flux is varied with respect to time by moving the magnetic bar back and forth.

Referring again to the iron-core inductor of Figure 1.29, if we assume now that the iron-core now has an air-core of the same dimensions, there will also be a time-varying magnetic flux upon a current flowing through the coil, but the intensity or strength of such magnetic flux will be considerably smaller. How much smaller depends on the specifics of the iron material and the air-core. The reason is that magnetic flux or magnetic line forces travel much more easily through a ferromagnetic material than through air. The property or parameter that characterizes magnetic materials is called *permeability*. In practical terms, *relative permeability* is more frequently used, and relative permeability is defined as the magnetic material permeability divided by the permeability of vacuum or free space. Relative permeability is dimensionless. Permeability of air is close to that of vacuum.

$$\mu = \mu_o \cdot \mu_r. \tag{1.99}$$

The permeability of vacuum also referred to as the permeability of free space is a physical constant, and it was determined experimentally to be

Permeability of Free Space

$$\mu_o = 4\pi 10^{-7}\ \text{Wb/A} \cdot \text{m}. \tag{1.100}$$

The relative permeability of free space is 1. The relative permeability of magnetic materials ranges from as little as 10 to as much as several hundreds of thousands. For example, the permeability of 3–6% FeSi ranges from 1000 to 10,000.

1.6.3 Magnetic Materials and Permeability

In relation to their capacity of allowing the passage of magnetic lines of force, materials can be classified into four categories:

- Nonmagnetic materials like vacuum, air, wood, paper, and plastic. These materials have no effect on the passage of magnetic lines of force.
- Diamagnetic materials which show a small opposition to magnetic lines of force. For most practical purposes are nonmagnetic materials. Examples are copper and silver. Their $\mu < \mu_o$ or their $\mu_r < 1$.
- Paramagnetic materials which somewhat assist the passage of magnetic lines of force. Examples are aluminum and platinum. Their permeability is $\mu > \mu_o$ or their relative permeability is $\mu_r > 1$.
- Ferromagnetic materials, sometimes simply called magnetic materials, which greatly assist the passage of magnetic lines of force through them. Some magnetic materials are iron, nickel, cobalt, steel (an iron alloy), and ferrites (ceramic composite materials). Ferromagnetic materials' permeability is usually much larger than that of free space μ_o. μ_r, the relative permeability of magnetic materials is usually several thousands

or hundreds of thousands larger than the relative permeability of free space. Note: By definition the relative permeability of free space or air is 1, because we are referring the relative permeability of free space to its permeability.

1.6.4 Electromagnetic Induction and Inductor Current–Voltage Relationship

We studied that an AC current induces a magnetic field. The converse is also true. An alternating magnetic field produces an AC current.

Michael Faraday* experimentally discovered electromagnetic induction. Given an electrical setup like the one shown in Figure 1.30, if we move a permanent magnet back and forth over the coil, an electromagnetic force (or *emf*) is generated across the coil terminals, and an electric current will be generated in the circuit with the coil and the resistor. It is very interesting to note that there is no physical or electrical contact between the magnetic bar and the coil. Alternatively, an *emf* can be generated by switching on and off the switch in the primary circuit as shown by Figure 1.31. The switching action will cause a time-varying magnetic flux produced by the iron-core inductor. This magnetic flux will be magnetically coupled onto the stationary secondary winding coil and an *emf* will be produced across the secondary inductor. Such *emf* will generate a current that will flow through resistor R.

Electromagnetic induction allows the generation of an electromagnetic force by the net movements of charges with respect to time.

In Figure 1.31, the opening and closure of the switch causes a time-varying current in the primary side of the circuit. This varying current produces a magnetic flux in the primary circuit loop. The varying magnetic flux due to the current in the primary coil induces an *emf* on the secondary coil. This *emf* produces a current in the secondary coil. The induced *emf* in the secondary coil is equal to the rate at which the flux changes with respect to time or in mathematical form,

$$emf = -d\Phi/dt. \tag{1.101}$$

Equation (1.101) is referred to as the Faraday–Lenz law, where Φ is the magnetic flux and t is the time variable. $d\Phi/dt$ is the derivative of Φ with respect to time.

The minus sign in front of $d\Phi/dt$ means that the induced *emf* opposes the change of flux. In other words, the induced current has a direction such that the magnetic field due to the current opposes the change in magnetic flux induced by the current in the primary coil.

* Michael Faraday was an English physicist and chemist (1791–1867). He is most known for having discovered electromagnetic induction and the laws of electrolysis.

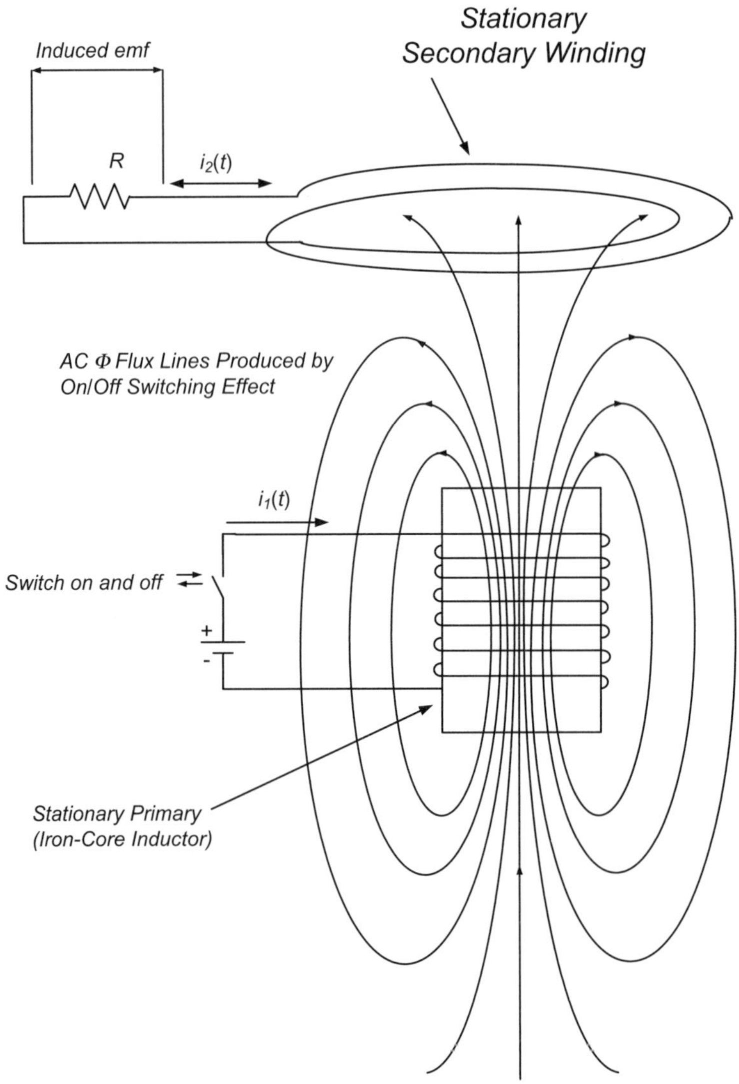

Figure 1.31 Magnetic induction generated electromotive force (*emf*) using an on/off switch.

The inductance (L) of a single turn coil is defined as

$$L = \Phi / i. \tag{1.102}$$

When the coil has N number of turns, the inductance L becomes

$$L = N\Phi / i. \tag{1.103}$$

In Equation (1.103), $N\,\Phi$ is defined as the magnetic flux linkage. The inductance of a coil is a measure of the flux linkage per unit of current. The unit of inductance is defined as a henry (H).

Rearranging terms in Equation (1.103), we obtain

$$N\Phi = Li. \tag{1.104}$$

Differentiating Equation (1.104) yields

$$d(N\Phi)/dt = Ldi/dt. \tag{1.105}$$

Since N, the number of inductor turns is constant, Equation (1.105) becomes

$$Nd\Phi/dt = Ldi/dt, \tag{1.106}$$

where $N\,d\Phi/dt$ is the voltage drop across the inductor L or $v_L(t)$. The magnitude of the voltage across an inductor depends on the rate of change of the current flowing through the inductor with respect to time.

Then, the fundamental voltage current relationship of an inductor is

$$v_L(t) = Ldi/dt. \tag{1.107}$$

The minus sign from Lenz law is dropped, since we are only interested in the magnitude of the voltage across the inductor. In circuit analysis, when a current enters the inductor, the voltage drop across the inductor has the polarity and direction shown based on the direction of the current as depicted in Figure 1.32.

Example 1.19 Determine the voltage waveform across an inductor knowing that its current waveform is 2 A DC and the inductance value is $L = 0.1$ mH.

Solution to Example 1.19

From Equation (1.107), the voltage–current relationship in an inductor is

$$v_L(t) = Ldi/dt.$$

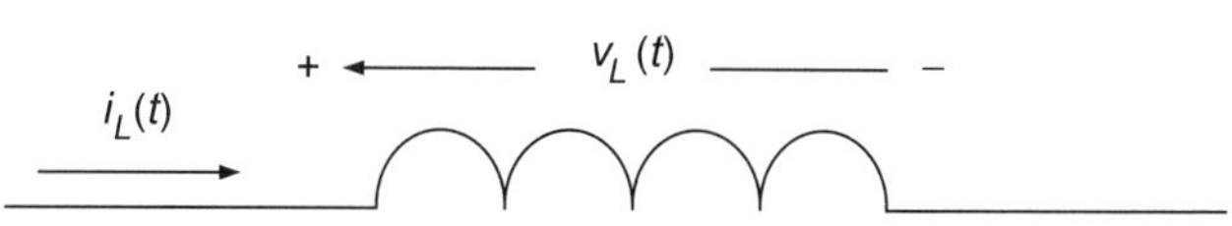

Figure 1.32 Inductor voltage drop polarity and current direction.

Since $i = 2$ A DC, the derivative of a constant is zero; thus, the voltage across the inductor is zero when a DC current flows through it. Idealized inductors have a zero-ohm winding resistance. Real inductors, however, have a winding resistance larger than zero ohm, whose value depends on the length of the wire used, its cross section, and the resistivity of the wire material used. (Refer again to Eq. 1.52 on resistivity.)

More generally, we can affirm that regardless of the inductance value, a DC current flowing through an inductor will develop 0 V of AC voltage across the inductor terminals.

Example 1.20 Determine the voltage developed across inductor terminals when a sinusoidal current that is a function of time (t) flows through it. The sinusoidal current is depicted in Figure 1.33, and three key parameters determine the sinusoidal current:

- Frequency (f): the inverse of the sinusoidal waveform period T.
- Phase angle (φ): the angle with respect to the origin of the time axis that the waveform is shifted.
- Amplitude: also called the peak value of the waveform.

Figure 1.33 depicts a sinusoidal current waveform that conforms to Equation (108):

$$i(t) = I_{peak} \sin(2\pi f t + \varphi). \tag{1.108}$$

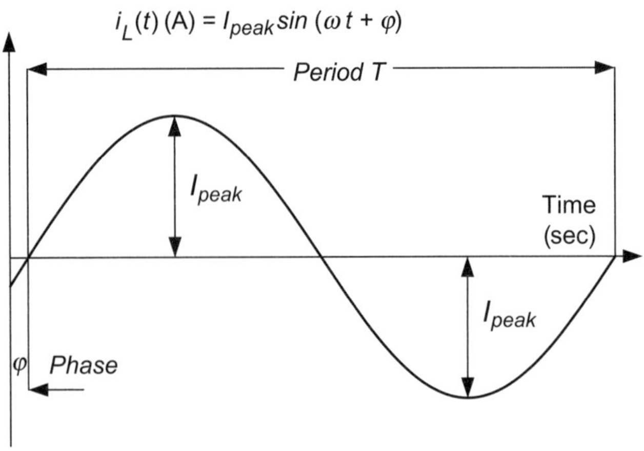

Figure 1.33 AC current waveform.

For this example we assume that

$$I_{peak} = 1\text{ A}, f = 1000\text{ Hz, and } \varphi \text{ is zero degrees.} \tag{1.109}$$

Where the unit of peak current is the ampere, the unit of frequency is second^{-1} (s^{-1}), also called hertz and abbreviated Hz. The unit of phase angle is either degrees or radians. When the phase is expressed in degrees, the "°" has to be explicitly shown next to the number of degrees. If radians are used, then no units are indicated next to the phase angle in radians. The radian is considered to be dimensionless. From high school geometry, let us remember that 2π radians or simply 2π equals 360°. So, for example, a 45° angle equals to π/4 (i.e., π/4 radians).

Rewriting Equation (1.108) with the parameters given by Equation (1.109) leads to

$$i(t) = 1\sin(2000\pi t + 0°)\text{ A}. \tag{1.110}$$

This can be simply stated as

$$i(t) = \sin(2000\pi t)\text{ A}. \tag{1.111}$$

The argument of the sinusoidal waveform in Equation (1.111), $2\pi ft$ is also referred to as ωt; where ω (the Greek letter omega) is called the sinusoidal waveform angular frequency or pulsation. Its units are hertz or s^{-1}. In some electrical engineering literature, the units of ω are also referred to as radians per second (rad/s). Since the radian is a dimensionless unit, hertz and rad/s are basically the same thing.

Figure 1.33 depicts the waveform described by Equation (1.111) with a zero phase angle.

Solution to Example 1.20

$$v_L(t) = L\,di/dt. \tag{1.112}$$

The voltage across the inductor produced by the sinusoidal current given by Equation (1.112) is easily calculated plugging Equation (1.111) into Equation (1.112).

Thus,

$$\begin{aligned} v_L(t) &= L\frac{d}{dt}[\sin(2000\pi t)] \\ &= v_L(t) = 2000\pi L\cos(2000\pi t). \end{aligned} \tag{1.113}$$

Figure 1.34 shows both the sinusoidal current through the inductor and the sinusoidal voltage developed across the inductor terminals. It is important to note that the voltage waveform leads the current waveform by 90°. Note: Cosine and sine waveforms are both referred to as sinusoidal waveforms in a general sense.

1.6.5 Inductors in Series

When inductors are connected in series, and there is no mutual coupling among the magnetic fields produced by each inductor in the series, the current flowing through the series of inductor is the same. Upon the series of inductors being excited by a time-varying current, each inductor develops a voltage across its terminals given by Equation (1.107), repeated here for the reader's convenience:

$$v_L(t) = Ldi/dt.$$

Referring to Figure 1.35, the current in series with the inductors is the same, and each inductor develops a voltage across its terminals given by Equation (1.107). This voltage may be different for every inductor since individual inductances may be different.

Given L_1, L_2, L_3, . . . and L_n, current $i(t)$ develops the following voltages across each inductor:

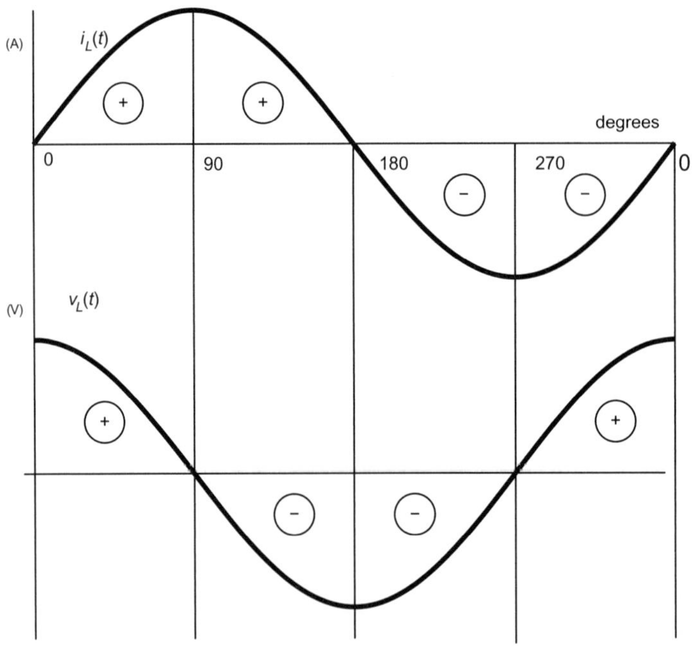

Figure 1.34 Sinusoidal current and voltage on an inductor.

Figure 1.35 Inductors in series.

$$v_{L1}(t) = L_1 di/dt \tag{1.114}$$

$$v_{L2}(t) = L_2 di/dt \tag{1.115}$$

$$v_{L3}(t) = L_3 di/dt \ldots \tag{1.116}$$

and

$$v_{Ln}(t) = L_n di/dt. \tag{1.117}$$

The sum of each inductor generated *emf* equals to the sum of the total *emf* applied across the complete series of inductors. Mathematically,

$$\begin{aligned} v_{total}(t) &= v_{L1}(t) + v_{L2}(t) + v_{L3}(t) + \ldots + v_{Ln}(t) \\ &= L_1 di/dt + L_2 di/dt + L_3 di/dt + \ldots + L_n di/dt, \end{aligned} \tag{1.118}$$

and since the current through the series of inductors is the same for all inductors, then

$$v_{total}(t) = (L_1 + L_2 + L_3 + \ldots + L_n) di/dt. \tag{1.119}$$

Thus, we can say that the equivalent series inductor equals the sum of each individual inductance in series:

$$L_{series\text{-}equivalent} = L_1 + L_2 + L_3 + \ldots + L_n. \tag{1.120}$$

Figure 1.35 depicts a series of inductors. Note that the current in the circuit is the same for every inductor.

Example 1.21 Inductors in series: Given three inductors $L_1 = 25$ nH, $L_2 = 75$ nH, and $L_3 = 50$ nH, determine the series equivalent inductor of L_1, L_2, and L_3. Assume that the magnetic field produced by each inductor does not couple with the magnetic field produced by any of the other inductors.

Solution to Example 1.21

From Equation (1.120) we can state that

$$L_{series\text{-}equivalent} = L_1 + L_2 + L_3. \tag{1.121}$$

Solving for the given values of inductance,

$$L_{series\text{-}equivalent} = 25\,\text{nH} + 75\,\text{nH} + 50\,\text{nH} = 150\,\text{nH}. \tag{1.122}$$

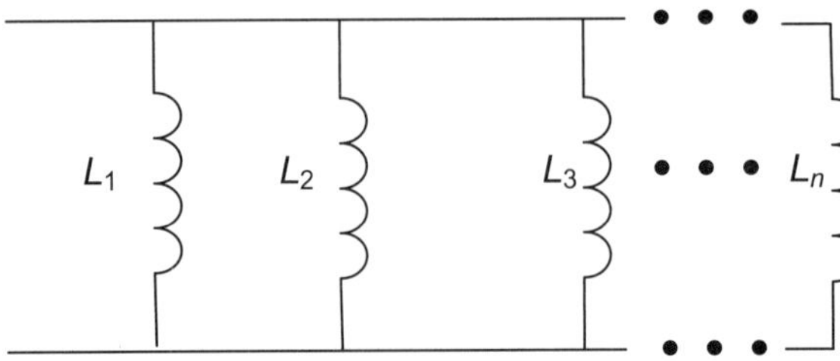

Figure 1.36 Inductors in parallel.

1.6.6 Inductors in Parallel

When inductors are connected in parallel, and there is no mutual coupling among the magnetic fields produced by each inductor in the parallel arrangement, the voltage across all the inductors in parallel is the same. Refer to Figure 1.36 for an arrangement of inductors in parallel. Upon the paralleled inductors being excited by a time-varying voltage, each inductor has across its terminals the same time-varying voltage. According to Equation (1.107), repeated here for convenience one more time,

$$v_L(t) = L di/dt.$$

Integrating the voltage across the inductor yields

$$i(t) = \frac{1}{L}\int v_L(t)dt. \tag{1.123}$$

Referring to Figure 1.36, the voltage is the same for all inductors, and since each inductance may be different (i.e., $L_1 \neq L_2 \neq \ldots \neq L_n$), the current through each inductor is

$$i_1(t) = \frac{1}{L_1}\int v_L(t)dt \tag{1.124}$$

$$i_2(t) = \frac{1}{L_2}\int v_L(t)dt \tag{1.125}$$

$$i_3(t) = \frac{1}{L_3}\int v_L(t)dt \ldots \tag{1.126}$$

$$i_N(t) = \frac{1}{L_N}\int v_L(t)dt. \tag{1.127}$$

From Equations (1.124) through (1.127), it can be seen that if we add the current of each of the inductors and name it $i_{total\text{-}parallel\text{-}equivalent\text{-}L}(t)$, there must exist a value of inductance that is equivalent to all of the inductances in parallel.

Thus, for

$$i_{total\text{-}parallel\text{-}equivalent\text{-}L}(t) = i_1(t) + i_2(t) + i_3(t) + \ldots + i_n(t). \tag{1.128}$$

There must exist a total parallel equivalent inductance such that

$$\begin{aligned} i_{total\text{-}parallel\text{-}equivalent\text{-}L}(t) &= \frac{1}{L_{parallel\text{-}equivalent}} \int v_L(t)dt \\ &= \frac{1}{L_1} \int v_L(t)dt + \frac{1}{L_2} \int v_L(t)dt \\ &+ \frac{1}{L_3} \int v_L(t)dt + \ldots + \frac{1}{L_n} \int v_L(t)dt. \end{aligned} \tag{1.129}$$

Now, since the voltage across each inductor is identical, we obtain from Equation (1.129) that

$$\frac{1}{L_{parallel\text{-}equivalent}} = \frac{1}{L_1} + \frac{1}{L_2} + \frac{1}{L_3} + \ldots + \frac{1}{L_n}. \tag{1.130}$$

Example 1.22 Given $L_1 = 5\ \mu\text{H}$ and $L_2 = 10\ \mu\text{H}$, determine the parallel equivalent inductor.

Solution to Example 1.22

Since

$$\begin{aligned} 1/L_{parallel\text{-}equivalent} &= 1/L_1 + 1/L_2, \\ L_{parallel\text{-}equivalent} &= (1/5\ \mu\text{H} + 1/10\ \mu\text{H})^{-1} = 3.33\ \mu\text{H}. \end{aligned} \tag{1.131}$$

Example 1.23 Given five identical inductors whose value is 45 nH, determine the parallel equivalent inductor of all five 45 nH inductors in parallel. Assume that the magnetic field produced by each inductor does not couple with the magnetic field produced by any of the other inductors.

Solution to Example 1.23

From Equation (1.130) we have that

$$1/L_{parallel\text{-}equivalent} = 1/L_1 + 1/L_2 + 1/L_3 + 1/L_4 + 1/L_5. \tag{1.132}$$

But since

$$L_1 = L_2 = L_3 = L_4 = L_5 = 45\,\text{nH}, \tag{1.133}$$

$$1/L_{parallel\text{-}equivalent} = (1/45\,\text{nH} + 1/45\,\text{nH} + 1/45\,\text{nH} + 1/45\,\text{nH} + 1/45\,\text{nH}), \tag{1.134}$$

from where it is immediate to find that

$$L_{parallel\text{-}equivalent} = 45\,\text{nH}/5 = 9\,\text{nH}. \tag{1.135}$$

1.6.7 Mutual Inductance

Given an N-turn inductor and an AC voltage source excitation across its terminals, we know from Faraday's law of induction that

$$Flux\ Linkage\ \lambda = N\Phi = Li(t) \tag{1.136}$$

and

$$v_L = Nd\Phi/dt = Ldi/dt, \tag{1.137}$$

where N is the number the inductor turns, Φ is the magnetic flux, and L is the self-inductance of the coil or simply the inductance. Refer to previously seen Figure 1.29 where the flux of the sole inductor is drawn. The term coil and inductor will be interchangeably used. However, an inductor is an ideal circuit element that only has inductance; a coil more commonly refers to a physical inductor, which is predominantly inductive, but it may also have some parasitic (usually undesirable) resistive and capacitive properties.

Applying the right-hand rule in Figure 1.37, we determine from the direction of the current flow the direction of the magnetic flux Φ. From Faraday's

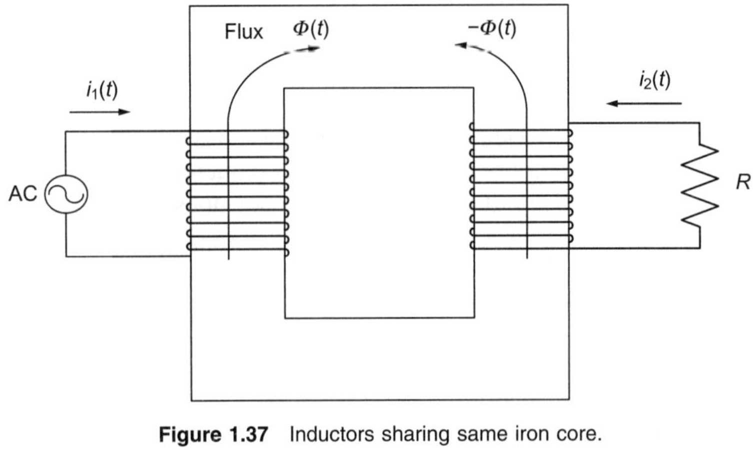

Figure 1.37 Inductors sharing same iron core.

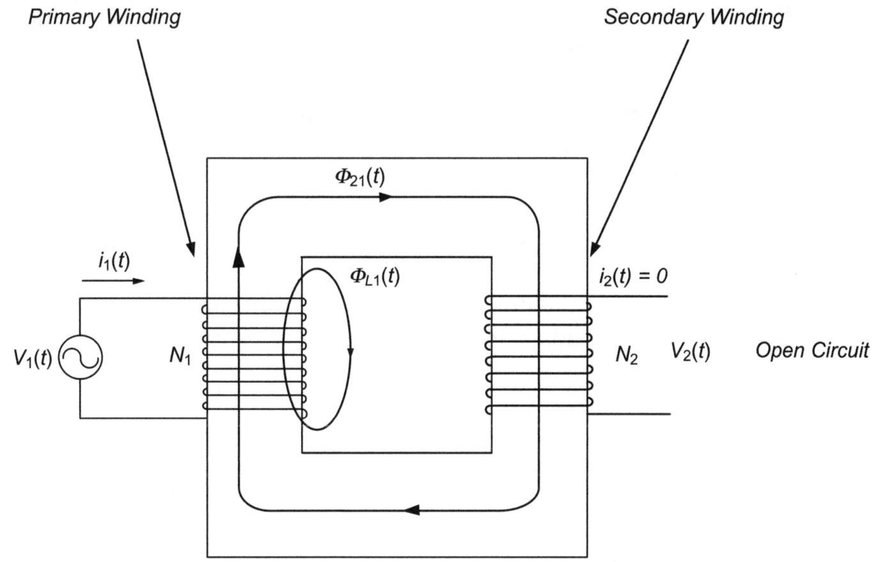

$\Phi_{21}(t)$ *Magnetic flux coupled to the secondary generated by the primary current.*

$\Phi_{L1}(t)$ *Magnetic flux in the primary generated by the primary current. This flux does not couple to the secondary, that is why it referred to as primary leakage flux.*

Figure 1.38 Magnetically coupled primary and secondary coils: primary excitation, open-circuited secondary.

law, a voltage is induced on the inductor only if the flux Φ varies with respect to time. This flux variation need not be a sinusoidal time-varying flux; it just needs to be a time-varying flux. If the voltage exciting the inductor is DC or constant with respect to time, no voltage is induced across the inductor, and the inductor behaves as a virtual short circuit to the DC source. Why is it a virtual short? Real inductors or coils have a finite amount of DC resistance due to the resistance of its winding. However, such resistance is relatively small, typically some small fraction of an ohm, and is thus a virtual short circuit to DC.

We will analyze the interaction of two magnetically coupled coils. One coil is referred to as the primary, and drawn on the left-hand side of the circuits; the right-hand side coil is referred to as the secondary in Figures 1.38 through 1.40. The primary is assumed to have N_1 turns and the secondary has N_2 turns. The coils are assumed to be physically located in close proximity and static with respect to each other. Three cases will be analyzed:

1. Primary is excited by a time-varying voltage source, secondary is open-circuited and there is no secondary current flow.
2. Primary is open-circuited, no primary current flows, and the secondary is excited by a time-varying voltage source.

3. Primary is excited by a time-varying voltage source, and current flows through the primary and secondary circuits.

Case 1 Let us analyze Case 1 by inspection of Figure 1.38.

The primary current $i_1(t)$ produced by voltage source $v_1(t)$ in Figure 1.38 generates a magnetic flux:

$$\Phi_{11} = \Phi_{Leakage1} + \Phi_{21}, \tag{1.138}$$

where $\Phi_{Leakage1}$ (Φ_{L1}) is the flux produced by primary current $i_1(t)$ that does not couple to the secondary inductor; it is also referred to as the leakage flux of the primary inductor. Φ_{21} is the flux produced by current $i_1(t)$ that couples between secondary inductor and the primary inductor. Φ_{21} is also called the mutual flux in the secondary due to the current in the primary. Since the secondary is an open circuit, there is no current in the secondary circuit; that is, $i_2(t) = 0$. Faraday's law gives the voltage induced in the secondary:

$$v_2(t) = d\lambda_2/dt = N_2 d\Phi_{21}/dt, \tag{1.139}$$

where λ_2 is the flux linkage on the secondary coil (refer to Figure 1.38).

If linearity* holds, then

$$N_2\Phi_{21} = M_{21}i_1(t), \tag{1.140}$$

where the term M_{21} is referred to as the mutual inductance between inductor 2 and inductor 1. The unit of mutual inductance is the henry, the same unit used for inductance and self-inductance. Now we can rewrite Equation (1.139) in terms of mutual inductance Equation (1.140) as follows:

$$v_2(t) = N_2 d\Phi_{21}/dt = M_{21}di_1(t)/dt. \tag{1.141}$$

The mutual inductance term M_{21} is a constant because the relative position of the primary and secondary inductors is fixed. The inductance or self-inductance of the primary winding is

$$L_1 i_1(t) = N_1\Phi_{11}, \tag{1.142}$$

* Linearity in magnetically coupled inductors holds when the inductors have an air-core. When the inductors have ferromagnetic cores, linearity holds when the cores are operated with currents well below from their core saturation (nonlinear) regions.

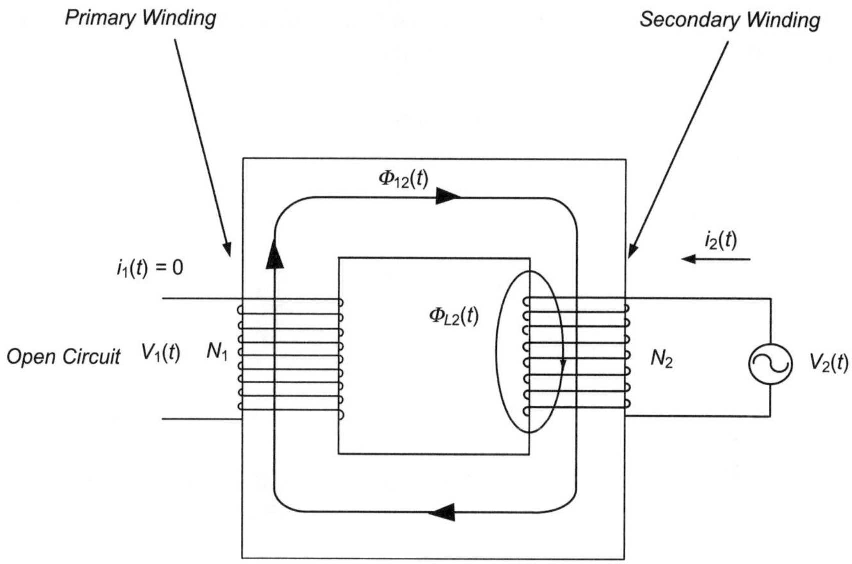

$\Phi_{12}(t)$ Magnetic flux coupled to the primary generated by the secondary current.

$\Phi_{L2}(t)$ Magnetic flux in the secondary generated by the secondary current. This flux does not couple to the primary, that is why is referred to as secondary leakage flux.

Figure 1.39 Magnetically coupled primary and secondary coils: secondary excitation and open-circuited primary.

where Φ_{11} is given by Equation (1.138).

Using Equation (1.137), the voltage on the primary is then

$$v_1(t) = L_1 di_1(t)/dt = N_1 d\Phi_{11}/dt. \tag{1.143}$$

Case 2 Let us analyze Case 2 by inspection of Figure 1.39.

We see that the primary is open-circuited and no current flows through this circuit, and the secondary is excited by a time-varying voltage source.

The secondary current $i_2(t)$ produced by voltage source $v_2(t)$ in Figure 1.39 generates a magnetic flux:

$$\Phi_{22} = \Phi_{Leakage2} + \Phi_{12}, \tag{1.144}$$

where $\Phi_{Leakage2}$ is the flux produced by secondary current $i_2(t)$ that does not couple to the primary inductor; it is also referred to as the leakage flux of the secondary inductor. Φ_{12} is the flux produced by current $i_2(t)$ that couples between primary inductor and the secondary inductor. Φ_{12} is also called the mutual flux in the primary due to the current in the secondary. Since the primary is open circuit, there is no current in the primary circuit; that is, $i_1(t) = 0$. The flux linkage of the primary circuit is

$$\lambda_1 = N_1 \Phi_{12}. \tag{1.145}$$

Using Equation (1.136) and Faraday's law lead to the voltage induced in the primary:

$$v_1(t) = N_1 d\Phi_{12}/dt. \tag{1.146}$$

If linearity holds then,

$$N_1 \Phi_{12} = M_{12} i_2(t), \tag{1.147}$$

where the term M_{12} is referred to as the mutual inductance between inductor 1 and inductor 2, due to current $i_2(t)$. The unit of mutual inductance is the henry, the same unit used for inductance and self-inductance. Now we can rewrite Equation (1.146) in terms of mutual inductance Equation (1.147) as follows:

$$v_1(t) = N_1 d\Phi_{12}/dt = M_{12} di_2(t)/dt. \tag{1.148}$$

The mutual inductance term M_{12} is a constant because the relative position of the primary and secondary inductors is fixed.
The inductance or self-inductance of the primary winding is

$$L_2 i_1(t) = N_2 \Phi_{22}, \tag{1.149}$$

where Φ_{22} is given by Equation (1.144).

Case 3 This is the most general case, that is, when there is nonzero current on both the primary and secondary circuits. We assume for this case that the primary is excited by a time-varying source, and a resistor or load in the secondary allows the induced voltage on the secondary to produce a current that flows through R.

From Equations (1.138) and (1.44), repeated here for the reader's convenience,

$$\Phi_{11} = \Phi_{Leakage1} + \Phi_{21}$$

$$\Phi_{22} = \Phi_{Leakage2} + \Phi_{12}.$$

Referring now to Figure 1.40, we find that the fluxes in the primary and secondary are respectively,

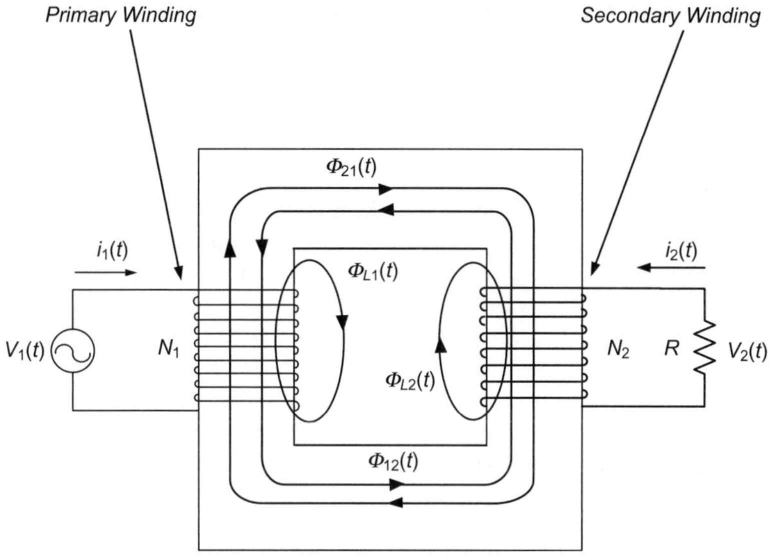

$\Phi_{12}(t)$ *Magnetic flux coupled to the primary generated by the secondary current.*

$\Phi_{21}(t)$ *Magnetic flux coupled to the secondary generated by the primary current.*

$\Phi_{L1}(t)$ *Magnetic flux in the primary generated by the primary current. This flux does not couple to the secondary, that is why it is referred to as leakage flux.*

$\Phi_{L2}(t)$ *Magnetic flux in the secondary generated by the secondary current. This flux does not couple to the primary, that is why it is referred to as leakage flux.*

Figure 1.40 Magnetically coupled primary and secondary coils. Primary excitation and secondary with resistive load.

$$\Phi_1 = \Phi_{Leakage1} + \Phi_{21} + \Phi_{12} = \Phi_{11} + \Phi_{12}. \tag{1.150}$$

$$\Phi_2 = \Phi_{Leakage2} + \Phi_{12} + \Phi_{21} = \Phi_{21} + \Phi_{12}. \tag{1.151}$$

Then the flux linkages for the primary and secondary are

$$\lambda_1 = N_1\Phi_{11} + N_1\Phi_{12}. \tag{1.152}$$

$$\lambda_2 = N_2\Phi_{21} + N_2\Phi_{22}. \tag{1.153}$$

Finally, differentiating the flux linkages, the complete primary and secondary voltages for the basic transformer are obtained:

$$v_1 = L_1 di_1(t)/dt \pm M_{12} di_2(t)/dt \tag{1.154}$$

$$v_2 = M_{21} di_1(t)/dt \pm L_2 di_2(t)/dt, \tag{1.155}$$

where $M_{12} = M_{21} = M$.

Equations (1.154) and (1.155) have been derived assuming that the windings directions are not known. It is not always possible to know the windings directions of a pair of mutually coupled inductors; consequently, the dot convention is used.

The Dot Rule for Coupled Inductors

It is not always possible or practical to know the directions of the windings of a pair of mutually coupled inductors. Dots are assigned in the following manner: pick one inductor, say the primary, and place a dot where the current to be injected enters the winding. Determine the flux created by such current using the right hand rule. The flux generated in the secondary inductor from Lenz law has to oppose the direction of the primary flux. Remember that the secondary current also has to meet with the right-hand rule with respect to the secondary flux. Now assume you would load the secondary, place the dot on the secondary terminal where this natural current leaves the secondary winding.

Summary of the dot rules:

1. When both primary and secondary currents enter (or leave) the dotted inductor terminals, the sign on the M (mutual inductance) terms shall have positive sign so that the equations are

$$v_1 = L_1 di_1(t)/dt + M_{12} di_2(t)/dt. \tag{1.156}$$

$$v_2 = M_{21} di_1(t)/dt + L_2 di_2(t)/dt. \tag{1.157}$$

2. When one current enters the dotted terminal of one inductor and leaves the dotted terminal of the other inductor, the signs on the M terms shall have negative signs so that the equations are

$$v_1 = L_1 di_1(t)/dt - M_{12} di_2(t)/dt. \tag{1.158}$$

$$v_2 = -M_{21} di_1(t)/dt + L_2 di_2(t)/dt. \tag{1.159}$$

The dots on the end of each coil are markings, typically made by the transformer manufacturer to indicate the relative polarities of the windings mutual voltages. Given that we find a dot on the primary and a dot on the secondary, the dot on the primary inductor indicates that a current entering this side of the inductor produces an induced voltage with its positive sign on the dotted side of the secondary inductor.

Figure 1.41 shows a transformer with its respective windings dots. In Figure 1.41 $v_1(t)$ produces a current into the dot, and induced voltage $v_2(t)$ produces a current with the direction shown by $i_2(t)$.

Figure 1.42 shows the dotted primary and secondary inductors. As usual, the primary current enters the dotted primary winding. The dotted secondary

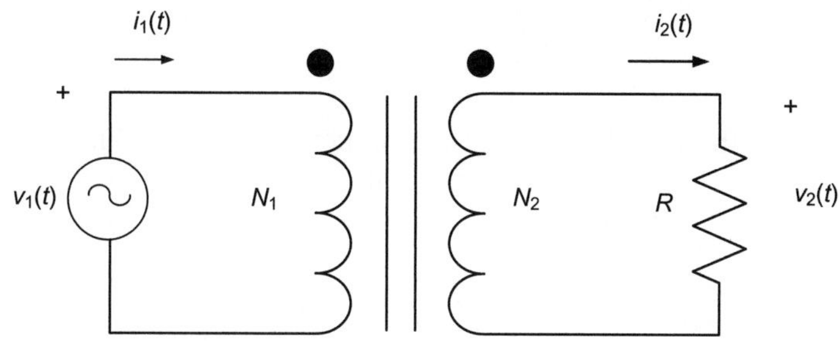

Figure 1.41 Dotted convention for two mutually coupled inductors (Case 1).

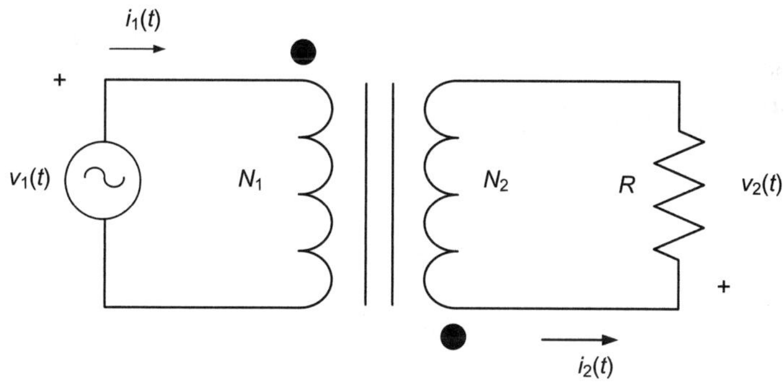

Figure 1.42 Dotted convention for two mutually coupled inductors (Case 2).

shows the positive polarity of the induced voltage and the direction of the secondary current [7].

1.6.8 Energy Stored by an Inductor

The energy held by an inductor between times t_0 and t_1 is given by

$$w_L(t) - w_L(0) = \int_{t0}^{t} v(t)i(t)dt, \tag{1.160}$$

where $v(t)$ is the voltage across the inductor, and $i(t)$ is the current through it. Since

$$v(i) = Ldi(t)/dt, \tag{1.161}$$

plugging Equation (1.161) in Equation (1.160) leads to

$$w_L(t) - w_L(0) = \int_{t0}^{t} Li(t)di(t) \tag{1.162}$$

$$w_L(t) - w_L(0) = \frac{1}{2} L i^2(t) \text{J}, \tag{1.163}$$

where $w_L(0)$ is the initial energy of the inductor, L is its inductance, and $i(t)$ the current flowing through it. In particular, for a pair of mutually coupled inductors (a transformer), the energy held by the transformer equals

$$w_L(t) - w_L(0) = \frac{1}{2} L_1 i_1^2(t) + \frac{1}{2} L_2 i_2^2(t) \pm M i_1(t) i_2(t) \tag{1.164}$$

where in Equation (1.164), the plus sign applies if both currents enter or leave the dotted marked inductor terminals, and it is minus if one current enters a dotted terminal, while the other current leaves its dotted terminal [7].

1.6.9 Inductor Nonlinearity

The magnetic flux $\Phi(t)$ of an inductor of inductance L is proportional to the current flowing through such inductor, provided that the core used does not get saturated by the amount of current flowing through the inductor winding:

$$\Phi(t) = L i_L(t). \tag{1.165}$$

Equation (1.165) remains linear if an air-core is used. Similarly, Equation (1.165) remains linear when a ferromagnetic core is used, and the current does not exceed the linear limits of the magnetic flux to current relationship. This means, as long as the ferromagnetic core does not get saturated. L is the constant of proportionality in Equation (1.165) and in most practical cases is time independent and strongly depends on the geometry of the coil winding and the type of core used.

Figure 1.43 depicts the nonlinear flux–current relationship that exists on a ferromagnetic core. The magnetic flux Φ, also equal to the magnetic inductance field B per unit area, maintains proportionality to the voltage applied to the inductor terminals. The magnetic field intensity H is proportional to the current flowing through the inductor times the number of turns.

Alternatively, some manufacturers present the flux-current ($\Phi - i$) characteristic curve instead of the B-H characteristic curve.

Figure 1.43 graphically depicts the hysteresis phenomenon. Let us start at $i = 0$ and $\Phi = 0$, the origin of coordinates; as current i is increased, flux Φ follows curve A until it reaches point $i = i_1$ and $\Phi = \Phi_1$. Note that the magnetic flux saturates the core at this point. Core saturation means that as the current continues to increase, the flux will no longer increase in a significant fashion. As current i decreases, tracing over curve B, flux Φ becomes zero at $i = -i_2$. If we continue pushing current in the minus direction toward $-i_3$, the core will eventually saturate (flattened curve at $i = -i_3$ and $\Phi = -\Phi_3$). Upon increasing the current toward the positive direction, the flux increases, but it is still negative. When the current reaches a value of i_4, the flux becomes zero, and it is no

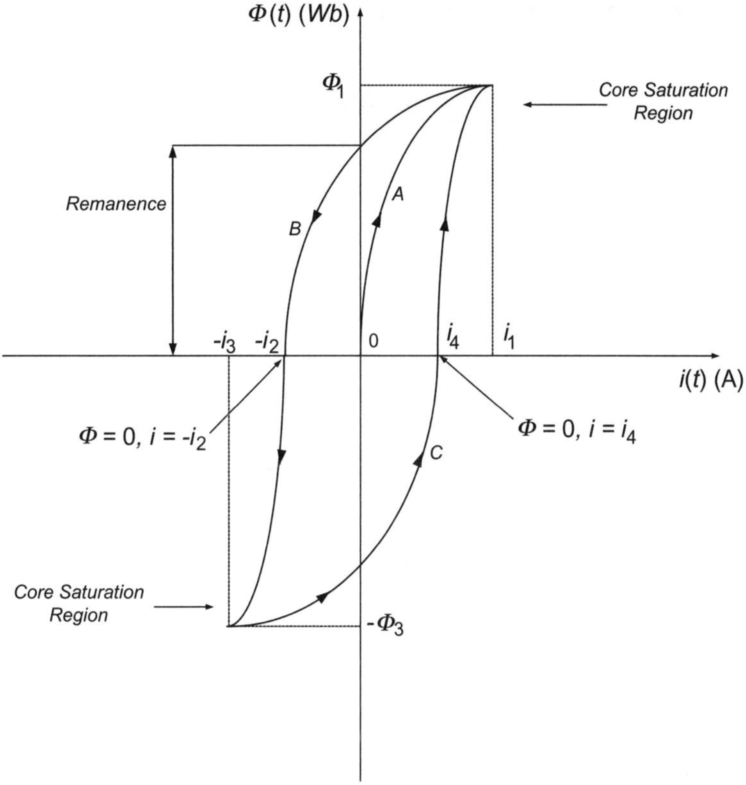

Figure 1.43 Magnetic characteristic (Φ-i) or B-H curve of an iron-core inductor.

longer negative (lower portion of curve C). If we continue to increase the current in the positive direction, curve C will finally meet point $i = i_1$ and $\Phi = \Phi_1$.

According to the best of our knowledge there are no close form equations describing the B-H ($\Phi - i$) characteristics of ferromagnetic materials. Core manufacturers empirically obtain such magnetic curves.

The slope of the B-H curve at each point determines the permeability μ of the ferromagnetic material:

$$\mu = \mu_R \mu_o = B/H.$$

When the relative permeability of the ferromagnetic material is almost constant for all points throughout the B-H curve, the material behaves linearly.

1.6.10 Inductor Component Selection

Inductor selection requires not only a good understanding of the desired circuit performance but also the data sheet information available on the inductor

Table 1.9 Typical printed circuit board inductor data sheet excerpt

Part Number	L ± 20%[a] (μH)	DCR (max) (Ω)	SRF (MHz)	I_{sat}[b] (A)	I_{RMS}[c] (A)
Value[d]	100	0.020	10	10	7

[a] Inductance tested at 100 kHz, 0.1 V_{RMS}.
[b] Inductance drop = 10% typ. at I_{sat}.
[c] For 40°C temperature rise typ. at I_{RMS}.
[d] All parameters tested at 25°C.

provided by the manufacturer. The purpose of this section is to present some general guidelines of the key parameters and factors that need to be taken into consideration when an inductor is selected.

For example, Table 1.9 shows an excerpt from a typical inductor data sheet. To use the data sheet parameters properly, one must have some understanding of what the parameters mean and how they were derived. Most manufacturers will not show all the performance parameters of their inductors under all possible and different sets of operational conditions. This would become practically prohibitive, since the testing and characterization cost of inductors would rise.

Following the entries of Table 1.9, after the part number entry, the inductance is provided. Inductance is the most basic parameter for selecting the inductor. This may be based on the energy storage capability required by the inductor or by the volt-second capacity, derived from equation $v_L = L\, di/dt$. Associated with the inductance value is the tolerance; it is common to see a wide spread tolerance, that is, ±20% is common.

The next parameter is the DC resistance (DCR) of the inductor wire. This resistance depends on the wire material, its gauge, and its length (or number of turns). DCR is almost always some fraction of an ohm for relatively small inductors used in the electronics industry. Industrial and heavy-duty inductors are not addressed by this example. The inductor's self-resonant frequency (SRF) is the frequency at which the inductor resonates naturally with its parasitic distributed capacitance. The resonating frequency is the frequency at which the inductive reactance equals the capacitive reactance. For all practical purposes, the inductor should be used at a frequency one order of magnitude lower than its self-resonating frequency. The topic inductive and capacitive reactances will be addressed in more depth in Chapter 2.

I_{sat} stands for the saturation current that the inductor can take just before its inductance goes down in value. Typically, manufacturers define I_{sat} as the current at which the inductance drops down by 10% of its nominal value at room temperature. Finally, the I_{RMS} current provides a measure of how much average current can continuously flow through the inductor while producing less than some specified temperature rise. This figure contains the ambient temperature at which it was measured and the temperature range at which the inductor will be used, taking into account self-heating effects of the inductor.

Shall any of the inductor manufacturers' parameters not be sufficiently explicit on the use of the inductor, it is the responsibility if the circuit designer to test the inductor and characterize its behavior for the desired operating conditions. As a conservative rule of thumb, it is wise to use an inductor up to some fraction, like 50% or 75% of the smaller of the two rated currents, that is, I_{sat} and I_{RMS}. The circuit designer should not be surprised that not all manufacturers specify both currents discussed.

1.7 KIRCHHOFF'S VOLTAGE LAW (KVL) AND KIRCHHOFF'S CURRENT LAW (KCL)

An electric circuit or network consists of a number of electric components, independent or dependent current, and voltage sources interconnected to each other. Figure 1.44a through f shows a variety of circuit topologies that are commonly seen in electrical and electronics engineering applications.

Note that the blank rectangles represent virtually any R, L, or C circuit series or parallel combination. However, in this section, and without loss of generality, we will apply Kirchhoff's laws when such rectangular shaped elements are resistors.

Note that when the Lattice Network of Figure 1.44f has an electrical element loading the two right-hand side terminals, the topologies of the Lattice and the Wheatstone bridge structures are identical.

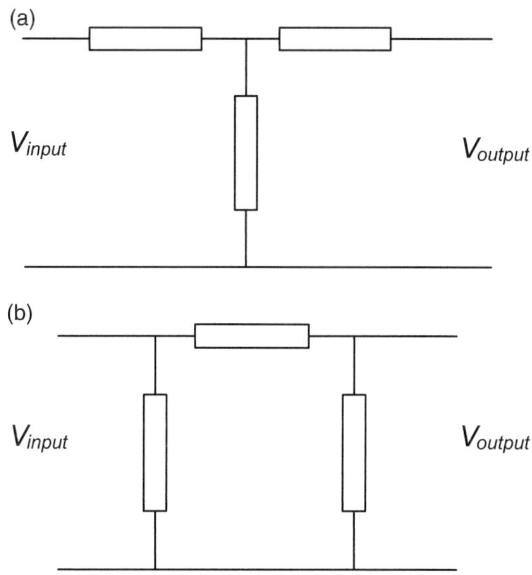

Figure 1.44 Various circuit topologies: (a) T-network or Y-network; (b) π-network or delta-network; (c) ladder network; (d) bridged-T-network; (e) bridge or Wheatstone bridge network; and (f) lattice network.

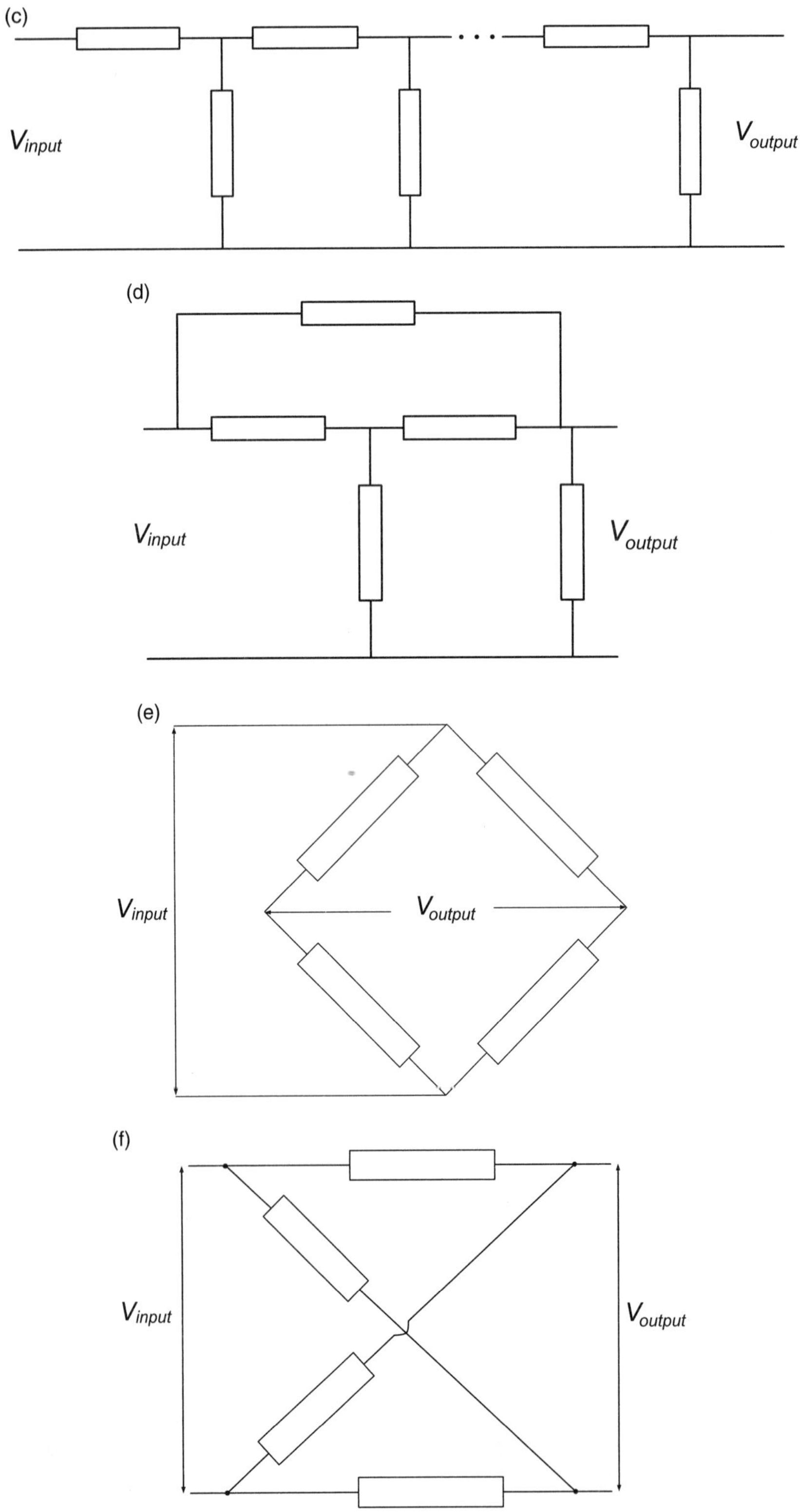

Figure 1.44 (*Continued*)

Student Exercise: Redraw a Lattice network with a resistor on its two right-hand side terminals and justify that it indeed matches the Wheatstone bridge topology.

Other circuit structures, which are common in engineering applications, are derivatives of the ladder networks. In general, they are referred to as window-pane topologies. Figure 1.45a through d shows a few examples of them.

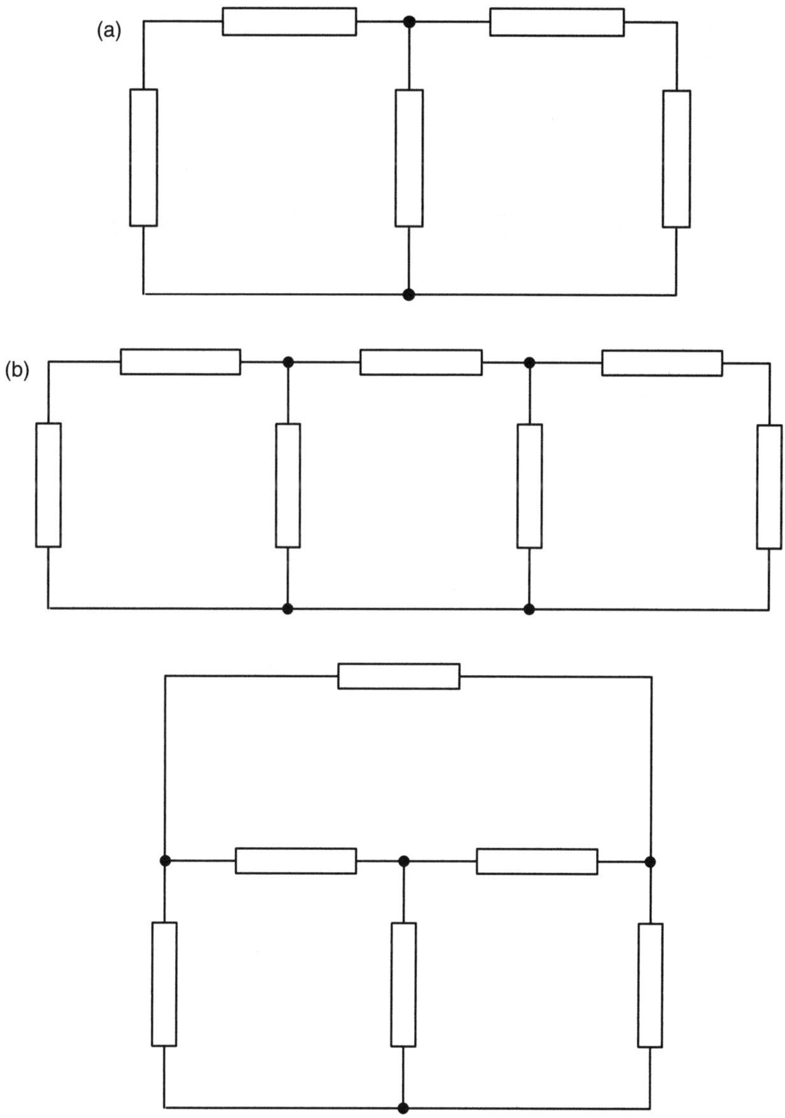

Figure 1.45 Window pane topologies networks: (a) 2-window; (b) 3-window; (c) 4-window; (d) 6-window.

(c)

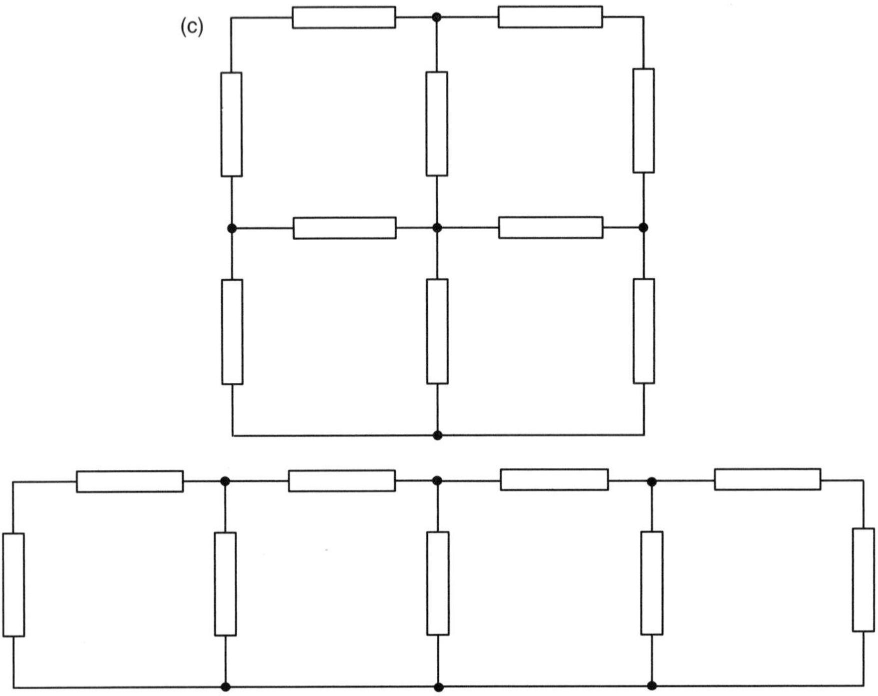

(d)

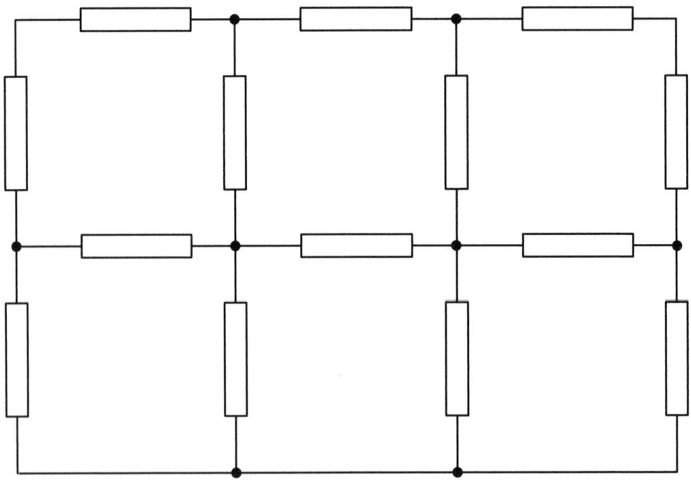

Figure 1.45 (*Continued*)

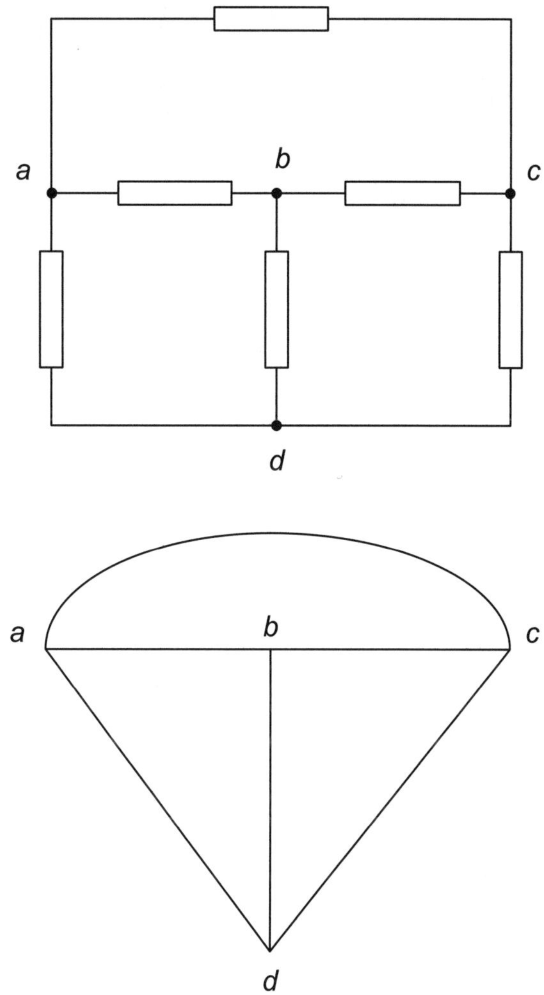

Figure 1.46 (a) 3-window network; (b) its topology.

The graph or the topology of a circuit or network consists in finding the *skeleton* of the circuit. Replacing every element in the circuit with a line does that. For example the graph or topology of one of the circuits of Figure 1.45b is shown in Figure 1.46.

Circuit Definitions: Branches, Nodes, and Loops

Referring to the circuit topology of Figure 1.47a, segments $\overline{ab}$, $\overline{bc}$, $\overline{cd}$, $\overline{da}$, $\overline{bd}$, and $\overline{ac}$ are the *branches* or links in the network. We will use the term *branch*. The junction of two or more branches is a *node*. A loop in a network is a closed path formed by a number of connected branches. For example, for the circuit

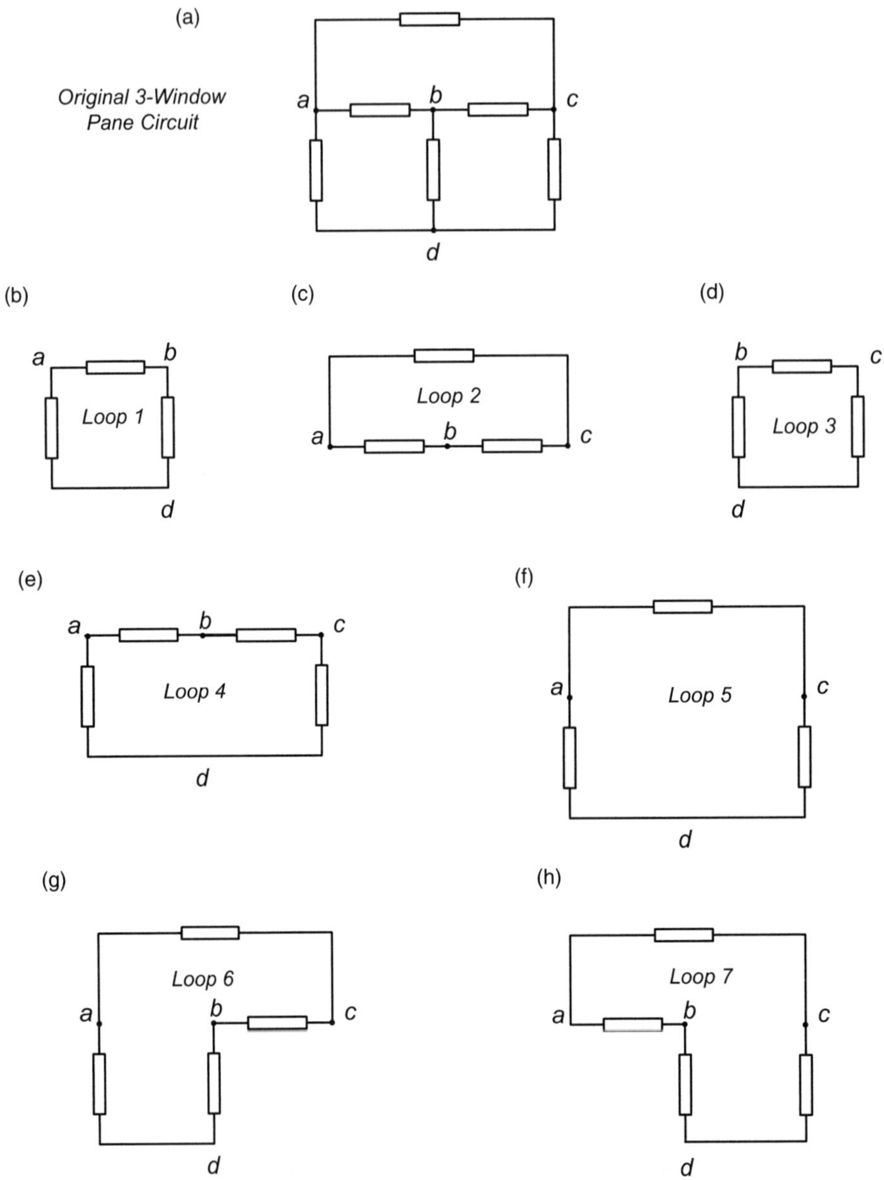

Figure 1.47 (a) Original 3-window pane circuit; (b) loop 1; (c) loop 2; (d) loop 3; (e) loop 4; (f) loop 5; (g) loop 6; (h) loop 7.

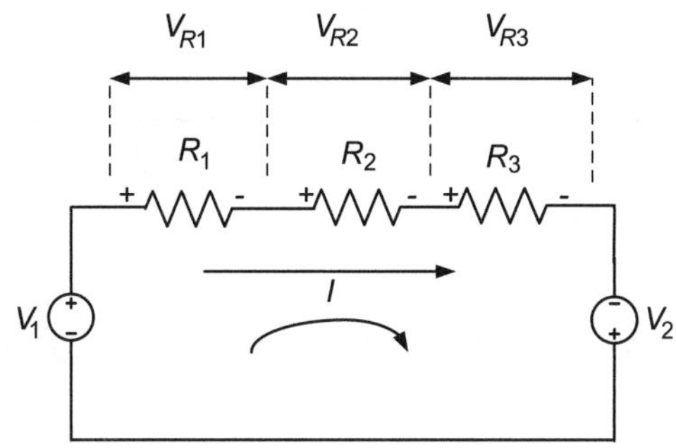

Figure 1.48 Single loop circuit, with two DC voltage sources.

whose topology is drawn in Figure 1.46, all the possible loops are shown in Figure 1.47.

Now let us assume a single loop circuit like the one depicted by Figure 1.48. V_1 and V_2 are DC voltage sources. R_1, R_2, and R_3 are all in series in the circuit shown.

KVL states that in any closed loop path in a network or circuit, the algebraic sum of all branch voltages equals to zero at all times.

An alternative way of stating KVL is

At any instant of time, in a closed loop path in a network, the sum of all the voltage rises must equal the sum of all the voltage drops.

V_1 and V_2 are both voltage rises, since they generate a rise in voltage. The current I developed in the series circuit produces voltage drops on each one of the resistors. $I\,R_1 = v_{R1}$ is the voltage drop across resistor R_1; $I\,R_2 = v_{R2}$ is the voltage drop across resistor R_2; and $I\,R_3 = v_{R3}$ is the voltage drop across resistor R_3.

Before using KVL, let us establish a direction we will be traveling around the loop. Arbitrarily, we will assume that we travel the loop in a clockwise direction. Voltages rises with their positive terminal before their negative terminal and in the direction of traveling the loop are negative. Whereas voltages rises with their negative terminal before their positive terminal and in the direction of traveling the loop are positive. Current I flows through the resistors from higher potentials to lower potentials. In this example, all three voltage drops have the same sign; however, in multiloop circuits, that may not be the case. Current I was arbitrarily assumed to flow from left to right as shown in the circuit of Figure 1.48. More examples will follow to explain this.

Applying KVL to the circuit of Figure 1.48, we obtain that

$$V_1 - V_2 - v_{R1} - v_{R2} - v_{R3} = 0, \qquad (1.166)$$

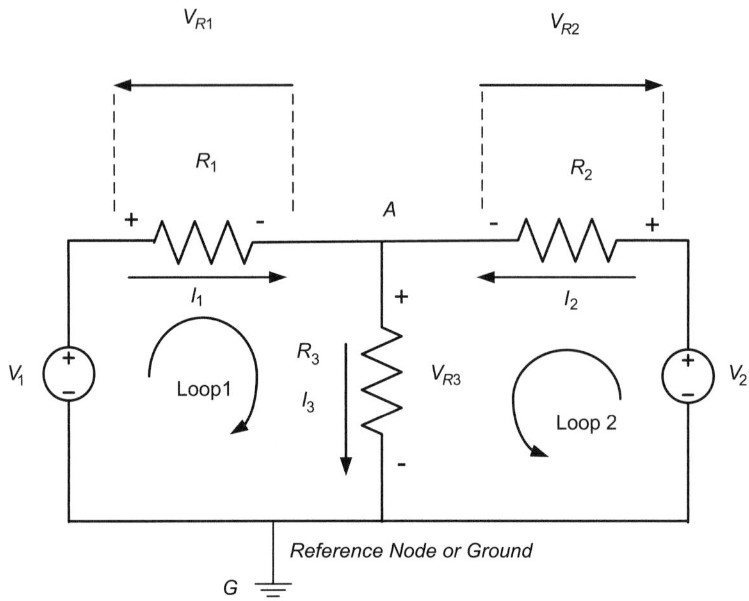

Figure 1.49 Circuit for KVL, Example 1.24.

Let us remember that an algebraic sum is a sum where every addend or sum member is taken into account with its respective sign. This is what Equation (1.166) shows.

Example 1.24 Given the circuit of Figure 1.49, determine the KVL equations for loops 1 and 2.

Solution to Example 1.24

Figure 1.49 indicates the two loops chosen. For the left-most loop, we have

$$V_1 = V_{R1} + V_{R3}. \tag{1.167}$$

And for the right-most loop, we have

$$V_2 = V_{R2} + V_{R3}. \tag{1.168}$$

Note that the loop directions (see arrows in Fig. 1.49) have been arbitrarily chosen.

Practice Problem Choosing different loop traveling directions, rewrite KVL for the circuit of Figure 1.49. Prove that your new equations are algebraically equivalent to Equations (1.167) and (1.168).

KCL states that the algebraic sum of all branch current at a node is zero at all instants of time.

Algebraic sum means to take into consideration the direction of flow of the current, that is, its sign. We can arbitrarily choose that any current entering a node is positive, and any current leaving a node is negative. The opposite can also be assumed. The important fact is to pick one current direction convention and keep such direction consistently throughout the solution of the complete circuit.

An alternative way of stating KCL is

The sum of currents entering a node must equal the sum of currents leaving such node at all instants of time.

Example 1.25 Apply KCL to the Two-Loop Circuit Given in Figure 1.49

Solution to Example 1.25

Let us apply KCL at node A of the circuit given in Figure 1.49. Node G (Ground) is assumed to be the reference node, which is the node with respect to which all other node voltages are referenced. Figure 1.49 shows branch current directions in the circuit, which can be arbitrarily assigned. Based on the current directions assumed for node A,

$$I_3 = I_1 + I_2. \tag{1.169}$$

Note that we may also write the KCL equations for node G and that leads to

$$I_1 + I_2 = I_3. \tag{1.170}$$

It should be clear that both Equations (1.169) and (1.170) are identical. Thus, just one of the equations is used.

Example 1.26 Using the circuit of Figure 1.49, assume that $R_1 = 6\ \Omega, R_2 = 3\ \Omega$, $R_3 = 10\ \Omega$, $V_1 = 5$ V, and $V_2 = 4$ V. Write the *KCL* equation for node A and *KVL* equations for loops 1 and 2 as shown on the figure.

For the reader's convenience Figure 1.49 is repeated here and referred to as Figure 1.50.

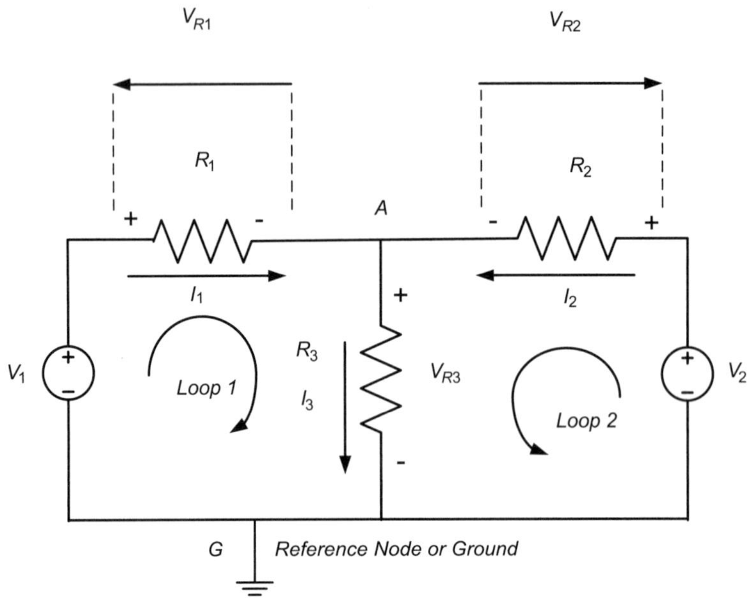

Figure 1.50 Circuit for KVL.

Solution to Example 1.26

By inspection of Figure 1.50 and applying KCL for node A,

$$I_1 + I_2 = I_3. \tag{1.171}$$

Applying KVL on loop 1,

$$V_1 = I_1 R_1 + I_3 R_3. \tag{1.172}$$

Applying KVL on loop 2,

$$V_2 = I_2 R_2 + I_3 R_3. \tag{1.173}$$

Equations (1.171) through (1.173) are three linearly independent equations, and since we have three unknowns which are I_1, I_2, and I_3, we obtain unique solutions for each branch current.

Plugging the resistor and voltage sources values given into Equations (1.172) and (1.173) we obtain

$$5 = I_1 6 + I_3 10. \tag{1.174}$$

$$4 = I_2 3 + I_3 10. \tag{1.175}$$

$$I_3 = I_1 + I_2. \tag{1.176}$$

The solving of the three simultaneous Equations (1.174) through (1.176) is left as an exercise to the reader. As promised on the Preface, this book covers hardware essentials, and it is not intended to be a math book.

The solutions for the three currents are

$$I_1 = 0.2315 \text{ A}. \tag{1.177}$$

$$I_2 = 0.1296 \text{ A}. \tag{1.178}$$

$$I_3 = 0.3611 \text{ A}. \tag{1.179}$$

All three numerical results were rounded to the fourth decimal place. Let us plug results given by Equations (1.177) through (1.179) into Equations (1.174) through (1.176):

$$5 = 0.2315 \times 6 + 0.3611 \times 10. \tag{1.180}$$

$$4 = 0.1296 \times 3 + 0.3611 \times 10. \tag{1.181}$$

$$0.3611 = 0.1296 + 0.2315. \tag{1.182}$$

It is also easy to verify that the voltage at node A (Fig. 1.50) is

$$V_A = I_3 R_3. \tag{1.183}$$

$$V_A = 0.36111 \text{ A} \times 10\ \Omega = 3.6111 \text{ V}. \tag{1.184}$$

It is also instructive to realize that node voltage V_A also equals from Figure 1.50 using KVL:

$$V_A = V_1 - I_1 R_1. \tag{1.185}$$

Using the values for V_1, I_1, *and* R_1 in Equation (1.185) we find that

$$V_A = 5 - 0.2315 \times 6 = 3.6111 \text{ V}. \tag{1.186}$$

Applying KVL to loop 2 of the circuit of Figure 1.50,

$$V_A = V_2 - I_2 R_2. \tag{1.187}$$

And using the values known for V_2, I_2, and R_2 yields

$$V_A = 4 - 0.1296 \times 3 = 3.6112 \text{ V}. \tag{1.188}$$

In summary, the nodal voltage V_A was found using one set of KVL Equations (1.174) and (1.175) and found to be the same when using an alternate set of

KVL equations given by Equations (1.185) through (1.187). Keep in mind that due to the use of finite precision (4 decimal places in our example), the numbers may not be 100% exact. This is just due to numerical round-off errors and not to Kirchhoff's laws.

Practice Problem 1.27: Using the numerical Example 1.26, find solutions for the KVL and KCL equations expressed with rational numbers, instead of rounded or truncated decimal numbers to prove that KVL and KCL are exact.

Example 1.27 Given the three-mesh circuit of Figure 1.51, state KVL equations for meshes: *ABG*, *BCG*, and *ABC*, and KCL equations for node *B*.

Solution to Example 1.27

By inspection of the circuit of Figure 1.51 we write the following KVL equations:

Mesh ABG:

$$V_1 = I_1R_1 + I_4R_4 \tag{1.189}$$

Mesh BCG:

$$V_2 = I_2R_2 + I_4R_4 \tag{1.190}$$

Mesh ABC:

$$0 = I_1R_1 - I_2R_2 + I_3R_3 \tag{1.191}$$

Node B:

$$I_1 + I_2 = I_4 \tag{1.192}$$

Equations (1.189) through (1.192) are a set of linearly independent simultaneous equations and four unknowns. The unknowns are: I_1, I_2, I_3, and I_4.

Practice Problem 1.28: Using the circuit provided by Figure 1.51, assume the following resistor and voltage values:

$$R_1 = 1\,\Omega,\ R_2 = 2\,\Omega,\ R_3 = 3\,\Omega,\ R_4 = 4\,\Omega,\ V_1 = 5, \text{and } V_2 = 6\text{ V}.$$

For the given numerical resistor and voltage values, find the numerical values of I_1 through I_4. Hint: Equations (1.189) through (1.191) constitute a set of linearly independent simultaneous equations. Solving, we obtain

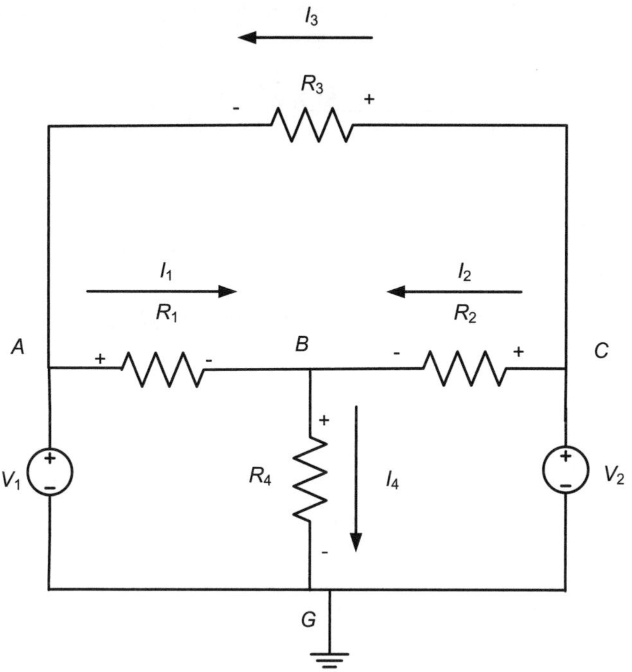

(Reference Node or Ground)

Figure 1.51 Three-mesh circuit with two voltage sources.

$$I_1 = 0.428571 \text{ A}. \tag{1.193}$$

$$I_2 = 0.714286 \text{ A}. \tag{1.194}$$

$$I_3 = 0.333333 \text{ A}. \tag{1.195}$$

$$I_4 = 1.14286 \text{ A}. \tag{1.196}$$

Using the circuit of Figure 1.51 is easy to verify that all currents comply with KCL.

State the remaining KCL equations not stated above (i.e., nodes A and C), and plug in the numerical values obtained in Equations (1.193) through (1.196) to validate KCL. Finally, using the found values of currents, validate KVL Equations (1.189) through (1.191).

Example 1.28 State the KVL and KCL of the loops and nodes found on the circuit of Figure 1.52. Assume that the value of each resistor (R_1 through R_4) is 1 Ω, V = 6 V, and I = 5 A.

Provide numerical answers for currents I_1 through I_4, node voltages V_B and V_C. Finally, determine the current that the voltage source V supplies, that is, I_{V1}

Writing KVL and KCL for above circuit,

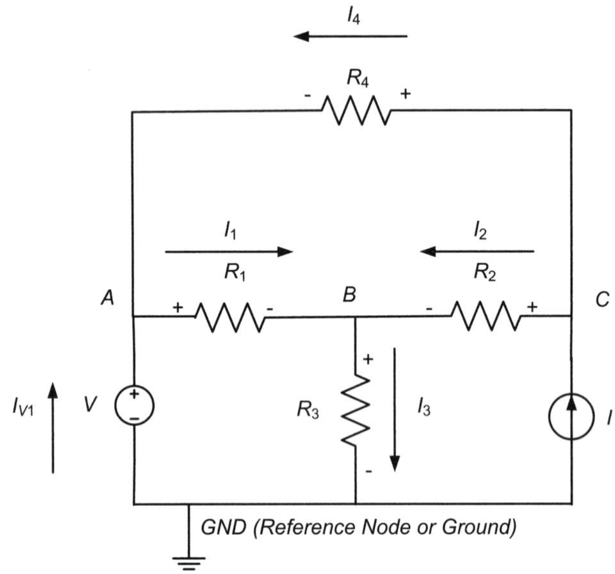

Figure 1.52 Circuit for Example 1.29: apply KVL and KCL.

Mesh ABG:

$$V = I_1R_1 + I_3R_3 \tag{1.197}$$

Mesh BCG:

$$V_{CG} = I_2R_2 + I_3R_3 \tag{1.198}$$

Mesh ABC:

$$0 = I_1R_1 - I_2R_2 + I_4R_4 \tag{1.199}$$

And for the nodes:
Node A:

$$I_{V1} + I_4 = I_1 \tag{1.200}$$

Node B:

$$I_1 + I_2 = I_3 \tag{1.201}$$

Node C:

$$I = I_2 + I_4 \tag{1.202}$$

Node G:

$$I_{V1} + I = I_3 \tag{1.203}$$

Solving a set of linearly independent equations we obtain

$$V_B = 4.6\text{ V.}$$

$$V_C = 7.8\text{ V.}$$

$$I_1 = 1.4\text{ A.}$$

$$I_2 = 3.2\text{ A.}$$

$$I_3 = 4.6\text{ A.}$$

$$I_4 = 1.8\text{ A.}$$

$$I_{V1} = -0.4\text{ A.}$$

Practice Problem 1.30: Verify that all answers given above meet Kirchhoff's Equations (1.197) through (1.203).

1.8 SUMMARY

This chapter covers the essentials of DC circuits. It starts with resistors, capacitors, and inductors, all three passive circuit elements, and voltage and current sources. The emphasis on this chapter is on DC or direct current circuits. The fundamental laws of circuit analysis are presented: Ohm's law and Kirchhoff's voltage and current laws are the pillars to solve simple as well as complicated circuits. Magnetics basics were also presented.

It is important that the reader works out as many problems as possible reading and reading the chapter as often as it is necessary.

FURTHER READING

1. Vishay Intertechnology, Inc., *Tech Note TN0015: Design Guide for Use of Precision Resistors*, 2005.
2. David Halliday, Robert Resnick, and Jearl Walker, *Fundamentals of Physics*, 7th ed., Wiley, Hoboken, NJ, 2004.
3. Paul Horowitz and Windfield Hill, *The Art of Electronics*, 2nd ed., Cambridge University Press, New York, 1989.
4. Sanjaya Maniktala, *Switching Power Supply Design & Optimization*, McGraw-Hill, New York, 2004.
5. Arthur F. Kip, *Fundamentals of Electricity and Magnetism*, McGraw-Hill Inc., New York, 1962.
6. David A. Bell, *Fundamentals of Electric Circuits*, 4th ed., Prentice Hall, Upper Saddle River, NJ, 1988.
7. Mahmood Navhi and Joseph Edminister, *Electric Circuits*, 4th ed., McGraw-Hill, New York, 2003.

PROBLEMS

1.1 What is electrical conductivity of a material?

1.2 Which atomic particles have zero charge?

1.3 Which atomic particles have a mass about 1800 times larger than the mass of an electron?

1.4 Name the four types of materials based on their electrical characteristics.

1.5 How is an electric current defined?

1.6 Assume you have an automobile battery that has a 12-V nominal output voltage and a nominal internal resistance of 20 mΩ. We safely apply a short circuit across the battery terminals. Assume that the internal resistance of the battery does not change in a significant manner during the first few seconds after applying the short circuit. What is the current that the battery will deliver within the first couple of seconds?

1.7 We are given two DC voltage sources; one of them has a 12-V open-circuit voltage and an internal resistance of 1 Ω, the second source also has a 12-V open-circuit voltage and an internal resistance of 0.5 Ω. Which of the two sources is capable of delivering more current? Calculate the short circuit current of each source.

1.8 A resistor has 150 V applied across it and the current through it is 10 A. Find the value of resistance.

1.9 Given the circuit of Figure 1.53, calculate the current through the resistor for Figure 1.53a–c. What conclusion can you make as the resistor across the DC voltage source increases? What is the voltage across the resistor for all three cases?

1.10 Given the circuit of Figure 1.54, calculate the voltage across the resistor for Figure 1.54a–c. What conclusion can you make as the resistor across the DC current source increases? What is the current through the resistor for all three cases?

1.11 Given 1 MΩ nominal valued resistor, of a ±5% accuracy rating and a temperature coefficient of ±300 ppm/°C, calculate the resistance range that the resistor will span for a temperature range of –50°C to +150°C. Assume that the resistor nominal value provided is such, at 25°C.

1.12 Repeat Problem 1.11 for 1 MΩ resistor of ±5% accuracy rating, with a ±50 ppm/°C temperature coefficient.

1.13 Referring to the circuits of Figure 1.55a through l, which circuits are meaningless from a circuit analysis perspective? For each case, explicitly state the reason why each circuit is meaningless or not. In all cases, make an attempt to calculate the voltage and/or current in all the circuit

(a)

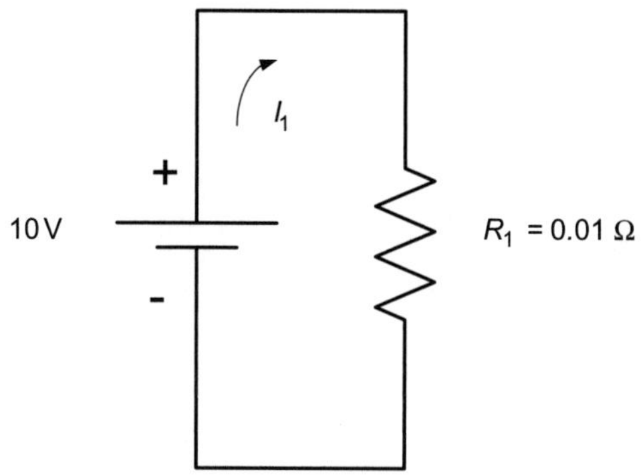

(b)

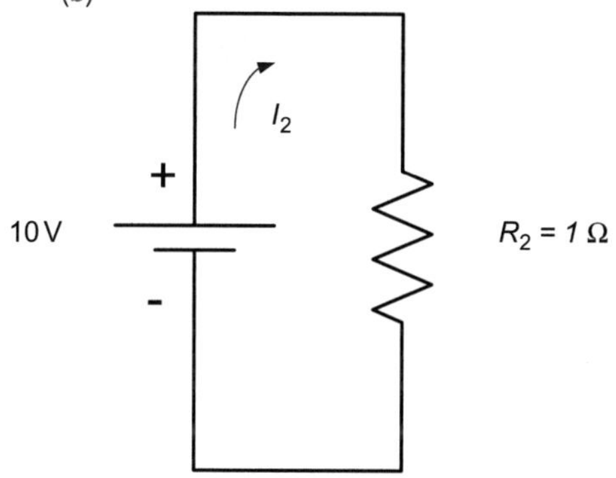

(c)

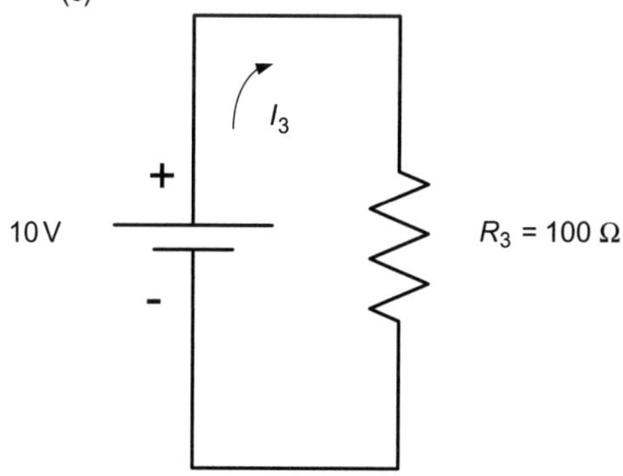

Figure 1.53 Circuit for Problem 1.9.

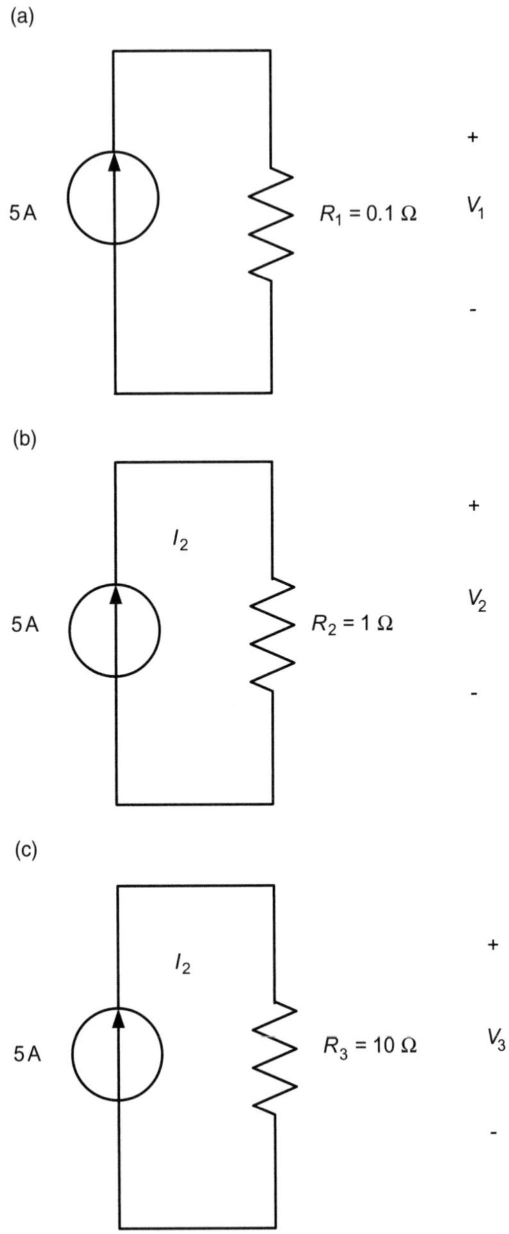

Figure 1.54 Circuit for Problem 1.10.

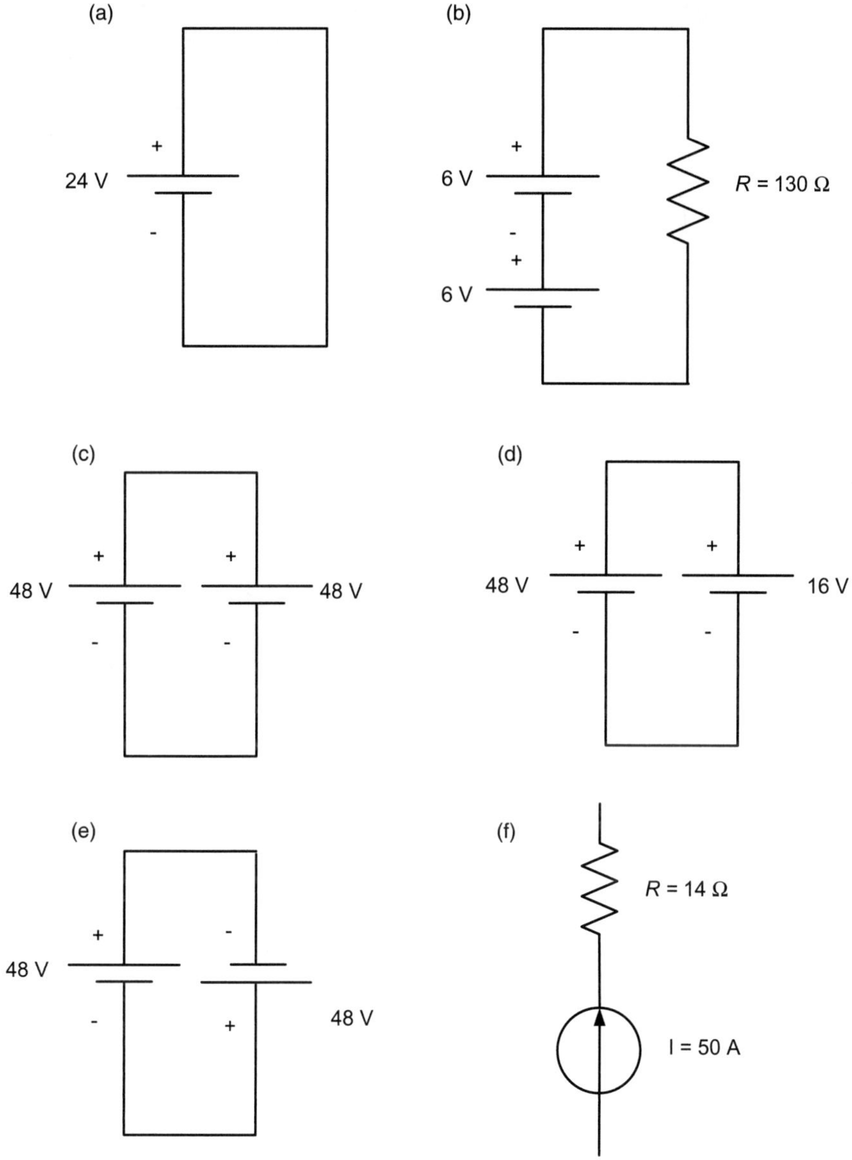

Figure 1.55 Circuit for Problem 1.13.

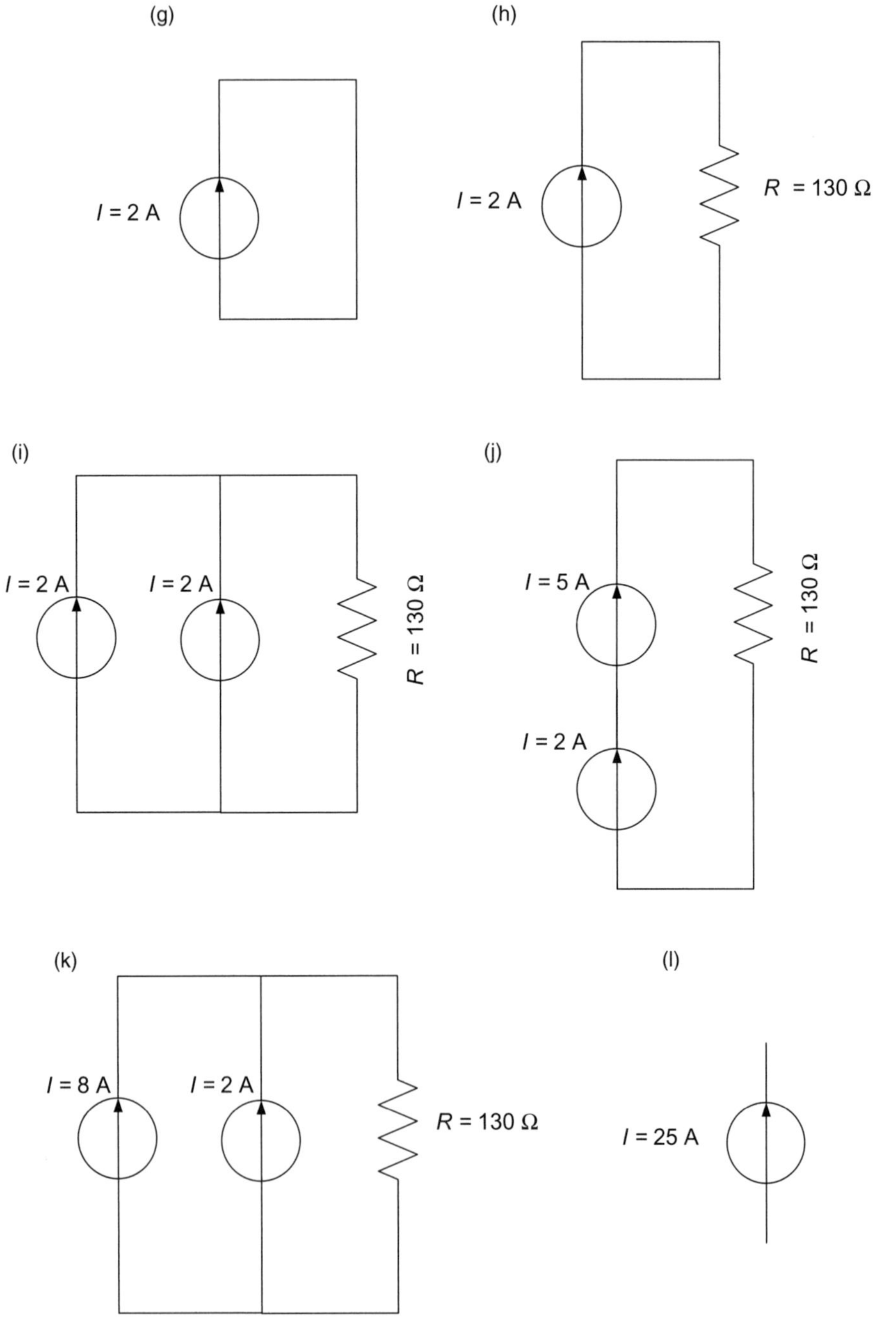

Figure 1.55 (*Continued*)

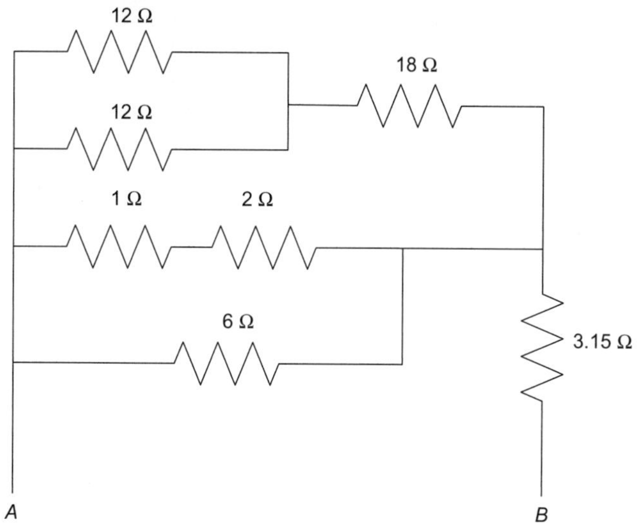

Figure 1.56 Circuit for Problems 1.14 through 1.18.

elements and sources whenever possible. If it is not possible, justify why it is not.

1.14 Given the circuit of Figure 1.56, calculate the equivalent resistance between points A and B.

1.15 Given the same circuit of Figure 1.56, assume that the series of the 1 Ω and 2 Ω resistors is replaced with a 0 Ω resistance. Calculate the equivalent resistance between points A and B.

1.16 Given the same circuit of Figure 1.56, assume that the 18 Ω resistor is removed from the circuit and replaced with a short circuit. Calculate the equivalent resistance between points A and B.

1.17 Given the same circuit of Figure 1.56, assume that only one of the 12 Ω resistors is replaced with a short circuit. Calculate the equivalent resistance between points A and B.

1.18 Given the same circuit of Figure 1.56, assume that the 3.15 Ω resistor is removed from the circuit and replaced with a short circuit. Calculate the equivalent resistance between points A and B.

1.19 Calculate the resistance of a 35-mm^2 cross-section copper wire that is 1 km long. Assume the room temperature is 20°C.

1.20 Repeat Problem 1.19 using silver for the wire material.

1.21 Repeat Problem 1.19 using gold for the wire material.

1.22 This question is not addressed by the material covered in this chapter. (Just for fun.) If the resistance of a gold wire is higher than that of a copper or a silver wire of the same cross section, length, and temperature, why do you believe that in some applications, for example, integrated circuit manufacturing, gold is used over copper and silver?

1.23 Which is the resistance value of the copper wire of Problem 1.19 if the ambient temperature of the wire is 100°C?

1.24 What is the resistance value of a 110 V, 100 W-rated lightbulb? At what temperature and current value do you believe that you have information to provide the answer? Justify your answers.

1.25 An ideal voltage source has an infinite capacity of generating current. Explain what is the point in paralleling two ideal voltage sources of the same voltage?

1.26 How should a real current source be connected so as not to damage the real device or create huge voltages?

1.27 Given two parallel-plate capacitors of the same plate area, and the same plate separation, assume that one has air as dielectric while the other one has Teflon. Which capacitor has a higher capacitance value?

1.28 If we apply a sinusoidal current source to a capacitor, draw the given current waveform and the voltage waveform developed across the capacitor.

1.29 A capacitor of value C is completely discharged. If a constant DC current source is indefinitely applied across the capacitor terminals, what does the voltage waveform as a function of time look like? Why?

1.30 Calculate the total equivalent capacitance between points A and B of the circuit of Figure 1.57.

1.31 Calculate the energy stored in a 100 μF capacitor with 1 kV applied across its terminals after a long period of time.

1.32 How much charge does the capacitor in Problem 1.31 have when it is fully charged?

1.33 Assume that we want to build a gigantic 1-F capacitor. We want to implement it with two square parallel metallic plates separated by a 10 mm air dielectric. Knowing that $\varepsilon_0 = 8.85 \times 10^{-12}$ F/m, determine the area of the plates needed. Translate your area result into acres. Ignore practical difficulties building the capacitor.

1.34 What can you conclude about the size of the capacitor from Problem 1.33?

1.35 Research problem: Using the World Wide Web, find the following features:

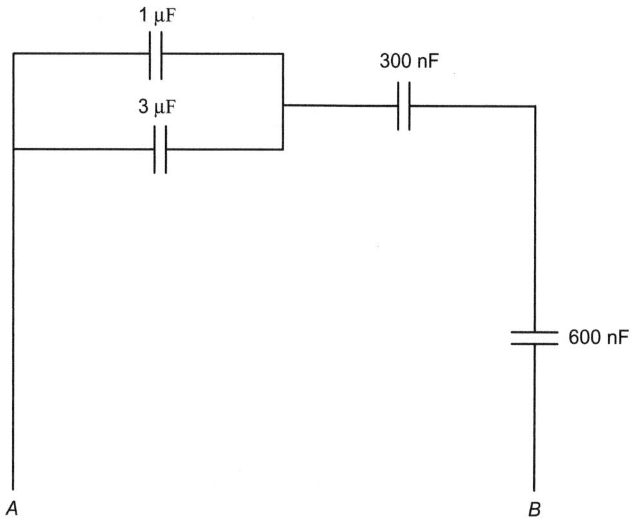

Figure 1.57 Circuit for Problem 1.30.

(1) Operational temperature range and (2) standard capacitance tolerances of the following capacitor dielectrics:

NPO/COG
X5R
X7R
Y5U
Z5U

1.36 Find at least five ferromagnetic materials, which can be either natural elements or man-made. Annotate the approximate relative permeability of each one. Hint: Access the World Wide Web.

1.37 Review the inductors section of this chapter; derive an equation for straight coil inductance which is a function of the core length, the loop effective area, the number of wire turns, and the core permeability.
Refer to Figure 1.58 for a graphical representation of the inductor.

1.38 Calculate the total equivalent inductance of the circuit of Figure 1.59 between points A and B.

1.39 Figure 1.41d depicts a transformer. Two coupled inductors constitute a transformer. The input inductor on the left is the primary; the output inductor on the right is the secondary. An ideal transformer receives AC current and voltage at a power level and produces at its output the same power level, but voltage and current are transformed by the following idealized equations:

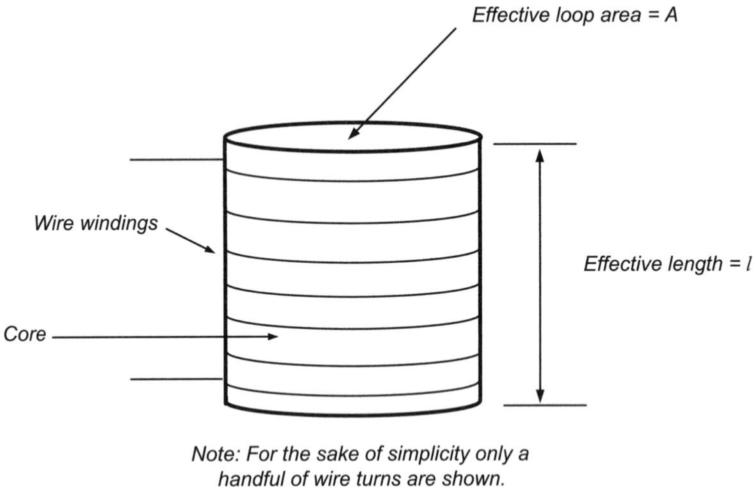

Figure 1.58 Inductor graphical representation.

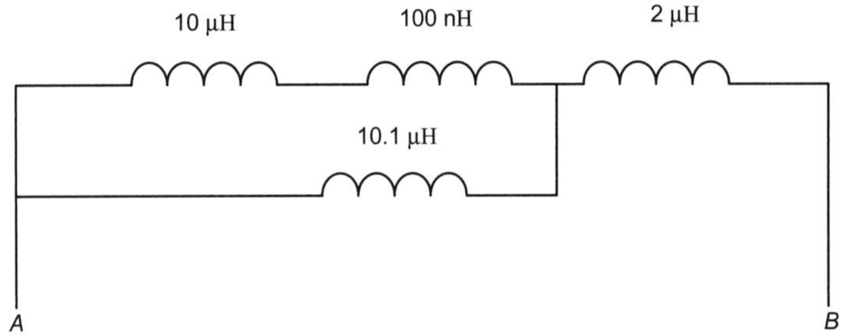

Figure 1.59 Inductor combination.

$$v_1 i_1 = v_2 i_2$$

$$v_1 N_2 = v_2 N_1$$

$$i_1 N_1 = i_2 N_2$$

In the given equations, v_1 is the primary voltage or excitation, i_1 is the primary current, v_2 is the output or secondary voltage, i_2 is the secondary current, N_1 is the primary number of wire turns, and N_2 is the secondary number of wire turns.

An ideal transformer differs from a real one in that the ideal transformer coupling between primary and secondary is assumed to be 100%; there is no leakage flux. High power, 50/60 Hz transformers work very close to such model. Radio-frequency (RF) transformers do not follow that closely the ideal transformer model.

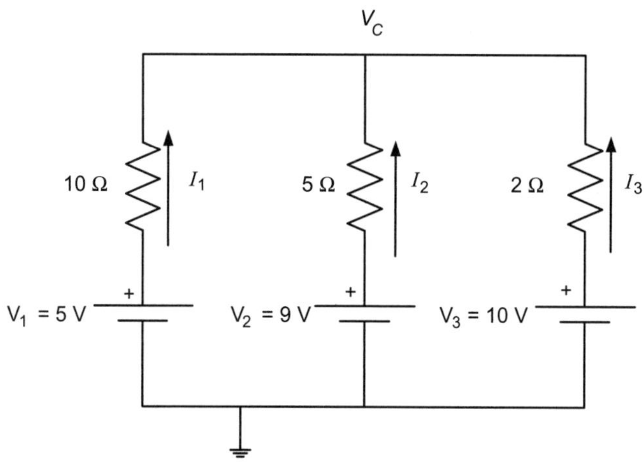

Figure 1.60 KVL/KCL circuit for Problem 1.42.

If the primary voltage is 240 V and the primary current is 10 A, determine the secondary voltage and secondary current. Assume that the transformer is ideal and its N_1/N_2 turns ratio is 4/1.

1.40 Based on your knowledge of an air-core inductor and an iron-core inductor, which one will store a larger amount of magnetic energy if the same current flows through both of them, and why?

1.41 An iron-core inductor becomes saturated because a current higher than its maximum saturation current is applied to it. (a) Which is the value of inductance L after the inductor is fully saturated? (b) Is the saturated inductor behaving in linear mode? (c) Which will be the current value through the inductor after it becomes fully saturated? (d) Explain one way by which the core saturation can be eliminated.

1.42 Using KVL and KCL equations for the circuit depicted in Figure 1.60, (a) Find the voltage and current on each resistor. (b) Find the current that each voltage source provides. (c) Verify that the sum of the powers produced by all sources equals to the power consumed by all resistors.

2

ALTERNATING CURRENT CIRCUITS

2.1 AC VOLTAGE AND CURRENT SOURCES, ROOT MEAN SQUARE VALUES (RMS), AND POWER

When we plug a toaster into an electrical outlet in our kitchen, insert a slice of bread into a slot, we notice that the toaster starts to get hot very quickly. If we take a peek in the slot where the slice of bread is, we can see that the internal wires in the toaster become red hot. The toaster-heating elements are approximately 1 to 2 kW rated resistors, depending on the toaster make and model, rated to operate at the household AC supply voltage. This is a simple example of an alternating current (AC) voltage source, supplying an AC current to the toaster-heating elements in operation. Both of these waveforms, voltage and current, vary sinusoidally with respect to time. The AC current, being "*pushed*" by the AC voltage source, is the cause of heat being produced in the immediate vicinity of the toaster-heating element. The outlet on the kitchen wall is the point where we connect the appliance to the AC voltage source. The AC voltage source from the electric utility company is usually located in a remote site, far away from the home. In most households in the Unites States, the standard AC voltage is 120 V. The 120 V refers to the root mean square (RMS) value of the sinusoidal waveform, where RMS is defined mathematically by the following equation:

Electrical, Electronics, and Digital Hardware Essentials for Scientists and Engineers, First Edition. Ed Lipiansky.

$$f_{RMS} = \sqrt{\frac{1}{T}\int_0^T [f(t)]^2\,dt}. \tag{2.1}$$

Equation (2.1) is the RMS value of waveform $f(t)$.

In Equation (2.1) T is the period of the waveform. The waveform $f(t)$ can be either a voltage or a current, and t is the time, the independent variable. RMS of a waveform is also referred to as the effective value of the waveform.

When $f(t)$ is a sinusoidal waveform such as $v(t) = V\sin(\omega t + \theta)$; V is the amplitude (or peak value) of the sinusoidal waveform in volts, ω is its angular frequency equal to $2\pi f$, where f equals the inverse of the sinusoid's period T or the sinusoid frequency, given in units of second^{-1} or hertz, and θ is the sinusoidal waveform phase shift. The units of the angular frequency ω are given in radians per second. Solving Equation (2.1) for a sinusoidal voltage, the RMS value of it is

$$V_{effective} = V_{RMS} = V/\sqrt{2} \cong 0.707\ \text{V}, \tag{2.2}$$

where V is the peak value or magnitude of the sinusoidal waveform. So the 120 V at the kitchen outlet is the RMS value of the sinusoidal waveform that the electric utility company provides to U.S. households. Also applying Equation (2.2) to a current waveform, $i(t) = I\sin(\omega t + \theta)$, we find that its RMS value is also

$$I_{effective} = I_{RMS} = 0.707\ I. \tag{2.3}$$

In Equation (2.3) I is the peak value or amplitude of the current waveform.

2.1.1 Ideal and Real AC Voltage Sources

An ideal AC voltage source is one that produces a sinusoidal voltage that varies with time. Most importantly, the amplitude and the RMS value of such voltage source does not vary based on how much current the load across the source terminals is drawing. This means that the internal resistance of an ideal AC voltage source is zero. So whether the voltage source supplies no current or very large currents, the voltage amplitude and RMS value remain constant. It is also true that the waveforms retain their sinusoidal shape and original frequency f and phase angle θ. On the other hand, a real AC voltage source amplitude does not remain constant with the level of current being supplied by the real AC voltage source. This concept is similar to that of ideal and real DC voltage sources. The real AC voltage source can be modeled as an ideal AC voltage source in series with its internal resistance, the real AC voltage source internal resistance is not zero, and it is a finite number as shown by Figure 2.1.

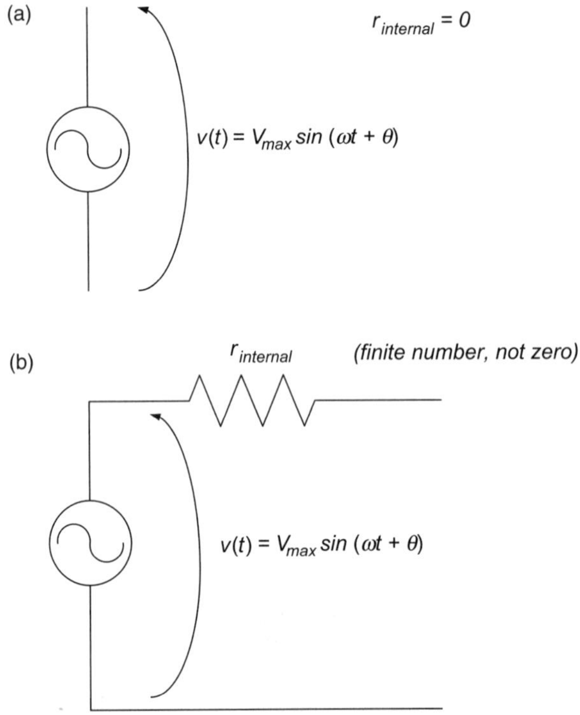

Figure 2.1 Representation of (a) ideal and (b) real AC voltage sources.

- *The internal resistance of an ideal AC voltage source is zero, which means that the source can supply an unlimited current to its load.*
- *In contrast, real AC voltage sources cannot provide infinite current when the source terminals are short-circuited.*
- *The internal resistance of a real AC voltage source is never zero and it is a finite number.*
- *The internal resistance of a real AC voltage source is always greater than 0 Ω.*
- *The internal resistance is an indicator of the current sourcing capability of the voltage source.*

Example 2.1 Ideal versus Real Voltage Sources

Let us assume that we have an ideal voltage source, this can be a DC or an AC source. An ideal voltage source has zero internal resistance. Thus, a load connected across the terminals of the voltage source can draw any amount of current dictated by the load value. For example, given an ideal 12-V DC source connected across a 10 Ω, 1 Ω, 0.1 Ω, 0.01 Ω, or any other resistor value (except for zero), the current is always determined by Ohm's law. A 10 Ω resistor draws 12 V/10 Ω = 1.2 A from the ideal 12-V source. The 1 Ω resistor draws 12 V/1 Ω = 12 A; a 0.1 Ω draws 12 V/0.1 Ω = 120 A and the 0.01 Ω resistor draws 12 V/0.01 Ω = 1200 A from the voltage source. Now what happens if the resistor placed across the ideal voltage source has a 0 Ω value? The current that the ideal voltage source would have to supply is infinite. So to be realistic with how much current an ideal voltage source can supply, it is fair to say that any amount of current desired can be provided by the source, but not an infinite current. Using circuit simulators, if we simulated a short-circuited voltage source with a 0-Ω internal resistor, it produces an indetermination.

For the ideal AC voltage source, there is conceptually no difference with respect to the ideal DC source. The key difference is that the AC source supplies a perfectly sinusoidal time varying waveform; which has a peak value or amplitude, a frequency, and a phase angle.

Sinusoidal Waveforms: A sinusoid, from basic trigonometry, is a periodic waveform that repeats itself with a period T; it is also positive half of the time, and it is negative the other half of the time. Figure 2.2 shows a sinusoidal voltage of frequency f, amplitude V, and phase θ. Figure 2.3 shows a sinusoidal source applied to a resistor.

When we discussed DC circuits we stated that $V = I\,R$ (Ohm's law), where I and V are DC values of current and voltage, respectively. For sinusoidal-varying waveforms $v(t)$ and $i(t)$, Ohm's law holds true as well:

Generalized Form of Ohm's Law

$$v(t) = i(t)R \tag{2.4}$$

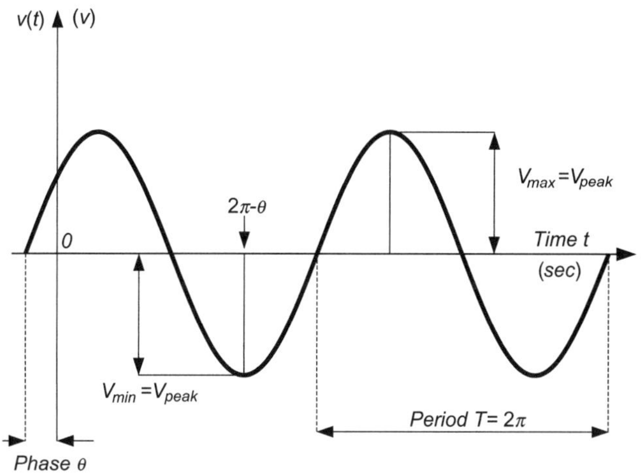

Figure 2.2 Sinusoidal voltage.

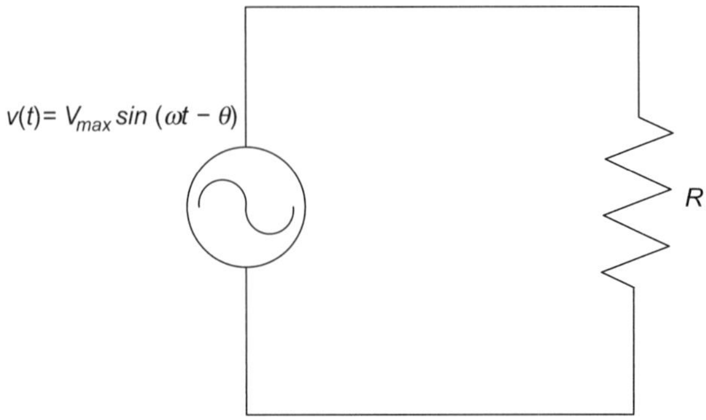

Figure 2.3 An AC voltage source applied across a resistor.

Example 2.2 Assume an ideal AC voltage source that generates a voltage equal to $v(t) = V_{peak} \sin(\omega t + \theta)$, $V_{peak} = 12$ V, $\omega = 2\pi\ 1$ rad/s, and $\theta = 0°$. Evaluate the current waveform obtained for the following resistive loads R_L: (a) 10 Ω; (b) 1 Ω; (c) 0.1 Ω, and (d) 0.01 Ω.

Answer to Example 2.2

From Equation (2.4), since $v(t) = i(t)\ R_L$, thus $i(t) = V_{peak}/R_L \sin(\omega t + \theta)$.
Thus, we obtain

(a) $i(t) = 1.2 \sin(2\pi t)$ for $R_L = 10\ \Omega$;
(b) $i(t) = 12 \sin(2\pi t)$ for $R_L = 1\ \Omega$;
(c) $i(t) = 120 \sin(2\pi t)$ for $R_L = 0.1\ \Omega$; and
(d) $i(t) = 1200 \sin(2\pi t)$ for $R_L = 0.01\ \Omega$.

All currents are given in amperes.

Equation (2.4) holds for all values of time such that $t \geq 0$. Moreover, Equation (2.4) is not limited to sinusoidal-varying waveforms but to any real-world time-varying currents and voltages that are functions of time. Finally, Equation (2.4) tells us that whatever the current as a function of time waveform is, the voltage developed across such resistor is proportional to the current waveform.

In particular, when $v(t) = V \sin(\omega t + \theta)$ and $i(t) = I \sin(\omega t + \theta)$,

$$V \sin(\omega t + \theta) = RI \sin(\omega t + \theta). \tag{2.5}$$

Figure 2.4 below depicts a plot of Equation (2.5). Note that both current and voltage waveforms are sinusoidal and proportional to each other. Resistance R is the constant of proportionality. It is also important to note that the angular frequency ω (or $2\pi f$) is the same for the current and voltage waveforms.

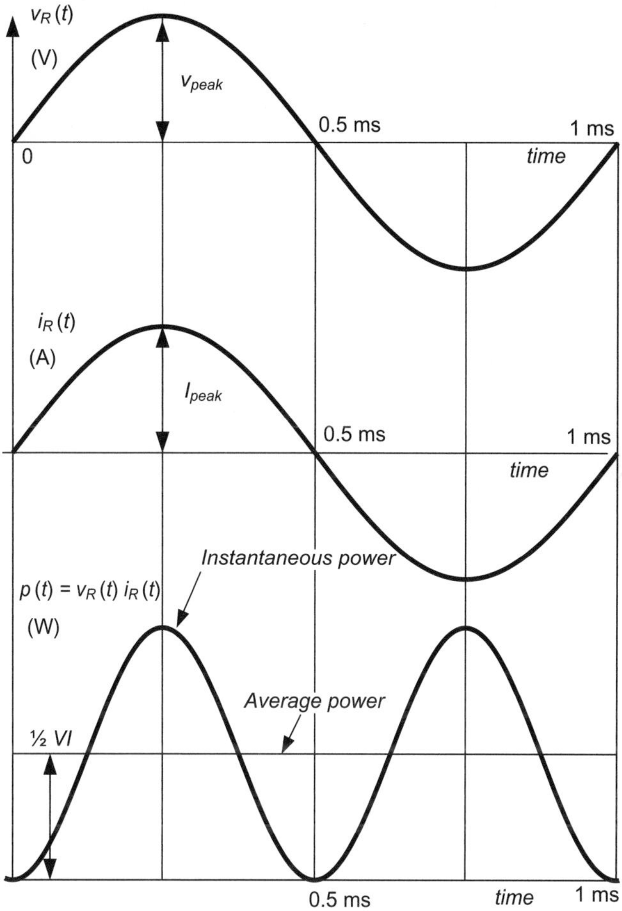

Figure 2.4 Resistor with sinusoidal voltage, current, and instantaneous power.

From Figure 2.4, we can see that both sine waves are exactly in-phase. This means that the voltage and current peak values occur at the same time, as well as their valleys (negative peaks), zero crossings, and so on.

Referring to the toaster example powered by a sinusoidal voltage source, we calculate that the instantaneous power consumed by the resistor is

$$p(t) = v(t)i(t). \tag{2.6}$$

In particular, when $v(t) = V\sin(\omega t + \theta)$ and $i(t) = V\sin(\omega t + \theta)$, then

$$p(t) = VI\sin^2(\omega t + \theta). \tag{2.7}$$

From the trigonometric equality,

$$\sin^2 x = \frac{1}{2}(1 - \cos 2x). \tag{2.8}$$

Using Equation (2.8) in Equation (2.7), we obtain

$$p(t) = \frac{1}{2}[VI - \cos(2\omega t + \theta)]. \tag{2.9}$$

Equation (2.9) is the instantaneous power on resistor R. Refer to Figure 2.4 which depicts, from top to bottom, voltage across the resistor, current through the resistor, instantaneous power consumed by the resistor, and the average power on the resistor.

Equation (2.9) has two terms, a constant power term equal to

$$\frac{1}{2}VI. \tag{2.10}$$

The second term varies with twice the original frequency and is given by Equation (2.11):

$$-\frac{1}{2}\cos(2\omega t + \theta). \tag{2.11}$$

The average power consumed by the resistor is given by

$$P_{average} = \frac{1}{T}\int_0^T p(t)dt. \tag{2.12}$$

Integrating Equation (2.12), where $p(t)$ is given by Equation (2.9), yields

$$P_{average} = \frac{1}{2}VI. \tag{2.13}$$

V and I, respectively, are the peak values of voltage and current.

The resistor will dissipate an amount of heat that is the average value of its instantaneous power. Again looking at Equation (2.9), the average value of $p(t)$ is

$$P_{average} = \frac{1}{T}\int_0^T VI \sin^2(\omega t + \theta)\, dt. \tag{2.14}$$

And solving the integral of Equation (2.14) yields

$$P_{average} = \frac{1}{2}VI. \tag{2.15}$$

V and I are respectively the amplitude (or peak values) of voltage and current.

Integrating the term

$$-\frac{1}{2}\cos(2\omega t + \theta) \tag{2.16}$$

between 0 and period T, yields zero.

Now from Equations (2.2) and (2.3), we know that for sinusoidal waveforms,

$$V_{RMS} = V/\sqrt{2} \approx 0.707\ V \tag{2.17}$$

and

$$I_{RMS} = I/\sqrt{2} \approx 0.707\ I. \tag{2.18}$$

Substituting the V_{RMS} and I_{RMS} values in Equation (2.15), we obtain that

$$P_{average} = I_{RMS}V_{RMS}. \tag{2.19}$$

The average power dissipated by a resistor when a sinusoidal current flows through it, developing a sinusoidal voltage across it, is the product of the RMS (or effective) values of such current and voltage.

The *RMS* values of current and voltage on the resistor are thermally equivalent to DC values of same current and voltage. The following example explains.

Example 2.3 Power Calculations on a Resistor Powered by an AC Voltage Given a 10 Ω resistor R, with an AC voltage source $v(t)$ applied across its terminals, where $v(t) = 25\sin(2\pi\ 60t)$, where $f = 60$ Hz, note that the phase θ, in this example has a value of zero. Note: The peak value of the sinusoidal waveform above is 25 V.

(a) Determine the value of the AC current developed through the resistor.

(b) Find the average AC power dissipated by the resistor finding the AC waveform corresponding RMS values.

(c) Find equivalent values of DC voltage and DC current that will produce the same power dissipation as the RMS values of the AC waveforms produce.

Solution to Example 2.3

(a) The current through the resistive circuit is

$$\begin{aligned} i(t) &= v(t)/R \\ &= 25/10\sin(120\pi t) \\ &= 2.5\sin(120\pi t)\ \text{A}. \end{aligned}$$

(b) Using Equations (2.17) and (2.18), we find the RMS values of voltage and current waveforms are

$$V_{RMS} = V/\sqrt{2} \cong 17.68\ \text{V}. \tag{2.20}$$

$$I_{RMS} = \text{I}/\sqrt{2} \cong 1.77\ \text{A} \tag{2.21}$$

Thus, the power dissipated by the resistor equals

$$P_{dissipated} = V_{RMS} I_{RMS} = 17.68\ \text{V}\ 1.77\ \text{A} = 31.29\ \text{W}.$$

(c) Since $V_{RMS} = 17.68$ V and $I_{RMS} = 1.77$ A, DC values of 17.68 V and 1.77 A will produce the same power dissipation of

$$\begin{aligned} P_{dissipated} &= V_{RMS} I_{RMS} = V_{DC} I_{DC} \\ P_{dissipated} &= 17.68 \times 1.77 = 31.29\ \text{W}. \end{aligned} \tag{2.22}$$

From a thermal perspective, the resistor sees no difference between the power produced by sinusoidal current and voltage or by equivalent DC values.

Example 2.4 Given a 1 Ω resistor and a 1 A DC current source, determine the peak value of an AC current source with a 1 Ω load, which produces the same power dissipation as the DC source. Hint: The resistor dissipates 1 W in DC and must also dissipate 1 W in AC.

Solution to Example 2.4

Given that the DC current value is 1 A, a sinusoidal AC current with an RMS current of 1 A will produce the same power dissipation as the DC value. Thus, the peak value of the sinusoidal current is $1 \times \sqrt{2} \approx 1.41$ A. Refer to Equation (2.18).

2.1.2 Ideal and Real AC Current Sources

An ideal AC current source is one that produces a current that varies in a sinusoidal fashion with respect to time. Most importantly, the amplitude of an ideal AC current source does not vary based on how much voltage gets developed across the current source, based on the load that it has across its terminals. So whether the current source supplies current to a short-circuit load or a very light load (resistor of high ohmic value), the current amplitude remains constant. The internal resistance of an ideal current source is infinitely large; this means that regardless of the load applied across its terminal, the current remains constant and the voltage is given by the current times the voltage across the load. The standby condition of a current source is obtained by short-circuiting the current source terminals. When ideal current source terminals are left *open-circuited*, the voltage developed across the current source approaches an infinitely large value. When we have a real current source and leave its terminals open-circuited, the voltage developed across the current source terminals is very large, and there is a great likelihood of damaging the current source. A real AC current source amplitude does not remain constant with the level of voltage being developed across the real AC current source. An ideal current source is depicted in Figure 2.5a. The real AC current source can be modeled as an ideal AC current source in parallel with its internal resistance, as shown by Figure 2.5b. Current source terminals, whether the source is real or ideal, must always be short-circuited when no load is

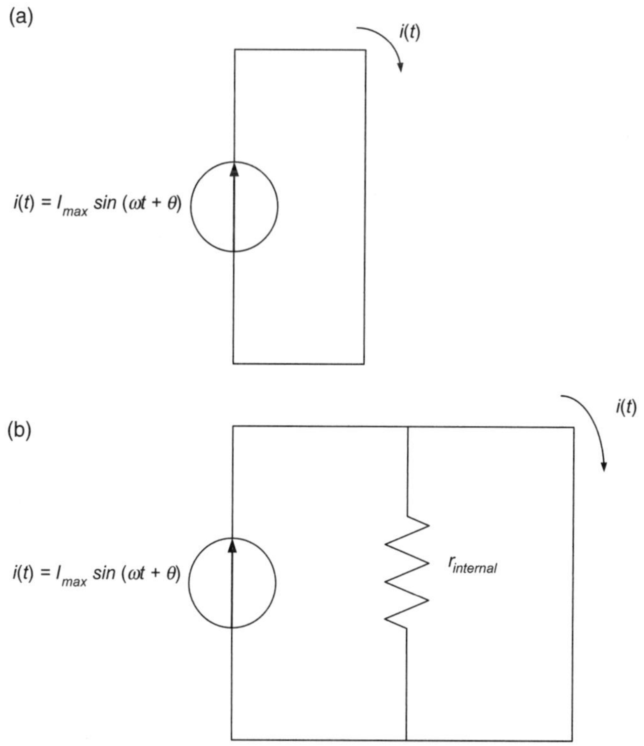

Figure 2.5 (a) Ideal and (b) real current source models in standby mode.

connected to its terminals (standby mode). Why? Because when the current source terminals are open-circuited, then the ideal and real current source voltage approaches a very large voltage value.

Just like with DC circuit voltage sources, an AC voltage source is in a standby mode when its terminals are in an open-circuit condition; its open-circuit voltage is read, but since there is no load applied across its terminals, no current is delivered by the voltage source.

An ideal or real AC current source in standby mode must have its terminals short-circuited, or a 0-Ω resistance across its terminals. A current source is in a benign state when its terminals are short-circuited. Figure 2.6a shows a basic voltage source with internal resistance and load resistance in series. Figure 2.6b depicts an ideal load line of an ideal 10-V voltage source with internal

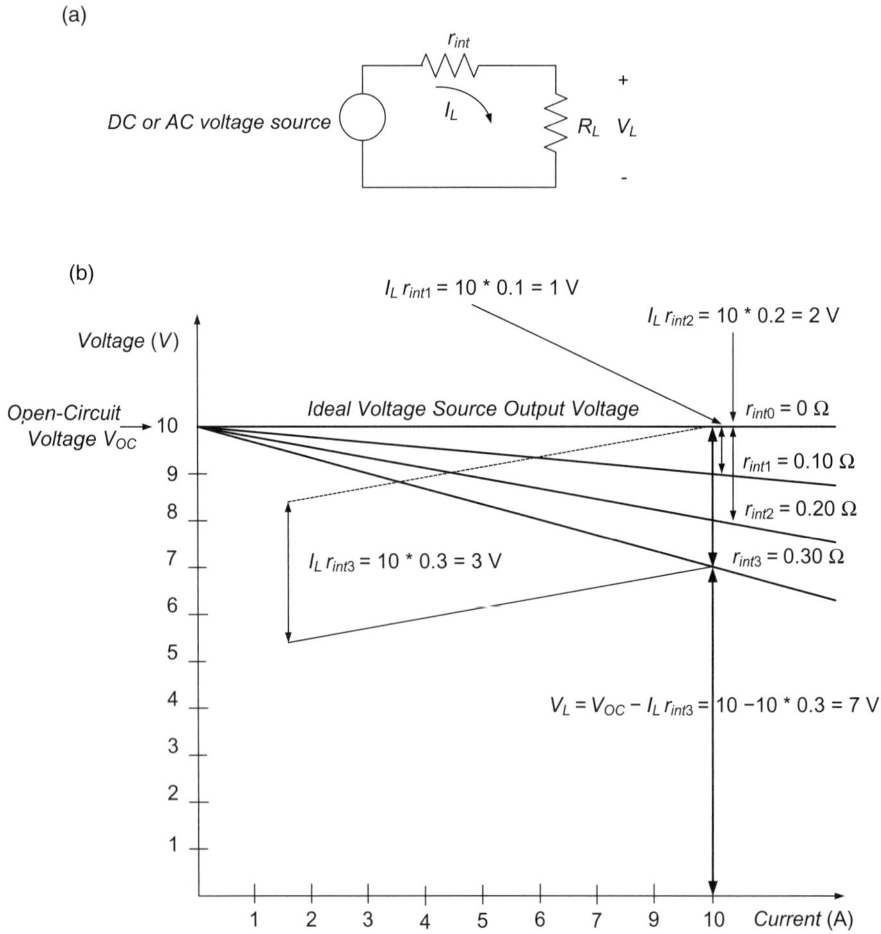

Figure 2.6 (a) Voltage source model with internal resistance; (b) different internal resistance voltage sources models under the same load condition.

resistance $r_{int0} = 0\ \Omega$, and three real voltage sources with internal resistances of $r_{int1} = 0.1\ \Omega$, $r_{int2} = 0.2\ \Omega$, and $r_{int3} = 0.3\ \Omega$. All four voltage sources have a 10-A current load. It is important to note that for equal current loading, the output voltage (load voltage V_L) of the source with the largest internal resistance (r_{int3}) is the lowest. The ideal source with zero internal resistance produces the highest possible voltage, which is 10 V. The load line equation is given from Kirchhoff's and Ohm's laws by

$$V_L = V_{oc} - r_{int} I_L. \tag{2.23}$$

In Equation (2.23) V_L is the voltage across the load (V_L). Refer again to Figure 2.6a.

The load voltages for each load line equation for r_{int0}, r_{int1}, r_{int2}, and r_{int3} for $V_{oc} = 10$ V and load current $I_L = 10$ A, respectively, are

$$V_L = V_{oc} - r_{int0} I_L = 10\ \text{V}. \tag{2.24}$$

$$V_L = V_{oc} - r_{int1} I_L = 10 - 0.1 \times 10 = 9\ \text{V}. \tag{2.25}$$

$$V_L = V_{oc} - r_{int2} I_L = 10 - 0.2 \times 10 = 8\ \text{V}. \tag{2.26}$$

$$V_L = V_{oc} - r_{int3} I_L = 10 - 0.3 \times 10 = 7\ \text{V}. \tag{2.27}$$

Exercise: For Equations (2.25), (2.26), and (2.27) determine the actual load resistance R_L at the given conditions.

For Equation (2.25), the output or load voltage is 9 V, and since the load current is 10 A, then $R_L = 9$ V/10 A = 0.9 Ω. Similarly for Equation (2.26), $R_L = 8$ V/10 A = 0.8 Ω and for Equation (2.27), $R_L = 7$ V/10 A = 0.7 Ω.

Figure 2.7 depicts an ideal load line of an ideal 10-A current source with internal resistance $r_{int0} \to \infty$ and three real current sources with internal resistances of $r_{int1} = 10\ \Omega$, $r_{int2} = 5\ \Omega$, and $r_{int3} = 3.333\ \Omega$. All four current sources have a load that causes the load voltage to be 10-V. It is important to note that for equal voltage at the load, the current of the source with the numerically smallest internal resistance (r_{int3}) produces the lowest load current. The goal is to obtain as much of the current source current to flow through the load. The ideal source with an infinite internal resistance produces the highest possible load current I_L, which is 10 A. For the example on hand a current source with a 10-Ω internal resistance and 10 V at the load produces 9 A through the load and 1 A through the internal resistance. A current source with a 5-Ω internal resistance and 10 V at the load produces 8 A through the load and 2 A through the internal resistance. Finally, a current source with a 3.3333-Ω internal resistance and 10 V at the load produces 7 A through the load and 3 A through the internal resistance. Refer to Equations 2.24 through 2.27.

The load line equation is given from Kirchhoff's and Ohm's laws by Equation (2.28):

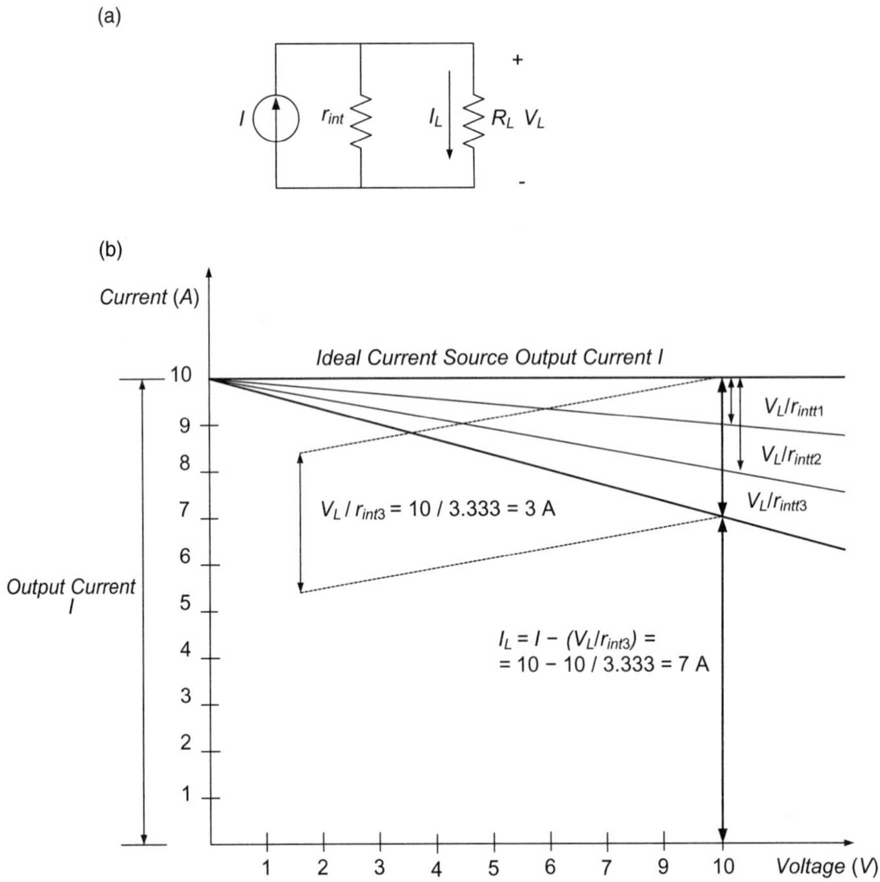

Figure 2.7 (a) Current source model with internal resistance; (b) different internal resistance current sources at the same load voltage.

$$I_L = I - \frac{V_L}{r_{int}}. \tag{2.28}$$

In Equation (2.28), I_L is the current through the load resistance R_L, I is the total current that the current source supplies, r_{int} is the current source internal resistance, and V_L is the load voltage or the voltage across R_L (R_L does not appear in Equation (2.28), refer to Figure 2.7a for the location of R_L). Thus, using Equation (2.28), the line load equations for r_{int0}, r_{int1}, r_{int2}, and r_{int3} for $I = 10$ A and load voltage $V_L = 10$ V, respectively, are

$$I_L = I = 10 \text{ A, because } r_{int0} \to \infty. \tag{2.29}$$

$$I_L = I - \frac{V_L}{r_{int1}} = 10 - (10/10) = 9 \text{ A}. \tag{2.30}$$

$$I_L = I - \frac{V_L}{r_{int2}} = 10 - (10/5) = 8 \text{ A}. \tag{2.31}$$

$$I_L = I - \frac{V_L}{r_{int3}} = 10 - (10/3.333) = 7 \text{ A}. \tag{2.32}$$

Independent current source I can be a *DC* current or an *AC* current source. When using a DC current source, I simply is the current DC value; when using an AC current source, I is typically the peak value of the sinusoidal current.

2.2 SINUSOIDAL STEADY STATE: TIME AND FREQUENCY DOMAINS

When sinusoidal voltage or current sources excite an *RLC* network, the sinusoidal voltage and current waveforms are of the same angular frequency ω in sinusoidal steady state. Sinusoidal steady state means that transient behavior is over. For the circuit given in Figure 2.8, which shows an AC voltage source in series with a resistor R, capacitor C, and inductor L, we can state the circuit equations using Kirchoff's voltage law (KVL) and Kirchoff's current law (KCL) directly in the time domain.

The time domain circuit equations for a resistor, capacitor, and inductor are summarized in Table 2.1 from previous sections of this chapter. The various scientists and engineers that developed basic circuit theory throughout most of the 19th century experimentally obtained such equations. It is important to state that the equations of Table 2.1 hold true regardless of the waveform that excites each element. For example, for a resistor, if its current $i_R(t)$ is a constant (DC), then its voltage $v_R(t)$ is a constant, since the voltage–current behavior is $v_R(t) = i_R(t)R$. Similarly, if the resistor current is a sinusoidal function of time, so will be the voltage across it. A second example for an inductor, if its current is a sinusoidal waveform with respect to time, the voltage developed across such inductor varies proportionally to the derivative of its current with respect to time. That is, $v_L(t) = Ldi(t)_L/dt$.

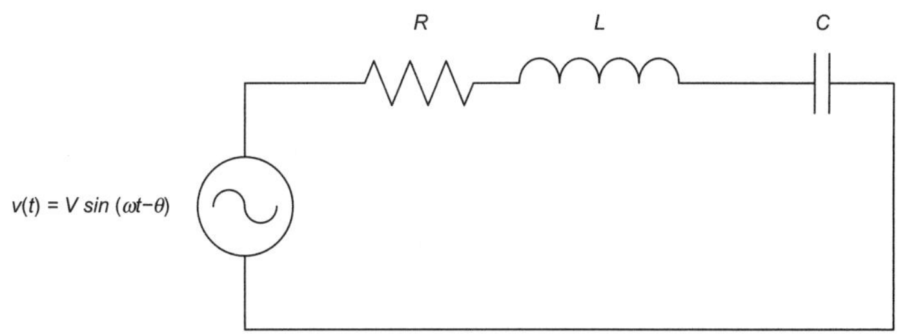

Figure 2.8 Series *RLC* circuit with sinusoidal voltage source.

Table 2.1 Voltage–current and current–voltage relationships for electric components (Universal time domain equations)

Circuit Element	Basic Voltage–Current Relationship	Basic Current–Voltage Relationship
R	$v_R(t) = i_R(t)\,R$	$i_R(t) = \dfrac{v_R(t)}{R}$
L	$v_L(t) = L\dfrac{di_L(t)}{dt}$	$i_L(t) = \dfrac{1}{L}\int_{-\infty}^{t} v_L(t)\,dt$
C	$v_C(t) = \dfrac{1}{C}\int_{-\infty}^{t} i_C(t)\,dt$	$i_C(t) = C\dfrac{dv_C(t)}{dt}$

2.2.1 Resistor under Sinusoidal Steady State

Based on the voltage–current relationships for R in Table 2.1, when

$$i_R(t) = I\sin(\omega t + \theta). \tag{2.33}$$

$$v_R(t) = i_R R = IR\sin(\omega t + \theta) = V\sin(\omega t + \theta). \tag{2.34}$$

V, *the peak voltage*, is defined as:

$$V = IR. \tag{2.35}$$

In Equation (2.35), I is the peak value of the current waveform and R is the resistor value.

Previously seen Figure 2.4 depicts the voltage and current waveform of a resistor with sinusoidal excitation. Figure 2.4 also shows the instantaneous power on the resistor and the average value of the power dissipated. Important facts to observe are that both voltage and current waveforms are exactly in phase; that is, they both have the same zero crossings, positive and negative peaks.

2.2.2 Inductor under Sinusoidal Steady State

Based on the voltage–current relationships for L in Table 2.1, when

$$i_L(t) = I\sin(\omega t + \theta). \tag{2.36}$$

$$v_L(t) = Ldi_L/dt = \omega LI\cos(\omega t + \theta) = V\cos(\omega t + \theta). \tag{2.37}$$

V *peak* is defined as

$$V = \omega LI, \tag{2.38}$$

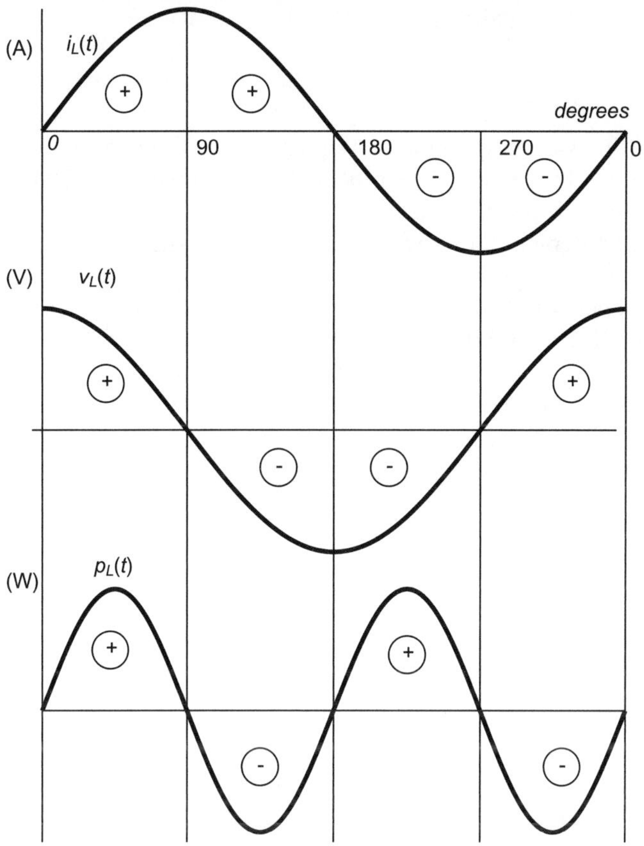

Figure 2.9 Inductor under sinusoidal voltage, current, and instantaneous power.

where ω is the angular frequency of the exciting current, I is the peak value of the current waveform, and L is the inductor value.

Figure 2.9 depicts the voltage and current waveform of an inductor with sinusoidal excitation. An important fact to observe is that the voltage waveform leads the current waveform by 90° (or $\pi/2$ radians). It is also interesting to note that the peak value of the voltage waveform (V) is a frequency-dependent term (recall that $\omega = 2\pi f$). We will discuss instantaneous power in the inductor shortly.

2.2.3 Capacitor under Sinusoidal Steady State

Based on the voltage–current relationships for C in Table 2.1, when

$$v_C(t) = V\sin(\omega t + \theta). \tag{2.39}$$

$$i_C(t) = C dv_C/dt = \omega CV\cos(\omega t + \theta) = I\cos(\omega t + \theta). \tag{2.40}$$

I, the peak current, is defined as:

$$I = \omega CV, \tag{2.41}$$

where ω is the waveform angular frequency, C is the capacitance value, V is the peak value of the voltage waveform; thus, I is the peak value of the current waveform through the capacitor.

Figure 2.10 depicts the current, voltage, and instantaneous power waveforms of a capacitor with sinusoidal excitation. An important fact to observe is that in a capacitor, the current waveform leads the voltage waveform by 90° (or $\pi/2$ rad). It is also interesting to note that the peak value of the current waveform (i_C) is an angular frequency-dependent term ($\omega = 2\pi f$). We will discuss instantaneous power in the capacitor shortly.

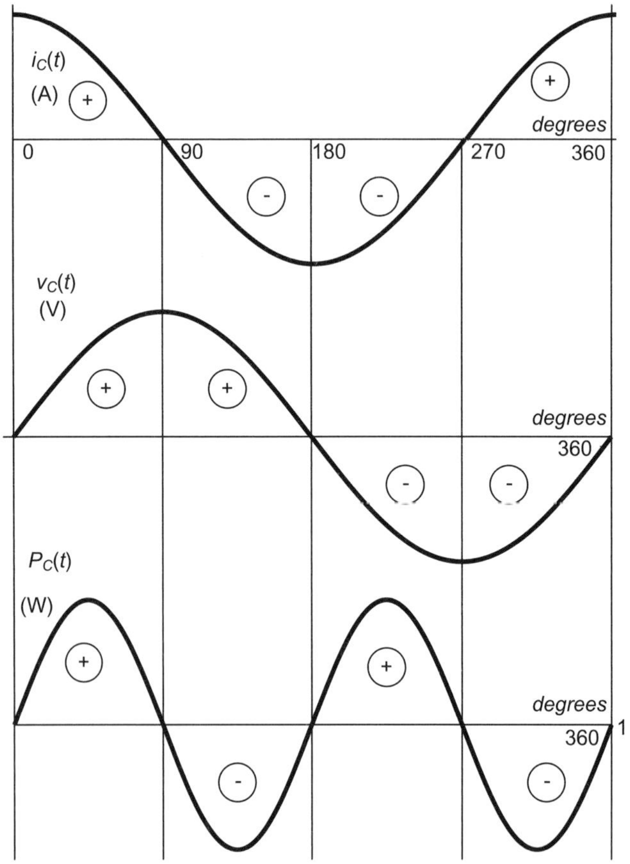

Figure 2.10 Capacitor under sinusoidal voltage, current, and instantaneous power.

Table 2.2 Time domain equations for *R*, *L*, and *C* with sinusoidal excitation

Electric Element	Voltage[a]	Current[a]	Voltage–Current Phase Relationship
Resistor	$v_R(t) = V \sin(\omega t + \theta)$	$i_R(t) = I \sin(\omega t + \theta)$	v_R and i_R are in-phase
Inductor	$v_L(t) = V \cos(\omega t + \theta)$	$i_L(t) = I \sin(\omega t + \theta)$	v_L leads i_L by 90°
Capacitor	$v_C(t) = V \sin(\omega t + \theta)$	$i_C(t) = I \cos(\omega t + \theta)$	i_C leads v_C by 90°

[a] All waveforms are referred to as sinusoidal, regardless whether they are expressed by a sine or a cosine function.

From Equations (2.33) through (2.41), Table 2.2 summarizes the results obtained.

Note: Figures 2.9 and 2.10 depict degrees in their horizontal axis; this is totally equivalent to display time, where 90° is 1/4 of a sinusoidal period, 180° is half-a-period, and so on.

When circuits operate in sinusoidal steady state, it is particularly useful to use complex numbers instead of manipulating time domain equations. When using time domain equations, differential equations need to be solved. When dealing with complex numbers, complex algebra manipulations are required instead of having to solve differential equations. This topic will be addressed further in the section about phasors.

2.2.4 Brief Complex Number Theory Facts

The purpose of this section is to provide a brief review on complex numbers and their basic operations.

Mathematically, "*i*" is the imaginary number unit; however, electrical engineers prefer to use "*j*" because the letter *i* is reserved for current.

Complex number theory begins with its fundamental assumption or definition:

$$j = \sqrt{-1} \tag{2.42}$$

A complex number z is a number of the form $a + jb$, where a is the real part of the complex number z, $\mathrm{Re}\{z\} = a$ and b is the imaginary part of z, $\mathrm{Im}\{z\} = b$.

Complex number:

$$z = a + jb \tag{2.43}$$

is said to be represented in *rectangular* form. Complex numbers can be represented on the complex plane. The horizontal axis of this plane is used to represent the real part of the complex number, and the vertical axis or the

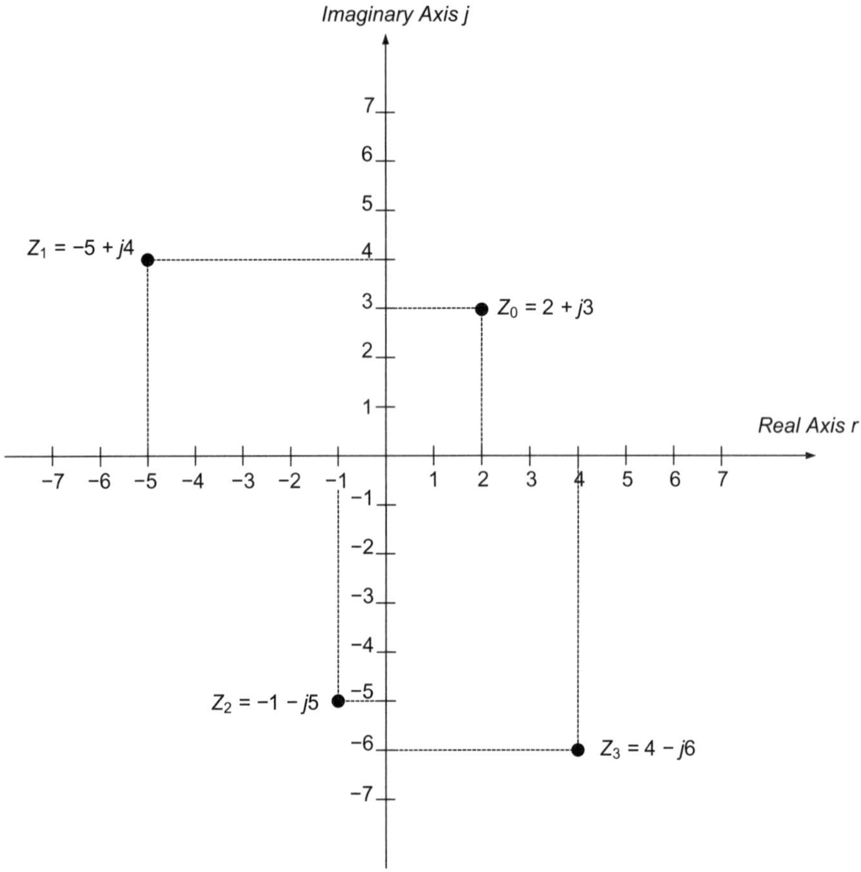

Figure 2.11 Complex plane showing complex numbers in rectangular form.

j-axis is used to represent the imaginary part of the complex number. Figure 2.11 shows the complex plane, and on the plane there are four examples of complex numbers.

In particular, a complex number with its zero real part is said to be a pure imaginary number. Conversely, a complex number with zero imaginary part is said to be a real number.

Examples of *pure imaginary numbers* in rectangular form are

$$0 + j3 = j3;$$

$$0 + j4.5 = j4.5;$$

$$0 + j1 = j;$$

$$0 + j\pi = j\pi.$$

Examples of *real numbers* in rectangular form are

$$+1 + j0 = 1;$$

$$\pi - j0 = \pi;$$

$$23.7 + j0 = 23.7;$$

$$1 + j0 = 1.$$

2.2.4.1 Complex Numbers in Polar Form Complex numbers can also be represented in *polar* form. Figure 2.12 shows a complex number with real part a, imaginary part b, and how it relates to its modulus or absolute value ρ (rho) and its phase angle θ (theta) with respect to the real axis.

From Figure 2.12 and trigonometric identities, one can see that the absolute value of the complex number is related to its rectangular component as follows:

$$\rho = \sqrt{a^2 + b^2}. \tag{2.44}$$

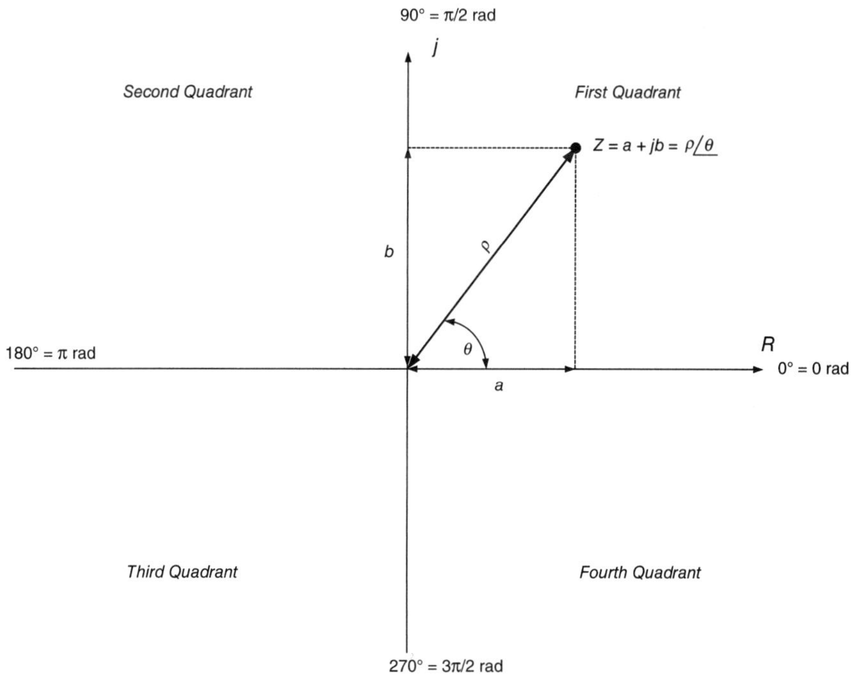

Figure 2.12 Complex numbers in rectangular and in polar forms.

The phase angle θ, also called the argument of z, is related to its *rectangular* components as follows:

$$\theta = \tan^{-1}\left(\frac{b}{a}\right), \tag{2.45}$$

where a and b are respectively the real and imaginary part of complex number z.

The complex number in polar form is represented as follows:

$$z = \rho\angle\theta. \tag{2.46}$$

Figure 2.12 also depicts the four quadrants within the trigonometric circle:

Quadrant I encompasses angles in the range: $90° < \theta < 0°$
Quadrant II encompasses angles in the range: $180° < \theta < 90°$
Quadrant III encompasses angles in the range: $270° < \theta < 180°$
Quadrant IV encompasses angles in the range: $360° < \theta < 270°$

where 0°, 90°, 180°, 270°, and 360° angles are the boundaries between quadrants.

It is also important to note that the following convention is also accepted:

Negative angles whose angle absolute value is within the range: $90° < |\theta| < 0°$ are in Quadrant IV.

Example 2.5 Negative angles whose angle absolute value is within the range: $90° < |\theta| < 0°$.

–30°, –5°, and –75° are all examples of angles that reside in Quadrant IV.

Example 2.6 Negative angles whose angle absolute value is within the range: $180° < |\theta| < 90°$.

–110°, –145°, and –175° are all examples of angles that reside in Quadrant III.

Example 2.7 Negative angles whose angle absolute value is within the range: $270° < |\theta| < 180°$.

–190°, –205°, and –265° are all examples of angles that reside in Quadrant II.

Example 2.8 Negative angles whose angle absolute value is within the range: $270° < |\theta| < 360°$.

–280°, –300°, and –334° are all examples of angles that reside in Quadrant I.

Example 2.9 Convert the following complex numbers from *rectangular* form to *polar* form:

(a) $z_0 = 2 + j3$
(b) $z_1 = -5 + j4$
(c) $z_2 = -1 - j5$
(d) $z_3 = 4 - j6$

Applying Equations (2.43) through (2.45) for (a) through (d) we obtain

(a) $z_0 = 2 + j3 = (2^2 + 3^2)^{½} \angle\tan^{-1}(3/2) = 3.606\angle 56.31°$
(b) $z_1 = -5 + j4 = [(-5)^2 + 4^2]^{½} \angle\tan^{-1}[4/(-5)] = 6.403\angle 141.34°$
(c) $z_2 = -1 - j5 = [(-1)^2 + (-5)^2]^{½} \angle\tan^{-1}[(-5)/(-1)] = 5.099\angle 258.69°$
(d) $z_3 = 4 - j6 = [4^2 + (-6)^2]^{½} \angle\tan^{-1}[(-6)/4] = 7.211\angle -56.31°$

2.2.4.2 *Complex Numbers in Euler's Form*

From Euler's identity,

$$z = \rho e^{j\theta} = \rho(\cos\theta + j\sin\theta), \tag{2.47}$$

where ρ is the modulus or amplitude of the complex number z and θ the angle that its module has with respect to the real axis; complex number z then is

$$z = \rho e^{j\theta} = \rho\angle\theta. \tag{2.48}$$

From Euler's equality, Equation (2.47), it can be seen by looking at the *rectangular* representation of a complex number, previously given by Equation (2.43), that

$$\mathrm{Re}\{z\} = a = \rho\cos\theta \tag{2.49}$$

and

$$\mathrm{Im}\{z\} = b = \rho\sin\theta \tag{2.50}$$

Equations (2.49) and (2.50) show a direct conversion of complex number from polar form into Euler form.

Example 2.10 Convert the following complex numbers from *polar* form to *Euler's* form:

$$z_1 = 3.606\angle 56.31°$$

$$z_2 = 6.403\angle 141.34°$$

$$z_3 = 5.099\angle 258.69^\circ$$

$$z_4 = 7.211\angle -56.31^\circ$$

The conversion from polar form is straightforward; it just uses the modulus and the phase angle in Euler's equation. Yielding

$$z_1 = 3.606\angle 56.31^\circ = 3.606e^{j56.31\circ}$$

$$z_2 = 6.403\angle 141.34^\circ = 6.403e^{j141.34\circ}$$

$$z_3 = 5.099\angle 258.69^\circ = 5.099e^{j258.69\circ}$$

$$z_4 = 7.211\angle -56.31^\circ = 7.211e^{-j56.31\circ}$$

2.2.4.3 Arithmetic Operations with Complex Numbers

2.2.4.3.1 Rectangular Form Addition/Subtraction

$$\text{Given } z_1 = a + jb \text{ and } z_2 = c + jd\text{, then } z_1 + z_2 = (a + c) + j(b + d). \quad (2.51)$$

$$\text{Given } z_1 = a + jb \text{ and } z_2 = c + jd\text{, then } z_1 - z_2 = (a - c) + j(b - d). \quad (2.52)$$

From Equations (2.51) and (2.52), it can be seen that for addition or subtraction in *rectangular* form, real parts get added or subtracted, and imaginary parts get added or subtracted.

Example 2.11 Given complex numbers z_0, z_1, z_2, and z_3 in rectangular form, perform the following operations: (a) $z_0 + z_1$; (b) $z_2 - z_3$; (c) $z_1 + z_2 - z_3$; and (d) $-z_0 - z_2$.

$$z_0 = 2 + j3$$

$$z_1 = -5 + j4$$

$$z_2 = -1 - j5$$

$$z_3 = 4 - j6$$

Solutions to Example 2.11

(a) $z_0 + z_1 = (2 - 5) + j(3 + 4) = -3 + j7$

(b) $z_2 - z_3 = (-1 - j5) - (4 - j6) = -1 - 4 - j5 + j6 = -5 + j$

(c) $z_1 + z_2 - z_3 = -5 + j4 - 1 - j5 - (4 - j6) = -10 + j5$

(d) $-z_0 - z_2 = -(2 + j3) - (-1 - j5) = -1 + j2$

2.2.4.3.2 Polar and Euler's Forms Addition/Subtraction To add or subtract complex numbers in *polar* or *Euler*'s forms, it is convenient to convert the complex numbers to *rectangular* form, do the addition (or subtraction), and convert the results back to *polar* or *Euler*'s form.

2.2.4.3.3 Rectangular Form Multiplication

$$\text{Given } z_1 = a + jb \text{ and } z_2 = c + jd, \text{ then } z_1 \times z_2 = (a + jb) \times (c + jd). \tag{2.53}$$

Performing the term-by-term multiplication of both complex numbers in rectangular form and taking into account that $j^2 = -1$, leads to

$$z_1 \times z_2 = (ac - bd) + j(ad + bc) \tag{2.54}$$

Example 2.12 Multiplication of complex numbers given in rectangular form.
Given $z_1 = 8 + j6$, $z_2 = 2 - j1$; find the product $z_1 \times z_2$ operating with both numbers in their given rectangular form.

Solution to Example 2.12

$$\begin{aligned} z_1 \times z_2 &= (8 + j6)(2 - j1) = 16 + (j6)(-j1) + j \times 6 - j1 \times 8 \\ &= 16 + 6 + j12 - j8 = 22 + j4 \end{aligned}$$

2.2.4.3.4 Euler's and Polar Forms Multiplication Given: $z_1 = \rho_1 e^{j\theta_1} = \rho_1 \angle \theta_1$ and $z_2 = \rho_2 e^{j\theta_2} = \rho_2 \angle \theta_2$, respectively in Euler's form and polar form.

The product is obtained by multiplying ρ_1 and ρ_2, and by adding their respective phase angles, $\theta_1 + \theta_2$, so that the final product is

$$z_1 \times z_2 = (\rho_1 + \rho_2) e^{j(\theta 1 + \theta 2)}. \tag{2.55}$$

Equation (2.55) is in Euler's form and similarly in polar form:

$$z_1 \times z_2 = (\rho_1 + \rho_2) \angle (\theta 1 + \theta 2). \tag{2.56}$$

Example 2.13 Find the product of z_1 and z_2. $z_1 = 12\angle 25°$ and $z_2 = 3\angle 60$.

Solution to Example 2.13

Applying Equation (2.55), we calculate the desired product:

$$z_1 \times z_2 = (12\angle 25° \times (3\angle 60°) = 12.3\angle(25° + 60°) = 36\angle 85°.$$

2.2.4.3.5 *Rectangular Form Division*

$$\text{Given } z_1 = a + jb \text{ and } z_2 = c + jd, \text{ then } z_2/z_1 = (c + jd)/(a + jb). \tag{2.57}$$

Multiplying the numerator and denominator by the complex conjugate of the denominator allows rationalizing the complex number. That is, it eliminates the imaginary part of the number of the denominator.

Since the denominator in the given case is

$$(a + jb),$$

its complex conjugate has the same real part but complementary imaginary part; that is,

$$\text{Complex Conjugate } (a + jb) = a - jb. \tag{2.58}$$

Since $z_1 = a + jb$, its complex conjugate is indicated as

$$z_1^* = a - jb. \tag{2.59}$$

Then,

$$z_2/z_1 = \frac{(c + jd)(a - jb)}{(a + jb)(a - jb)} \tag{2.60}$$

$$\frac{(c + jd)(a - jb)}{(a^2 + b^2)} = \frac{(ac + bd) + j(ad - bc)}{(a^2 + b^2)}. \tag{2.61}$$

Example 2.14 Division of complex numbers given in rectangular form.

Given $z_2 = 8 + j6$, and $z_1 = 2 - j1$, find the quotient z_2/z_1 using both numbers in rectangular form.

Solution to Example 2.14

$$\begin{aligned} z_2/z_1 &= (8 + j6)/(2 - j1) = \frac{(8 + j6)(2 + j1)}{(2 - j1)(2 + j1)} = \frac{16 - 6 + j(12 + 8)}{(2^2 + 1^2)} \\ &= \frac{10 + j20}{5} = 2 + j4. \end{aligned} \tag{2.62}$$

2.2.4.3.6 *Polar and Euler's Forms Division* Given: $z_1 = \rho_1 e^{j\theta_1} = \rho_1\angle\theta_1$ and $z_2 = \rho_2 e^{j\theta_2} = \rho_2\angle\theta_2$, where we define z_2 as the dividend and z_1 as the divisor.

The quotient of z_2/z_1 is obtained by dividing the modulus of the dividend by the modulus of the divisor (ρ_2/ρ_1), and by subtracting the divisor phase angle θ_1 from the dividend phase angle θ_2 so that the final quotient is

$$z_2 / z_1 = (\rho_2 / \rho_1) e^j (\theta_2 - \theta_1) \tag{2.63}$$

in Euler's form and similarly in polar form:

$$z_2 \times z_1 = (\rho_2 / \rho_1) \angle(\theta_2 - \theta_1). \tag{2.64}$$

Example 2.15 Find the quotient of z_2/z_1, where $z_2 = 3\angle 60°$ and $z_1 = 12\angle 25°$.

Solution to Example 2.15

Applying Equation (2.64) we calculate the desired quotient:

$$z_2 / z_1 = (3\angle 60°) \times (12\angle 25°) = 3/12\angle(60° - 25°) = 0.25\angle 35°. \tag{2.65}$$

2.3 TIME DOMAIN EQUATIONS: FREQUENCY DOMAIN IMPEDANCE AND PHASORS

The basic equations describing the voltage–current relationships, where voltage and current are functions of time in resistors, capacitors, and inductors (see Table 2.1), are referred to as the time domain equations of those electric components. Those equations were experimentally determined. In the particular case that we need to deal with sinusoidal steady-state regime, current and voltage waveforms have a single frequency, and they vary sinusoidally with respect to time; it is possible to manipulate the waveform with phasors instead of the time domain differential or integral equations.

We will address phasors shortly, but the main advantage of using phasors, provided that the circuit is in sinusoidal steady state, is that the voltage and current calculations need not be in the time domain; consequently, no differential equations need to be solved. Phasors allow current and voltage calculations to be made with simple arithmetic equations. The catch is that such arithmetic is complex arithmetic; real and imaginary numbers are involved.

2.3.1 Phasors

A sinusoidal voltage or current waveform varying with respect to time, such as

$$v(t) = V \sin(\omega t + \theta)$$

can also be described with a phasor of amplitude or peak value V, rotating at a constant angular frequency ω, and θ its phase shift with respect to zero degrees. Figure 2.13 depicts a phasor rotating in counterclockwise direction generating as it rotates each ordinate or sine value of our sinusoid.

For Figure 2.13 above, the phase angle θ is assumed to be zero, which is the reason why the sine wave in the time domain begins at the origin of the time axis.

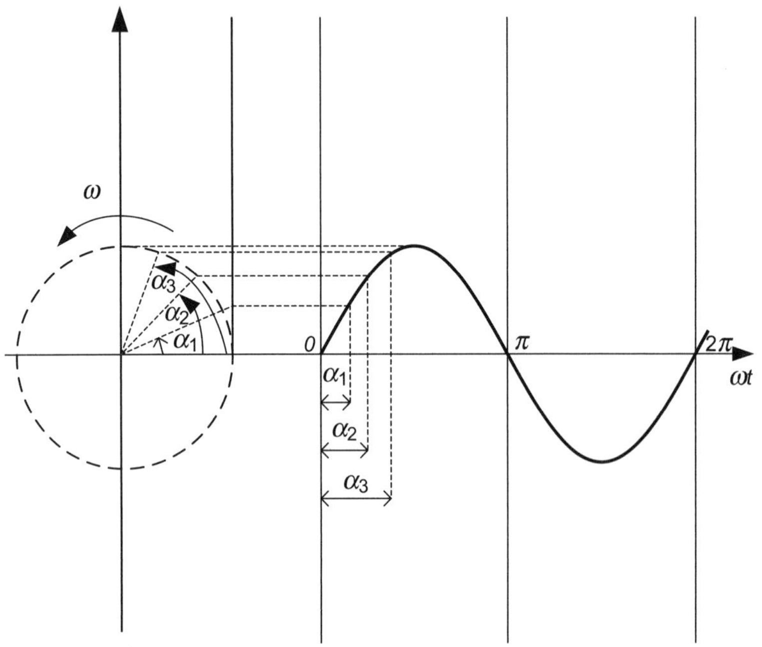

Figure 2.13 Phasor-generated sine wave with zero-phase angle.

Example 2.16 Determine the phasor of the following sinusoidal, time domain equation.

Given: $v(t) = V\sin(\omega t + \theta)$, where $V = 20$ V, $\omega = 60$ rad/s, and $\theta = 45°$.

Rewriting the sinusoidal waveform with the given numerical values results in

$$v(t) = 20\sin(60t - 45°).$$

We can represent the sinusoidal waveform with its generating phasor instead of using the time-varying sine function. The phasor is: $20\angle 45°$. Figure 2.14 depicts this phasor.

Example 2.17 Phasor of sinusoidal waveform $i(t) = 5\sin(60t + 30°)$.

$i(t)$ is a current waveform, of a 5 A peak amplitude, 60 rad/s angular frequency ω, and a 30° phase angle θ. The phasor is: $5\angle 30°$.

Figure 2.15 depicts such current phasor.

2.3.2 The Impedance Concept

The impedance of a circuit element, where a circuit element can be a resistor, an inductor or a capacitor, is defined as the ratio of its voltage phasor **V** divided by its current phasor **I**. So referring to the time domain equations for R, L, and C circuit elements (Table 2.1) with sinusoidal excitation, we will find their equivalent voltage and current phasor to determine what their impedance is.

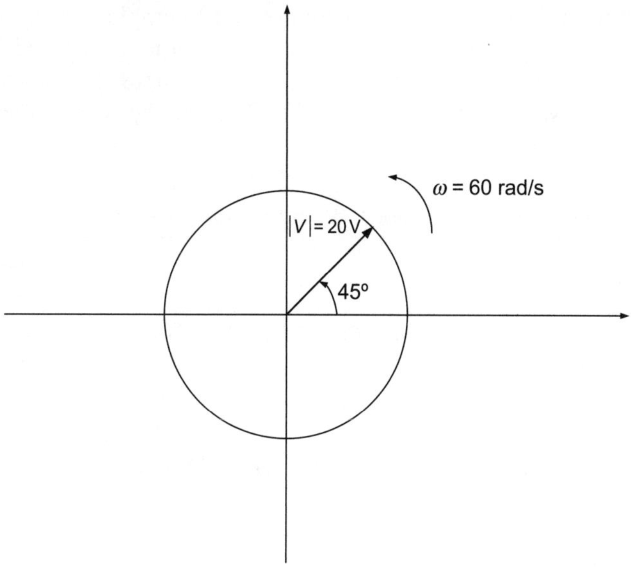

Figure 2.14 Phasor of sinusoidal waveform: $v(t) = 20\sin(60t - 45°)$.

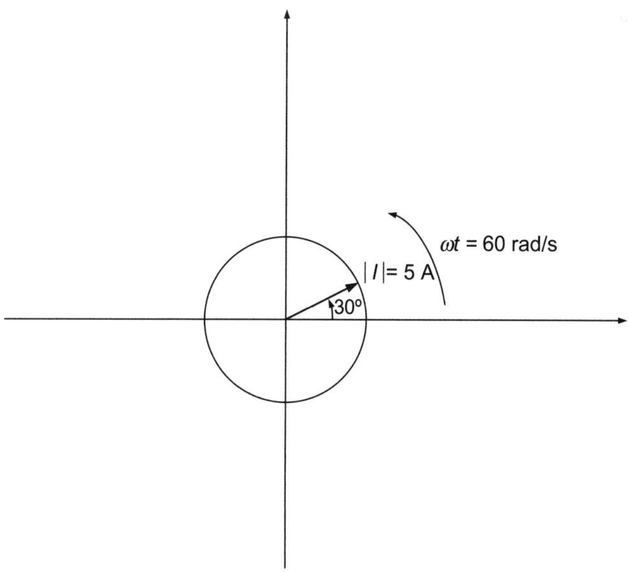

Figure 2.15 Current phasor for current waveform $i(t) = 5\sin(60t - 30°)$.

Before proceeding much further, it is important to emphasize that the impedance concept is only meaningful at one angular frequency and when the voltage and current waveforms are sinusoidal. To evaluate the phasors for each circuit element, we will make use of Table 2.2. For the reader's convenience, Table 2.2 is repeated here:

Table 2.2 Time domain equations for *R*, *L*, and *C* with sinusoidal excitation

Electric Element	Voltage[a]	Current[a]	Voltage–Current Phase Relationship
Resistor	$v_R(t) = V\sin(\omega t + \theta)$	$i_R(t) = I\sin(\omega t + \theta)$	v_R and i_R are in-phase
Inductor	$v_L(t) = V\cos(\omega t + \theta)$	$i_L(t) = I\sin(\omega t + \theta)$	v_L leads i_L by 90°
Capacitor	$v_C(t) = V\sin(\omega t + \theta)$	$i_C(t) = I\cos(\omega t + \theta)$	i_C leads v_C by 90°

[a] All waveforms are referred to as sinusoidal, regardless whether they are expressed by a sine or a cosine function.

2.3.3 Purely Resistive Impedance

For a resistor from Table 2.2 we have that both sinusoidal voltage and current are in phase, so the impedance of a *pure** resistor is a real number, expressed by Equation (2.66):

$$\mathbf{Z_R} = \mathbf{V_R}/\mathbf{I_R}. \tag{2.66}$$

In Equation (2.66), $\mathbf{V_R}$ is the voltage phasor that corresponds to the time-varying sinusoidal voltage developed across the resistor. $\mathbf{I_R}$ is the current phasor that corresponds to the time-varying current through the resistor.

That is, $v_R(t) = V\sin(\omega t + \theta)$. $\mathbf{I_R}$ is the current phasor of the sinusoidal time-varying waveform that flows through the resistor. $\mathbf{Z_R}$ denotes impedance and, in a general sense, **V**, **I,** and **Z** are complex numbers (actually referred to as phasors). However, because both voltage and current phasors are always in phase on a resistor, the actual impedance for a *pure* resistor is a real number. Often times, the impedance of a pure resistor is referred to as simply *R*, the resistance itself.

Example 2.18 Given a sinusoidal voltage and current equal to

$$v_R(t) = 120\sin(2\pi 60t - 45°) \tag{2.67}$$

and

$$i_R(t) = 20\sin(2\pi 60t - 45°), \tag{2.68}$$

* A *pure resistor*, also called an *ideal* resistor, means within this context, that the resistor exclusively has resistive properties and has no parasitic inductive or capacitive characteristics.

where $v_R = 120$ V is the voltage peak value, $\omega = 2\pi\ 60$ rad/s, ($f = 60$ Hz), and phase angle θ is 45° for the resistor voltage waveform of Equation (2.67). Similarly for the current waveform, $i_R = 20$ A is the current peak value, $\omega = 2\pi\ 60$ rad/s, ($f = 60$ Hz), and phase angle θ is 45° (Eq. 2.68).

Determine the voltage and current phasors on the resistor and the resistor value.

Solution to Example 2.18

From Equation (2.66) we can see that the voltage phasor corresponding to such time domain waveform is

$$\mathbf{V_R} = 120\angle 45° \tag{2.69}$$

and the current phasor is

$$\mathbf{I_R} = 20\angle 45°. \tag{2.70}$$

Thus, $\mathbf{Z_R} = 120/20 = 6\ \Omega$, a real number, which means that the impedance is purely resistive in this case.

Note that the resistive impedance turns out to be a real number after all. This will not happen with inductors and capacitors. In general, impedance phasors will always be of the complex form, with nonzero real and imaginary parts, when a circuit contains resistance, plus inductance and/or capacitance.

Graphical interpretation of phasors $\mathbf{V_R} = 120\angle 45°$ and $\mathbf{I_R} = 20\angle 45°$

Both phasors $\mathbf{V_R}$ and $\mathbf{I_R}$ rotate at a constant angular frequency $\omega = 2\pi\ 60$ rad/s = 376.98 rad/s, and since both phasors are in phase, their phase difference is zero. Figure 2.16 is a representation of both phasors in the complex plane.

2.3.4 Inductive Impedance: Inductive Reactance

For an inductor, from Table 2.2, we have that the sinusoidal voltage across the inductor leads the sinusoidal current through the inductor by 90°. The impedance of a *pure** inductor is

$$\mathbf{Z_L} = \mathbf{V_L}/\mathbf{I_L}. \tag{2.71}$$

In Equation (2.71), $\mathbf{V_L}$ is the voltage phasor that corresponds to the time-varying sinusoidal voltage developed across the inductor. That is, $v_L(t) = V\cos(\omega t + \theta)$. $\mathbf{I_L}$ is the current phasor of the sinusoidal time-varying waveform that flows through the inductor. $\mathbf{Z_L}$ denotes impedance and in a general sense,

* A *pure inductor* also called an ideal inductor means, in this context, that the inductor exclusively has inductive properties and no parasitic resistive or capacitive characteristics.

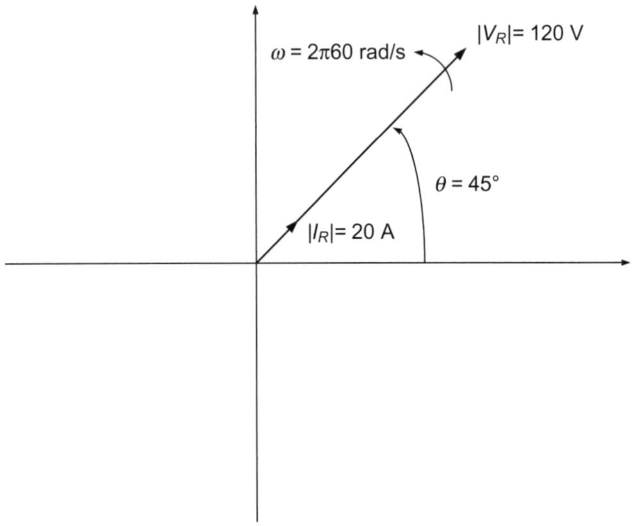

Figure 2.16 Resistor voltage and current phasors.

$\mathbf{V_L}$, $\mathbf{I_L}$, and $\mathbf{Z_L}$ are complex numbers (they are also referred to as phasors). However, because on a pure inductor its voltage phasor always leads the current phasor by 90°, the actual impedance of a *pure* inductor is a pure imaginary number (i.e., has zero real part).

Example 2.19 Given: a sinusoidal voltage and current equal to

$$v_L(t) = 45\cos(2\pi 100 + 20°) \tag{2.72}$$

and

$$i_L(t) = 5\sin(2\pi 100t + 20°). \tag{2.73}$$

From trigonometry we know that

$$\cos x \text{ leads } \sin x \text{ by } 90°. \tag{2.74}$$

Thus

$$v_L(t) \text{ leads } i_L(t) \text{ by } 90° \tag{2.75}$$

We can now proceed and determine that the respective phasors for $v_L(t)$ and $i_L(t)$ are

$$\boldsymbol{V_L} = 45\angle 110° \tag{2.76}$$

and

$$\boldsymbol{I_L} = 5\angle 20°, \tag{2.77}$$

where $|\boldsymbol{V_L}| = 45$ V, $\omega = 2\pi\ 100$ rad/s, ($f = 100$ Hz), and phase angle θ is 110° for the inductor voltage waveform of Equation (2.61). Similarly for the current waveform, $\boldsymbol{I_L} = 5$ A, $\omega = 2\pi\ 100$ rad/s, ($f = 100$ Hz), and the phase angle θ is 20°.

Using Equation (2.71) with Equations (2.76) and (2.77), we get that $\boldsymbol{Z_L} = \boldsymbol{V_L}/\boldsymbol{I_L} = 9\Omega\angle 90° = j9\Omega$, an imaginary number.

Example 2.20 Determine the voltage and current phasors and the impedance of the pure inductor from Example 2.19.

Equations (2.76) and (2.77) are repeated here for the reader's convenience:

$$\boldsymbol{V_L} = 45\angle 110°. \tag{2.78}$$

$$\boldsymbol{I_L} = 5\angle 20°. \tag{2.79}$$

From Equations (2.78) and (2.79), since we know that the impedance of an inductor is the ratio of voltage and current phasors, this leads to

$$\begin{aligned} \mathbf{Z}_L &= 45/5\angle(110° - 20°) \\ &= \mathbf{Z}_L = 9\ \Omega\angle 90° \end{aligned} \tag{2.80}$$

or simply

$$\mathbf{X}_L = 9\ \Omega\angle 90°\ \text{(in polar form)} \tag{2.81}$$

or

$$\mathbf{X}_L = j9\ \Omega\ \text{(in rectangular form)}, \tag{2.82}$$

where $\mathbf{X}_L$ is defined as the *reactive inductance* of the given inductor. The *reactive inductance*, Equation (2.81), represents a pure imaginary number as predicted earlier.

We can further look at equations for v_L and i_L from Table 2.2 since

$$v_L(t) = V\cos(\omega t + \theta) \tag{2.83}$$

and

$$i_L(t) = I\sin(\omega t + \theta), \tag{2.84}$$

also remembering from Table 2.2 that $v_L(t) = Ldi_L(t)/dt$. Using this equation into Equations (2.83) and (2.84) we obtain for

$$v_L(t) = Ldi_L(t)/dt = Ld[I\sin(\omega t + \theta)]/dt \tag{2.85}$$

$$v_L(t) = \omega LI\cos(\omega t + \theta), \tag{2.86}$$

where the term (ωLI) is the peak voltage V of Equation (2.86) for $v_L(t)$:

$$V = V_{peak} = \omega LI = |\mathbf{X_L}|, \tag{2.87}$$

where $|\mathbf{X_L}|$ is the absolute value of the inductive reactance given in Equation (2.71). The absolute value of the inductive reactance equals the absolute value of the inductor impedance, because its impedance real part is zero.

Again identifying the phasors for time domain Equations (2.85) and (2.86), we get that for an inductor,

$$\mathbf{V_L} = \mathbf{I_L X_L}. \tag{2.88}$$

Equation (2.88) is very important because it describes Ohm's law in phasor form or for an inductor when used in sinusoidal steady state. Note that all three terms, $\mathbf{V_L}$, $\mathbf{I_L}$, and $\mathbf{X_L}$, are complex numbers, and in a general sense they have magnitude and phase.

From Equations (2.78) and (2.79), the impedance of a pure inductor in rectangular form is

$$\mathbf{Z_L} = \mathbf{X_L} = \mathbf{V_L}/\mathbf{I_L} = j\omega L \quad \text{(for an inductor, } \mathbf{Z_L} = \mathbf{X_L}\text{)}. \tag{2.89}$$

Graphical interpretation of inductor phasors $\mathbf{V_L} = 45\angle 110°$ and $\mathbf{I_L} = 5\angle 20°$
Both phasors $\mathbf{V_L}$ and $\mathbf{I_L}$ are rotating at a constant angular frequency $\omega = 2\pi\ 100$ rad/s = 628.30 rad/s; both phasors are separated by a fixed 90° phase difference, where $\mathbf{V_L}$ leads $\mathbf{I_L}$ by 90°. Figure 2.17 is a representation of both $\mathbf{V_L}$ and $\mathbf{I_L}$ inductor phasors in the complex plane. Note that the initial phase

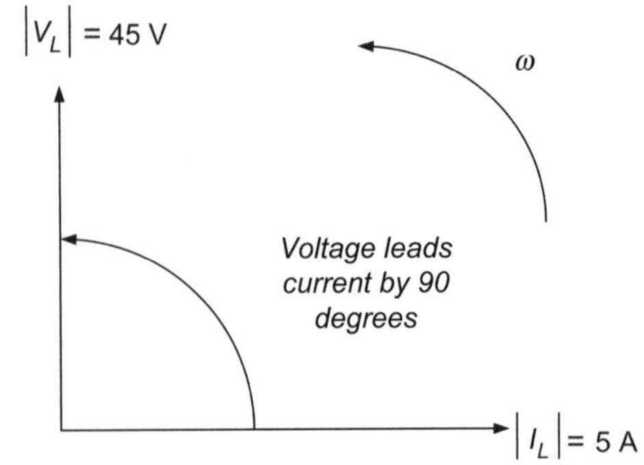

Figure 2.17 Inductor voltage and current phasors.

angle from both the inductor voltage and capacitor current were eliminated from Figure 2.17.

2.3.5 Purely Capacitive Impedance: Capacitive Reactance

For a capacitor from Table 2.2 we have that the sinusoidal current through the capacitor leads the sinusoidal voltage drop across the capacitor by 90°. The impedance of a *pure** capacitor is

$$\mathbf{Z}_C = \mathbf{V}_C / \mathbf{I}_C, \tag{2.90}$$

where $\mathbf{V}_C$ is the voltage phasor that corresponds to the time-varying sinusoidal voltage developed across the inductor, i.e., $v_C(t) = V \sin(\omega t + \theta)$. $\mathbf{I}_C$ is the current phasor of the sinusoidal time-varying waveform that flows through the capacitor. $\mathbf{Z}_C$ denotes impedance, and in a general sense, $\mathbf{V}_C$, $\mathbf{I}_C$, and $\mathbf{Z}_C$ are complex numbers (actually referred to as phasors).

However, because on a pure capacitor its current always leads the voltage phasor by 90°, the actual impedance of a *pure* capacitor is a pure imaginary number (has zero real part).

Example 2.21 Given a capacitor with a sinusoidal voltage and current equal to

$$v_C(t) = 14\sin(2\pi\ 2\ \text{MHz}\ t + 45°) \tag{2.91}$$

and

$$i_C(t) = 2\cos(2\pi\ 2\ \text{MHz}\ t + 45°). \tag{2.92}$$

Thus

$$v_C(t) = 14\sin(4\pi \times 10^6 t + 45°) \tag{2.93}$$

and

$$i_C(t) = 2\cos(4\pi \times 10^6 t + 45°) \tag{2.94}$$

We can see from (2.93) and (2.94) that the capacitor current leads the capacitor voltage by 90°. Please also refer to Table 2.2.

$$\textit{Capacitor phasor } I_C \textit{ leads capacitor phasor } V_C \textit{ by } 90° \tag{2.95}$$

We can now proceed and determine that the respective phasors for $v_C(t)$ and $i_C(t)$ are

$$V_C = 14\angle 45° \tag{2.96}$$

* A *pure capacitor*, also called an ideal capacitor, means in this context, that the capacitor exclusively has capacitive properties and has no parasitic resistive or inductive characteristics.

and since $\boldsymbol{I_C}$ leads $\boldsymbol{V_C}$ by 90°

$$\boldsymbol{I_C} = 2\angle 135°, \tag{2.97}$$

where $\boldsymbol{V_C} = 14$ V peak voltage, $\omega = 12.57$ Mrad/s, ($f = 2$ MHz), and phase angle θ is 45° for the capacitor voltage waveform of Equation (2.91). Similarly for the current waveform, $\boldsymbol{I_C} = 2$ A peak current, $\omega = 12.57$ Mrad/s, ($f = 2$ MHz), and phase angle θ is 135°.

Determine the voltage and current phasors and the impedance of the pure capacitor.

Solution to Example 2.21

From Equation (2.91), we can see that the voltage phasor corresponding to such time domain waveform is

$$V_C = 14\angle 45°. \tag{2.98}$$

And from Equation (2.95), the current phasor is

$$I_C = 2\angle 135°. \tag{2.99}$$

From Equations (2.98) and (2.99), we know that since the impedance of a capacitor is the ratio of voltage and current phasors, this leads to

$$\begin{aligned} \mathbf{Z_C} &= 14/2\angle(45° - 135°) \\ &= \mathbf{Z}_C = 7\ \Omega\angle -90° \end{aligned} \tag{2.100}$$

or simply

$$\mathbf{X_C} = 7\ \Omega\angle -90°\ (\textit{in polar form}). \tag{2.101}$$

Or

$$\begin{aligned} \mathbf{X_C} &= -j7\ \Omega\ (\textit{in rectangular form}) \\ &= 1/j(7\ \Omega)\ (\textit{also in rectangular form, without rationalizing } j). \end{aligned} \tag{2.102}$$

In Equation (2.102), $\mathbf{X_C}$ is defined as the *reactive capacitance* of the given capacitor. The *reactive capacitance*, Equation (2.102), represents a pure imaginary number as predicted earlier.

We can further look at equations for v_C and i_C from Table 2.2, and since

$$v_C(t) = V\sin(\omega t + \theta) \tag{2.103}$$

and

$$i_C(t) = I\cos(\omega t + \theta), \tag{2.104}$$

also remember from Table 2.1 that $i_C(t) = Cdv_C(t)/dt$. Using this expression into Equations (2.103) and (2.104) we obtain for

$$v_C(t) = Cdv_C(t)/dt = Cd[V\sin(\omega t + \theta)]/dt \tag{2.105}$$

$$v_C(t) = \omega CV\cos(\omega t + \theta), \tag{2.106}$$

where the term $\omega\, C\, V$ is the peak voltage V of Equation (2.106):

$$V = V_{peak} = \omega CV. \tag{2.107}$$

Again identifying the phasors for time domain Equations (2.104) and (2.106), we get that for a capacitor,

$$\mathbf{V_C} = \mathbf{I_C X_C}. \tag{2.108}$$

Equation (2.108) is very important because it describes Ohm's law in phasor form, for a capacitor when used in sinusoidal steady state. Note that all three terms, $\mathbf{V_C}$, $\mathbf{I_C}$, and $\mathbf{X_C}$ are complex numbers that have magnitude and phase.

In general, for a pure capacitor, the reactive capacitance is given by

$$\mathbf{Z_C} = \mathbf{X_C} = \frac{1}{j\omega C} = -j\frac{1}{\omega C}. \tag{2.109}$$

Note that $|\mathbf{X_C}| = 1/\omega C$ is the absolute value of the capacitive reactance.

Graphical interpretation of inductor phasors $\mathbf{V_C} = 14\angle 45°$ and $\mathbf{I_C} = 2\angle 135°$
Both phasors $\mathbf{V_C}$ and $\mathbf{I_C}$ are rotating at a constant angular frequency $\omega = 2\pi$ 2 Mrad/s = 12.57 Mrad/s; both phasors are separated by a fixed 90° phase difference, where $\mathbf{V_C}$ lags $\mathbf{I_C}$ by 90°. Figure 2.18 is a representation of both $\mathbf{V_C}$ and $\mathbf{I_C}$ capacitor phasors in the complex plane.

2.3.6 R, L, and C Impedances Combinations

From the previous three sections we can summarize that the impedances for R, L, and C elements are

$$\mathrm{Z_R} = \mathrm{V_R}/\mathrm{I_R} = R \tag{2.110}$$

$$\mathrm{Z_L} = \mathrm{X_L} = \mathrm{V_L}/\mathrm{I_L} = j\omega L \tag{2.111}$$

$$\mathbf{Z_C} = \mathbf{X_C} = \frac{1}{j\omega C} = -j\frac{1}{\omega C}. \tag{2.112}$$

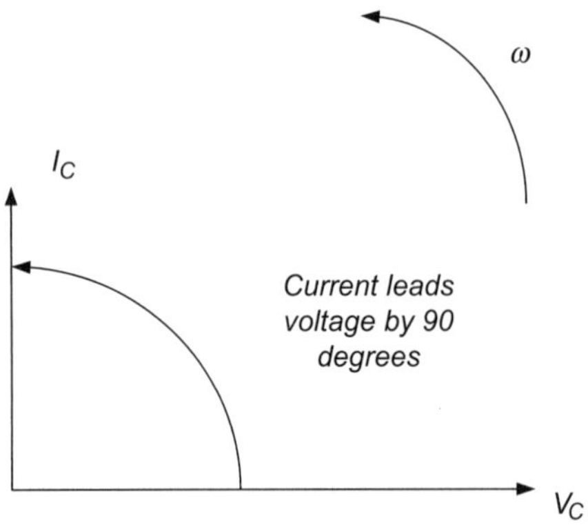

Figure 2.18 Capacitor voltage and current phasors.

Given any R, L, C circuit series and parallel combinations of impedance are handled similarly to how serial and parallel combinations of resistors are calculated. In DC circuit calculations with resistors, all the operations are done with real numbers. In AC, sinusoidal steady-state analysis impedances are in general complex numbers and in phasor form. Note: The concept of impedance and phasors is defined for the frequency domain. Impedance and phasors do not make any sense in the time domain.

Example 2.22 Compute the series equivalent impedance of $\mathbf{Z_R}$, $\mathbf{Z_C}$, and $\mathbf{Z_L}$.

$$\mathbf{Z}_{\textbf{series-equivalent}} = \mathbf{Z_R} + \mathbf{Z_C} + \mathbf{Z_L}. \tag{2.113}$$

Given that $\mathbf{Z_R} = 10\ \Omega$, $\mathbf{Z_L} = j\,60\ \Omega$ and $\mathbf{Z_C} = -j30\ \Omega$, calculate the series equivalent impedance.

Since

$$\mathbf{Z}_{\textbf{series-equivalent}} = \mathbf{Z_R} + \mathbf{Z_C} + \mathbf{Z_L}, \tag{2.114}$$

using the given values leads to

$$\begin{aligned} \mathbf{Z}_{\textbf{series-equivalent}} &= 10\ \Omega - j30\ \Omega + j60\ \Omega \\ &= 10\ \Omega + j30\ \Omega \text{ (in rectangfular form)} \end{aligned} \tag{2.115}$$

and

$$\begin{aligned} &= (10^2 + 30^2)^{1/2} \angle{-\arctan(+30/10)} \\ &= 31.62\angle 71.57^\circ \text{ (in polar form).} \end{aligned} \tag{2.116}$$

Example 2.23 For the previous example, calculate the values of capacitance and inductance assuming that the angular frequency is 1 Mrad/s.

Solution to Example 2.23

From Equations (2.111) and (2.112) we know that

$$\mathbf{Z_L} = \mathbf{X_L} = \mathbf{V_L}/\mathbf{I_L} = j\omega L. \tag{2.117}$$

$$\mathbf{Z_C} = \mathbf{X_C} = \frac{1}{j\omega C} = -j\frac{1}{\omega C}. \tag{2.118}$$

Since $\omega = 1$ Mrad/s and since $|\mathbf{Z_L}| = |j\omega L| = \omega L$,

$$L = |\mathbf{Z_L}|/1\ \text{Mrad/s} = 60\ \Omega/1\ \text{Mrad/s} = 60\ \mu\text{H} = 60 \times 10^{-6}\ \text{H},$$

and for C, using Equation (2.118),

$$C = 1/\omega|\mathbf{Z_C}| = 1/1\ \text{Mrad/s} \times 30\ \Omega = 33.33\ \text{nF} = 33.33 \times 10^{-9}\ \text{F}.$$

Example 2.24 Parallel Equivalent Impedance

Given that $\mathbf{Z_R} = 10\ \Omega$, $\mathbf{Z_L} = j60\ \Omega$ and $\mathbf{Z_C} = -j30\ \Omega$, calculate the parallel equivalent impedance in rectangular, polar, and Euler's forms.

Similarly to what we did with resistors, we do it with impedances, but remembering that impedances are phasors, or complex numbers with magnitude and phase, then,

$$\frac{1}{\mathbf{Z_{parallel\text{-}equivalent}}} = \frac{1}{\mathbf{Z_R}} + \frac{1}{\mathbf{Z_L}} + \frac{1}{\mathbf{Z_C}}. \tag{2.119}$$

It is convenient to transform the given impedance into their polar forms, which are $\mathbf{Z_R} = 10\ \Omega\angle 0°$, $\mathbf{Z_L} = 60\Omega\angle{+90°}$, and $\mathbf{Z_C} = 30\Omega\angle{-90°}$.

Then using Equation (2.119) and using the impedances in polar form we obtain

$$\mathbf{1/Z_{parallel\text{-}equivalent}} = 1/10\ \Omega\angle 0° + 1/60\ \Omega\angle{-90°} + 1/30\ \Omega\angle{+90°}. \tag{2.120}$$

After converting each impedance term on the right-hand side of Equation (2.120) from polar form into rectangular form, adding the three of them and then obtaining the inverse of the addition, leads to

$$\mathbf{1/Z_{parallel\text{-}equivalent}} = 1/10 + 1/j60 - j30 \tag{2.121}$$

$$\mathbf{Z_{parallel\text{-}equivalent}} = 9.729 - j1.622, \tag{2.122}$$

where Equation (2.122) is the equivalent impedance in rectangular form.

$$\mathbf{Z}_{\text{parallel-equivalent}} = 9.863\angle{-9.465^\circ} \tag{2.123}$$

And finally,

$$\mathbf{Z}_{\text{parallel-equivalent}} = 9.863e^{-j9.465^\circ} \tag{2.124}$$

where Equation (2.122) is in rectangular form, Equation (2.123) is in polar form, and Equation (2.124) is in Euler's form.

2.4 POWER IN AC CIRCUITS

Circuits on sinusoidal steady state draw power from their sinusoidal power source. When R, L, C circuits are connected to a sinusoidal power source, some of the power drawn by the circuit is consumed by it; this is called real, true, active, or average power. *Active* power is measured in Watts (W). Active power is power that the load takes from the source to perform useful work. Active power gets converted into heat on the resistive part of the impedance. The capacitors and/or inductors present in the circuit cause the source to produce some additional power that is not consumed by the load. In the capacitor case, this power establishes the electric field on the capacitor itself; in the inductor case, this power establishes the magnetic field on the inductor. The power drawn from the power source by the capacitive and inductive elements does not produce any active power. This capacitive and inductive power is referred to as *reactive* power, and it is measured in *reactive volt-amperes* (VAR). The total power taken by a load from the AC power source is some combination of the total active power plus the total reactive power. This total power is referred to as the *apparent* power (S), measured in volt-amperes (VAs). So what is the relationship between apparent (S), active (P), and reactive (Q) powers?

We will answer this question soon, but first let us study the instantaneous power drawn by a resistor, a capacitor, and an inductor when they are fed by an AC source.

Let us go over active, reactive, and apparent power one more time. Active or real power is the easiest to understand. And it is the total energy absorbed by the resistive component of the load during each sinusoidal cycle. Energy is measured in units of power (W) multiplied by units of time, for example, watt-seconds or watt-hours. Real or active power is measured in watts.

The physical meaning of reactive power is not as easy or intuitive to understand. Reactive power, denoted by Q, refers to the maximum value of instantaneous power absorbed by the reactive component of the load. The instantaneous reactive power is alternatively positive and negative, twice per sinusoidal cycle. For an inductor, refer to Figure 2.9, and for a capacitor, refer to Figure 2.10. Note that the instantaneous power in a reactive element (i.e., either an inductor or a capacitor) is positive for the first quarter of the sinusoidal cycle, and then it becomes negative during the second quarter of the cycle, positive on the third quarter, and negative on the final quarter cycle. Refer to Figures 2.9 and 2.10. Positive instantaneous power means that the generator provides power to the reactive load; negative instantaneous power means to the load returns the power back to the source. Note that the average or active power consumed by a reactive element is zero on a cycle per cycle basis. Active and reactive powers are related, and the combination of both is referred to as apparent power measured in VAs. In the next several sections we will discuss instantaneous power in resistors, inductors, and capacitors. This will lead to active, reactive, and apparent powers and their relationship which is explained by means of the *triangle of powers*.

2.4.1 AC Instantaneous Power Drawn by a Resistor

From Table 2.1, when a sinusoidal current and voltage are produced on a resistor, we know that both waveforms are in phase. And from Equation (2.9), repeated here for the reader's convenience, the instantaneous power on the resistor is

$$p_R(t) = 1/2VI - 1/2VI\cos(2\omega t + \theta), \tag{2.125}$$

where V and I are respectively peak values of voltage and current.

We also have seen (from Eqs. 2.12 through 2.19) that the average power consumed by the resistor is evaluated as follows:

$$\mathrm{P}_{\text{average-resistor}} = 1/T\int_0^T [I_{RMS}V_{RMS} - V_{RMS}I_{RMS}\sin(2\omega t)]\,dt, \tag{2.126}$$

which leads to

$$\mathrm{P}_{\text{average}} = 1/2VI = I_{RMS}V_{RMS} \tag{2.127}$$

because the term $V_{RMS}I_{RMS}\sin 2\omega t$ average value is zero.

Earlier in this Chapter, Figure 2.4 shows the sinusoidal current, voltage on a resistor, the instantaneous power, and the average power consumed by the resistor. It is important and interesting to observe that the average power consumed by the resistor always flows from the AC power source into the resistor. Such average power is always positive.

2.4.2 AC Instantaneous Power Drawn by a Capacitor

The instantaneous power drawn by a capacitor is the product of its instantaneous voltage and current. From previous sections we know that the instantaneous voltage across the capacitor lags the instantaneous current waveform by 90°. That is to say,

$$V_C = -jX_C I_C \text{ in phasor form or the frequency domain} \tag{2.128}$$

and

$$I_C \cos(\omega t + \theta);\, V_C \sin(\omega t + \theta) \text{ in the time domain,} \tag{2.129}$$

where in both Equations (2.128) and (2.129), I_C and V_C are respectively the peak values of AC current and AC voltage on the capacitor. As usual, ω is the *angular frequency* and θ is an arbitrary phase angle. Note that θ shows up on both AC current and voltage. For simplicity and without loss of generality we will assume that θ is zero.

The product of its AC voltage and current gives the instantaneous power on the capacitor;

$$p_C(t) = v_C(t) \times i_C(t), \tag{2.130}$$

where we substitute the waveforms from Equation (2.129) into Equation (2.130) and obtain

$$p_C(t) = V_C \sin(\omega t) \times I_C \cos(\omega t) \tag{2.131}$$

$$= V_C I_C \sin(\omega t)\cos(\omega t). \tag{2.132}$$

In Equation (2.132), V_C and I_C are respectively the peak voltage and current values.

Using the following trigonometric identity in Equation (2.132),

$$\sin 2x = 2 \sin x \cos x \tag{2.133}$$

leads to

$$= 1/2 V_C I_C \sin 2\omega t \tag{2.134}$$

$$= V_{RMS} I_{RMS} \sin 2\omega t, \tag{2.135}$$

where $V_{RMS} = V_C/\sqrt{2}$ and $I_{RMS} = I_C/\sqrt{2}$ are the RMS values of voltage and current.

$$\mathrm{P}_{\text{average-capacitor}} = 1/T \int_0^T V_{RMS} I_{RMS} \sin 2\omega t \, dt \tag{2.136}$$

$$\mathrm{P}_{\text{average-capacitor}} = 0. \tag{2.137}$$

From Equation (2.137) we observe that the average power consumed by the capacitor during an AC period is zero. Referring to the double frequency instantaneous power waveform of Figure 2.10, it is possible to see that the integral of the instantaneous power waveforms between an integral number of cycles T is zero.

Again referring to Figure 2.10, it can be seen that the instantaneous power drawn by the capacitor is alternatively positive and negative every quarter of a period of the original voltage and current waveforms. When the instantaneous power is positive, it means that the source is providing instantaneous power to the capacitor; when the instantaneous power is negative, the capacitor is returning power to the source. This is what originates the *capacitive reactive* power in a capacitor, and it is sometimes called as the *entertaining power* between the source and the capacitor.

2.4.3 AC Instantaneous Power Drawn by an Inductor

The instantaneous power drawn by the inductor is the product of its instantaneous voltage and current. From previous sections we know that the instantaneous voltage across the inductor leads the instantaneous current waveform by 90°. That is to say,

$$V_L = jX_L I_L \text{ in phasor form or the frequency domain} \tag{2.138}$$

and

$$V_L \cos(\omega t + \theta);\ I_L \sin(\omega t + \theta) \text{ in the time domain,} \tag{2.139}$$

where in both equations above, V_L and I_L are respectively the peak values of AC voltage and AC current. As usual, ω is the *angular frequency* and θ is the phase angle.

Note that θ shows up on both AC voltage and current. For simplicity and without loss of generality we will assume that θ is zero.

The product of its AC voltage and current gives the instantaneous power on the inductor.

$$p_L(t) = v_L(t) \times i_L(t), \tag{2.140}$$

where we substitute the waveforms from 2.139 into Equation (2.140) and obtain

$$p_L(t) = V_L \cos \omega t \times I_L \sin \omega t \tag{2.141}$$

$$= V_L I_L \cos \omega t \sin \omega t \tag{2.142}$$

Using the following trigonometric identity in Equation (2.142),

$$\sin 2x = 2 \sin x \cos x \tag{2.143}$$

leads to

$$= 1/2 V_L I_L \sin 2\omega t \tag{2.144}$$

$$= V_{RMS} I_{RMS} \sin 2\omega t, \tag{2.145}$$

where $V_{RMS} = V_L/\sqrt{2}$ and $I_{RMS} = I_L/\sqrt{2}$ are the RMS values of voltage and current:

$$\mathrm{P}_{\text{average-inductor}} = 1/T \int_0^T V_{RMS} I_{RMS} \sin 2\omega t \, dt. \tag{2.146}$$

Evaluating the integral

$$\mathrm{P}_{\text{average-inductor}} = 0. \tag{2.147}$$

From Equation (2.147) we can see that the average power consumed by the inductor during a sinusoidal AC period is zero. Refer to previously seen Figure 2.9. From this figure it can be seen that the instantaneous power drawn by the inductor is alternatively positive and negative every quarter of a period. A period refers to the voltage or current period. Both voltage and current waveforms on an inductor have the same frequency when the excitation is sinusoidal.

When the instantaneous power is positive, it means that the source is providing instantaneous power to the inductor; when the instantaneous power is negative, the inductor is returning power to the source. This is what originates the *inductive reactive* power in an inductor, and it is sometime called as the *entertaining power* between the source and the inductor.

2.4.3.1 AC Power Triangle Active, capacitive reactive, inductive reactive, and apparent powers are geometrically related by the power triangle. When an impedance **Z** has all three electric components (*R*, *L*, and *C*), the active power, dissipated on the resistive part of the impedance, is drawn horizontally and is denoted as *P*. The inductive reactive power is represented vertically and pointing toward the positive side of the complex plane. It is denoted as *Q* inductive and has a positive sign. The capacitive reactive power is represented vertically and pointing down toward the negative side of the complex plane. It is denoted as *Q* capacitive and has a negative sign. The active and net reactive power (*inductive reactive power or capacitive reactive power*) are related to each other by the Pythagorean relationship, from phasor analysis:

$$S^2 = P^2 + Q^2, \tag{2.148}$$

where *S* in VA is the apparent power; *P* is the active power in watts consumed by the resistive part of the impedance; *Q* is the net reactive power in VAR (reactive volt-ampere).

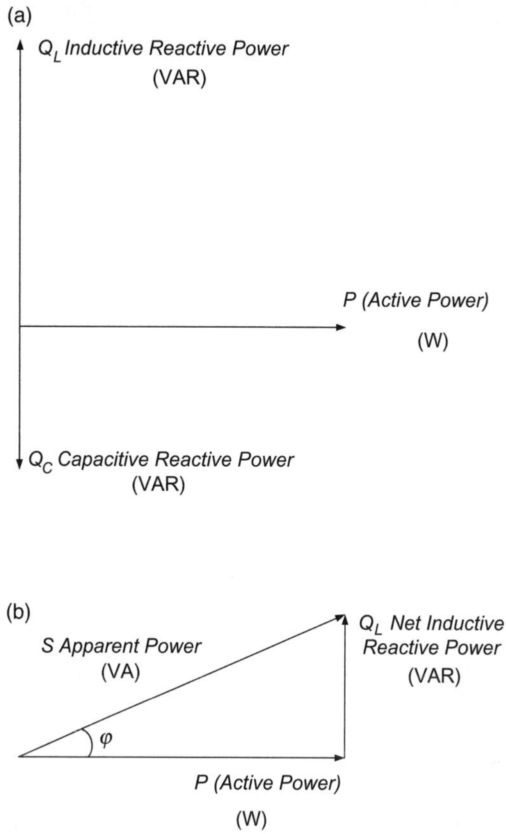

Figure 2.19 Power triangle with net inductive reactive power.

Figure 2.19 depicts active power with net inductive reactive power, and Figure 2.20 depicts active power with net capacitive reactive power.

From basic trigonometry it can be observed that

$$P = S\cos\varphi \tag{2.149}$$

and

$$Q = S\sin\varphi, \tag{2.150}$$

where φ is defined as the *power factor* for sinusoidal steady state.

Thus,

$$\text{Power Factor} = \text{PF} = \cos\varphi. \tag{2.151}$$

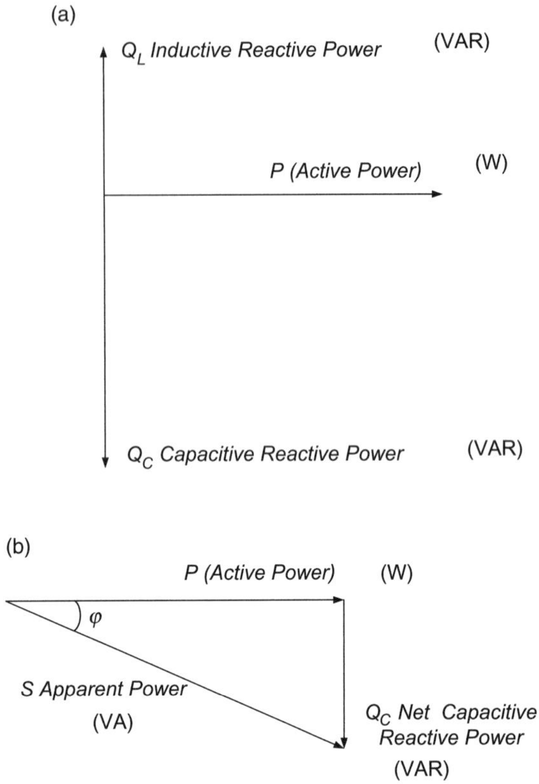

Figure 2.20 Power triangle with net capacitive reactive power.

For sinusoidal steady state, it can also be observed that

$$\mathrm{PF} = P/S. \tag{2.152}$$

Power factor is an important figure of merit that electric utility companies observe closely. The utility company does not want its customer's electrical loads to demand too much reactive power. Why? Because the electric generators need to produce an excess power (reactive power) that does not end up as useful work developed at the load. Remember that reactive power is power that is supplied by the generator to the load and returned back from the load to the generator on a cyclical basis. When a capacitor's electric fields and inductor's magnetic fields are created, they cause for the existence of capacitive and inductive reactive power respectively. Ideally, the electric utility company wants that reactive power to be zero, or in other words, they want to see a very close to unity power factor (PF = 1). From Equation (2.151) for PF to be one, φ has to be zero. For inductive loads, the current through the inductor lags the voltages across it, and the power factor is said to be *lagging*. For capacitive loads, the current through the capacitor leads the voltage across it, and the power factor is said to be *leading*.

After all being said, why is it that important that the power factor of an electrical load be one or very close to one? The reasons are that if power factor is smaller or much smaller than one, the utility company electric generator has to generate excessive power that will not end up being used by the load as active power. Let us remember that the load produces useful work consuming active power. The power distribution wiring needs to be thicker if the power factor is smaller than one. The dimensioning of the power distribution wiring must be made based on the apparent power drawn by the load. This ensures that the AC power distribution wires to the load are appropriately sized.

Example 2.25 Determine the total apparent, active, and reactive power that a 2 Ω resistive load with a unity power factor draws from an AC 220 V_{RMS} voltage generator.

Solution to Example 2.25

Since

$$S^2 = P^2 + Q^2 \tag{2.153}$$

where

$$P = S\cos\varphi \tag{2.154}$$

and

$$Q = S\sin\varphi \tag{2.155}$$

where $\cos\varphi$ equals one, as stated by the problem assumption, thus, φ equals 0° and $\sin\varphi = 0$.

From Equations (2.154) and (2.155),

$$P = S \tag{2.156}$$

and

$$Q = 0 \tag{2.157}$$

where

$$P = S = V_{RMS}^2/R = 220^2/2 = 24.2\ \text{kW} = 24.2\ \text{kVA}$$

In this example the apparent power equals the active power dissipated by the load, and there is zero reactive power between the generator and the load.

Example 2.26 Determine the total apparent, active, and reactive powers that an impedance of an absolute value of 11 Ω, and an inductive power factor of 0.8. The impedance draws 20 A_{RMS} from a 220 V_{RMS} AC voltage generator. Also determine the impedance real and imaginary parts.

Solution to Example 2.26

Since apparent power

$$S = I_{RMS} V_{RMS} \tag{2.158}$$

$$S = 20 \text{ A} 220 \text{ V} = 4400 \text{ VA}.$$

And since

$$S^2 = P^2 + Q^2 \tag{2.159}$$

where

$$P = S \cos \varphi \tag{2.160}$$

and

$$Q = S \sin \varphi \tag{2.161}$$

$$P = 4400 \times 0.8 = 3520 \text{ W}$$

and where $\cos \varphi$ equals 0.8, as stated by the problem assumption, thus, $\varphi =$ 36.87° and $\sin \varphi = 0.6$.

Then,

$$Q = 4400 \times 0.6 = 2640 \text{ VAR}.$$

$$R = |Z| \cos 36.87° = 8.8 \ \Omega \tag{2.162}$$

and

$$X_L = |Z| \sin 36.87° = 6.6 \ \Omega \tag{2.163}$$

where X_L is inductive.

2.5 DEPENDENT VOLTAGE AND CURRENT SOURCES

Dependent sources produce either a voltage or a current, where such voltage or current depends on either a voltage or a current on some other part of the circuit or network.

Dependent sources are widely used to model active circuits like operational amplifiers, transistor-based amplifiers, and transistors such as bipolars and MOSFETs.

There are four basic kinds of dependent sources; two dependent voltage sources and two dependent current sources. Within each type there are current- and voltage-controlled sources.

The four types of dependent sources are listed below:

1. Voltage-controlled dependent voltage source or VCVS
2. Current-controlled dependent voltage source or CCVS
3. Voltage-controlled dependent current source or VCCS
4. Current-controlled dependent current source or CCCS

2.5.1 Voltage-Controlled Voltage Source (VCVS)

The voltage-controlled voltage source is a dependent source that allows us to model a voltage amplifier. Without knowing yet about the internals of a voltage amplifier, we can define such a circuit element as a two-port device. One input port that receives and input voltage V_{in} and one output port that generates an output voltage which is a magnification of the input voltage V_{in} by some constant A, where A stands for *amplification factor* or simply its *amplification.*

Figure 2.21 depicts the symbol diagram of a VCVS which is very appropriate to model the behavior of a voltage amplifier, such as the one just described.

Figure 2.22 depicts the use of a VCVS in a circuit example. Note that $V_{out} = A\,V_{in}$; thus, A has to be dimensionless because $A = V_{out}/V_{in}$, and its units are then volts/volts.

Note that in the VCVS circuit example, the voltage source of value $V_{out} = A\,V_{in}$ produces an output voltage that depends on the value of input voltage V_{in},

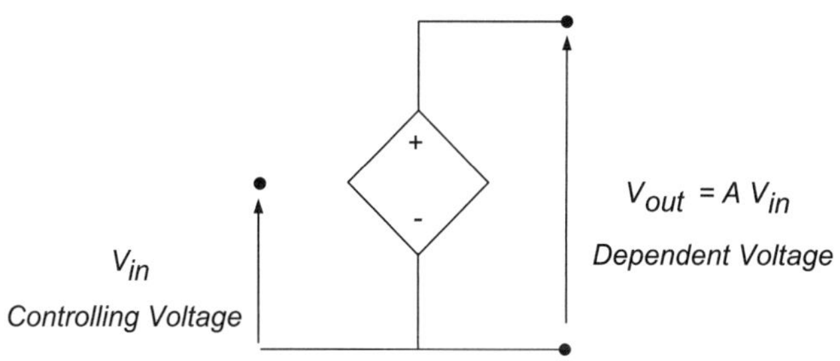

Figure 2.21 Voltage-controlled voltage source.

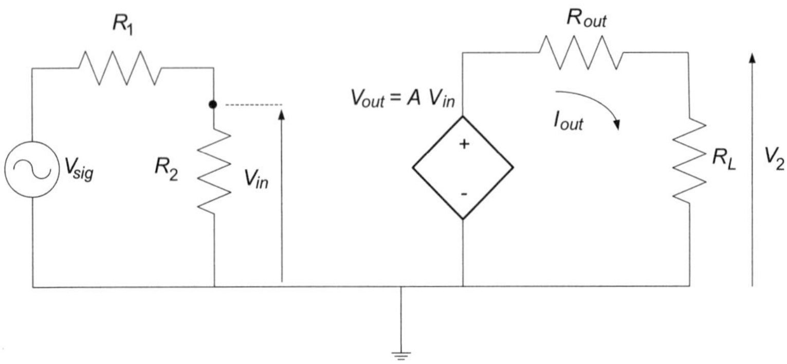

Figure 2.22 Use of a VCVS in a circuit example.

which is shown across resistor R_2. Just to wrap up this example, let us evaluate the overall output voltage of the complete circuit, V_2 as a function of V_{sig}.

From Figure 2.22, using KVL, we can write for the left-hand side loop that

$$V_{in} = \frac{R_2}{R_1 + R_2} V_{sig}. \tag{2.164}$$

Using KVL for the right-hand side loop we get

$$AV_{in} = I_{out}(R_{out} + R_L). \tag{2.165}$$

Combining Equation (2.164) with Equation (2.165), and since $V_2 = I_{out} \times R_L$, it yields

$$V_2 = \frac{A}{R_{out} + R_L} R_L \frac{R_2}{R_1 + R_2} V_{sig}. \tag{2.166}$$

2.5.2 Current-Controlled Voltage Source (CCVS)

A current-controlled voltage source is a dependent source that allows us to model a trans-resistance amplifier. Without knowing yet about the internals of a trans-resistance amplifier, we can define such a circuit element as a two-port device. One input port that receives and input current I_{in} and one output port that generates an output voltage which is a magnification of the input current I_{in} by some constant Γ (rho), where Γ stands for *trans-resistance amplification factor* or simply its *amplification* Γ. Note that the units of Γ are ohms.

Figure 2.23 depicts the symbol diagram of a CCVS which is very appropriate to model the behavior of a trans-resistance amplifier. Figure 2.24 depicts the use of a CCVS in a circuit example. Note that $V_{out} = \Gamma\, I_{in}$; thus, Γ is in ohms, because $\Gamma = V_{out}/I_{in}$ units are volts/ampere.

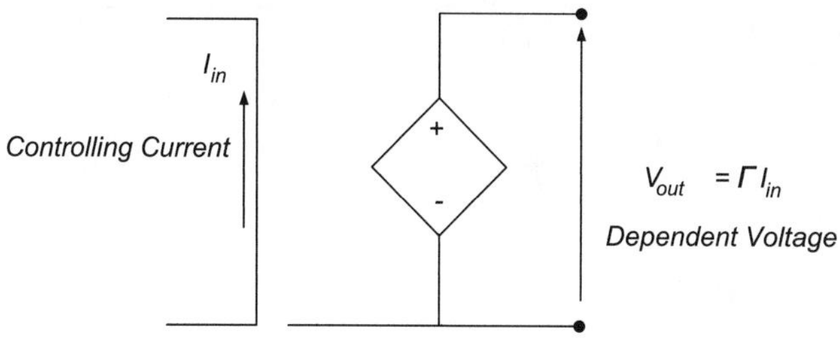

Figure 2.23 Current-controlled voltage source.

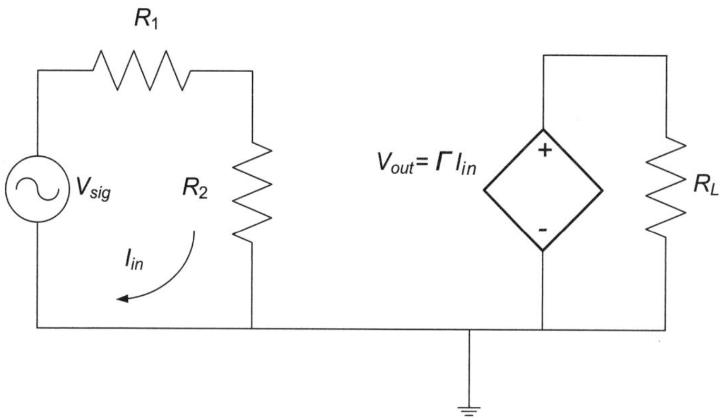

Figure 2.24 Use of a CCVS in a circuit example.

Note that in the CCVS circuit example of Figure 2.24, the voltage source of value $V_{out} = \Gamma\ I_{in}$ produces an output voltage that depends on the value of input current I_{in}, which flows in the circuit of R_1 and R_2 and excited by V_{sig}.

2.5.3 Voltage-Controlled Current Source (VCCS)

A voltage-controlled current source is a dependent source that allows us to model a trans-conductance amplifier. Without knowing yet about the internals of a trans-conductance amplifier, we can define such a circuit element as a two-port device. One input port that receives and input voltage V_{in} and one output port that generates an output current which is a magnification of the input voltage V_{in} by some constant G, where G stands for *trans-conductance amplification factor* or simply its *amplification* G, where G has conductance units (Ω^{-1}).

Figure 2.25 depicts the symbol diagram of a VCCS which is very appropriate to model the behavior of a trans-conductance amplifier, such as the one just described.

Figure 2.26 depicts the use of a VCCS in a circuit example. Note that I_{out} = G V_{in}.

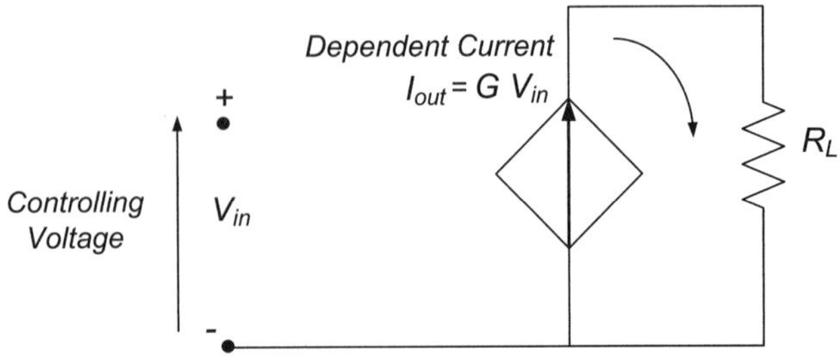

Figure 2.25 Voltage-controlled current source.

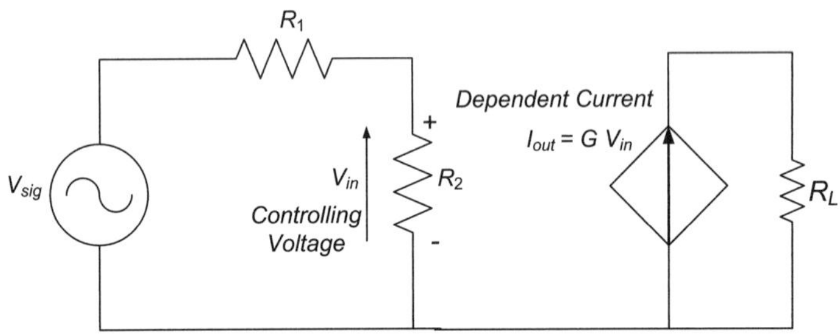

Figure 2.26 Use of a VCCS in a circuit example.

Note that in the VCCS circuit example the current source of value: $I_{out} = G\,V_{in}$ produces an output current that depends on the value of input voltage V_{in}, which is shown across resistor R_2.

2.5.4 Current-Controlled Current Source (CCCS)

A current-controlled current source is a dependent source that allows us to model a current amplifier. Without knowing yet about the internals of a current amplifier, we can define such a circuit element as a two-port device. One input port that receives and input current I_{in} and one output port that generates an output current which is a magnification of the input current I_{in} by some constant β, where β stands for *current amplification factor* or simply its *amplification* β. Note that β has no dimensions since it is obtained as the ratio of two currents. A current amplifier is also referred to as a buffer. We will see in later chapters that buffers can be implemented with transistors or with operational amplifiers.

Figure 2.27 depicts the symbol diagram of a CCCS which is very appropriate to model the behavior of a current amplifier, such as the one just described.

Figure 2.28 depicts the use of a CCCS in a circuit example.

Note that in the CCCS circuit example the voltage source of value $I_{out} = \beta I_{in}$ produces an output current that depends on the value of input current I_{in}, which is shown in the circuit of Figure 2.28.

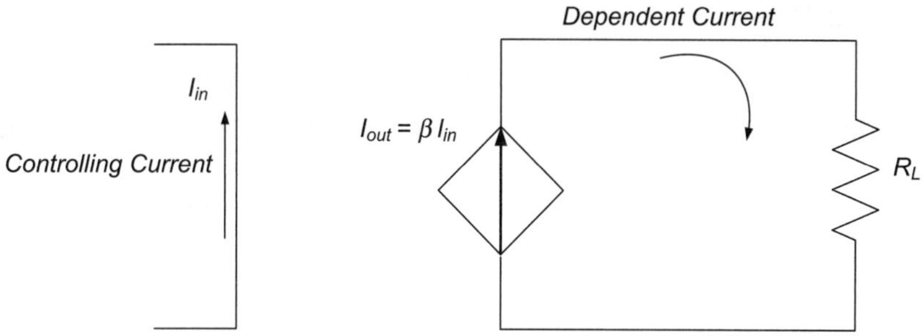

Figure 2.27 Current-controlled current source.

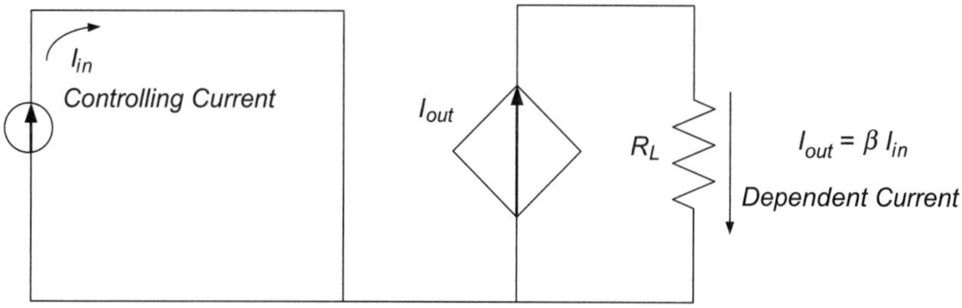

Figure 2.28 Use of a CCCS in a circuit example.

2.6 SUMMARY OF KEY POINTS

This chapter covers the fundamentals of AC circuits. It is important to understand the concept of effective value (RMS) of voltage and current and the role they play on R, L, and C elements and how they produce different kinds of AC power: active, reactive, and apparent. It is also of great interest to know how to manipulate circuit equations in the time domain as well as in the frequency domain. In the time domain, derivatives and integrals of current or voltage usually apply. In the frequency domain, phasors replace the tedious-to-deal-with differential equations. Phasor diagrams make AC circuit calculations easier. The catch is that this method works when the voltage and current frequencies are the same. "Dependent sources" is a topic of great interest to model electronic devices or active devices that have gain. More on this subject is covered in Chapters 5 and 6.

FURTHER READING

1. Charles Alexander and Matthew Sadiku, *Fundamentals of Electric Circuits*, 2nd ed., McGraw Hill, New York, 2004.
2. Charles Monier, *Electric Circuit Analysis*, Prentice Hall, Upper Saddle River, NJ, 2001.
3. David Bell, *Fundamentals of Electric Circuits*, 4th ed., Prentice Hall, Upper Saddle River, NJ, 1988.

PROBLEMS

2.1 A toaster is rated at 1 kW and for an AC voltage of 120 V at 60 Hz.

(a) Determine the resistance of the toaster, before its temperature increases. Assume that the resistance is at room temperature.

(b) Determine the RMS value of current flowing through the toaster when it is dissipating 1 kW.

(c) Determine the peak value of the current through the toaster when it is dissipating 1 kW.

(d) If the resistance of the toaster has +/–10% tolerance, calculate the minimum and maximum power that the toaster will consume under the two extremes of resistance values.

2.2 Evaluate the RMS value of the voltage waveform drawn in Figure 2.29. Assume that the peak amplitude of the waveform is 1 V, its period T is 1 msec, and 50% duty cycle.

2.3 Evaluate the average DC voltage waveform for the double rectified sine-wave waveform depicted in Figure 2.30. Analytically, the waveforms can be described as follows:

$$v(t) = \sin \omega t; \text{ for: } \pi/2 \le t \le 0$$

$$v(t) = -\sin \omega t; \text{ for: } \pi/2 \le t \le \pi$$

This alternatively can be expressed as

$$v(t) = |\sin \omega t|.$$

2.4 Calculate the RMS value of a 10 A DC current.

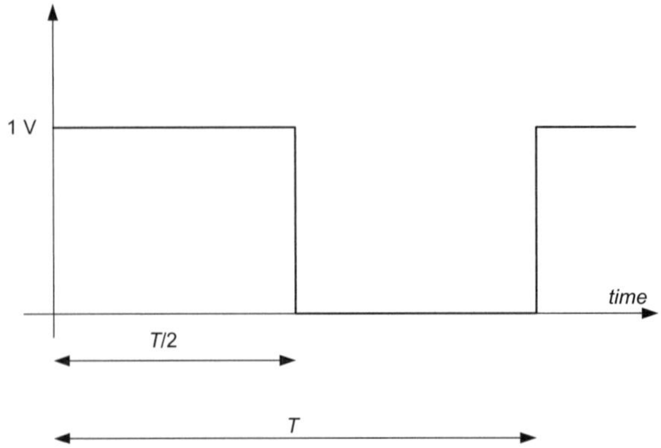

Figure 2.29 50% duty cycle square-wave voltage waveform for Problem 2.2.

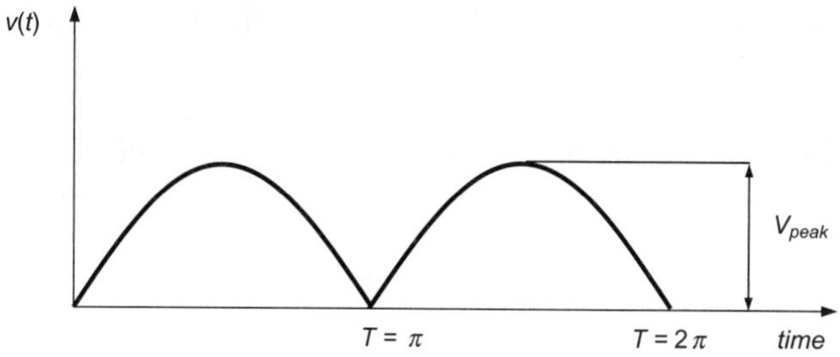

Figure 2.30 Double rectified sine-wave waveform for Problem 2.3.

2.5 Given an *RLC*-series circuit, where $R = 10\ \Omega$, $L = 320$ nH, $C = 100\ \mu$F, find the absolute value of the impedance of the circuit at the following frequencies:

(a) 1 Hz
(b) 10 Hz
(c) 100 Hz
(d) 1 kHz
(e) 10 kHz
(f) 100 kHz
(g) 1 MHz, and
(h) 10 MHz.

2.6 For an RLC-series circuit, where $R = 10\ \Omega$, $L = 320$ nH, $C = 100\ \mu$F, calculate the absolute value of inductive reactance and the capacitive reactance at the following frequencies:

(a) 1 Hz
(b) 10 Hz
(c) 100 Hz
(d) 1 kHz
(e) 10 kHz
(f) 100 kHz
(g) 1 MHz, and
(h) 10 MHz.

2.7 (a) For the circuit given in Problem 2.5, find the frequency at which the absolute value of the inductive reactance equals the absolute value of the capacitive reactance (i.e., resonance condition). (b) At this frequency find the peak value of current for a 1-V peak sinusoidal voltage at the resonant frequency.

2.8 The circuit of Problem 2.5 is said to be at its resonant frequency when the absolute value of its inductive reactance equals the absolute value of its capacitive reactance. The resonant frequency was calculated in Problem 2.7. Assuming a 1-V peak sinusoidal voltage, (a) find the value of current in the circuit at a frequency equal to 10 times the resonant frequency of the circuit, and (b) find the value of current in the circuit at a frequency equal to one-tenth of the resonant frequency of the circuit.

Draw conclusions from the numerical answers that you obtain for this problem.

2.9 An impedance of value $Z = (400 + j350)\ \Omega$ is connected to a sinusoidal voltage of 416 V RMS. (a) Compute the apparent, active, and reactive powers that the impedance absorbs from the AC generator. (b) Determine the power factor of the circuit.

2.10 Establish the time domain equations (i.e., differential equations) of an *RLC*-series circuit powered by a sinusoidal voltage source $v(t) = V_{peak}\sin(\omega t + \theta)$. Hint: The final equation is a second-order differential equation with constant coefficients.

2.11 Establish the time domain equations (i.e., differential equations) of a parallel *RLC* circuit powered by a sinusoidal current source $i(t) = I_{peak}\sin(\omega t + \theta)$.

Hint: The final equation is a second-order differential equation with constant coefficients.

2.12 Given a 10 Ω resistor in series with a 10 μF capacitor, and an AC voltage source of $V_{in} = 100\ \text{V}\ e^{j0}$, of a 1 kHz frequency, determine: (1) if the current through the circuit leads or lags the input voltage V_{in} across the *RC* series; (2) the phase angle between the input voltage and the circuit series current), and (3) the phase angle between the voltage source and the voltage across the capacitor.

2.13 Given an *RLC* series circuit, where $R = 100\ \Omega$, $L = 1\ \mu\text{H}$, and $C = 10\ \mu\text{F}$, determine the frequency at which the circuit goes into resonance.

2.14 Express the series impedance given in Problem 2.12 in complex notation. Hint: $Z(j\omega)$.

2.15 Given that $Z_1 = (30 + j25)\ \Omega$ and $Z_2 = (20 - j15)\ \Omega$, calculate: (1) the series combination of both impedances, and (2) the parallel combination of both impedances.

2.16 Given impedance $(Z_1 = 30 + j25)\ \Omega$, find the value of inductance of its inductive reactance at a frequency of 1 kHz.

2.17 Given impedance $(Z_2 = 20 - j15)\ \Omega$, find the value of the capacitive reactance of the impedance at a frequency of 1 kHz.

2.18 Given an *RLC* series circuit, where $R = 5\ \Omega$, the reactive reactance is $+j18\ \Omega$, and the capacitive reactance is $-j10\ \Omega$ connected to a sinusoidal

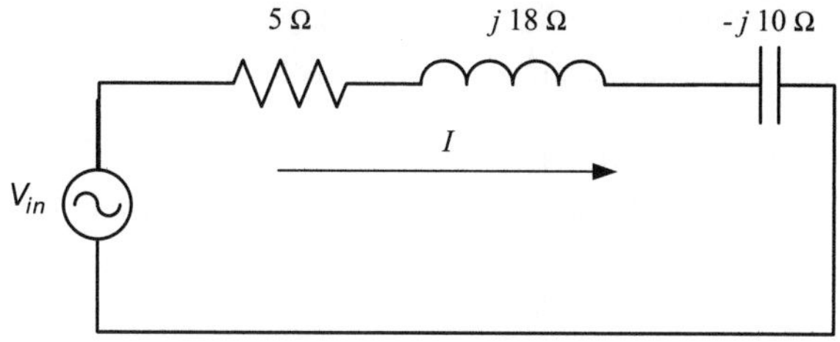

Figure 2.31 *RLC* series circuit for Problem 2.18.

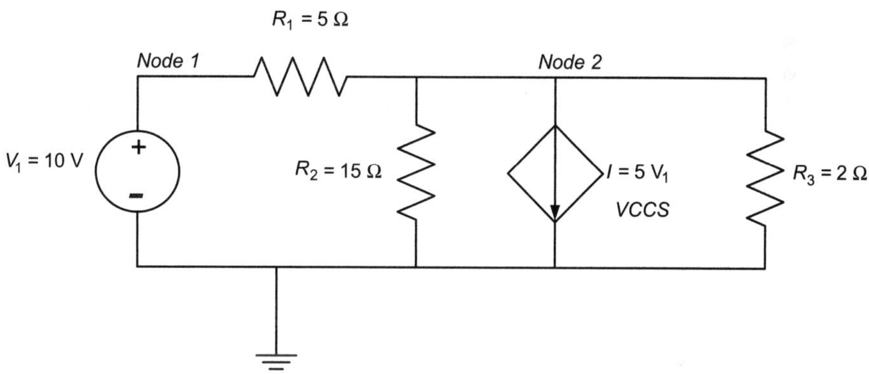

Figure 2.32 Circuit for Problem 2.20.

voltage source V_{in} of *RMS* value of 100 V, determine the circuit complete phasor diagram. The following phasors must be shown: (1) circuit current phasor, $\boldsymbol{I}$, (2) resistor voltage phasor, $\boldsymbol{V_R}$, (3) capacitive reactance voltage phasor, $\boldsymbol{V_C}$, and (4) inductive reactance voltage phasor, $\boldsymbol{V_L}$. Find and show on the phasor diagram all the numerical phase angle values between the current and the three voltages (Fig. 2.31).

2.19 Given an RLC series circuit with an impedance $Z = 100 - j45$ at 60 Hz, assume that the circuit is energized by a 240 V 60 Hz sinusoidal voltage generator. (1) Calculate the real, apparent, and reactive power of the circuit, and (2) calculate the circuit power factor.

2.20 Given the circuit of Figure 2.32, note that $I = 5\ V_1$, between node 2 and ground, is a voltage-controlled current source (VCCS), whose output current value is $I = 5\ V_1$, and the control voltage is $V_1 = 10$ V. Calculate the voltage at every node with respect to ground and the currents through every resistor, the independent voltage source, and the VCCS.

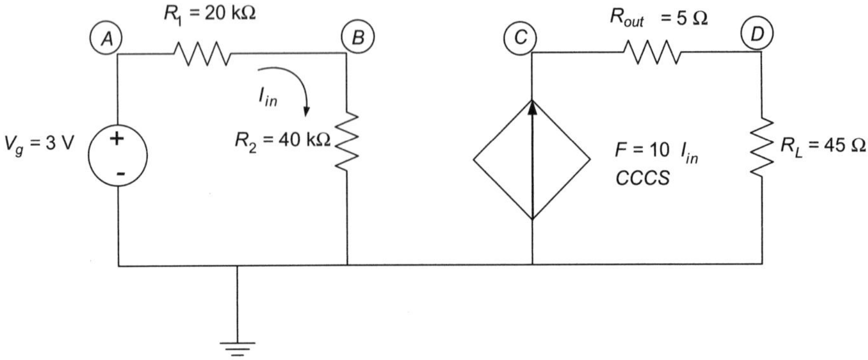

Figure 2.33 Circuit for Problem 2.21.

2.21 Given the circuit of Figure 2.33, note that the element between node C and ground is a current-controlled current source (CCCS), whose output current is 10 I_{in}, and the control current is I_{in}. Calculate the voltage at every node (A through D) with respect to ground and the currents through every resistor, the independent voltage source, and the VCCS dependent source.

3

CIRCUIT THEOREMS AND METHODS OF CIRCUIT ANALYSIS

3.1 INTRODUCTION

Circuit analysis is finding the current and voltage on every element of the circuit being analyzed. In previous chapters we addressed solving circuits using Ohm's and Kirchhoff's laws. This chapter will enhance your portfolio of circuit-solving techniques by introducing new circuit methods of analysis. The methods covered in this chapter are superposition, Thévenin's, Norton's, Mesh, and Nodal methods. But why do we need so many more methods? The answer is an issue of practicality. Solving a circuit becomes easier with more knowledge of different methods. This helps the person solving a circuit in several ways. Many times the number of variables in a circuit is too large, and thus difficult to solve by hand. If we have a computer to solve the circuit, why do I care about the number of variables? Well so far in our world, computers are faster but may not always generate the correct answer. It is important that we, as circuit analysis engineers, have at least a rough idea if the numerical answer that the computer will provide is within reasonable expectations. It is always important to be able to do a rough analysis to understand if computer findings are at least meaningful for the given circuit and within the expected range. All the following methods will ultimately allow us to use a *checks and balances* approach to circuit solving.

Electrical, Electronics, and Digital Hardware Essentials for Scientists and Engineers, First Edition. Ed Lipiansky.

3.2 THE SUPERPOSITION METHOD

The superposition method is applicable and valid for solving circuits if the circuit is linear. In a general sense, a function $f(x)$ is said to be linear if-and-only-if* the following conditions are met:

1. Function $f(x)$ domain and its range are linear spaces over the same scalar field.
2. *Homogeneity Property*: For all values of x in the function domain and every scalar α then

$$f(\alpha x) = \alpha f(x). \tag{3.1}$$

3. *Additivity Property:* For every pair of element domains x_1 and x_2, the following holds:

$$f(x_1 + x_2) = f(x_1) + f(x_2). \tag{3.2}$$

It can be observed that Equation (3.1) holds when $f(x)$ is a linear function.

In general a function $f(x)$, that has the form:

$$f(x) = ax + b, \tag{3.3}$$

where a is the slope of the line and b is its y-intercept, is said to be linear only if its y-intercept is zero. That is,

$$f(0) = 0. \tag{3.4}$$

Important Points:

$$f(\mathrm{x}) = ax + b, \tag{3.5}$$

for $b \neq 0$, *is NOT a linear function.* However, Equation (3.5) is still the equation of a straight line. The function

$$f(x) = ax \tag{3.6}$$

is linear for any value of a and x [1].

Graphically we state that a line that goes through the origin of coordinates is a linear function. However, a line, whose equation does not go through the

* The symbol "⇔" stands for "if-and-only-if," meaning that logically, it is a necessary and sufficient condition.

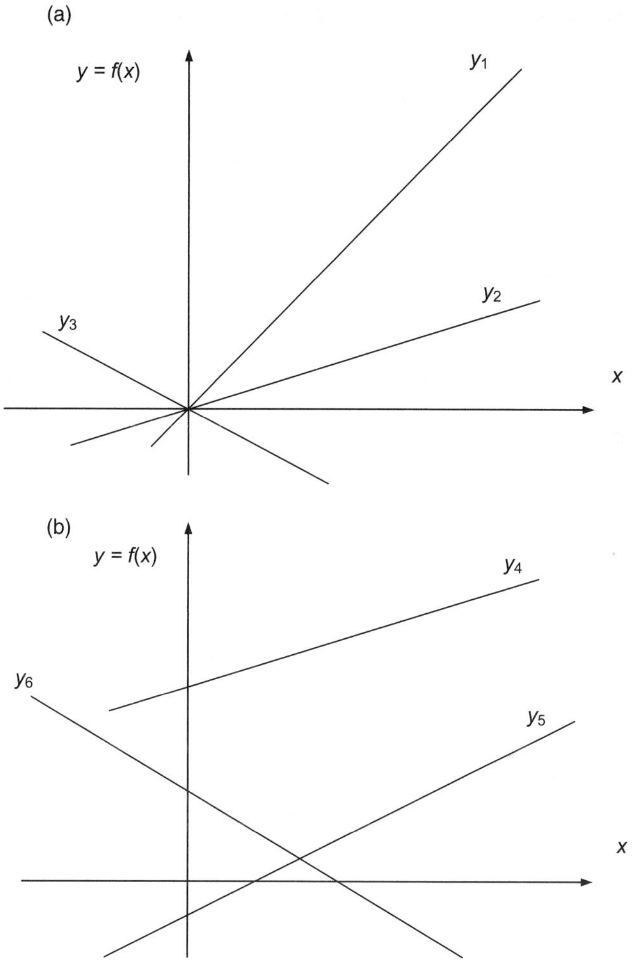

Figure 3.1 (a) Lines that are linear functions; (b) lines that are *not* linear functions.

origin of coordinates, is *not* a linear function. Figure 3.1a,b show lines that are linear functions and lines that are not linear functions.

Let us present an example of the homogeneity property given that

$$f(x) = 3x \tag{3.7}$$

and $\alpha = 4.5$, we need to verify that Equation (3.1) holds for all values of x when applied to Equation (3.7):

$$f(4.5x) = 3(4.5x). \tag{3.8}$$

To prove that the homogeneity property holds, let us present Table 3.1.

Table 3.1 Table used to exemplify numerically the homogeneity property

Col 1	Col 2	Col 3	Col 4	Col 5
x	$f(x) = 3x$	α	$\alpha f(x)$	$f(\alpha x) = 4.5 \times (3x)$
0	$f(0) = 3 \times (0) = 0$	4.5	$4.5 \times (0) = 0$	$4.5 \times [3 \times (0)] = 0$
1	$f(1) = 3 \times (1) = 3$	4.5	$4.5 \times (3) = 13.5$	$4.5 \times [3 \times (1)] = 13.5$
2	$f(2) = 3 \times (2) = 6$	4.5	$4.5 \times (6) = 27$	$4.5 \times [3 \times (2)] = 27$
3	$f(3) = 3 \times (3) = 9$	4.5	$4.5 \times (9) = 40.5$	$4.5 \times [3 \times (3)] = 40.5$
n	$f(\text{n}) = 3 \times (\text{n}) = 3\text{ n}$	4.5	$4.5 \times (3\text{ n}) = 13.5\text{ n}$	$4.5 \times [3 \times (\text{n})] = 13.5\text{ n}$

Note: Col stands for Column.

The entries of this table are:

Column 1: x
Column 2: $f(x) = 3x$
Column 3: α
Column 4: $\alpha f(x)$ and
Column 5: $f(\alpha x)$

By inspection of Table 3.1's columns 4 and 5, it is clear that Equation (3.1) holds. Without loss of generality we can state Equation (3.1) will hold for the infinitely many values of x for the given $f(x)$ and for any given value of α.

In reference to the additivity property we will prove that it is met by a linear function in a graphical way (Fig. 3.2).

By inspection of Figure 3.2 we can see that Equation (3.2) holds, repeated for the reader's convenience

$$f(x_1 + x_2) = f(x_1) + f(x_2).$$

Homogeneity and additivity properties together are completely equivalent to stating that a linear function complies with the *superposition property*:

$$f(\alpha_1 x_1 + \alpha_2 x_2) = \alpha_1 f(x_1) + \alpha_2 f(x_2). \tag{3.9}$$

So when a function complies with Equation (3.9), it is said to be linear. Conversely, when a function is linear, it complies with Equation (3.9).

Logically, the above is stated as follows:

Given $f(x)$, a function whose domain is x, and its range $f(x)$ is a

$$\textit{Linear function} \Leftrightarrow f(\alpha_1 x_1 + \alpha_2 x_2) = \alpha_1 f(x_1) + \alpha_2 f(x_2).$$

Example 3.1 Prove that the equation of a straight line $f(x) = 4x + 7$, not passing through the origin of coordinates (i.e., $b \neq 0$), is not a linear function of x.

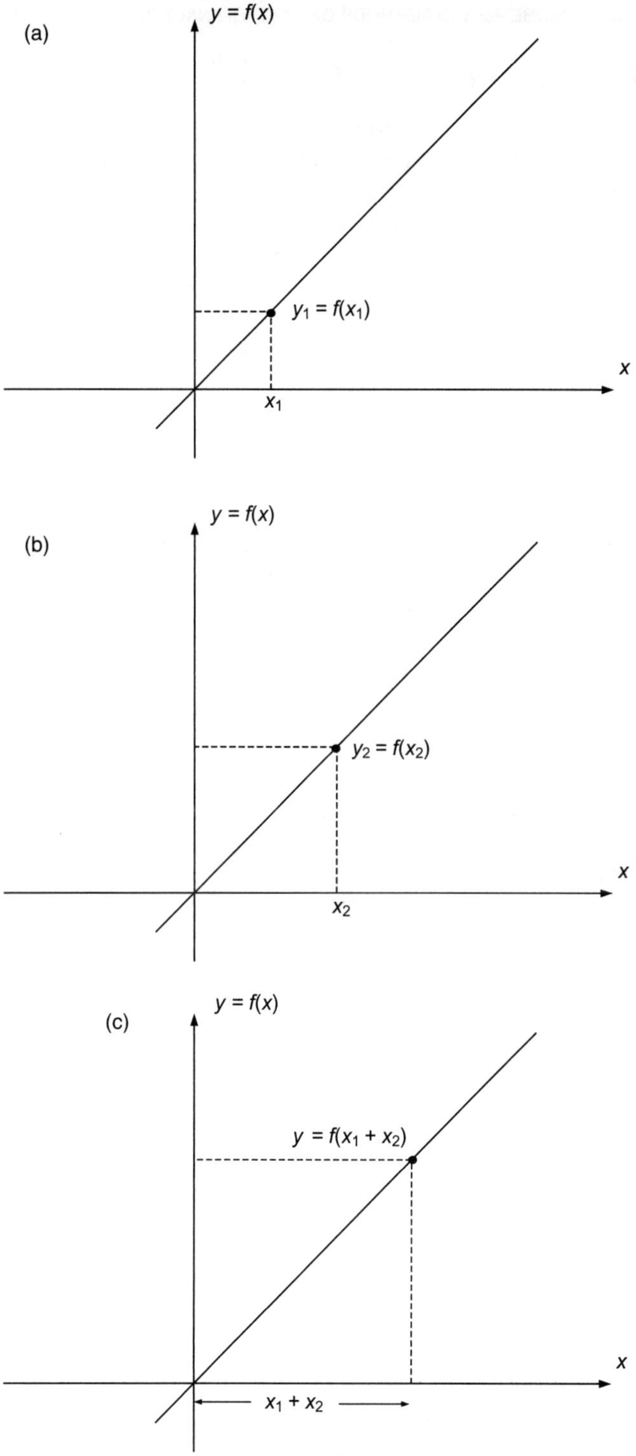

Figure 3.2 The validity of the additivity property for a linear function. (a) Linear function evaluated at x_1: $f(x_1)$; (b) linear function evaluated at x_2: $f(x_2)$; (c) linear function evaluated at $x_1 + x_2$: $f(x_1 + x_2) = f(x_1) + f(x_2)$.

Solution to Example 3.1

Simply using the homogeneity property, Equation (3.1), $f(\alpha x) = \alpha f(x)$.

It can be seen that $f(x) = 4x + 7$ is not a linear function because

$$f(\alpha x) = 4\alpha x + 7 \tag{3.10}$$

and

$$\alpha f(x) = \alpha(4x + 7) = 4\alpha x + 7\alpha. \tag{3.11}$$

From Equations (3.10) and (3.11) we see that

$$f(\alpha x) \neq \alpha f(x). \tag{3.12}$$

Thus, function $f(x) = 4x + 7$, the equation of a straight line, is *not* a linear function from the standpoint that it does not comply with Equation (3.9). Nonetheless, $f(x) = 4x + 7$ is the equation of a straight line.

3.2.1 Circuits Superposition

Let us now apply the superposition property to electric circuits. Assume that we are given an electrical circuit that can contain any number of resistors, in the black box represented in Figure 3.3. Two external voltage sources are applied to the circuit. We also refer to these two voltage sources as the circuit excitations. The output of the circuit is referred to as the circuit response.

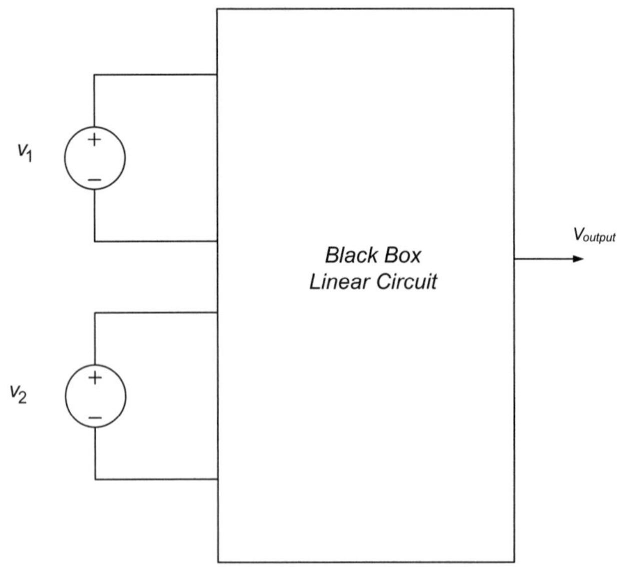

Figure 3.3 Electrical linear circuit with two external voltage sources: v_1 and v_2.

If we have a linear circuit where x is the excitation and $y = f(x)$ is its response, the superposition property tells us that

Given:

$y_1 = f(v_1)$, where y_1 is the response of the circuit due to excitation v_1 and
$y_2 = f(v_2)$, where y_2 is the response of the circuit due to excitation v_2.

The sum of the circuit responses $y_1 + y_2 = f(v_1) + f(v_2)$ equals the response of the sum of the circuit excitations $y_1 + y_2 = f(v_1 + v_2)$.

Moreover, thanks to the linearity of the circuit, we can also calculate the response of the circuit to excitation v_1 while excitation v_2 is inhibited. This yielding the response y_1 for v_2 = inhibited. Similarly, we can calculate the response of the circuit y_2 when excitation v_1 is inhibited. Finally, adding the individually found responses we obtain

$$y_{1(\text{for } v2=\text{inhibited})} + y_{2(\text{for } v1=\text{inhibited})}. \tag{3.13}$$

Equation (3.13) provides the complete response of the circuit due to noninhibited excitations or the response of the circuit due to both excitations applied simultaneously.

When the excitation is a voltage source v, inhibiting the excitation means to replace the voltage source with a short circuit ($v = 0$). When the excitation is a current source i, inhibiting the excitation means to remove the current source from the circuit, or open-circuiting the current source.

To follow up with the circuit given in Figure 3.3, we can solve the circuit by superposition, which means by applying one excitation at a time, while inhibiting the other one. The complete response of the circuit is obtained by adding each of the individual responses as Figure 3.4a,b show. So the total response of the circuit is

$$V_{output} = V_{output\text{-}due\text{-}to\text{-}v1} + V_{output\text{-}due\text{-}to\text{-}v2}. \tag{3.14}$$

So now we may ask the question, why is it better to use superposition to solve a circuit, if it seems that the number of steps grows in the process? So let us address this question with an example.

Example 3.2 Given the circuit of Figure 3.5, find the current I_2 through resistor R_2 using superposition.

Now let us apply superposition. Calculate the current through R_2 but only due to the presence of the $V = 12$ V excitation, removing or open-circuiting current source I. We obtain the circuit shown in Figure 3.6, which clearly is simpler to solve than the original circuit of Figure 3.5. By inspection of circuit in Figure 3.6, the 12 V source is applied directly across R_1, thus the current through R_1 is easily calculated using Ohm's law:

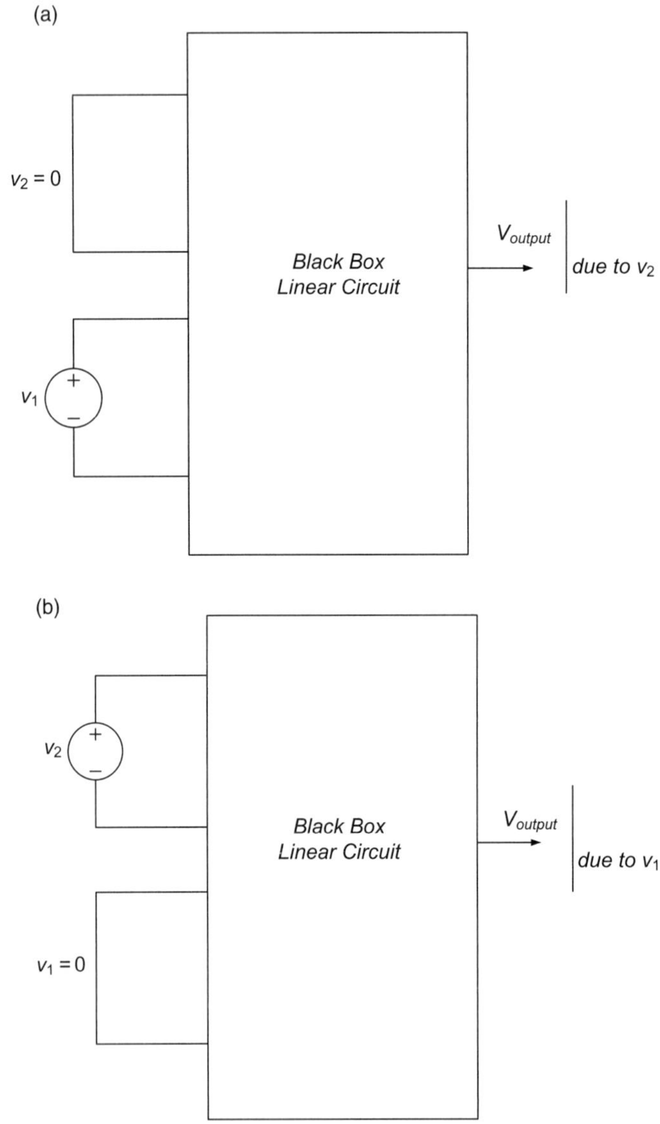

Figure 3.4 Application of superposition to the circuit of Figure 3.3: (a) circuit response due to v_1 when $v_2 = 0$; (b) circuit response due to v_2 when $v_1 = 0$.

$$I_1 = 12\ \text{V}/6\ \Omega. \tag{3.15}$$

Now let us calculate the current through the series of R_2 and R_3. Thus, current I_{23} is

$$I_{23} = \text{V}/(\text{R}_2 + \text{R}_3). \tag{3.16}$$

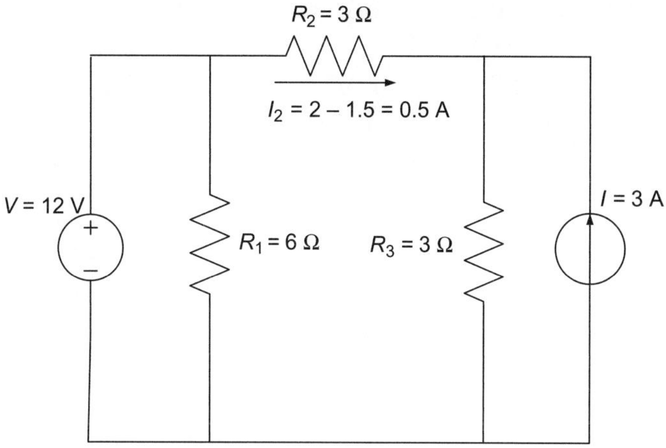

Figure 3.5 Circuit for Example 3.2.

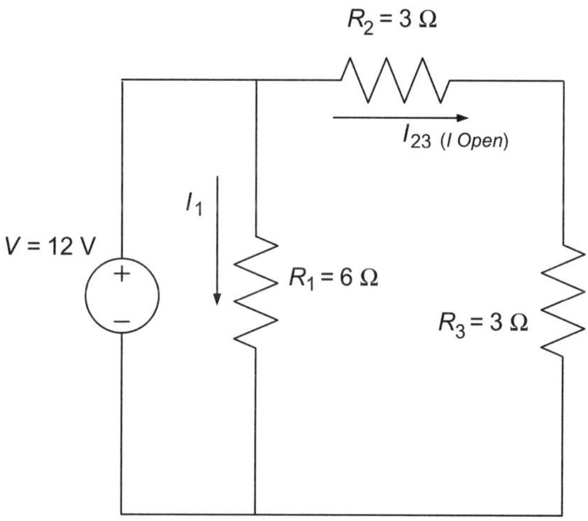

Figure 3.6 Example 3.2: Removing current source *I* and applying superposition under the effect of *V*.

Using the numerical values from Figure 3.6 leads to

$$I_{23} = 12\ \text{V}/(3+3)\ \Omega = 2\ \text{A}. \tag{3.17}$$

Now we need to calculate the current that flows through R_2 when excitation V = 12 V is replaced by a short circuit. We present this circuit in Figure 3.7.

By close examination we can see that R_1 is short-circuited, so basically only R_2 and R_3 are in parallel with the 3 A current source. This circuit is shown in Figure 3.8.

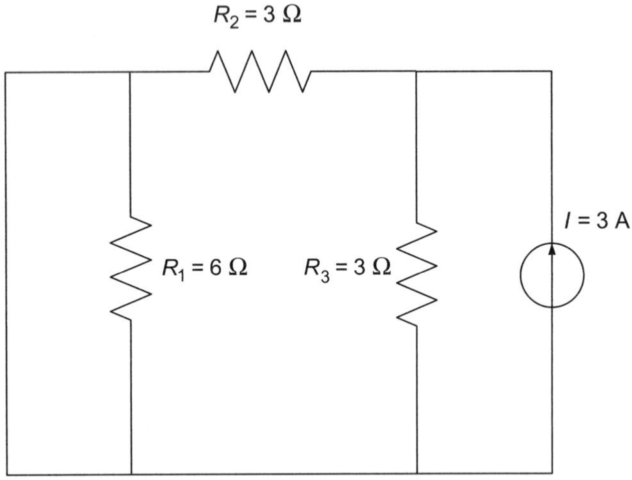

Figure 3.7 Removing voltage source *V* and applying superposition under the effect of current *I*.

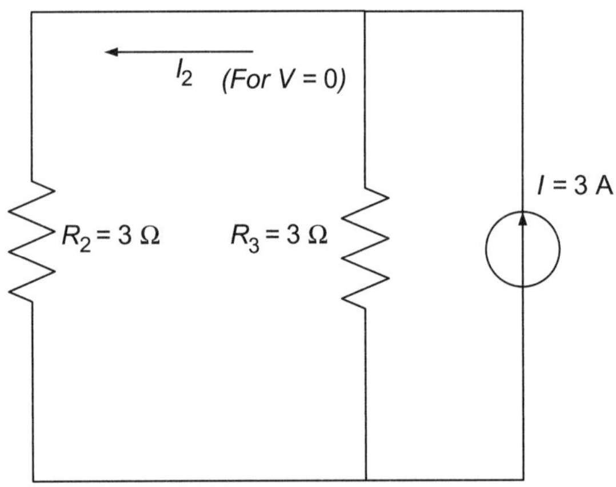

Figure 3.8 Example 3.2: Eliminating the short circuited resistor R_1.

By inspection of Figure 3.8 note that the circuit was redrawn eliminating the presence of R_1, because it was short-circuited (see Fig. 3.7). Clearly we see now that the 3-A current source *I* delivers current to two equal valued resistors in parallel R_2 and R_3. From Kirchoff's current law (KCL) we know that the current has to be divided equally between R_2 and R_3. Thus, the current flowing through R_2, after the elimination of the 12-V voltage source, is 1.5 A.

Finally, to complete the application of superposition for Example 3.2, we present the previously obtained currents that flow through R_2. When the 3-A current source was removed, the current through R_2 flowed from left to right, as shown in Figure 3.9, under $I_{due\ to\ V=12\ V}$. When the 12-V voltage source was

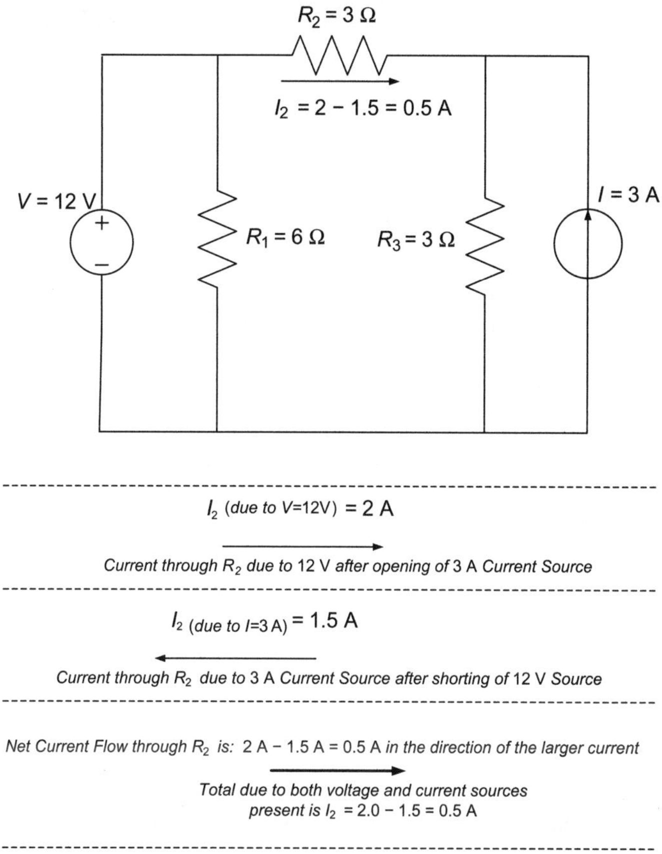

Figure 3.9 Net current flowing through resistor R_2.

removed, the current through R_2 flowed from right to left, as shown in Figure 3.9, under $I_{due\ to\ I=3\ A}$.

The net resulting current of 0.5 A, flows through R_2 due to the simultaneous effect of both the voltage and the current sources in the direction on the larger current of 2 A.

3.3 THE THÉVENIN METHOD

The Thévenin method is very powerful since it allows one to replace a large linear circuit with a voltage source and a resistor in series. Such voltage is referred to as the Thévenin voltage and the resistor is called the Thévenin resistor. If we are dealing with an AC circuit, the term resistor is replaced with impedance. From Chapter 2, impedance is a combination of R, L, and C circuit elements.

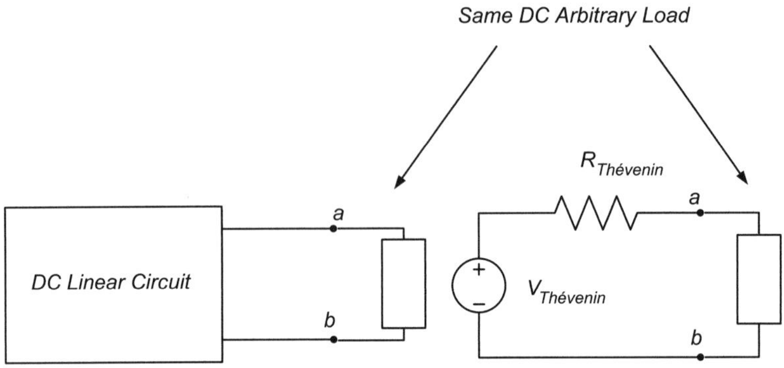

Figure 3.10 DC Thévenin equivalent of a DC circuit.

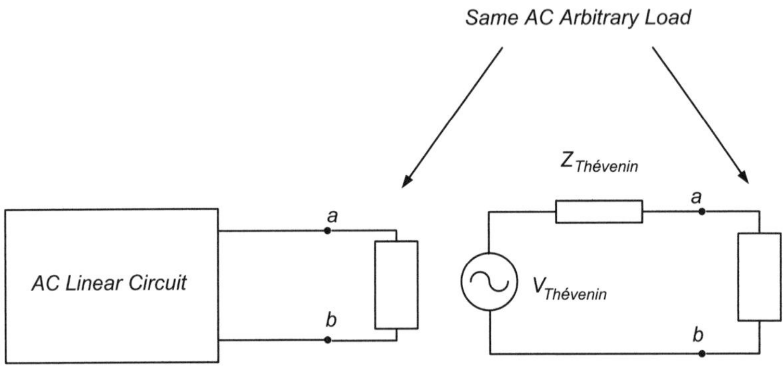

Figure 3.11 AC Thévenin equivalent of an AC circuit.

The linear circuit, to be replaced, can contain any number of circuit elements*, independent and/or dependent voltage, and current sources. It is important to note that the controlling voltage or current of the dependent sources need to reside within the same linear circuit that is to be replaced with the Thévenin equivalent circuit.

Figure 3.10 shows a DC linear circuit and its Thévenin equivalent circuit. The Thévenin equivalent voltage is a DC source for the DC case. Figure 3.11 shows an AC linear circuit and its Thévenin equivalent. The Thévenin equivalent voltage is an AC source for the AC case.

In both Figures 3.10 and 3.11, the load can even be a nonlinear load, it is not required for it to be linear, as the circuit that will be replaced with its Thévenin equivalent does. The arbitrary load may also have dependent voltage or current sources. Their controlling voltage or current shall be within the arbitrary load circuit itself.

* Circuit elements refer to resistors in DC circuits; but it refers to resistors, inductors, and capacitors in AC circuits.

3.3.1 Application of the Thévenin Method

Methodology to find the Thévenin equivalent circuit: (as it applies to Example 3.3, Fig. 3.12)

1. First slice the portion of the circuit that we want to find the Thévenin equivalent circuit at two nodes only. *This circuit has to be linear.*

   In our example we have drawn a dotted line *a-b*. The portion of the circuit that we want to *Thévenize* is on the left-hand side of the dotted line. Ensure that if dependent sources are present on the circuit to be *Thévenized*, the slicing of the circuit must not separate the dependent sources from their respective controlling variables.
2. The Thévenin voltage is calculated as follows: separate the circuit to be *Thévenized* at the terminals *a* and *b* from the associated circuit that is on the right-and side of the dotted line. The Thévenin voltage ($V_{Thévenin}$) is calculated as the *open-circuit voltage* across terminals *a* and *b*. In particular, this is the voltage across resistor R_3 for our example of Figure 3.12a.
3. To calculate the Thévenin resistance for DC circuits or the Thévenin impedance in AC circuits, inhibit in the linear circuit to be *Thévenized* all the independent voltage and current sources. To inhibit a voltage source, each source should be replaced with a short circuit. Open-circuiting or simply removing the current source from the circuit inhibits a current source. No action needs to be taken with dependent voltage or current sources. However, it is appropriate to remind the reader that any dependent sources in the linear circuit to be *Thévenized* must have their controlling variable within the same circuit to be *Thévenized.* Once all independent sources have been inhibited, calculate the resistance in the DC case (or the impedance in the AC case) seen across terminals *a* and *b*. The result is what we call the Thévenin resistance (or impedance) referred to as either R_{Th} (or Z_{Th}).
4. The entire circuit on the left of the dotted line can now be replaced by the series of the Thévenin voltage source and the Thévenin resistance (or impedance).

   Example 3.3, which follows, goes explicitly over a numerical application of Thévenin.

Example 3.3 Given the circuit of Figure 3.12a, find the Thévenin equivalent circuit at the left of the a-b dotted line. Use the Thévenin equivalent circuit found to calculate the current of the original circuit.

Referring to Figure 3.12a, we want to find the Thévenin equivalent of the circuit to the left of the a-b dotted line. Separating this circuit from the rest of the circuit on the right side of the a-b dotted line, we next find the open-circuit voltage across resistor R_3. This is the Thévenin voltage, and it is calculated as follows applying Kirchoff's voltage law (KVL):

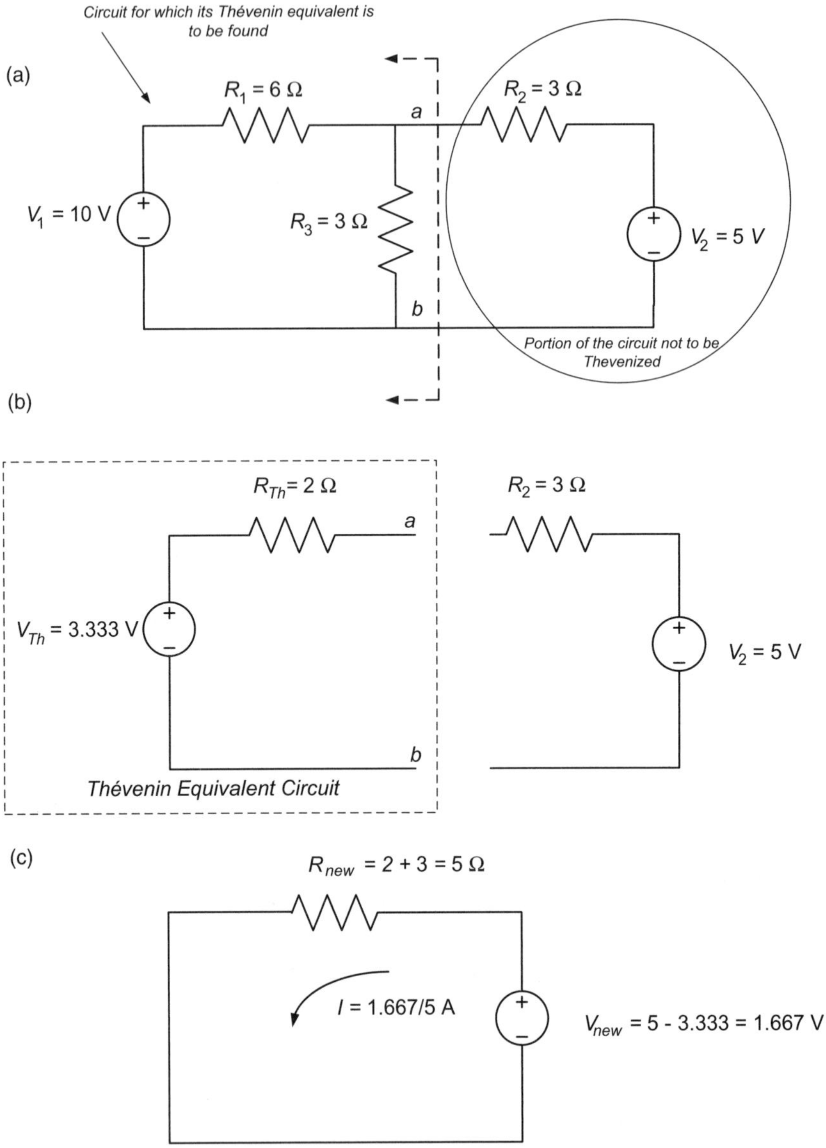

Figure 3.12 Circuit to apply the Thévenin method: (a) original circuit; (b) Thévenin equivalent on left of portion not to be Thévenized; (c) merging of the found Thévenin equivalent with the non-Thévenized portion of the circuit.

$$V_{Th} = V_1 R_3/(R_1 + R_3) \tag{3.18}$$

$$V_{Th} = 10 \times 3/(6+3) = 3.333 \text{ V}. \tag{3.19}$$

The Thévenin resistance R_{Th} is calculated by inhibiting (short-circuiting) the 10-V voltage source and calculating the resistance seen to the left of the *a-b* dotted line.

This yields the parallel of R_1 and R_3:

$$R_{Th} = (R_1 \times R_3)/(R_1 + R_3) = 2\ \Omega. \tag{3.20}$$

Next we substitute the circuit to the left of the *a-b* dotted line with its Thévenin equivalent, which is a 3.333-V DC source, from Equation (3.19) in series with $R_{Th} = 2\ \Omega$, from Equation (3.20). Finally, in Figure 3.12c we merged the Thévenin equivalent circuit with the rest of the untouched right-hand side original circuit. The resulting circuit is almost trivial and allows us to calculate the current in the circuit by Ohm's law in a straightforward fashion. We connect the found Thévenin equivalent circuit with the right-hand side circuit of Figure 3.12b. Combining the two voltage sources into V_{new}, we obtain Figure 3.12c:

$$I = (5 - 3.333)\text{ V}/5\ \Omega = 0.3334 \text{ A}. \tag{3.21}$$

Example 3.4 Given the circuit of Figure 3.13a, find the Thévenin equivalent circuit of the circuit to the left of the a-b dotted line. Reattach the equivalent circuit to resistor R_3 to calculate the current through R_3.

Figure 3.13a shows the originally given circuit, while Figure 3.13b shows the circuit to-be-Thévenized not connected to its load R_3. At this point we calculate the Thévenin voltage; the open-circuit voltage across terminals *a* and *b* needs to be determined. To proceed with this calculation, we will use the superposition method; refer to the circuit of Figure 3.13b. We will split the problem into two easier problems to solve. We will compute the voltage V_{ab} due to the effect of 8-V voltage source *V*, open-circuiting the current source *I*. This will yield $V_{Th\ due\ to\ 8\,V}$. On the second step we calculate V_{ab} due to the effect of the 1-A current source *I*, short-circuiting the 8-V voltage source. This will yield $V_{Th\ due\ to\ 1\,A}$. Upon obtaining those two partial voltages, the total voltage, which is the Thévenin voltage, is the algebraic sum of $V_{\text{Th due to 8 V}}$ and $V_{\text{Th due to 1 A}}$. An algebraic sum refers to performing a sum taking into account the signs or polarities of the voltages involved. In our particular case, both polarities are positive.

Let us refer to Figure 3.14a and b to see how we partition the originally given circuit (Fig. 3.13a) into two separate circuits, each of which will be driven by one of the sources while the other source is *inhibited*.

Referring to Figure 3.14a we calculate $V_{Th\ due\ to\ 8\,V}$ as follows:

$$V_{Th\ due\ to\ 8\,V} = VR_2/(R_1 + R_2). \tag{3.22}$$

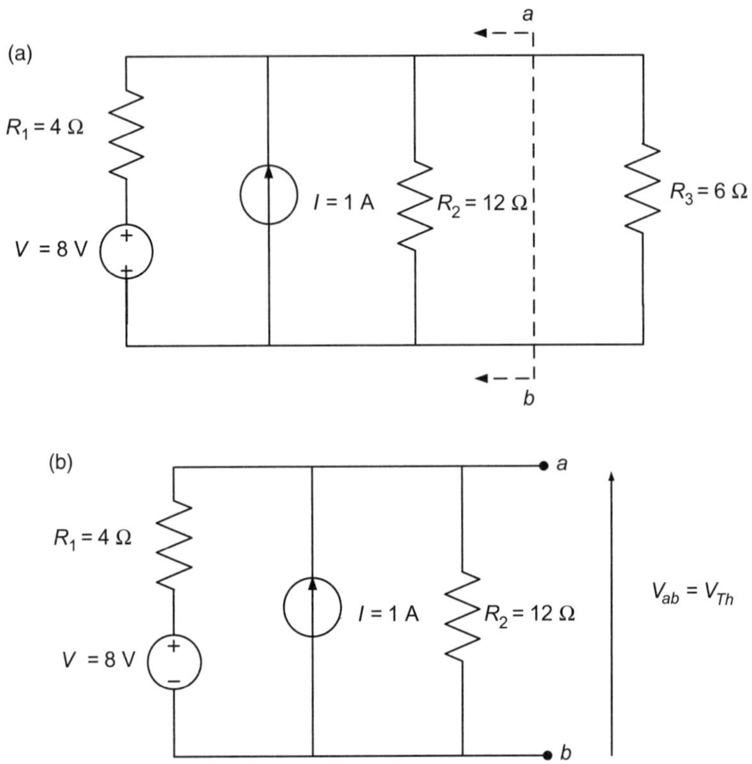

Figure 3.13 Circuit for Example 3.4: (a) original circuit; (b) sliced circuit to which Thévenin is applied.

Using the respective values in Equation (3.22) from Figure 3.14a, we obtain

$$V_{Th\ due\ to\ 8\,\mathrm{V}} = 8 \times 12/(4+12) = 6\ \mathrm{V}. \tag{3.23}$$

Now referring to Figure 3.13b, we calculate $V_{Th\ due\ to\ 1\,\mathrm{A}}$ as follows:

By KCL we can see by inspection that

$$1 = (V_{ab}/R_1) + (V_{ab}/R_2), \tag{3.24}$$

where: V_{ab}/R_1 and V_{ab}/R_2 are respectively the currents through resistors R_1 and R_2.

Using the component values from Figure 3.13b into Equation (3.24) yields

$$1 = V_{ab}(1/4 + 1/12). \tag{3.25}$$

$$V_{Th\ due\ to\ 1\,\mathrm{A}} = V_{ab} = 3\ \mathrm{V}. \tag{3.26}$$

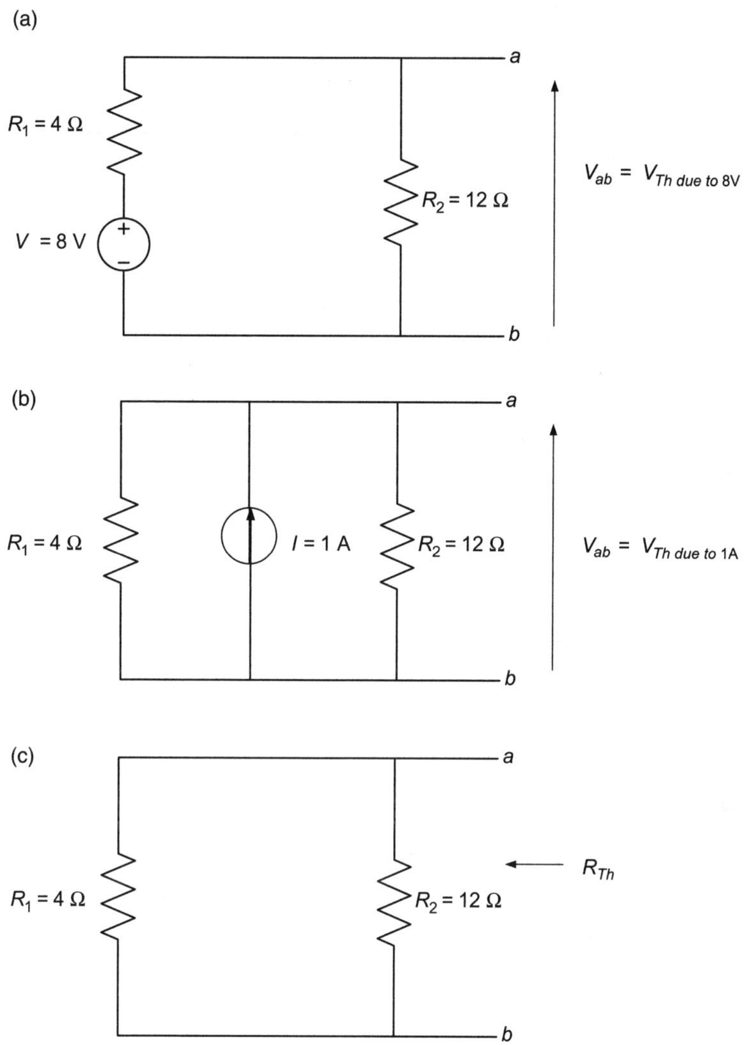

Figure 3.14 Superposition method for Thévenin Example 3.4: (a) effect of $V = 8$ V voltage source, 1 A current source open-circuited; (b) effect of $I = 1$ A current source, 8 V voltage source short-circuited; (c) elimination of all sources to compute the Thévenin resistance.

Now from the results of Equations (3.23) and (3.26) we add both voltages leading to

$$V_{Th\ due\ to\ 8\,\mathrm{V}} + V_{Th\ due\ to\ 1\,\mathrm{A}} = 6\ \mathrm{V} + 3\ \mathrm{V} = 9\ \mathrm{V}. \tag{3.27}$$

Now we need to calculate the Thévenin resistance (refer to Figure 3.14c). The voltage source is replaced by a short circuit and the current source is open-circuited. R_{Th} is simply the parallel of resistors R_1 and R_2. So now we have all

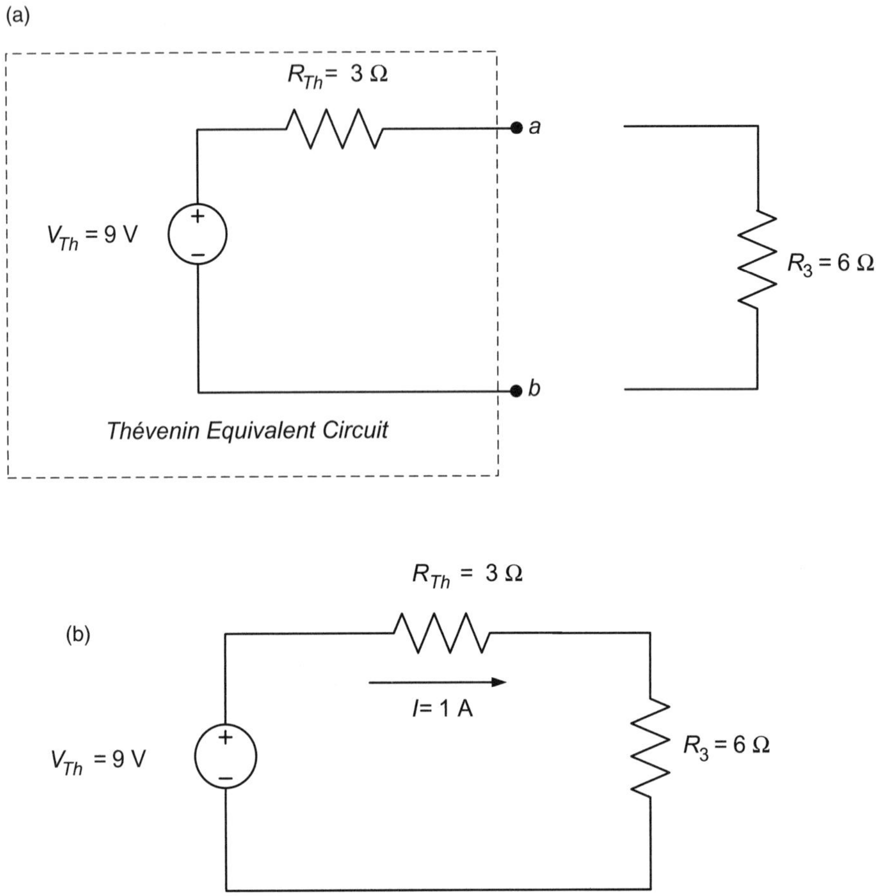

Figure 3.15 (a) Thévenin equivalent of Example 3.4; (b) Thévenin equivalent merged with the resistive load.

the elements of the Thévenin equivalent circuit. These are shown in Figure 3.15a. In Figure 3.15b, the Thévenin equivalent circuit is joined to resistor R_3. Finally, the current through R_3 is simply calculated using Ohm's law, leading to

$$I = V_{Th}(R_{Th} + R_3) = 9/(3+6) = 1\ \text{A}. \tag{3.28}$$

3.4 NORTON'S METHOD

Several decades after the invention of the Thévenin method, American engineer *Edward Norton* invented an analysis method which bears his name today. Norton's method of analysis is the dual of Thévenin's method. Duality, in

Table 3.2 Some dual pairs

Resistance	Conductance
Inductance	Capacitance
Voltage	Current
Voltage source	Current source
Node	Mesh
Open circuit	Short circuit
KVL	KCL
Thévenin	Norton
Elements in series	Elements in parallel

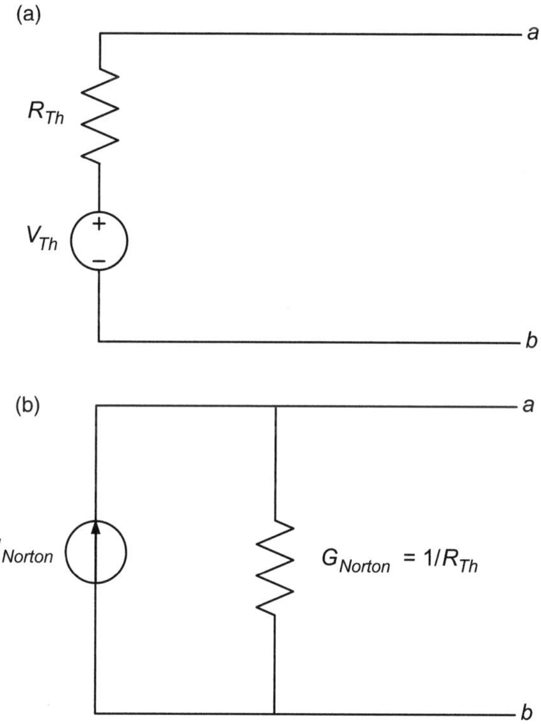

Figure 3.16 (a) Thévenin equivalent circuit; (b) Norton equivalent circuit.

circuit analysis, refers to circuits that can be described by the same set of equations and solutions, except that certain elements are interchanged.

Table 3.2 lists the dual-pair elements.

The dual of a Thévenin equivalent circuit is its Norton's equivalent.

Figure 3.16 shows the dual of Thévenin equivalent. Thévenin is a series of two elements; Norton is transformed by duality into a parallel of two elements. The Thévenin voltage source becomes a Norton current source. The Thévenin resistance becomes a Norton conductance. For more details on *duality in circuit theory*, the reader is referred to the Bibliography at the end of the chapter.

In Figure 3.16b the value of the Norton current source is given by

$$I_{Norton} = V_{Th}/R_{Th} \tag{3.29}$$

The Norton current is obtained short-circuiting the Thévenin equivalent circuit and calculating the current that flows through the short. So the Norton current source (I_{Norton}) is the short-circuit current in the Thévenin equivalent circuit.

And the Norton resistance is

$$R_{\mathrm{N}} = R_{Th}. \tag{3.30}$$

Note the Norton equivalent resistance is identical to the Thévenin equivalent resistance. For the sake of simplicity we will continue to use R_{Th} whether we use the Thévenin or the Norton equivalent circuits.

3.4.1 Source Transformations

Every voltage source in series with a resistance or impedance can be converted into a parallel equivalent of a current source in parallel with a conductance or admittance.

Equations (3.31) and (3.32) address the source transformations between Thévenin and Norton equivalent circuits. The equations below address the source transformation when we have an AC Thévenin source and in series with a Thévenin impedance.

$$\mathbf{I}_{\mathbf{Norton}} = \mathbf{V}_{\mathbf{Th}}/\mathbf{Z}_{\mathbf{Th}} \tag{3.31}$$

and

$$\mathbf{Z}_{\mathrm{N}} = \mathbf{Z}_{\mathbf{Th}}. \tag{3.32}$$

It is important to understand that the current, voltage, and impedance in Equations (3.31) and (3.32) are all phasors. Some textbooks define the Norton admittance $\mathrm{Y_N} = 1/\mathrm{Z_N}$, which is consistent with Equations (3.31) and (3.32).

Example 3.5 Given the Thévenin equivalent circuit of Figure 3.16a, find the Norton equivalent circuit. Assume that $V_{Th} = 24$ V and $R_{Th} = 4\ \Omega$.

Using the source transformation from Equations (3.31) and (3.32), the Norton current source is calculated as follows:

$$I_{Norton} = V_{Th}/R_{Th} = 24/4 = 6\ \mathrm{A}. \tag{3.33}$$

The Norton equivalent resistance equals the Thévenin resistance:

$$R_{Th} = R_N = 4\,\Omega. \tag{3.34}$$

The source transformation method allows the conversion of a Thévenin equivalent circuit into a Norton equivalent circuit and vice versa. There is a direct way to obtain the Norton equivalent circuit from the given circuit, without previously finding its Thévenin equivalent. This is the topic of the following section.

3.4.2 Finding the Norton Equivalent Circuit Directly from the Given Circuit

The procedure follows:

1. Separate the circuit for which the Norton equivalent circuit is to be found from the rest of the circuit or its load. If there are any dependent voltage or current sources in the circuit for which the equivalent circuit is to be found, the dependent source and its control variable must reside within such circuit.
2. To find the Norton resistance, calculate it exactly as the Thévenin resistance was calculated. Inhibit all voltage and current sources in the circuit. That is, open-circuit all current sources, and short-circuit all voltage sources.
3. Calculate the Norton equivalent current source by shorting terminals a and b of the circuit whose Norton equivalent circuit is to be found. Refer to Figure 3.17. The Norton current source I_N is the calculated short-circuit current that flows through shorted terminals a and b.
4. The Norton equivalent of the original circuit is the parallel of I_N and R_{Th} (remember that R_{Th} and R_N are always identical).

Example 3.6 Find the Norton equivalent circuit of the circuit of Figure 3.17b. Use the found Norton equivalent circuit to calculate the load current I_L through resistor R_L, refer to Figure 3.17a.

By inspection of Figure 3.17b, inhibiting all voltage and current sources, we find that the Norton resistance equals the series of R_1 and R_2 in parallel with R_3. Thus,

$$R_N = R_{Th} = (R_1 + R_2)R_3/(R_1 + R_2 + R_3). \tag{3.35}$$

Using the given values for the resistors,

$$R_N = R_{Th} = (3+3)6/(3+3+6) = 3\,\Omega. \tag{3.36}$$

Let us calculate the Norton current source using Equation (3.33). Prior to this calculation, let us simplify the circuit from Figure 3.17b a little further.

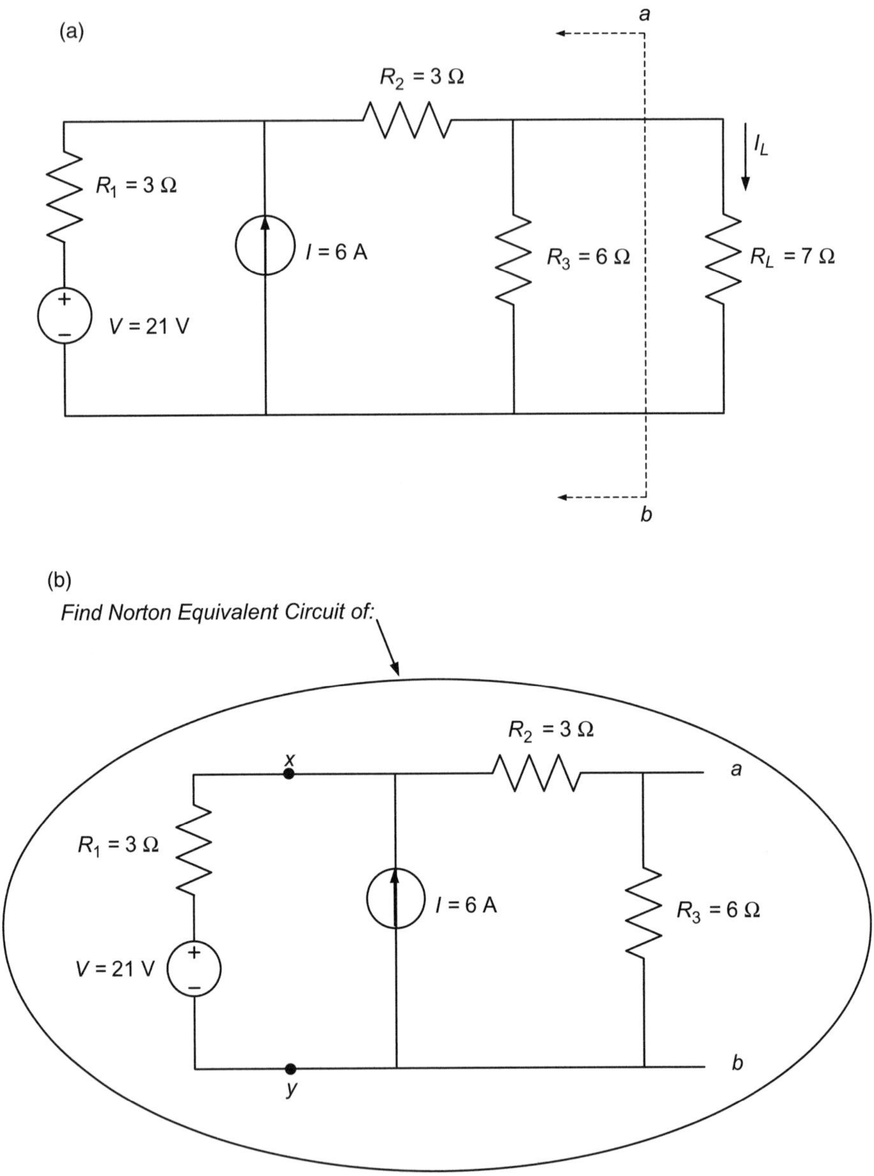

Figure 3.17 Finding the Norton equivalent circuit for Example 3.6: (a) original circuit; (b) circuit from which to find Norton's equivalent circuit.

By inspection of the circuit in Figure 3.17b note that V and R_1 are in series, on the left side of nodes x and y. We can do a source transformation of V and R_1 to convert them into a current source in parallel with a resistor.

$$I_{short\text{-}circuit} = I_{Norton} = V_{Th}/R_{Th} = V/R_1. \tag{3.37}$$

$$I_{short\text{-}circuit} = I_{Norton} = I_{ST} = 21/3 = 7 \text{ A}. \tag{3.38}$$

The result of this source transformation is provided in Figure 3.18b. Note that after such source transformation, the 3 Ω resistor R_1 appears in Figure 3.18b renamed as R_{ST}. The Norton current calculated in the source transformation is presented as $I_{ST} = 7$ A. Note: The subscript ST stands for *source transformation*. Then, using the Norton equivalent circuit for the voltage source transformation and substituting it into the original circuit of Figure 3.17b, we obtain Figure 3.19a.

So let us combine the two current sources into one, thus 6 A + 7 A = 13 A and name this current I_{Combined} (refer to Figure 3.19b). Now let us short-circuit the terminals a and b and calculate the Norton short circuit current. This will be done in four steps, with Equations (3.39) through (3.42).

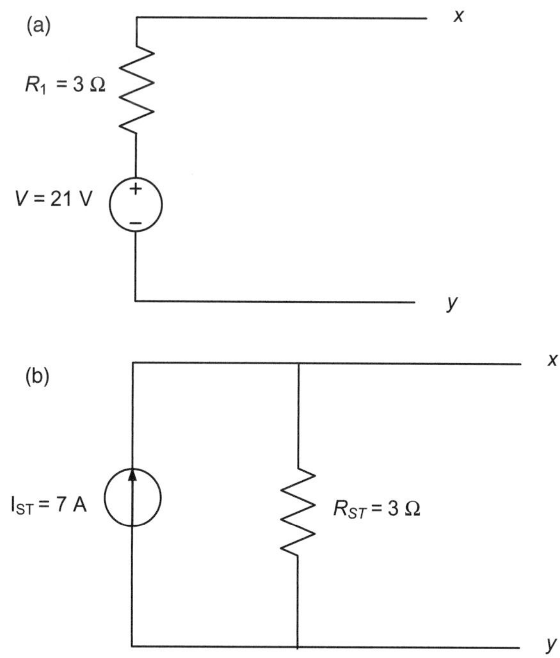

Figure 3.18 Source transformation as an interim step toward finding the Norton equivalent: (a) Thévenin equivalent; (b) Norton equivalent.

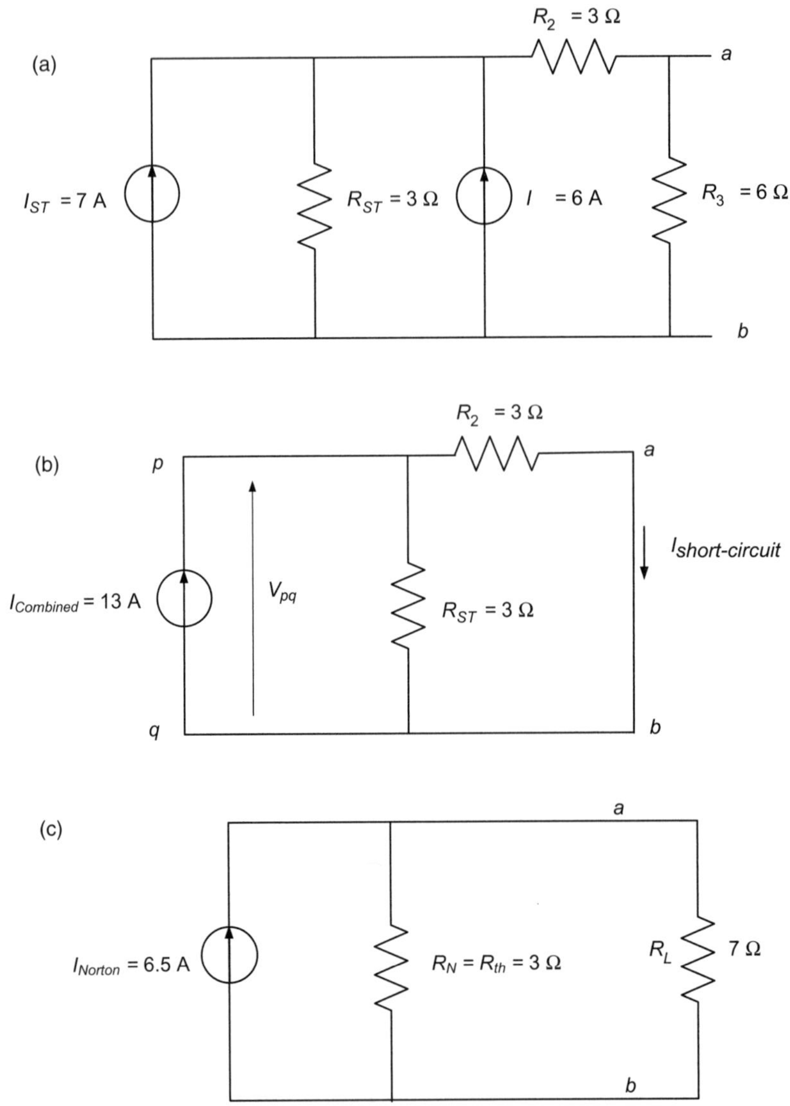

Figure 3.19 (a) Circuit of Example 3.6 after a source transformation; (b) circuit of Example 3.6 after combination of the two current sources; (c) final Norton equivalent circuit: Example 3.6.

From Figure 3.19b and since R_2 and R_{ST} are in parallel, we get

$$V_{pq} = I_{Combined} R_2 R_{ST} / (R_2 + R_{ST}), \tag{3.39}$$

where V_{pq} is the voltage across the terminals of the (13 A) $I_{Combined}$ current source.

Note: In Figure 3.19b, the shorting of terminals a and b, essentially eliminates resistor R_3 from the circuit seen in Figure 3.19a. Since this voltage V_{pq} is the same as the voltage across the parallel of R_{ST} and R_2, by Ohm's law:

$$I_{short\ circuit} = [I_{Combined} R_2 R_{ST} / (R_2 + R_{ST})] / R_2 \tag{3.40}$$

Eliminating R_2 from numerator and denominator:

$$I_{short\ circuit} = I_{Combined} R_{ST} / (R_2 + R_{ST}) \tag{3.41}$$

Now using the values for Equation (3.41) from Figure 3.19b:

$$I_{short\ circuit} = 13\ 3/(3+3) = 6.5\ \text{A}. \tag{3.42}$$

This short circuit current of 6.5 A will be the Norton equivalent current source of the originally provided circuit (Fig. 3.17b). The 3-Ω resistor R_{ST} from Equation (3.36) is the Thévenin resistor of the equivalent model, as it can be seen in Figure 3.19c.

The final step, Figure 3.19c, is to attach the 7-Ω load resistor R_L to the Norton equivalent circuit and calculate the current through R_L.

Thus, we obtain

$$V_{ab} = I_{Norton}[R_N R_L / (R_N + R_L)]. \tag{3.43}$$

Thus, the current through R_L is:

$$I_L = V_{ab} / R_L. \tag{3.44}$$

Plugging the value of V_{ab} from Equation (3.44) into Equation (3.43)

$$I_L = I_{Norton}[R_N / (R_N + R_L)]. \tag{3.45}$$

$$I_L = 1.95\ \text{A}. \tag{3.46}$$

3.5 THE MESH METHOD OF ANALYSIS

The mesh method of circuit analysis is based on KVL. It provides a more effective way of deriving circuit equations virtually by quick inspection of the circuit. The mesh method is more suitable and intuitive when the circuit contains independent voltage sources. The method is somewhat less intuitive when current sources are also included and probably the least intuitive when dependent voltage and current sources are also present. The mesh method solves for mesh currents as opposed to finding individual branch currents for every circuit branch. This is advantageous because the number of unknowns is somewhat reduced.

We will address the methodology of writing mesh equations for various circuits via examples that will grow in complexity.

The following assumptions when using the mesh method are made:

1. *All circuits that we will analyze are planar.* The mesh method does not work for non-planar circuits.
2. *A mesh is a closed loop that does not contain other loops within it.*

Planar circuits are those circuits that can be drawn on a plane without its branches crossing each other. Nonplanar circuits are those that cannot be drawn or redrawn without one or more branches crossing. Figure 3.20 presents examples of planar and nonplanar circuits. Nonplanar circuits are outside the scope of this book and are studied in advanced circuit analysis courses. Careful observation of Figure 3.20a reveals that the circuit is actually planar; however, at first sight it initially may appear nonplanar. Circuit Figure 3.20b is the exact same circuit as that in a. Finally, c is a true nonplanar circuit.

Figure 3.21 shows a simple circuit with two meshes: one of them is: a-b-c-d-a, the second one is b-e-f-c-b. It is important to observe that a-b-e-f-c-d-a is a loop and not a mesh, because it includes one previously defined mesh.

3.5.1 Establishing Mesh Equations. Circuits with Voltage Sources

Let us assume we have a two-mesh circuit as the one shown in Figure 3.22. Note that the circuit has three independent voltage sources, three resistors and two meshes. We also introduce the concept of mesh currents. Mesh current I_I is the current in the mesh formed by elements V_1, R_1, R_2, and V_2. Mesh current I_{II} is the current in the mesh formed by elements V_2, R_2, R_3, and V_3. The branch currents are I_{b1}, I_{b2}, and I_{b3}. It is important to see that mesh currents are not in general the same as branch currents. Note that mesh currents are named with Roman numeral subscripts in this example, whereas branch currents are named with regular number subscripts.

Branch current I_{b1} is the current that flows through the branch that contains voltage source V_1 and resistor R_1. Similarly, branch current I_{b2} is the current that flows through the branch that contains V_2 and R_2; and branch current I_{b3} is the branch current that flows through elements R_3 and V_3. So let us look into the relationship that exists between branch currents and mesh currents.

In particular for the circuit shown in Figure 3.22, the following are how the branch and mesh currents relate to each other:

$$I_I = I_{b1} \tag{3.47}$$

$$I_I - I_{II} = I_{b2} \tag{3.48}$$

$$I_{II} = I_{b3}, \tag{3.49}$$

where in Equations (3.47) through (3.49) the currents in the left-hand side of the equal signs are mesh currents. The currents on the right-hand side of the equal sign are branch currents. Once we establish the mesh equations for the

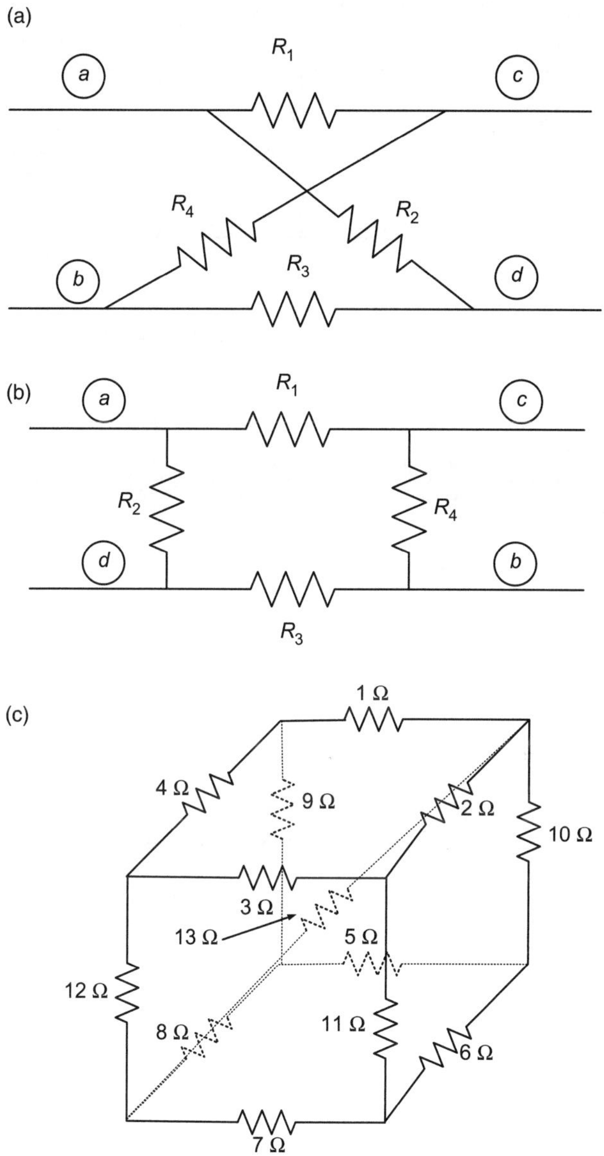

Figure 3.20 Planar and nonplanar circuits: (a) nonplanar circuit, that is, planar circuit in disguise; (b) same planar circuit redrawn; (c) true nonplanar circuit.

given circuit, the mesh currents are the unknowns in the mesh method of analysis. Since we have two meshes, we will be able to obtain two mesh equations and solve for the unknown mesh currents I_I and I_{II}. The branch currents in each specific circuit element are calculated using Equations (3.47) through (3.49) after the mesh currents are found.

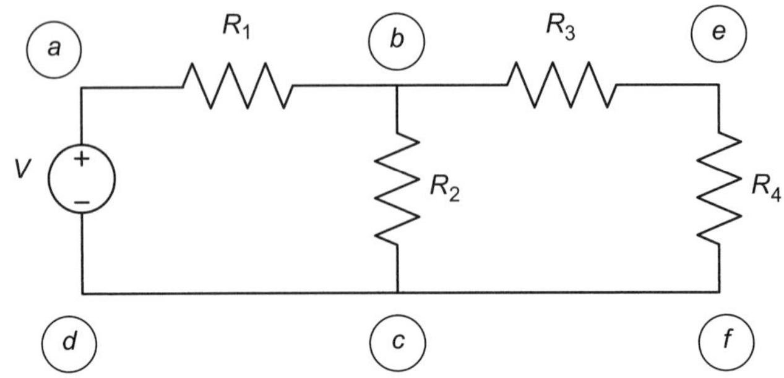

Figure 3.21 Meshes and loops in a circuit.

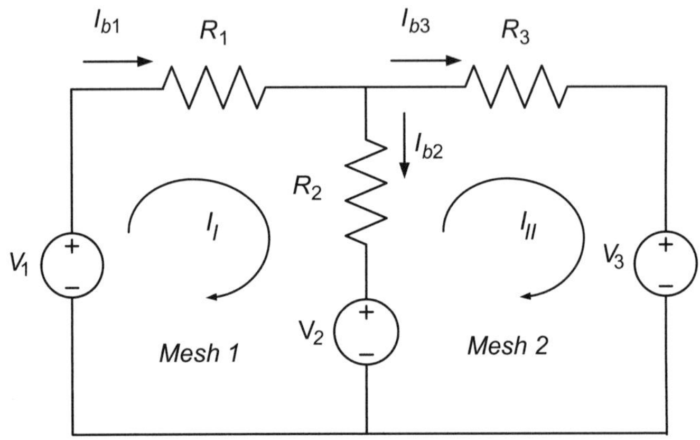

Figure 3.22 Two-mesh equations for Example 3.7.

Initially we will apply KVL to each mesh, but work with mesh currents instead of branch currents. So for

$$\text{Mesh 1: } V_1 - V_2 = I_I R_1 + (I_I - I_{II}) R_2. \tag{3.50}$$

$$\text{Mesh 2: } V_2 - V_3 = (I_{II} - I_I) R_2 + I_{II} R_3. \tag{3.51}$$

Note that the direction of the mesh currents was arbitrarily chosen to be clockwise. When applying KVL to each mesh we travel each mesh in the clockwise direction too. It is usually a headache to the reader, understanding why is that currents directions and the direction of traveling the meshes are picked arbitrarily? The simple answer to this is that as long as the voltage rises and voltage drops signs are respected in a consistent manner, the numerical answer will provide a positive sign when such current direction was assigned in the way in which it was assumed; or it will provide a negative sign if the current actually flows in the direction opposite to the one assumed. Do not get hung up on this; solving problems will clarify these apparently confusing arbitrary choices.

So now let us regroup the terms of Equations (3.50) and (3.51) with respect to the mesh currents and obtain

$$\text{Mesh 1: } V_1 - V_2 = I_I(R_1 + R_2) - I_{II}R_2. \tag{3.52}$$

$$\text{Mesh 2: } V_2 - V_3 = -I_I R_2 + I_{II}(R_2 + R_3). \tag{3.53}$$

Equations (3.52) and (3.53) are a system of simultaneous linear equations that allows us to find the two unknown mesh currents I_I and I_{II}.

We can also rewrite the system of simultaneous equations in matrix form as follows:

$$\begin{vmatrix} V_1 - V_2 \\ V_2 - V_3 \end{vmatrix} = \begin{vmatrix} R_1 + R_2 & -R_2 \\ -R_2 & R_2 + R_3 \end{vmatrix} \begin{vmatrix} I_I \\ I_{II} \end{vmatrix}. \tag{3.54}$$

Example 3.7 Now let us consider a numerical example using the circuit of Figure 3.22 and mesh equations in matrix form, from Equation (3.54), assuming the following component values:

$$R_1 = 3\,\Omega,\ R_2 = 1\,\Omega,\ R_3 = 3\,\Omega. \tag{3.55}$$

$$V_1 = 2\text{ V},\ V_2 = 1\text{ V},\ V_3 = 2\text{ V}. \tag{3.56}$$

Using the values from Equations (3.53) and (3.54) into the mesh equation obtained in Equation (3.52) we obtain

$$\begin{bmatrix} 2-1 \\ 1-2 \end{bmatrix} = \begin{bmatrix} 3+1 & -1 \\ -1 & 3+1 \end{bmatrix} \begin{bmatrix} \mathrm{I_I} \\ \mathrm{I_{II}} \end{bmatrix}. \tag{3.57}$$

Solving matrix Equation (3.57) we obtain

$$I_I = 0.2\text{ A}. \tag{3.58}$$

$$I_{II} = -0.2\text{ A}. \tag{3.59}$$

Using the mesh to branch currents relationships from Equations (3.47), (3.48), and (3.49) we obtain

$$I_I = I_{b1} = 0.2\text{ A} \tag{3.60}$$

$$I_I - I_{II} = I_{b2} = 0.2\text{ A} - (-0.2\text{ A}) = 0.4\text{ A} \tag{3.61}$$

$$I_{II} = I_{\mathrm{b}3} = -0.2\text{ A}. \tag{3.62}$$

Refer to Figure 3.23 to see the original circuit from Figure 3.22 with the added mesh and branch currents values found in the above calculations.

To verify the correctness of the numerical results, work out the mesh equations of circuit of Figure 3.23, using the results of Equations (3.58) through (3.62). Make sure that KCL for all nodes and KVL for all meshes are met.

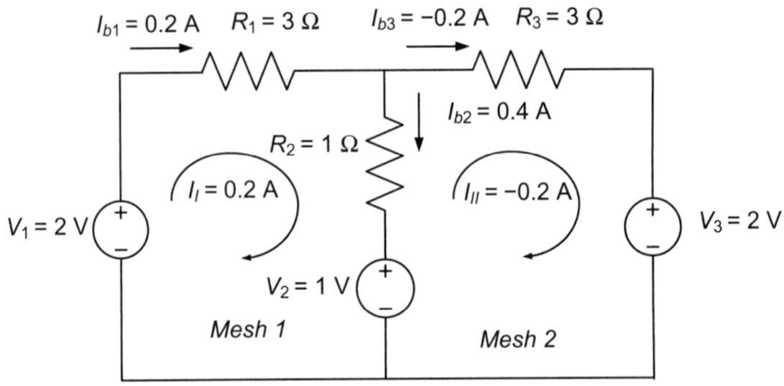

Figure 3.23 Two-mesh equations solutions for Example 3.7.

Example 3.8 Given the circuit of Figure 3.24, derive the mesh equations; find all the branch currents as functions of the mesh currents and the *voltage at node A with respect to ground.* Provide numerical answers for all of the currents and voltages requested.

By inspection of the circuit in Figure 3.24 we can write the mesh equations in the same way we did it for the previous problem, using KVL:

$$\text{Mesh 1: } V_1 - V_2 = (I_I - I_{III})R_1 + (I_I - I_{II})R_2. \tag{3.63}$$

$$\text{Mesh 2: } V_2 - V_3 = (I_{II} - I_I)R_2 + (I_{II} - I_{III})R_3. \tag{3.64}$$

$$\text{Mesh 3: } V_4 = I_{III}R_4 + (I_{III} - I_{II})R_3 + (I_{III} - I_I)R_1. \tag{3.65}$$

Regrouping Equations (3.63) through (3.65) based on each of the mesh currents, we obtain

$$\text{Mesh 1: } V_1 - V_2 = I_I(R_1 + R_2) - I_{II}R_2 - I_{III}R_1. \tag{3.66}$$

$$\text{Mesh 2: } V_2 - V_3 = -I_IR_2 + I_{II}(R_2 + R_3) - I_{III}R_3. \tag{3.67}$$

$$\text{Mesh 3: } V_2 - V_3 = -I_IR_1 - I_{II}R_3 + I_{III}(R_1 + R_3 + R_4). \tag{3.68}$$

We will come back to Equations (3.66) through (3.68) when we will cover finding out the mesh equations simply by inspection of the circuit; eliminating the steps where we applied KVL, Equations (3.63) through (3.65).

Now rewriting Equations (3.66) through (3.68) in matrix form we get

$$\begin{vmatrix} V_1 - V_2 \\ V_2 - V_3 \\ V_4 \end{vmatrix} = \begin{vmatrix} R_1 + R_2 & -R_2 & -R_1 \\ -R_2 & R_2 + R_3 & -R_3 \\ -R_1 & -R_3 & R_1 + R_3 + R_4 \end{vmatrix} \begin{vmatrix} I_I \\ I_{II} \\ I_{III} \end{vmatrix}. \tag{3.69}$$

Using the numerical values from Figure 3.24 into matrix Equation (3.69), we obtain

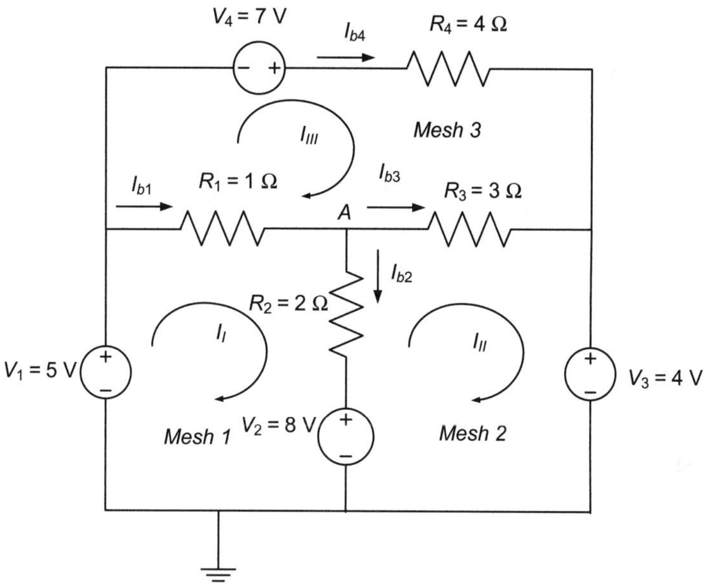

Figure 3.24 Mesh analysis for circuit for Example 3.8.

$$\begin{vmatrix} 5-8 \\ 8-4 \\ 7 \end{vmatrix} = \begin{vmatrix} 1+2 & -2 & -1 \\ -2 & 2+3 & -3 \\ -1 & -3 & 1+3+4 \end{vmatrix} \begin{vmatrix} I_I \\ I_{II} \\ I_{III} \end{vmatrix}. \tag{3.70}$$

Solving the matrix above, it yields:

$$I_I = 1.36364 \text{ A}. \tag{3.71}$$

$$I_{II} = 2.54545 \text{ A}. \tag{3.72}$$

$$I_{III} = 2 \text{ A}. \tag{3.73}$$

By inspection of the circuit of Figure 3.24 we see that the branch to mesh current relationships are

$$I_{b1} = I_I - I_{III}. \tag{3.74}$$

$$I_{b2} = I_I - I_{II}. \tag{3.75}$$

$$I_{b3} = I_{II} - I_{III}. \tag{3.76}$$

$$I_{b4} = I_{III}. \tag{3.77}$$

Plugging the values of I_I (Eq. 3.71) through I_{III} (Eq. 3.73) into Equations (3.74) through (3.77) yields:

$$I_{b1} = -0.63644 \text{ A}. \tag{3.78}$$

$$I_{b2} = -1.18182 \text{ A}. \tag{3.79}$$

$$I_{b3} = 0.54545 \text{ A}. \tag{3.80}$$

$$I_{b4} = 2 \text{ A}. \tag{3.81}$$

By inspection of the branch currents in Figure 3.24 and the results of Equations (3.78) through (3.81) we can see that results for currents I_{b1} and I_{b2} produced negative results. This means that if we go back to Figure 3.24, currents I_{b1} and I_{b2} actually flow in the opposite direction as that shown in the picture.

Finally, it is easy to see that voltage at node *A with respect to ground* (or V_A) equals

$$V_A = V_1 - I_{b1}R_1. \tag{3.82}$$

Plugging the given and the calculated values into Equation (3.77) we obtain

$$V_A = 5 - (-0.63644)1 = 5.63643 \text{ V}. \tag{3.83}$$

As an additional exercise to the reader, verify that all the branch currents, given by Equations (3.74) through (3.77), numerically comply with KCL. The reader should also verify numerically that all KVL Equations (3.66) through (3.68) hold. Hint: Use the calculated values and plug them into the appropriate circuit equations.

3.5.2 Establishing Mesh Equations by Inspection of the Circuit

From the example problems already addressed, we notice that we have been working with circuits that only have voltage sources. Because of this, it is more suitable and also straightforward to derive mesh equations using KVL around each mesh. What we will do in this section is to skip the writing of the circuit equations using KVL, as we did for Example 3.7; see Equations (3.52) and (3.53).

Example 3.9 Writing Mesh Equations by Inspection of the Circuit
Let us start first with the circuit of Example 3.7, Figure 3.22. We repeat this circuit for the reader's convenience in Figure 3.25.

Let us study the circuit diagram carefully. Mesh 1 contains voltage sources V_1 and V_2 and resistors R_1 and R_2. Mesh current I_I is defined to travel mesh 1 in the clockwise direction. Mesh 2 contains voltage sources V_2 and V_3 and resistors R_2 and R_3. Mesh current I_{II} is defined to travel mesh 2 also in a clockwise direction.

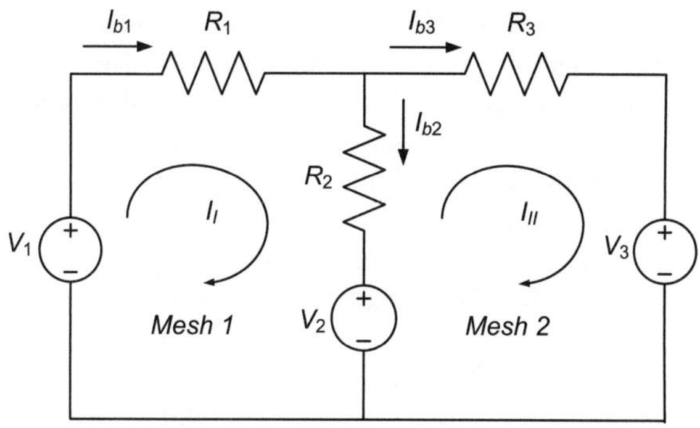

Figure 3.25 Circuit for Example 3.9 finding mesh equations by circuit inspection.

Referring to the previously obtained mesh equations in matrix form, refer to Equation (3.54), we repeat them here again for the reader's convenience:

$$\begin{vmatrix} V_1 - V_2 \\ V_2 - V_3 \end{vmatrix} = \begin{vmatrix} R_1 + R_2 & -R_2 \\ -R_2 & R_2 + R_3 \end{vmatrix} \begin{vmatrix} I_I \\ I_{II} \end{vmatrix} \tag{3.84}$$

So now referring to matrix Equation (3.84), note that the vector column of voltages has 2 × 1 dimensions and elements:

$$v_1 = V_1 - V_2. \tag{3.85}$$

and

$$v_2 = V_2 - V_3. \tag{3.86}$$

The 2 × 2 resistance matrix has elements

$$a_{11} = R_1 + R_2 \quad a_{12} = -R_2 \tag{3.87}$$

$$a_{21} = -R_2 \quad a_{22} = R_2 + R_3 \tag{3.88}$$

The 2 × 1 vector column of currents contains mesh currents I_I and I_{II}. Usually the mesh currents are the unknowns to be found.

We can also express the mesh equations with Ohm's law in matrix form:

$$|V| = |R||I|, \tag{3.89}$$

where $|V|$ is a 2 × 1 voltage column, $|R|$ is a 2 × 2 resistance matrix, and $|I|$ is a 2 × 1 current column.

For the construction of the mesh equations in matrix form, we make the following observations:

The top element of the voltage column v_1 equals the algebraic sum (taking into account the sign of each source) of the voltage sources in mesh 1, traveling mesh 1 in the clockwise direction. That is $V_1 - V_2$ from Equation (3.85).

The bottom element of the voltage column v_2 equals the algebraic sum of the voltage sources of mesh 2 traveling mesh 2 in the clockwise direction. That is $V_2 - V_3$ from Equation (3.86)

Now for the resistive matrix, element a_{11} will always have the sum of all the resistive elements in mesh 1. Note that this sum will always be a sum of positive numbers.

For the resistive matrix, element a_{22} will always have the sum of all the resistive elements in mesh 2. Note that this sum will always be a sum of positive numbers.

Let us concentrate on the a_{12} term of the resistive matrix. We see (Fig. 3.25) that resistor R_2 is a common element between meshes I and II. And since mesh current I_I flows in the clockwise direction, while mesh current I_{II} flows through R_2 in the opposite direction; the contribution of the $I_{II}\ R_2$ term will have a negative sign. Note that if both mesh currents had been chosen such that they both flowed through the common element in the same direction, then the sign of term $I_{II}\ R_2$ would have been positive.

Finally, for the resistance matrix terms a_{21} is the term in mesh 2 that is common to mesh 1. For the same reason, since mesh current I_{II} flows in the opposite direction of mesh current I_I, the term a_{21} will have a negative sign.

The column of mesh currents simply contains the unknown mesh currents to be found which are I_I and I_{II}. Let us also observe that the resistance matrix will always have positive elements on its main diagonal: that is, elements a_{11} and a_{22}. The reciprocal terms (a_{12} and a_{21}) may be both positive and both negative depending on the directions chosen for the mesh currents, as explained earlier. Finally, if the circuit is passive, that means it does not contain any dependent sources, elements a_{12} and a_{21} are identical in sign and magnitude. This is to say that the resistance matrix is symmetrical.

Important Points: Deriving Mesh Equations by Circuit Inspection:
Resistance Matrix: Main diagonal always contains positive elements, none of which can be zero.

If the circuit is passive the resistance matrix is symmetrical (*i.e.*, $a_{12} = a_{21}$).

A passive circuit only contains resistors, inductors, or capacitors, but it cannot contain dependent voltage or dependent current sources. Some examples using dependent sources will be given in Chapter 6.

Drill Problem 3.10: Using the previously seen methodology, find out by inspection the mesh equations for the circuit of Figure 3.26 (this is the same circuit used for Example 3.8).

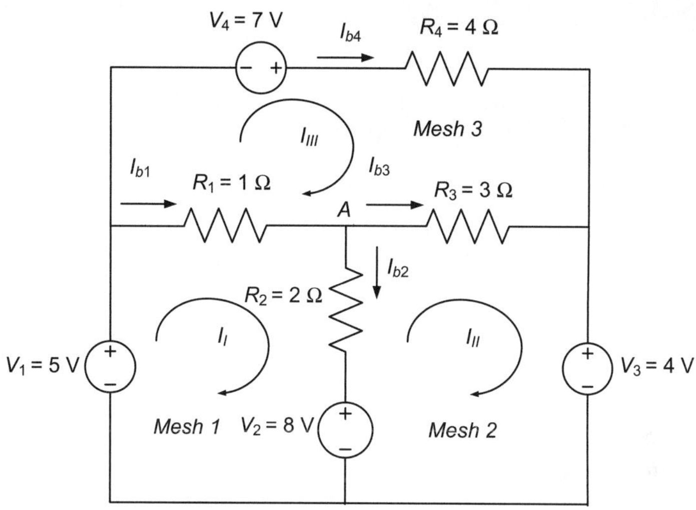

Figure 3.26 Mesh equations by circuit inspection for Drill Problem 3.10.

3.5.3 Establishing Mesh Equations When There Are also Current Sources

The mesh method of analysis is very straightforward when the circuit contains voltage sources. However, if the circuit in addition to containing voltage sources contains current sources, some changes will occur in the mesh equations. To better understand the differences, let us address this with Example 3.11.

Example 3.11 Refer to the circuit of Figure 3.27 to work on this example.

Some Important Notations Pertaining to Circuits:

Referring to the circuit of Figure 3.27: Note that the voltage across resistor R_1 is $V_A - V_{REF} = V_A - 0 = V_A$

The voltage across resistor R_2 assumes the highest voltage at node B with respect to node A is denoted V_{BA}, which is also equal to $V_B - V_A$. It also means that node B is more positive than node A. Note that if the voltage at node B is less positive than the voltage of node A, then $V_B - V_A$ is a negative number. For example, If $V_B = 4$ V and $V_A = 5$ V, then $V_B - V_A = -1$ V.

On the other hand, if we want to talk about the voltage across resistor R_2, where the higher voltage is assumed to be at node A, and the lower voltage is at node B, the voltage across R_2 is V_{AB} which is also equal to $V_A - V_B$. Also note that $V_{AB} = -V_{BA}$.

The voltage across current source I_{S1} is V_{DA}. The voltage at node D is assumed to be larger than the voltage at node A. That is: $V_{DA} = V_D - V_A$.

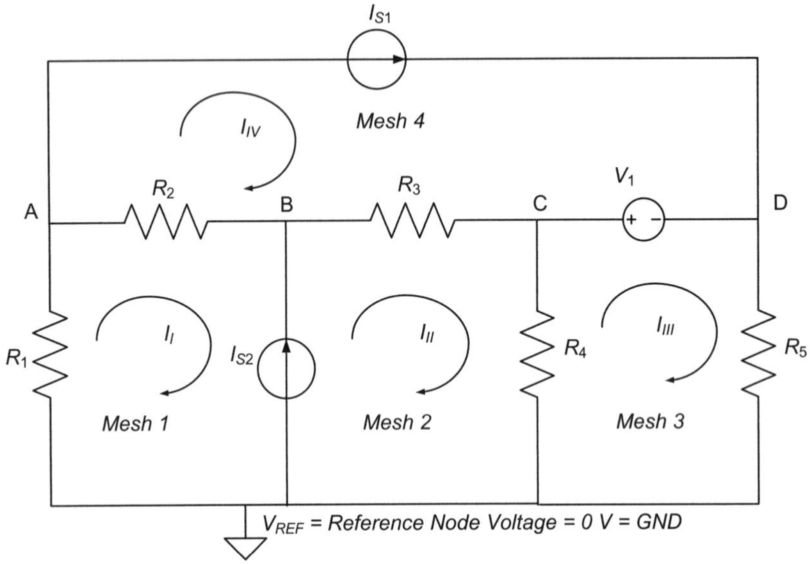

Figure 3.27 Mesh equations of circuits with voltage and current sources for Example 3.11.

Similarly, the voltage across current source I_{S2} is V_B. The voltage at node B is assumed to be higher than the voltage at node V_{REF}; note that V_{REF} was defined as our reference or zero-volt ground node.

So now by inspection of circuit of Figure 3.27 we see four meshes, that is, I, II, III, and IV. Starting with mesh I, if we want to establish the mesh equations for this mesh, we cannot use the value of the current source I_{S2} in the KVL equations since the voltage across current source I_{S2} is V_B. We can then write mesh I equation as follows:

$$\text{Mesh } I \quad V_B = (I_{IV} - I_I)R_2 - I_I R_1 \tag{3.90}$$

$$\text{Mesh } II \quad V_B = (I_{II} - I_{IV})R_3 + (I_{II} - I_{III})R_4 \tag{3.91}$$

$$\text{Mesh } III \quad -V_1 = I_{III}R_5 + (I_{III} - I_{II})R_4 \tag{3.92}$$

It is not necessary to write the equation for *Mesh IV* since mesh current *IV* (I_{IV}) is known numerically. That is,

$$I_{IV} = I_{S1}. \tag{3.93}$$

The last equation we need is current source I_{S2} which equals the differences between mesh currents I_{II} and I_I. That is,

$$I_{S2} = I_{II} - I_I. \tag{3.94}$$

Now if we subtract Equation (3.91) from Equation (3.90), the unknown voltage V_B is eliminated from the result and we obtain

$$(I_{IV}-I_I)R_2 - I_I R_1 + (I_{IV}-I_{II})R_3 + (I_{III}-I_{II})R_4 = 0. \qquad (3.95)$$

Reordering Equation (3.95) grouping by mesh currents yields

$$-I_I(R_1+R_2) - I_{II}(R_3+R_4) + I_{III}R_4 + I_{IV}(R_2+R_3) = 0. \qquad (3.96)$$

The elimination of voltage V_B from Equations (3.95) and (3.96) is equivalent to thinking as merging meshes *I* and *II*; this new merged mesh is called a *super-mesh*. This super-mesh consists of elements R_1, R_2, R_3, and R_4 after the elimination of current source I_{S2}.

Refer to Figure 3.28, the super-mesh just described is shown after the physical removal of current source I_{S2}. Remember this step is justified by Equations (3.95) and (3.96).

Referring to Figure 3.28, note that independent current source I_{S1} is identical to the selected mesh current I_{IV}. In a typical circuit, like the one of Figure 3.27, the independent current and voltage sources are known; the same goes for the resistors. Generally, the mesh currents are unknown. But let us talk about how many mesh current equations we need and how many mesh currents are unknown. We have a total of four mesh currents I_I, I_{II}, I_{III}, and I_{IV}.

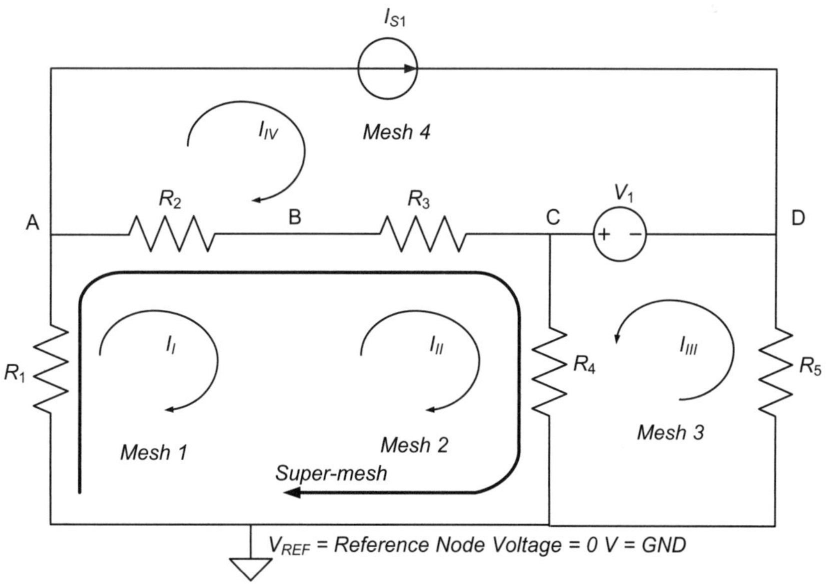

Figure 3.28 The creation of the super-mesh for Example 3.11.

We already mentioned that I_{IV} is numerically known because it is equal to the value of independent current source I_{S1}, see Figure 3.27. Thus, the only unknown mesh currents are: I_I, I_{II}, and I_{III}. To find the three unknown mesh currents we need three linearly independent equations. The first one is Equation (3.96) and the other two are Equations (3.92) and (3.94). We repeat these three key equations here for the reader's convenience. Since we also know from Equation (3.93) that mesh current I_{IV} is known and equals I_{S1}, we replace I_{IV} with I_{S1} in Equation (3.96) and obtain

$$-I_I(R_1+R_2)-I_{II}(R_3+R_4)+I_{III}R_4+I_{S1}(R_2+R_3)=0. \quad (3.97)$$

$$-V_1=I_{III}R_5+(I_{III}-I_{II})R_4. \quad (3.98)$$

where $I_{S2}=I_{II}-I_I$

In a typical problem all resistors R_1 through R_5, the two current sources I_{S1} and I_{S2}, and voltage source V_1 are numerically known.

What we just did mathematically with Equation (3.97) is the following:

The voltage V_B across current source I_{S2} is not initially known and Equation (3.96) eliminates V_B. This merges or creates a so called *super-mesh* with meshes *I* and *II*. Two meshes that share a current source are referred to as an essential mesh. So we re-draw the circuit of Figure 3.27 showing the newly formed super-mesh and it is shown in Figure 3.28. On the other hand, mesh *IV* is nonessential because its current source I_{S1} is not shared with any other mesh. Thus, we eliminate (or open circuit) current source I_{S1}. These steps along with the super-mesh are both shown in the circuit of Figure 3.29.

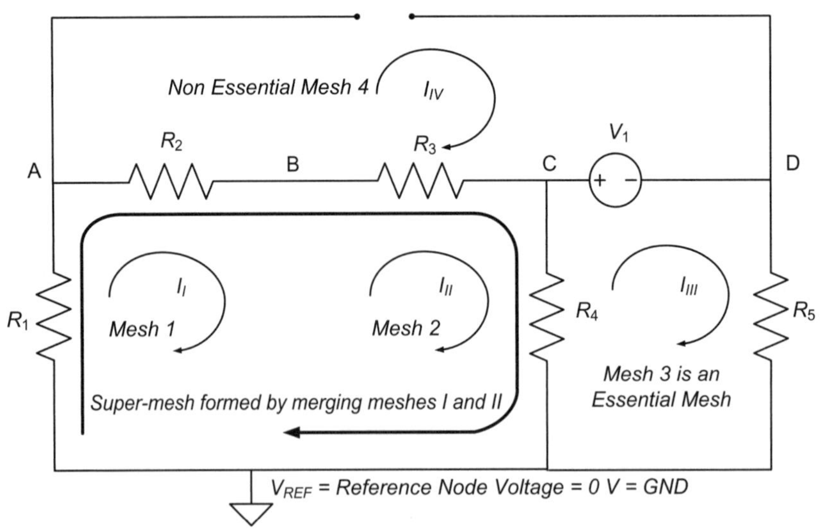

Figure 3.29 Circuit for Example 3.11 after the elimination of all independent current sources.

It is important to state that a super-mesh does not have a current of its own. Note that the original mesh currents I_I through I_{IV} continue to flow through the elements of the newly formed *super-mesh* in Figure 3.29. This is certainly a requirement which was derived by Equation (3.96). We will show next a simpler method using the super-mesh concept of deriving Equation (3.96), without having to write the individual equations for meshes *I* and *II* as we did previously.

Following the *super-mesh* of Figure 3.29, travel the super-mesh in the direction indicated by the heavy arrow accounting for all voltage drops and rises. In our example, the super-mesh does not have any voltage rises (i.e., voltage sources) of its own; however, it may have them in other examples.

$$I_I R_1 + (I_I - I_{IV})R_2 + (I_{II} - I_{IV})R_3 + (I_{II} - I_{III})\mathrm{R}_4 = 0. \tag{3.99}$$

Since mesh current I_{IV} equals the current I_{S1} (Fig. 3.27),

$$I_{IV} = I_{S1}. \tag{3.100}$$

After regrouping terms in Equation (3.99) around, the mesh current becomes:

$$-I_I(R_1 + R_2) - I_{II}(R_3 + R_4) + I_{III}R_4 + I_{S1}(R_2 + R_3) = 0. \tag{3.101}$$

Note that Equation (3.101) is identical to Equation (3.97).

The fourth and last equation is for essential mesh *III*. This is probably the simplest equation to write since it contains only a voltage source and we need to write the mesh equation using KVL. Note: An essential mesh is one that has current sources, and it is not a super mesh.

$$\textit{Mesh III} - V_1 = (I_{III} - I_{II})R_4 + I_{III}R_5. \tag{3.102}$$

Solving the four equations, which we rewrite below for the reader's convenience, all mesh currents, are numerically obtained.

$$-I_I(R_1 + R_2) - I_{II}(R_3 + R_4) + I_{III}R_4 + I_{S1}(R_2 + R_3) = 0. \tag{3.103}$$

$$-V_1 = I_{III}R_5 + (I_{III} - I_{II})R_4. \tag{3.104}$$

$$I_{S2} = I_{II} - I_I. \tag{3.105}$$

A system of three simultaneous linear equations with three unknowns; Equations (3.103) through (3.105), is solved to obtain mesh currents I_I, I_{II}, and I_{III}.

Remember that I_{IV} is already known by inspection of the circuit of Figure 3.27, Equation (3.100).

By inspection of the circuit of Figure 3.30 we can find the branch currents on every resistor as a function of their mesh currents.

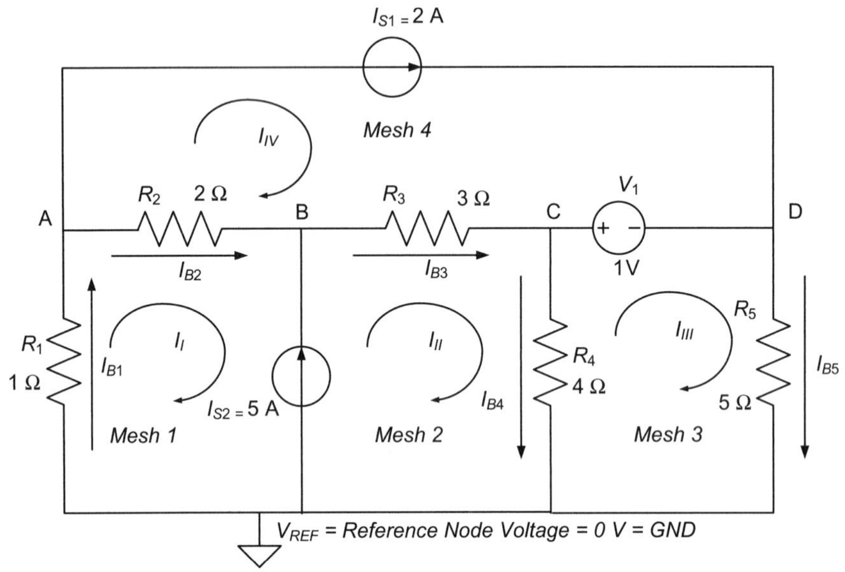

Figure 3.30 Circuit to solve by mesh analysis method for Example 3.12.

$$\textit{Branch current through resistor } R_1 = I_{B1} = I_I. \tag{3.106}$$

$$\textit{Branch current through resistor } R_2 = I_{B2} = I_I - I_{IV}. \tag{3.107}$$

$$\textit{Branch current through resistor } R_3 = I_{B3} = I_{II} - I_{IV}. \tag{3.108}$$

$$\textit{Branch current through resistor } R_4 = I_{B4} = I_{II} - I_{III}. \tag{3.109}$$

$$\textit{Branch current through resistor } R_5 = I_{B5} = I_{III}. \tag{3.110}$$

Example 3.12 Using the circuit of Figure 3.27, and assuming the element values given by Equations (3.111) through (3.118), calculate the values of all four mesh currents. Hint: Use Equations (3.103) through (3.105).

Once the mesh currents are obtained, calculate the branch currents through resistors R_1 through R_5. Hint: Use Equations (3.106) through (3.110). Then, calculate the voltages at nodes A, B, C, and D with respect to ground.

$$R_1 = 1\,\Omega, \tag{3.111}$$

$$R_2 = 2\,\Omega, \tag{3.112}$$

$$R_3 = 3\,\Omega, \tag{3.113}$$

$$R_4 = 4\,\Omega, \tag{3.114}$$

$$R_5 = 5\ \Omega, \tag{3.115}$$

$$I_{S1} = 2\ \text{A}, \tag{3.116}$$

$$I_{S2} = 5\ \text{A}, \tag{3.117}$$

$$V_1 = 1\ \text{V}. \tag{3.118}$$

The circuit is presented again for the reader's convenience in Figure 3.30.

Using the circuit values given by Equations (3.111) through (3.118), plugging them into mesh Equations (3.103) through (3.105), we obtain the following mesh currents:

The author assumes that the reader can solve a system of linear simultaneous equations. Bear in mind that one of this book's goals is to learn circuit analysis; however, it is not the main goal of this book to walk the reader through solving algebraic equations.

Then,

$$I_I = -2.01351\ \text{A}. \tag{3.119}$$

$$I_{II} = 2.98649\ \text{A}. \tag{3.120}$$

$$I_{III} = 1.21622\ \text{A}. \tag{3.121}$$

$$I_{IV} = 2\ \text{A}\ (\textit{found by circuit inspection}). \tag{3.122}$$

Now, using Equations (3.106) through (3.110) to calculate the branch currents, we obtain

$$I_{B1} = I_I = -2.01351\ \text{A}. \tag{3.123}$$

$$I_{B2} = I_I - I_{IV} = -4.01351\ \text{A}. \tag{3.124}$$

$$I_{B3} = I_{II} - I_{IV} = 0.98649\ \text{A}. \tag{3.125}$$

$$I_{B4} = I_{II} - I_{III} = 1.77027\ \text{A}. \tag{3.126}$$

$$I_{B5} = I_{III} = 1.21622\ \text{A}. \tag{3.127}$$

Refer one more time to Figure 3.30 to see the branch current directions and compare them with the signs of Equations (3.123) through (3.127).

Note that branch current I_{B1} was defined in the same direction as mesh current I_I; however, the numerical result of $I_{B1} = -2.01351$ A means branch current I_{B1} actually flows from node A to the reference node. It is also true that branch current I_{B2} defined to flow from node B into node C, because of the negative sign of its result, actually flows from C to B.

By inspection of Figure 3.30 we can easily find the corresponding nodal voltages as function of their branch currents and their respective branch resistors.

$$V_A = -I_{B1}R_1. \tag{3.128}$$

$$V_A - V_B = I_{B2}R_2. \tag{3.129}$$

$$V_B - V_C = I_{B3}R_3. \tag{3.130}$$

$$V_C = I_{B4}R_4. \tag{3.131}$$

$$V_D = I_{B5}R_5. \tag{3.132}$$

Now plugging the values of branch current and resistors into (we find the nodal voltages)

$$V_A = 2.013521 \text{ V}. \tag{3.133}$$

$$V_B = 10.0405 \text{ V}. \tag{3.134}$$

$$V_C = 7.08108 \text{ V}. \tag{3.135}$$

$$V_D = 6.08108 \text{ V}. \tag{3.136}$$

Let us note that from Equation (3.133), V_A is a positive voltage with respect to ground, which means that $V_{GND} - V_A = 0 - V_A = -V_A = I_{B1} \times R_1$, which is consistent with the direction which branch current I_{B1} has in Figure 3.30 and its negative result given by Equation (3.123).

Similarly note that nodal voltage V_A is positive but smaller than the nodal voltage at V_B (i.e., $V_A < V_B$ or 2.01351 V < 10.0405 V). That explains why based on the direction defined for branch current I_{B2} (Fig. 3.30), the numerical result is negative; that is, from Equation (3.124)

$$I_{B2} = -4.01351 \text{ A}.$$

On a final note on this example, mesh currents are defined currents just for the mesh method of analysis. Mesh currents are not currents that can be directly measured, like branch currents can.

3.5.4 Establishing Mesh Equations When There Are also Dependent Sources

In this section we will address a circuit with an independent voltage source and also a dependent current source. We will see that the mesh equations can be stated simply starting with the circuit KVL equations. The fact that there is a dependent source does not change significantly how KVL equations need to be written. *We will find then that a constraint equation links the dependant source output (a current $4I_A$ in our next example) and its independent variable (I_A).* Finally, we will see that the matrix mesh equations lead to a *nonsymmetrical* matrix, because a dependent source represents an active device. More on dependent sources will be covered on the chapter on transistors.

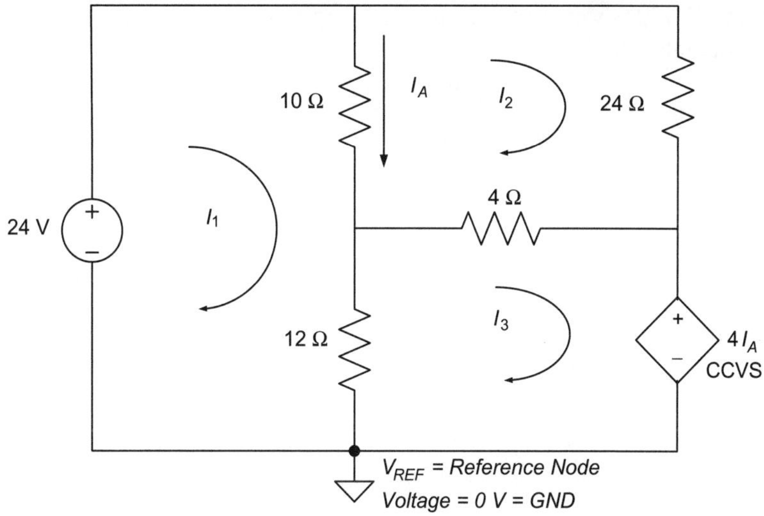

Figure 3.31 Establishing mesh equations for circuits with a dependent source.

Example 3.13 Establish the mesh equations starting with *KVL:* Let us refer to the circuit of Figure 3.31. There is a dependent voltage source, whose output voltage is, $4I_A$, which is referred to as the dependent variable or the voltage source output. The current I_A is the control variable of our dependent source. This current I_A is defined to be the current that flows through the 10-Ω resistor, with the direction shown in Figure 3.31. Keep in mind that dependent sources (current of voltage types) are mathematical models to represent devices that have gain. You cannot buy a dependent source in a battery store or anywhere else; a dependent source is a circuit-modeling concept. We will address the meaning of gain when studying operational amplifiers and transistorized circuits. Finally, we add that in cases that have dependent sources, the author prefers not to address a *by-inspection* method, because its rules are more complex than those for the straightforward cases of mesh equations with just independent voltage sources.

Using the already predefined mesh current of Figure 3.31, we can write for each mesh their respective mesh equations using KVL around each mesh:

$$\textit{Mesh 1: } 24 = 10(I_1 - I_2) + 12(I_1 - I_3). \quad (3.137)$$

$$\textit{Mesh 2: } 0 = 10(I_2 - I_1) + 24I_2 + 4(I_2 - I_3). \quad (3.138)$$

$$\textit{Mesh 3: } -4I_A = 12(I_3 - I_1) + 4(I_3 - I_2). \quad (3.139)$$

Note that in Mesh 3 (3.139), term $-4I_A$ is a voltage not a current; refer again to Figure 3.31.

By inspection of the circuit in Figure 3.31, it is easy to see that

$$I_A = I_1 - I_2. \quad (3.140)$$

Equation (3.139) shows branch current I_A expressed as a function of the circuit mesh currents. Now regrouping terms in Equations (3.137) through (3.139) and using Equation (3.140) to eliminate the use of I_A, we obtain the following mesh equations:

$$\textit{Mesh 1:}\ 24 = 22I_1 - 10I_2 - 12I_3. \tag{3.141}$$

$$\textit{Mesh 2:}\ 0 = -10I_1 + 38I_2 - 4I_3. \tag{3.142}$$

$$\textit{Mesh 3:}\ 0 = -8I_1 - 8I_2 + 16I_3. \tag{3.143}$$

Dividing by two on both sides of the equal sign Equations (3.141) and (3.142), dividing (3.143) by eight, and rewriting them in their matrix form yields:

$$\begin{bmatrix} 12 \\ 0 \\ 0 \end{bmatrix} = \begin{bmatrix} 11 & -5 & -6 \\ -5 & 19 & -2 \\ -1 & -1 & 2 \end{bmatrix} \begin{bmatrix} I_1 \\ I_2 \\ I_3 \end{bmatrix} \tag{3.144}$$

Solving the matrix system, one obtains that

$$I_1 = 2.25\ \text{A}. \tag{3.145}$$

$$I_2 = 0.75\ \text{A}. \tag{3.146}$$

$$I_3 = 1.5\ \text{A}. \tag{3.147}$$

And using Equation (3.140) for I_A,

$$I_A = 2.25 - 0.75 = 1.5\ \text{A}. \tag{3.148}$$

The actual matrix solving is left as an exercise to the reader.

Notice that as predicted, the resistance matrix in Equation (3.144) is not symmetrical, that is, because there was a dependent source in the circuit. That is, $a_{23} = -2$ is not equal to $a_{32} = -1$. Question to the reader: Which other elements of matrix (3.144) prove that the matrix is not symmetrical?

3.5.4.1 *Commentary on Mesh Analysis* Note that given a circuit with only voltage sources and "n" meshes, there are n mesh currents that can be defined. This yields a system of n independent equations with n unknowns.

However, if there are any current sources in a mesh, each current source reduces the number of linearly independent equations by one per current source per mesh. Finally, if the circuit contains at least one dependent source, the resistance matrix will not be symmetrical like it is in the case of a passive circuit. A passive circuit only contains resistors (additionally capacitors and inductors if it is an AC circuit) and independent voltage and/or current sources.

3.6 THE NODAL METHOD OF ANALYSIS

The nodal method of circuit analysis is based on KCL. It provides a more effective way of deriving circuit equations virtually by quick inspection of the circuit. The nodal method is more suitable and intuitive when the circuit contains independent current sources. The method is somewhat less intuitive when voltage sources are also included and probably the least intuitive when dependent voltage and current sources are present. For a circuit that contains *n* nodes, one of the nodes is arbitrarily chosen as the reference node or ground, and the remaining "*n* – *1*"nodal voltages of the circuits are typically the unknowns. We will address the methodology of writing KCL equations for various circuits via examples that will grow in complexity.

Unlike the mesh method, the nodal method works for *planar* and *nonplanar* circuits. It is commonly the method of choice of some electric and electronic circuit simulation programs.

3.6.1 Establishing Nodal Equations: Circuits with Independent Current Sources

Let us assume that we have a circuit such as the one presented in Figure 3.32. By inspection we see that the circuit has four nodes. The reference node is usually chosen to be at the bottom of the circuit. Additionally, the nonreference nodes are: *A*, *B*, and *C*.

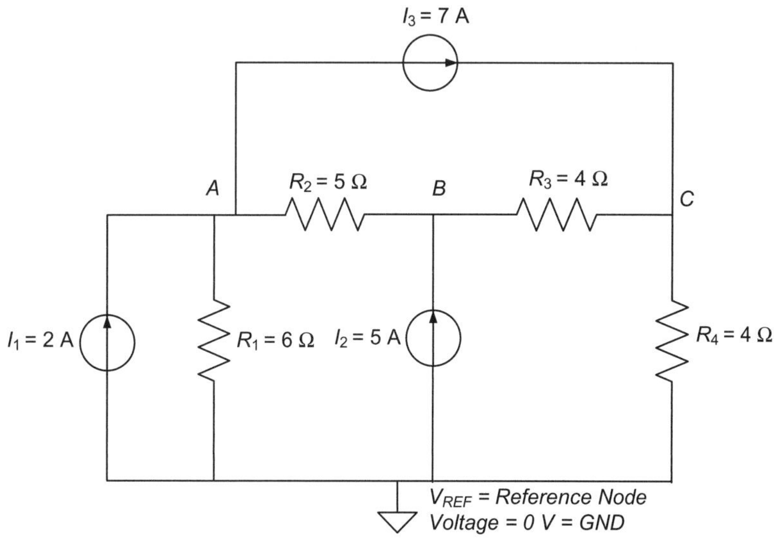

Figure 3.32 Circuit with current sources to establish nodal equations.

Using KCL at each of the nodes we obtain

$$\textit{Node A: } I_1 - I_3 = V_A/R_1 + (V_A - V_B)/R_2. \quad (3.149)$$

$$\textit{Node B: } I_2 = (V_B - V_A)/R_2 + (V_B - V_C)/R_3. \quad (3.150)$$

$$\textit{Node C: } I_3 = V_C/R_4 + (V_C - V_B)/R_3. \quad (3.151)$$

Regrouping Equations (3.149) through (3.151) around their nodal voltages we obtain

$$\textit{Node A: } I_1 - I_3 = V_A(1/R_1 + 1/R_2) - V_B(1/R_2). \quad (3.152)$$

$$\textit{Node B: } I_2 = -V_A(1/R_2) + V_B(1/R_2 + 1/R_3) - V_C(1/R_3). \quad (3.153)$$

$$\textit{Node C: } I_3 = -V_B(1/R_3) + V_C(1/R_3 + 1/R_4) \quad (3.154)$$

In the above three equations we have three unknowns, the nodal voltages V_A, V_B, and V_C. Once those voltages are found, the branch currents in every branch element can easily be calculated using Ohm's law.

Let us make a notation simplification, remembering that the inverse of a resistance R is its conductance G, where $G = 1/R$.

We can re-write Equations (3.152) through (3.154) in matrix form and they become:

$$\begin{bmatrix} I_1 - I_3 \\ I_2 \\ I_3 \end{bmatrix} = \begin{bmatrix} G_1 + G_2 & -G_2 & 0 \\ -G_2 & G_2 + G_3 & -G_3 \\ 0 & -G_3 & G_3 + G_4 \end{bmatrix} \begin{bmatrix} V_A \\ V_B \\ V_C \end{bmatrix}, \quad (3.155)$$

where [I] is a 3 × 1 column of current sources. The 3 × 3 matrix in Equation (3.155) is referred to as the conductance matrix [G]. The vector of nodal voltages contains all nonreference node voltages, V_A, V_B, and V_C. Equation (3.155) can be written in a more compact form and that is

$$[I] = [G][V]. \quad (3.156)$$

Equation (3.156) is another matrix form of Ohm's law using the *conductance* matrix G.

Example 3.14 Find the nodal voltages V_A, V_B, and V_C and currents on resistors R_1 through R_4 in the circuit of Figure 3.32. Hint: Use Equations (3.152) through (3.154) or matrix system (Eq. 3.155).

Referring again to the circuit of Figure 3.32 and using the corresponding values for the independent current sources and resistors we rewrite Equation (3.155) as follows:

$$\begin{bmatrix} 2-7 \\ 5 \\ 7 \end{bmatrix} = \begin{bmatrix} 1/6+1/5 & -1/5 & 0 \\ -1/5 & 1/5+1/4 & -1/4 \\ 0 & -1/4 & 1/4+1/4 \end{bmatrix} \begin{bmatrix} V_a \\ V_b \\ V_c \end{bmatrix}. \quad (3.157)$$

Solving the linear system of three equations with three unknowns we obtain for the nodal voltages:

$$V_A = 0.947368 \text{ V}. \tag{3.158}$$

$$V_B = 26.7368 \text{ V}. \tag{3.159}$$

$$V_C = 27.3684 \text{ V}. \tag{3.160}$$

Having obtained the nodal voltages we can use them to calculate the branch currents for resistors R_1 through R_5. Thus:

$$I_{B1} = V_A / R_1 = 0.947368/6 = 0.157895 \text{ A} \tag{3.161}$$

$$I_{B2} = (V_B - V_A)/R_2 = (26.7368 - 0.947368)/5 = 5.15789 \text{ A} \tag{3.162}$$

$$I_{B3} = (V_C - V_B)/R_3 = (27.3684 - 26.7368)/4 = 0.157895 \text{ A} \tag{3.163}$$

$$I_{B4} = V_C / R_4 = 27.3684/4 = 6.84211 \text{ A}, \tag{3.164}$$

where, in Equation (3.161), current I_{B1} flows from node A to ground; in Equation (3.162) I_{B2} flows from node B into node A; in (3.163) I_{B3} flows from node C into node B. Finally, in (3.164) I_{B4} flows from node C to ground.

Drill Problem 3.15: Using the circuit of Figure 3.32 and the found branch currents given by Equations (3.161) through (3.163), check that KCL is met at nodes A, B, C, and V_{REF} (the grounded node).

3.6.2 Establishing Nodal Equations by Inspection: Circuits with Current Sources

The nodal equations of circuits with current sources and a relatively small number of nodes are quite easy to determine. Let us start working on the circuit of the next example.

Example 3.16 Refer to the circuit of Figure 3.33. On first sight the circuit appears to be complicated; however, after looking at it for some time and understanding its topology, it is not so.

Although the circuit contains six resistors and six independent current sources, its topology is not that much different from that of the circuit of Figure 3.32.

Note that our circuit, Figure 3.33, still has three nonreference nodes and a reference node (ground). For simplicity and convenience, the nonreference nodes are named A, B, and C.

So although this circuit has many more resistors and current sources than the one of Figure 3.32, it is still a circuit whose nodal equations matrix is 3×3, that is, three rows by three columns. Let us assume that the elements of the

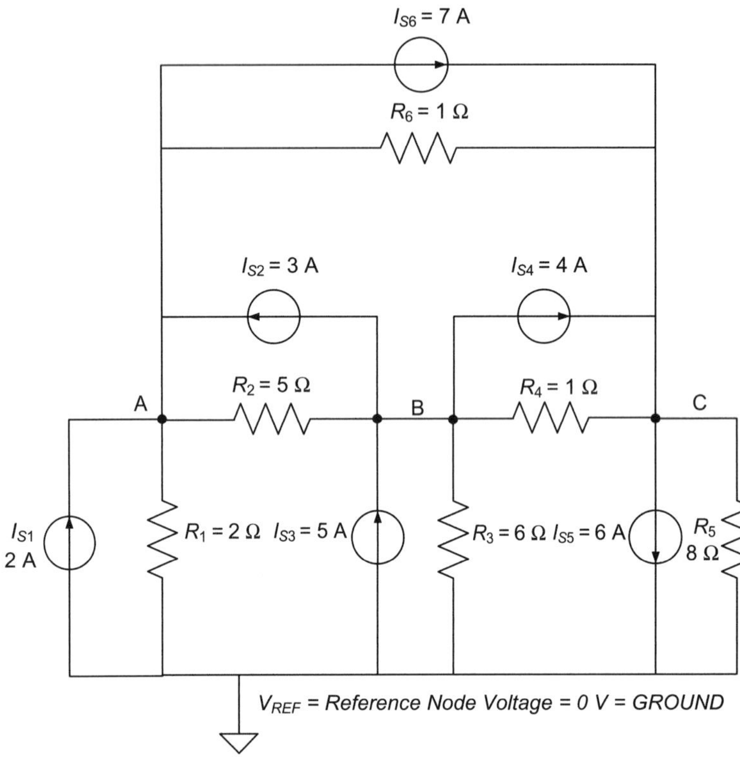

Figure 3.33 Circuit for Example 3.14: establishing nodal equations by inspection.

first row of our matrix 3 × 3 are: a_{11}, a_{12}, a_{13}. Second row elements are: a_{21}, a_{22}, a_{23}; and the third row elements are: a_{31}, a_{32}, a_{33}. The system of three nodal equations and three unknowns is provided first. Then we will look at every equation term and identify how it is associated to the given circuit diagram.

Again referring to Figure 3.33 by inspection of the circuit we have

$$\text{Node } A\text{: } I_{S1} + I_{S2} - I_{S6} = (G_1 + G_2 + G_6)V_A - G_2V_B - G_6V_C. \tag{3.165}$$

$$\text{Node } B\text{: } I_{S3} - I_{S2} - I_{S4} = -G_2V_A + (G_2 + G_3 + G_4)V_B - G_4V_C. \tag{3.166}$$

$$\text{Node } C\text{: } I_{S4} + I_{S6} - I_{S5} = -G_6V_A - G_4V_B + (G_4 + G_5 + G_6)V_C. \tag{3.167}$$

A general inspection of Equations (3.165) through (3.167) should make it clear that each equation is written for every nonreference node and the effects of the other nonreference node over the node in question. Equation (3.165) corresponds to node A, Equation (3.166) corresponds to node B, and Equation (3.167) corresponds to node C. In particular in Equation (3.165), we observe that it has a current term on the left-hand side of the equal sign. The term $(G_1 + G_2 + G_6)V_A$ expresses the effect of all conductances connected to node

A; the term: $-G_2V_B$ expressed the effect of adjacent node B over node A. Finally, the last term: $-G_6V_C$ expresses the effect of adjacent node C over node A.

Referring to Figure 3.33 let us go over every term of Equation (3.165) one more time. The left hand term of Equation (3.165) is the algebraic sum of the currents at node A. Currents flowing into the node are positive, while currents leaving the node are negative. Thus, the term: $I_{S1} + I_{S2} - I_{S6}$. The first term to the right of the equal sign: $(G_1 + G_2 + G_6)V_A$ consists of the sum of all the conductances directly connected to node A, that is, conductances G_1, G_2, and G_6. Because they are all connected to node A, they need to be multiplied by V_A. The next term of Equation (3.165), that is, $-G_2V_B$, contains a term that is minus the conductance between nodes A and B times nodal voltage V_B. G_2 is also referred to as the *shared* conductance between nodes A and B. Finally, the last term of Equation (3.165), that is, $-G_6V_C$, is minus the shared conductance between nodes A and C ($-G_6$) times the nodal voltage V_C. Similarly, we can go over Equation (3.166). In the current term $I_{S3} - I_{S2} - I_{S4}$, note that I_{S3} enters node B, that is why I_{S3} has a positive sign, while currents I_{S2} and I_{S4} leave node B, thus their negative sign. Continuing with Equation (3.166), the term: $-G_2V_A$ denotes the influence of node A over node B. The next term, $(G_2 + G_3 + G_4)V_B$ shows the effect of the actual node in question, that is, node B, and the sum of all conductances connecting to such node times its nodal voltage V_B. Finally, term: $-G_4V_C$ expresses the influence of node C over node B.

Last, let us briefly describe Equation (3.167). The current term: $I_{S4} - I_{S5} + I_{S6}$ is the sum of the currents entering node C minus the currents leaving node C. Term: $-G_6V_A$ shows the impact of node A over node C through their shared conductance G_6. Term: $-G_4V_B$ shows the impact of node B over node C through their shared conductance G_4. Finally, term: $(G_4 + G_5 + G_6)V_C$ is the effect of all conductances connecting to node C, times its nodal voltage V_C.

Now that we have gone over the terms of Equations (3.165) through (3.167), it is easy to convert such equation into their matrix form:

$$\begin{bmatrix} I_{S1}+I_{S2}-I_{S6} \\ I_{S3}-I_{S2}-I_{S4} \\ I_{S4}-I_{S5}+I_{S6} \end{bmatrix} = \begin{bmatrix} G_1+G_2+G_6 & -G_2 & -G_6 \\ -G_2 & G_2+G_3+G_4 & -G_4 \\ -G_6 & -G_4 & G_4+G_5+G_6 \end{bmatrix} \begin{bmatrix} V_A \\ V_B \\ V_C \end{bmatrix} \quad (3.168)$$

In the above matrix equations we have that

$$a_{11} = G_1 + G_2 + G_6;\ a_{12} = -G_2;\ a_{13} = -G_6. \quad (3.169)$$

$$a_{21} = -G_2;\ a_{22} = G_2 + G_3 + G_4;\ a_{23} = -G_4. \quad (3.170)$$

$$a_{31} = -G_6;\ a_{32} = -G_4;\ a_{33} = G_4 + G_5 + G_6. \quad (3.171)$$

Note all the main diagonal elements of the $[G]$ matrix, a_{11}, a_{22}, and a_{33} are always nonzero and positive. Additionally, if the circuit matrix has no dependent sources, that is to say, the circuit is passive and thus its matrix is symmetrical.

A symmetrical matrix is that whose elements that are mirrored around its main diagonal are identical. In general $a_{ij} = a_{ji}$ for *i and j* from 1 to 3, but $i \neq j$. In particular for our example this means that

$$a_{12} = a_{21}. \tag{3.172}$$

$$a_{13} = a_{31}. \tag{3.173}$$

$$a_{23} = a_{32}. \tag{3.174}$$

We will provide the method without applying KCL, as it was done in the previous section and stating directly the procedural steps:

1. For arithmetic convenience, convert every resistor into its equivalent conductance. That is, $G = 1/R$.
2. Identify all nonreference nodes and the reference node. This step is already given in our example.
3. Determine the conductance matrix $[G]$ which will have dimensions of $(number\ of\ nodes - 1) \times (number\ of\ nodes - 1)$, in our example: 3×3.
4. Determine the independent current sources column vector *[I]*.
5. Write the matrices in the form: $[I] = [G][V]$,

where $[V]$ is the column of nodal voltages or the unknowns. In our example:

$$[V] = \begin{bmatrix} V_A \\ V_B \\ V_C \end{bmatrix}$$

Example 3.17 Again, using the circuit of Figure 3.33, calculate the numerical values of nodal voltages V_A, V_B, and V_C. After you find the three nodal voltages, find the currents through every resistor. Hint: Use Equations (3.165) through (3.167) or solve the matrix system given by Equation (3.168).

For this example, all we have to do is to use the values given for the independent current sources and the resistors and solve Equation (3.168).

So plugging in the appropriate values into matrix Equation (3.168) we obtain

$$\begin{bmatrix} 2+3-7 \\ 5-3-4 \\ 4-6+7 \end{bmatrix} = \begin{bmatrix} 0.5+0.2+1 & -0.2 & -1 \\ -0.2 & 0.2+0.1667+1 & -1 \\ -1 & -1 & 1+0.125+1 \end{bmatrix} \begin{bmatrix} V_A \\ V_B \\ V_C \end{bmatrix}. \tag{3.175}$$

Remember that the 3×3 matrix coefficients are formed by sum of the conductances (not the resistances) given by the circuit of Figure 3.33.

Solving Equation 3.175 we obtain the following nodal voltages:

$$V_A = 0.84812\ \text{V}. \tag{3.176}$$

$$V_B = 1.02857\ \text{V}. \tag{3.177}$$

$$V_C = 3.23609\ \text{V}. \tag{3.178}$$

The current for every resistor are simply given by Ohm's law:

$I_{R1} = V_A/R_1 = 0.84812\ \text{V}/2\ \Omega = 0.42406\ \text{A}$ (I_{R1} flows from node A to ground)

$I_{R2} = (V_B - V_A)/R_2 = (1.02857 - 0.84812)\ \text{V}/5\ \Omega$
$= 0.0360902\ \text{A}$ (I_{R2} flows from B to A)

$I_{R3} = V_B/R_3 = 1.02857\ \text{V}/6\ \Omega = 0.171429\ \text{A}$ (I_{R3} flows from B to ground)

$I_{R4} = (V_C - V_B)/R_4 = (3.23609 - 1.02857)\ \text{V}/1\ \Omega$
$= 2.20752\ \text{A}$ (I_{R4} flows from C to B)

$I_{R5} = V_C/R_5 = 3.23609\ \text{V}/8\ \Omega = 0.404511\ \text{A}$ (I_{R5} flows from C to ground)

$I_{R6} = (V_C - V_A)/R_6 = (3.23609 - 0.84812)\ \text{V}/1\ \Omega$
$= 2.38797\ \text{A}$ (I_{R6} flows from C to A)

3.6.3 Establishing Nodal Equations When There Are also Voltage Sources

The nodal method is more suitable and straightforward, when the circuit only contains current sources, because the method is derived using KCL at each nonreference node. It is also possible to apply the nodal method when, in addition to current sources, voltage sources are present. This, even though it is somewhat less intuitive, it simplifies the nodal equations by one equation per voltage source that constitutes a super node.

Example 3.18 Refer to Figure 3.34 to solve the three-node circuit that contains two independent current sources and two independent voltage sources. The nonreference nodes are clearly labeled: A, B, and C. We cannot apply nodal analysis as usual because of the presence of the voltage sources. It is not possible to know the currents through the independent voltage sources, before making any calculations. Inspecting the circuit of Figure 3.34 carefully, we can see that even though there are three nodes in the circuit, the voltage at node A is already known, it actually is $V_1 = 4$ V. We can also see by inspection that the difference of nodal voltages V_B and V_C equals the value of the voltage source $V_2 = 3$ V. That is, $V_2 = V_B - V_C = 3$ V.

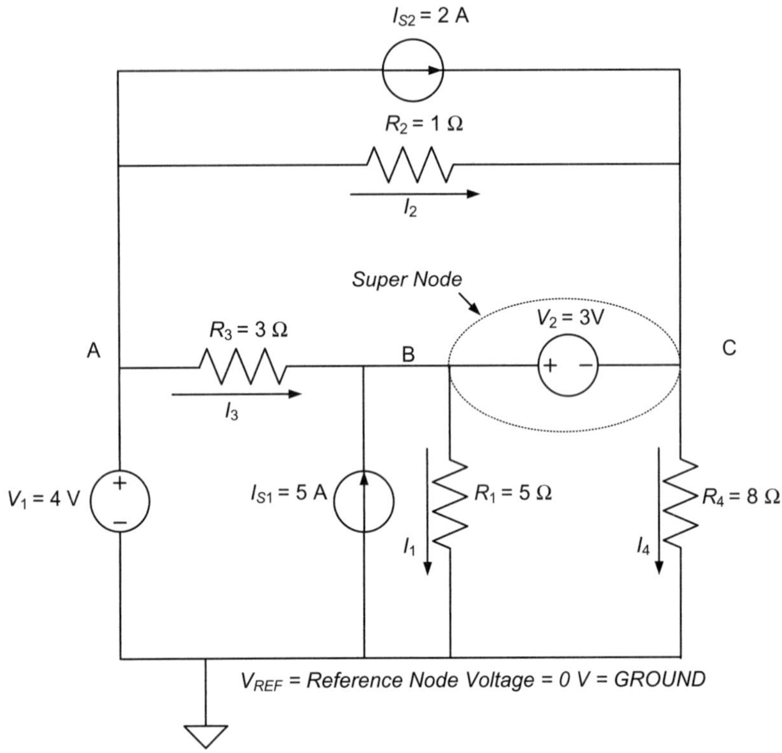

Figure 3.34 Circuit to be solved by the nodal method.

So there are two constraint equations:

$$V_2 = V_B - V_C = 3\text{ V}. \tag{3.179}$$

$$V_1 = V_A = 4\text{ V}. \tag{3.180}$$

Nodes B and C are encircled in Figure 3.34, and they are defined as a *super-node.*

Note that at the indicated *super-node*, KCL also applies. So the sum of all *super-node* entering current equals the sum of all super node leaving currents, that is,

$$I_{S2} + I_2 + I_3 + I_{S1} = I_1 + I_4. \tag{3.181}$$

Since $I_{S1} = 5$ A and $I_{S2} = 2$ A, then Equation (3.181) becomes

$$I_2 + I_3 + 7 = I_1 + I_4. \tag{3.182}$$

Additionally, by inspection of the circuit of Figure 3.34, we see that

$$I_1 = V_B/R_1 = V_B/5. \tag{3.183}$$

$$I_2 = (V_A - V_C)/R_2 = (V_A - V_C)/1. \tag{3.184}$$

$$I_3 = (V_A - V_B)/R_3 = (V_A - V_B)/3. \tag{3.185}$$

$$I_4 = V_C/R_4 = V_C/8. \tag{3.186}$$

Using Equations (3.183) through (3.186) in Equation (3.182) we obtain

$$(V_A - V_C)/1 + (V_A - V_B)/3 + 7 = V_B/5 + V_C/8. \tag{3.187}$$

Solving Equation (3.187) with constraint Equations (3.181) and (3.182), we obtain the values for V_B and V_C.

$$V_B = 9.47236 \text{ V}. \tag{3.188}$$

$$V_C = 6.47236 \text{ V}. \tag{3.189}$$

Now since all the nodal voltages are known, we can easily find the branch currents through all the resistors using Equations (3.183) through (3.186).

$$I_1 = V_B/R_1 = V_B/5 = 1.89447 \text{ A}. \tag{3.190}$$

$$I_2 = (V_A - V_C)/R_2 = (V_A - V_C)/1 = -2.47236 \text{ A}. \tag{3.191}$$

$$I_3 = (V_A - V_B)/R_3 = (V_A - V_B)/3 = -1.82412 \text{ A}. \tag{3.192}$$

$$I_4 = V_C/R_4 = V_C/8 = 0.809045 \text{ A}. \tag{3.193}$$

To determine the currents of each of the independent voltage sources, it is a simple application of KCL. This is left as an exercise to the reader.

3.6.4 Establishing Nodal Equations When There Are Dependent Sources

Let us analyze a circuit that contains two independent current sources, an independent voltage source and a dependent voltage source. An example of this nature is likely as complex as it can be, to solve by hand.

Example 3.19 Using the circuit of Figure 3.35 establish the nodal equations for the circuit using KCL.

By inspection of the circuit of Figure 3.35 we can see that the 10-V dependent voltage source has a control voltage V, which is the voltage developed across the independent 10-A current source, with the shown polarity. Nodes C and D are super-nodes. The voltage at node D with respect to ground is

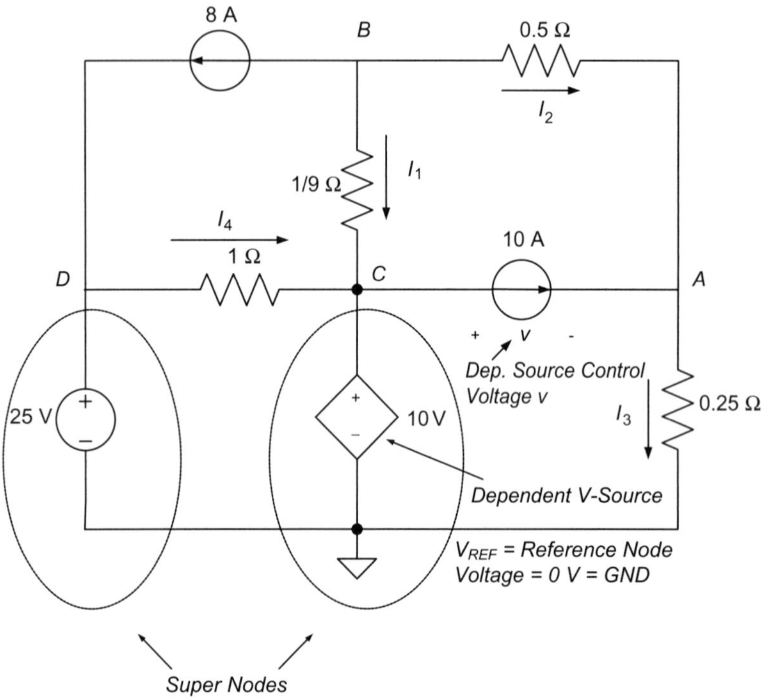

Figure 3.35 Circuit for Example 3.19: nodal method containing dependent sources.

25 V. The only nodal equations that we need to establish are those at nodes A and B.

One more time referring to the circuit in Figure 3.35, let us note that the currents entering node A are 10 A (an independent current source), and branch current I_2, while branch current I_3 leaves node A. Additionally, we observe that for node B, the 8 A-independent current source and both branch currents I_1 and I_2 leave node B.

Since that branch current I_1 flows from node B to node C, thus

$$I_1 = (V_B - V_C)/(1/9). \tag{3.194}$$

Branch current I_2 flows from node B to node A. Thus,

$$I_2 = (V_B - V_A)/0.5. \tag{3.195}$$

Branch current I_3 flows from node A to ground, and

$$I_3 = V_A/0.25, \tag{3.196}$$

we can establish the two nodal equations:

at Node A:

$$10 = V_A/0.25 - (V_B - V_A)/0.5 \tag{3.197}$$

and

at Node B:

$$-8 = (V_B - V_C)/1/9 + (V_B - V_A)/0.5. \tag{3.198}$$

Now, regrouping equations and finding the inverses of the resistances in Equations (3.197) and (3.198), we obtain the nodal equations for the circuit:

$$\textit{Node A: } 10 = (4+2)V_A - 2V_B. \tag{3.199}$$

$$\textit{Node B: } -8 = -2V_A + (9+2)V_B - 9V_C \tag{3.200}$$

Note that Equations (3.199) and (3.200) are two equations with three unknown nodal voltages: V_A, V_B, and V_C. Again by inspection of the circuit of Figure 3.35 we can write one constraint equation that relates nodal voltage V_C to the dependent source voltage V:

By inspection of the circuit we can see that

$$10\text{ V} = V_C. \tag{3.201}$$

$$V_A = 10\text{ V} - \text{V} = 9\text{ V}. \tag{3.202}$$

And

$$V = V_C - V_A. \tag{3.203}$$

Using Equations (3.201) and (3.202) into Equations (3.199) and (3.200) and solving for the nodal voltages we get

$$V_A = 2.2381\text{ V}. \tag{3.204}$$

$$V_B = 1.71429\text{ V}. \tag{3.205}$$

$$V_C = 2.4867\text{ V}. \tag{3.206}$$

And since $V = V_C - V_A$ from Equation (3.203),

$$\text{V} = 2.48677 - 2.2381 = 0.24867\text{ V}. \tag{3.207}$$

Now that we have all the nodal voltages, the calculation of the branch currents easily follows:

$$I_1 = -6.95238 \text{ A}. \tag{3.208}$$

$$I_2 = -1.04762 \text{ A}. \tag{3.209}$$

$$I_3 = 8.95238 \text{ A}. \tag{3.210}$$

$$I_4 = 22.5132 \text{ A}. \tag{3.211}$$

The current through the 25-V independent voltage source is: 14.5132 A *and the current through the voltage-controlled voltage source* 10 V *is* 5.56085 A.

The reader is strongly encouraged to apply KCL to each node of the circuit of Figure 3.35 to validate that the calculated currents are correct.

3.6.4.1 Commentary on Nodal Analysis Note that given a circuit with only current sources and "n" nodes, there are "$n - 1$" necessary and sufficient linearly independent nodal equations required to find the n nodal voltages. However, if there are any voltage sources in the circuit, this creates so-called super-nodes. Each super-node voltage source reduces the number of nodal voltages by one per source. Finally, if the circuit contains at least one dependent source, the resistance matrix will not be symmetrical like it is in the case of a passive circuit. A passive circuit only contains resistors (additionally capacitors and inductors if it is an AC circuit) and independent sources.

3.7 WHICH ONE IS THE BEST METHOD?

We looked at the following circuit theorems and methods of analysis:

1. Superposition theorem
2. Thévenin theorem
3. Norton theorem
4. Source transformations
5. Mesh method of analysis
6. Nodal method of analysis

It is strictly the user, you, who has to make a decision of what method to use. Not everyone will necessarily agree that one method is better than another one. However, it is true, if one is comfortable using any of the methods presented, equally well, sometimes using one method instead of another one can really speed up the circuit solving process, particularly when this is done by hand.

3.7.1 Superposition Theorem Highlights

It is a divide-and-conquer approach. Given a circuit with multiple independent current or voltage sources, one is able to calculate the effect of all sources by enabling one source at a time while inhibiting all other ones. This is repeated until every source was enabled in a mutually exclusive fashion. The algebraic sum of the individual effects leads to the composite result, that is, as if all the sources were applied simultaneously. This method only works with linear circuits, usually the majority of circuits that we will be dealing with, but be careful it is not all of them (e.g., a diode is a nonlinear device). Superposition of power quantities *does not* apply even if the circuit is linear. The applicability of superposition applies to the currents and voltages of linear circuits. The most interesting feature of superposition is that one solves a larger number of problems, where each problem is easier to solve. Or at least that should be the idea when applying this method.

It works with independent and dependent sources, AC and DC voltages and currents. However, it is important to note that when applying superposition, the dependent sources *must not* be inhibited like the independent sources are (i.e., one at a time); the dependent sources must be left alone. Some Problems at the end of the chapter will allow you to practice solving circuits by superposition with dependent and independent sources.

3.7.2 Thévenin Theorem Highlights

Thévenin theorem allows one to replace a piece of a circuit that we choose, with a Thévenin voltage source in series with a Thévenin resistance (or impedance). Many times this simplifies solving the circuit. Thévenin applies to linear circuits with independent and dependent voltage and current sources and AC and DC voltages and currents. When dealing with dependent sources, just like when applying superposition, we do not touch (or inhibit) the dependent sources. The reason is that dependent sources have their own control variable. Finally, upon slicing a circuit to find its Thévenin equivalent, if the sliced circuit contains a dependent voltage or current source, you must make sure that the control variables of such sources *do not* get separated from the circuit being Thévenized.

3.7.3 Norton's Theorem Highlights

Norton's theorem allows one to replace a piece of a circuit that we choose, with a Norton current source in parallel with a Norton resistance (or impedance). The Norton resistance is calculated in exactly the same way as the Thévenin resistance. Norton only applies to linear circuits with independent and dependent voltage and current sources and AC and DC voltages and currents. When dealing with dependent sources, just like when applying superposition, we do not touch (or inhibit) the dependent sources. The reason is that dependent sources have their own control variable. Finally, upon slicing a circuit to find its Norton equivalent, if the sliced circuit contains a dependent voltage or current source, you must make sure that the control variables of such sources *do not* get separated from the circuit whose Norton equivalent is being sought.

3.7.4 Source Transformations Highlights

This method basically derives from the convertibility that exists between independent voltage into independent current sources, by applying Thévenin and Norton Theorems. An independent voltage source in series with a resistor (or impedance) can be converted into an independent current source in parallel with the same resistor (or impedance).

So upon solving a circuit with a large number of mixed current and voltage sources, it may become convenient to either transform all the current sources into voltage sources or vice-versa in order to apply the by-inspection mesh or nodal methods. So whether we apply Thévenin, Norton, or source transformations, it is important to note that the Thévenin resistance (or impedance) is identical to the Norton equivalent resistance (or impedance).

3.7.5 Mesh Method of Analysis Highlights

The mesh method applies to circuits that are planar. Recall that a planar circuit is one that can be drawn without any branches crossing any other. Beware that there are circuits that may appear to be nonplanar; however, redrawing them reveals that they are actually planar. This method is more appealing when we have voltage sources in the meshes, because it uses KVL as its main method of analysis.

This method seems more intuitive and easier to approach with the "*by inspection method*" earlier described, but only when voltage sources are present. When current sources are present, it reduces the number of mesh equations by one per current source. The mesh method with both voltage and current sources also has a by inspection method, but we do not cover this on this text. The by-inspection method with voltage and current sources is somewhat more complicated to memorize. The mesh method applies when there are dependent sources as well. Finally, it can be said that the mesh method is usually an attractive choice when the number of meshes is small.

3.7.6 Nodal Method of Analysis Highlights

The nodal method is a more general method than the mesh method. The nodal method applies to planar and nonplanar circuits. This method is more appealing to use when the circuit contains current sources because it fundamentally uses KCL for the analysis. The method, however, is also applicable when there are voltage sources. This creates a reduction in the number of nodal equations of one by voltage source. There is also a *by inspection nodal method* which is intuitive and easier to apply when only current sources are present. The *by*

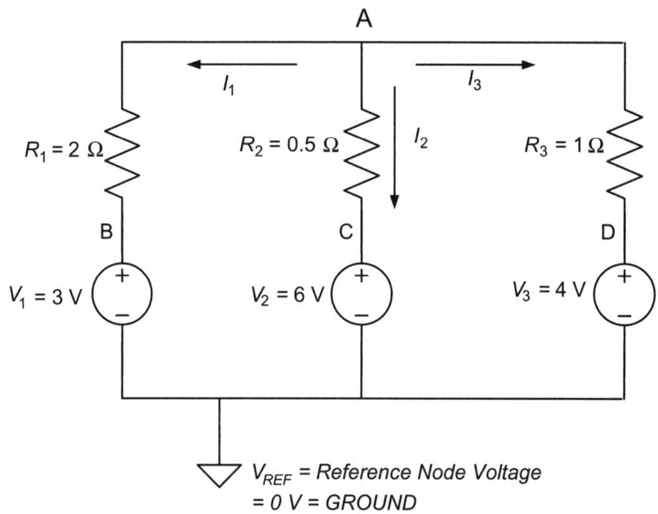

Figure 3.36 Circuit example to be solved by six different methods.

inspection nodal method also exists for current and voltage sources but it is less intuitive to apply, thus we do not cover it.

3.8 USING ALL THE METHODS

To help determine which method is more effective, the circuit of Figure 3.36 will be solved by all six methods covered throughout this chapter. These are: superposition, Thévenin, Norton, source transformations, mesh, and nodal.

3.8.1 Solving Using Superposition

Let us present Example 3.20 to appreciate the different approaches to solving circuits using the different methods presented in this chapter. Refer to the circuit of Figure 3.36. The problem to solve in this example is to find the value of the nodal voltage A with respect to ground.

Example 3.20 Solving the Circuit of Figure 3.36 by Superposition
This is a circuit with three independent voltage sources and three resistors. Solving this problem by superposition takes three steps to disable one source at a time and calculating the value of voltage A. In a fourth and last step, we calculate the composite solution by obtaining the algebraic sum of the three previous results.

Proceeding with this problem, we break this circuit down into three simpler circuits. Refer to Figure 3.37 parts *a*, *b*, and *c*.

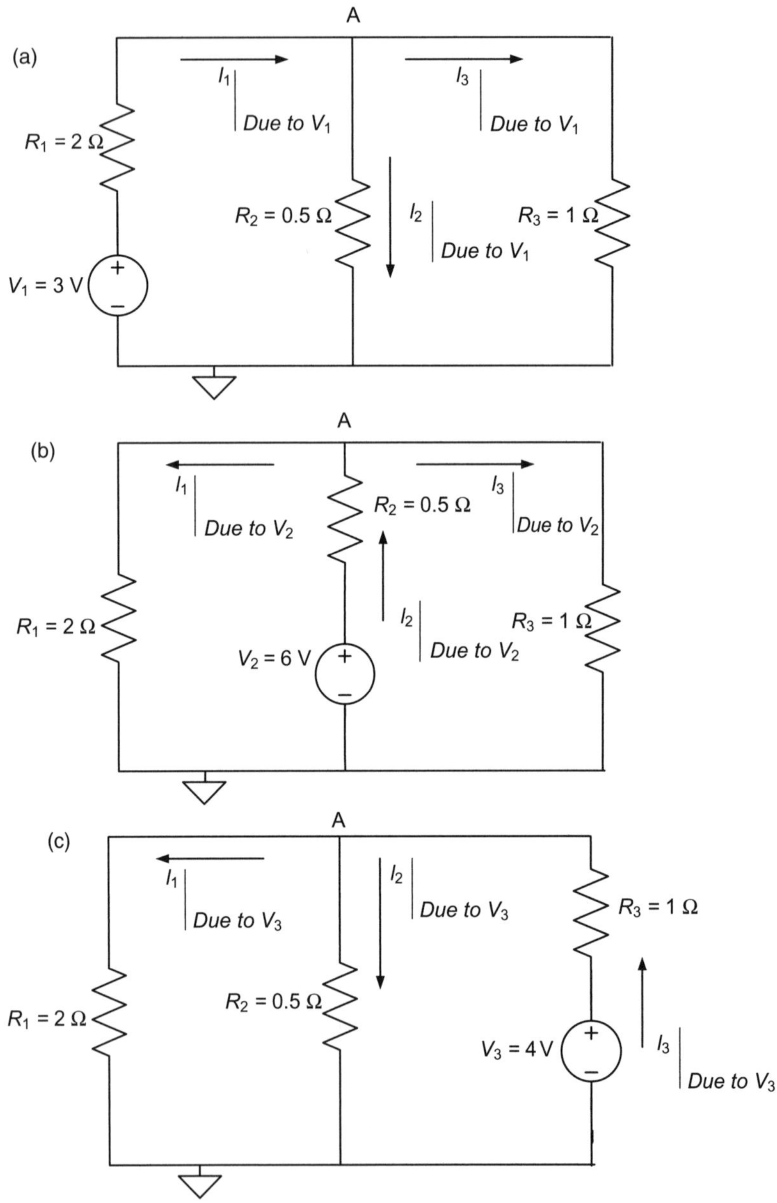

Figure 3.37 Solving circuit of Example 3.20 by superposition: (a) only source V_1 is present; (b) only source V_2 is present; (c) only source V_3 is present.

Circuit of Figure 3.37a is the original circuit powered by source V_1, while sources V_2 and V_3 are inhibited (replaced with short circuits). The circuit of b is powered by source V_2 while V_1 and V_3 are inhibited. Finally, the circuit of c is powered by source V_3, while V_1 and V_2 are inhibited.

So for the circuit of Figure 3.37a we have

$$V_{A|due\ to\ V1} = [V_1/(R_1 + R_2//R_3)](R_2//R_3). \tag{3.212}$$

Equation (3.212) tells us that voltage V_A is the voltage across the parallel of resistors R_2 and R_3. The current through the circuit is the voltage V_1 divided by the series combination of R_1 and ($R_2//R_3$), thus we arrive at Equation (3.212). $R_2//R_3$ is short-hand notation for the calculation of R_2 in parallel with R_3. Plugging in the component values from Figure 3.37a into Equation (3.212) we get

$$V_{A|due\ to\ V1} = [3/(2+0.5//1)](0.5//1) = 3/7\ \text{V}. \tag{3.213}$$

For the circuit of Figure 3.37b, the reasoning is similar as the case before. Thus,

$$V_{A|due\ to\ V2} = [V_2/(R_2 + R_1//R_3)](R_1//R_3) \tag{3.214}$$

Plugging in the component values from Figure 3.37*b* into Equation (3.214) we get:

$$V_{A|due\ to\ V2} = [6/(0.5+2//1)](2//1) = 24/7\ \text{V}. \tag{3.215}$$

For the circuit of Figure 3.37c, the reasoning is similar as the case before. Thus,

$$V_{A|due\ to\ V3} = [V_3/(R_3 + R_1//R_2)](R_1//R_2). \tag{3.216}$$

Plugging in the component values from Figure 3.37c into Equation (3.216) we get

$$V_{A|due\ to\ V3} = [4/(1+2//0.5)](2//0.5) = 8/7\ \text{V}. \tag{3.217}$$

$$V_A = 3/7\ \text{V, due to excitation } V_1 \text{ when } V_2 = 0 \text{ and } V_3 = 0. \tag{3.218}$$

$$V_A = 24/7\ \text{V, due to excitation } V_2 \text{ when } V_1 = 0 \text{ and } V_3 = 0. \tag{3.219}$$

$$V_A = 8/7\ \text{V, due to excitation } V_3 \text{ when } V_1 = 0 \text{ and } V_2 = 0, \tag{3.220}$$

Adding all three voltages at node A, due to voltage excitations V_1, V_2, and V_3 we obtain

$$V_{A|due\ to\ V1} + V_{A|due\ to\ V2} + V_{A|due\ to\ V3} = 3/7 + 24/7 + 8/7 = 5\ \text{V}. \tag{3.221}$$

Note: In this particular case, because all the voltages are positive, there is no difference between their sum and their algebraic sum.

Since V_A is known, it is now easy to find branch currents I_1, I_2. and I_3.

$$I_1 = (V_1 - V_A)/R_1 = (3-5)/2 = -1 \text{ A}. \quad (3.222)$$

$$I_2 = (V_2 - V_A)/R_2 = (6-5)/0.5 = 2 \text{ A}. \quad (3.223)$$

$$I_3 = (V_3 - V_A)/R_3 = (4-5)/1 = -1 \text{ A}. \quad (3.224)$$

Referring back to the circuit of Figure 3.36 note that current I_2 is positive; thus, it flows in the same direction assumed in the circuit. Currents I_1 and I_3 flow in the opposite direction than the one originally assumed. This is consistent with the fact that $V_A = 5$ V, which is smaller than V_2 but it is higher than V_1 and V_3. Remember, current flows from high voltages to lower voltages, and its flow is considered positive.

3.8.2 Example 3.21: Solving the Circuit of Figure 3.36 by Thévenin

Now let us analyze the same circuit of Figure 3.36 using Thévenin's theorem. We refer to the step-by-step procedure following the circuits in Figures 3.38 and 3.39. So the first thing we do is to slice the original circuit such that one piece is to be Thévenized while the remaining circuit will not be touched, initially.

We do exactly that if Figure 3.38a–c, respectively, show the circuits to calculate the Thévenin voltage and resistance.

Figure 3.38a shows the calculated Thévenin voltage and resistance, of 16/3 V and 1/3 Ω, respectively.

The calculation of V_{Th} simply is the open-circuit voltage at terminals A and B of the circuit of Figure 3.38a. Note that this circuit is a single mesh circuit and applying KVL to it we obtain

$$V_2 - V_3 = I_{mesh}(R_2 + R_3) \quad (3.225)$$

and

$$V_{Th} = V_2 - I_{mesh}R_2. \quad (3.226)$$

Solving Equations (3.225) and (3.226) we obtain that

$$I_{mesh} = (6-4)/(0.5+1) = 4/3 \text{ A}. \quad (3.227)$$

Thus, from Equation (3.226),

$$V_{Th} = 6 - (4/3)(1/2) = 6 - 2/3 = 16/3 \text{ V}. \quad (3.228)$$

From Figure 3.38c we see that R_{Th} is the parallel of R_2 and R_3; that is, $R_{Th} = 1/3\ \Omega$.

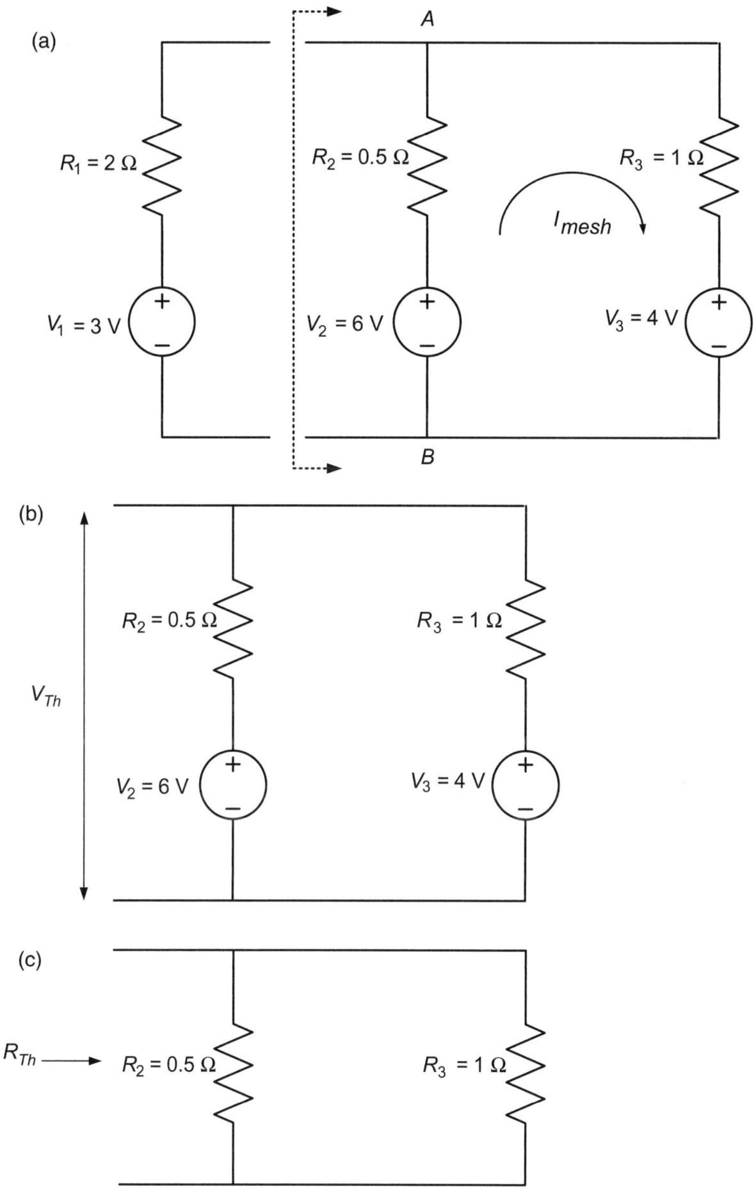

Figure 3.38 Example 3.21: (a) circuit to be Thévenized; (b) circuit used to find V_{Th}; (c) circuit used to find R_{Th}.

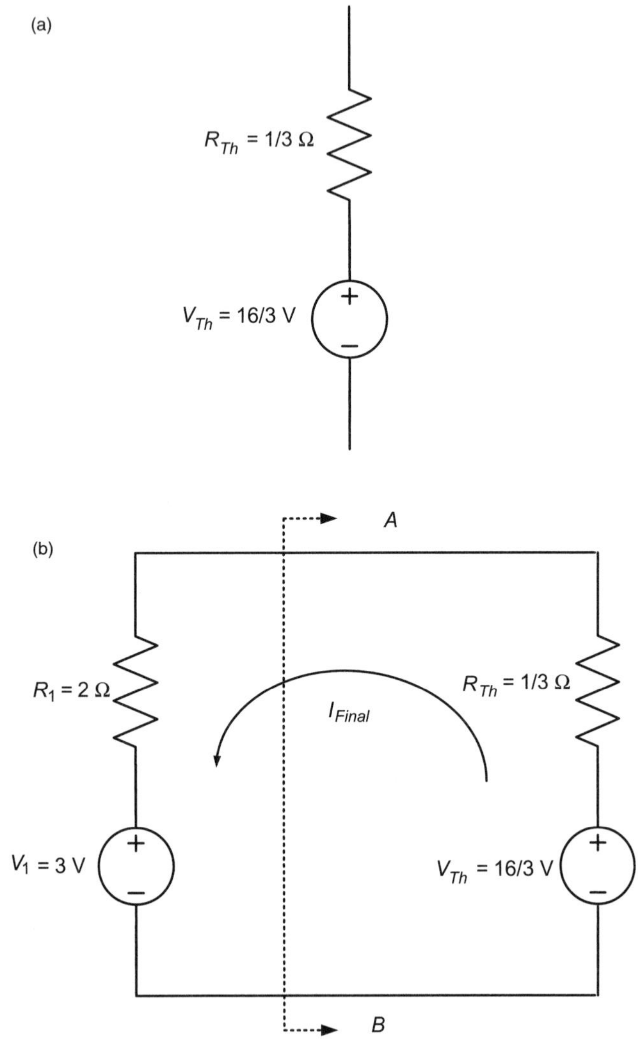

Figure 3.39 (a) The Thévenin equivalent; (b) the Thévenin equivalent circuit reattached to the left side of the sliced circuit, which was not Thévenized.

Figure 3.39 shows the final steps, the Thévenin equivalent circuit, and the attachment of the Thévenized circuit to the portion of the circuit that was not Thévenized.

The only things left are to apply KVL to the circuit of Figure 3.39b which is

$$V_{Th} - V_1 = I_{Final}(R_{Th} + R_1), \tag{3.229}$$

and calculate the voltage across nodes A and B.

$$V_{AB} = V_{Th} - I_{Final}R_{Th}. \tag{3.230}$$

Now using the numerical values for the terms of Equation (3.230) inspecting Figure 3.39b it yields

$$I_{Final} = 1\ \text{A}. \tag{3.231}$$

and

$$V_{AB} = 5\ \text{V}. \tag{3.232}$$

3.8.3 Example 3.22: Solving the Circuit of Figure 3.36 by Norton

In a similar fashion, the original circuit of Figure 3.36 can be solved using Norton's theorem. We use Figures 3.40 and 3.41 to follow the step-by-step process. We slice the circuit to which we will apply Norton in Figure 3.40a. The circuit to which Norton will be applied is on the right-hand side of the dotted line. Next we compute the short-circuit current, that is, the Norton current I_N seen on the circuit of Figure 3.40b, and calculate the Norton resistance with the circuit of Figure 3.40c. The Norton current of Figure 3.40b can be calculated using any other method of choice. You may choose to use KCL and state that the Norton or short-circuit current I_N of Figure 3.40b is

$$I_N = V_2/R_2 + V_3/R_3, \tag{3.233}$$

which numerically yields

$$I_N = 16\ \text{A}. \tag{3.234}$$

Note that the above takes into account that the voltage at node A is identical to the voltage at Ground or 0 V. Why? Because nodes A and *Ground* are tied together by a wire of zero resistance, thus both nodes are really the same node and they are at the same voltage level. Now looking at Figure 3.40c, the Norton resistance is the parallel equivalent of R_2 and R_3 and it is $R_N = 1/3\ \Omega$. Recall that to calculate the Norton resistance, we short-circuit voltage sources and open-circuit current sources. Our example only has two voltage sources V_1 and V_2, which are short-circuited to calculate the Norton resistance, Figure 3.40c.

Moving now to Figure 3.41 we draw the Norton equivalent circuit attached to the left-hand side of the original circuit that was left alone at slicing time, see Figure 3.40a. We use KCL at node A and state that

$$(V_1 - V_A)/R_1 + 16 = V_A/R_N. \tag{3.235}$$

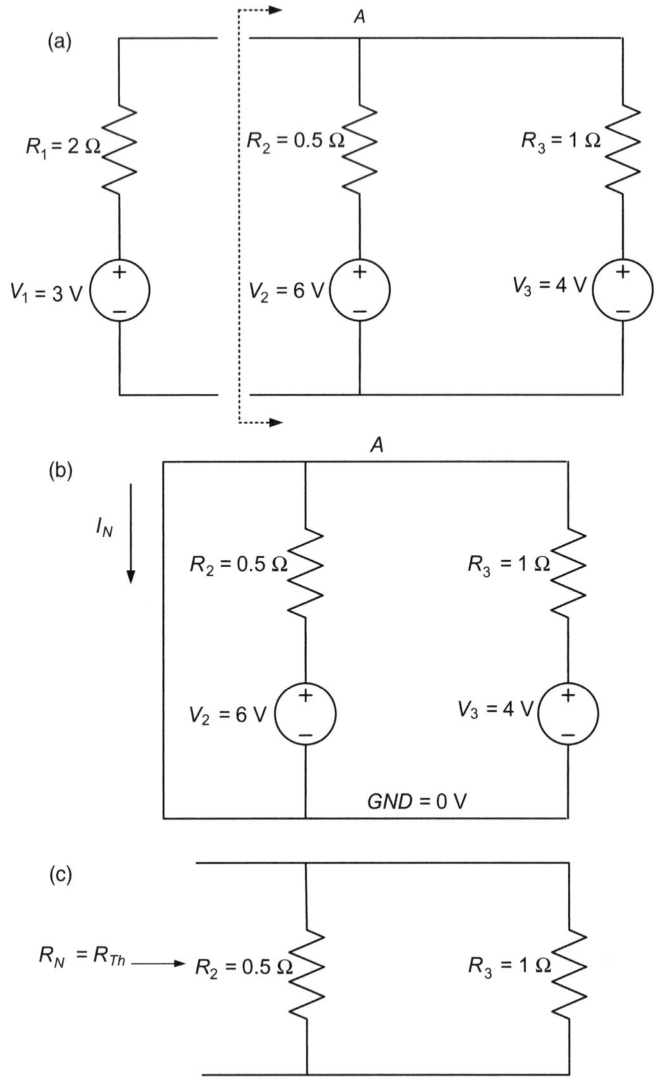

Figure 3.40 Example 3.22: (a) slicing the circuit; (b) calculating the Norton current; (c) calculating the Norton resistance.

Plugging numerical values into Equation (3.235) it yields that

$$V_A = 5\ \text{V}. \tag{3.236}$$

Note that in the circuit of Figure 3.40, V_A is now zero because node A is grounded; refer to Figure 3.40b.

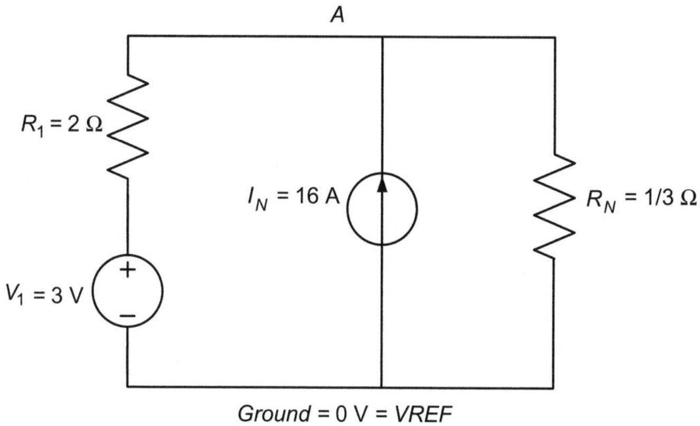

Figure 3.41 Example 3.22: Norton equivalent circuit attached to original circuit left out.

As a final step, we can apply a source transformation to the series circuit formed by V_1 and resistor R_1 of Figure 3.41. This last step, not shown in Figure 3.40, will convert the entire circuit into a single current source in parallel with a single resistor.

Drill Problem: Derive the above final step in Example 3.22, using Figure 3.41.

3.8.4 Example 3.23: Solving the Circuit of Figure 3.36 Using Source Transformations

Referring again to the original circuit of Figure 3.36, it is easy to see that we have three voltage sources in series with a resistor. Applying Thévenin's to Norton's source transformation to each source and its resistor in series, we convert them into a current source in parallel with a resistor.

The equations to do this are:

$$I_{N1} = V_1/R_1 = 3/2 = 1.5 \text{ A.} \tag{3.237}$$

$$I_{N2} = V_2/R_2 = 6/0.5 = 12 \text{ A.} \tag{3.238}$$

$$I_{N3} = V_3/R_3 = 4/1 = 4 \text{ A.} \tag{3.239}$$

The resistances in parallel with each one of the Norton current sources are the same resistors that were in series with each voltage source.

From Figure 3.42a we see that the parallel independent current sources equal to their algebraic sum of currents. The resistors in parallel are combined into a single parallel equivalent resistor, shown in Figure 3.42b

Thus, the total resulting current is obtained adding Equations (3.237) through (3.239), and this is

$$I_{N1} + I_{N2} + I_{N3} = I_{Ntotal} = 17.5 \text{ A.} \tag{3.240}$$

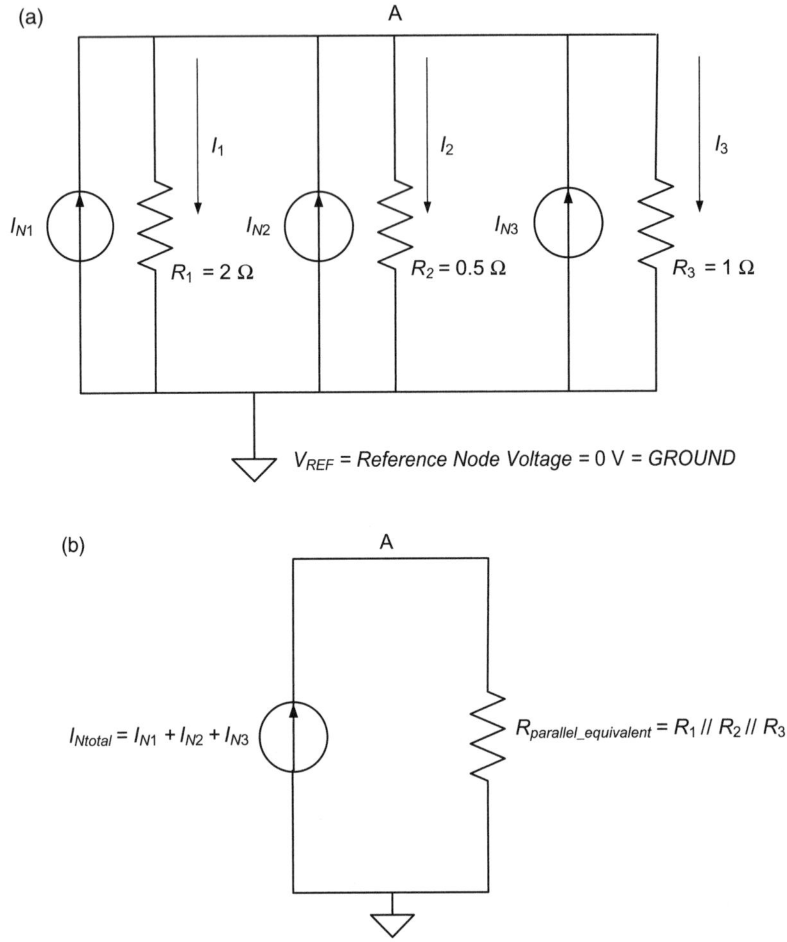

Figure 3.42 (a) Voltage source to current source transformation of circuit of Figure 3.36; (b) Norton resistance.

The total equivalent parallel resistance, that is, 2 Ω in parallel with 0.5 Ω and in parallel with 1 Ω equals

$$R_{par\text{-}equiv} = R_1 // R_2 // R_3 = 2//0.5//1 = 2/7\ \Omega \tag{3.241}$$

and the nodal voltage,

$$V_A = I_{Thtotal} R_{par\text{-}equiv} = 5\ \text{V}. \tag{3.242}$$

Note: The operator "//" stands for parallel equivalent resistor: so that *a*//*b* is equal to: $(ab)/(a + b)$.

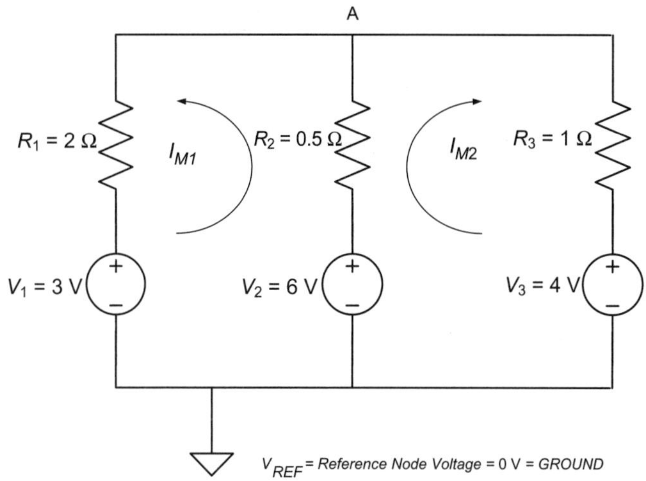

Figure 3.43 Circuit of Figure 3.36 solved by the mesh method.

3.8.5 Example 3.24: Solving the Circuit of Figure 3.36 Using the Mesh Method

The circuit of Figure 3.36 is repeated in Figure 3.43. In addition to the original information two mesh currents I_{M1} and I_{M2} have been arbitrarily defined. Note that I_{M1} was chosen to flow in the counter clockwise direction in mesh 1. I_{M2} was chosen to flow in the clockwise direction in mesh 2. Now we can start writing the mesh equations for this circuit.

$$\text{Mesh 1:} \quad V_2 - V_1 = (I_{M1} + I_{M2})R_2 + I_{M1}R_1 \tag{3.243}$$

$$\text{Mesh 2:} \quad V_2 - V_3 = (I_{M2} + I_{M1})R_2 + I_{M2}R_3. \tag{3.244}$$

Distributing and regrouping Equations (3.243) and (3.244) by mesh currents we obtain

$$\text{Mesh 1:} \quad V_2 - V_1 = I_{M1}(R_1 + R_2) + I_{M2}R_2 \tag{3.245}$$

$$\text{Mesh 2:} \quad V_2 - V_3 = I_{M1}R_2 + I_{M2}(R_2 + R_3). \tag{3.246}$$

Important observation: In Equations (3.245) and (3.246) the $I_{M2}R_2$ and $I_{M1}R_2$ terms have a positive sign because both mesh currents I_{M1} and I_{M2} flow through R_2, the common element between both meshes, in the same direction. We could have also obtained Equations (3.245) and (3.246) directly by inspection of the circuit in Figure 3.43.

We plug into Equations (3.245) and (3.246) the values from Figure 3.43 and get

$$6-3 = I_{M1}(2+0.5)+I_{M2}0.5. \quad (3.247)$$

$$6-4 = I_{M1}0.5+I_{M2}1.5. \quad (3.248)$$

Solving Equations (3.247) and (3.248) for I_{M1} and I_{M2} we obtain

$$I_{M1} = 1\text{ A}. \quad (3.249)$$

$$I_{M2} = 1\text{ A}. \quad (3.250)$$

Finally, by inspection of the circuit of Figure 3.43 we see that

$$V_A = V_2 - (I_{M1}+I_{M2})R_2. \quad (3.251)$$

And again plugging the values from Figure 3.43 and from Equations (3.249) and (3.250) we obtain

$$V_A = 5\text{ V}. \quad (3.252)$$

3.8.6 Example 3.25: Solving the Circuit of Figure 3.36 Using the Nodal Method

Figure 3.44 addresses the solving of this circuit by the nodal method. Note that the circuit of Figure 3.44 has actually four nodes (A, B, C, D, and V_{REF}). But fortunately, the nodal voltages at nodes B, C, D with respect to V_{REF} are known. That is,

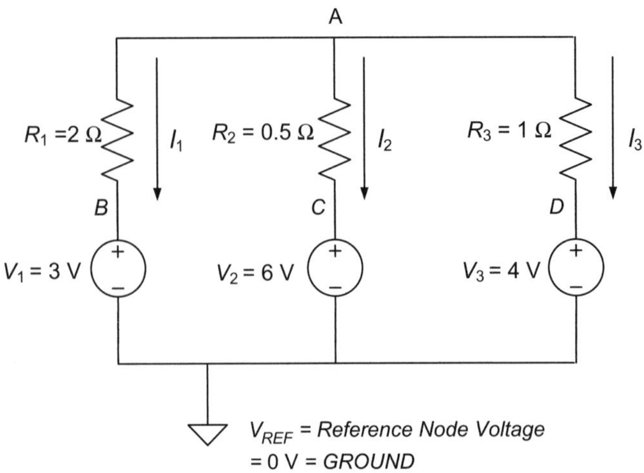

Figure 3.44 Solving the circuit of Figure 3.36 using the nodal method.

$$V_B = V_1 = 3\text{ V}. \tag{3.253}$$

$$V_C = V_2 = 6\text{ V}. \tag{3.254}$$

$$V_D = V_3 = 4\text{ V}. \tag{3.255}$$

So in actuality there is one nonreference node and an unknown nodal voltage V_A.

Nodes B, C, D are super nodes, and we can state the single nodal equation needed to solve V_A. By inspection of the circuit of Figure 3.44, the single nodal equation is

$$(V_A - V_1)/R_1 + (V_A - V_2)/R_2 + (V_A - V_3)/R_3 = 0. \tag{3.256}$$

Now plugging the values from Figure 3.44 into Equation (3.256) it yields

$$(V_A - 3)/2 + (V_A - 6)/0.5 + (V_A - 4)/1 = 0. \tag{3.257}$$

Solving for V_A we obtain that

$$V_A = 5\text{ V}. \tag{3.258}$$

3.9 SUMMARY AND CONCLUSIONS

After all six methods have been used, the reader is encouraged to go over them at least one more time to understand each of the techniques used. Leaving personal preferences aside, the nodal and the source transformation methods are quite brief and powerful. For example, look at the single Equation (3.257) used with the nodal method. The nodal method allows one to solve the problem with a single nodal equation because of the presence of super nodes. The source transformation method allows solving the problem with simple arithmetic. Now whether one can state that these two are the easiest methods is a different story. Clearly, superposition breaks down a single problem into three simpler ones. In some ways this introduces more opportunity to make an arithmetic error. Thévenin and Norton are not so bad. Finally, the mesh method allows us to solve for voltage V_A writing two equations. Ultimately, it is the reader who can clearly state which is the easiest and fastest method to apply for him or her, this becoming more of a personal preference and not an absolute fact.

FURTHER READING

1. Charles A. Desoer and Ernest S. Kuh, *Basic Electric Circuit Theory*, McGraw-Hill, New York, 1969.

2. Thomas Kailath, *Linear Systems*, Prentice Hall, Upper Saddle River, NJ, 1980.
3. Mahmood Nahvi and Joseph Edminister, *Electric Circuits*, 4th ed., McGraw-Hill, New York, 2003.
4. Charles K. Alexander and Matthew N. O. Sadiku, *Fundamentals of Electric Circuits*, 2nd ed., McGraw-Hill, New York, 2004.
5. Davis E. Johnson, Johnny R. Johnson, John L. Hilburn, and Peter D. Scott, *Electric Circuit Analysis*, 3rd ed., Prentice Hall, Upper Saddle River, NJ, 1997.

PROBLEMS

3.1 For the circuit in Figure 3.45 determine using the superposition method: (1) Current delivered by the current source I, (2) voltage across resistor R, (3) current through resistor R, (4) voltage across current source I, (5) power delivered by voltage source V, (6) power delivered by current source I, (7) power consumed by resistor R.

3.2 For the circuit in Figure 3.46 determine using the superposition method the power consumed by resistors R_1, R_2, and R_3, and the power delivered by voltage sources V_1, V_2, and V_3.

3.3 For the circuit given in Figure 3.47, find the voltage value at node A using superposition. Note that the circuit has two independent voltage sources and one voltage-controlled voltage source (VCVS). Hint: When you apply superposition, eliminate the independent sources one at a time. Never eliminate the dependent source and its control voltage.

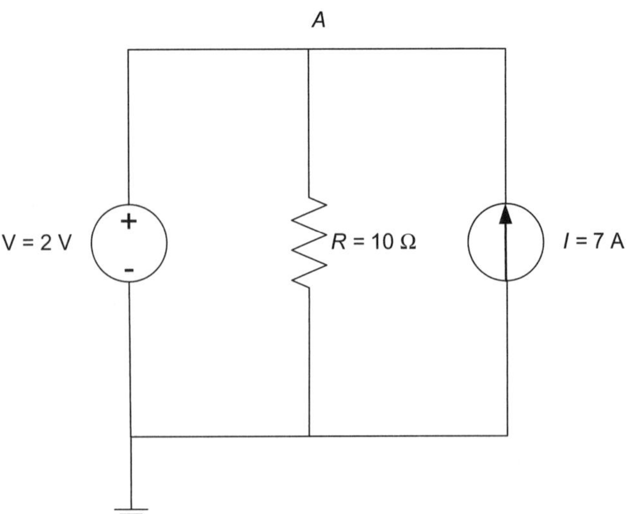

Figure 3.45 Circuit for Problem 3.1.

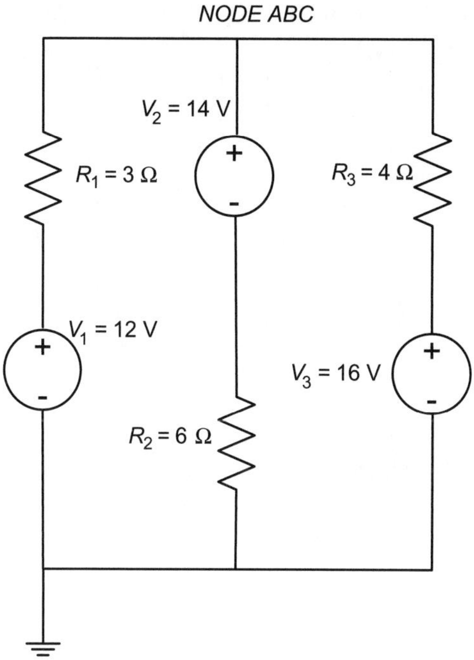

Figure 3.46 Circuit for Problem 3.2.

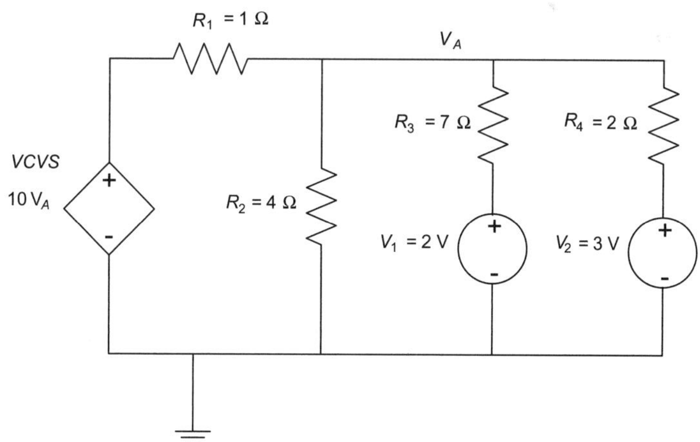

Figure 3.47 Circuit for Problem 3.3.

3.4 For the circuit given in Figure 3.48, find the voltage at node A using superposition. Note that the circuit has two independent sources and one VCVS. Hint: When you apply superposition, eliminate the independent sources one at a time. Never eliminate the dependent source and its control voltage or current.

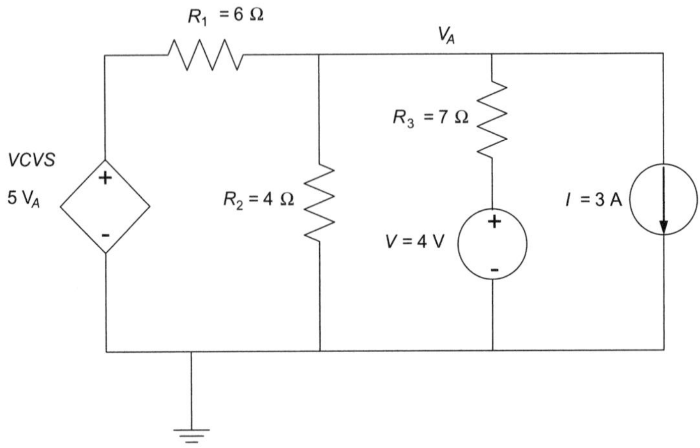

Figure 3.48 Circuit for Problem 3.4.

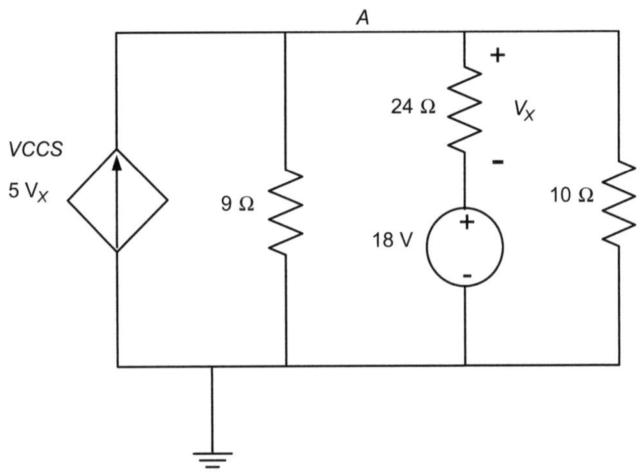

Figure 3.49 Circuit for Problem 3.5.

3.5 For the circuit given in Figure 3.49, find the voltage between node A and ground. Apply any circuit solving method of your preference.

3.6 For the circuit given in Figure 3.50, calculate the power consumed by load resistor $R_L = 2\ \Omega$.

3.7 Using the circuit of Figure 3.46 and assuming that $V_2 = 0$ V, recalculate using any circuit analysis method the power consumed by resistors R_1, R_2, and R_3, and the power delivered by voltage sources V_1, and V_3.

3.8 Using the circuit of Figure 3.51, apply any circuit analysis method to determine current I_3 through resistor R_3. Hint: This problem is much simpler than what it appears to be.

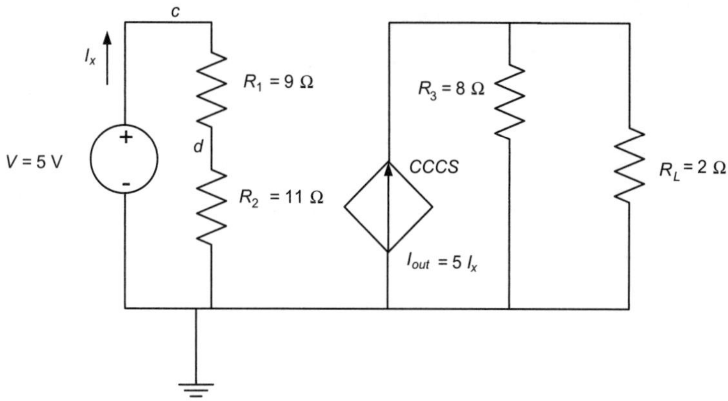

Figure 3.50 Circuit for Problem 3.6.

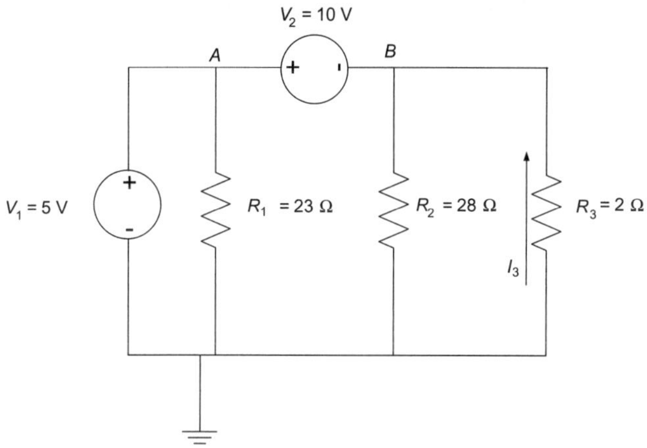

Figure 3.51 Circuit for Problem 3.8.

3.9 Using the circuit of Figure 3.52, apply Thévenin's method as many times as needed, to find current I_4 through resistor R_4.

3.10 Using the circuit of Figure 3.52, find current I_4 through resistor R_4 transforming all voltage sources to current sources and applying KCL.

3.11 Using the circuit of Figure 3.52, find current I_4 through resistor R_4 using superposition.

3.12 Using the circuit of Figure 3.52, find current I_4 through resistor R_4 using source transformations. Hint: Convert voltage sources into current sources and apply KCL.

3.13 Using the circuit of Figure 3.53, find the voltage across and current through every resistor, that is, R_1 through R_4.

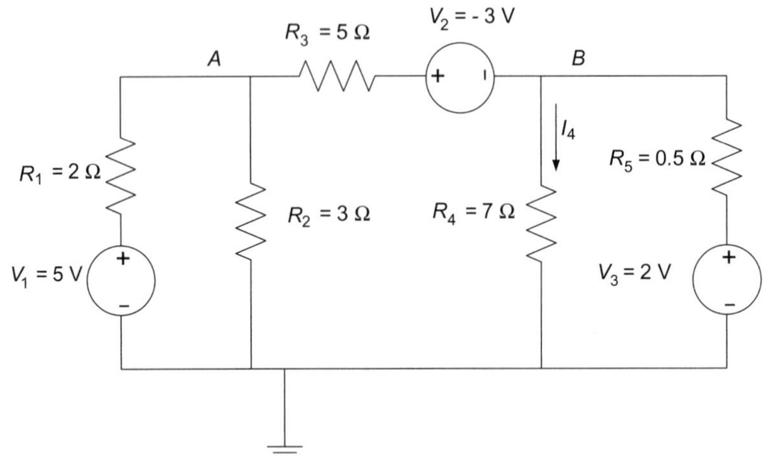

Figure 3.52 Circuit for Problems 3.9 through 3.12.

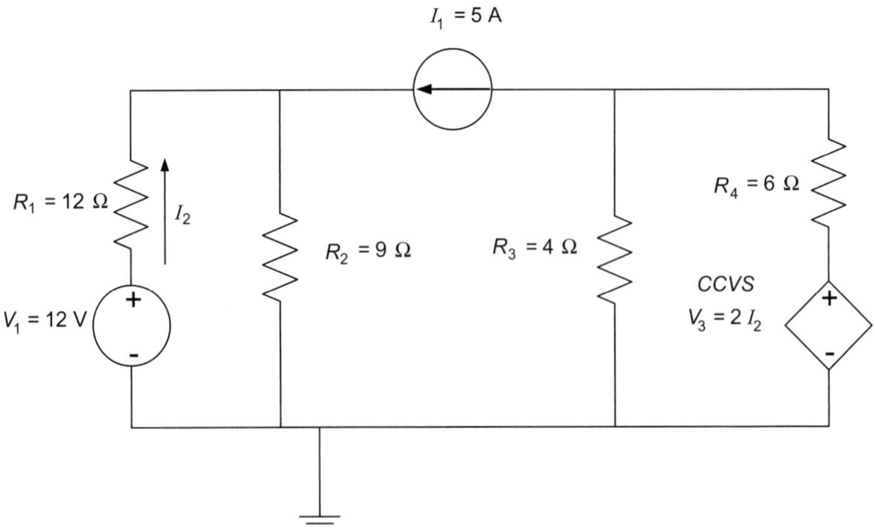

Figure 3.53 Circuit for Problem 3.13.

3.14 Calculate the total power delivered by all sources and consumed by all resistors in the circuit of Figure 3.54. Hint: There may be situations when a source does not deliver power, because it is being *charged* by some other source in the circuit. Question: Is that the case for this example? Justify your answer.

3.15 Find branch currents I_1, I_2, and I_3 of the circuit of Figure 3.55.

3.16 Apply mesh analysis to calculate branch currents I_1, I_2, and I_3 of the circuit of Figure 3.56.

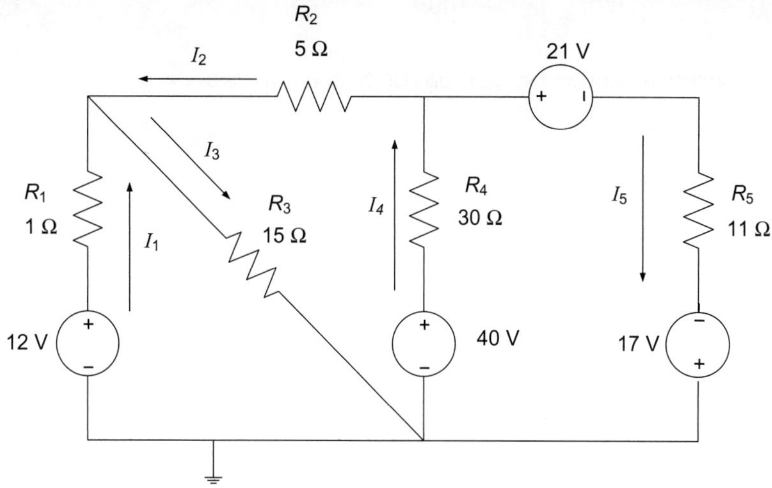

Figure 3.54 Circuit for Problem 3.14.

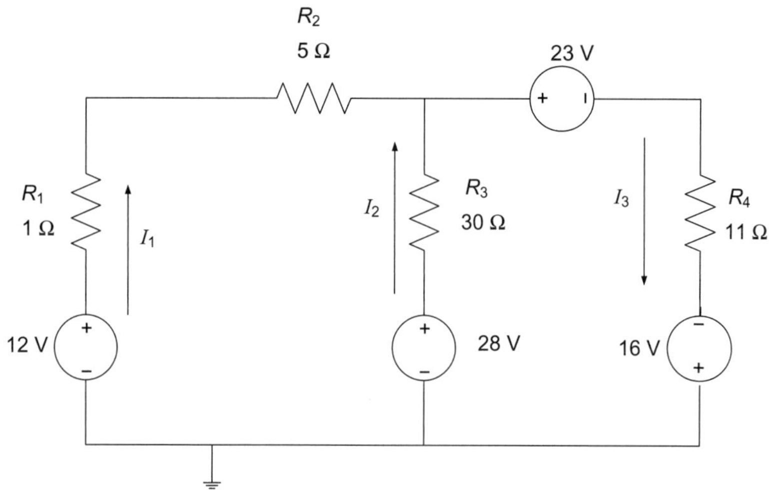

Figure 3.55 Circuit for Problem 3.15.

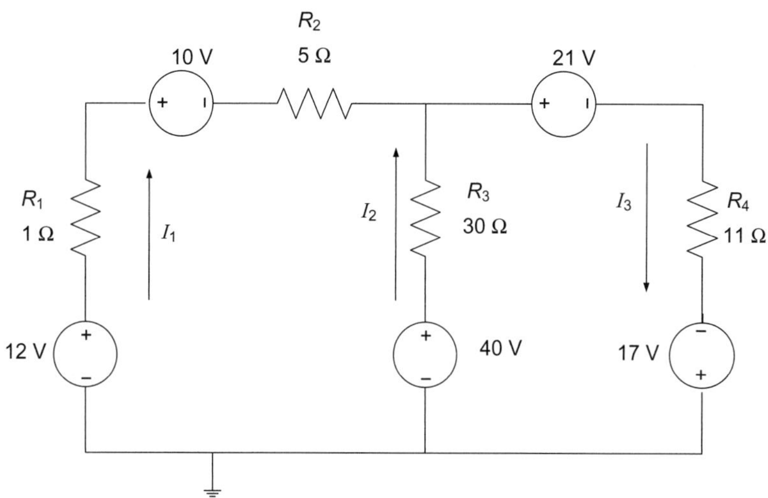

Figure 3.56 Circuit for Problem 3.16.

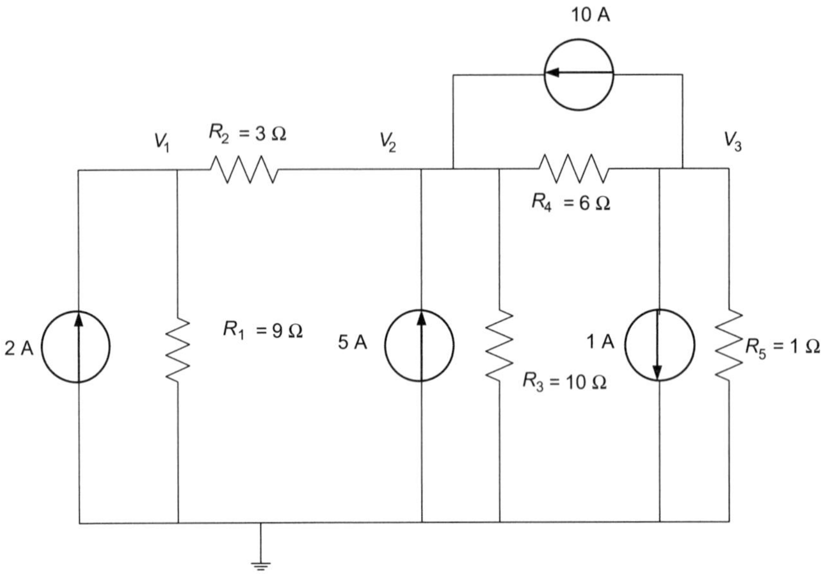

Figure 3.57 Circuit for Problem 3.17.

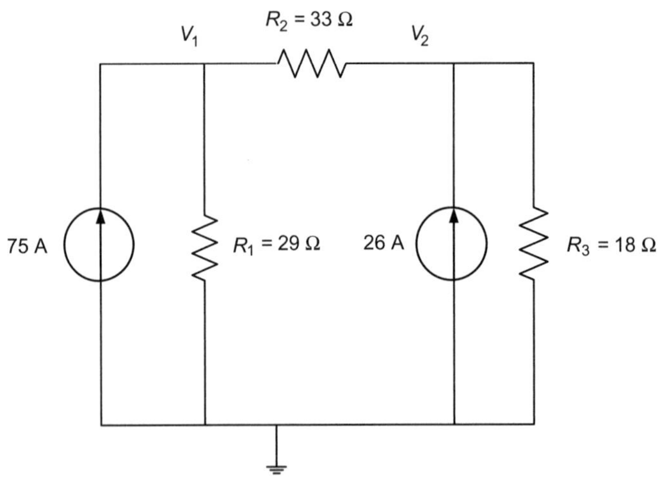

Figure 3.58 Circuit for Problems 3.18, 3.19, and 3.20.

3.17 Apply nodal analysis to calculate node voltages V_1, V_2, and V_3 of the circuit of Figure 3.57.

3.18 Apply nodal analysis to calculate node voltages V_1 and V_2 of the circuit of Figure 3.58.

3.19 Apply Norton's analysis to calculate node voltages V_1 and V_2 of the circuit of Figure 3.58.

3.20 Applying superposition calculate node voltages V_1 and V_2 of the circuit of Figure 3.58.

4

FIRST- AND SECOND-ORDER CIRCUITS UNDER SINUSOIDAL AND STEP EXCITATIONS

4.1 INTRODUCTION

First-order circuits are very important in electrical and electronic engineering. Many higher order circuits can be reduced to a first-order circuit. Analyzing the behavior either in the time or in the frequency domains of a first-order circuit is unquestionably simpler than analyzing that of a higher order circuit. Essentially, first-order circuits have a single energy storage device. Such devices can be either a capacitor or an inductor. Examples of circuits that can be reduced to first-order circuits under certain conditions are electronic amplifiers, operational amplifiers, servomechanisms, electric motors, and other control networks.

Let us present an example of a first-order circuit.

Example 4.1 RL Series First-Order Circuit

Given a circuit that contains one resistor in series with an inductor, such as the one shown in Figure 4.1, we can calculate the output voltage to input voltage ratio as a function of frequency. Such ratio of voltages in the frequency domain is commonly referred to as $H(j\omega)$, where $H(j\omega)$ is called the circuit transfer function.

$$H(j\omega) = \mathbf{V}_{out}(j\omega)/\mathbf{V}_{in}(j\omega). \tag{4.1}$$

Electrical, Electronics, and Digital Hardware Essentials for Scientists and Engineers, First Edition. Ed Lipiansky.

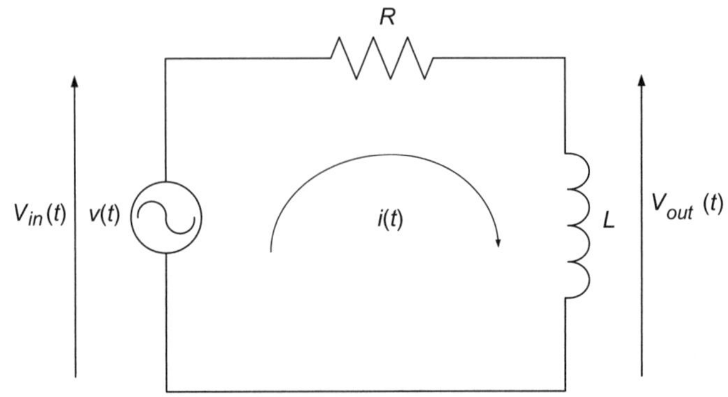

Figure 4.1 Circuit for Example 4.1, a first-order series RL circuit.

Previously reviewing the material from Chapter 3 on AC analysis we can easily calculate $H(j\omega)$ for the circuit in Figure 4.1.

The current in the RL circuit is calculated as follows:

$$\boldsymbol{I} = V_{in}/(R + j\omega L), \tag{4.2}$$

where $\boldsymbol{I}$ and $\boldsymbol{V_{in}}$ are respectively the current and voltage phasors of the circuit.

Now the output voltage $\boldsymbol{V_{out}}$ is calculated by multiplying the circuit current times the impedance or reactance of inductor L. Thus, we obtain

$$\boldsymbol{V_{out}}(j\omega)/\boldsymbol{V_{in}}(j\omega) = j\omega L/(R + j\omega L). \tag{4.3}$$

And finally, our transfer function is

$$H(j\omega) = \mathbf{V}_{out}(j\omega)/\mathbf{V}_{in}(j\omega) = j\omega L/(R + j\omega L). \tag{4.4}$$

Furthermore, rationalizing the denominator, that is, multiplying numerator and denominator of Equation (4.4) by the complex conjugate of the denominator, $(R - j\omega L)$, we obtain

$$\boldsymbol{H}(j\omega) = \frac{\omega^2 L^2 + j\omega RL}{R^2 + (\omega L)^2} = \tag{4.5}$$

$$= \frac{\omega^2 L^2}{R^2 + (\omega L)^2} + j\frac{\omega RL}{R^2 + (\omega L)^2} \tag{4.6}$$

Equation (4.6) is of the form $a + jb$ where the terms a and jb are frequency dependent. Additionally term jb is of inductive nature.

We can also write the time domain circuit equation for the circuit of Figure 4.1; this leads to

$$v_{in}(t) = i(t)R + L\frac{di(t)}{dt}, \tag{4.7}$$

where $v_{in}(t)$ is the excitation, $i(t)$ is the current through the circuit, $i(t)R$ is the voltage drop across the resistor, and $L\ d\ i(t)/dt$ is the voltage drop across the inductor.

Equation (4.7) is the time domain first-order differential equation that describes the circuit on hand.

The highest derivative in a differential equation determines the *order* of the differential equation.

4.2 THE FIRST-ORDER *RC* LOW-PASS FILTER (LPF)

Let us investigate the *RC* circuit of Figure 4.2. This circuit is excited by a sinusoidal voltage waveform. Elements *R* and *C* are in series, the input voltage is applied to the two elements in series, the output is taken across the capacitor terminals.

4.2.1 Frequency Domain Analysis

Let us calculate the transfer function of this circuit:

$$\mathrm{H}(j\omega) = \mathbf{V}_{out}(j\omega)/\mathbf{V}_{in}(j\omega). \tag{4.8}$$

The impedance of the resistor and capacitor in series is

$$Z_{series}(j\omega) = R + 1/j\omega C. \tag{4.9}$$

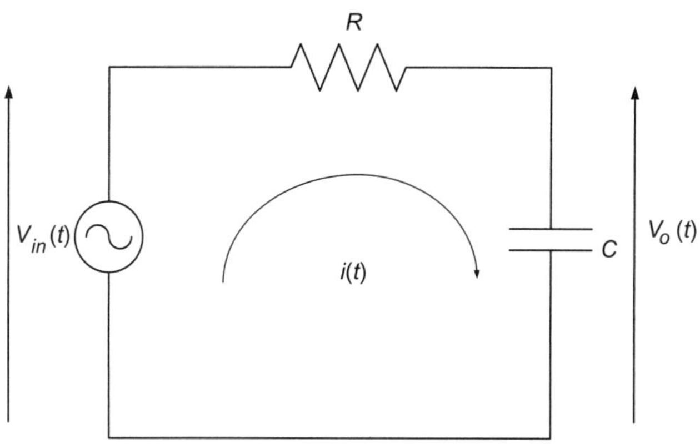

Figure 4.2 First-order *RC* low-pass filter.

Then the current through the circuit is

$$\mathbf{V}_{in}(j\omega)/Z_{series}(j\omega). \tag{4.10}$$

The above current times the impedance or reactive capacitance of the capacitor $1/j\omega C$ equals the output voltage V_{out} $(j\omega)$, thus we get

$$\mathrm{H}(j\omega)=\frac{V_{out}(j\omega)}{V_{in}(j\omega)}=\frac{1}{1+j\omega RC}. \tag{4.11}$$

Remember that from Chapter 3, ω is the angular frequency, which equals $2\pi f$, where f is the frequency of the sinusoidal waveform in hertz.

Equation (4.11) describes the ratio of output voltage to input voltage of the *RC* circuit given by Figure 4.2, again referred to as the circuit transfer function in the frequency domain.

So let us now construct a table to plot the values of the transfer function $\mathrm{H}(j\omega)$ for a given R and a given C. The product

$$RC \tag{4.12}$$

is referred to as the circuit time constant. We will plot the transfer function of Equation (4.11) using two separate plots. One plot is for its magnitude, and the second plot for its phase, both are functions of frequency. Note that transfer function of Equation (4.11) is a complex quantity and as such, it has magnitude and phase.

4.2.2 Brief Introduction to Gain and the Decibel (dB)

It is common in circuit theory to refer to the output to input voltage ratio as the *gain* of the circuit. In the case of a first-order *RC* LPF circuit, such *gain* is always one* or less than one. Active circuits are those circuits containing operational amplifiers or transistors that usually are designed to have a gain larger than one. More on active circuits will be covered in Chapter 6.

When the *gain* is greater than *1*, it is referred to as *gain*, but when the gain is less than *1*, it is sometimes referred to as *attenuation*, or simply as *a less-than-one gain*. Passive first-order circuits have gains that are one or strictly less than 1.

When we talk about *gain* without any units associated to it, it is simply a ratio of voltages, thus it has no units because *gain* units are volts/volts. In circuit analysis, it is very common to define a new unit for *gain* and *attenuation* called the decibel (dB).

* Mathematically speaking, the gain of an *RC* circuit may be very close to *1*, but it is never exactly *1*. In practical terms the gain is *1* for frequencies one-tenth of the cutoff frequency and below.

The decibel is defined as follows for a ratio of voltages:

$$Gain\ in\ dB = 20log_{10}|V_{out}/V_{in}|. \tag{4.13}$$

It is important to mention that the argument of a decimal log must be a positive number. The log of a number less than or equal to zero is undefined.

So note that the *gain* (without units or in volts/volts), is also referred to, as the *linear gain* of a circuit. For example, given a circuit with a *linear gain* of *10*, by virtue of Equation (4.13) the *gain* in dB becomes:

$$20log_{10}(10/1) = 20\text{ dB}. \tag{4.14}$$

For a circuit whose *linear gain* is *0.1*, or an *attenuation* of *10*, the gain in dB becomes:

$$20log_{10}(1/10) = -20\text{ dB}. \tag{4.15}$$

So from Equations (4.14) and (4.15) larger than one *gains* have units of positive decibels, while *attenuations* are always given in negative decibels. Note that a linear gain of *1* is:

$$0\text{ dB} = 20log_{10}(1/1). \tag{4.16}$$

When we refer to a circuit with a *gain* of *1* or a *gain* of *zero dB*, we are talking about the exact same thing.

To summarize the relation between linear gain and logarithmic or gain in decibel we develop Table 4.1.

Positive linear gains are above 1 and negative linear gains are below 1. On the other hand, following Table 4.1, a –20 dB and a –40 dB gains can also be referred to as 20 dB and 40 dB attenuations, respectively.

Table 4.1 Relationship between linear gain or attenuation and gain in decibel

Linear Gain (V/V)	Gain (dB)
0.01	–40
0.1	–20
1	0
10	20
100	40
1,000	60
10,000	80
100,000	100
1,000,000	120

4.2.3 *RC* LPF Magnitude and Phase Bode Plots

Since *gains* may range from very small to very large values, plotting the gain in decibel is very advantageous. A very large range of *gains*, for example from 0.01 to 100,000 V/V, looks somewhat cumbersome when plotted in a linear scale. The gains are extremely compressed at low gain values and extremely expanded for large gain values.

To more evenly distribute the gain along the height of the *y*-axis the magnitude of the gain is plotted in decibel. Since the decibel is a logarithmic function, plotting gain in decibel is effectively *gain* in a logarithmic scale. Plotting the gain values in decibel (e.g., –40 dB for a *linear gain* of 0.01 and +100 dB for a *linear gain* of 100,000) allows the plot to have the same amount of vertical space allocated to display all the values of *gain*.

The horizontal axis of the magnitude Bode plot is frequency. Similarly to what is done to gain, frequency is plotted in logarithmic scale. So if we are interested in plotting the gain as a function of frequencies from 0.01 Hz to 100 kHz, the *x*-axis is scaled logarithmically. Doing this allows us to see the frequency range in a decompressed fashion. Plotting the frequency as a linear quantity would make the frequency axis very compressed at low frequencies and greatly expanded at high frequencies. So for all practical purposes, a Bode magnitude plot displays decibel linearly and frequency logarithmically; thus, it is a semi-log plot; but since the decibel is a logarithmic function, the magnitude plot is a log-log plot for gain in decibel versus frequency in hertz.

The phase portion of the Bode plot of a transfer function displays degrees in the vertical axis and logarithmic frequency along the *x*-axis. Degrees are always plotted in a linear scale, because the range of degrees is in general not very large as gains or frequencies are.

Example 4.2 Bode Plots of an *RC* LPF Transfer Function

Continuing with the circuit given in Figure 4.2 with transfer function given by Equation (4.11), the transfer function is repeated below for the reader's convenience:

$$H(j\omega) = \frac{V_{out}(j\omega)}{V_{in}(j\omega)} = \frac{1}{1 + j\omega RC}. \quad (4.17)$$

We define a new term, the cutoff angular frequency ω_0 in radians per second and the *cutoff frequency* f_0 in hertz, as

$$\omega_0 = 1/RC, \quad (4.18)$$

where:

$$\omega_0 = 2\pi f_0. \quad (4.19)$$

Now let us assume that our filter has a cutoff frequency f_0 of 1 KHz, thus from combining Equations (4.18) and (4.19) we obtain

$$f_0 = 1/2\pi RC \tag{4.20}$$

and from Equation (4.20)

$$RC = 1/2\pi f_0 = 159.155\ \mu\text{s}. \tag{4.21}$$

From Equation (4.18) the term RC is known as the circuit *time constant* τ (Greek letter tau) and its units are ohms multiplied by farads, which lead to time in seconds in the SI system of units (see Chapter 1).

For our particular example,

$$\tau = 159.155\ \mu\text{s}. \tag{4.22}$$

Important Point

The time constant τ of the circuit, that is, the RC product, determines the circuit frequency behavior.

The cutoff frequency of the circuit f_0 is a very important characteristic of a circuit, as we will see shortly.

Now we can rewrite Equation (4.17) using ω_0 and this becomes

$$H(j\omega) = \frac{V_{out}(j\omega)}{V_{in}(j\omega)} = \frac{1}{1 + j\omega/\omega_0} = \frac{1}{1 + jf/f_0}.$$

Since the cutoff frequency of the circuit is 1 kHz, we will pick to start plotting the Bode plots from frequencies much smaller than the cutoff frequency. In our case we will start at 1 Hz, somewhat arbitrarily we pick a high end frequency of 1 MHz.

Let us construct a table listing frequency on the leftmost column followed by the linear magnitude of our transfer function and a third and last column with the phase angle of our transfer function. Remember that the transfer function of interest given by Equation (4.11) is a complex number that has a magnitude and a phase.

The magnitude is

$$|\mathrm{H}(j\omega)| = \left|\frac{1}{1 + j\omega RC}\right| = \frac{1}{|1 + j\omega RC|} = \frac{1}{\sqrt{1 + (\omega RC)^2}} \tag{4.23}$$

and since $1/RC = \omega_0$ from Equation (4.18), simply becomes

$$|\mathrm{H}(j\omega)| = \frac{1}{\sqrt{1 + (\omega/\omega_0)^2}}. \tag{4.24}$$

Table 4.2 ***RC*** **LPF transfer function: magnitude and phase as a function of frequency**

Frequency (Hz)	Linear Gain (V/V)	Phase (Degrees)
1.000	0.999999500	–0.05729576
10.000	0.999950007	–0.57293870
100.000	0.995037481	–5.71059313
1,000.000	0.707117209	–44.99999998
10,000.000	0.099506625	–84.28940686
100,000.000	0.009999795	–89.42706130
1,000,000.000	0.001000029	–89.94270424

The phase of Equation (4.11) is

$$\angle H(j\omega) = -\arctan(\omega/\omega_0). \tag{4.25}$$

So our Table 4.2 follows:

Table 4.2 lists frequencies in the leftmost column, the magnitude as a dimensionless number (volts/volts) in the center column and the phase in degrees in the rightmost column. In electronics the preferred way of displaying magnitude is in decibel. The output to input voltage ratio, which we will refer to as a *gain*, is usually expressed in decibels. A decibel was defined by Equation (4.13), which we repeat for the reader's convenience.

$$V_{out}/V_{in} \; in \; dB = 20log_{10}|V_{out}/V_{in}|. \tag{4.26}$$

We can easily verify using Equation (4.26) that for a gain ratio of 1 the gain in decibel equals 0 dB; for a gain ratio of 10, the gain equals 20 dB; a gain ratio of 100, the gain equals 40 dB; and a for gain ratio of 1000 equals 60 dB. So for every order of magnitude that the gain goes up, the gain in decibel goes up by 20 dB. On the other hand, for a gain of 0.1, the gain in decibel equals –20 dB, for 0.01 it equals –40 dB, for 0.001 it equals –60 dB, and so on.

Figure 4.3 depicts the magnitude and phase Bode plots of the *RC* LPF tabulated in Table 4.2.

Important Points

For a first-order RC LPF circuit, the gain is close to 0 dB at frequencies below one-tenth of the cutoff frequency f_0.

For a first-order RC LPF circuit the gain at the cutoff frequency f_0 *is –3.01 dB.*

For a first-order RC LPF, the circuit the gain drops at a rate of –20 dB per decade from its cutoff frequency. This is to say that the gain drops by 20 dB for a frequency 10 times f_0, *it drops another 20 dB for a frequency 100 times* f_0, *another 20 dB for a frequency 1000 times* f_0, *and so on. (Refer to Table 4.3.)*

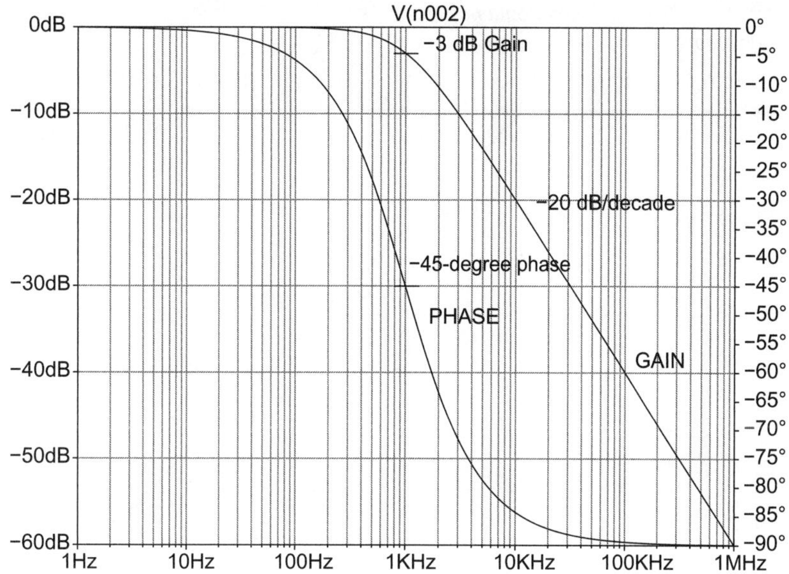

Figure 4.3 Exact magnitude and phase Bode plots of the first-order circuit of Example 4.2.

The frequency axis does not have a zero or origin of frequencies because log of zero is nonexistent. The lowest frequency can be represented with a value as small as we desire, but not with zero.

Frequency is represented logarithmically for both magnitude and phase plots.

The magnitude or gain in decibel is represented linearly. There is a 0 dB origin for the vertical axis because the scale is linear in decibels.

The phase in degrees is represented linearly on the vertical axis.

The phase of an RC LPF is approximately 0° at frequencies below 1/10 of f_0*. The phase of an RC LPF is approximately –90° at frequencies larger than 10* f_0*. At* f_0 *the phase equals –45°.*

Although the definition of the decibel may initially seem capricious, it is actually a better way that allows us to visualize the growth or the decay of the gain in a magnitude plot.

From the calculation in Table 4.2 we will build Table 4.3 that will contain the magnitude in decibels. Frequency, magnitude (linear gain), and phase are shown with a generous number of decimal places.

4.2.4 *RC* LPF Drawing a Bode Plot Using Just the Asymptotes

To draw the asymptotes of the Bode plots of our circuit (Fig. 4.2) we will normalize frequency. Instead of listing on the frequency axis the actual cutoff

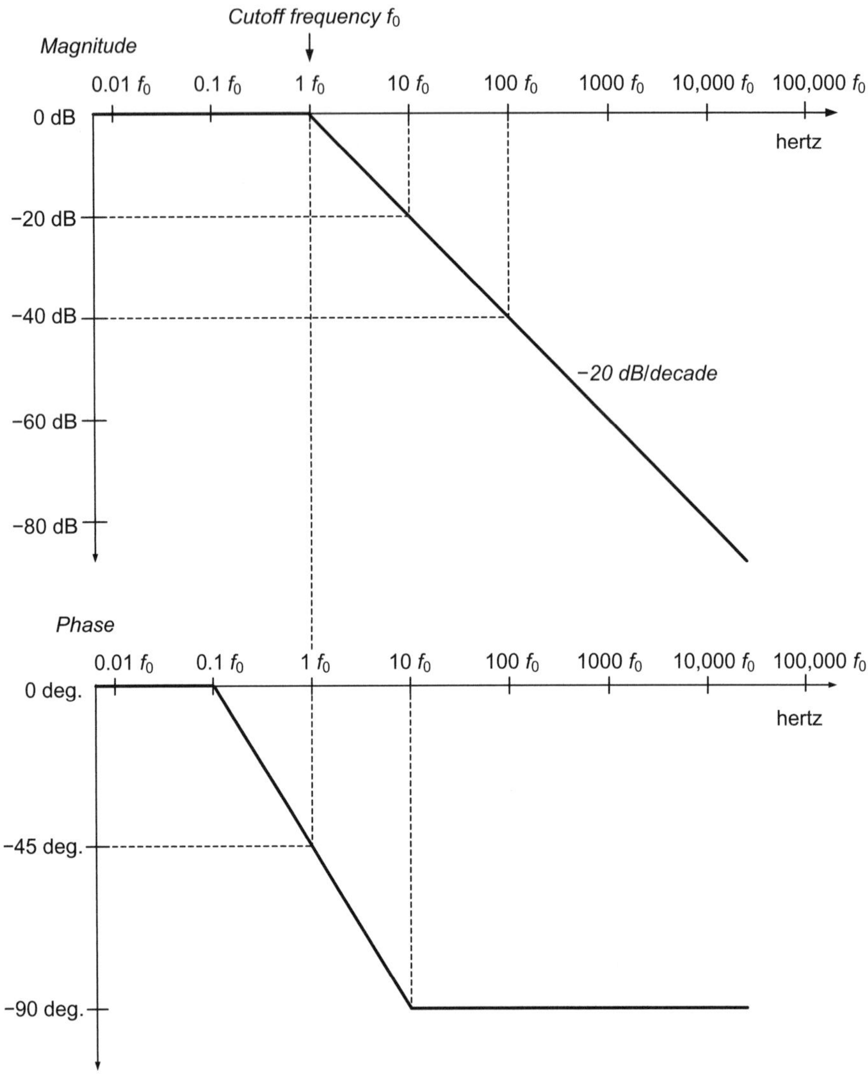

Figure 4.4 First-order *RC* LPF asymptotic Bode plots: magnitude and phase.

frequency value of 1 KHz, we will denote this frequency as f_0, so lower frequencies will be a tenth, a one-hundredth, and so on of f_0. Similarly, frequencies above f_0 are 10 times, 100 times, and so on of f_0. Figure 4.4 shows these normalized frequencies on the horizontal axis.

From Table 4.3 and Figure 4.3 we see that the magnitude asymptotically approaches 0 dB from the cutoff frequency f_0 to smaller frequencies. Also from the cutoff frequency to higher frequencies, the gain drops at a constant rate

Table 4.3 *RC* LPF transfer function: magnitude in *dB* and phase as a function of frequency

Frequency (Hz)	Normalized Frequency f/f_0	Linear Gain (V/V)	Gain in dB	Phase (Degrees)
1.000	0.001 f_0	0.999999500	0.00	−0.05729576
10.000	0.01 f_0	0.999950007	0.00	−0.57293870
100.000	0.1 f_0	0.995037481	−0.04	−5.71059313
1,000.000	1 f_0	0.707117209	−3.01	−44.99999998
10,000.000	10 f_0	0.099506625	−20.04	−84.28940686
100,000.000	100 f_0	0.009999795	−40.00	−89.42706130
1,000,000.000	1,000 f_0	0.001000029	−60.00	−89.94270424

of −20 dB per decade. But at the cutoff frequency, the gain is approximately −3 dB. In linear terms, this means that the amplitude of the sinusoidal waveform that excites the *RC* circuit becomes attenuated to about 70% from its original value. Referring one more time to Table 4.3, we can see that at the cutoff frequency f_0 the output voltage magnitude is 0.707 of the original input, which has a magnitude of 1. That is, the output magnitude is approximately 70.7% of the input magnitude.

In a similar fashion, we can see that the phase is −45° at f_0. One more time looking at the phase in Table 4.3, we see that the phase at about one-tenth of f_0 is −5.7°. And for even lower frequencies, the phase asymptotically approaches 0°. On the other hand, at a frequency of 10 times f_0, the phase is approximately 5.7° below −90°.

So when Bode plots need to be drawn by hand, drawing its asymptotes is the preferred and quickest way of constructing magnitude and phase plots. This process not only saves a tremendous amount of number crunching but also makes the understanding of the plots more clear. The magnitude and phase plots are commonly referred to as the frequency response of the circuit. Figure 4.4 shows the asymptotic Bode plots for the circuit of Figure 4.2. Note that the gain is 0 dB flat from very low frequencies approximately up to the cutoff frequency f_0. The second magnitude asymptote simply decays from f_0 at a rate of −20 dB *per decade*. Once the magnitude asymptotes are drawn one can fill in by hand, the approximated gain curves. It is important to realize that the gain at f_0 is −3 dB and not 0 dB as Table 4.3 shows.

For the phase, we can also draw its plot using the phase asymptotes. The phase is a little bit more involved than the magnitude at least initially. Let us start with frequencies well below f_0, up to one-tenth of f_0, we draw a straight line at zero degrees from low frequencies all the way up to 1/10 of f_0. At a frequency of 10 times f_0 the frequency asymptote is a horizontal line at −90 degrees, starting at 10 f_0 continuing at −90° into higher frequencies. From Table 4.3, we know that at the cutoff frequency f_0, the phase is −45°. Looking at Figure 4.4, we now draw a straight line of a phase angle of 0 degrees at 1/10 of f_0 all the way to –90° at 10 f_0, such that this straight line passes through a –45° phase at f_0. In this way the phase asymptotes are drawn. Now one can draw by hand the approximate phase curves. Note that at 1/10 of f_0 and at 10 f_0 the phase is about 5.7° below zero degrees and 5.7° above –90°. Finally, it is important to note that the phase curve has an inflexion point at –45°, see Figure 4.3.

4.2.5 Interpretation of the *RC* LPF Bode Plots in the Time Domain

We will use the asymptotically drawn Bode plots to explain the meaning of the Bode magnitude and phase plots in terms of sinusoidal inputs applied to the *RC* LPF circuit. The same concepts can be extended to the actual (or exact) Bode plots that are tabulated in Table 4.3. So referring to Figure 4.4, the asymptotic Bode plots tell us the following:

First let us assume that a sinusoidal waveform of 1 V peak amplitude and a frequency of 0.01 f_0 is applied to the input of the first-order *RC* LPF. The output voltage waveform that will be observed across the capacitor terminals is for all practical purposes equal in magnitude to the input waveform and equal in frequency with no phase shift with respect to the input. That is, both inputs and outputs have the same magnitude: 1; and both input and output sinusoidal waveforms are in phase (phase = 0°). The lower the input waveform frequencies with respect to the cutoff frequency of the RC filter, the more accurate the preceding statement is. Refer to the numerical values of linear gain and gain in *dB* for frequencies much smaller than f_0 in Table 4.3.

Assume now that a sinusoidal input waveform of 1 V peak-amplitude and of a frequency f_0 (equal to the circuit cutoff frequency) is applied to the input of the RC circuit. The output voltage across the capacitor will be a sinusoidal waveform of peak amplitude 30.3% smaller than the input amplitude; however, it will still be of the same frequency, as the input waveform; but the output will be lagging the input by a *45* degree-phase. *Forty five degrees* is an eight of a full sinusoidal cycle.

Let us consider now that a sinusoidal input waveform of 1 V peak-amplitude and of a frequency 10 times larger than f_0, which is applied to the input of the *RC* circuit. The output voltage across the capacitor will be a sinusoidal waveform of peak amplitude 10 times smaller (20 dB) than the input amplitude; however, it will still be of the same frequency, as the input waveform; but the output will be lagging the input by about –84.3 degrees (almost –90 degrees). Clearly examining again the exact plots of Figure 4.3, the higher is the frequency of the input waveform with respect to the circuit cutoff frequency, the closer the output to input phase will be to –90 degrees. If the input waveform frequency is 100 times f_0 the output amplitude, while the input amplitude is always 1 V, the output waveform amplitude will be 100 times smaller (40 dB) than the input waveform amplitude. This behavior goes on and on, for every time the frequency goes up by a factor of 10 from the gain decays another 20 dB. (Refer again to Figs. 4.3 and 4.4.)

4.2.6 Why Do We Call This Circuit a LPF?

From the Bode plots just presented in Figures 4.3 and 4.4, it is clear to see that frequencies well below the cutoff frequency f_0 do not get attenuated, they just pass through the circuit with a *0 dB* gain (*linear gain of 1*), and a zero-degree phase shift. Frequencies well above the cutoff frequency become attenuated. We also see that the higher the frequency is above f_0, the higher the attenuation. The attenuation grows by 20 dB for every order of magnitude that the frequency grows above f_0. Alternatively, the gain decreases at a rate of 20 dB per decade of frequency.

In summary, the *RC* circuit just analyzed allows low frequency signals to go through the circuit without attenuation and without phase-shift, whereas the high frequencies become progressively attenuated as the input signal frequency goes up. At frequencies beyond 10 times the cutoff frequency, output signals exhibit a phase shift of approximately –90°. In reference to our *RC* LPF, which are low and which are high frequencies? The reference frequency is f_0 the filter cutoff frequency. One-tenth below f_0, the frequency is considered low. Ten times above f_0, a frequency is considered high.

4.2.7 Time Domain Analysis of the *RC* LPF

Now let us analyze the time domain equations of the low-pass *RC* circuit. Referring one more time to the circuit of Figure 4.2, it is possible to establish the differential equation for such circuit. Let us apply Kirchoff's voltage law (KVL) for the series of elements.

$$v_{in}(t) = i(t)R + v_o(t),$$

where $v_{in}(t)$ is the excitation or the circuit input voltage.

$i(t)$ is the current through the circuit, thus $i(t)\ R$ is the voltage drop across the resistor and $v_o\ (t)$ is the voltage across the capacitor:

We will define a unit-step function excitation $u(t)$ as follows:

$$u(t) = 1 \text{ for } t \geq 0$$

and

$$u(t) = 0 \text{ for } t < 0.$$

So applying the unit step function $u(t)$ *to* the input of *RC* LPF we have that

$$v_{in}(t) = u(t)$$

and substituting v_{in} with $u(t)$ into $v_{in}(t) = i(t)\ R + v_o(t)$, yields:

$$u(t) = i(t)R + v_o(t). \tag{4.27}$$

Since the current through the resistor and the capacitor have the same value,

$$i(t) = Cdv_o(t)/dt. \tag{4.28}$$

Plugging Equation (4.28) into Equation (4.27) leads to

$$u(t) = RC\frac{dv_o(t)}{dt} + v_o(t), \tag{4.29}$$

where Equation (4.29) is a first-order differential equation.

From calculus considerations to solve differential equations, the solution of Equation (4.29) has the form

$$v_o(t) = A_1 + A_2e^{-t/\tau} \tag{4.30}$$

where A_1 and A_2 are two constants and τ is the circuit time constant RC. A_1 is the steady-state value of the output voltage since

$$\begin{aligned} &for\ t \to \infty, \\ &v_o(t) \to A_1. \end{aligned} \tag{4.31}$$

This means that A_1 is the final value of output voltage $v_o(t)$after the transient is over. We call this final value V_{final}.

Thus,

$$V_{\text{final}} = A_1. \tag{4.32}$$

Also note that the initial value of the output voltage $v_o(t)$is found by making $t = 0$. Thus, from Equation (4.30),

$$= V_{\text{initial}} = v_o(0) = A_1 + A_2. \tag{4.33}$$

Finally plugging Equations (4.32) and (4.33) in Equation (4.30) we obtain that

$$v_o(t) = V_{final} + (V_{initial} - V_{final})e^{(-t/\tau)}. \tag{4.34}$$

Equation (4.34) is a general solution for a first-order circuit or a circuit with a single-time constant. We will use Equation (4.34) several times throughout this text and the homework problems.

Example 4.3 Given the single-time constant circuit of Figure 4.2, assuming that the input voltage is a unit-step function $u\ (t - 1)$, $R = 1\ \text{M}\Omega$, $C = 1\ \mu\text{F}$, calculate the final value of the output voltage $v_o(t)$ across the capacitor. Plot the output voltage waveform for positive values of time. Figure 4.5 displays

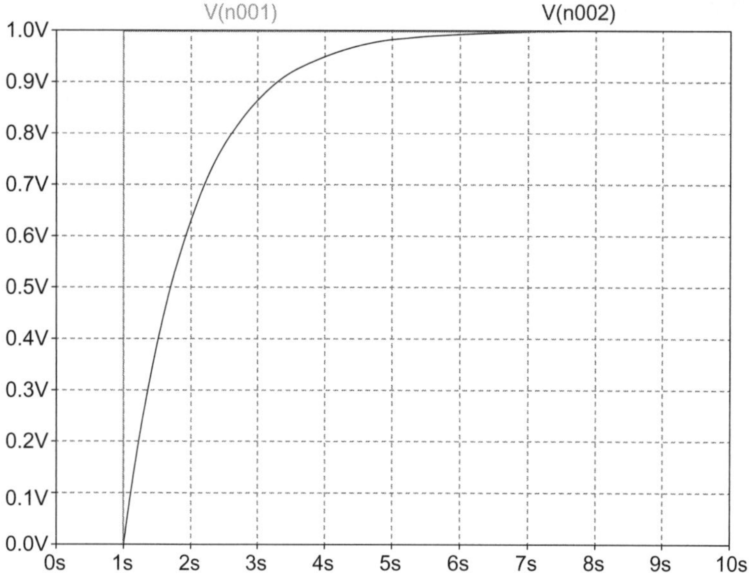

Figure 4.5 Unit step response of first-order *RC* LPF.

the input step function and the output as a function of time. Note that $u\,(t-1)$ is displaced or delayed from the origin of time by 1 second.

Now, since $R = 1\ \text{M}\Omega$ and $C = 1\ \mu\text{F}$, the time constant τ is 1 second. Now using Equation (4.34) and knowing that $V_{initial} = 0$ V, $V_{final} = 1$ V we obtain

$$v_o(t) = 1 - e^{-(t-1)/\tau}. \tag{4.35}$$

Equation (4.35) is plotted in Figure 4.5 from time 0 to 10 seconds. Note that at time $t = 2$ seconds the output voltage $v_o(t)$ has risen to

$$v_o(2) = 1 - e^{-(2-1)/\tau} = 0.6321. \tag{4.36}$$

Observing the waveform $v_o(t)$ in Figure 4.5, Equation (4.36) tells us that after one time constant the output reaches approximately 63% of its final value. After five time constants the output voltage across the capacitor reaches approximately 99% of its final value. Figure 4.6 displays several step input responses to circuits that have a range of time constants from 0.1, 0.5, 1, 2, 5, and 10 seconds. The shorter the time constant of the circuit, the faster the output voltage will approach its final value. So for our Figure 4.6, the circuit with $\tau = 0.1$ second has the fastest response of all the waveforms displayed. On the other hand, the circuit with $\tau = 10$ seconds has the slowest response. Notice that this waveform (10-second time constant) just reaches 63% of its final value after *one* time constant. The trajectory of the output waveform for a 10-second time constant is only shown for two time constants (20 seconds). If we had plotted the Figure 4.6 up to $t = 50$ seconds, the waveform would have reached 99% of 1 V.

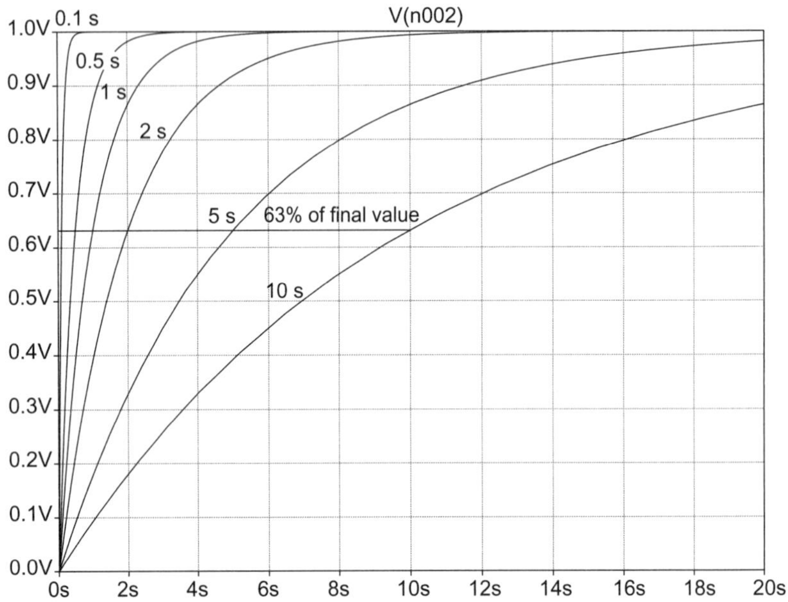

Figure 4.6 Unit step responses of *RC* LPF of time constants 0.1, 0.5, 1, 2, 5, and 10 seconds.

4.2.8 First-order *RC* LPF under Pulse and Square-Wave Excitation

We defined a unit step excitation $u(t)$ using Equation (4.29) in the previous section.

Let us combine to unit step functions such that the first one $u(t)$ is added, to a second $u(t)$ that is delayed by t_p seconds and inverted, so that:

$$u(\mathrm{t}) - u(t - t_p). \qquad (4.37)$$

Equation (4.37) is the expression of a single pulse of width t_p.

Applying such pulse to an *RC* LPF and its responses are shown in Figure 4.7 for a number of time constants. Note that for small time constants like 0.01 second and 0.05 second the output voltage waveform resembles the input more closely that the larger time constant curves. As the circuit time constant increases, the output waveform looks less exponential and more linear (2, 5, and 10-second time constants).

Now let us consider a square-wave input, as the one shown in Figure 4.8. Such waveform is a continuous train of pulses that swings between 0 V and 1 V with a 50% duty cycle. The waveform starts at 1 V at zero time for 1 second, at this time it drops very quickly to 0 V for another second. After this last second at 0 V, the earlier described process repeats itself indefinitely. Note that the period T of this pulse train is 2 seconds.

Let us apply such excitation to the input of a first-order *RC* LPF. We will look at the responses of several *RC* LPF with time constants equal to 0.01,

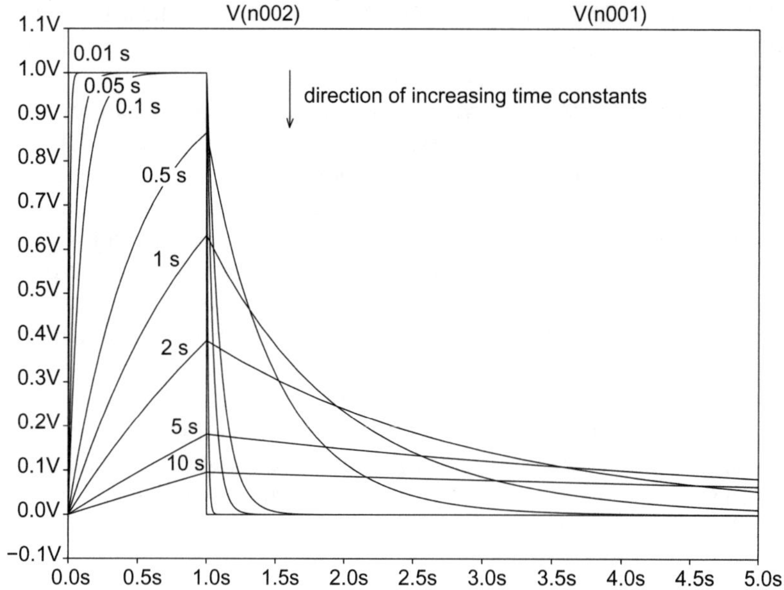

Figure 4.7 First-order *RC* LPF and pulse responses for various time constants.

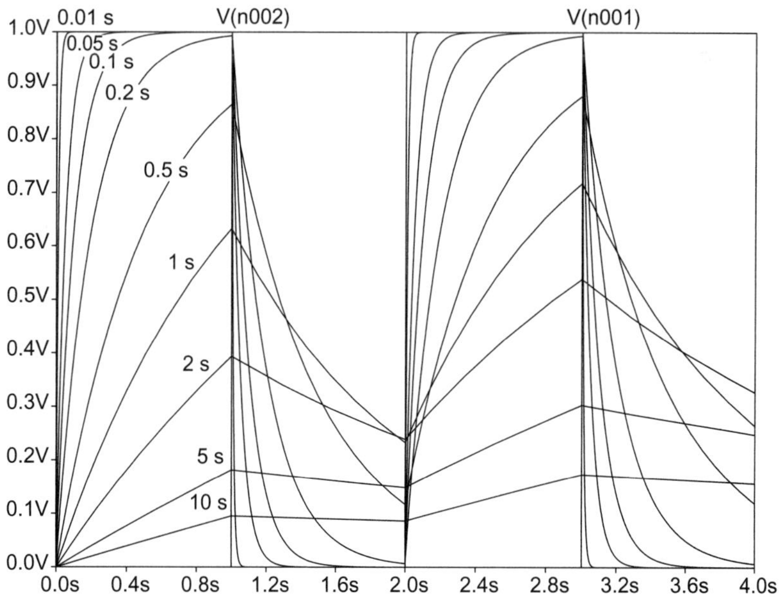

Figure 4.8 The outputs of all nine different time constants *RC* LPF circuits.

0.05, 0.1, 0.2, 0.5, 1, 2, 5, and 10 seconds. Note that the smallest time constant is 1/200th of the 2-second period of the excitation. The longest time constant is five times the period of the excitation.

Figure 4.8 shows 50% duty cycle square-wave driving *RC* LPF of nine different time constants, for two full excitation periods (i.e., 4 seconds).

The shorter is the time constant with respect to the excitation period, the faster the output of each *RC* circuit follows the 50% duty cycle square-wave input. For example, referring to Figure 4.8, the output of the circuit whose time constant is 0.01 seconds closely follows the square-wave input, some rounding is seen at the end of the rising and falling edges of the output response. Additionally, this behavior of reaching fairly quickly a steady state is reached virtually from the first excitation period. Now let us concentrate on the slowest time constant circuit of 10 seconds. Note that because this time constant is actually larger than the excitation period, it takes some time for the output of the 10-second time constant circuit to reach a steady-state value. Within this context, a steady-state value refers to the waveform moving up and down with time such that its average value settles down to a constant value, and it does not change significantly anymore. Figure 4.9 depicts the similar waveforms of those of Figure 4.8 display but for a much longer period of time, that is, 30 seconds. Note that Figure 4.9 only displays responses for nine different circuit of time constants equal to: 0.01, 0.05, 0.1, 0.2, 0.5, 1, 2, 5, and 10 seconds. Carefully observing the 10-second time constant circuit response, we see that during the first 20 seconds starting at zero time, the output little by little rises as

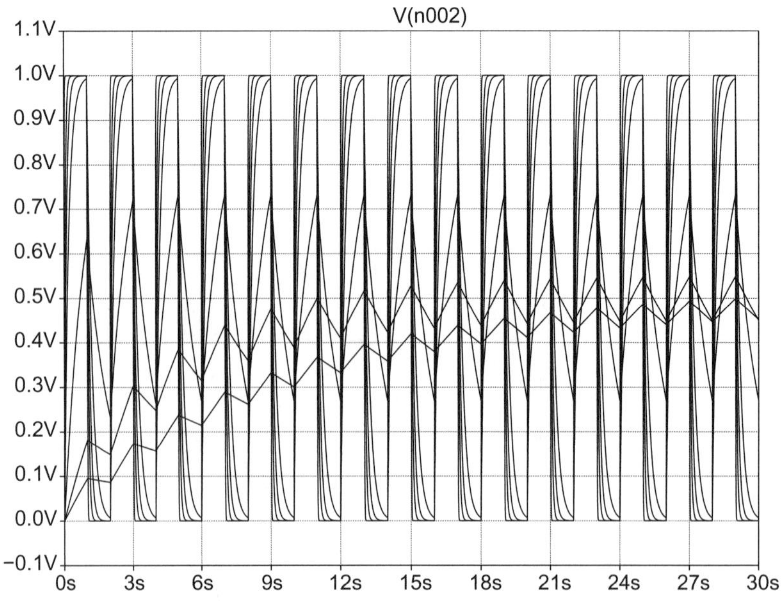

Figure 4.9 Square-wave responses of *RC* LPF of various time constants.

seconds go by. Then somewhere in the neighborhood of 30 seconds, the output settles around the mean value of the square-wave input. Since our input has a 50% duty cycle and swings between 0 and 1 V, its mean value is exactly 0.5 V. The average value to which the output voltage will settle is 0.5 V. Figure 4.9 does not quite show when the 10 seconds time constant circuit settles to 0.5 V because a few more seconds should have been plotted. Because the drawing becomes too busy and for a longer time, the author determined that 30-second was a better time frame to display. Looking at Figure 4.9 once more, the second slowest 5-second time constant response, second waveform from the time axis, more clearly reaches 0.5 V. On the other hand, note that those responses whose circuits have very fast time constant relative to the period of the square-wave excitation, simply follow relatively closely their input. Such responses will never settle to an average value of 0.5 V of the input waveform.

4.2.9 The *RC* LPF as an Integrator

When a first-order *RC* LPF has a large time constant in comparison with the time that it takes for the input signal to make an appreciable change, the voltage drop across the output capacitor is small compared to the drop across the resistor. Referring again to Figure 4.2, the current through the circuit is

$$i(t) = CdV_o(t)/dt \tag{4.38}$$

and since V_o is small compared to the voltage across resistor R then,

$$i(t) = V_{in}/R. \tag{4.39}$$

Combining Equations (4.38) and (4.39) we obtain

$$CdV_o(t)/dt = V_{in}/R, \tag{4.40}$$

which, after some algebraic manipulation and integration on both sides of the equal sign, it becomes

$$V_o = 1/RC\int V_{in}dt. \tag{4.41}$$

Equation (4.41) states that the output voltage of our *RC* LPF circuit is proportional to the integral of the input voltage. The constant of proportionality is $1/RC$.

Referring again to the responses of Figures 4.8 and 4.9, it is clear to note that the shorter the time constant of the circuit with respect to the period of the square-wave excitation, the output signal tends to follow the input waveform. This is noted for time constants of 0.01, 0.05, and 0.2 seconds. For longer time constants such as 5 and 10-second, the circuit behaves more like an integrator. Notice that the integral of a (constant) horizontal line is a ramp. Indeed waveforms responses for 5- and 10-second time constants like fairly linear ramps and not so much exponential as described by Equation (4.34).

Summary of Important Points about RC LPFs in the Frequency Domain and Integrators in the Time Domain

A first-order RC LPF circuit allows sinusoidal frequencies smaller than one order of magnitude of its cutoff frequency to go through the circuit with little attenuation and no significant change in phase with respect to the input sinusoidal.

Sinusoidal frequencies of one order of magnitude higher that the cutoff frequency of the circuit become attenuated by 20 dB.

Sinusoidal frequencies of at least one order of magnitude higher that the cutoff frequency of the circuit or higher, approach a –90-degree phase shift with respect to the sinusoidal input.

The same first-order RC circuit performs time integration of the signals that are at least one order of magnitude higher in frequency than the filter cutoff frequency.

A practical limitation of the integrator implemented with a first-order RC LPF circuit is that the integrated output signal is attenuated, while other lower frequency signals below the cutoff frequency pass through the filter practically unaltered. We will see how to overcome these problems using an operational amplifier in Chapter 5.

4.3 THE FIRST-ORDER *RC* HIGH-PASS FILTER (HPF)

Let us investigate the *RC* circuit of Figure 4.10. This circuit is excited by a sinusoidal voltage waveform. Elements *R* and *C* are in series, the input voltage

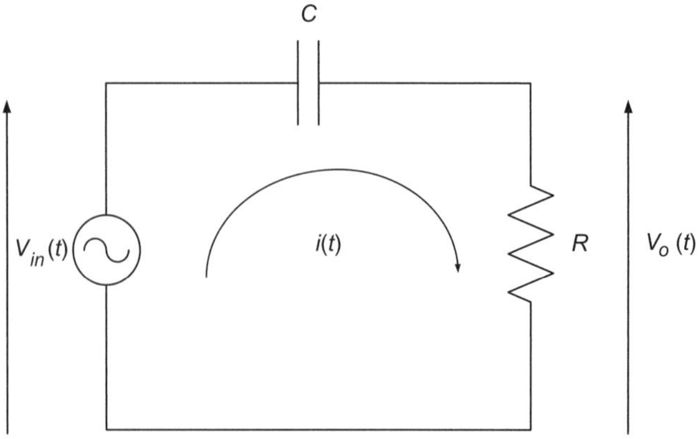

Figure 4.10 First-order *RC* high-pass filter (HPF).

is applied to the two elements in series, the output is taken across the resistor terminals.

4.3.1 *RC* HPF Frequency Domain Analysis

Let us calculate the transfer function of this circuit:

$$\mathrm{H}(j\omega) = \mathbf{V}_{out}(j\omega)/\mathbf{V}_{in}(j\omega). \tag{4.42}$$

The impedance of the resistor and capacitor in series is

$$Z_{series}(j\omega) = R + \frac{1}{j\omega C}. \tag{4.43}$$

Then, the current through the circuit is

$$\mathbf{V}_{in}(j\omega)/Z_{series}(j\omega). \tag{4.44}$$

The above current times the resistance equals the output voltage $\boldsymbol{V}_{out}(j\omega)$, thus we get

$$\mathrm{H}(j\omega) = \frac{V_{out}(j\omega)}{V_{in}(j\omega)} = \frac{R}{R + 1/j\omega C} = \tag{4.45}$$

After some algebraic manipulations,

$$= \frac{j\omega RC}{1 + j\omega RC}. \tag{4.46}$$

Remember that from Chapter 2, ω is the angular frequency, which equals $2\pi f$, where f is the frequency of the sinusoidal waveform in hertz. We also define the angular cutoff frequency:

$$\omega_0 = 1/RC, \tag{4.47}$$

and the *RC* HPF cutoff frequency,

$$f_0 = 1/2\pi RC. \tag{4.48}$$

Equation (4.46) describes the ratio of output voltage to input voltage of the *RC* circuit given by Figure 4.10, again referred to as the circuit transfer function in the frequency domain. Using the definition for f_0 by Equation (4.48), we can rewrite the transfer function of the circuit as follows:

$$H(j\omega) = \frac{j\omega/\omega_0}{1 + j\omega/\omega_0} \tag{4.49}$$

$$|H(j\omega)| = \frac{\omega/\omega_0}{\sqrt{1+(\omega/\omega_0)^2}} \tag{4.50}$$

$$\angle H(j\omega) = \pi/2 - arctan(\omega/\omega_0), \tag{4.51}$$

where Equation (4.50) is the linear magnitude of Equation (4.49). Equation (4.51) is the phase in radians of Equation (4.49).

So let us now construct a table to plot the values of the transfer function $H(j\omega)$ for a given *RC* HPF with a cutoff frequency f_0. Assuming a 1 kHz cutoff frequency f_0,

$$RC = 1/2\pi 1000 = 159.155\ \mu s. \tag{4.52}$$

We will plot the transfer function of Equation (4.49) using two separate plots. One plot is for its magnitude and the second plot for its phase, both as functions of frequency. Note that transfer function Equation (4.49) is a complex quantity and as such, it has magnitude and phase. Repeating the procedure that we previously used for the *RC* LPF we will tabulate the magnitude and phase values for the *RC* HPF. We show the outcome of such calculations in Table 4.4.

The gain and phase values of Table 4.4 have been calculated finding the magnitude and phase of the complex expression given by Equation (4.49).

Figure 4.11 depicts the exact magnitude and phase Bode plots for the *RC* HPF, generated using the values of Table 4.4.

4.3.2 Drawing an *RC* HPF Bode Plot Using Just the Asymptotes

To draw the asymptotes of the Bode plots of our first-order *RC* HPF (Fig. 4.10), we will normalize the frequency axis. Instead of listing on the frequency axis the actual cutoff frequency value of 1 kHz, we will denote this frequency

Table 4.4 ***RC*** **HPF transfer function: magnitude and phase as functions of frequency**

Frequency (Hz)	Normalized Frequency f/f_0	Linear Gain (V/V)	Gain in dB	Phase (Degrees)
1.000	0.001 f_0	0.001000253	–60.00	89.94270424
10.000	0.01 f_0	0.010002036	–40.00	89.42706130
100.000	0.1 f_0	0.099528982	–20.04	84.28940687
1,000.000	1 f_0	0.707296534	–3.01	45.00000002
10,000.000	10 f_0	0.995318596	–0.04	5.710593139
100,000.000	100 f_0	1.000233088	0.00	0.572938695
1,000,000.000	1,000 f_0	1.000282601	0.00	0.057295758

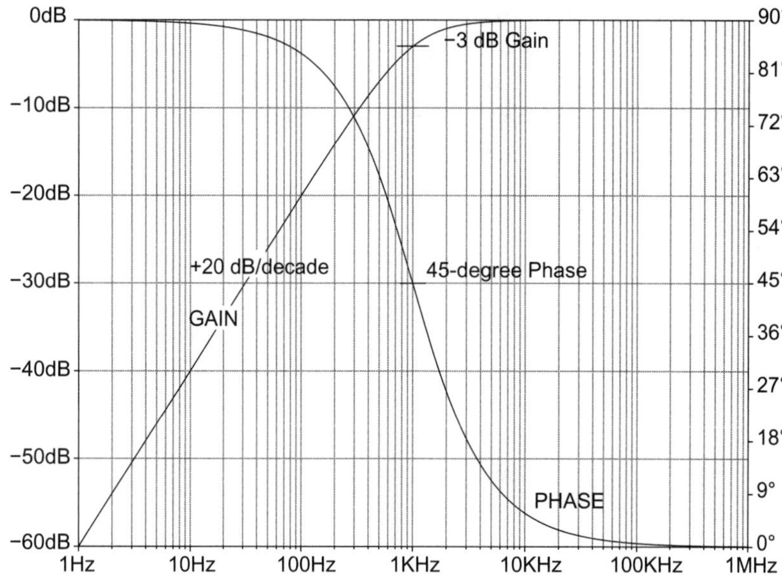

Figure 4.11 First-order *RC* HPF exact Bode plots.

as f_0, so lower frequencies will be a tenth, a one-hundredth, and so on of f_0. Similarly, frequencies above f_0 are 10 times, 100 times, and so on of f_0. Figure 4.11 shows these frequencies on the horizontal axis.

From Table 4.4 we see that the magnitude asymptotically approaches 0 dB from the cutoff frequency f_0 to higher frequencies. Also from the cutoff frequency to lower frequencies, the gain drops at a constant rate of 20 dB per decade. But at the cutoff frequency the gain is approximately –3 dB. In linear terms, this means that the amplitude of the sinusoidal waveform of frequency equal to f_0, the cutoff frequency of our circuit that excites the *RC* circuit, becomes attenuated down to about 70.7% from its original value. Referring one more time to Table 4.4 we can see that at the cutoff frequency f_0, the output voltage magnitude is 0.707 of the original input, which has a magnitude of 1. That is, the output magnitude is approximately 70.7% of the input magnitude.

In a similar fashion we can see that the phase is +45° at f_0. One more time looking at the phase in Table 4.4, we see that the phase at one-tenth of f_0 is about +84.3°. And for even lower frequencies, the phase asymptotically approaches +90°. On the other hand, at a frequency 10 times f_0, the phase is approximately 5.7° *more* than the high frequency value of the phase, which is 0°. For all frequencies higher than about 10 times f_0, the phase of the HPF is approximately 0 degrees (Figure 4.11).

We can also draw the HPF phase Bode plot using the phase asymptotes. Let us start with frequencies well below f_0, up to one-tenth of f_0, we draw a straight line at +90 degrees from low frequencies all the way up to 1/10 of f_0. At a frequency of 10 times f_0, the phase asymptote is a horizontal line at 0 degrees, continuing into higher frequencies. From Table 4.4, we know that at the cutoff frequency f_0, the phase is +45°. Looking at Figure 4.12, we now draw a straight line of a phase angle of 90° at 1/10 of f_0 all the way to 0° at 10 f_0,

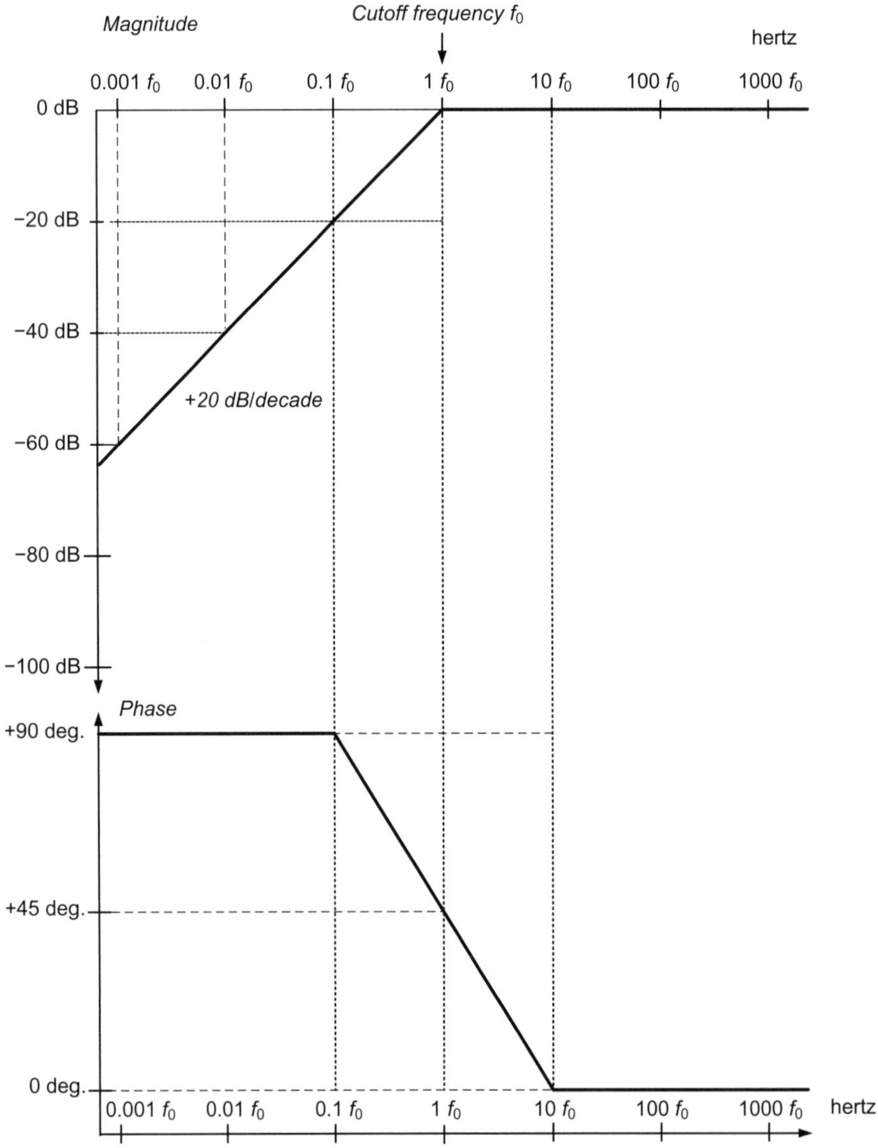

Figure 4.12 First-order *RC* HPF Bode plots asymptotes: magnitude and phase.

such that this straight line passes through a +45° phase at f_0. In this way, the phase asymptotes are drawn. Now one can draw by hand the approximate phase curves. Note that at 1/10 of f_0 and at 10 f_0, the phase is about 5.7° below +90° and about 5.7° above 0°, respectively. Finally, it is important to note that the phase curve has an inflexion point at 45°, see Figure 4.12. Figure 4.12 depicts the asymptotes of the magnitude and phase Bode plots for a first-order *RC* HPF.

Important Points about the First-Order RC HPF

For a first-order RC HPF the circuit gain at the cutoff frequency f_0 *is –3.01 dB.*

For a first-order RC HPF the gain is practically 0 dB at 10 f_0 *all the way to larger frequencies.*

From f_0 *down in frequency the gain starts decreasing at a 20 dB per decade rate. So that at 0.1* f_0 *the gain is 20 dB below 0 dB. At 0.01* f_0 *the gain is another 20 dB below, or 40 dB below the 0 dB gain line. At 0.001* f_0 *the gain is another 20 dB below the preceding decibel at the previous decade in frequency or 60 dB below the 0 dB gain line. This gain behavior continues to drop 20 dB as the frequency decreases by an order of decimal magnitude.*

The frequency axis does not have a zero or origin of frequencies because log of zero is nonexistent. The lowest frequency can be represented with a value as small as we desire, but not with zero.

Frequency is represented logarithmically for both magnitude and phase plots.

The magnitude or gain in dB is represented linearly. There is a 0 dB origin for the vertical axis because the scale is linear in dB.

The phase in degrees is represented linearly on the vertical axis.

The phase of an RC HPF is approximately +90° at frequencies below 1/10 of f_0*. The phase of an RC HPF is 0° at frequencies larger than 10* f_0*. At* f_0 *the phase equals +45°.*

4.3.3 Interpretation of the *RC* HPF Bode Plots in the Time Domain

We will use the asymptotically drawn Bode plots to explain the meaning of the Bode magnitude and phase plots in terms of sinusoidal inputs applied to the first-order *RC* HPF circuit (Fig. 4.12). We will explain this section at a faster pace because of the similarity that exists with first-order *RC* LPF, Section 4.2.

First let us assume that a sinusoidal waveform of 1 V peak amplitude and a frequency of 0.01 f_0 is applied to the input of the first-order *RC* HPF. The output voltage waveform that will be observed across the resistor terminals is 40 dB (a factor of 100) smaller than the 1 V input. At 0.1 f_0, the output waveform is 20 dB (factor of 10) smaller than the 1 V input. At the cutoff frequency f_0, the output magnitude is 3 dB below the 1 V input, meaning that the output is 70.7% of 1 V.

For the phase of the *RC* HPF, there is 90° phase shift for frequencies below 1/10th of f_0. A phase of +45° exists at the cutoff frequency. Finally, the phase becomes close to 0° (actually 5.7°) for 10 times f_0 and practically 0° at 100 f_0 frequencies and above.

4.3.4 Why Do We Call This Circuit an HPF?

From the Bode plots just presented in Figures 4.11 and 4.12 it is clear to see that frequencies below 1/10th of the cutoff frequency f_0 get attenuated. Frequencies above 10 times the cutoff frequency pass through the circuit with little or no attenuation. In summary, the *RC* circuit just analyzed allows high frequency signals to go through the circuit without attenuation, whereas the low frequencies become progressively attenuated as the input signal frequency goes below 1/10th of f_0. In summary, our first-order *RC* HPF greatly attenuates low frequencies and passes high frequencies without any significant attenuation. As usual, low frequencies are those that are smaller than 1/10th of f_0, and high frequencies are those that are larger than 10 times f_0. It is also interesting to notice that at frequencies equal to 10 f_0 and above, the region of frequency at which the gain is 0 dB, the phase shift is also 0°. The range of frequencies starting at 10 f_0 and going to larger frequencies is the pass-band frequency range of the filter. Within such range, signals pass through the filter unaltered in magnitude and in phase.

4.3.5 Time Domain Analysis of the *RC* HPF

Now let us analyze the time domain equations of the high-pass *RC* circuit. Referring one more time to the circuit of Figure 4.10, it is possible to establish the differential equation for such circuit. Simply applying KVL for the series of elements,

$$v_{in}(t) = v_{cap}(t) + i(t)R, \tag{4.53}$$

where $v_{in}(t)$ is the excitation or the circuit input voltage, $i(t)R$ equals the output voltage $v_o(t)$ and $v_{cap}(t)$ is the voltage across the capacitor of our *RC* HPF.

From Figure 4.10 we see that

$$v_{cap}(t) = v_{in}(t) - v_o(t). \tag{4.54}$$

Also,

$$i(t) = C\frac{d[v_{in}(t) - v_o(t)]}{dt} \tag{4.55}$$

because $[v_{in}(t) - v_o(t)]$ is the voltage across the capacitor.

Since the current times the resistor R is the output voltage $v_o(t)$, then,

$$v_o(t) = RCdv_{in}(t)/dt - RCdv_o(t)/dt. \tag{4.56}$$

Rearranging terms, Equation (4.56) becomes

$$\frac{dv_0(t)}{dt} + \frac{1}{RC}v_o(t) = \frac{dv_{in}(t)}{dt}, \tag{4.57}$$

where Equation (4.57) is a first-order differential equation. When input $v_{in}(t)$ is a step function $u(t)$, the solution is given by Equation (4.34), repeated here for the reader's convenience.

$$v_o(t) = V_{final} + (V_{initial} - V_{final})e^{(-t/\tau)}. \tag{4.58}$$

In particular for our *RC* HPF, Figure 4.10, we calculate the values of the initial and final values from circuit boundary considerations.

The initial value of the output waveform is 1 V, the magnitude of our step input excitation function $u(t)$. Why? Because upon impressing the 1-V pulse at the input of the circuit, assuming that the capacitor is initially discharged, the capacitor behaves like a short circuit to the 1-V edge. The final value of the output voltage after the transient behavior of the output is 0 V. Note that the capacitor has a *blocking* effect to the DC value of the step input. The output waveform will have no average value.

Now using Equation (4.54) and knowing that $V_{initial} = 1$ V, $V_{final} = 0$ V we obtain

$$v_o(t) = e^{-\frac{t}{\tau}}. \tag{4.59}$$

Equation (4.59) is plotted in Figure 4.13 with six different time constant values: 0.1, 0.5, 1, 2, 5, and 10 seconds from time 0 to 20 seconds. By observation of the response curve for time constant 10, note that its value is down to 36.8% from its original value of 1 V after 10 seconds from the origin of time.

Note that an *RC* HPF with a time constant of 10 seconds, at time $t = 10$ seconds, the output voltage $v_o(t)$ decays from its initial value of 1 V to

$$v_o(10) = e^{-\frac{10}{10}} = 0.368 \text{ V}. \tag{4.60}$$

The same is true for all other waveforms, as an example the response curve for 1-second time constant is down to 36.8% of its original value of 1 V after 1 second.

Back to Equation (4.58), we can verify that after five time constants, the response of the *RC* HPF will be down from its initial value of 1 V to 1% of 1 V.

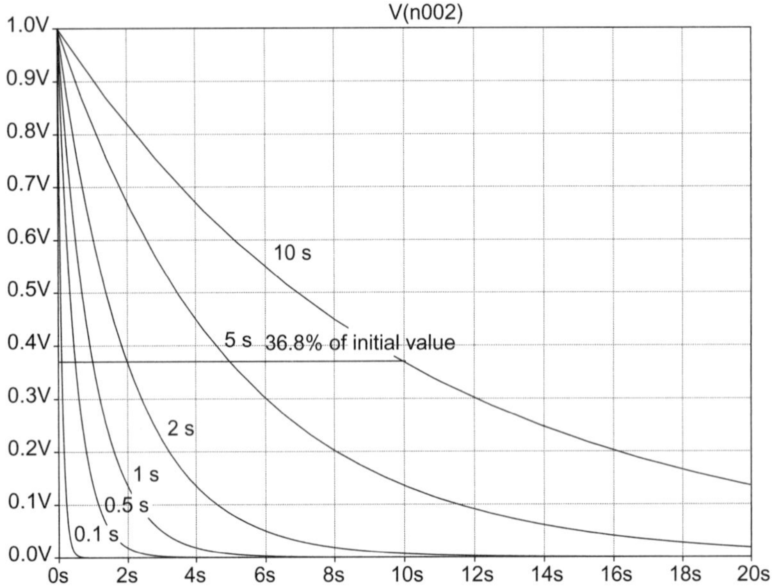

Figure 4.13 Unit step response of an *RC* HPF for six different time constants.

4.3.6 First-Order *RC* LPF under Pulse and Square-Wave Excitation

Let us apply a pulse of unity magnitude and 1-second duration to a first-order *RC* HPF with a 1-second time constant. The excitation and the corresponding response can be seen in Figure 4.14. The positive portion of the response is not new to us, since this is what we previously obtained in Figure 4.13 upon applying a unit-step. The difference in this example is that we are not applying a step, but a pulse. The 1-second pulse shown on the top of Figure 4.14 can be thought as the sum of a unit step at the origin, plus a 1-second delay inverted unit step. The equation for such step follows:

$$u(t) - u(t-1). \tag{4.61}$$

The positive portion of the response to such unit pulse is shown on the bottom section of Figure 4.14, and it is very much what we obtained for a unit step. This positive portion of the response is exponential and follows Equation (4.58). The difference in this case is that we are applying a pulse. The pulse cuts short the step at 1 second. So at such time, a negative 1-V step is applied to the to the circuit input. Note that the response of the circuit due to this negative 1-V step applied at *time* = 1 second, will also be exponential but will start at $t = 1$ second and 1 volt below the voltage of magnitude V_p in the first exponential in Figure 4.14. From that point on, the pulse over the exponential response from the negative portion continues to decay

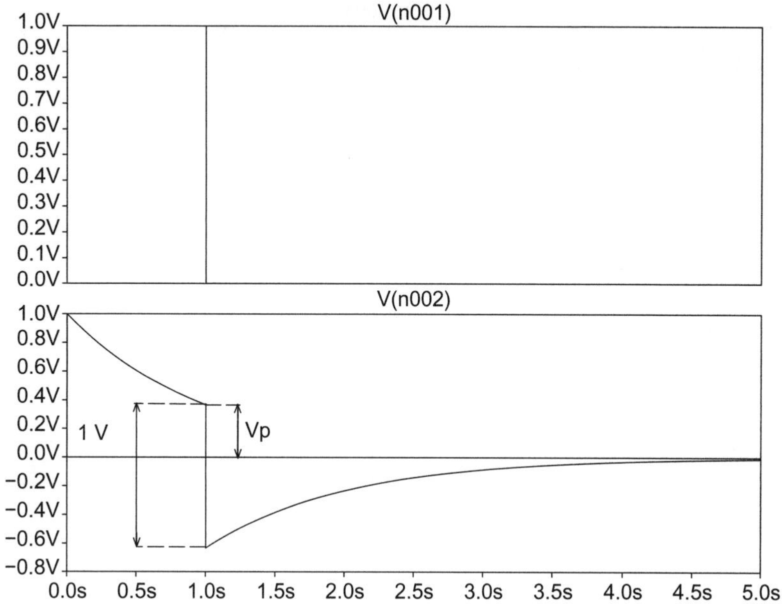

Figure 4.14 First-order *RC* HPF input pulse response.

exponentially (actually to increase exponentially) toward 0 V. After several time constants, the response reaches a zero value. Let us ask ourselves, what is the voltage V_p at time $t = 1$ second? Using Equation (4.58) and knowing that the circuit time constant is 1 second, we obtain

$$V_p = e^{-1/1} = +0.368 \text{ V}. \tag{4.62}$$

The voltage at which the negative portion of the exponential response begins at $t = 1$ second is:

$$0.368 - 1 = -0.632 \text{ V}. \tag{4.63}$$

It is important to observe that the average value of the complete response to the pulse, that is, the positive and the negative exponentials, have an average value of 0 V. In other words, the response has no *DC* component. Another way of saying this is that the area under the positive exponential equals the area above the negative exponential with respect to the time axis in both cases.

Now let us consider a square-wave input, as the one shown in Figure 4.15. Such waveform is a continuous train of pulses that swings between 0 V and 1 V with a 50% duty cycle. The waveform starts at 1 V at zero time for 1 second, at this time it drops very quickly to 0 V for another second. After this last second at 0 V, the earlier described process repeats itself indefinitely. Note that the period T of this square wave is 2 seconds.

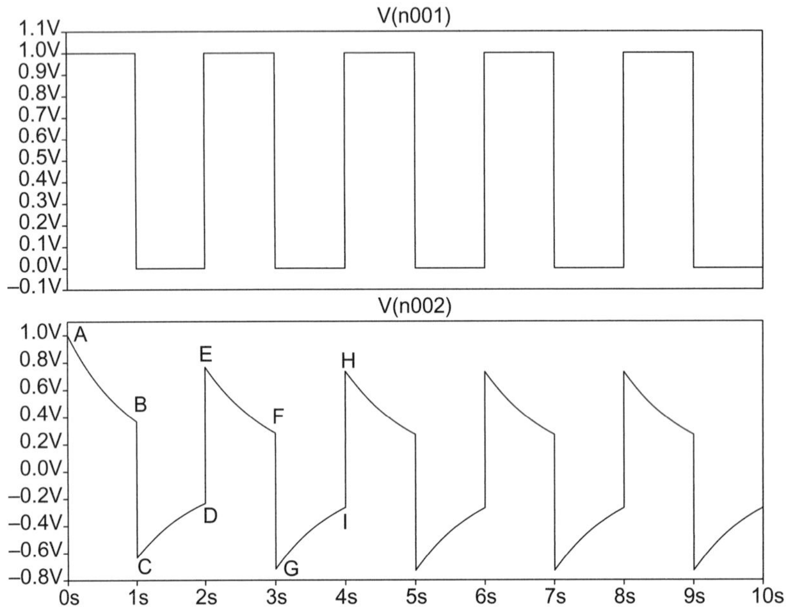

Figure 4.15 50% duty cycle square-wave driving *RC* HPF.

Let us apply such excitation to the input of a 1-second time constant first-order *RC* HPF. Note that during the first period of the square-wave input the response of the HPF settles down to a periodic response. The waveform indicated by points A, B, C, and D is a transient waveform; the second portion of the response, D, E, F, G, H, and I becomes the waveform that will be repeated over and over as long as the excitation is applied to the input. In our next example, Figure 4.16, the excitation input is the same as in the previous example, 0 V to 1 V swing, 50% duty cycle square wave, period $T = 2$ seconds. But now the time constant of the HPF is very small compared to the period of the excitation, that is, τ is 0.1 seconds. The RC circuit is much faster that the period of the excitation; this is the cause why the response attains steady-state value within the first period of the excitation.

Our next and final example of an *RC* HPF response to a square-wave input is applied to a circuit with a 100-second time constant. This is slower than the 2-second period of the excitation. Referring to Figure 4.17, it is clear to see that it takes in the order of 300 seconds (or three time constants) for the response to attain its steady state.

It is important and interesting to observe from Figures 4.15–4.17 that regardless of the *RC* circuit time constant, once the response attains a steady state, the average value or DC component of the response is zero. Let us remember that this occurs because of the DC blocking capacitor in the circuit. That is to say the output waveform has a zero average or zero DC value after the output transient behavior is over.

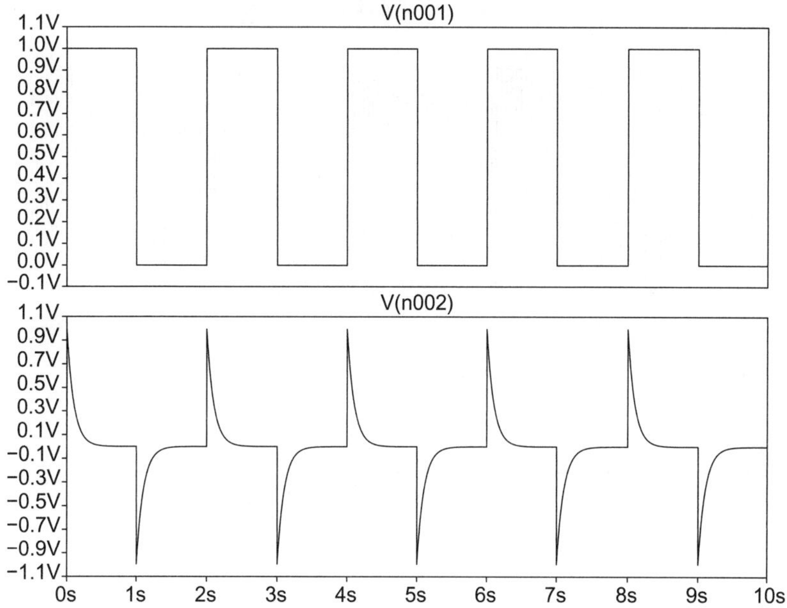

Figure 4.16 Square-wave excitation applied to a 0.1-second time-constant *RC* HPF.

4.3.7 The *RC* HPF as a Differentiator

From Equation (4.57), repeated here for the reader's convenience,

$$\frac{dv_0(t)}{dt} + \frac{1}{RC} v_o(t) = \frac{dv_{in}(t)}{dt}. \tag{4.64}$$

When the *RC* time constant and the output voltage are small, Equation (4.64) becomes

$$v_o(t) = RC \frac{dv_{in}(t)}{dt}. \tag{4.65}$$

Equation (4.65) shows that under the conditions previously stated, the output voltage is proportional to the derivative of the input voltage.

As an example of differentiation, let us look back at Figure 4.16, the *RC* HPF has a time constant of 0.1 second, which is smaller than the excitation 2-second period. Note that the circuit produces the derivative of the input waveform; the positive going transitions of the square-wave input become positive spikes, the negative going transitions become negative spikes upon the square wave becoming differentiated. Note that the constant levels of the square wave are zero, because the derivative of any constant is zero. As a counterexample of what is not a differentiator, refer this time to Figure 4.17,

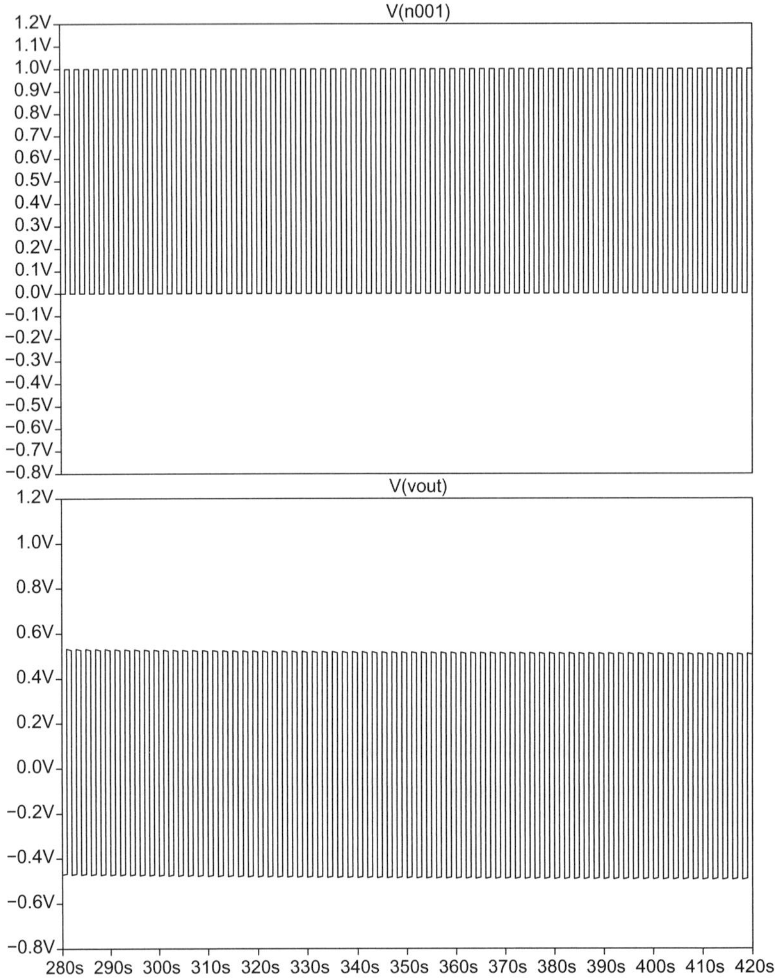

Figure 4.17 Second square-wave excitation applied to a 100-second time-constant *RC* HPF, in steady state.

the time constant of this *RC* HPF is 100 seconds (a large number), while the period of the excitation waveform is still 2 seconds as in Figures 4.15 and 4.16.

The plot shown presents the signals after its transient portion, in other words, in steady state condition. Note that the input square wave of 50% duty cycle, period of 2 seconds, swings between 0 V and 1 V. This input waveform contains a non-zero DC component of 0.5 V. The *RC* HPF allows the waveform to pass straight through with little attenuation, but notice that its DC component of 0.5 V has been removed by the filter. This is noticed by the fact that the output now swings between –0.5 V and +0.5 V, its peak to peak amplitude is still 1 V, no change with respect to the input. One more time referring to Figure 4.17, looking closely at the output waveform of the filter, we notice a slight slope on the top and the bottom of the output waveform positive and negative cycles. The reason for this is that the filter passes through high frequencies; however, it changes the phase of each frequency component by a positive 90° phase. Figure 4.17 displays the HPF response 280 seconds after the excitation was applied at the origin of time, 0 second.

Summary of Important Points about RC HPFs in the Frequency Domain and Differentiators in the Time Domain

A first-order RC HPF circuit allows sinusoidal waveforms of frequencies larger than one order of magnitude of its cutoff frequency to go through the circuit with little attenuation and with a 0° phase shift with respect to the sinusoidal input.

Sinusoidal waveforms whose frequencies are one order of magnitude lower than the cutoff frequency of the circuit are blocked by the RC HPF by being attenuated by 20 dB. Frequencies two orders of magnitude smaller than the cutoff are attenuated 40 dB. This goes on at a rate of 20 dB attenuation per decade. The phase of all frequencies at least one order of magnitude lower than f_0 *experience an approximate +90-degree phase shift.*

The same first-order RC circuit performs time differentiation of the signals that have frequencies at least one order of magnitude lower than than the filter cutoff frequency.

A practical limitation of the differentiator implemented with a first-order RC HPF circuit is that the differentiated output signal is attenuated, while other higher frequency signals above the cutoff frequency pass through the filter practically unaltered. We will see how to overcome these problems using an operational amplifier in Chapter 5.

4.4 SECOND-ORDER CIRCUITS

Second-order circuits are described by second-order ordinary differential equations with constant coefficients. Refer to Equation (4.66) to observe a second-order circuit differential equation:

$$a_0 \frac{d^2 f(t)}{dt^2} + a_1 \frac{df(t)}{dt} + a_2 f(t). \tag{4.66}$$

In Equation (4.66), $f(t)$ usually is $i(t)$ or $v(t)$, respectively current or voltage varying with respect to time. a_0, a_1, and a_2 are the constant coefficients, typically real numbers. t is time, the independent variable.

Equation (4.66) may be equated to zero or to a constant or to a function of time. Equation (4.66) equates the differential equation to zero, thus Equation (4.67):

$$a_0 \frac{d^2 f(t)}{dt^2} + a_1 \frac{df(t)}{dt} + a_2 f(t) = 0. \quad (4.67)$$

The differential equation of Equation (4.67) describes a second-order circuit without any external excitation. When Equation (4.66) is equated to a constant, it is usually when the second-order circuit is excited by a step. From now on, we will refer to a circuit described by a differential equation of the form given by Equation (4.66) simply as a second-order circuit. Second-order circuits have one inductor, one capacitor, and they may or may not have a resistor. When the second-order circuit does not have any resistors, it is said to be lossless.

4.5 SERIES RLC SECOND-ORDER CIRCUIT

We will analyze now a series *RLC* circuit, with a step input. Figure 4.18 depicts such a circuit.

From the circuit of Figure 4.18 we can derive the time domain equations. We obtain

$$L\frac{di(t)}{dt} + i(t)R + v_C = V_{step}. \quad (4.68)$$

In Equation (4.68), the first term on the left is the voltage drop on the inductor, the voltage drop on the resistor follows, and v_C is the drop across the capacitor. The differential equation, that is, Equation (4.68) is equated to V_{step}, assumes that a step input is applied to the circuit at time $t = 0$.

Since

$$i_C = C\frac{dv_C(t)}{dt}, \quad (4.69)$$

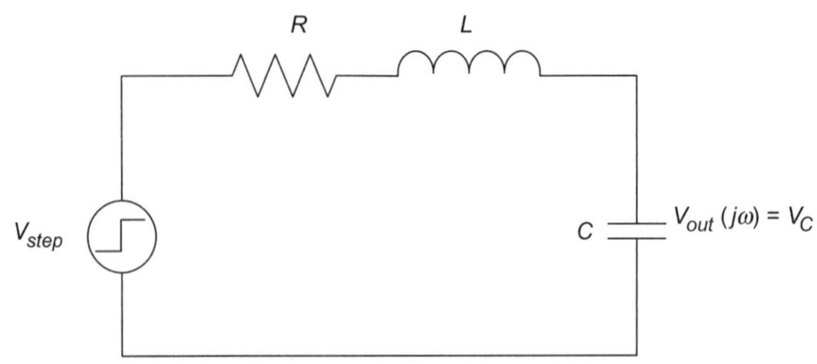

Figure 4.18 RLC series circuit with a step input excitation.

Then

$$v_C = \frac{1}{C}\int i_C(t)dt = \frac{1}{C}\int i(t)dt, \tag{4.70}$$

because $i_C(t)$ equals $i(t)$. We are analyzing a series circuit so the current through any one of its elements is the same current in the circuit.

We plug Equation (4.70) into Equation (4.68):

$$L\frac{di(t)}{dt} + i(t)R + \frac{1}{C}\int i(t)dt = V_{step}. \tag{4.71}$$

Differentiating Equation (4.71) and rearranging terms, we obtain

$$\frac{d^2i(t)}{dt^2} + \frac{R}{L}\frac{di(t)}{dt} + \frac{1}{LC}i(t) = 0. \tag{4.72}$$

In Equation (4.72) $i(t)$ is the current through the capacitor which is the same as the current through the series circuit. When we solve, or find the solutions for differential Equation (4.72), we are finding the value of i as a function of time.

To find the solution of differential Equation (4.72), we will always end up with solutions that are functions of the following form:

$$i_1(t) = k_1e^{s_1t}. \tag{4.73}$$

$$i_2(t) = k_2e^{s_2t}. \tag{4.74}$$

Note that two solutions are found because it is a second-order system.

Equations (4.73) and (4.74) are solutions of Equation (4.72), and this means that if we plug each of the solutions into the differential equation, the solution will satisfy the mathematical operations of differential Equation (4.72). In Equations (4.73) and (4.74), k_1, s_1, k_2, and s_2 are constants, which can be real, imaginary, or complex. The differential equation solutions will determine three classic behaviors of second-order systems. These are

1. *Overdamped,*
2. *Critically damped, and*
3. *Underdamped.*

The reader is encouraged to plug Equation (4.73) into Equation (4.72) and validate the equation; similarly with Equation (4.74). So let us now plug a generic solution of the form of Equation (4.73) to our differential Equation (4.72):

$$\frac{d^2(ke^{st})}{dt^2}+\frac{R}{L}\frac{d(ke^{st})}{dt}+\frac{ke^{st}}{LC}=0. \tag{4.75}$$

Computing the derivatives of Equation (4.75) we obtain

$$s^2ke^{st}+\frac{R}{L}ske^{st}+\frac{1}{LC}ke^{st}=0. \tag{4.76}$$

Since e^{st} can never be zero for any finite time t, we can eliminate the instances of ke^{st} from Equation (4.76) and obtain

$$s^2+\frac{R}{L}s+\frac{1}{LC}=0. \tag{4.77}$$

Equation (4.77) is called the *characteristic equation* of our differential equation.

Now, finding the roots for Equation (4.77) yields

$$s_1, s_2=-\frac{R}{2L}\pm\frac{1}{2}\sqrt{(R/L)^2-4\frac{1}{LC}}. \tag{4.78}$$

The roots of the characteristic equation are of three possible types:

1. *Both roots are real and different, or*
2. *Both roots are real and equal, or*
3. *Both roots are complex conjugates**

The solutions of differential Equation (4.72) have one of the following forms:

$i(t)=k_1e^{s_1t}+k_2e^{s_2t}$: both roots are real and different: overdamped (4.79)

$i(t)=(k_1+k_2t)e^{\alpha t}$: both roots are real and identical: critically damped (4.80)

$i(t)=(k_1\cos\omega t+k_2\sin\omega t)e^{-\alpha t}$: roots are complex conjugates: underdamped (4.81)

When the roots of the characteristic equation are complex conjugates, the roots have the following complex notation:

$$s_1, s_2=-\alpha\pm j\omega. \tag{4.82}$$

* There is a fourth case when the roots are complex conjugate but pure imaginary. However, this is a special case of Equation (4.81).

In Equations (4.80) and (4.81),

$$\alpha = -\zeta\omega_n, \tag{4.83}$$

where ξ is defined as the damping factor, and ω_n is the natural or undamped frequency.

ω in Equation (4.84) is called the damped frequency, equal to

$$\omega = \omega_n\sqrt{1-\zeta^2}. \tag{4.84}$$

Based on Equation (4.72), repeated here for the reader's convenience,

$$\frac{d^2i(t)}{dt^2} + \frac{R}{L}\frac{di(t)}{dt} + \frac{1}{LC}i(t) = 0. \tag{4.85}$$

We can find the relationship between the damping factor ξ, the damped frequency ω, and the undamped or natural frequency ω_n with the circuit R, L, and C components.

From Equation (4.72) R/L is defined as $2\xi\omega_n$. $1/LC$ is ω_n^2, that is, the square of the circuit natural frequency. The notation using ξ and ω_n is commonly used in control theory. Given those new defined parameters, we can rewrite Equation (4.77) as follows:

$$s^2 + \frac{R}{L}s + \frac{1}{LC} = s^2 + 2\zeta\omega_n s + \omega_n^2. \tag{4.86}$$

In reference to Equations (4.77) through (4.79), constants k_1 and k_2 are evaluated for a specific problem like ours, by the knowledge of the circuit *initial conditions*. Exponents s_1 and s_2 are the roots of the characteristic equation. Referring to Equation (4.80), α is the real part of the s_1 and s_2 roots of our system. And $\pm\, \omega$ is the imaginary part of the complex conjugate roots, also called the damped frequency. Referring once more to Figure 4.18 at time $t = 0$ when the voltage step is applied, the current in the circuit cannot change instantaneously, because the inductor is initially opposed to any current changes. Thus, $i(0+) = 0$. This means that the second and third voltage terms of Equation (4.68) are zero. The iR term is zero because $i(0+) = 0$ and v_C because the initial voltage across the capacitor is zero. Equation (4.68) is reduced to

$$\frac{di(0^+)}{dt} = \frac{V_{step}}{L}. \quad \text{ampère per second}$$

Example 4.4 Using the circuit of Figure 4.18, assume the following circuit components parameters:

$$R = 5\,\Omega,\ \mathrm{L} = 1\,\mathrm{H},\ \mathrm{C} = \frac{1}{6}\mathrm{F} \tag{4.87}$$

And a step input of 1 V at t_0+.

Note: The large values of inductance and capacitance are simply used to simplify the arithmetic of the problem.

Referring to Equation (4.72), repeated here for the reader's convenience, thus,

$$\frac{d^2i(t)}{dt^2}+\frac{R}{L}\frac{di(t)}{dt}+\frac{1}{LC}i(t)=0. \tag{4.88}$$

Using the values provided by (4.87) in Equation (4.88) yields

$$\frac{d^2i_C(t)}{dt^2}+5\frac{di_C(t)}{dt}+6i_C=0. \tag{4.89}$$

The characteristic equation of Equation (4.89) is

$$s^2+5s+6=0. \tag{4.90}$$

The roots of Equation (4.84) are

$$s_1=-2;\ s_2=-3. \tag{4.91}$$

Thus, the solution of Equation (4.88) is of the form

$$i_C(t)=k_1e^{s_1t}+k_2e^{s_2t}. \tag{4.92}$$

Using the roots of the characteristic equation, Equation (4.92) becomes

$$i_C(t)=k_1e^{-2t}+k_2e^{-3t}. \tag{4.93}$$

Let us determine constants k_1 and k_2 based on the problem initial conditions.

For $t = 0$, Equation (4.93) becomes:

$$0=k_1+k_2. \tag{4.94}$$

Now taking the derivative of Equation (4.92) yields

$$\frac{di(t)}{dt}=-2k_1e^{-2t}-3k_2e^{-3t}. \tag{4.95}$$

$$\frac{di(t0^+)}{dt}=\frac{V}{L}=1. \tag{4.96}$$

Because the current in the series circuit is zero, the inductor current cannot instantaneously change at $t = t_{0+}$. The voltages across the capacitor and resistor are zero.

As previously explained, using Equation (4.95) with the numerical values on-hand we obtain

$$1 = -2k_1 - 3k_2. \tag{4.97}$$

Solving the system of simultaneous Equations (4.94) and (4.97) yields

$$k_1 = 1; k_2 = -1. \tag{4.98}$$

Using the values of k_1 and k_2 from Equations (4.98) and (4.93), we obtain the complete current response:

$$i_C(t) = e^{-2t} - e^{-3t} \quad \textit{<<< overdamped case} \tag{4.99}$$

This example had a characteristic equation with two real and distinct roots; this is an overdamped-type response. In the next example we will study the response of the same second-order *RLC* circuit but with characteristic equation roots that are real and both are identical to each other. Since Example 4.4 was covered in great detail, the next two examples will be dealt without that many steps.

Example 4.5 Using the circuit of Figure 4.18, assume the following circuit component parameters:

$$R = 4\,\Omega, L = 1\,\text{H}, C = \frac{1}{4}\,\text{F} \tag{4.100}$$

and a 1-V step input. Derive an equation for the transient response of the circuit current, *i*(*t*).

From Equation (4.88),

$$\frac{d^2i(t)}{dt^2} + \frac{R}{L}\frac{di(t)}{dt} + \frac{1}{LC}i(t) = 0. \tag{4.101}$$

Equation (4.101) holds because we are dealing with the same series *RLC* circuit. Using the value given by Equation (4.100), Equation (4.101) becomes

$$\frac{d^2i_C(t)}{dt^2} + 4\frac{di_C(t)}{dt} + 4i_C = 0. \tag{4.102}$$

From Equation (4.102) the characteristic function is

$$s^2 + 4s + 4 = 0. \tag{4.103}$$

The roots of Equation (4.103) are $s_1, s_2 = -2$; that is, -2 is a double root of characteristic Equation (4.103).

The solution will have the form of Equation (4.80), repeated below for the reader's convenience:

$$i_C(t) = (k_1 + k_2 t)e^{\alpha t}. \tag{4.104a}$$

Same as before for t_0^+, when the step is applied to the circuit, since the inductor will not allow an instantaneous current change, $i_C(t) = 0$ at the initial time t_{0+}.

Thus,

$$i_C(t_0^+) = 0 = k_1. \tag{4.104b}$$

Differentiating Equation (4.104a) after we substitute k_1 with 0, we obtain

$$\frac{di_C(t)}{dt} = \alpha k_2 t e^{\alpha t}. \tag{4.105}$$

Evaluating Equation (4.105) at time t_{0+}, yields

$$\frac{di_C(t_0^+)}{dt} = \alpha k_2 t e^{\alpha t} = 1. \tag{4.106}$$

Since $\alpha = -2$, then

$$k_2 = -\frac{1}{2}. \tag{4.107}$$

Using Equations (4.104) and (4.107), the solution is

$$i_C(t) = \frac{1}{2} t e^{-2t} \quad \text{<<< \textbf{\textit{critically damped case}}} \tag{4.108}$$

Example 4.6 This example will address the series *RLC* circuit, with a 1-V step input when the roots of the characteristic equation are complex conjugates: $R = 2\ \Omega$, $L = 1$ H, $C = 1/2$ F.

The second-order differential equation that describes such system is

$$\frac{d^2 i_C(t)}{dt} + 2\frac{di_C(t)}{dt} + 2i_C = 0. \tag{4.109}$$

The circuit characteristic equation is

$$s^2 + 2s + 2 = 0. \tag{4.110}$$

The roots of Equation (4.109) are

$$s_1, s_2 = -1 \pm j1. \tag{4.111}$$

The general solution of Equation (4.109) is of the form

$$\begin{aligned} i_C(t) &= k_1 e^{(-1+j1)t} + k_2 e^{(-1-j1)t} = \\ &= e^{-t}(k_1 e^{jt} + k_2 e^{-jt}) \end{aligned} \tag{4.112}$$

which is also of the general form previously shown by Equation (4.81):

$$i_C(t) = e^{-t}(k_3 \cos t + k_4 \sin t). \tag{4.113}$$

Note: The mathematical equivalence between Equations (4.112) and (4.113) is justified with Euler's identity; that is,

$$e^{\pm j\omega t} = \cos \omega t \pm j \sin \omega t. \tag{4.114}$$

Repeating Equation (4.81) for the reader's convenience,

$$i_C(t) = (k_1 \cos \omega t + k_2 \sin \omega t)e^{-\alpha t} \tag{4.115}$$

and now we equate Equations (4.113) and (4.115):

$$e^{-t}(k_3 \cos t + k_4 \sin t) = e^{-t}(k_1 e^{jt} + k_2 e^{-jt}). \tag{4.116}$$

From Euler's Equation (4.114) it can be shown that

$$\cos \omega t = \frac{e^{j\omega t} - e^{-j\omega t}}{2} \tag{4.117}$$

and

$$\sin \omega t = \frac{e^{j\omega t} + e^{-j\omega t}}{2j}. \tag{4.118}$$

Expanding the right-hand side term of Equation (4.116) using Euler's identities we obtain

$$e^{-t}(k_3 \cos t + k_4 \sin t) = e^{-t}[k_1(\cos t - j \sin t) + k_2(\cos t - j \sin t)]. \tag{4.119}$$

Rearranging terms on the right-hand side of Equation (4.119) and comparing them against the left-hand side of Equation (4.119) we obtain that

$$k_3 = k_1 + k_2 \quad \text{and} \quad k_4 = j(k_1 - k_2). \tag{4.120}$$

The initial conditions for this problem are exactly the same as what they were for Examples 4.4 and 4.5.

$$i_C(t0+) = 0 \quad \text{and} \quad \frac{di_C(0+)}{dt} = 1. \tag{4.121}$$

We evaluate the left-hand side term of Equation (4.119) at time $t = 0+$ and get

$$i_C(0+) = e^{-0}(k_3 \cos 0 + k_4 \sin 0) = 0 = k_3. \tag{4.122}$$

Now since k_3 is 0,

$$\frac{di_C(t)}{dt} = \frac{d}{dt}(e^{-t}k_4 \sin t). \tag{4.123}$$

Thus,

$$\frac{di_C(t)}{dt} = k_4(e^{-t}\cos t - e^{-t}\sin t). \tag{4.124}$$

Since the initial condition $\frac{di_C(0+)}{dt} = 1$ from Equation (4.121), we evaluate Equation (4.124) at time $t = 0+$

And this yields

$$\frac{di_C(0)}{dt} = k_4(e^{-0}\cos 0 + e^{-0}\sin 0) = 1 \tag{4.125}$$

$$k_4 = 1. \tag{4.126}$$

Now we are ready to find our particular solution for

$$i_C(t) = e^{-t}(k_3 \cos t + k_4 \sin t). \tag{4.127}$$

Recall that $k_3 = 0$ and $k_4 = 1$. Thus,

$$i_C(t) = e^{-t}\sin t \quad \text{<<<} \ \textbf{\textit{underdamped case}} \ \text{(Table 4.5)} \tag{4.128}$$

Exercise for the reader: Technically speaking there is a fourth case, when the roots are pure imaginary and conjugate. Find the series *RLC* circuit voltage response across the capacitor due to a 1-V step voltage. Hint: Assume that the characteristic equation is $s^2 + 1 = 0$. Determine the values of all three circuit components for the given characteristic equation.

Table 4.5 Time domain step-input responses

Case	Type of Roots	Time Domain Response
Overdamped	Negative real and distinct	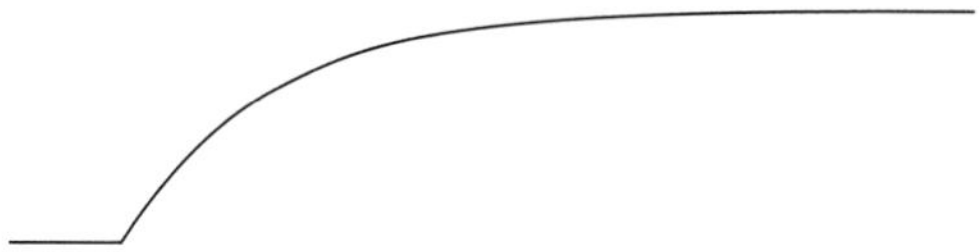
Critically damped	Negative real and equal	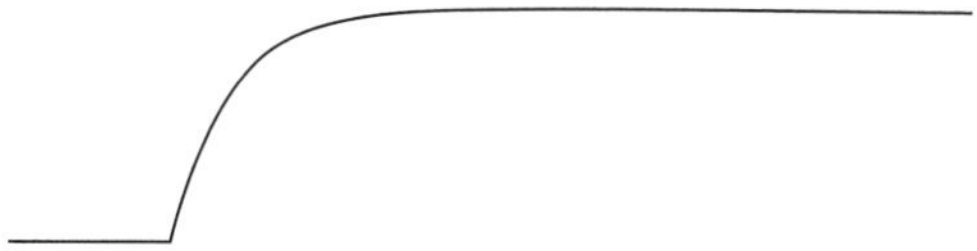
Underdamped	Complex conjugates with negative real parts	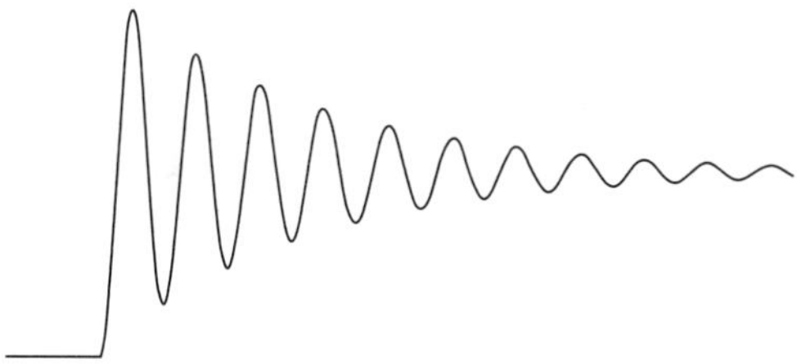

4.6 SECOND-ORDER CIRCUIT IN SINUSOIDAL STEADY STATE: BODE PLOTS

In this section we will observe the behavior of a second-order circuit in the frequency domain. That is to say we will look at its magnitude in decibels and its phase in degrees.

The circuit of Figure 4.18 depicts a second-order *RLC* circuit. We are interested in the voltage across the capacitor. Let us apply an AC voltage source to the input of the series.

Now we can calculate the ratio of the output voltage over the input voltage of the circuit.

The total impedance seen by the AC source is the series of the *R*, *L*, and *C* circuit elements. That is,

$$Z(j\omega) = R + j\omega L + \frac{1}{j\omega C}. \tag{4.129}$$

We are interested in the output voltage, which is the voltage across the capacitor. If we think of the R, L series as one impedance, say we call it Z_1, and we think of the capacitor as being impedance which we call Z_2, we have then

$$Z_1(j\omega) = R + j\omega L \tag{4.130}$$

and

$$Z_2(j\omega) = \frac{1}{j\omega C}. \tag{4.131}$$

The output voltage is calculated as if the impedances worked as resistor dividers.

Thus,

$$\frac{V_{out}}{V_{in}} = \frac{Z_2}{Z_1 + Z_2}. \tag{4.132}$$

However, it is important to understand that all the voltages and impedances in Equation (4.132) are complex numbers, because they are representing components operating at a the same sinusoidal frequency.

Plugging the values from Equations (4.130) and (4.131) into Equation (4.132), replacing the variable $j\omega$ with the operator s yields after doing some arithmetic:

$$\frac{V_{out}}{V_{in}} = \frac{1}{LC}\left(\frac{1}{s^2 + s\frac{R}{L} + \frac{1}{LC}}\right) \tag{4.133}$$

Note: The s operator is called the Laplace variable or operator. We simply used the operator as a substitute for the complex number $j\omega$. A whole entire course can be taken on the Laplace transforms and its applications. Certainly, this is not the book to read about Laplace transforms.

Equation (4.133) is also referred to as the circuit or system transfer function.

This is the transfer function that we will plot to understand the magnitude and the phase behavior with respect to frequency.

The denominator of Equation (4.133) is a second-order equation (nothing new here). This denominator can be factored as $(s - root_1).(s - root_2)$, where $root_1$ and $root_2$ are the denominator roots.

For the sake of simplicity and a clear presentation, we will assume the following numerical values for R, L, and C.

Assume that: $L = 1$ H, $C = 1$ F, and we plot 10 magnitude and 10 phase plots for the following values of R in ohms: 0.1, 0.3, 0.6, 0.9, 1, 2, 3, 4, 5, 10. Figure 4.19 is a computer generated Bode plot (magnitude and phase) for the transfer function given by Equation (4.133).

The purpose of this demonstration is to reveal the most important characteristics that a second-order system transfer function Bode plot has. Also compare those against the first-order Body plots at the beginning of this chapter.

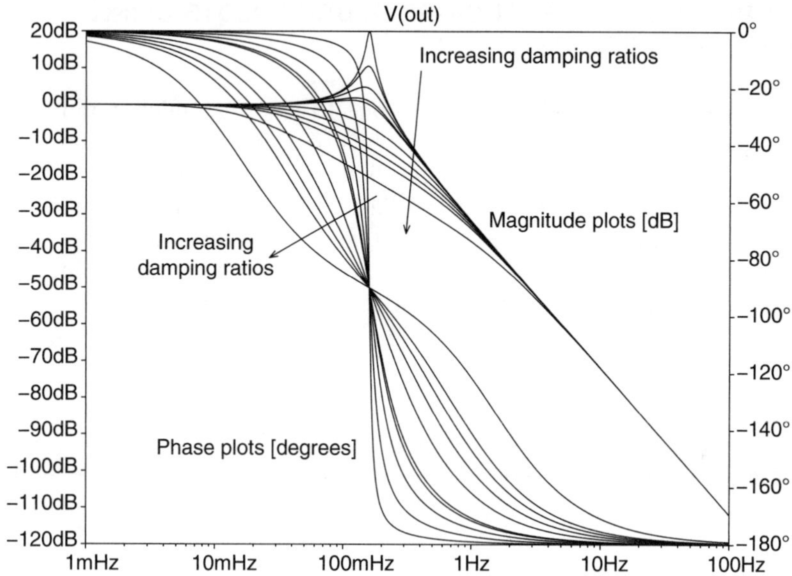

Figure 4.19 Magnitude and phase Bode plots of *RLC* circuit.

Figure 4.19 shows magnitude and phase plots on the same sheet. The unit of magnitude is the decibel, the unit of phase is the degree.

Magnitude characteristics:

10 different magnitude plots are shown for the 10 given values of R.

The natural frequency f_n, according to Equation (4.86), is $\omega_n^2 = 1/LC$, where $\omega_n = 2\pi f_n$, which for $L = 1$ H and $C = 1$ F, $f_n = 0.15924$ Hz. By inspection of Figure 4.19 we see that the magnitude peaks for small damping rations, and as the damping factor increases, the magnitude becomes less "peaky." It is also of importance to mention that the magnitudes peak at the natural frequency of the circuit:

$$f_n = \frac{1}{2\pi\sqrt{LC}} \text{ [hertz]}. \tag{4.134}$$

The negative slope of the magnitude plots are −40 dB per decade. Once the magnitude is at a frequency greater than or equal to 10 times the natural frequency, the slope is −40 dB/dec regardless of the damping ratio of the circuit.

Phase characterisitics:

The phase changes from 0 degrees to −180 degrees in approximately two decades of frequency. This statement is more accurate for lower damping ratios. The natural frequency is the crossover point for all phase plots. All phase plots will cross over at the f_n regardless of the value of the damping ratio. The phase crossover point for the second-order system is −90 degrees.

4.7 DRAWING THE SECOND-ORDER BODE PLOTS USING ASYMPTOTIC APPROXIMATIONS

The approximate methodology allows one to get very quickly approximate magnitude and phase plots. The Bode plots of a second-order system can be constructed as the composite plots of 2 first-order Bode plots.

In a generic way, assume that the natural frequency is f_n. Following the asymptotic magnitude plot of Figure 4.20, we see that for frequencies less than or equal to 1/10 f_n the magnitude plot is approximated by a 0 dB line.

From frequency f_n, we draw a line with a –40 dB/dec slope.

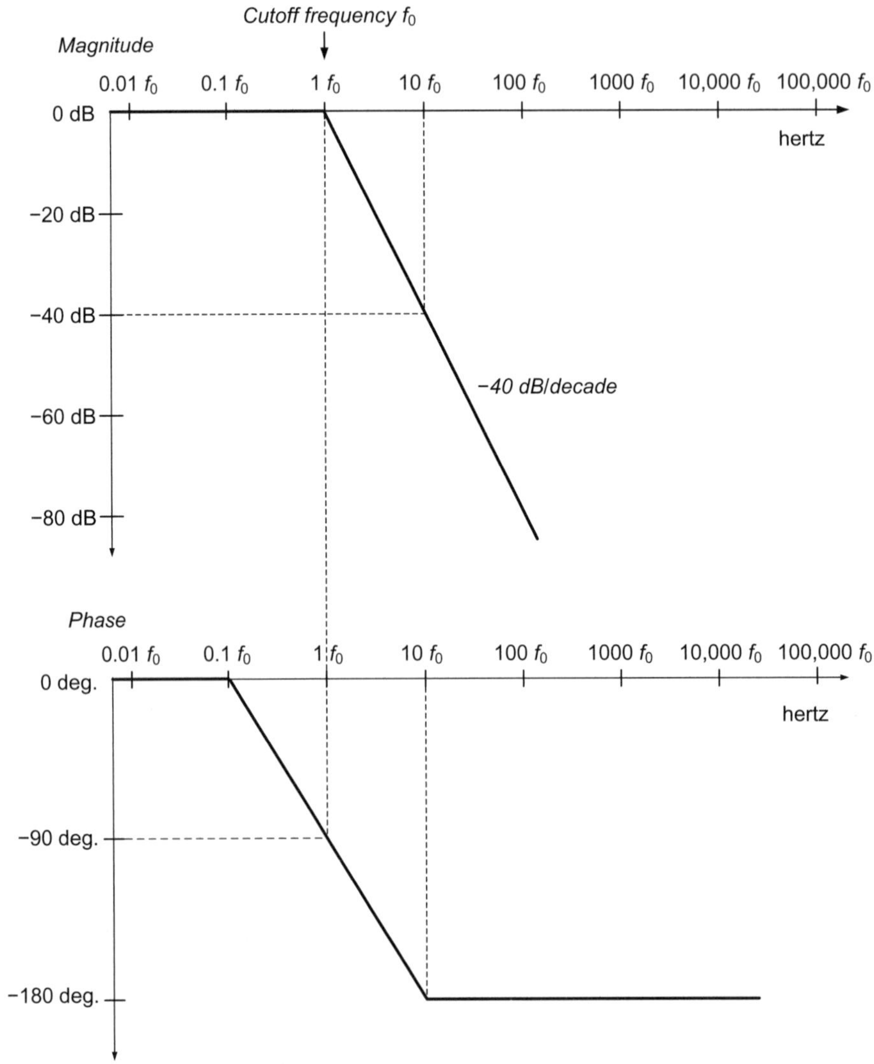

Figure 4.20 Asymptotic method for a second-order system Bode plots.

For the phase, we approximate it with a zero degree phase line at frequencies less than or equal to $1/10 \times f_n$. We then draw a horizontal phase line starting at a frequency greater or equal to $10 \times f_n$. Finally and once more referring to Figure 4.20, we draw a line for the phase from 0 degrees at 1/10 of f_n all the way through the −180 degree point at a frequency 10 times f_n.

4.8 SUMMARY

We looked at two of the most fundamental circuits in electrical and electronics engineering, the first-order *RC* LPF and HPF. They are first-order circuits because they have a single energy storage element, a capacitor. Their time domain equations are first-order differential equations. The circuits are fully characterized; that is, their time behavior as well as their frequency behavior are completely known by their *RC* time constant.

The *RC* LPF as its name states allows low frequencies to pass through it unaltered. The *RC* HPF allows high frequencies to pass through it, unaltered.

The behavior of an *RC* LPF can be that of an LPF or that of an integrator at frequencies well above the filter cutoff frequency. The behavior of an *RC* HPF can be that of an HPF or that of a differentiator at frequencies well below the filter cutoff frequency.

The *RC* LPF integrates when the frequency of the signal to be integrated is at least 10 times f_0 or more. The *RC* HPF differentiates when the frequency of the signal to be differentiated is at most 0.1 times f_0 or less. Recall that f_0 is for both, HPF and LPF, their cutoff frequency.

Second-order circuits, have one capacitor and one inductor in addition to some resistance. Those two energy-storing circuit elements are what cause the overshooting and undershooting of the second-order time response, of course depending on the damping ratio. The larger the damping ratio, the smoother the response and no overshoot/undershoot will be observed. The smaller the damping ratio, the larger the overshoots and undershoots will be. Overshooting and undershooting are phenomena not observed in first-order circuits. For overshooting and undershooting to occur, the circuit has to be a second-order system or higher.

FURTHER READING

1. M. E. Van Valkenburg, *Analog Filter Design*, HRW, New York, 1982.
2. M. E. Van Valkenburg, *Network Analysis*, 3rd ed., Prentice Hall, Englewood Cliffs, NJ, 2006.
3. Sergio Franco, *Design with Operational Amplifiers and Analog Integrated Circuits*, McGraw-Hill Book Company, New York, 1988.
4. Mahmood Nahvi and Joseph Edminister, *Electric Circuits*, 4th ed., Schaum's Outline Series, McGraw-Hill, New York, 2003.

PROBLEMS

4.1 Given an RC low-pass filter circuit, like the one shown in Figure 4.2, assume that $R = 1$ kΩ and $C = 1$ μF. (a) Determine the filter cutoff frequency, (b) determine the time constant of the circuit, and (c) draw the magnitude and phase asymptotic Bode plots of such filter for the following frequencies: 0.01 f_0, 0.1 f_0, 1 f_0, 10 f_0, 100 f_0, where f_0 refers to the cutoff or corner frequency. Make sure to use semi-log paper to draw the Bode plots.

4.2 For an RC low-pass filter with $R = 1$ kΩ and $C = 1$ μF, determine the steady-state output $v_{out}(j\omega)$ magnitude and phase when the a sinusoidal voltage $v_{in}(j\omega)$ is applied at the input. Tabulate magnitude and phase for the following frequencies: 0.01 f_0, 0.1 f_0, 1 f_0, 10 f_0, 100 f_0, where f_0 refers to the cutoff or corner frequency. Note: $v_{out}(j\omega)$ is the voltage across the capacitor.

4.3 For the circuit given in Figure 4.22, initially the capacitor is completely discharged. Determine the voltage that the capacitor will get charged up to, after the switch is closed instantaneously at time t_o and waiting for two circuit time constants.

4.4 Recall the current–voltage relationship of the voltage across a capacitor and the current flowing through it, is given by: $i_C(t) = CdV_C/dt$. (a) Calculate the voltage developed across an initially discharged 1 μF capacitor when a DC current source is applied as shown by Figure 4.21. (b) Justify your answer based on the capacitor current–voltage relationship.

4.5 Using the circuit depicted by Figure 4.23, (a) draw the current through the 10 nH inductor when the square wave shown is applied to the inductor for two complete periods; (b) determine the current numerical value at $t = 1$ μs; (c) determine the current numerical value at $t = 2$ μs.

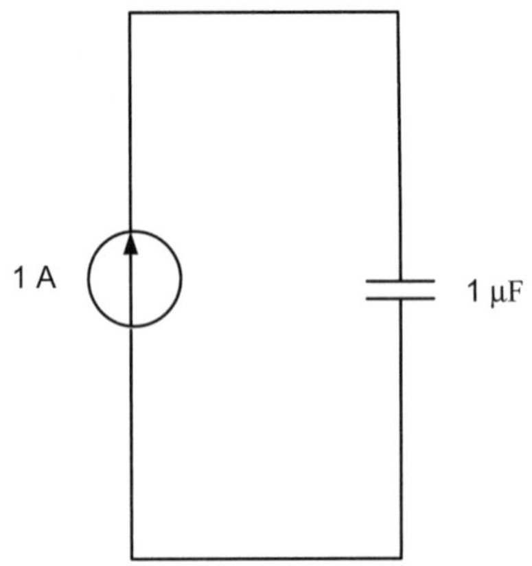

Figure 4.21 Circuit for Problem 4.4.

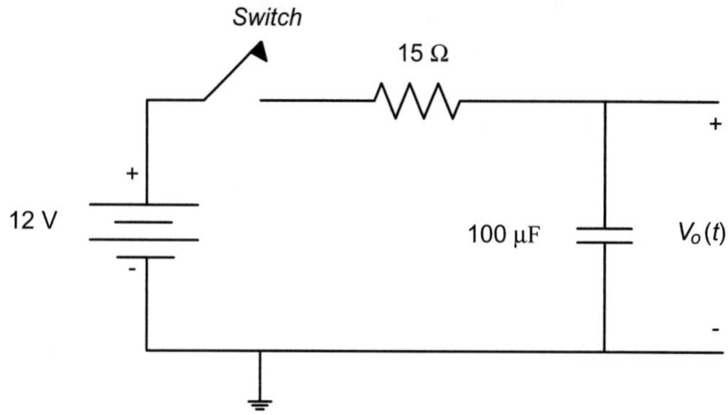

Figure 4.22 Circuit for Problem 4.3.

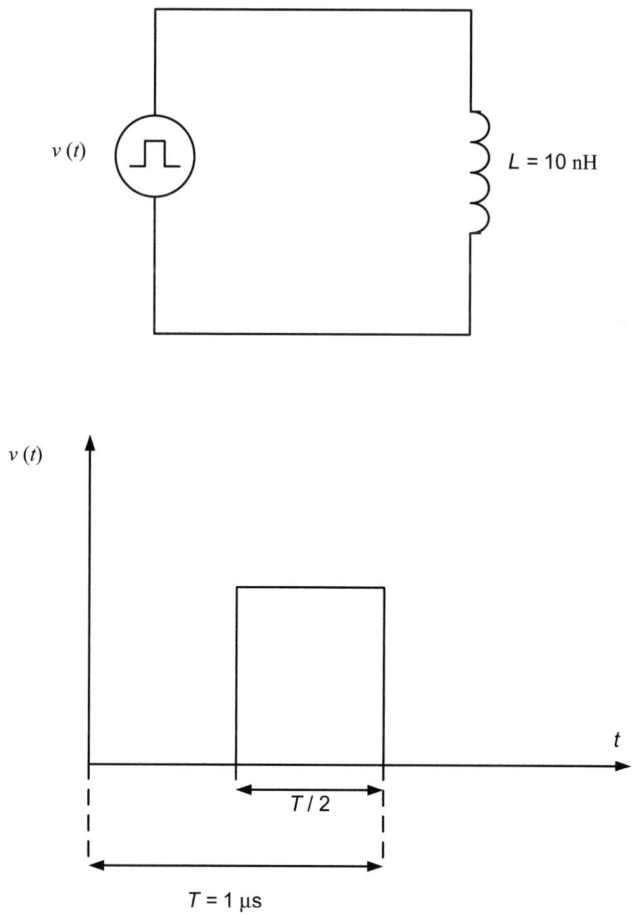

Figure 4.23 Circuit for Problem 4.5.

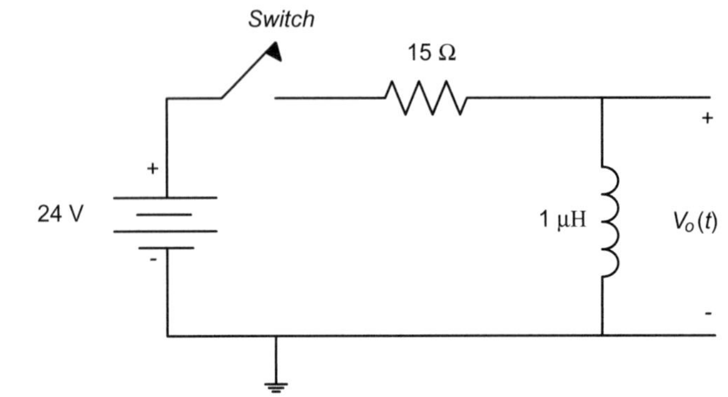

Figure 4.24 Circuit for Problem 4.6.

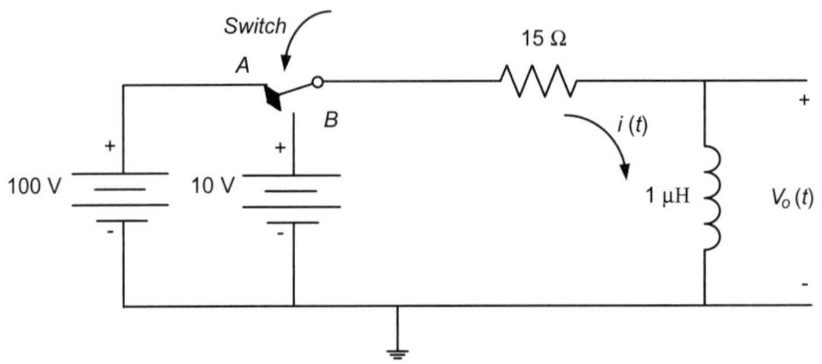

Figure 4.25 Circuit for Problem 4.7.

4.6 Using the circuit depicted by Figure 4.24, (a) determine the output voltage $V_o(t)$ equation as a function of time, when the switch is quickly closed, (b) draw the output voltage $v_o(t)$, (c) determine the output voltage after 10 μs from closing the switch, and (d) Determine the output voltage after one second from closing the switch.

4.7 The switch in Figure 4.25 has been closed for a very long time in position *A*. At time $t = 0$, the switch is quickly moved to position *B*. (a) Determine the equation of $i(t)$ for $t > 0$; (b) draw current $i(t)$ for $t > 0$.

4.8 The switch in Figure 4.26 has been closed for a very long time in position *A*. At time $t = 0$, the switch is quickly moved to position *B*. (a) Determine the equation of $i(t)$ for $t > 0$; (b) draw current $i(t)$ for $t > 0$.

4.9 Given the circuit of Figure 4.27, (a) determine the circuit time constant, (b) determine the circuit cutoff frequency f_0, and (c) construct the magnitude and phase Bode plots using the asymptotic method for the transfer function: $V_{out}(j\omega)/V_{in}(j\omega)$. Use as frequency range, 2 decades below cutoff frequency f_0 up to 2 decades above f_0.

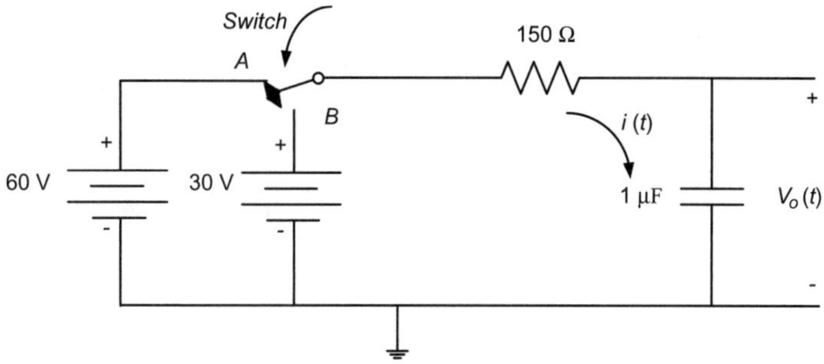

Figure 4.26 Circuit for Problem 4.8.

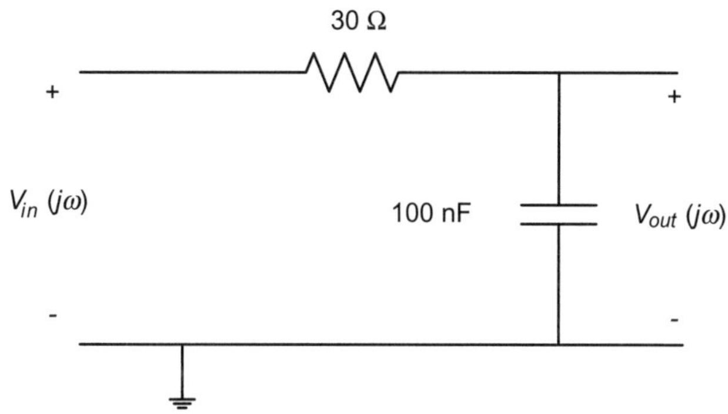

Figure 4.27 Circuit for Problem 4.9.

4.10 For the *RL* series circuit of Figure 4.28, (a) determine the transfer function of the circuit, that is, $V_{out}(j\omega)/V_{in}(j\omega)$; (b) determine the circuit cutoff frequency f_0; (c) determine and draw the asymptotic Bode plots for the magnitude and the phase of the transfer function; the frequency range used should be from 0.01 f_0 to 100 f_0. This is a total of four decades of frequency; (d) which type of filter this circuit represents?

4.11 The capacitor in the circuit of Figure 4.29 is charged up to 50 V DC when the switch is open. Upon closing the switch very quickly, determine the transient current as a function of time that will flow through the circuit. Note: The circuit that initially charged the capacitor is not shown.

4.12 For the circuit of Figure 4.30, (a) calculate the circuit transfer function as a function of $j\omega$; (b) calculate the cutoff or corner frequency of the circuit; (c) draw the asymptotic magnitude and phase Bode plots.

4.13 For the circuit of Figure 4.30, determine the current transient response for a step input voltage of 0 to 1.

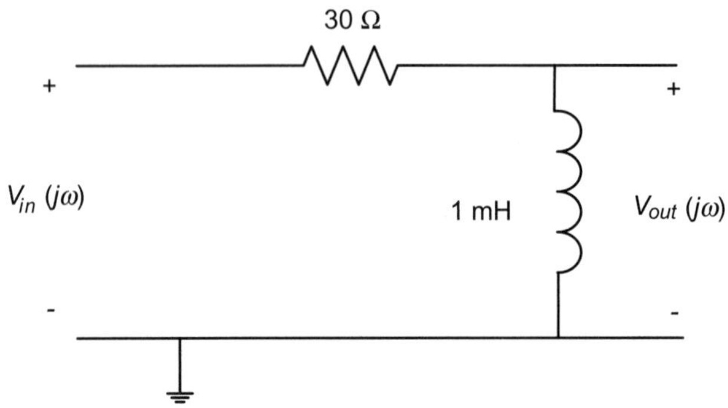

Figure 4.28 Circuit for Problem 4.10.

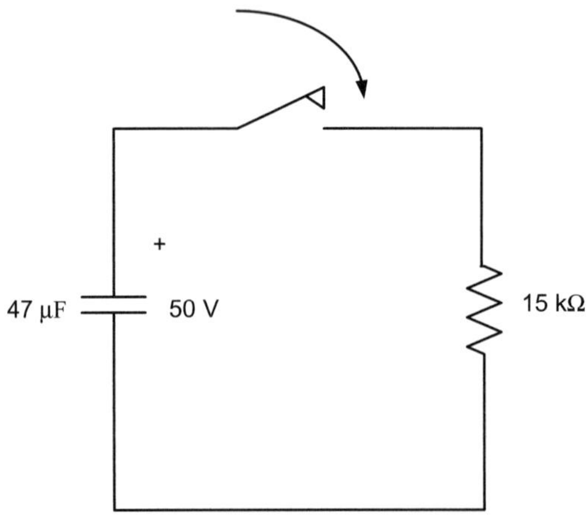

Figure 4.29 RC circuit for Problem 4.11.

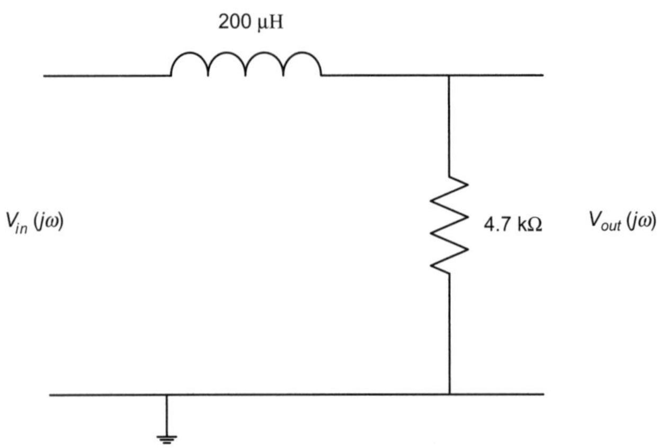

Figure 4.30 Circuit for Problems 4.12 and 4.13.

4.14 For an RC high-pass filter, such as the one shown in Figure 4.10, if $R = 100\ \Omega$ and $C = 0.159\ \mu F$, (a) derive the transfer function of the circuit $V_{out}(j\omega)/V_{in}(j\omega)$; (b) draw the magnitude and phase Bode plots from 2 frequency decades below the corner frequency up to 2 decades above the corner frequency.

4.15 Assume that you are given an RC low-pass filter, whose corner frequency is 10 kHz. Calculate the exact magnitude in decibels and phase in degrees at 100 Hz, 10 kHz, and 100 kHz.

4.16 For the filter shown in Figure 4.31, assume a 1-V step is applied to the input. (a) Derive a time domain equation of the current through the circuit, (b) calculate the circuit time constant, and (c) plot the current response for two time constants. Hint: Apply Thévenin to simplify the problem.

4.17 Refer to the circuit of Figure 4.32. Determine the time domain equation of the current as a function of time. Make sure that you find all the initial conditions of the circuit. Assume that the capacitor is initially discharged. Hint: Apply Thévenin to the left-hand side of the 47 μF capacitor to simplify the problem.

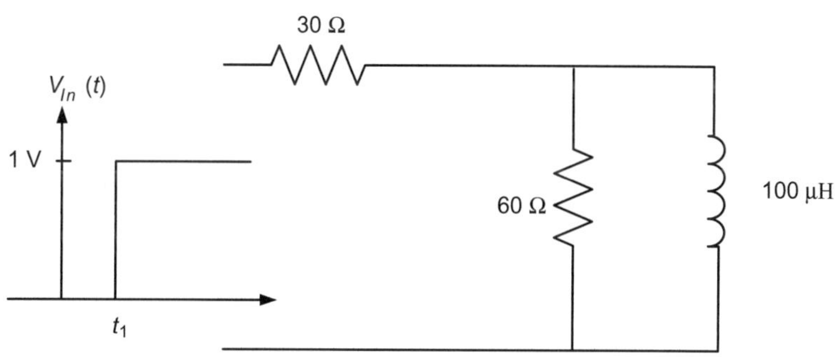

Figure 4.31 Circuit for Problem 4.16.

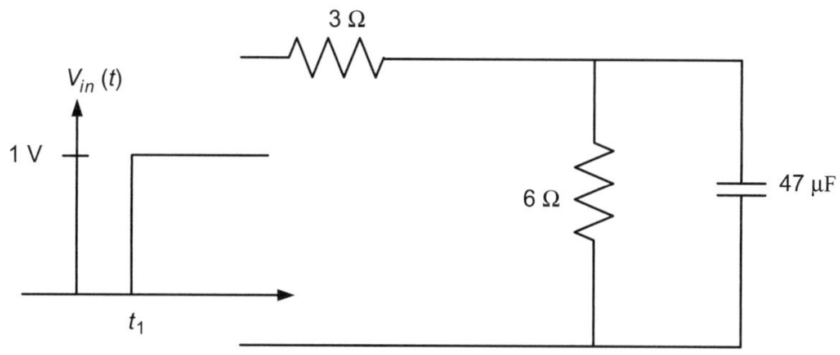

Figure 4.32 Circuit for Problem 4.17.

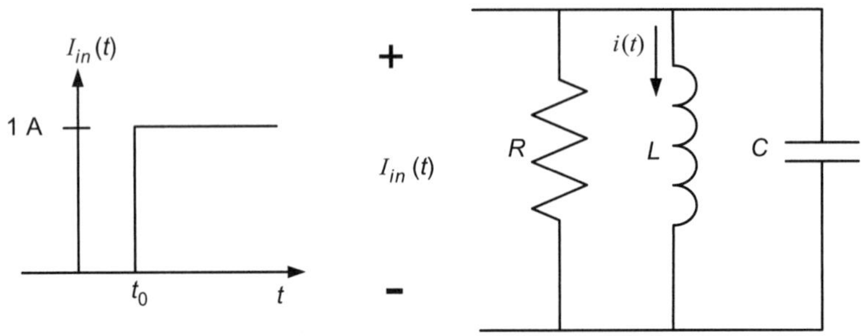

Figure 4.33 Circuit for Problem 4.19.

4.18 Refer to the second-order *RLC* series circuit of Figure 4.18. A 1-V step input voltage is applied at time t_0, all the circuit initial conditions are zero.

(a) State the time domain equation of the circuit; ensure that you show the equation as a second-order system equation. (b) In a general fashion, explain the consequences when the roots are

(i) negative real and different, (ii) negative real and equal, and (iii) when the root are complex conjugates.

4.19 Refer to the second-order *RLC* parallel circuit of Figure 4.33. Assume that a 1-A step input current is applied at time t_0, all the circuit initial conditions are zero.

(a) State the time domain equation of the circuit inductor current; ensure that you show the equation as a second-order system equation. (b) In a general fashion, explain the consequences when the roots are

(i) negative real and different, (ii) negative real and equal, and (iii) when the roots are complex conjugates.

5

THE OPERATIONAL AMPLIFIER AS A CIRCUIT ELEMENT

5.1 INTRODUCTION TO THE OPERATIONAL AMPLIFIER

The operational amplifier, more commonly known as *op amp,* is an analog circuit. Op amps perform many arithmetic functions, linear and nonlinear operations in the analog or continuous domain. Op amps are also used in several kinds of analog amplifiers and active filters. They are also used to implement nonlinear circuits such as voltage comparators and continue to have a widespread use in the field of analog electronics. In its very early years of electronics, the beginning of the twentieth century, the first op amps were implemented with vacuum tubes, later on with transistors, and most currently, op amps are available in a single (or monolithic) integrated circuit (IC) device. That is to say that the transistors that implement the op amp itself reside within the IC. Why, if the basic components of an op amp are transistors, do we choose to cover op amps prior to the introductory chapter on electronic devices (*diodes, bipolar*, and *MOSFET transistors*)? The reason is simple and justifiable: op amps can be dealt with as circuit elements without necessarily knowing all the details of their internals. In this chapter we will start using dependent sources, to model the operation of op amps. Moreover, the op amp can perform a variety of functions that can be easily understood without initially having the knowledge of how the actual integrated circuit is designed. Finally, an op amp can be effectively used as a circuit element knowing the behavior of its

Electrical, Electronics, and Digital Hardware Essentials for Scientists and Engineers, First Edition. Ed Lipiansky.

inputs and output terminals and knowing its parameters from the manufacturer's data sheet. Op amp parameters are indicators of how much a real op amp deviates from an idealized model of an op amp. It is in the interest of the circuit designer using op amps to establish when the op amp-based circuit behaves as if its op amps were ideal and when they are not. When op amp-based circuits have nonidealities, the circuit designer has to take such deviations from the ideal op amp into account to predict their circuit behavior more accurately.

5.2 IDEAL AND REAL OP AMPS

It is extremely useful to use a model of an idealized op amp. In many applications, as we will see later throughout this chapter, the idealized behavior is just a first-order approximation of the way the op amp works as a circuit element. Later on, we will add the influence of the real op amp parameters that may not let an op amp-based circuit to always be analyzed as an idealization. Our goal in this chapter is to understand the ideal op amp, understand how its inputs and output work. It is also our goal to know when and what to take into account from the manufacturer's data sheet, which otherwise would not be correct with the idealized model. Finally, one of our goals is to address and analyze the most important and useful linear and nonlinear applications using op amps.

The most basic symbol of an op amp is given in Figure 5.1. It has two inputs, a noninverting or positive input and an inverting or negative input. It also has a single output. The most generic way of representing an op amp, whether it is a real one or an idealized one, is the one seen in Figure 5.1a. A more complete graphical representation is to draw the power terminals that provide positive and negative power sources to the real op amp internals, Figure 5.1b. *Warning* to the reader: some technical publications, data sheets, or textbooks draw the positive input at the top left of the op amp symbol, some others draw the negative input at the top left. However, many other publications interchange the location of the positive and negative inputs. Thus, the reader has to be very cautious and find out which are the noninverting and the inverting inputs of the op amp. Confusing the correct input may mislead one into a completely incorrect interpretation of the function of an op amp-based circuit.

Some of the most basic ideal op amp characteristics are

(a) Open-loop gain is infinite: ∞, or $A_{OL} \to \infty$

(b) The noninverting and inverting terminals do not draw or source any current from or into the op amp. $Z_{input} = Z_i = \infty$, means that its input impedance is infinitely high.

(c) The output of the op amp can provide an infinitely large current. In other words, the op amp output impedance is zero ($Z_{output} = 0$ or $Z_o = 0$).

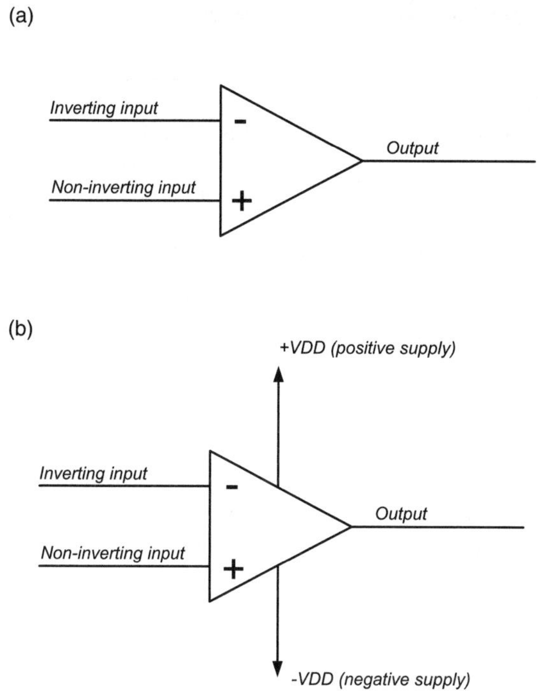

Figure 5.1 Graphic representation of an op amp (a) idealized or real op amp without power terminals, (b) real op amp showing its power terminals.

(d) When the op amp is operated in linear mode using negative feedback, the voltage difference between the noninverting and inverting input is infinitesimally small.

$$\Delta V = V^{+} - V^{-} \rightarrow 0$$

We will come back to the negative feedback concept shortly.

(e) Bandwidth of an ideal amplifier is infinite, because the ideal amplifier can react to signals of any frequency equally well. Bandwidth in a real op amp refers to small signal bandwidth, that is, signals whose peak-to-peak amplitudes are a small fraction of the op amp power supply rail. For example, signals of a 1 V for a ±15-V powered op amp are considered to be small signal amplitudes.

(f) Slew-rate: For large signal behavior, that is, for signals that are comparable to the power supply rail magnitude in a real op amp, slew-rate is a finite and nonzero number. This is because real op amps take time to react to large voltage swings. Typically, slew-rates are expressed in volts per microsecond.

Table 5.1 Some operational amplifiers idealizations

Parameter	Ideal Value
Open-loop gain (A_{OL})	Infinite
Input resistance R_i (more generically input impedance)	Infinite
Output resistance R_o (more generically output impedance)	Zero
Voltage difference (ΔV) between noninverting and inverting inputs, when negative feedback path exists. That is, the op amp is working in a linear application. **Note:** This is not true when the op amp operates in open-loop mode or with positive feedback.	Zero
Bandwidth (refers to small signal response capability)	Infinite
Slew-rate (refers to large signal response capability)	Infinite
Offset voltage	Zero
Bias current	Zero
Offset current	Zero
Common Mode Rejection Ratio (CMMR)	Infinite

(g) Offset voltage: Zero for the ideal case. This value is nonzero for real op amps since their negative and positive inputs cannot be perfectly matched.

(h) Bias current: Zero for an ideal op amp. This value is nonzero for real op amps since their negative and positive inputs cannot be perfectly matched.

(i) Offset current (difference of bias currents at positive and negative inputs): Zero for the ideal op amp. This value is nonzero for real op amp because bias currents, even when finite, cannot be perfectly matched.

(j) Common Mode Rejection Ratio (CMRR): The ratio of differential mode gain and the common mode gain, usually expressed in dBs.

Table 5.1 summarizes the idealizations made above for an ideal op amp.

As we discuss more op amp-based circuits, we will introduce some more of the ideal op amp characteristics. We will be able to go a long way using the top four characteristics mentioned in Table 5.1. For now, in what way real op amps parameters differ from the idealizations of Table 5.1? Without getting into much detail this early in the chapter, we will just say that for a real op amp, none of the characteristics listed in Table 5.1 is true. Upon studying real op amp data sheet parameters, we will expand what that means in a more qualitative manner.

5.3 BRIEF DEFINITION OF LINEAR AMPLIFIERS

Let us study the basic linear amplifiers that are available, before we zoom into the operational amplifier-based circuits. Op amp-based circuits are usually special cases of the most generic cases of the four types of amplifiers that we will cover in this section.

An amplifier is a two-port device that receives an input signal and produces and output that is proportional to some constant, that we call the amplifier *gain* or *A,* which stands for amplification factor. In general, the *gain* of an amplifier (unlike the gain of an ideal op amp) is finite. Generally when we talk about amplifiers, we will always refer to linear amplifiers, unless it is otherwise stated. For example, multipliers are nonlinear amplifiers, whereas adders, subtractors, inverters, buffers, and difference amplifiers all are linear amplifiers.

Four key types of linear amplifiers exist from the point of view of the kind of input and output signals that they involve:

(a) Voltage amplifier
(b) Current amplifier
(c) Trans-conductance amplifier
(d) Trans-resistance amplifier

A voltage amplifier receives an input voltage and produces an amplified output voltage. A current amplifier receives an input current and produces an amplified output current. A trans-resistance amplifier receives an input current and produces an amplified output voltage. A trans-conductance amplifier receives an input voltage and produces an amplified output current.

It is common practice to cascade amplifiers. That means connecting the output of one into the input of the next one. It is common to cascade two or three stages of amplifiers in that way. The input stage of an amplifier typically loads the output of an amplifier that precedes it. Figure 5.2 presents three cascaded amplifiers.

Figure 5.3 depicts the four amplifiers types described above.

The voltage amplifier of Figure 5.3a has an ideal infinite input resistance, a zero output resistance, and a voltage gain of

$$
\begin{aligned}
&R_i \to \infty \\
&R_o = 0 \\
&A_v = v_o / v_i. \qquad (5.1)
\end{aligned}
$$

The current amplifier of Figure 5.3b has an ideal zero input resistance, an infinite output resistance, and a current gain of

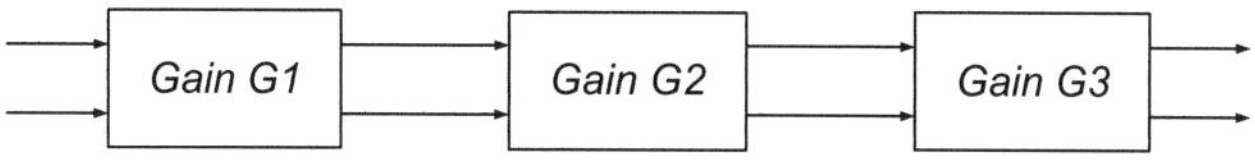

Figure 5.2 Cascaded amplifiers.

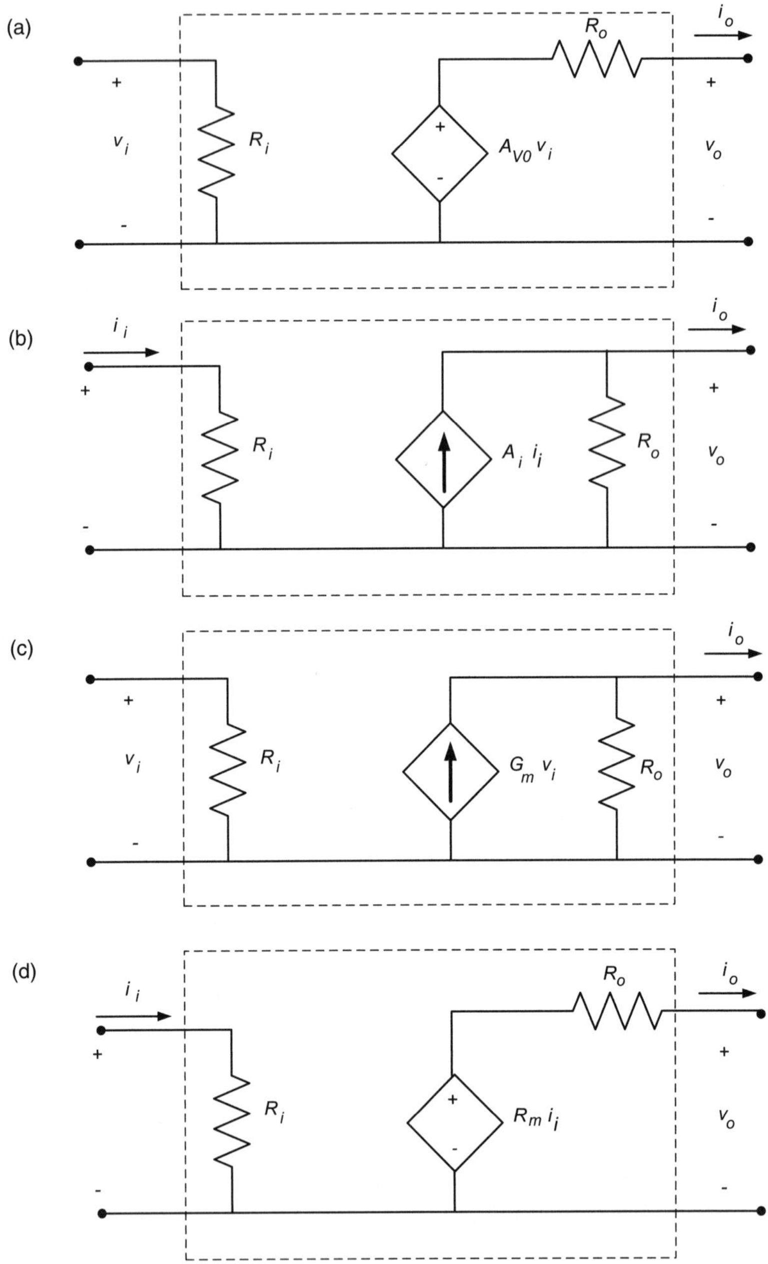

Figure 5.3 Amplifier types: (a) voltage, (b) current, (c) trans-conductance, (d) trans-resistance.

$$R_i = 0$$
$$R_o \to \infty$$
$$A_I = i_o / i_i. \quad (5.2)$$

The trans-conductance amplifier of Figure 5.3c has an ideal infinite input resistance, an infinite output resistance, and a trans-conductance gain of

$$R_i \to \infty$$
$$R_o \to \infty$$
$$A_G = i_o / v_i. \quad (5.3)$$

The trans-resistance amplifier of Figure 5.3d has an ideal zero input resistance, a zero output resistance, and an open circuit trans-resistance gain of

$$R_i = 0$$
$$R_o = 0$$
$$A_R = v_o / i_i. \quad (5.4)$$

But what do Equations (5.1) through (5.4) mean? Let us start with Equation (5.1) on the voltage amplifier and let us refer to Figure 5.3a. The voltage amplifier is modeled with a voltage-controlled voltage source (VCVS). A voltage amplifier has an input resistance (in more general terms we say that it is an input impedance), and from an ideal point of view, we do not want the input stage of the amplifier to load the source that is driving it. So that is the reason why, ideally speaking, a voltage amplifier should have an infinite input resistance. This in effect means that the amplifier does not draw any current from its driving source. A finite output resistance of a voltage amplifier is what would actually limit a real amplifier from driving current to a load connected across its output terminals. Since ideally one wants the amplifier to have no current sourcing limitation, thus we say that the ideal voltage amp should have a zero output resistance.

For the current amplifier of Figure 5.3b the amp is modeled by a current-controlled current source (CCCS). The controlling input is the input current i_i; we want the amplifier to be controlled by the current and not by a voltage developed across its input resistance. Thus, in this case, the ideal current amplifier should have a zero input resistance, since the input current has to enter the amplifier for control purposes. For the same amplifier, notice that the output resistance is in parallel with the output current source $A_i\, i_i$; we certainly do not want all the output current to be drained or consumed by its output resistance, we want the output current to go to the load. Thus, the output resistance of an ideal current amplifier wants to be infinite. The trans-conductance amplifier of Figure 5.3c is modeled with a voltage-controlled current source (VCCS). The controlling input is voltage v_i, so we want the ideal trans-conductance amplifier not to draw any current from its driving input source; thus, the need for an infinite input resistance. The infinite output resistance of the trans-conductance amp is justified in the same way as the output resistance of the current amplifier was. The trans-resistance amp is modeled with a current-controlled voltage source (CCVS); the input is controlled by current i_i, thus we want a zero input resistance for the trans-resistance amplifier, just like we have it for the current amplifier. The output of the trans-resistance amplifier of Figure 5.3d behaves just like the output of the voltage amplifier, thus we want the ideal trans-resistance amplifier to have a zero output resistance to drive any load. Finally, let us note that the gain of the trans-resistance (A_R) and trans-conductance (A_G) amplifiers have respectively units of ohms and ohms^{-1}. Current and voltage amplifiers have dimensionless amplification factors.

5.4 LINEAR APPLICATIONS OF OP AMPS

Linear applications of op amps refer to those circuits that have a linear relationship between output and input, whereas nonlinear applications do not. The most common linear applications are inverting amplifiers, noninverting amplifiers, buffers, and difference amplifiers. Other interesting op amp-based linear circuits are integrators and differentiators.

5.4.1 The Inverting Amplifier

Let us now look at op amp-based circuit show of Figure 5.4. This circuit is called an inverting amplifier configuration. To analyze how this circuit works, we will assume that the op amp of the inverting amplifier is ideal. Let us carefully describe by inspection of Figure 5.4 how this circuit is connected. First,

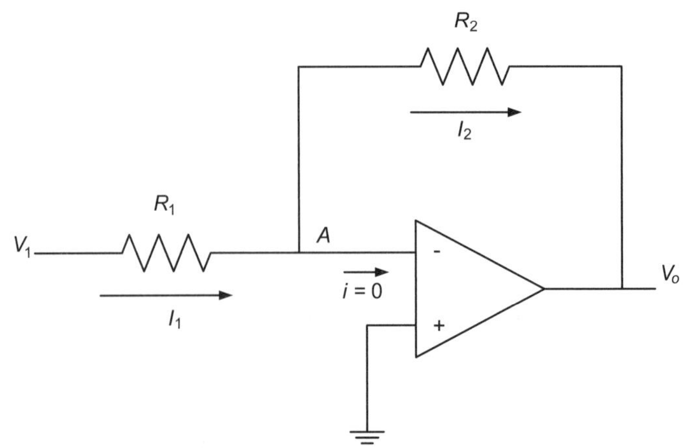

Figure 5.4 Op amp-based inverting amplifier configuration.

besides the op amp, we have two external resistors. Resistor R_1 is connected from the input source V_1 to the inverting input terminal of the op amp. Resistor R_2 connects the output of the op amp back to the inverting input of the op amp. This path from the output to the inverting input of the op amp is referred to as a negative feedback path. We will see that all linear circuits implemented with op amps have a negative feedback path. Finally, the noninverting input of the op amp is tied to reference ground.

Example 5.1 Calculate the output voltage to input voltage ratio, V_o/V_i or the voltage gain of the circuit of Figure 5.4, assuming that the op amp is ideal.

Because the op amp has a negative feedback path, we can assume that the voltage difference between inverting and noninverting inputs is zero (see Table 5.1). Now because the noninverting input is tied to ground, then the voltage at the inverting input is referred to as being virtually grounded. People refer to as this node as being "virtual ground." The closer is the op amp to the idealization, the truer that statement becomes. Now we can state a Kirchoff's current law (KCL) equation at virtual ground node A:

$$I_1 = I_2 \tag{5.5}$$

because $i = 0$.

But since

$$I_1 = \frac{V_1}{R_1} \tag{5.6}$$

because node A is virtually grounded, and since

$$I_2 = -\frac{V_A - V_o}{R_2} = -\frac{V_0}{R_2}, \tag{5.7}$$

V_A is zero because it is *virtually grounded.*

At this point, let us refer to Figure 5.4 one more time. Note that the entire current I_1 that flows through R_1 also flows through R_2, because there is no current at all going into or out of the inverting terminal of the ideal op amp, $i = 0$, Equation (5.5).

So now combining Equations (5.6) and (5.7) yields:

$$\frac{V_o}{V_1} = -\frac{R_2}{R_1}. \tag{5.8}$$

Equation (5.8) is the approximated closed-loop voltage gain of the inverting amplifier, which assumes an infinite op amp A_{OL}. Equation (5.8) is also commonly referred to as the inverting amplifier closed-loop gain (CLG). Such

CLG is approximate because the ideal op amp parameters have been assumed (Table 5.1). Note that the absolute magnitude of this gain is the ratio of R_2 and R_1, whereas its sign is negative. It is important to observe that the CLG of the amplifier, Equation (5.8), depends only on the external resistors, and it is independent of the op amp, as long as the op amp open-loop gain is "large enough" and other op amp idealizations are met (Table 5.1).

5.4.1.1 Effects of Finite Op Amp Open-Loop Gain in the CLG

Now let us investigate what happens when the open-loop gain of the op amp is large, but not quite as large as it would have to be. We will see next how to quantify when the open-loop gain of the op amp is large enough for Equation (5.8) to be accurate. We will calculate the error we make using Equation (5.8) when the open-loop gain is finite. Figure 5.5 shows an op amp model where the open-loop gain is no longer infinite ($A_{OL} < \infty$). The output is modeled with a VCVS that depends on ΔV, which is the difference between the noninverting and the inverting input voltages.

Upon making the above assumptions, not only A_{OL} is finite but also ΔV is no longer zero. Figure 5.5 shows the usage of the model of a configured as an inverting amplifier with two external resistors, R_1 and R_2.

Let us assume that the one nonideality of the op amp model that we are interested in is its finite open-loop gain. By inspection of the circuit of Figure 5.5 we can state that

$$I_1 = I_2.$$

Because the op amp model still assumes that the input resistance of the op amp is infinite, thus

$$i = 0.$$

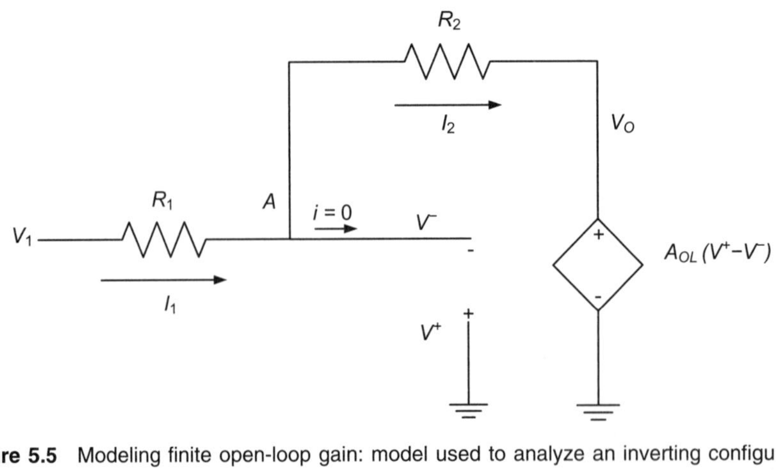

Figure 5.5 Modeling finite open-loop gain: model used to analyze an inverting configuration CLG with a finite-open-loop gain op amp.

Additionally,

$$(V_1 - V_A)/R_1 = (V_A - V_o)/R_2. \tag{5.9}$$

Since the open-loop gain is assumed to be finite, then

$$|V_A| = |V_o / A_{OL}|.$$

And since the positive input is grounded (0 V), the voltage at node A has to be

$$V_A = -V_o / A_{OL}. \tag{5.10}$$

Using Equation (5.10) in Equation (5.9) and doing a little bit of algebra we arrive at

$$V_o / V_1 = -\frac{R_2 / R_1}{1 + \frac{1}{A_{OL}}(1 + R_2 / R_1)}. \tag{5.11}$$

Equation (5.11) shows the CLG of the inverting amplifier of Figure 5.4 when the open-loop gain is finite. We will also refer to this as the true value CLG. Note that if

$$A_{OL} \to \infty,$$

then Equation (5.11) becomes

$$V_o / V_1 = -R_2 / R_1. \tag{5.12}$$

Equation (5.12) is the CLG of the inverting amplifier with an infinite A_{OL} op amp or simply the estimated or approximated CLG.

Example 5.2 Assume a ratio of $R_2/R_1 = 1$, using Equation (5.11), evaluate V_o/V_1 for the following values of A_{OL}: 1, 10, 10^2, 10^3, 10^4, 10^5, 10^6, and 10^7. Determine the error that exists between the more accurate close loop gain of Equation (5.11) with respect to the approximated CLG given by Equation (5.12).

Let us define the absolute value of the error between the two CLGs as the difference of the absolute values of Equations (5.12) and (5.11):

$$Abs\_Error = Estimated\_Value - True\_Value$$

$$Abs\_Error = |-R_2 / R_1| - \left| -\frac{R_2 / R_1}{1 + \frac{1}{A_{OL}}(1 + R_2 / R_1)} \right| \tag{5.13}$$

And the relative error:

$$Relative\_Error = (Abs\_Error / True\_Value) \times 100 \; [\%]$$

$$Relative\_Error = \left| \frac{\left(R_2 / R_1 - \frac{R_2 / R_1}{1 + \frac{1}{A_{OL}}(1 + R_2 / R_1)} \right)}{\frac{R_2 / R_1}{1 + \frac{1}{A_{OL}}(1 + R_2 / R_1)}} \right| \times 100. \tag{5.14}$$

Table 5.2 depicts an inverting amplifier CLG absolute and relative errors for various finite values of A_{OL} for a CLG of 1. Table 5.3 depicts the same values as Table 5.2 but for a CLG of 10.

It is interesting and important to note that for small CLG of 1 of an inverting amplifier, for a finite op amp open-loop gains (A_{OL}) of 60 dB and higher (Table 5.2), the relative error that exists between the CLG assuming an op amp with an infinite open-loop gain versus the CLG with a finite op amp open-loop gain is just 0.2%. This is quite a small error for the CLG equation. Since most op amps today have gains of at least 80 dB, the closed-loop error gain of the inverting amplifier with finite A_{OL} is practically the same as the

Table 5.2 Inverting amplifier closed-loop gain (CLG) errors for finite A_{OL} and CLG of 1

Finite Op Amp Open-Loop Gain (A_{OL})	Op Amp A_{OL} in dB	Closed-Loop Gain of 1 (CLG = 1) with Finite A_{OL}	CLG = 1 Absolute Error	CLG = 1 Relative Error
(V/V)	(dB)	(V/V)	(V/V)	(%)
1	0	0.333333	0.666667	200.000000
10	20	0.833333	0.166667	20.000000
100	40	0.980392	0.019608	2.000000
1,000	60	0.998004	0.001996	0.200000
10,000	80	0.999800	0.000200	0.020000
100,000	100	0.999980	0.000020	0.002000
1,000,000	120	0.999998	0.000002	0.000200
10,000,000	140	1.000000	0.000000	0.000020

Table 5.3 Inverting amplifier closed-loop gain (CLG) errors for finite A_{OL} and CLG of 10

Finite Op Amp Open-Loop Gain (A_{OL}) (V/V)	Op Amp A_{OL} in dB (dB)	Closed-Loop Gain of 10 (CLG = 10) with Finite A_{OL} (V/V)	CLG = 10 Absolute Error (V/V)	CLG = 10 Relative Error (%)
1	0	0.833333	9.166667	1,100.000000
10	20	4.761905	5.238095	110.000000
100	40	9.009009	0.990991	11.000000
1,000	60	9.891197	0.108803	1.100000
10,000	80	9.989012	0.010988	0.110000
100,000	100	9.998900	0.001100	0.011000
1,000,000	120	9.999890	0.000110	0.001100
10,000,000	140	9.999989	0.000011	0.000110

gain with an infinite op amp open-loop gain (just a 0.02% error; again refer to Table 5.2). Now when the CLG of the inverting amplifier is higher than 1, in our example of Table 5.3, we assume a CLG of 10, note that for a finite open-loop gain of 60 dB, the CLG relative error of the amplifier is 1.1% (Table 5.3).

We can generalize and state that the larger the op amp A_{OL}, the more accurate is the approximated CLG of Equation (5.8). When the op amp A_{OL} is not as large, then Equation (5.11) should used for better accuracy. However, once we have selected "an" op amp, A_{OL} is fixed. So under these conditions, as presented by the Tables 5.2 and 5.3, the larger the CLG that is desired, the less accurate it will be when compared to another CLG that is smaller. For example, this is to say that given an op amp with a 100 dB A_{OL}, implementing an inverting configuration of a CLG of 1 will be more accurate than an inverting configuration of a CLG of 10. Once more looking at Tables 5.2 and 5.3, we have an error of 0.002% for a CLG of 1 and an error of 0.011% for a CLG of 10, in both cases for an A_{OL} of 100 dB.

5.4.1.1.1 Effect of Op Amp Output Swing Due to Saturation A real op amp has to receive power from positive and negative power supplies. Some op amps are designed to operate off a unipolar power supply; we will in general assume that the op amps we use require plus and minus power supplies unless it is otherwise stated. So we need to ask ourselves the following question, what can the maximum output of an op amp be? Regardless of whether the op amp is used in a linear or a nonlinear application, closed-loop or open-loop (will cover open-loop applications when talking about comparators), the highest and lowest output of the amplifier cannot exceed its power supply rails minus a saturation voltage (V_{SAT}) imposed by the op amp. For example, if an op amp is operated from a +15-V and −15-V power supplies, if the op amp

V_{SAT} is 2 V *below the positive rail* and 2 V *above the negative rail*, its output shall always be within a voltage range of –13 V to +13 V. From now on, even though we will continue to deal with ideal op amps, we will assume that our ideal op amp requires positive and negative power, and its output shall be required to stay away from the saturation limits.

Example 5.3 Design an inverting amplifier with resistors and assume you have an ideal op amp. Let us assume that we want a CLG of –4. (a) Determine some possible resistor value pairs. (b) Determine the maximum and minimum input signal values not to saturate the op amp. Assume the op amp is powered from +15 V/–15 V supplies and for the output not to saturate the op amp is allowed excursions from +13 V to –13 V.

Solution to Example 5.3

(a) the simplest combination of values that come to mind are 4 kΩ and 1 kΩ resistors. Other combinations of values such as 80 kΩ and 20 kΩ, or 400 kΩ and 100 kΩ are possible. This should pose a question on our mind: how large or how small such resistor values can be. Could we use a 400 MΩ and a 100 MΩ? Since we are assuming that we are dealing with an ideal op amp even 400 MΩ and 100 MΩ are fine to use because their ratio provides a CLG of –4 in an inverting amplifier configuration. *Later on we will see that it may not be possible to use arbitrarily large or arbitrarily small resistor values when we use real-world op amps. We will see that there are upper limits as well as lower limits for the resistor values we can choose.*

(b) Now, since the op amp output swing cannot exceed +13 V and cannot be under –13 V, the largest possible input signal magnitude is determined by dividing:

$$13/4 = 3.25\ \text{V}. \tag{5.15}$$

So the input signal should not exceed +3.25 V and should not be under –3.25 V for the inverting amplifier not to saturate. Why? Because if the input signal is +3.25 V, then

$$3.25\ \text{V}.(-4) = -13\ \text{V}, \tag{5.16}$$

and when the input signal is

$$-3.25\ \text{V}.(-4) = +13\ \text{V}. \tag{5.17}$$

So keeping the input within –3.25 V and +3.25 V voltage range will prevent the op amp from saturating. This is sometimes referred to as *the op amp hitting or exceeding the power rails*.

What happens, with the circuit of Example 5.3, when we saturate the op amp-based inverting amplifier? If the input is outside of the voltage range discussed, that is, ±3.25 V, the output of the op amp will not try to go beyond its positive output saturation voltage (V_{SAT}) or below its negative saturation voltage ($-V_{SAT}$). Figure 5.6a,b,c present three examples of what occurs when the inverting amplifier op amp output becomes saturated. In essence, the op amp-based circuit ceases to work as the inverting amplifier that we were trying

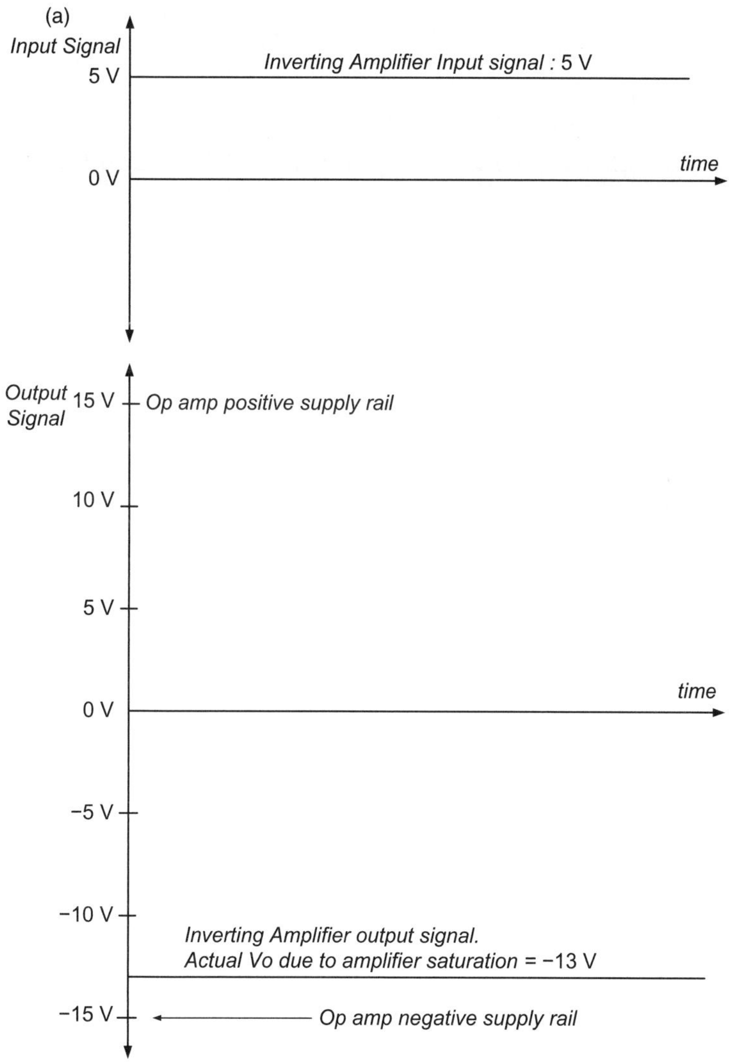

Figure 5.6 Inverting amplifier (*gain* = –4) circuit with signals exceeding the output saturation limits. (a) 5 V DC input, (b) –5 V DC input, and (c) 5 sin (2π 1 kHz) sinusoidal input.

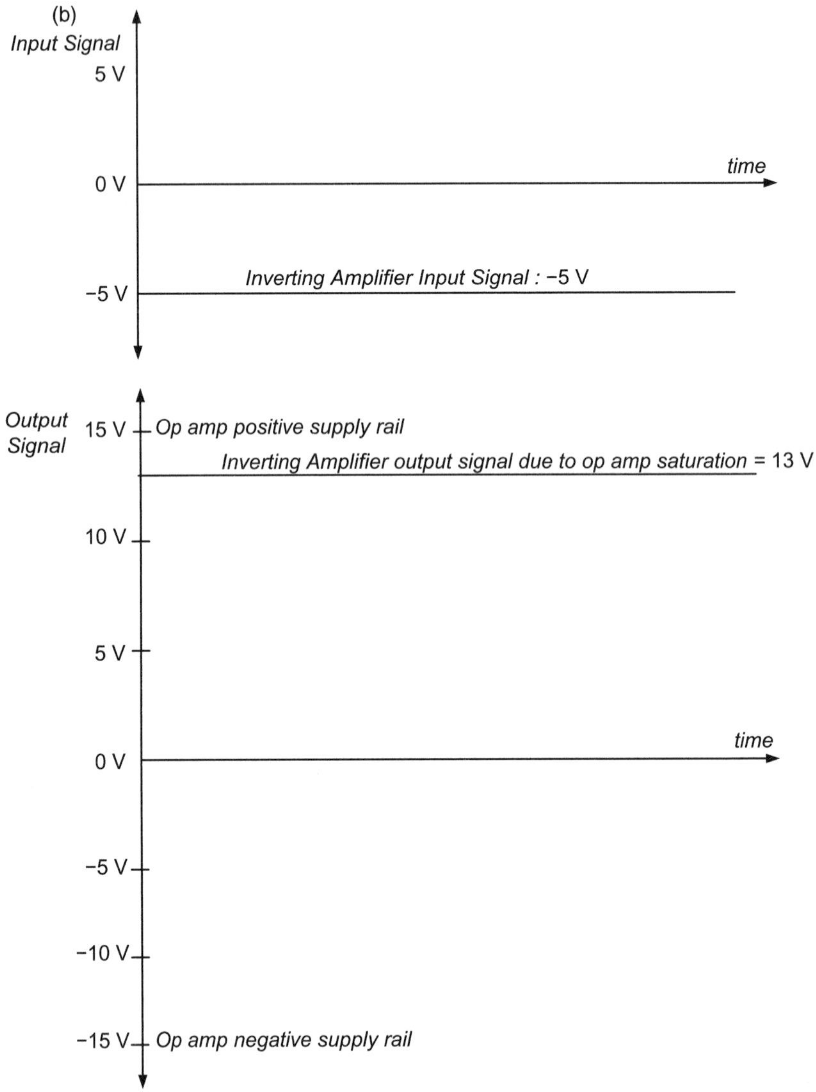

Figure 5.6 (*Continued*)

to implement. When the input times the amplifier gain exceeds the maximum voltage that the op amp can produce at its output, which is +15 V – 2 V = 13 V for positive outputs and when the op amp output is less than –15 V – (–2 V) = –13 V, the output of the op amp is clipped. This means that the output will never be over +13 V or be below –13 V. Referring again to Figure 5.6c, note that the clipped sinusoidal output waveform is a sinusoidal up until the clipping voltage limits are reached by the output. Such limits are

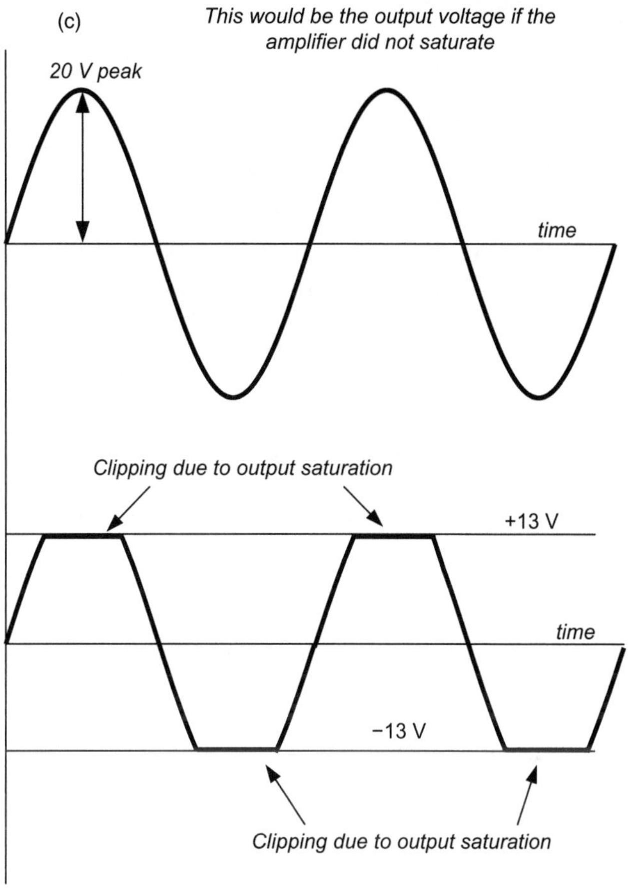

Figure 5.6 (*Continued*)

±13 V. Note that the top part of Figure 5.6c depicts the output as if the inverting amplifier could reproduce it with a linear gain of –4. The lower portion of Figure 5.6c shows what actually happens to the output because of the op amp saturation. This effect is called clipping and causes a usually undesired nonlinearity. Note: Figure 5.6c ignores the inverting sign of the CLG for simplicity.

Note that in all three cases, the maximum signal value times –4, the inverting gain of the amplifier, leads to a voltage of –20 V DC for case (a), +20 V DC for case (b), and –20 sin (2π 1 kHz) for (c). However, in all three cases, the inverting amplifier op amp output cannot go beyond ±13 V.

Figure 5.7 shows the same inverting amplifier of a gain of –4 V, when a sinusoidal signal of 3 sin (2π 1 kHz) gets well amplified by a factor of –4 without any clipping or distortion.

The output waveform in Figure 5.7 does not saturate on either positive or negative cycles.

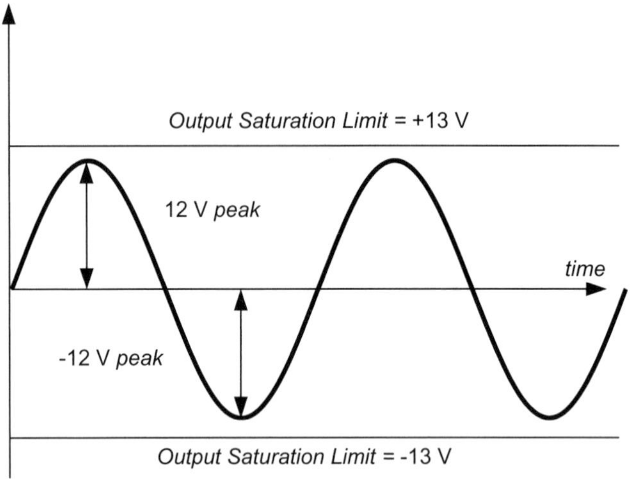

Figure 5.7 Inverting amplifier with a sinusoidal input that does not cause the op amp to saturate.

The 3 V peak input sinusoidal waveform is not shown in Figure 5.7. Since the inverting amp gain is –4 V, the magnitude of the positive and negative peaks of the output do not exceed ±12 V.

5.4.1.1.2 Powering and Decoupling the Op Amp-Based with Positive and Negative Supplies Most op amps require two-polarity power supplies. Only special op amps require a single polarity supply. Figure 5.8 shows the interconnection required for two 15-V power supplies for the op amp to effectively see a +15 V at its +VDD power pin and –15 V at its –VDD power pin. In addition to the power supplies, the op amp requires what is usually referred to as decoupling capacitors. Decoupling capacitors need to be placed very near the op amp power pins, to eliminate any stray inductance on the wire leads. It is the decoupling capacitors that keep the voltage across the op amp power pins constant and without electrical noise. When the output or inputs of the op amp make transitions, the power supply cannot instantaneously supply the ±15-V power that the op amp requires at all times to operate correctly. The instantaneous voltage during such time is supplied for a short time by the decoupling capacitors, until the power supplies have time to respond to the transient.

What value of decoupling capacitor we need? As usual we need to make some assumptions about the situation, if that is not already given to us. Assume the power supplies can respond to the changes in power supply current demand in 1 ms, but no sooner than that. Then we have to size the $\pm V_{DD}$ decoupling capacitors to hold voltage level of the supplies constant for at least 1 ms at not less than the normal V_{DD} value minus 100 mV. Also assume that the operational amplifier requires at most 10 mA of current of each of its supplies. However, the capacitor will not be able to maintain strictly a constant voltage.

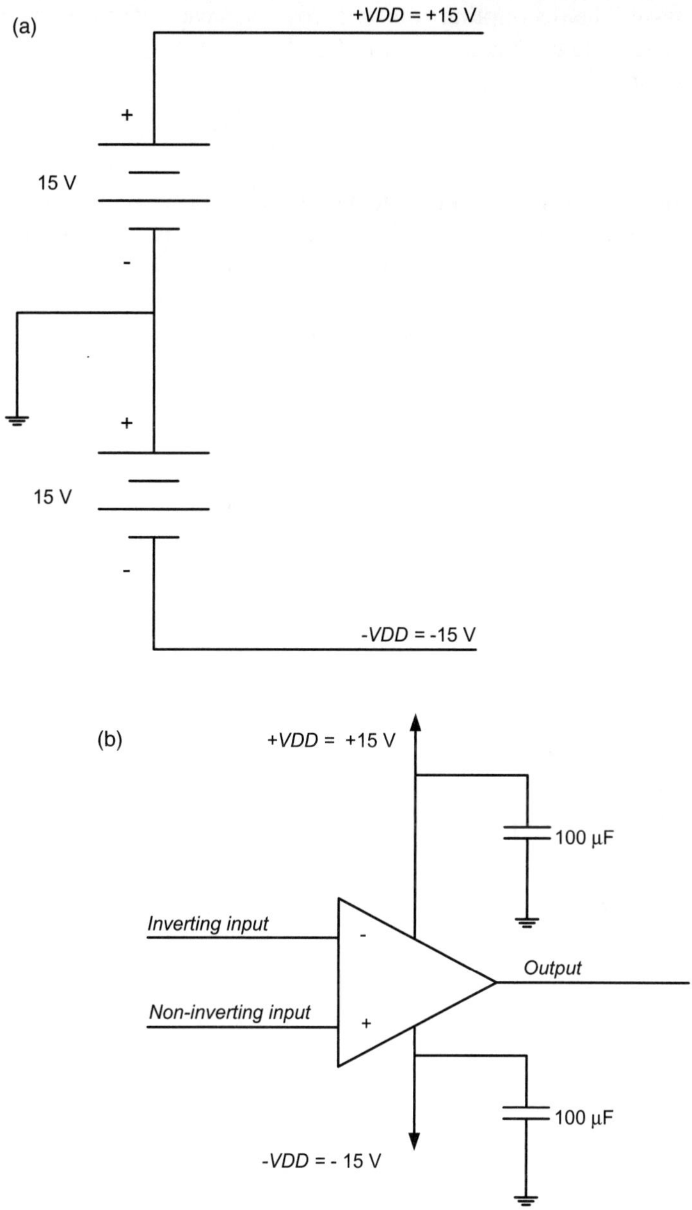

Figure 5.8 Op amp power (a) $+V_{DD}$ and $-V_{DD}$, (b) decoupling capacitors.

As time passes, the decoupling voltage across each capacitor will droop a little bit. Why? The voltage–current relationship that governs the electrical behavior of a capacitor is

$$i(t) = C\,dv(t)/dt. \tag{5.18}$$

So let us use Equation (5.18) to calculate a capacitor value that will hold the voltage to 100 mV or less for 1 ms. Plugging in the numbers into Equation (5.18) and replacing differentials by finite increments we obtain

$$i(t) = C\,\Delta v(t)/\Delta t, \tag{5.19}$$

where

$$C = i(t)\Delta t/\Delta v(t) \tag{5.20}$$

$$C = 0.010 \times 0.001/0.1 = 100\ \mu\text{F}. \tag{5.21}$$

Figure 5.8a depicts the generation of $+V_{DD}$ and $-V_{DD}$, and Figure 5.8b shows the 100 μF decoupling capacitors just calculated, connected to the op amp power supply pins.

Note that the other terminal of each capacitor not connected to a power supply rail, connects to ground.

After the charge of the capacitors is depleted, so that they would not be able to continue to supply the 15 V less 100 mV, the power supply is ready to supply current again and not only powers the op amp at this time, but also recharges the decoupling capacitors.

5.4.2 The Noninverting Amplifier

An op amp-based noninverting amplifier circuit is shown in Figure 5.9.

Let us point out the similarities and differences that exist between this circuit and the inverting amplifier of Figure 5.4. Both amplifiers have negative feedback. Note that on both amplifiers, the output voltage is sampled and fed back into the inverting input. The noninverting amplifier, however, has the input voltage V_1 applied to the positive or noninverting input; this is not the case for the inverting amplifier. Assuming that the op amp is ideal, we will calculate the output to input voltage ratio or the closed-loop transfer function of the noninverting amplifier. Remember that our ideal amplifier, however, requires power, and it will saturate if the output gets too close to either supply rail.

Node A is virtually close to input voltage V_1, because the amplifier uses negative feedback and $\Delta V$is practically zero; the voltage at node A is V_1. So, by inspection of Figure 5.9, we have

$$I_1 = -V_1/R_1 \tag{5.22}$$

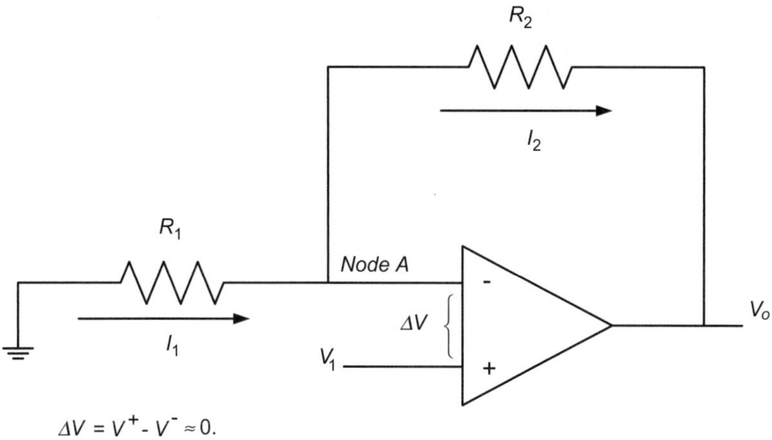

Figure 5.9 Noninverting amplifier.

and

$$V_1 - V_o = I_2 R_2. \tag{5.23}$$

Now, since there is no current entering or exiting the op amp inverting terminal,

$$I_1 = I_2. \tag{5.24}$$

Combining Equations (5.22) through (5.24) yields

$$\frac{V_o}{V_1} = 1 + \frac{R_2}{R_1}. \tag{5.25}$$

Equation (5.25) is the approximated transfer function of the closed loop gain of the noninverting amplifier. It is approximated because ideal op amp parameters have been assumed (Table 5.1). Note that the sign of the output matches the sign of the input waveform. Additionally, it is important to note that the CLG ($1 + R_2/R_1$) is always strictly greater than one if both resistors are greater than 0 Ω. The noninverting amplifier ±input signal times the CLG ($1 + R_2/R_1$) must be less than the absolute value of the power supply rail.

Example 5.4 Given a noninverting amplifier like the one shown in Figure 5.9, assuming that the op amp is ideal, but with an op amp $V_{SAT} = \pm 13$ V, and input V_1 is ±1 V maximum, determine the maximum CLG of the noninverting amplifier that will not allow the output to become saturated. Assume ±15 V power supplies.

Solution: Since the closed loop of the noninverting amplifier is: $(1 + R_2/R_1)$, the supply rail is +15 V, and $V_{SAT} = 13$ V, the maximum positive and negative swings that the amplifier can have is ±13 V. Note that this amplifier will only swing to the negative voltage rail if the input is negative.

$$\pm 13\text{ V} / 1\text{ V} = \pm 13. \tag{5.26}$$

Then using Equation (5.26) with Equation (5.25) we have

$$\pm V_o / V_1 = 1 + \frac{R_2}{R_1} = \pm 13. \tag{5.27}$$

From Equation (5.17) we determine that the R_2/R_1 ratio must equal 12. So for example $R_2 = 12\ k\Omega$ and $R_1 = 1\ k\Omega$.

5.4.2.1 Effects of Finite Op Amp Open-Loop Gain in the Closed-Loop Equation of the Noninverting Amplifier

In a very similar manner as it was done with the inverting amplifier configuration assuming a finite open-loop gain op amp model, we arrive at the noninverting amplifier CLG which is:

$$V_o / V_1 = \frac{1 + R_1 / R_2}{1 + \frac{1}{A_{OL}}(1 + R_1 / R_2)}. \tag{5.28}$$

The op amp circuit model used to derive Equation (5.28) is given by the circuit of Figure 5.10 where A_{OL} is assumed to be finite.

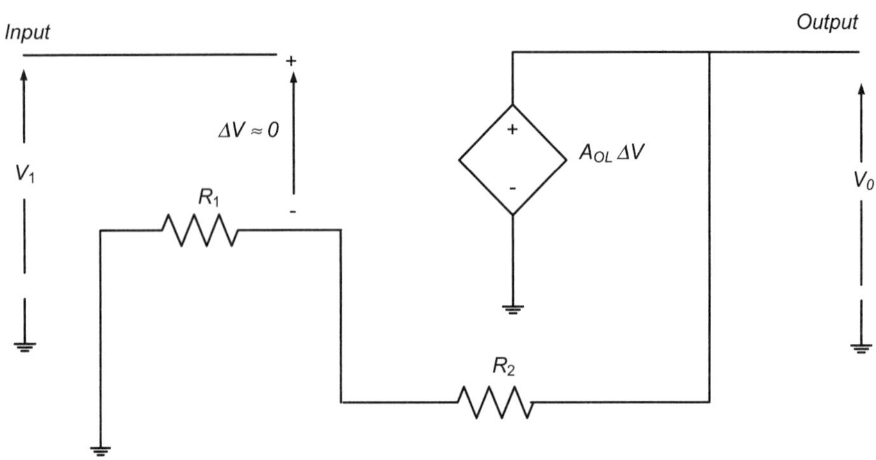

Figure 5.10 Model of noninverting amplifier with finite A_{OL}.

Table 5.4 Noninverting amplifier closed-loop gain (CLG) errors for finite A_{OL}: for an estimated CLG of *2*

Finite Op Amp Open-Loop Gain (A_{OL})	Op Amp A_{OL} in dB	Closed-Loop Gain of 2 (CLG = 2) with Finite A_{OL}	CLG = 2 Absolute Error	CLG = 2 Relative Error
(V/V)	(dB)	(V/V)	(V/V)	(%)
1	0	1.333333	0.666667	50.000000
10	20	1.833333	0.166667	9.090909
100	40	1.980392	0.019608	0.990099
1,000	60	1.998004	0.001996	0.099900
10,000	80	1.999800	0.000200	0.009999
100,000	100	1.999980	0.000020	0.001000
1,000,000	120	1.999998	0.000002	0.000100
10,000,000	140	2.000000	0.000000	0.000010

Table 5.5 Noninverting amplifier closed-loop gain (CLG) errors for finite A_{OL}: for an estimated CLG of 20

Finite Op Amp Open-Loop Gain (A_{OL})	Op Amp A_{OL} in dB	Closed-Loop Gain of 20 (CLG = 20) with Finite A_{OL}	CLG = 20 Absolute Error	CLG = 20 Relative Error
(V/V)	(dB)	(V/V)	(V/V)	(%)
1	0	0.952381	19.047619	2,000.000000
10	20	6.666667	13.333333	200.000000
100	40	16.666667	3.333333	20.000000
1,000	60	19.607843	0.392157	2.000000
10,000	80	19.960080	0.039920	0.200000
100,000	100	19.996001	0.003999	0.020000
1,000,000	120	19.999600	0.000400	0.002000
10,000,000	140	19.999960	0.000040	0.000200

Using Equation (5.25) as the estimated value of the CLG and Equation (5.28) as the true value of the CLG for an op amp with finite open-loop gain A_{OL}, we can calculate the absolute and relative errors of the noninverting amplifier CLG equation when the op amp A_{OL} is finite.

Tables 5.4 and 5.5 show the errors that exist for a noninverting amplifier configuration for CLGs of 2 and 20 for various values of A_{OL}.

5.4.3 The Buffer or Noninverting Amplifier of Unity Gain

The buffer amplifier is a special case of the noninverting amplifier. Referring to the circuit in Figure 5.9, if R_2 approaches 0 and R_1 approaches infinity, then $V_o = V_1$.

For the reader's convenience, we repeat the gain equation of the noninverting amplifier given by Equation (5.25):

$$\frac{V_o}{V_1} = 1 + \frac{R_2}{R_1}. \tag{5.29}$$

Expressing V_o as a function of V_1 and resistors R_1 and R_2, and taking the limit of the expression for $R_1 \to \infty$ and $R_2 \to 0$, we obtain

$$\lim_{\substack{R_1 \to \infty \\ R_2 \to 0}} V_o = (1 + R_2 / R_1) V_1 = V_1. \tag{5.30}$$

Figure 5.11a depicts a unity gain noninverting amplifier or simply a buffer.

So what is the meaning of $V_o = V_1$? Literally, it means that the output voltage is equal to the input voltage. If we connect just a wire between V_1 and V_0 without having any op amp in between, we get the same voltage relationship. To explain the importance of the circuit of Figure 5.11a we need to assume that the idealized input resistance of the op amp is actually finite and not infinite like it is for the ideal op amp model. Additionally, we will assume that the op amp A_{OL} is also finite. This op amp model described with finite input resistance and open-loop gain is presented in Figure 5.11b and the buffer amplifier, using the op amp model from Figure 5.11b, is shown in Figure 5.11c

Writing the Kirchhoff equations for the circuit of Figure 5.11c and realizing that

$$V_1 - V_o = V_1 - A_{OL}(V_1 - V_o),$$

we find that

$$A_{CL} = V_o / V_1 = A_{OL} / (A_{OL} + 1) \approx 1 \tag{5.31}$$

$$R_{inCL} = (A_{OL} + 1) R_{in} \approx A_{OL} R_{in}. \tag{5.32}$$

A_{CL} is the CLG of the buffer amplifier and R_{inCL} is the closed-loop input resistance of the buffer amplifier. Because for a buffer amplifier the input voltage is quite approximately equal to the output voltage, Equation (5.31), this amplifier is also called a *voltage follower*.

R_{in} is the op amp input resistance, usually of several mega-ohms. A_{OL} is the op amp open-loop gain. R_{inCL} stands for the closed-loop input resistance of the buffer amplifier, not the op amp input resistance.

From Equation (5.32) it is clear to see that since R_{in} is in the order of several mega-ohms, A_{OL} is in the order of 10^6 V/V, the closed-loop input resistance of the buffer amplifier configuration (see Figure 5.11c) is in the order of 10^{12} Ω. The closed-loop input resistance of the buffer amplifier is the effective resistance that the input V_1 sees at the noninverting input terminal of the amplifier.

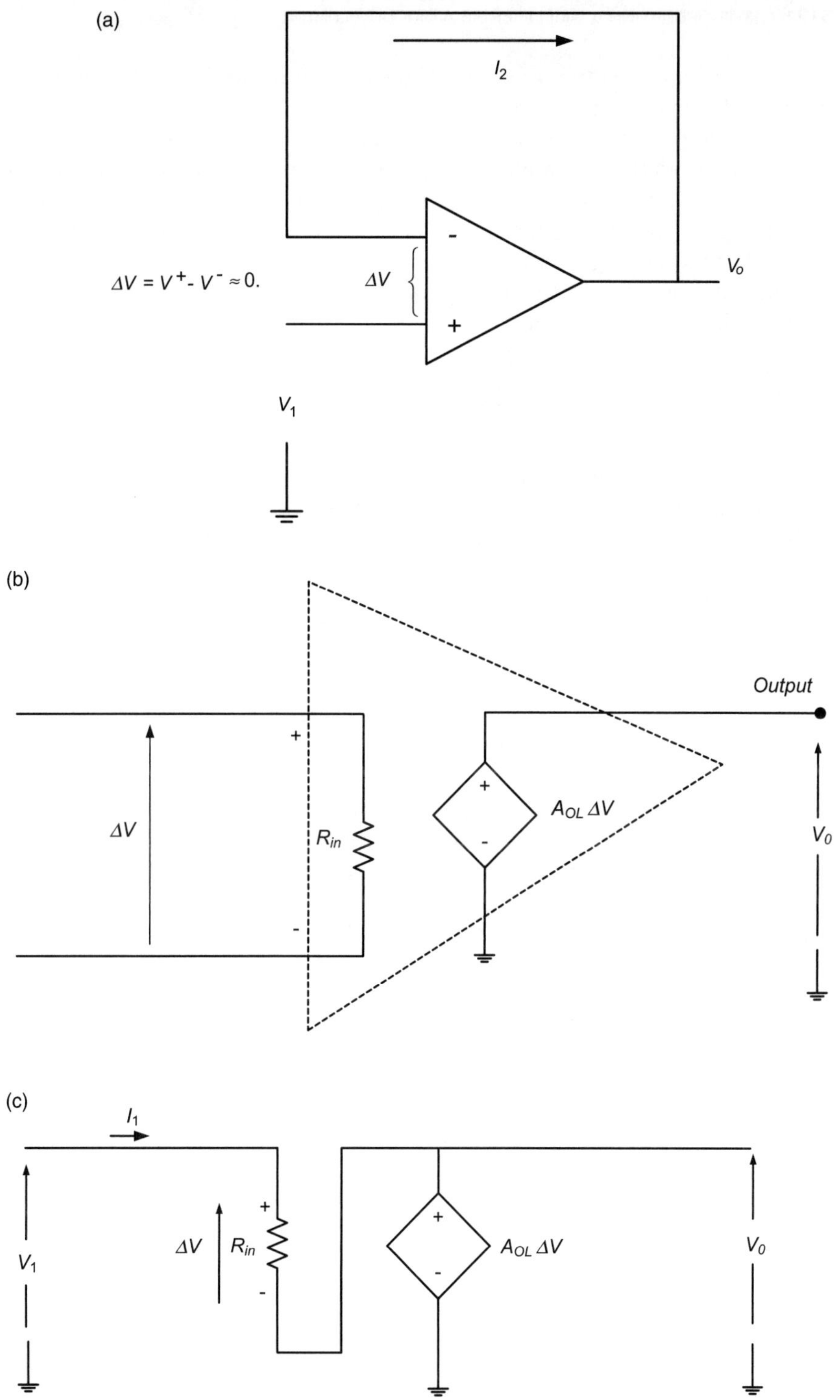

Figure 5.11 (a) Buffer amplifier, (b) op amp with model finite R_{in} and A_{OL}, (c) actual buffer amplifier circuit using the model from part (b).

So for input signals in the order of 1 V, the current drawn by the buffer amplifier is in the order of pico amps (10^{-12} A)!

Based on what we just discussed, the buffer amplifier, which operates as a unity gain noninverting amplifier, is in effect a current amplifier. The output of the buffer circuit is converted to a voltage with much higher drive capability than the pico amp input current. It can also be proven from the circuit of Figure 5.11c by adding a finite (nonzero) R_{out} between the plus sign of the A_{OL} Δ_V-controlled source and the output voltage V_o, that the effective output resistance of the buffer amplifier is

$$R_{outCL} = \frac{R_{out}}{(1 + A_{OL})}, \tag{5.33}$$

where R_{out} is the op amp output resistance, which we have assumed to be zero in the ideal op amp model. In a real op amp, R_{out} is larger than zero and typically is a small fraction of an ohm to a few ohms. R_{outCL} is the effective output resistance that the buffer amplifier presents to a load that can be connected at its output. Let us consider the following numerical example to quantify the significance of finite R_{in} and finite A_{OL}.

Example 5.5 The circuit of Figure 5.12a consists of a resistor divider formed with R_1 and R_2. The intent of this resistor divider is to provide 5 V to any load. This will only work with some accuracy if the load resistance is much higher than 1 kΩ. But in the circuit example the load resistance is also 1 kΩ. The load has a loading effect on the 5 V at node A of the resistor divider. After stating the circuit Kirchhoff equations and solving them we determine that the voltage across R_{LOAD} is 3.333 V, this is considerably lower than the desired 5 V. Figure 5.12b shows the circuit divider circuit to which a buffer amplifier configuration was added to right side of node A. The output of the buffer drives R_{LOAD}. Since the buffer does not draw any significant current from the resistor divider, the buffer supplies to R_{LOAD} practically the 5 V at the input of the buffer to R_{LOAD}. The advantage of this circuit (Fig. 5.12b) is that for load resistances much larger than 1 kohm, the op amp-based buffer can supply the current required by the load. To continue to work with this example, consider the circuit of Figure 5.12a and calculate the voltage at R_{LOAD}, when R_{LOAD} = 1 Ω, 100 Ω, 10 kΩ, 100 kΩ, and 1 MΩ, refer to Table 5.6 for the numerical answers of this example.

By Kirchhoff, the voltage at node A is

$$V_A = \frac{V_{input} \cdot (R_2 // R_{LOAD})}{R_1 + (R_2 // R_{LOAD})}. \tag{5.34}$$

Plugging into Equation (5.34) 1 Ω, 100 Ω, 10 kΩ, 100 kΩ, and 1 MΩ we obtain Table 5.6.

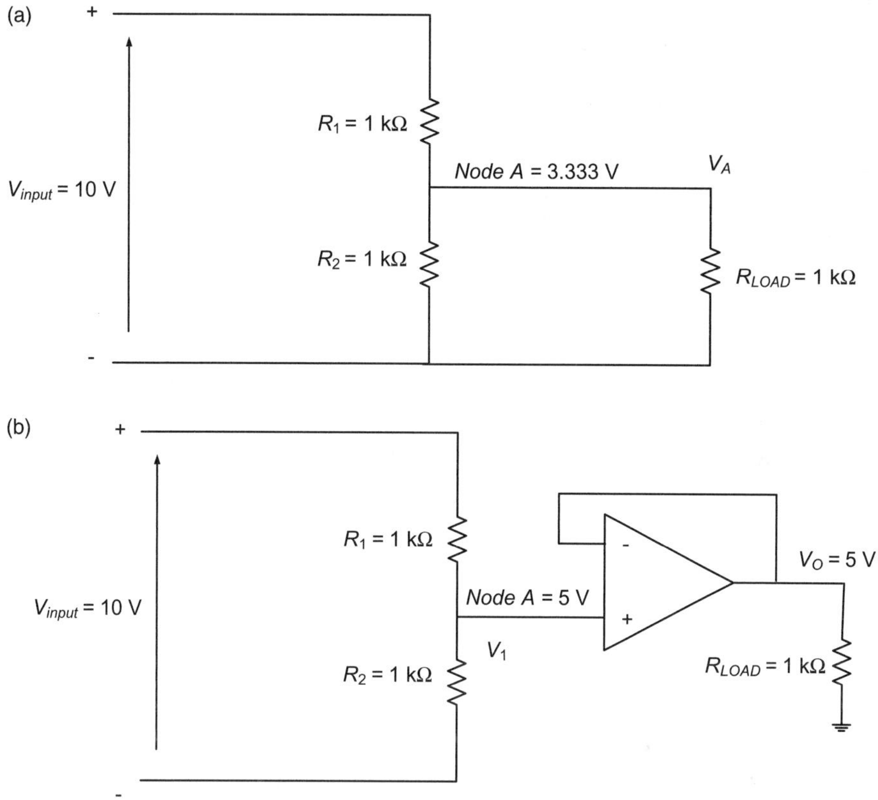

Figure 5.12 (a) Resistor divider loaded with a resistor at node *A* and no buffer amplifier, (b) resistor divider load with a resistor after buffering node *A*, with a voltage follower.

Table 5.6 Load effect on circuit without and with buffering op amp

R_{LOAD} (Ω)	V_A (V); Figure 5.12a without Op Amp	V_o (V); Figure 5.12b with Op Amp
1	0.00998004	5.0
100	0.098039216	5.0
10,000	4.761904762	5.0
100,000	4.975124378	5.0
1,000,000	4.997501249	5.0

It is interesting to note that the larger the load resistance value is with respect to the values of the resistor dividers, when no buffer amplifier is used, then the better is the output voltage accuracy. For instance, for 1 MΩ load resistance, without using a buffer amp, the voltage across the load is quite close to 5 V. For 1 Ω, that is not the case; that is, 0.00998004 V is much lower when compared against the 5 V obtained using the buffer amplifier.

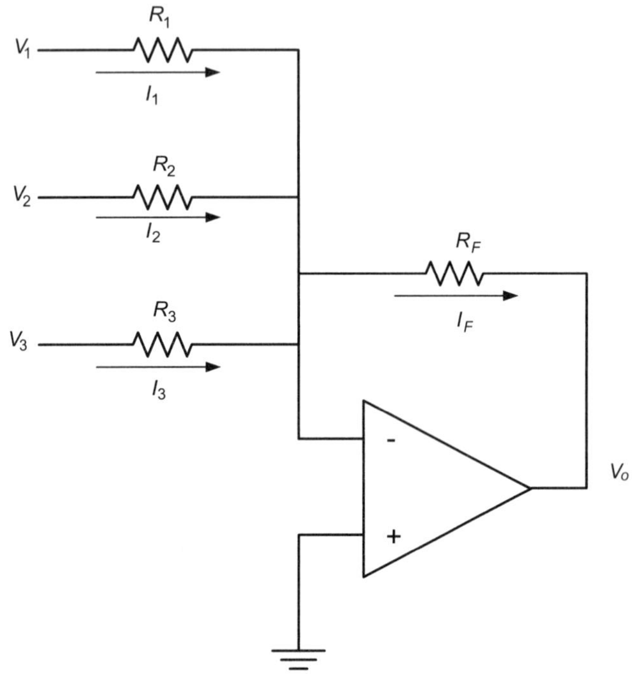

Figure 5.13 Three-input inverting adder amplifier.

5.4.4 The Inverting Adder

Figure 5.13 shows the circuit topology of an inverting adding amplifier. We only show a three-input circuit; however; the circuit can easily be generalized to n-inputs, where n is an integer. Let us analyze this circuit to determine the output voltage as a function of all the circuit resistors and the input voltages. We will do the analysis assuming that the op amp is ideal.

Since the op amp is ideal and the circuit has negative feedback, the voltage difference across the inverting and noninverting inputs of the op amp is zero ($\Delta V = V^+ - V^- = 0$).

Additionally, since the input resistance of the op amp is infinite and using KCL we have that

$$I_1 + I_2 + I_3 = I_F. \tag{5.35}$$

Now since the negative input of the op amp is virtually grounded, we have using Equation (5.35) and applying Ohms law to each circuit branch:

$$\frac{V_1}{R_1} + \frac{V_2}{R_2} + \frac{V_3}{R_3} = -\frac{V_o}{R_F}. \tag{5.36}$$

Doing some algebra on Equation (5.36) we obtain

$$V_o = -R_F\left(\frac{1}{R_1}V_1 + \frac{1}{R_2}V_2 + \frac{1}{R_3}V_3\right). \tag{5.37}$$

We usually prefer to make $R_1 = R_2 = R_3 = R$ so that Equation (5.37) becomes

$$V_o = -\frac{R_F}{R}(V_1 + V_2 + V_3). \tag{5.38}$$

Equation (5.38) is also referred to as the inverting adder with constant gain output voltage, since the ratio R_F/R is the same for all the inputs. When the input signals are audio frequency signals, the circuit is also called an *audio mixer*.

Example 5.6 Derive the output voltage equation of an n-input inverting adder with constant gain circuit. Assume the op amp is ideal.

Proceeding just like we did to obtain Equations (5.36)–(5.38), we simply generalize them to have n-inputs and obtain the following:

$$\frac{V_1}{R_1} + \frac{V_2}{R_2} + \frac{V_3}{R_3} + \cdots + \frac{V_n}{R_n} = -\frac{V_o}{R_F} \tag{5.39}$$

$$V_o = -R_F\left(\frac{1}{R_1}V_1 + \frac{1}{R_2}V_2 + \frac{1}{R_3}V_3 + \cdots + \frac{1}{R_n}V_n\right). \tag{5.40}$$

We usually prefer to make $R_1 = R_2 = R_3 = \cdots = R_n = R$ so that Equation (5.40) becomes

$$V_o = -\frac{R_F}{R}(V_1 + V_2 + V_3 + \cdots + V_n). \tag{5.41}$$

Important note about inverting adders of any number of inputs:

In order for the inverting adder circuit to operate linearly and without saturation, it is required that

$$|V_{SAT}| > \left|R_F\left(\frac{1}{R_1}V_1 + \frac{1}{R_2}V_2 + \frac{1}{R_3}V_3 + \cdots + \frac{1}{R_n}V_n\right)\right|. \tag{5.42}$$

And when $R_1 = R_2 = R_3 = \cdots = R_n = R$, then

$$|V_{SAT}| = \left|\frac{R_F}{R}(V_1 + V_2 + V_3 + \cdots + V_n)\right|.$$

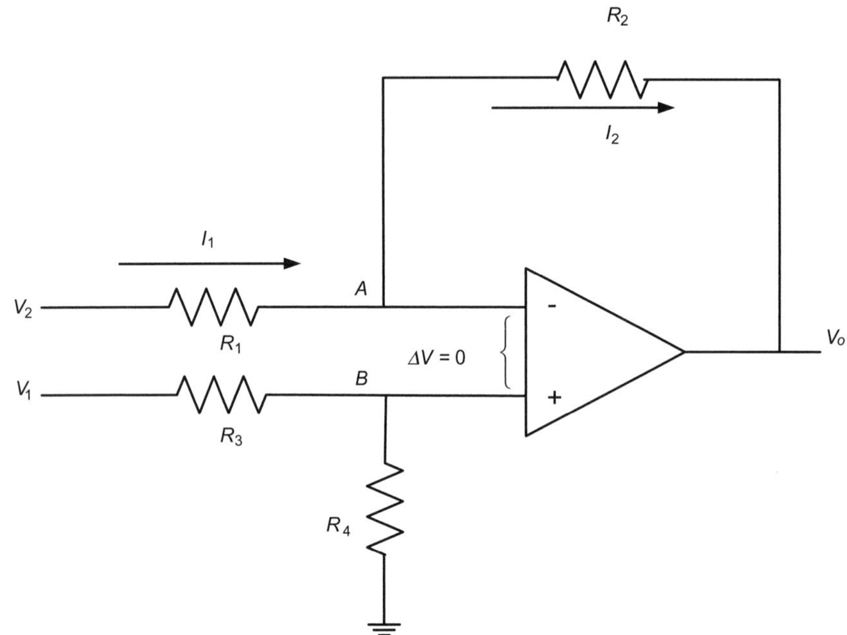

Figure 5.14 Op amp-based difference amplifier.

5.4.5 The Difference Amplifier

Figure 5.14 shows the circuit of an op amp-based difference amplifier. Let us analyze the circuit to calculate the output to inputs relationship of this amplifier.

First, let us note that the op amp has negative feedback, like all previous configurations did. The feedback is negative because the output is sampled and injected back into the inverting terminal of the op amp. Then assuming our op amp is ideal and since it has negative feedback, the ΔV or the voltage difference between the noninverting and inverting inputs is infinitely small or practically zero.

By inspection of Figure 5.14 we see that ΔV, which is the difference between node voltages B and A, is practically zero. So the voltage at node A is identical to the voltage at node B. We will refer to this voltage as V_A. Moreover, it is easy to see that

$$V_A = V_B = \frac{R_4}{R_3 + R_4} V_1. \tag{5.43}$$

The current that flows through resistor R_1 is

$$I_1 = (V_2 - V_A) / R_1. \tag{5.44}$$

Plugging the value of V_A from Equation (5.43) into Equation (5.44) we obtain

$$I_1 = \frac{V_2 - \frac{R_4}{R_3 + R_4} V_1}{R_1}. \tag{5.45}$$

Current I_2 through resistor R_2 is

$$I_2 = (V_A - V_o) / R_2. \tag{5.46}$$

Again plugging the value of V_A from Equation (5.43) into Equation (5.46) we obtain

$$I_2 = \frac{\frac{R_4}{R_3 + R_4} V_1 - V_0}{R_2}. \tag{5.47}$$

Since $I_1 = I_2$, equating Equations (5.45) and Equation (5.47) we get

$$\frac{V_2 - \frac{R_4}{R_3 + R_4} V_1}{R_1} = \frac{\frac{R_4}{R_3 + R_4} V_1 - V_0}{R_2}. \tag{5.48}$$

Carefully doing the algebra on Equation (5.48) we obtain the output of the difference amplifier as a function of its two input voltages and resistors R_1 through R_4:

$$V_o = \left(\frac{R_1 + R_2}{R_1}\right)\left(\frac{R_4}{R_3 + R_4}\right) V_1 - \frac{R_2}{R_1} V_2. \tag{5.49}$$

Equation (5.49) is the difference amplifier transfer function when all resistors (R_1 through R_4) are different in value. Additionally, Equation (5.49) is the expression of the difference amplifier gain when the op amp is assumed to be ideal.

Ideal difference amplifiers only amplify the difference of the two input signals and fully reject the average of the sum of the two input signals, also referred to as the common mode input signal. This means that Equation (5.50) is nonzero and Equation (5.51) is zero for an ideal op amp-based difference amplifier.

$$\textit{Differential input signal}: V_{idiff} = V_1 - V_2 \tag{5.50}$$

$$\textit{Common mode input signal}: V_{icm}\ ½(V_1 + V_2). \tag{5.51}$$

However, in the real world, difference amplifiers will not only amplify the differential mode input signal, but also the common mode signal to some extent.

The following equation is an expression of the output voltage produced by the presence of both types of gains, the differential and the common mode:

$$V_o = A_{diff} V_{idiff} + A_{cm} V_{icm}. \tag{5.52}$$

It is important to note that the difference amplifier of Figure 5.14 is nonideal even if it is implemented with an ideal operational amplifier. Why is this so? Because a real difference amplifier, regardless of using an ideal or a real op amp, has both differential and common mode gains that are not zero. The common mode gain is different from zero due greatly to the resistor inaccuracies. Using the circuit of Figure 5.14 we can calculate the value of the common mode gain by injecting an input of the same polarity to both the inverting and noninverting inputs of the difference amplifier. Figure 5.15 depicts the combined input signal to determine the difference amplifier common mode gain.

If the op amp used for the difference amplifier is ideal, again we have that the ΔV is zero, that is, node A and node B are at the same voltage level; however, the voltage at node A (and B) is not zero. Again making the usual assumptions about ideal op amps, we calculate the voltage at node B, as the one produced by the common mode input voltage V_{icm} and the resistor divider formed by R_3 and R_4 (Fig. 5.15):

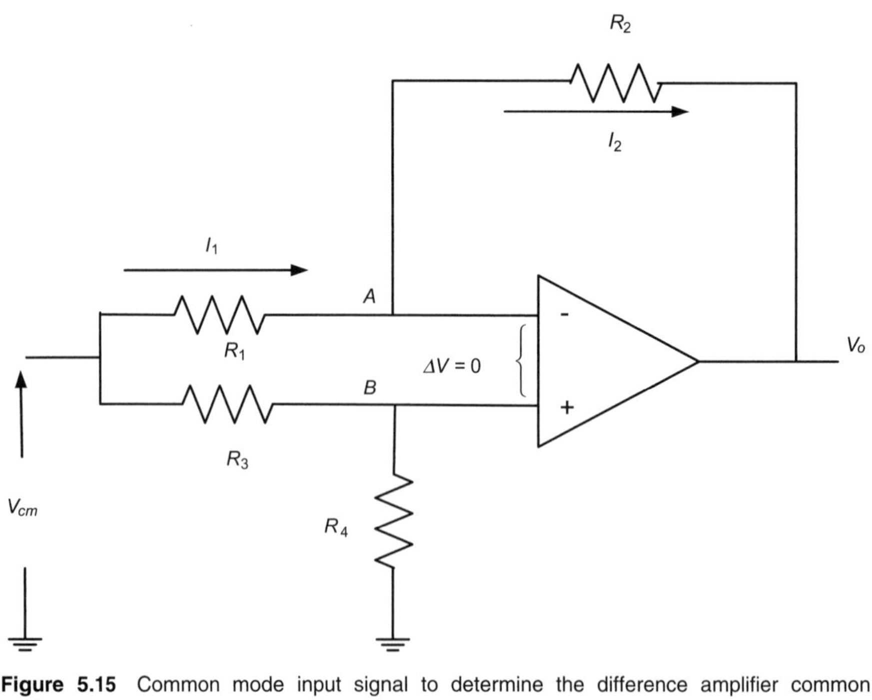

Figure 5.15 Common mode input signal to determine the difference amplifier common mode gain.

$$V_A = V_B = V_{cm}\left(\frac{R_4}{R_3 + R_4}\right). \tag{5.53}$$

From the circuit of Figure 5.15 we see that

$$I_1 = \frac{V_{cm} - V_A}{R_1}. \tag{5.54}$$

Plugging Equation (5.53) into Equation (5.54) we obtain

$$I_1 = \frac{V_{cm} - \dfrac{R_4}{R_3 + R_4}V_{cm}}{R_1}. \tag{5.55}$$

We also have that

$$I_2 = \frac{\dfrac{R_4}{R_3 + R_4}V_{cm} - V_o}{R_2}. \tag{5.56}$$

And since $I_1 = I_2$, we get

$$\frac{V_{cm} - \dfrac{R_4}{R_3 + R_4}V_{cm}}{R_1} = \frac{\dfrac{R_4}{R_3 + R_4}V_{cm} - V_o}{R_2}. \tag{5.57}$$

Doing some algebra on Equation (5.57) and expressing everything in terms of V_o/V_{cm} we get

$$\frac{V_o}{V_{cm}} = \frac{R_3}{R_3 + R_4}\left(1 - \frac{R_2}{R_1}\frac{R_3}{R_4}\right). \tag{5.58}$$

Equation (5.58) is the equation of the common mode gain of the difference amplifier.

Since we want a difference amplifier to have a zero common mode gain, so that it amplifies only differential signals and it eliminates common mode input signals. For the common mode gain to be zero, the right-hand side term of Equation (5.58) needs to be zero. This is achieved by having the following resistor ratios:

$$\frac{R_2}{R_1} = \frac{R_4}{R_3}. \tag{5.59}$$

When Equation (5.59) is met. the common mode gain of the difference amplifier with an ideal op amp becomes zero. Thus:

$$A_{cm} = \frac{V_o}{V_{cm}} = 0. \tag{5.60}$$

Example 5.7 Given a difference amplifier just like the one in Figure 5.14, determine the value of the differential mode gain, using Equation (5.52) as the condition that inhibits the common mode gain. Assume that the op amp is ideal.

So using the general expression for the difference amplifier gain from Equation (5.49), which we repeat here for the reader's convenience is

$$V_o = \left(\frac{R_1 + R_2}{R_1}\right)\left(\frac{R_4}{R_3 + R_4}\right)V_1 - \frac{R_2}{R_1}V_2. \tag{5.61}$$

Equation (5.61) consists of differential and common mode gains. Doing some algebraic manipulations on Equation (5.61) we obtain

$$V_o = \left(1 + \frac{R_2}{R_1}\right)\frac{1}{R_3}\frac{R_4}{1 + \frac{R_4}{R_3}}V_1 - \frac{R_2}{R_1}V_2. \tag{5.62}$$

Then, plugging the condition given by Equation (5.59) into Equation (5.62), we get an expression for the differential amplifier gain, rid of any common mode gain, thus:

$$V_o = \frac{R_2}{R_1}(V_1 - V_2). \tag{5.63}$$

Referring once more to Figure 5.14 note that V_1 is the input to the noninverting side of the difference amplifier, while V_2 is the input to the inverting side of the amplifier.

A closer look at Equation (5.63) reveals that the difference between V_1 and V_2 is amplified by the ratio of R_2/R_1. This ratio is called the difference amplifier gain. Remember that Equation (5.63) is valid when the condition given by Equation (5.59) is met.

Additionally, if

$$R_1 = R_2 = R_3 = R_4, \tag{5.64}$$

Equation (5.63) becomes

$$V_o = V_1 - V_2. \tag{5.65}$$

Equation (5.65) is the expression of the difference amplifier output voltage strictly as a function of the difference between voltages V_1 and V_2.

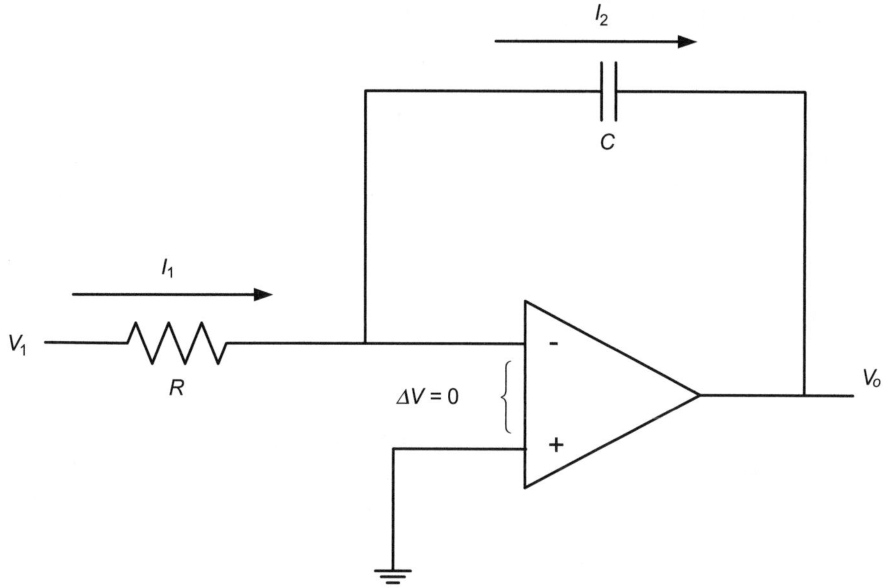

Figure 5.16 Pure integrator circuit using an ideal op amp.

5.4.6 The Inverting Integrator

If we have an inverting amplifier configuration and replace the resistor in the feedback loop with a capacitor, the circuit obtained is an integrator. Figure 5.16 depicts an integrator using an op amp. The output to input voltage ratio in the frequency domain is then

$$V_o / V_1 = -\frac{1/j\omega C}{R} = -\frac{1}{j\omega RC}. \tag{5.66}$$

In the time domain and referring to Figure 5.16 we have that

$$I_1 = I_2, \tag{5.67}$$

where

$$I_1 = V_1 / R \tag{5.68}$$

and

$$I_2 = -C\frac{dV_o}{dt}. \tag{5.69}$$

Combining Equations (5.67) through (5.69 yields

$$\frac{V_1}{R} = -C\frac{dV_o}{dt}. \tag{5.70}$$

Integrating Equation (5.70) and expressing the output voltage V_o, the following is obtained:

$$V_o = -\frac{1}{RC}\int V_1 dt. \tag{5.71}$$

The Bode plots of an integrator given by Equation (5.66) are shown in Figure 5.17.

The integrator design frequency is $f_o = 1/2\pi\ RC$ where R and C are the values of the resistor and capacitor of the integrator circuit shown in Figure 5.16. It is important to notice the difference between the op amp-based integrator of Figure 5.16 and the first order *RC* LPF from Chapter 4, Figure 4.4. The op amp-based integrator has an infinite gain at DC or at zero frequency. The gain or magnitude decreases at a constant rate of 20 dB *per decade*. The first-order *RC* LPF integrator circuit, which has no op amp, has a 0 dB gain at frequencies below the integrator f_0 cutoff frequency and the gain decays at a constant rate of 20 dB per decade above f_0. Refer to this previously discussed circuit Bode plots in Figure 4.4. So what does this all mean in terms of practical operation of the integration? The op amp-based integrator has an infinite gain at DC, so if the input signal to be integrated is constant, the op amp will saturate. That is the reason why op amp integrators implemented like in Figure 5.16 are not quite that practical. One needs to add a semiconductor switch in parallel with the capacitor to reset the op amp output by discharging the capacitor, to let the integration restart. Another issue with the op amp-based integrator of Figure 5.16 is that any noisy signals below the integrator f_0 frequency become amplified more than the integrated frequency. We will shortly address a practical integrating op amp-based circuit that performs better than the one being presented and does not have the deficiencies just mentioned. It is also important to mention that the circuit of Figure 5.16 has a constant phase shift of –90° for all frequencies, whereas the first-order *RC* integrator does not (refer to Figure 4.4).

5.4.7 The Inverting Differentiator

If we have an inverting amplifier configuration and replace the resistor in the input path with a capacitor, the circuit obtained is a differentiator. Figure 5.18 depicts a differentiator using an op amp. The output to input voltage ratio in the frequency domain is then

$$V_o / V_i = -j\omega RC. \tag{5.72}$$

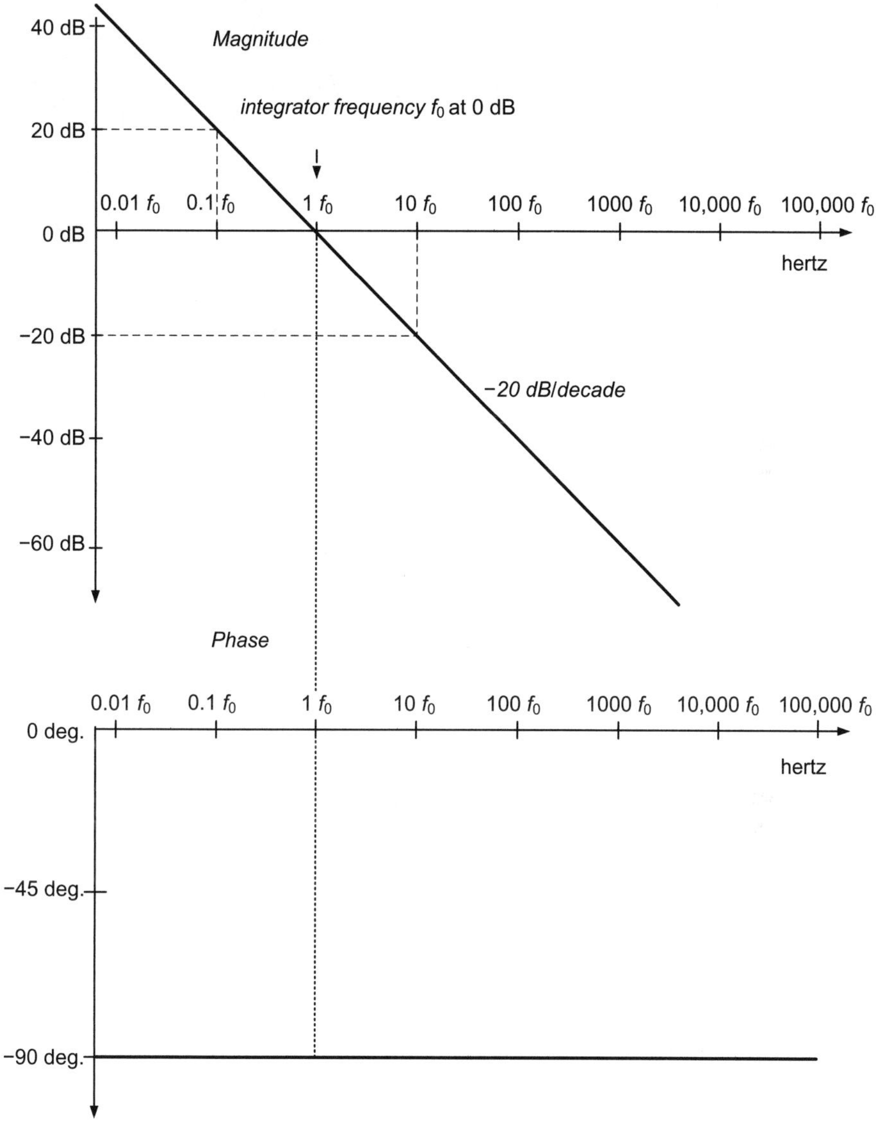

Figure 5.17 Op amp-based pure integrator asymptotic Bode plots.

In the time domain and referring to the differentiator of Figure 5.18 we have that

$$I_1 = I_2, \tag{5.73}$$

where

$$I_1 = \frac{V_1}{\dfrac{1}{j\omega C}} \tag{5.74}$$

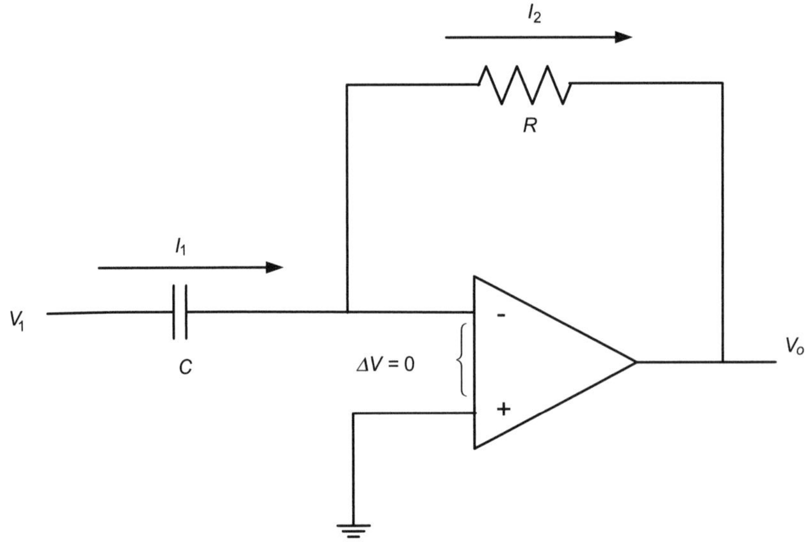

Figure 5.18 Pure differentiator circuit using an ideal op amp.

and

$$I_2 = -\frac{V_o}{R}. \tag{5.75}$$

Combining Equations (5.74) and (5.75) according to Equation (5.73) it yields

$$\frac{V_o}{V_1} = -j\omega RC. \tag{5.76}$$

Equation (5.76) is the expression of a differentiator transfer function in the frequency domain. We will also determine the time domain equation of the differentiator output voltage. Again by referring to Figure 5.18 we see that $I_1 = I_2$. Since current I_1 through the capacitor is

$$I_1 = CdV_1 / dt \tag{5.77}$$

and

$$I_2 = -V_o / R. \tag{5.78}$$

Combining Equations (5.77) and (5.78) we obtain the expression of the output voltage for the inverting differentiator:

$$V_o = -RC\frac{dV_1}{dt}. \tag{5.79}$$

The Bode plots of a differentiator derived from the frequency domain Equation (5.76) are shown in Figure 5.19.

The differentiator frequency is $f_0 = 1/2\pi\, RC$ where R and C are the values of the resistor and capacitor of the differentiator circuit shown in Figure 5.18. It is important to notice the difference between the op amp-based differentiator of Figure 5.18 and the first-order RC HPF differentiator from Chapter 4,

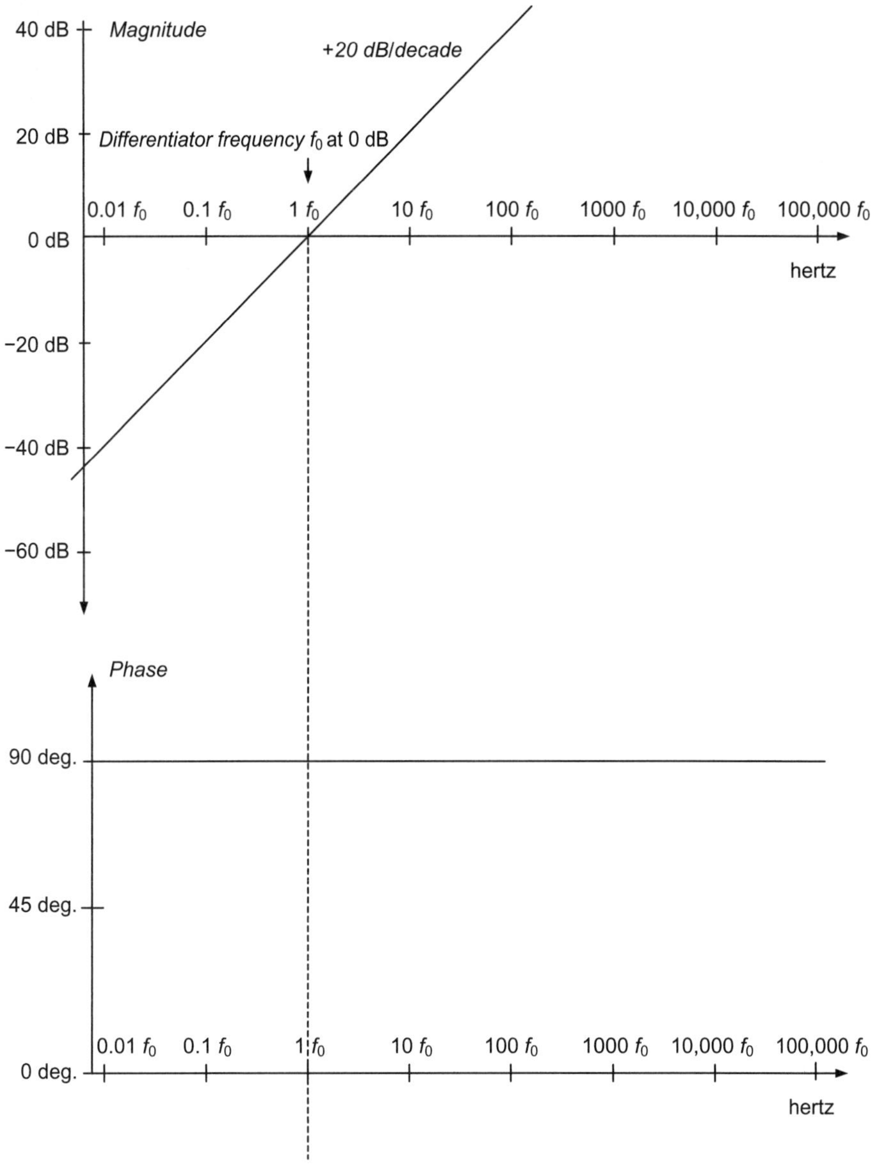

Figure 5.19 Asymptotic Bode plots of a pure differentiator.

Figure 4.10. The op amp-based differentiator has a negative infinite gain at DC. The gain or magnitude increases at a constant rate of 20 dB per decade of frequency. The first-order *RC* HPF differentiator circuit (Chapter 4, Fig. 4.10), which has no op amp, has a negative gain that increases at 20 dB per frequency decade from very low frequencies up until the filter cutoff frequency. Above this frequency, the op amp-less differentiator gain is 0 dB. Refer to the previously discussed circuit in Chapter 4, Figures 4.10 and 4.12. So what does this all mean in terms of practical operation of the differentiator? The op amp-based differentiator of Figure 5.18 actually will not work properly. Why? Because the gain at higher frequencies becomes extremely large; in fact, this high frequency gain is so large that any noisy or unwanted signals of a frequency higher than that of the signal that we intend to differentiate overwhelms the output of the op amp, effectively saturating it, and the op amp may even oscillate from rail to rail. To make the circuit of Figure 5.18 work, one has to limit the gain of the differentiator at very high frequencies. The implementation of such circuit, which will be used as a practical differentiator as well as an integrator, is discussed in the next section.

5.4.8 A Practical Integrator and Differentiator Circuit

We already discussed the reasons why the integrator circuit of Figure 5.16 and the differentiator circuit of Figure 5.18 will not operate properly. We summarize those results again. The integrator has an infinite gain at DC and thus it saturates the op amp output when a DC level is integrated for some finite time. The differentiator has an infinite gain for infinitely high frequencies, thus causing any high frequency unwanted signals to saturate the op amp and masking the differentiated signals of interest.

A practical solution to mitigate both of those problems is to limit the gain of the integrator at low frequencies and to limit the gain of the differentiator at very large frequencies. The circuit that does just that is depicted in Figure 5.20. Its corresponding asymptotic and exact Bode plots are presented in Figures 5.21 and 5.22, respectively.

The circuit in Figure 5.20 shapes the gain characteristics at low frequencies and at high frequencies. Refer once more to Figures 5.21 and 5.22. These figures show a fairly constant band-pass gain characteristic of 20 dB between cutoff frequencies f_2 and f_3. It behaves as a differentiator for all those signals of frequencies at f_1 and below, and it behaves as an integrator for those frequencies at f_4 and above. The cut-off frequencies are the –3 dB gain frequencies, for this particular example $f_2 = 100\,\text{Hz}$ and $f_3 = 10\,\text{kHz}$. Frequency range f_2 through f_3 defines the circuit mid-frequency band.

In other words our circuit combines the differentiating and integrating characteristics of both pure differentiator and pure integrator, and provides a flat gain band-pass characteristic at mid frequencies. Frequencies below f_1 are attenuated at a rate of +20 dB/decade. Frequencies above f_4 are attenuated at

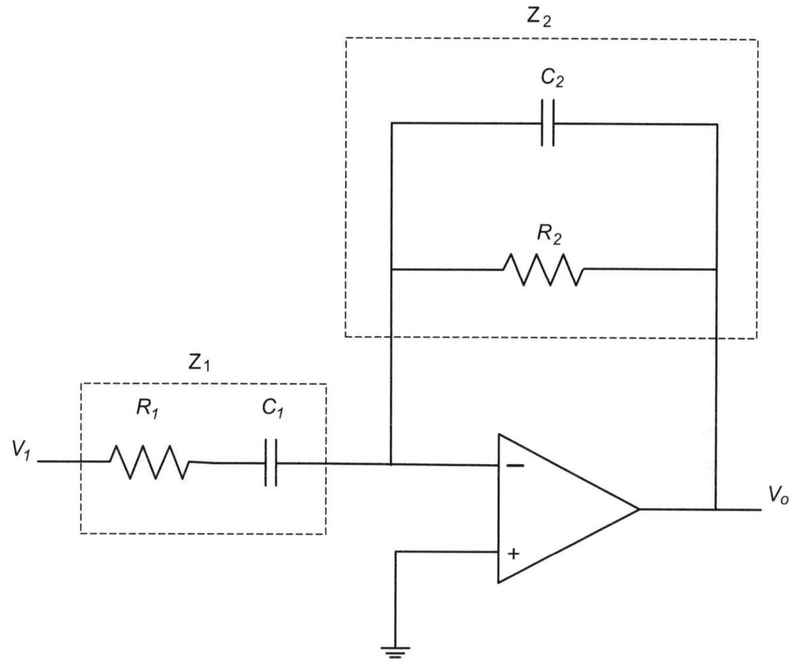

Figure 5.20 A practical differentiator, integrator, and band-pass filter circuit schematics.

a rate of –20 dB/decade. At mid-band frequencies the gain is quite constant and it is 20 dB. Such gain is mainly determined by the ratio of R_2 over R_1.

In a general sense our practical circuit is designed such that at frequencies f_2 through f_3 the ratio of the impedance module of R_2 in parallel with C_2 over the impedance module of the series of R_1 and C_1 is 10, thus a 20 dB gain. Capacitor C_1 and resistor R_2 determine f_1 at 0 dB gain, the high end of the differentiating frequency range. C_2 and R_1 determine f_4 at 0 dB, and it is the low end or the beginning of the integrating frequencies.

Our practical circuit phase behavior ranges from –90° at low frequencies to +90° at high frequencies. Now since the op amp produces a –180° phase shift, the overall circuit phase spans from –90° (+90° to 180°) from very low frequencies down to –270° (–90° to 180°) at high frequencies. The –180° phase shift is caused by the negative sign of the op amp-based inverting configuration output voltage over input voltage transfer function, that is, $V_o/V_{in} = -R_2/R_1$.

We can see that the topology of the circuit of Figure 5.20 is that of an inverting amplifier where

$$Z_2 = \frac{1}{\dfrac{1}{R_2} + j\omega C_2} \tag{5.80}$$

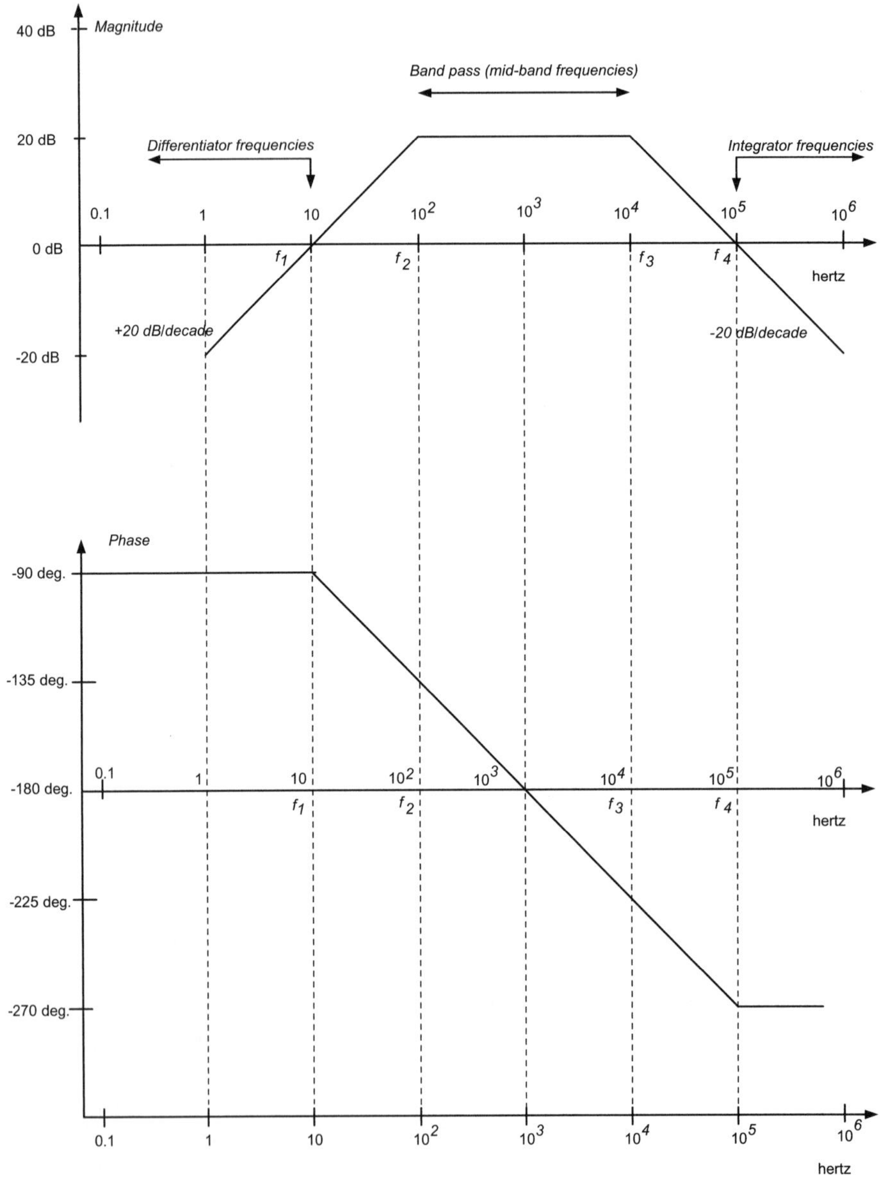

Figure 5.21 Asymptotic Bode plots of a practical differentiator, integrator, and band-pass filter.

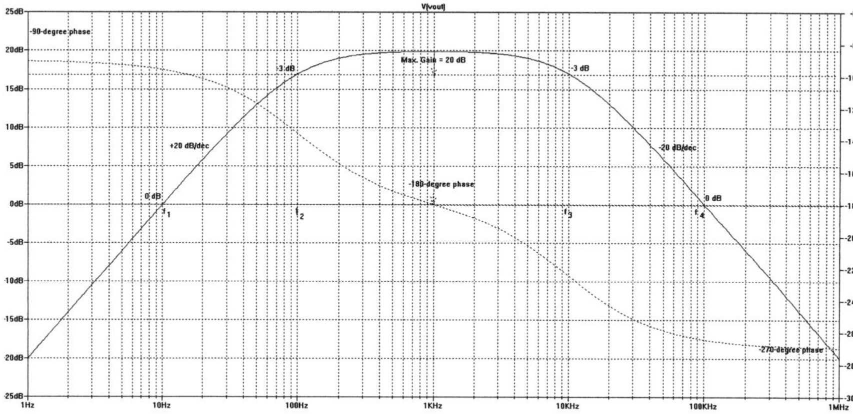

Figure 5.22 Exact Bode plots of a practical differentiator, integrator, and band-pass filter. See the Appendix to this chapter for a larger version.

and

$$Z_1 = R_1 + \frac{1}{j\omega C_1}. \tag{5.81}$$

The transfer function of the circuit of Figure 5.20 is

$$\frac{V_o}{V_1} = -\frac{Z_2}{Z_1}. \tag{5.82}$$

Using Equations (5.80) and (5.81) in Equation (5.82) we obtain

$$\frac{V_{out}}{V_{in}} = -\frac{\dfrac{1}{\dfrac{1}{R_2} + j\omega C_2}}{R_1 + \dfrac{1}{j\omega C_1}}. \tag{5.83}$$

Doing some algebraic manipulations with Equation (5.83) we arrive at

$$V_o / V_{in} = -\frac{j\omega R_2 C_1}{(1 + j\omega R_2 C_2)(1 + j\omega R_1 C_1)}. \tag{5.84}$$

Equation (5.84) has one zero at the origin in the numerator and two zeros on the denominator. Numerator zeros are simply referred to as zeros of the transfer function. Denominator zeros are referred to as poles of the transfer function.

We can rewrite Equation (5.84) using the following definitions:

$$\omega_1 = \frac{1}{R_2 C_1} \tag{5.85}$$

$$\omega_2 = \frac{1}{R_1 C_1} \tag{5.86}$$

$$\omega_3 = \frac{1}{R_2 C_2} \text{ and } \omega_4 = \frac{1}{R_1 C_2} \tag{5.87}$$

where $\omega_1 = 2\pi f_1$, $\omega_2 = 2\pi f_2$, $\omega_3 = 2\pi f_3$, and $\omega_4 = 2\pi f_4$ thus,

$$f_1 = 1/2\pi R_2 C_1 \tag{5.88}$$

$$f_2 = 1/2\pi R_1 C_1 \tag{5.89}$$

$$f_3 = 1/2\pi R_2 C_2 \text{ and } f_4 = 1/2\pi R_1 C_2 \tag{5.90}$$

We can plot Equation (5.84) either by writing a computer program that calculates the magnitude and the phase of Equation (5.84) or using the asymptotic Bode plot methodology. We have to keep in mind that Equation (5.88) is the frequency at which the transfer function magnitude has 0 dB gain and the differentiation in the time domain ends. Equation (5.89) is where the first pole of the transfer function is placed, and f_3 (in Eq. ((5.90)) is where the second pole of the transfer function is placed. Finally, f_4 is the frequency at which the gain is 0 dB and integration begins.

We will go over a numerical example to clarify the generation of the transfer function given by Equation (5.84).

Example 5.8 Design a pass band amplifier that has the following characteristics:

(a) Pass band gain = 20 dB
(b) 0 dB gain and end of differentiation at $f_1 = 10$ Hz
(c) Low cutoff frequency $f_2 = 100$ Hz
(d) High cut-off frequency $f_3 = 10$ kHz
(e) 0 dB gain and beginning of integration at 100 kHz

Assume you can use an ideal op amp.

The circuit topology is just like the circuit shown in Figure 5.20.

From Equations (5.88)–(5.90) and the given characteristics of the desired band pass amplifier with combined differentiating and integrating properties, we have the following:

$$f_1 = 1/2\pi R_2 C_1 = 10 \text{ Hz} \tag{5.91}$$

$$f_2 = 1/2\pi R_1 C_1 = 100 \text{ Hz} \tag{5.92}$$

$$f_3 = 1/2\pi R_2 C_2 = 10 \text{ kHz and } f_4 = 1/2\pi R_1 C_2 = 100 \text{ kHz.} \tag{5.93}$$

The mid-band frequency gain of our amplifier has to be 20 dB. This means that the amplifier closed loop gain has to be 10. Arbitrarily, we can choose $R_2 = 10\ \text{k}\Omega$, thus R_1 must be $1\ \text{k}\Omega$, thus $20\log_{10}(10) = 20$ dB. Using these values for R_1 and R_2 in Equations (5.91)–(5.93) we obtain

$$C_1 = 1.59\ \mu\text{F} \tag{5.94}$$

$$C_2 = 1.59\ \text{nF}. \tag{5.95}$$

Also remember that

$$R_1 = 1\ \text{k}\Omega \tag{5.96}$$

and

$$R_2 = 10\ \text{k}\Omega. \tag{5.97}$$

Figure 5.22 shows exact Bode magnitude and phase plots of the transfer function given by Equation (5.84) using the numerical values given by Equations (5.94) through (5.97).

Important Points

The differentiating frequencies of the amplifier are at and below f_1 (10 Hz and below).

The band pass flat gain of 20 dB is between $f_2 = 100$ Hz and $f_3 = 10$ kHz.

The integrating frequencies of the amplifier are at and above f_4 (100 kHz and above).

The current circuit topology will work as a differentiator, a band pass amplifier, and an integrator in the three different frequency ranges discussed above (see Figs. 5.21 and 5.22).

5.5 OP AMPS NONLINEAR APPLICATIONS

The most significant and widely used nonlinear application using op amps is when op amps operate in open loop. The op amp under such operation is referred to as a comparator, and op amp manufacturers optimize op amp parameters to operate them as comparators. So it is common for IC manufacturers to sell op amps and comparators. So what is a comparator? A comparator is designed to operate in open loop; its output swings between specified upper and lower limits. A very desirable feature is that the comparator wants to be fast to swing its output upon a detected voltage difference at its inputs. Usually comparators do not have internal compensating capacitors. On the other hand, op amps have internal compensating capacitors. The lack of compensation capacitors allows comparators to be faster than op amps. Op amps are designed to be accurate and stable; op amps have good DC and AC behavior. On a final note, most comparators have an open collector or open drain output; this means that its output can be connected to supply levels that may not necessarily be those of the comparator power supplies. This flexibility allows an easier interface of comparators to digital circuits.

5.5.1 The Open-Loop Comparator

The comparator is typically used in open-loop mode to compare when the signal level at the one of the inputs is greater or smaller that the signal level at the other input terminal. Since a comparator gain is very high, just like that of an open-loop op amp, upon detecting a difference between its inputs, the output will swing to the positive rail when V+ > V−. The output will swing to the negative rail when V− > V+. V+ refers to the input signal at the noninverting input of the comparator. V− is the signal at the inverting input of the comparator. Figure 5.23 shows a comparator operating with its inverting input tied to ground and an arbitrary waveform at its noninverting input. Note that when the input signal at the positive input is above zero the comparator output swings to its positive rail.

The output swings to the negative rail when the noninverting input signal is negative or below ground.

Figure 5.23 assumes that the comparator output used is very fast when reacting to a change of the noninverting input. Such circuit is called a noninverting zero-crossing detector. Note that since the inverting input of the comparator is grounded, it is at 0 V; every time the noninverting input is above zero, the output of the comparator saturates toward the positive rail, otherwise it saturates to the negative rail. We can easily design an inverting zero-crossing detector by swapping the inputs to the comparator of Figure 5.23. This means tying the ground to the noninverting input and connecting the input signal V_{in} to the inverting input.

5.5.2 Positive and Negative Voltage-Level Detectors Using Comparators

5.5.2.1 Positive Level Detectors Let us assume that we want to detect when a signal V_{in} is above a positive DC voltage level, which we will refer to as V_{REF}. Let us also assume that every time signal V_{in} is above V_{REF}, we want the output of the comparator to indicate this with a high output level at V_{out}. Figure 5.24 shows a possible implementation of such circuit.

Note that the reference voltage V_{REF} is connected to the inverting input of the comparator. The input signal V_{in} is tied to the noninverting input of the comparator. The comparator shows its positive and negative power supply levels. For simplicity, the decoupling capacitors are not shown. Let us see how this circuit works; each time V_{in} is greater than V_{REF}, the comparator output will saturate to the positive $+V_{SAT}$ level. When V_{in} is less than V_{REF} the comparator output will saturate at the negative $-V_{SAT}$ level. Although Figure 5.24 shows a sinusoidal waveform for V_{in}, there is no restriction on waveform V_{in},

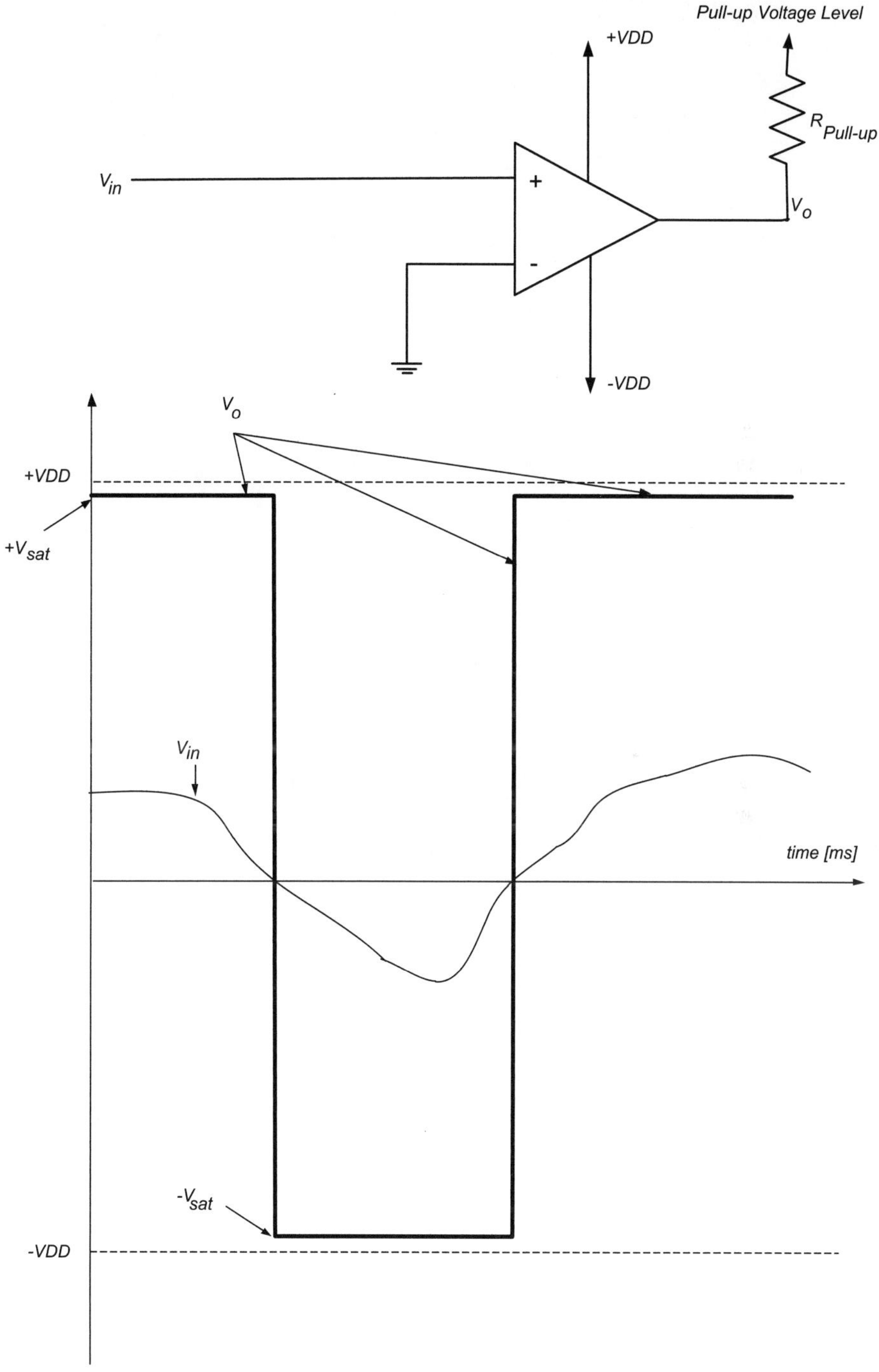

Figure 5.23 Open-loop comparator operation with grounded inverting input.

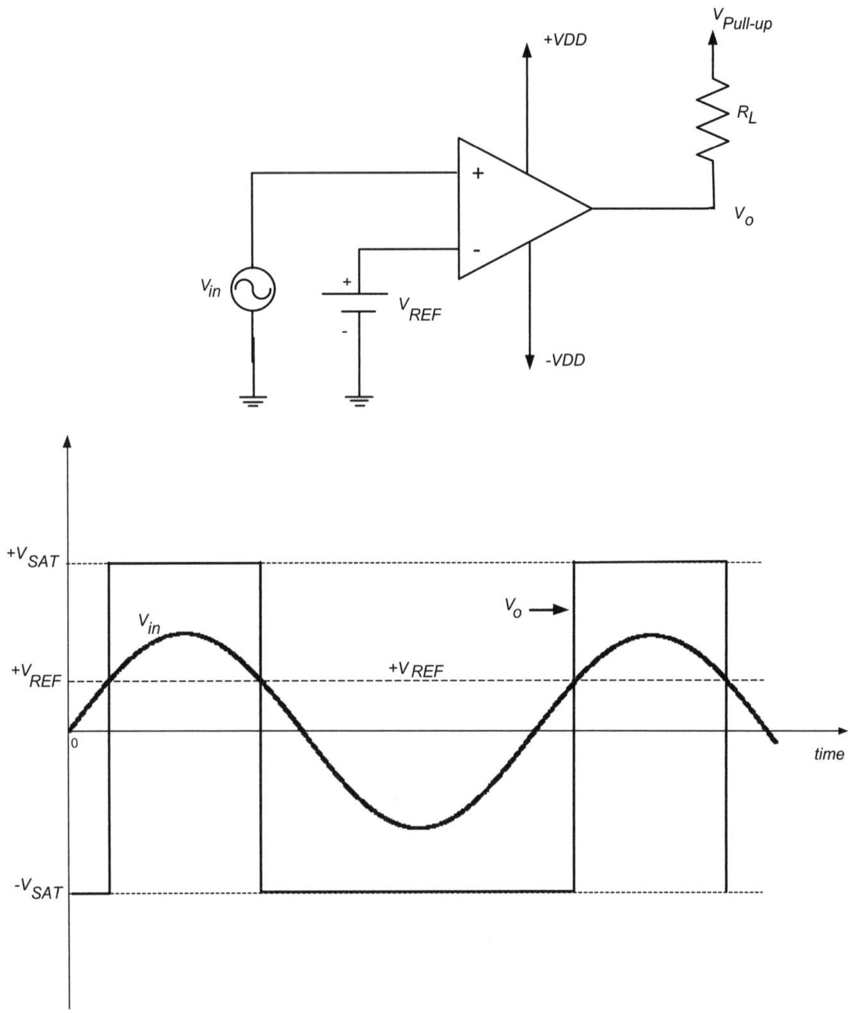

Figure 5.24 Noninverting positive-level detector.

it can be a saw-tooth, an exponential or even a piece-wise linear waveform. The positive-level detector of Figure 5.24 is referred to as a noninverting detector because the comparator output signals with a high level V_{out} at $+V_{SAT}$ when the input signal V_{in} is greater than the reference voltage. This is the simplest and most straightforward variation of the four types of level detectors that we will study in this section. When confused with the subsequent detector, make sure that you always come back to the one described by Figure 5.24. This will let one understand the concepts more easily than any of the other three will. The inverting positive-level detector is implemented in Figure 5.25. Note that the input signal is now applied to the inverting input of the comparator, while the reference voltage is applied to the noninverting input.

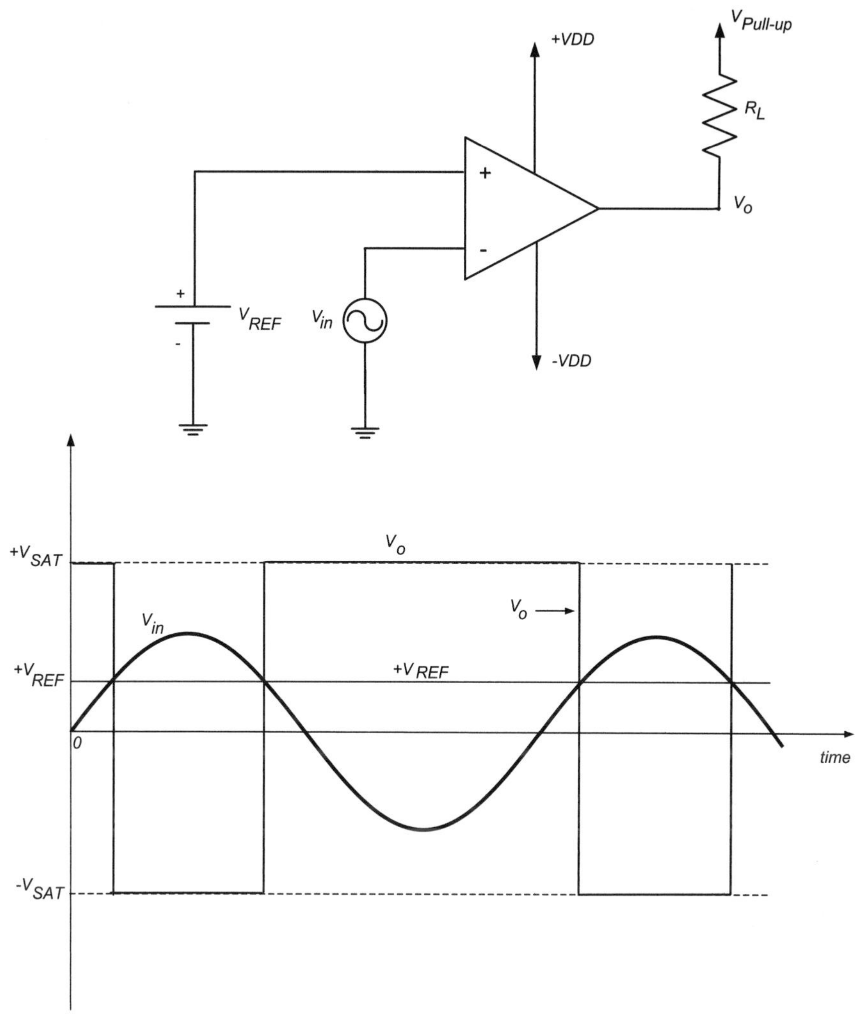

Figure 5.25 Inverting positive-level detector.

5.5.2.2 Negative-Level Detectors Just like there are two polarities of positive-level detectors, that is, inverting and noninverting, there are also two kinds of negative-level detectors. Negative level refers to the sign of the voltage, to which the input signal is being compared. Figure 5.26 depicts a noninverting negative-level voltage detector. V_{in} the input signal is applied to the positive input of the comparator, while the negative input of the comparator has a negative reference voltage. Note that V_{REF} negative terminal is connected to the comparator inverting input, while the positive terminal of V_{REF} is grounded.

Finally, for the inverting level negative edge voltage level detector, V_{REF} and V_{in} are swapped. Figure 5.27 depicts the inverting negative voltage-level detector. Note that when V_{in} is above the negative reference level ($-V_{REF}$), the output of the comparator is at $-V_{SAT}$.

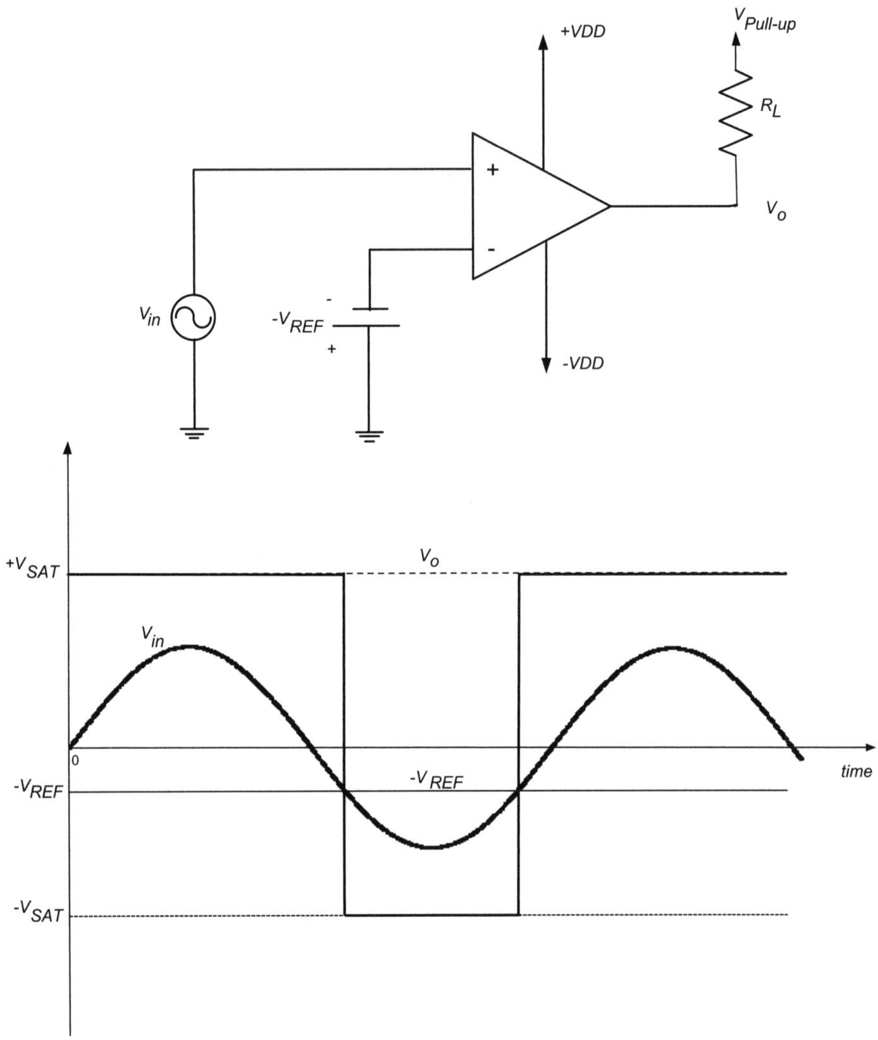

Figure 5.26 Noninverting negative voltage-level detector.

Summary of Voltage Level Detectors:

1. Noninverting positive-level detector (see Fig. 5.24)
2. Inverting positive-level detector (see Fig. 5.25)
3. Noninverting negative-level detector (see Fig. 5.26)
4. Inverting negative-level detector (see Fig. 5.27)

5.5.3 Comparator with Positive Feedback (Hysteresis)

Comparators work well in open-loop mode if the signal V_{in} at its input varies rapidly and it is not noisy. However, if a noisy input signal V_{in} is present, there is

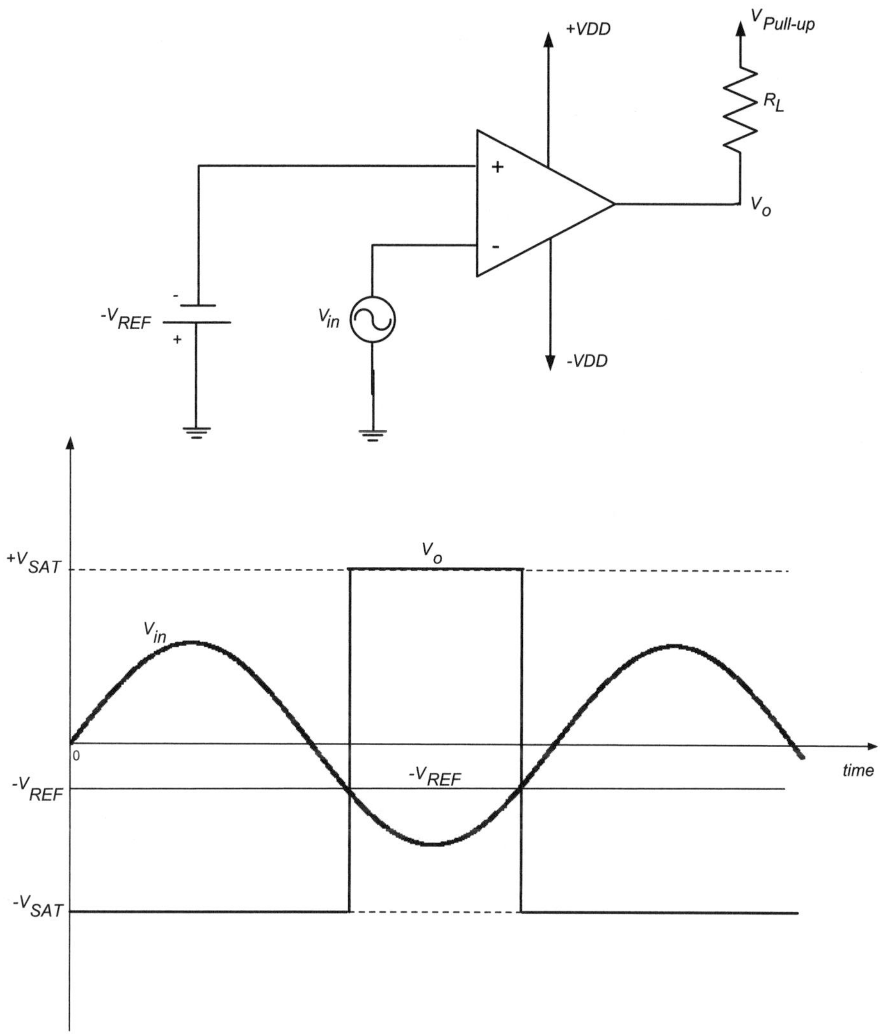

Figure 5.27 Inverting negative voltage-level detector.

a good opportunity for the open-loop comparator to oscillate once or more times before settling on its steady state either $+V_{SAT}$ or $-V_{SAT}$ output voltage level. Figure 5.28 shows a comparator operating in open-loop; the input signal has on top of it a noisy signal of much higher frequency than the input. We can observe that before the output settles to its final and correct value, the comparator output oscillates momentarily; this is referred to as *chattering*. A straightforward way of significantly reducing or eliminating this problem is to provide a little bit of positive feedback from the comparator output back into its positive input.

Taking a fraction of the output voltage and feeding it back to the noninverting input of the comparator provides positive feedback into the comparator.

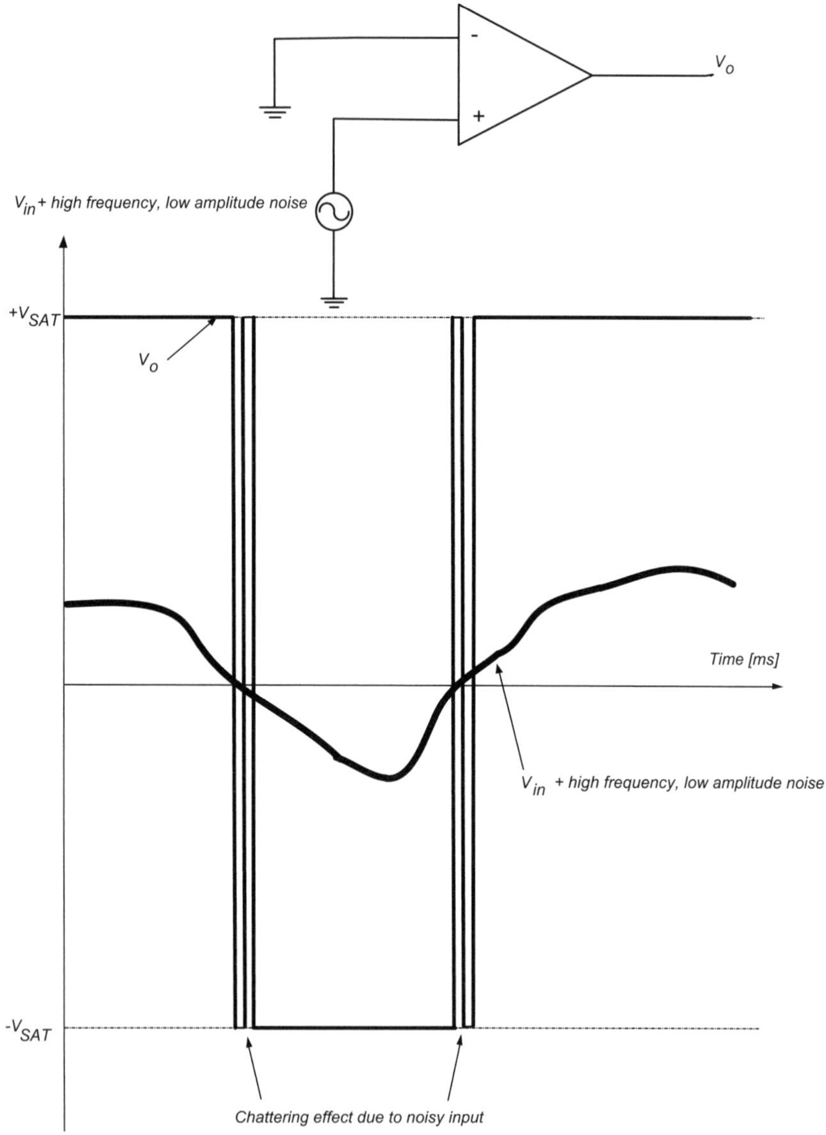

Figure 5.28 Comparator operating in open-loop mode that exhibits chattering.

Figure 5.29 shows a comparator with resistors R_1 and R_2 that provide positive feedback. The circuit is configured as a noninverting zero voltage-level detector. Let us understand how the resistors provide hysteresis to the circuit.

The upper threshold voltage is the output voltage times the resistor divider formed by R_1 and R_2. Thus,

$$V_{UT} = \frac{R_2}{R_1 + R_2} V_{SAT}. \tag{5.98}$$

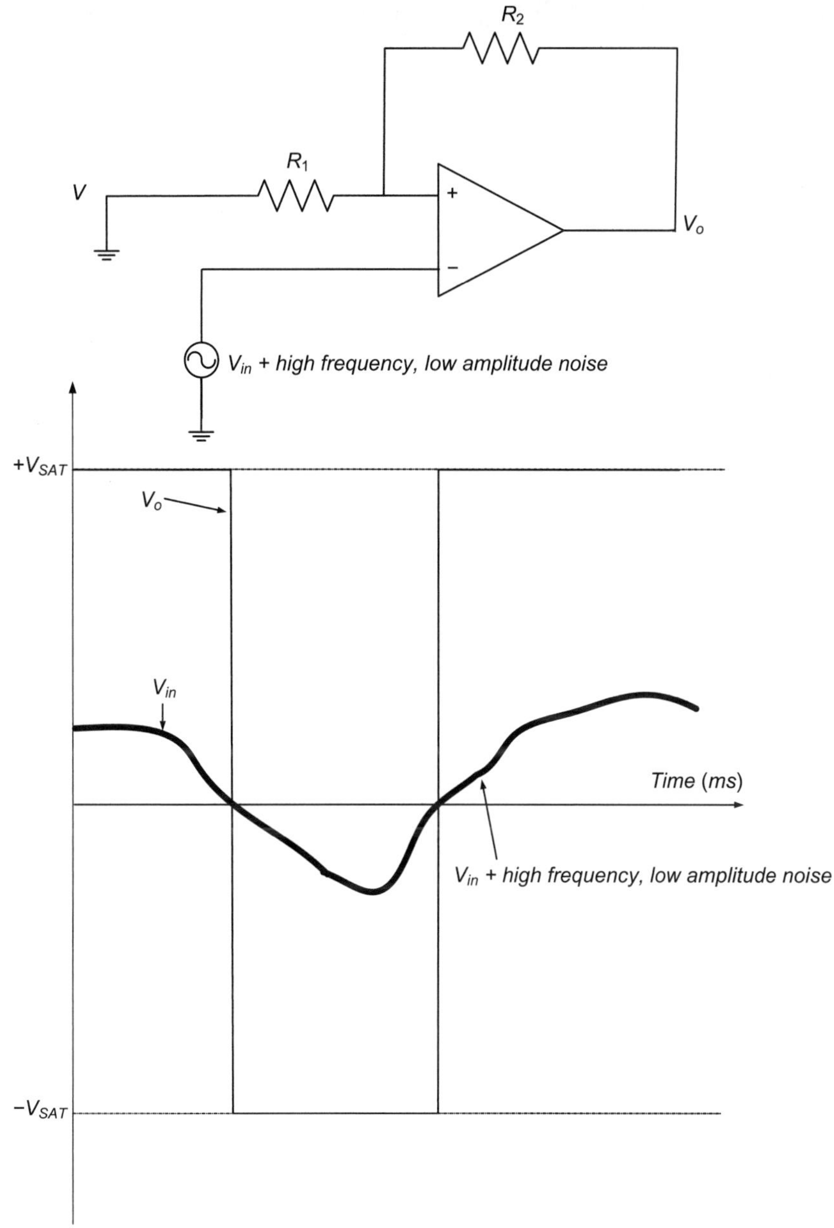

Figure 5.29 Comparator with hysteresis eliminates chattering.

When the comparator output voltage is equal to $+V_{SAT}$ the upper threshold voltage given by Equation (5.98) is a fraction of V_{SAT}. If V_{in} plus any noisy and unwanted components are present and are in absolute magnitude smaller that V_{UT}, the comparator snaps to $+V_{SAT}$ and will get locked at $+V_{SAT}$. As V_{in} decreases below 0 V (the ground level) $+V_{SAT}$ will eventually snap out of $+V_{SAT}$

and will switch to $-V_{SAT}$. Now $-V_{SAT}$ produces a voltage called the lower threshold, which is equal to

$$V_{LT} = -\frac{R_2}{R_1 + R_2} V_{SAT}. \tag{5.99}$$

The noise on top of V_{in} is still smaller in magnitude that the V_{LT}. The comparator output locks up at $-V_{SAT}$ now. As the input continues to decrease well below ground, the output of the comparator stays at $-V_{SAT}$. Not until V_{in} grows above ground the whole cycle repeats itself indefinitely. Every time the comparator output locked up to either $+V_{SAT}$ or $-V_{SAT}$, the undesirable chattering seen in the open-loop case of the previous section has been eliminated. Figure 5.29 shows a comparator with hysteresis.

Example 5.9 Given a 100 Hz triangular waveform, and assuming that noise signals of ±10 mV can be riding on the signal, design the hysteresis circuit divider to avoid chattering of the combined signal plus noise on a noninverting zero voltage detector. Let us assume that the positive and negative supplies of the comparator are respectively + and –15 V. Also assume that you checked the comparator data sheet and the + and $-V_{SAT}$ voltages are respectively + and –14 V.

Choosing $R_2 = 100\ k\Omega$ and $R_1 = 100\ \Omega$, using these values in Equations (5.98) and (5.99) we find that

$$V_{UT} = \frac{R_2}{R_1 + R_2} V_{SAT} = \frac{100k\Omega}{100k\Omega + 100\Omega} 14\ \text{V} \cong 14\ \text{mV} \tag{5.100}$$

and

$$V_{LT} = -\frac{R_2}{R_1 + R_2} V_{SAT} = -\frac{100k\Omega}{100k\Omega + 100\Omega} 14\ \text{V} \cong -14\ \text{mV}. \tag{5.101}$$

The total hysteresis voltage, or the voltage range for which the error on the signal is eliminated, is given by

$$V_H = V_{UT} - V_{LT} = 28\ \text{mV}. \tag{5.102}$$

Note that when the input signal goes above the upper threshold, the output voltage drops down to $-V_{SAT}$. The peak-to-peak noise voltage on the input signal has to be greater than or equal to 28 mV (the hysteresis voltage given by Equation (5.102)), to pull the input signal below the lower threshold and cause a false zero crossing. So as long as the peak-to-peak noise on top of the input signal does not exceed the hysteresis voltage range (V_H), the false crossing will not occur. Since for this example we are told that the peak-to-peak noise can be up to 20 mV and since the comparator hysteresis was designed to tolerate 28 mV of peak-to-peak noise, we still have a positive margin over the required immunity to noise.

5.6 OPERATIONAL AMPLIFIERS NONIDEALITIES

The actual electrical characteristics of real op amps are more complicated than those of the idealized op amp. The op amp is implemented with either bipolar or JFET transistors. The basic op amp consists of three main stages cascaded one after the other. The input stage is a transistorized differential stage, a level shifting stage follows the first stage and the last stage is an output stage that performs a differential to single ended output conversion. Differential signals come in pairs; there is a noninverted signal and an inverted signal to carry information over two separate wires. This provides to the signal better noise immunity; this is to say that the signals are more protected against noise or unwanted signals. Single ended signals are referenced with respect to ground and do not have as good noise immunity as differential signals do.

The differential input stage of the op amp has imperfections due to the fact that it is impossible to fabricate perfectly matched transistors. The currents in or out of the inverting and noninverting inputs are not really zero. They are nonzero and but small in magnitude, and they are referred to as the bias currents. There is an I_{B+} (positive input bias current) and an I_{B-} (negative input bias current). We will model this behavior of the input stage of the op amp with to current sources of values I_{B+} and I_{B-}. Figure 5.30 shows the model of a real op amp with the imperfections that we describe throughout this entire section. It is important to say that such bias currents are not only nonzero, but also they are not equal to each other and may flow in or out of their respective op amp input. The magnitude of the difference of the positive and negative bias current is referred to as the offset current (I_{OS}). The numerical value of the offset current generally is a fraction of the bias current. A voltage source models the offset voltage of the amplifier. The offset voltage (V_{OS}) of the op amp is the voltage that is required to apply across its inputs to obtain a 0 V output with the op amp in an open-loop condition. Differential impedance appears across both op amp inputs and common mode impedance appears between the inputs and ground. This impedance models the finite common mode the real op amp has. Figure 5.30 depicts an op amp model with some real electrical parameters.

The input stage has two AC current sources (I_N) and an AC voltage source (V_N) that model noise components that unavoidably exist in the op amp. Finally, for the input imperfections, we have to mention that both bias currents and offset voltages vary or drift with temperature variations. Manufacturers usually specify these parameters at one temperature, for example, at 25°C and at a range of temperatures, such as −40°C to 85°C.

Let us now talk about the output stage of the real op amp. The output of the op amp has a nonzero and finite output resistance (R_o), which is modeled

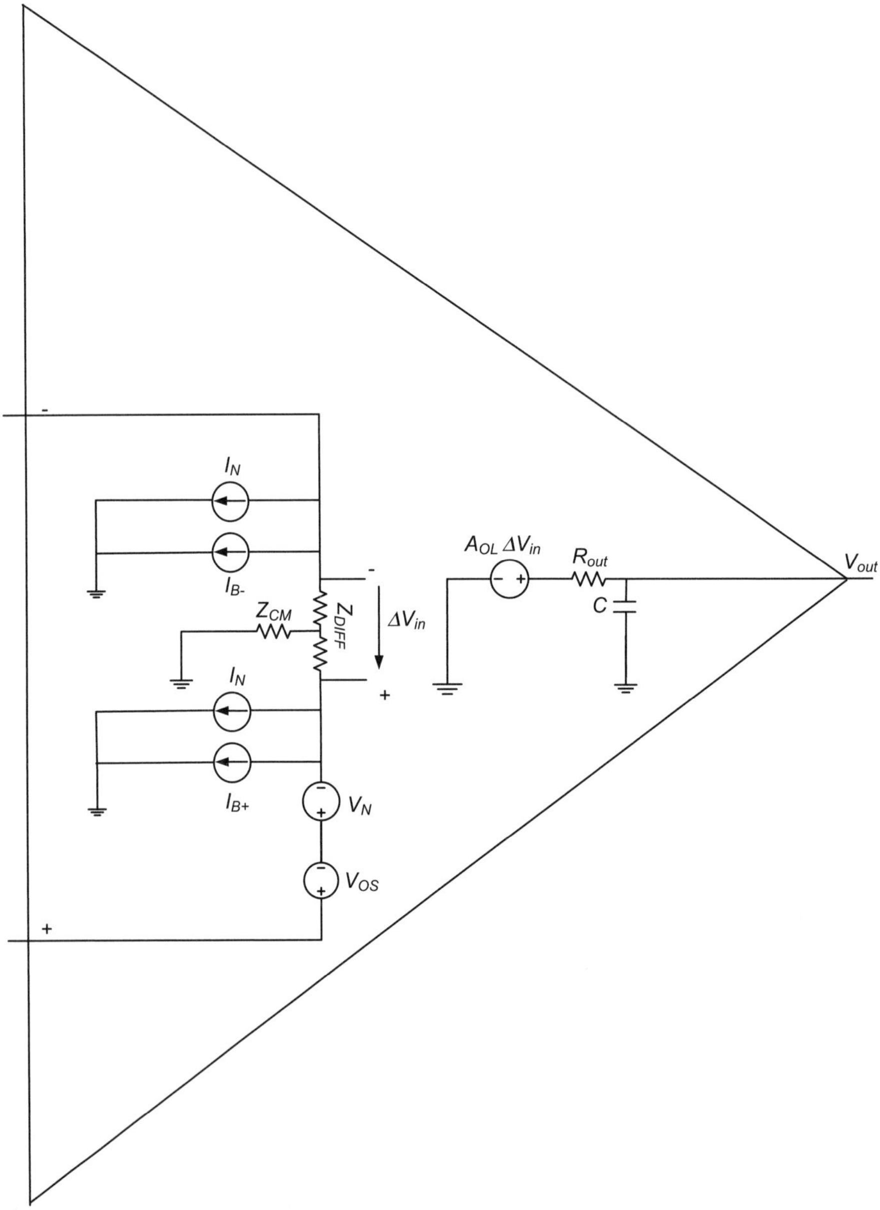

Figure 5.30 Model of a real operational amplifier.

in series with the output generator A_{OL} (Fig. 5.30). Generator A_{OL} models the finite nature of the op amp open-loop gain. In actuality A_{OL} is really large, but it is not infinite. The op amp open-loop gain is also a function of frequency, and it is usually modeled with a capacitor hanging from the output of the op amp to ground. This RC network at the output of the op amp is also referred to as the op amp single pole approximation. The op amp manufacturer specifies the device behavior with frequency with the open-loop gain bandwidth and with the op amp gain-bandwidth product. Both of these parameters model the frequency behavior of the op amp under small signal excitation. A small signal for an op amp is a signal whose amplitude is about one order of magnitude smaller that its supply voltage. For large signal operation, the slew-rate of the op amp determines how fast the op amp can react to changes. Slew is usually specified in volts per microsecond. A slow op amp may have a 0.5 V/μs slew-rate, while a high speed one may have a 6000 V/μs slew-rate.

There are a few more real parameters of an op amp such as common mode rejection ratio (CMRR), which provides information as to how much common mode signals become attenuated, by the op amp. CMRR is also a frequency-dependent parameter. Power supply rejection ratio specifies how sensitive the op amp operation is to variation of it power supply rails. This section intends to cover some of the most fundament imperfections that most op amp manufacturers specify. Beware that this list is not complete and there are manufacturers that provide more or less complete data sheet specifications for their devices. Table 5.7 summarizes some commonly specified op-amp parameters.

5.7 OP AMP SELECTION CRITERIA

Op amps are generally classified by the following characteristics:

1. DC parameters (offset voltage, bias and offset currents, and their drift with temperature) as precision op amps,
2. Low noise,
3. Speed (gain-bandwidth product and slew-rate),
4. Single or dual rail power supply,
5. Single, dual, or quad (1, 2, or 4 amplifiers) in a package,
6. Rail-to-rail output voltage swing,
7. Maximum common mode input voltage, low power consumption. Ultimately most manufacturers provide online tools to their customers to select any of their op amps by a selection of any number and combination of parameters offered by the manufacturer. Other characterizations exist; just a few of the most popular ones were listed above.

Why can't we have all of the parameters optimized and not worry about selecting op amps? The reason is that improving some op amp parameters imposes

Table 5.7 Real op amp parameters (National LM741, reproduced with permission of Texas Instruments Incorporated)

Absolute Maximum Ratings (Note 2)

If Military/Aerospace specified devices are required, please contact the National Semiconductor Sales Office/ Distributors for availability and specifications.

(Note 7)

	LM741A	LM741	LM741C
Supply Voltage	±22V	±22V	±18V
Power Dissipation (Note 3)	500 mW	500 mW	500 mW
Differential Input Voltage	±30V	±30V	±30V
Input Voltage (Note 4)	±15V	±15V	±15V
Output Short Circuit Duration	Continuous	Continuous	Continuous
Operating Temperature Range	−55°C to +125°C	−55°C to +125°C	0°C to +70°C
Storage Temperature Range	−65°C to +150°C	−65°C to +150°C	−65°C to +150°C
Junction Temperature	150°C	150°C	100°C
Soldering Information			
N-Package (10 seconds)	260°C	260°C	260°C
J- or H-Package (10 seconds)	300°C	300°C	300°C
M-Package			
Vapor Phase (60 seconds)	215°C	215°C	215°C
Infrared (15 seconds)	215°C	215°C	215°C
See AN-450 "Surface Mounting Methods and Their Effect on Product Reliability" for other methods of soldering surface mount devices.			
ESD Tolerance (Note 8)	400V	400V	400V

Electrical Characteristics (Note 5)

Parameter	Conditions	LM741A			LM741			LM741C			Units
		Min	Typ	Max	Min	Typ	Max	Min	Typ	Max	
Input Offset Voltage	$T_A = 25°C$										
	$R_S \leq 10\ k\Omega$					1.0	5.0		2.0	6.0	mV
	$R_S \leq 50\Omega$		0.8	3.0							mV
	$T_{AMIN} \leq T_A \leq T_{AMAX}$										
	$R_S \leq 50\Omega$			4.0							mV
	$R_S \leq 10\ k\Omega$						6.0			7.5	mV
Average Input Offset Voltage Drift				15							μV/°C
Input Offset Voltage Adjustment Range	$T_A = 25°C$, $V_S = \pm 20V$	±10				±15			±15		mV
Input Offset Current	$T_A = 25°C$		3.0	30		20	200		20	200	nA
	$T_{AMIN} \leq T_A \leq T_{AMAX}$			70		85	500			300	nA
Average Input Offset Current Drift				0.5							nA/°C
Input Bias Current	$T_A = 25°C$		30	80		80	500		80	500	nA
	$T_{AMIN} \leq T_A \leq T_{AMAX}$			0.210			1.5			0.8	μA
Input Resistance	$T_A = 25°C$, $V_S = \pm 20V$	1.0	6.0		0.3	2.0		0.3	2.0		MΩ
	$T_{AMIN} \leq T_A \leq T_{AMAX}$, $V_S = \pm 20V$	0.5									MΩ
Input Voltage Range	$T_A = 25°C$							±12	±13		V
	$T_{AMIN} \leq T_A \leq T_{AMAX}$				±12	±13					V

Table 5.7 (*Continued*)

Electrical Characteristics (Note 5) (Continued)

Parameter	Conditions	LM741A			LM741			LM741C			Units
		Min	Typ	Max	Min	Typ	Max	Min	Typ	Max	
Large Signal Voltage Gain	T_A = 25°C, $R_L \geq 2\ k\Omega$										
	$V_S = \pm20V$, $V_O = \pm15V$	50									V/mV
	$V_S = \pm15V$, $V_O = \pm10V$				50	200		20	200		V/mV
	$T_{AMIN} \leq T_A \leq T_{AMAX}$,										
	$R_L \geq 2\ k\Omega$,										
	$V_S = \pm20V$, $V_O = \pm15V$	32									V/mV
	$V_S = \pm15V$, $V_O = \pm10V$				25			15			V/mV
	$V_S = \pm5V$, $V_O = \pm2V$	10									V/mV
Output Voltage Swing	$V_S = \pm20V$										
	$R_L \geq 10\ k\Omega$	±16									V
	$R_L \geq 2\ k\Omega$	±15									V
	$V_S = \pm15V$										
	$R_L \geq 10\ k\Omega$				±12	±14		±12	±14		V
	$R_L \geq 2\ k\Omega$				±10	±13		±10	±13		V
Output Short Circuit	T_A = 25°C	10	25	35		25			25		mA
Current	$T_{AMIN} \leq T_A \leq T_{AMAX}$	10		40							mA
Common-Mode	$T_{AMIN} \leq T_A \leq T_{AMAX}$										
Rejection Ratio	$R_S \leq 10\ k\Omega$, $V_{CM} = \pm12V$				70	90		70	90		dB
	$R_S \leq 50\Omega$, $V_{CM} = \pm12V$	80	95								dB
Supply Voltage Rejection	$T_{AMIN} \leq T_A \leq T_{AMAX}$,										
Ratio	$V_S = \pm20V$ to $V_S = \pm5V$										
	$R_S \leq 50\Omega$	86	96								dB
	$R_S \leq 10\ k\Omega$				77	96		77	96		dB
Transient Response	T_A = 25°C, Unity Gain										
Rise Time			0.25	0.8		0.3			0.3		μs
Overshoot			6.0	20		5			5		%
Bandwidth (Note 6)	T_A = 25°C	0.437	1.5								MHz
Slew Rate	T_A = 25°C, Unity Gain	0.3	0.7			0.5			0.5		V/μs
Supply Current	T_A = 25°C					1.7	2.8		1.7	2.8	mA
Power Consumption	T_A = 25°C										
	$V_S = \pm20V$		80	150							mW
	$V_S = \pm15V$					50	85		50	85	mW
LM741A	$V_S = \pm20V$										
	$T_A = T_{AMIN}$			165							mW
	$T_A = T_{AMAX}$			135							mW
LM741	$V_S = \pm15V$										
	$T_A = T_{AMIN}$					60	100				mW
	$T_A = T_{AMAX}$					45	75				mW

Note 2: "Absolute Maximum Ratings" indicate limits beyond which damage to the device may occur. Operating Ratings indicate conditions for which the device is functional, but do not guarantee specific performance limits.

Table 5.7 (*Continued*)

Electrical Characteristics (Note 5) (Continued)

Note 3: For operation at elevated temperatures, these devices must be derated based on thermal resistance, and T_j max. (listed under "Absolute Maximum Ratings"). $T_j = T_A + (\theta_{jA} P_D)$.

Thermal Resistance	Cerdip (J)	DIP (N)	HO8 (H)	SO-8 (M)
θ_{jA} (Junction to Ambient)	100°C/W	100°C/W	170°C/W	195°C/W
θ_{jC} (Junction to Case)	N/A	N/A	25°C/W	N/A

Note 4: For supply voltages less than ±15V, the absolute maximum input voltage is equal to the supply voltage.

Note 5: Unless otherwise specified, these specifications apply for V_S = ±15V, −55°C ≤ T_A ≤ +125°C (LM741/LM741A). For the LM741C/LM741E, these specifications are limited to 0°C ≤ T_A ≤ +70°C.

Note 6: Calculated value from: BW (MHz) = 0.35/Rise Time(μs).

Note 7: For military specifications see RETS741X for LM741 and RETS741AX for LM741A.

Note 8: Human body model, 1.5 kΩ in series with 100 pF.

Schematic Diagram

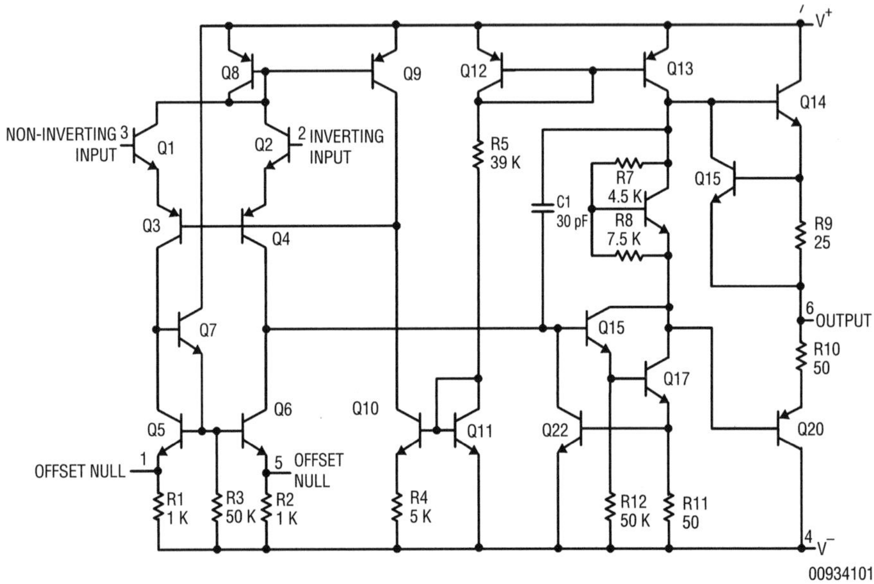

technological trade-offs. For example, if we want a high-precision op amp with excellent DC parameters and low drift, it will usually not be as fast as an op amp optimized for speed. Unfortunately, it is beyond the scope of this book to get into the op amp integrated circuit design techniques that address such issues in detail.

It is important for the reader to be aware that not all manufacturers publish the exact same list of parameters. There are various reasons for that; one of them is the cost of testing the op amps to guarantee every parameter published. Also not all manufacturers abbreviate or use the same nomenclature for the op amp parameters. It is the job of the op amp user to read very carefully how each manufacturer defines their parameters. For example, for some op amps, you may have a list of the most important parameter like input offset voltage, bias current, offset current, open-loop-gain, and so on, defined with typical, minimum, and maximum values at room temperature, most commonly 25°C/77°F and for *+/–15 V* power supplies. Then they publish the exact same list of parameters previously mentioned which are valid for their entire operating temperature range, for example, from –40°C to +85°C for +/–15-V power supplies. The important message here is that when one compares op amp or comparator parameters from different manufacturers, it is imperative to read each manufacturer's data sheet very carefully, and become aware of the conditions under which such parameters are being specified. Table 5.7 reproduces an LM741 op amp real data-sheet.

5.8 SUMMARY

This chapter is an overview of some of the most commonly used op amp-based circuits. Generally speaking, circuit designers use the op amp transfer functions derived, assuming the op amp is an ideal element. Most times that is correct for most routine applications. A routine application is one that is not high-precision demanding, or very high speed, or extremely low noise for example. However, when one designs highly sophisticated op amp applications, more care has to be paid and use the op amp model with the parameters that are most important for such application. One should model op amps introducing most important electrical parameters for the application. As an example, if one is interested in a high DC precision application, offset voltage, current, and bias currents should be introduced into the model; however, parameters like the gain-bandwidth product is certainly not important because the signals we are dealing with are DC signals. Conversely, if we have a very high speed, AC-coupled op amp application, certainly gain-bandwidth product is of utmost importance, while DC parameters are not important. Why are DC parameters not important? Because if an op amp is AC-coupled, that means that the input signals into the op amp as well as the output signal are capacitively coupled to their respective front-end and back-end stages, thus blocking the DC and AC errors due to offsets.

FURTHER READING

1. Sergio Franco, *Design with Operational Amplifiers and Analog Integrated Circuits*, McGraw-Hill, New York, 1988.
2. Walter G. Jung, ed., *Op Amp Applications*, Analog Devices, Norwood, MA, 2004.
3. Robert F. Coughlin and Frederick F. Driscoll, *Operational Amplifiers and Linear Integrated Circuits*, Prentice Hall, Englewood Cliffs, NJ, 1991.
4. Hank Zumbahlen, ed., *Basic Linear Design*, Analog Devices, Norwood, MA, 2007.

PROBLEMS

5.1 Mention five ideal operational amplifier parameters.

5.2 Using linear-dependent current or voltage-controlled sources, establish models for the following linear amplifiers:

(a) Voltage amplifier

(b) Current amplifier

(c) Trans-conductance amplifier

(d) Trans-resistance amplifier

5.3 Explain in your own words why a voltage amplifier features $R_i \to \infty$, $R_o = 0$, and $A_v = v_o / v_i$.

5.4 Explain in your own words why a current amplifier features $R_i = 0$, $R_o \to \infty$, and $A_I = i_o / i_i$.

5.5 Explain in your own words why a trans-conductance amplifier features $R_i \to \infty$, $R_o \to \infty$, and $A_G = i_o / V_i$.

5.6 Explain in your own words why a trans-resistance amplifier features $R_i = 0$, $R_o = 0$, and $A_R = v_o / i_i$.

5.7 Design an ideal op amp-based amplifier that has a gain of –2; draw the circuit.

5.8 Implement an ideal op amp-based circuit that produces the following arithmetic operation:

$$\mathrm{V}_{\mathrm{out}} = -\mathrm{V}_1 + 3\,\mathrm{V}_2. \tag{5.103}$$

In Equation (5.103) V_1 or $-\mathrm{V}_1$ can be used as an input voltage, so is V_2, but not ($3\,\mathrm{V}_2$). Implement the circuit with the smallest number of op

Table 5.8 Table for Problem 5.9: inverting amplifier with finite A_{OL}

Open-Loop Gain (A_{OL})	Ideal Closed-Loop Gain ($A_{OL} \to \infty$) (dB)	Closed-Loop Gain (CLG) Accounting for Finite A_{OL}	Absolute Error = CLG – ICLG	Relative Error = (CLG – ICLG) × 100/CLG (%)
1000	–2			
10,000	–2			
100,000	–2			
1,000,000	–2			
1,000	–20			
10,000	–20			
100,000	–20			
1,000,000	–20			

amps and explain why you can, or cannot implement it differently (i.e., with more or less op amps).

5.9 Assume that you have an op amp-based inverting amplifier; assume that the op amp used is ideal except for its gain. Complete Table 5.8 knowing the op amp open-loop gain and the ideal op amp-based inverting amplifier CLG. Refer to Table 5.8's first and second columns from the left.

5.10 For the circuit given in Figure 5.31, calculate the value of the output voltage V_{OUT}. Assume an ideal op amp.

5.11 For the circuit given in Figure 5.32, calculate the value of the output voltage V_{OUT}. Assume an ideal op amp.

5.12 Assume that you have an op amp in buffer amplifier configuration. The buffer CLG is 1 assuming an ideal op amp. Now assume the open-loop gain of the op amp (A_{OL}) is finite. For finite open-loop gains of 10^3, 10^4, 10^5, and 10^6 determine the CLG of the buffer amplifier and the relative error in ppm (parts per million).

5.13 Implement with the smallest possible number of ideal op amps the following analog expression:

$$V_{out} = K_1 V_1 + K_2 \int V_2 dt + K_3 \frac{dV_3}{dt}, \tag{5.104}$$

where K_1, K_2, and K_3 in Equation (5.103) are arbitrary constants. V_1, V_2, and V_3 are input voltages that vary with respect to time, and V_{out} is the total output voltage as shown by Equation (5.104). Hint: It is fine to use resistors and capacitors in addition to the op amps.

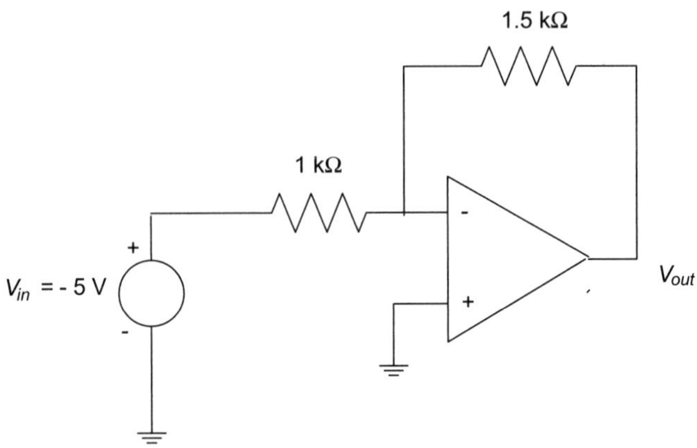

Figure 5.31 For Problem 5.10.

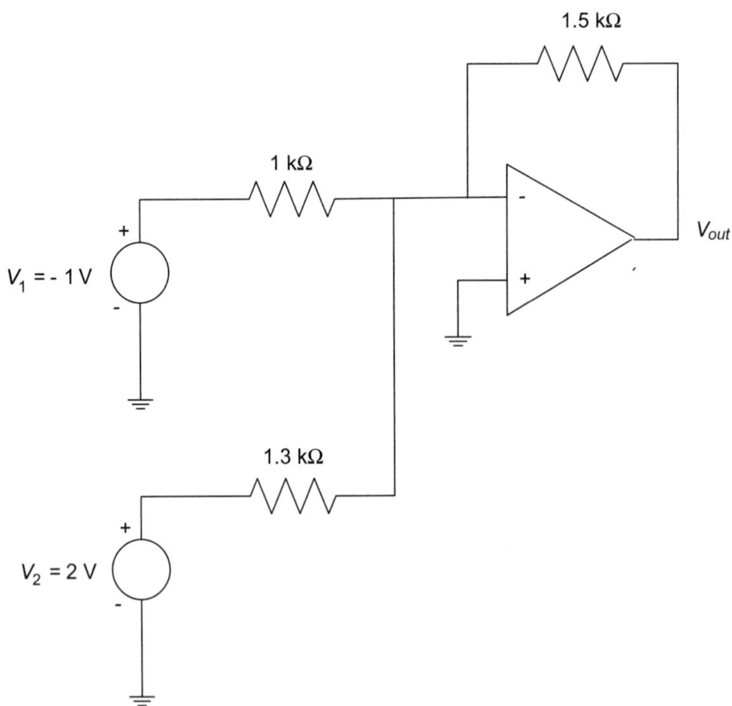

Figure 5.32 For Problem 5.11.

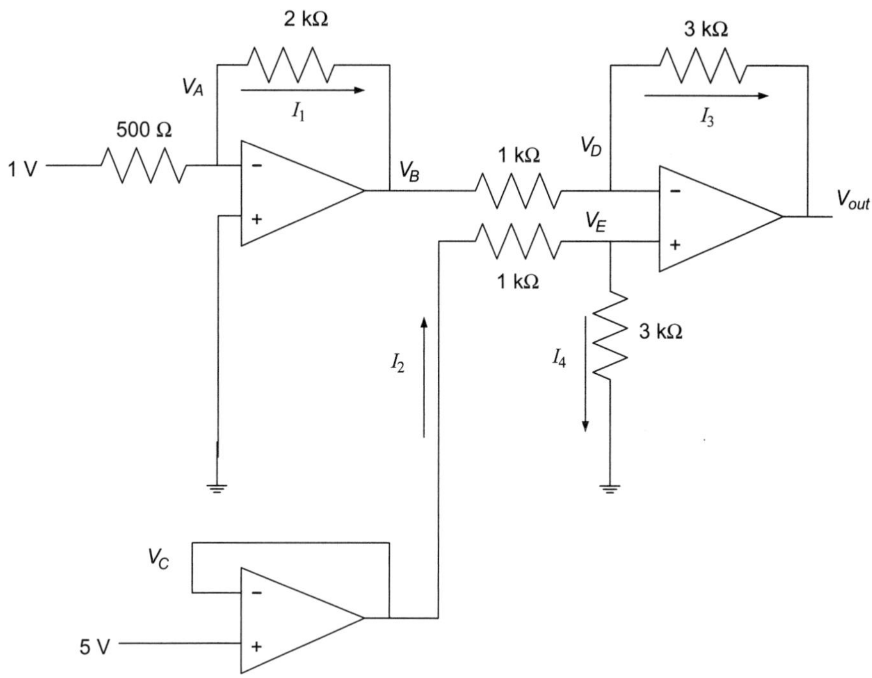

Figure 5.33 For Problem 5.15.

5.14 Find an alternate manner, also using ideal op amps, of implementing Equation (5.104). Hint: It is fine to use resistors and capacitors in addition to the op amps.

5.15 Given the circuit of Figure 5.33 calculate (a) V_{out}, (b) V_A, (c) V_B, (d) V_C, (e) V_D, (f) V_E, (g) I_1, (h) I_2, (i) I_3, and (j) I_4. Assume ideal op amps. Do not change the assumed directions for the currents.

5.16 Given the circuit of Figure 5.34 determine the circuit transfer function: $V_{out}(j\omega)/V_{in}(j\omega)$. Note: Express the final transfer function as a ratio of binomials with zeros and poles. Zeros are referred to as the numerator roots or zeros. Poles are referred to as the denominator roots or zeros.

5.17 For the circuit of Figure 5.34 assume the following component values: $R_1 = 1\ k\Omega$, $C_1 = 1\ \mu F$, $R_2 = 10\ k\Omega$, and $C_2 = 100$ pF. Draw the asymptotic magnitude and phase Bode plots from 1 Hz to 1 MHz.

5.18 For the circuit of Figure 5.34 assume the following component values: $R_1 = 1\ k\Omega$, $C_1 = 1\ \mu F$, $R_2 = 10\ k\Omega$, and $C_2 = 100$ pF. Draw the exact magnitude and phase Bode plots from 1 Hz to 1 MHz. Only calculate magnitude and phase values for 1 Hz, 10 Hz, . . . , 1 MHz.

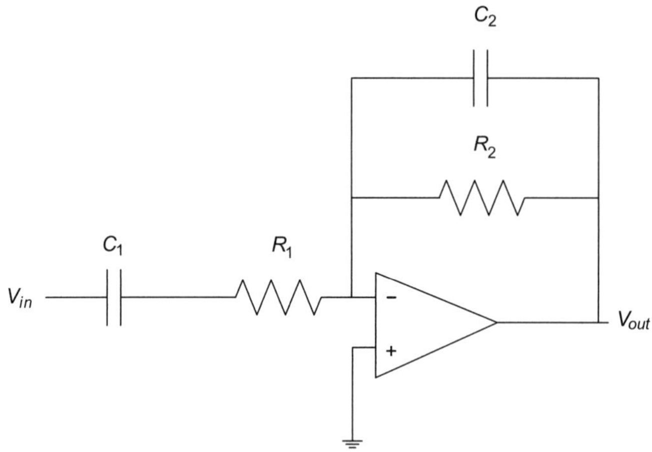

Figure 5.34 For Problems 5.16, 5.17, and 5.18.

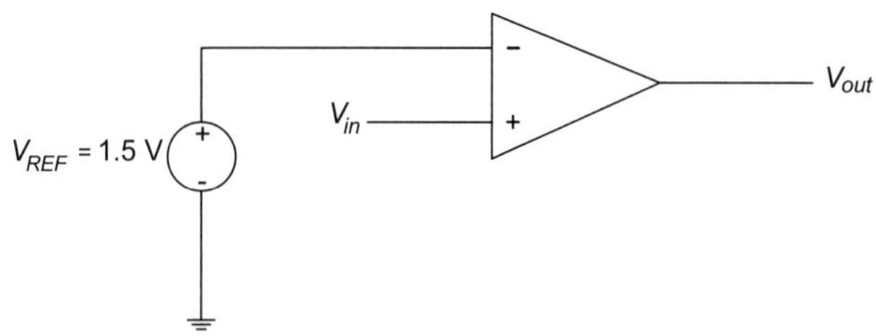

Figure 5.35 For Problem 5.19.

5.19 For the circuit of Figure 5.35 determine (a) V_{out} for $V_{in} = -1$ V, (b) V_{out} for $V_{in} = 0$ V, (c) V_{out} for $V_{in} = 1$ V, (d) V_{out} for $V_{in} = 2$ V, (e) V_{out} for $V_{in} = 3$ V. (f) Also for all five cases (a) through (e), calculate the voltage difference: $V_{in} - V_{REF}$

5.20 Research problem: Using Web sites of some op amps and comparator manufacturers, determine the fundamental differences between real-life operational amplifiers and comparators. Examples of some manufacturers are: http://www.linear.com, www.intersil.com, http://www.ti.com, and http://www.analog.com

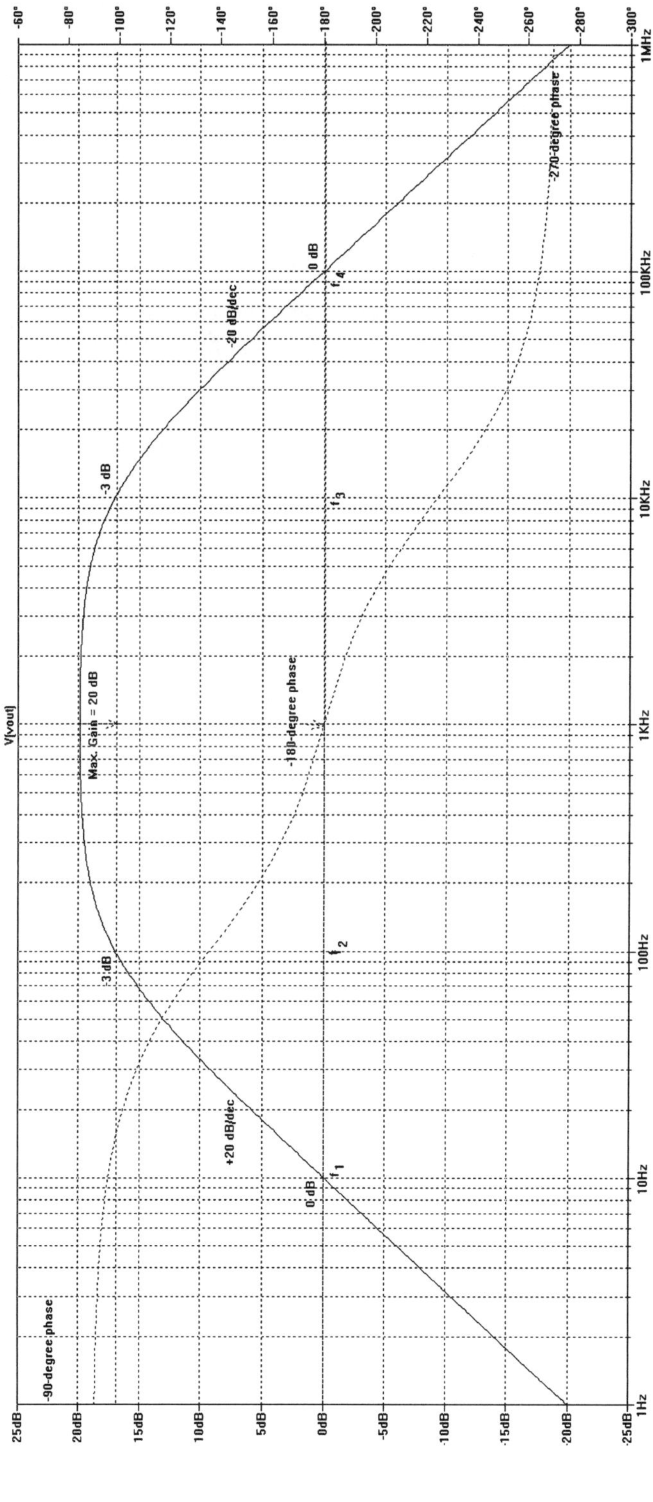

Figure 5.22 Exact Bode plots of a practical differentiator, integrator, and band-pass filter.

6

ELECTRONIC DEVICES: DIODES, BJTs, AND MOSFETs

6.1 INTRODUCTION TO ELECTRONIC DEVICES

Diodes, bipolar junction transistors (BJTs), and metal oxide field effect transistors or MOSFETs are the most common and most important electronic devices in use today. The junction field effect transistor or JFET is the predecessor of the MOSFET. Due to space reasons we will not address the JFET. This chapter presents to the reader the most basic and important considerations from a circuit analysis point of view of how to deal with diodes and transistors in the most common hardware applications. Some of such applications are: how do basic single-stage amplifiers work; how can we calculate the correct value of a resistor to correctly turn on a light-emitting-diode or LED; what are open collector and open drain outputs; why do we need them; how do we calculate the correct pull-up resistor value; and how do we increase the current drive capability of an integrated circuit output that needs to drive multiple chips or other discrete circuits. This chapter focuses on the characteristics and use of the three devices of our chapter title, their behavior, and their use in the most basic circuits that scientists and engineers need to know. It is not within the scope of this book to study electronic devices semiconductor physics concepts. Only a minimum of semiconductor physics will be addressed in particular when we cover MOSFETs. We generally take an approach of studying the

Electrical, Electronics, and Digital Hardware Essentials for Scientists and Engineers, First Edition. Ed Lipiansky.

electronic device, knowing its current–voltage relationship and its small signal model for AC analysis.

6.2 THE IDEAL DIODE

The diode is the simplest and most fundamental electronic device. Its idealized model describes the diode as a two-terminal device that acts like a mechanical switch. When the switch is closed, the ideal diode has a *zero-ohm* forward resistance. When the switch is open, the ideal diode has an infinitely large reverse resistance. Figure 6.1a shows the electronic symbol of a diode, indicating its two terminals, the *anode* (A) and the *cathode** (K). When the current through the diode flows from *anode* to *cathode*, the diode is said to be forward biased or *ON*. When a diode is forward biased, an external element in the circuit, such as a series resistor, is responsible for limiting the amount of current that flows through the turned-on diode. When the current through the diode intends to flow from *cathode* to *anode*, the diode is said to be reversed biased or *OFF*. Figure 6.1b shows the model of an ideal diode turned *ON*. While Figure 6.1c shows the model of an ideal diode turned *OFF*. Figure 6.1d depicts the ideal diode model *I-V* (current–voltage) characteristics. The bold positive current shown in Figure 6.1d represents the diode operating in a forward biased mode, that is the current flows through the diode from *anode* to *cathode*; this means that the *anode* is at a *higher* potential than the cathode. The external circuit must limit the forward current that flows through a forward biased diode; otherwise the diode would get destructed if such current was not limited. This portion of the forward biased region also indicates to us that the ideal diode forward voltage drop is zero. When the diode is in reverse biased mode, no current flows through the diode. The external circuit also limits the voltage across a reverse biased diode.

The diode *I-V* characteristic is nonlinear, thus the diode is a nonlinear element. Figure 6.1d shows the nonlinear characteristic formed by two linear segments; it is also appropriate to refer to such characteristic as piecewise linear. When the diode is used at a particular operating point, although the diode is a nonlinear device, it is possible and accurate to analyze the diode as a linear circuit element for such application.

Figure 6.2a shows a forward biased ideal diode. The current through this ideal diode equals the voltage applied divided by the series resistance, in our case 10 V/100 Ω = 1 mA. The forward biased diode has a zero *ON* or *forward resistance*.

The reverse biased diode of Figure 6.2b does not allow any current flow, and it has a –10 V reverse voltage applied across its *cathode* and *anode*. The

* Cathode derives from the Greek word "Kathodes", that is why it is traditionally abbreviated with the letter "K."

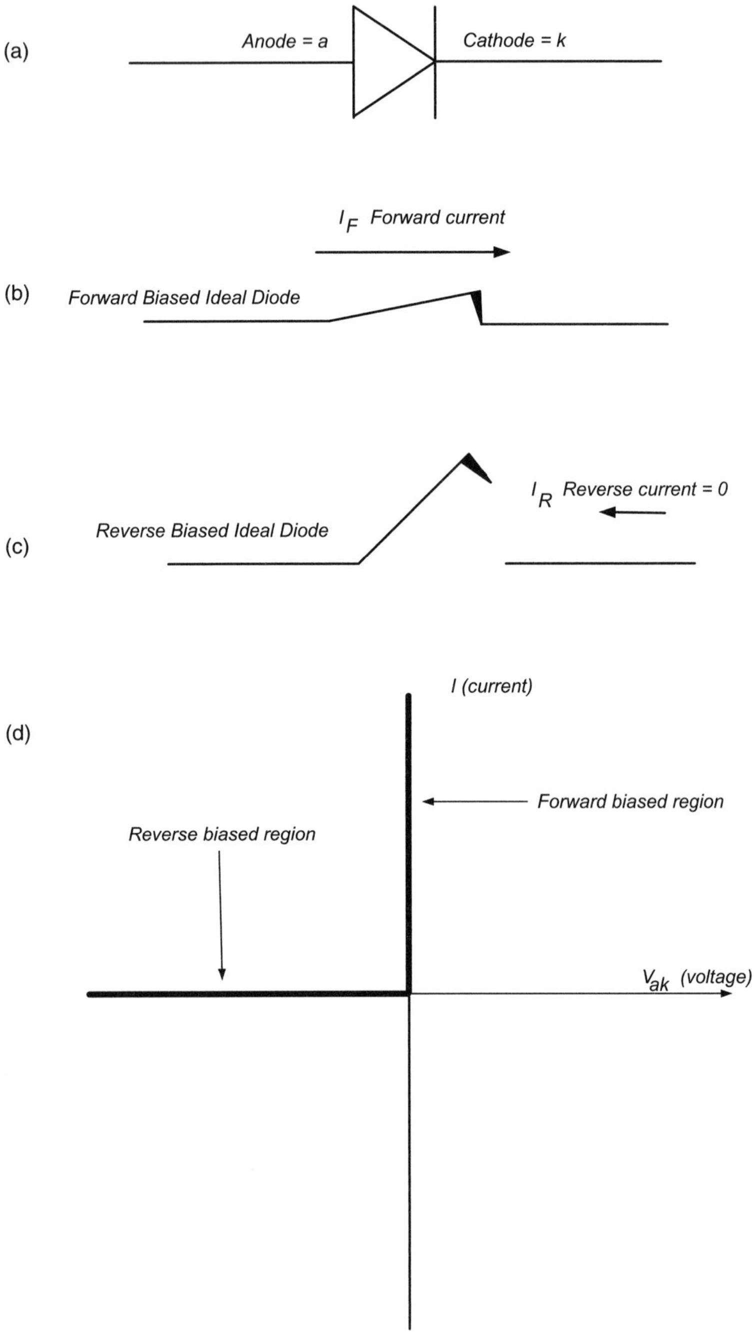

Figure 6.1 The ideal diode: (a) diode circuit symbol; (b) ideal diode turned *ON*; (c) ideal diode turned *OFF*; (d) ideal diode I-V characteristic.

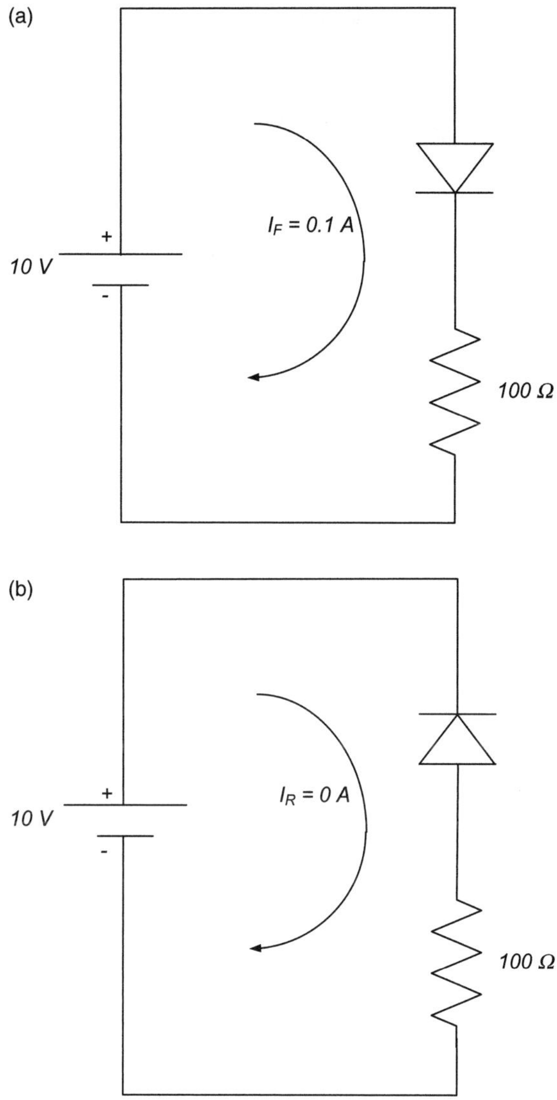

Figure 6.2 (a) Forward biased ideal diode; (b) reverse biased ideal diode.

reverse biased diode has an infinite resistance so that no current can flow through the circuit under those conditions.

6.2.1 The Half-Wave Rectifier

A half-wave rectifier consists of a diode and a resistor in series; a sinusoidal input waveform is applied between the diode anode and the free end of the resistor. The output voltage is taken across the resistor. Figure 6.3a shows

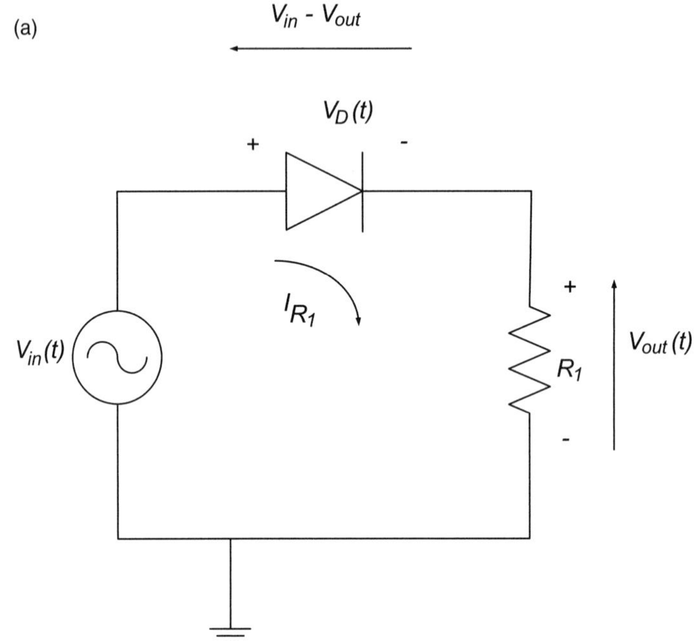

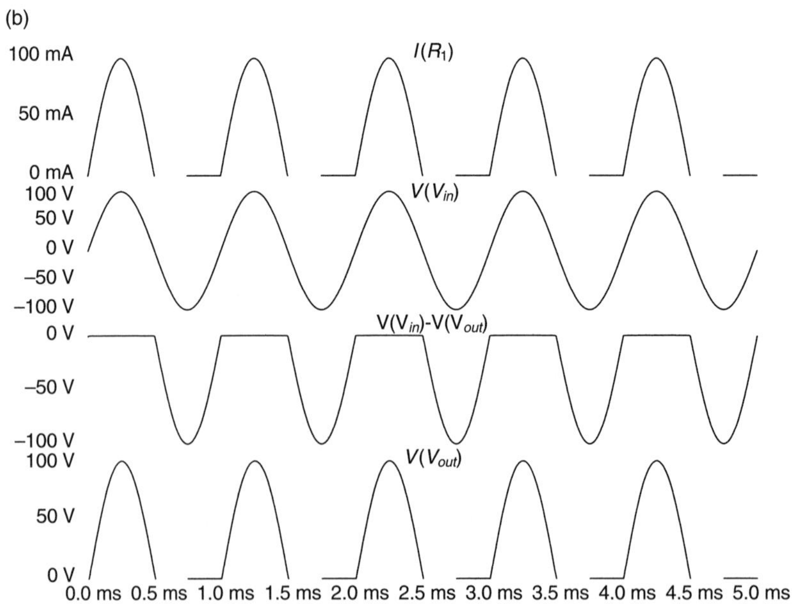

Figure 6.3 (a) Half-wave rectifier; (b) half-wave rectifier waveforms: $I_R(t)$, $V_{in}(t)$, $V_{out}(t)$, and $V_D(t) = V_{in}(t) - V_{out}(t)$.

a half-wave rectifier circuit. Figure 6.3b depicts the current $i(t)$ through the circuit, the sinusoidal input voltage waveform $V_{in}(t)$; the voltage $V_D(t)$ the *anode-to-cathode* voltage across the diode; the voltage $V_R = V_{out}(t)$ across the resistor

Let us describe how the circuit of Figure 6.3a works. Let us refer to the waveforms of Figure 6.3b. When the sinusoidal input voltage $V_{in}(t)$ is greater than zero, the diode becomes forward biased; this occurs for input phase angle ranges: 0 to π, 2π to 3π, 4π to 5π, which according to Figure 6.3b respectively correspond from 0 to 0.5 ms, 1 to 1.5 ms, 2 to 2.5 ms, and so forth. Note that for our specific waveform, its period 2π equals 1 ms. When the ideal diode is forward biased, the voltage drop across it is zero; refer to waveform $V_D(t) = V_{in} - V_{out}$, for time segments from 0 to 0.5 ms, 1 to 1.5 ms, 2 to 2.5 ms, and so forth. When $V_{in}(t)$ is negative, the diode is an open circuit because it becomes reverse biased; the voltage $V_D(t)$ across the diode is the negative cycle of the input sine-wave for phase angles of 1π to 2π, 3π to 4π, 5π to 6π (equivalently 0.5 to 1 ms, 1.5 to 2 ms, 2.5 to 3 ms), and so forth. The output voltage $V_{out}(t)$ across resistor R is equal to the positive half cycle of the input for phase angle ranges: 0 to π, 2π to 3π, 4π to 5π (equivalently for 0 to 0.5 ms, 1 to 1.5 ms, 2 to 2.5 ms), and so forth. For phase angles: 1 to 2π (0.5 to 1 ms), etc., the voltage across resistor R is zero. Finally, the current through the circuit has to have the same shape as the voltage across the resistor, because the resistor is a linear component. Now let us recall Kirchhoff's laws from earlier chapters. By inspection of the circuit of Figure 6.3a, Kirchoff's voltage law (KVL) affirms that

$$V_{in}(t) = V_D(t) + V_R. \tag{6.1}$$

Carefully examining the waveforms of Figure 6.3b, it is easy to verify that Equation (6.1) is met. This tells us that both linear and nonlinear circuits meet KVL. This is a very powerful statement, particularly since diodes did not exist during Kirchhoff's times.

Once more note the shapes of the current $i(t)$ and the voltage $V_{out}(t)$'s first and last waveforms of Figure 6.3b have the same shape. Current $i(t)$ is unidirectional and positive. If we eliminated the diode with a short-circuit current $i(t)$ would be sinusoidal, because only the source and the resistor are present in the circuit. We are just one simple next step away from obtaining a DC voltage level across resistor R. This is achieved by placing a capacitor of the appropriate value across load resistor R. Figure 6.4 shows the output waveform across the load resistor for capacitance values of: 5 μF, 10 μF, and 50 μF. Note that the larger the capacitor, the smoother is the DC waveform obtained. In all cases, the DC waveform is not a perfect horizontal straight line, it is rather a waveform that exhibits ripple. One can reduce the ripple by incrementing the capacitor size; however, the ripple cannot be completely eliminated.

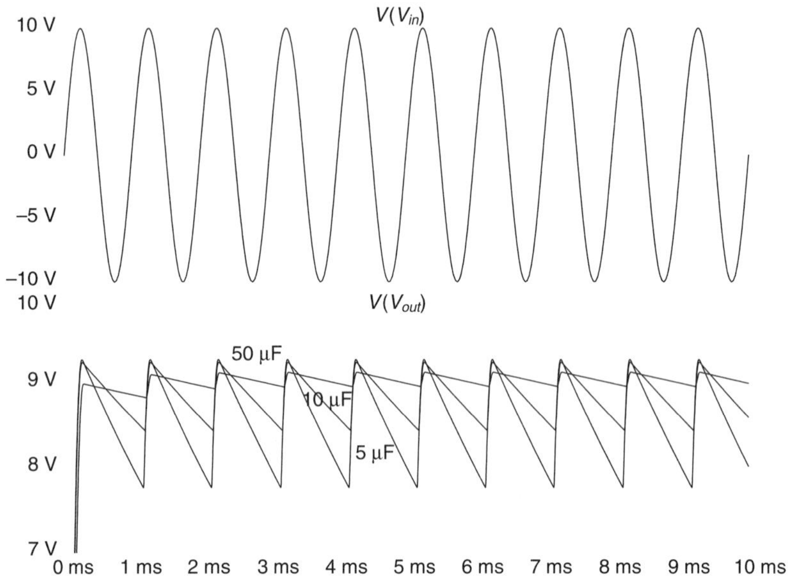

Figure 6.4 Half-wave rectifier with smoothing capacitor to reduce ripple.

6.2.2 The Full-Wave Bridge Rectifier

The circuit of Figure 6.3a rectifies only half of the sinusoidal input waveform. This means that only the positive half of the sinusoidal cycle is preserved, the negative half cycle becomes zero at the load resistor; this is assuming that there is no filtering capacitor across resistor R. The circuit of Figure 6.5a shows a four-diode bridge excited by a sinusoidal input V_{in} and the load resistor R across terminals V_{out} and V_{out_Ret}.

The circuit works as follows: when V_{in} is positive, signs in Figure 6.5a apply, diodes D_1 and D_2 conduct current through the circuit established between a positive V_{in}, D_1, R, and D_2. D_1 and D_2 are at this time forward biased, since for ideal diodes, the forward voltage drop is zero, drops across D_1 and D_2 are zero during this first half-period. Diodes D_3 and D_4 remain reversed biased at this time. The current through resistor R flows from terminal V_{out} to V_{out_Ret}, thus V_{out} is at a higher potential than V_{out_Ret}. The difference V_{out}-V_{out_Ret} is another form of referring to the voltage across resistor R where node V_{out} is more positive than node V_{out_Ret}. When V_{in} is negative, signs are opposite to those shown in Figure 6.5a, diodes D_3 and D_4 are forward biased, while D_1 and D_2 are reversed biased. The current flows from V_{in}, through D_3, R, and D_4. Note that this time, the current through R also flows from node V_{out} to V_{out_Ret}. The waveform obtained across resistor R terminal is referred to as a full-wave rectified sine wave. Figure 6.5b shows waveforms V_{in} and V_{out}-V_{out_Ret} the voltage across resistor R.

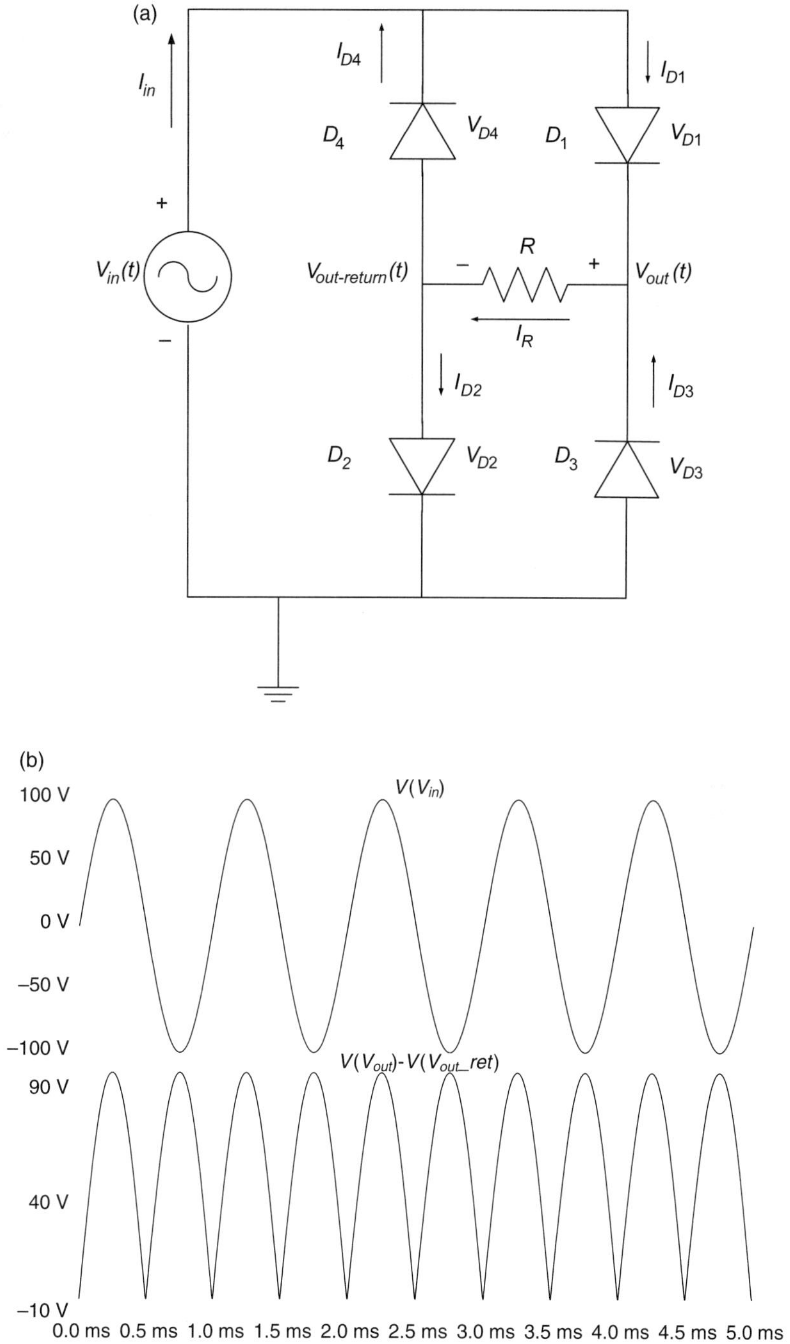

Figure 6.5 (a) Full wave rectifier; (b) voltage waveforms: V_{in} and $V_D = V_{out} - V_{out\_Ret}$; (c) current waveforms through four diodes: I_{D1}, I_{D2}, I_{D3}, I_{D4} and load current I_R.

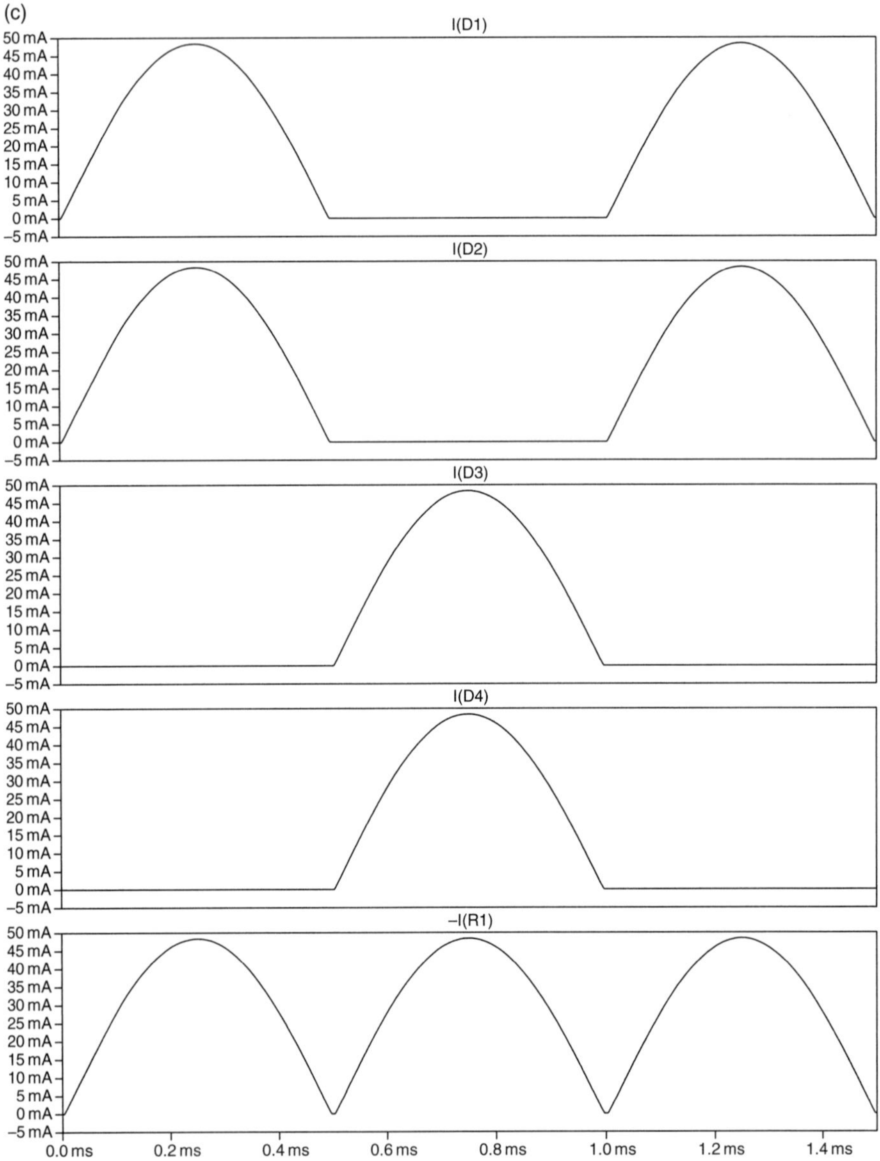

Figure 6.5 (*Continued*)

Continuing to look at the four-diode bridge circuit, we now take a look at the currents that flow through everyone of the circuit elements, the input voltage source, each of the four diodes, and load resistor R. Figure 6.5c shows the current waveforms through each diode I_{D1}, I_{D2}, I_{D3}, and I_{D4}, and the current waveform on load resistor R, I_R. It is very interesting again to find that all the currents comply with Kirchoff's current law (KCL). Similarly, it can be shown that the voltages across every one of the circuit elements also comply with KVL.

Carefully observe the current waveforms of Figure 6.5c: I_{D1}, I_{D2}, I_{D3}, I_{D4}, and I_R. It is important to refer to the circuit of Figure 6.5a to observe the direction of every current every diode and load resistor. Plot the current waveform I_{in} of the sinusoidal input source V_{in}.

Example 6.1 Draw all the voltage waveforms for the full wave rectifier circuit of Figure 6.5a. Such waveforms are: V_{in}, V_{D1}, V_{D2}, V_{D3}, V_{D4}, and V_R. Figure 6.6 presents all the waveforms requested. Note that at every node on the circuit, KVL is met. For example:

$$V_{in} = V_{D1} - V_{D3}$$

which is also equivalent to:

$$V_{in} = V_{D1} + V_R + V_{D2}.$$

Note that in Figure 6.6:

$V_{D1} = V_{in} - V_{out}$: is the voltage across diode 1
$V_R = \mathrm{V_{out}} - \mathrm{V_{out\_ret}}$: is the voltage across load resistor R
$V_{D2} = V_{out\_ret} - \mathrm{GND} = V_{out\_ret}$: is the voltage across diode 2
$V_{D3} = -V_{out}$: is the voltage across diode 3
$V_{D4} = V_{out\_ret} - V_{in}$: is the voltage across diode 4

Voltages across diodes have been defined as the difference between the anode and the cathode voltages.

The reader is encouraged to graphically add the appropriate waveforms to verify the validity of KVL. Beware of the plotted waveform polarites.

6.2.3 The Real Silicon Diode *I-V* Characteristics: Forward-Bias, Reverse-Bias, and Breakdown Regions

A real diode has an *I-V* characteristic as depicted by Figure 6.7. On the first quadrant, for positive anode-to-cathode voltage and current, the diode does not start conducting in its forward biased region until the forward voltage exceeds approximately 700 mV. After this voltage level is exceeded the diode conducts, and the diode external circuit only limits the current through it. In many cases, this external circuit is a resistor in series with the diode. The equation that models the behavior of the current through a diode as a function of its voltage is

$$\mathrm{i}_D(t) = I_S(\mathrm{e}^{V_D/nV_T} - 1) \tag{6.2}$$

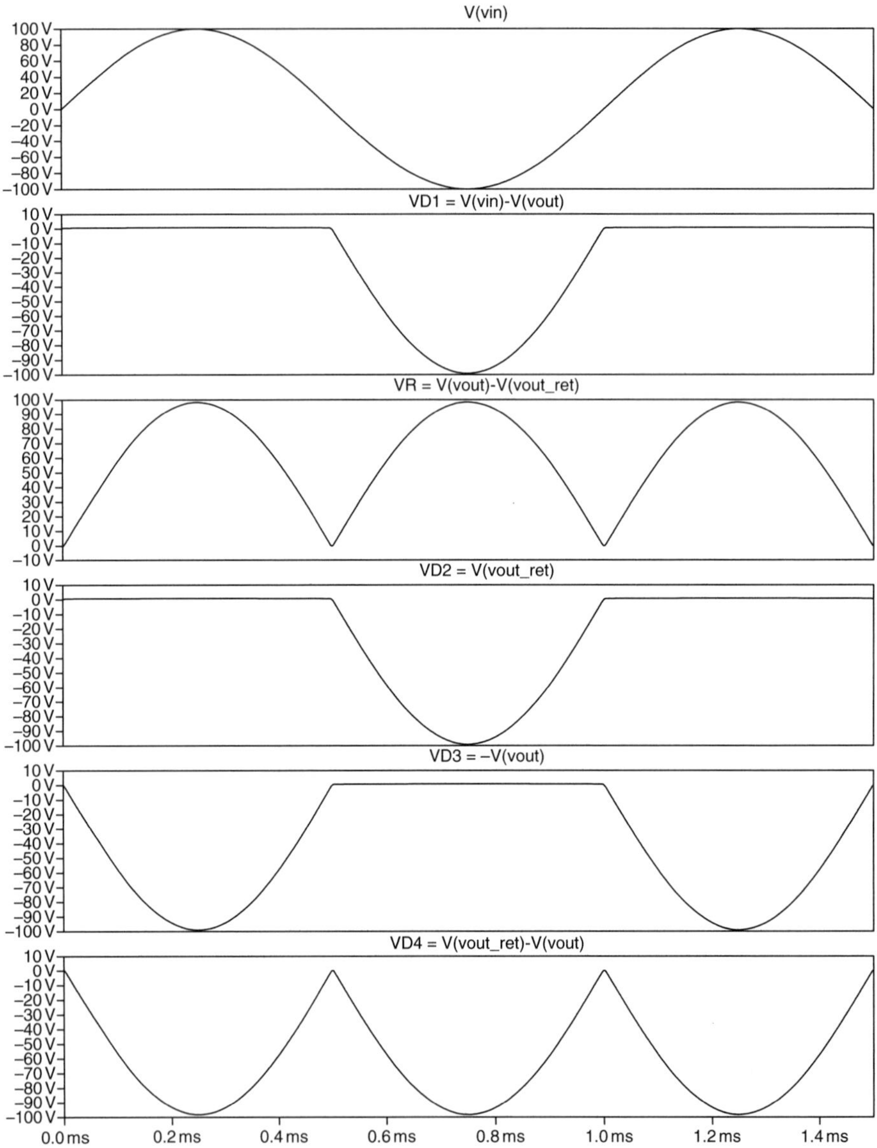

Figure 6.6 Voltage waveforms in full-wave rectifier: input voltage, voltages across four diodes, and voltage across the load.

Equation (6.2) describes the current through the diode. I_S is the diode saturation current, V_D is the anode-to-cathode voltage across the diode, n, the emission coefficient is a constant that is usually 1 for integrated circuit diodes and $n = 2$ for discrete diodes. V_T, the thermal voltage is defined from physical considerations as:

$$V_T = \frac{kT}{q} \tag{6.3}$$

where

k = Boltzmann's constant = 1.38×10^{-23} J/K
T = the absolute temperature in kelvin = 273 + temperature in °C
q = the magnitude of electronic charge = 1.60×10^{-19} C.

The thermal voltage V_T is approximately 26 mV at a room temperature of 300 °K.

The diode forward voltage drop has a negative temperature coefficient, which is

$$\frac{\Delta V_D}{V_T} = -2 \text{ mV} / °\text{C}. \tag{6.4}$$

Equation (6.4) tells us that for every degree C in temperature rise of the diode junction, its forward voltage drop decreases by 2 mV. This property of Silicone diode junctions is typically used to know the internal temperature of integrated circuits.

Referring again to Figure 6.7, we see that the forward current of the diode is negligible for voltages below about 0.5 V.

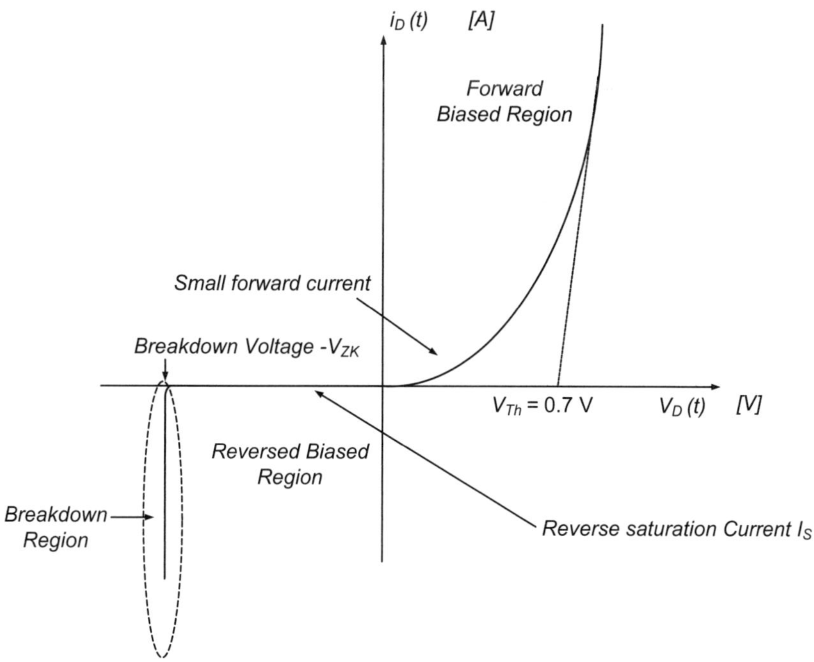

Figure 6.7 Diodes: three regions of their *I-V* characteristics:
For $V_D > 0$ the diode is in the forward-bias region, conducts significant current once $V_D > V_{Th}$
For $-V_{ZK} < V_D < 0$ the diode is in the reverse-bias region
For $V_D < -V_{ZK}$ the diode is in the breakdown region.

The current starts having significant value once a diode forward voltage threshold of approximately 0.6 V–0.7 V is reached.*

When the diode voltage V_D is negative the exponential term in Equation (6.2) becomes very small compared to the constant 1, thus Equation (6.2) becomes

$$i_D(t) \cong -I_S \tag{6.5}$$

where I_S is also referred to as the reverse bias saturation current. Current I_S has a positive temperature coefficient and it approximately doubles for every 10°C of temperature rise. This current is quite constant as the reverse bias voltage varies, and the temperature does not change (see Fig. 6.7, reverse-bias region). I_S can be in the order of 10^{-14} to 10^{-15} A. As V_D continues to decrease (i.e., becomes larger in magnitude but negative in sign) the diode enters the breakdown region. The voltage $-V_{ZK}$ is denoted the Zener voltage knee. Regular diodes are not designed to be operated in the breakdown region; however, another type of device, the Zener diode, is purposely designed to operate in the breakdown region. At a reverse voltage of $-V_{ZK}$, the diode characteristic is a virtual straight line (Fig. 6.7) so that means that within a range of reverse current, the voltage remains practically within a very small variation. Zener diodes are particularly used in voltage regulators.

Example 6.2 A silicon diode has the following characteristics:

Reverse saturation current at 20 V and 25°C = 25 nA.

Using Equation (6.2), determine the diode forward voltage drop for forward diode currents of (a) 5 mA, (b) 10 mA, (c) 20 mA, (d) 100 mA, and (e) 300 mA. This diode is fabricated so that it can conduct safely a current as high as 300 mA. Assume that the diode emission coefficient n is 2.

Using Equation (6.2) and a thermal voltage of 26 mV at 25°C we find:

(a) At 5 mA, $V_D = 2 \times 0.026 \ln [5 \times 10^{-3}/25\ 10^{-9}] = 0.630$ V
(b) At 10 mA, $V_D = 2 \times 0.026 \ln [10 \times 10^{-3}/25\ 10^{-9}] = 0.670$ V
(c) At 20 mA, $V_D = 2 \times 0.026 \ln [20 \times 10^{-3}/25\ 10^{-9}] = 0.710$ V
(d) At 100 mA, $V_D = 2 \times 0.026 \ln [100 \times 10^{-3}/25\ 10^{-9}] = 0.790$ V
(e) At 300 mA, $V_D = 2 \times 0.026 \ln [300 \times 10^{-3}/25\ 10^{-9}] = 0.850$ V

Note that for forward currents of 5 mA up to 300 mA, the increase in current is by a factor of 60 or 6000%, the diode forward voltage drop just increases by 220 mV or by 35%.

Why? The answer is to be provided by the reader.

* Power diodes, which are larger than general purpose and signal diodes, may have significantly higher forward drop voltages. It is not uncommon to see 1 V or more of forward voltage drop on a power diode.

6.2.4 Two More Realistic Diode Models

We have seen that the *I-V* characteristic of the diode is very steep when the diode forward voltage drops exceeds 600 mV. Thus, a better model than the ideal diode model can be produced taking into account the 600 mV forward diode drop. Figure 6.8a shows a circuit that models the diode with more realism than the idealized model of Figure 6.1. The 600 mV DC source in series with the diode, opposes the flow of current until the diode anode voltage is strictly greater than 600 mV with respect to the negative terminal of the 600 mV DC source. Figure 6.8b depicts the current–voltage characteristics of such diode model. This model is better than the ideal diode model because it takes into consideration a voltage forward drop of 600 mV. Let us remember that one of the most realistic equations used to model the real diode was the one given by Equation (6.2). Such equation takes into account thermal voltage, diode forward voltage drop, emission coefficient n, reverse saturation current I_S, which is used for the forward and reverse-bias regions.

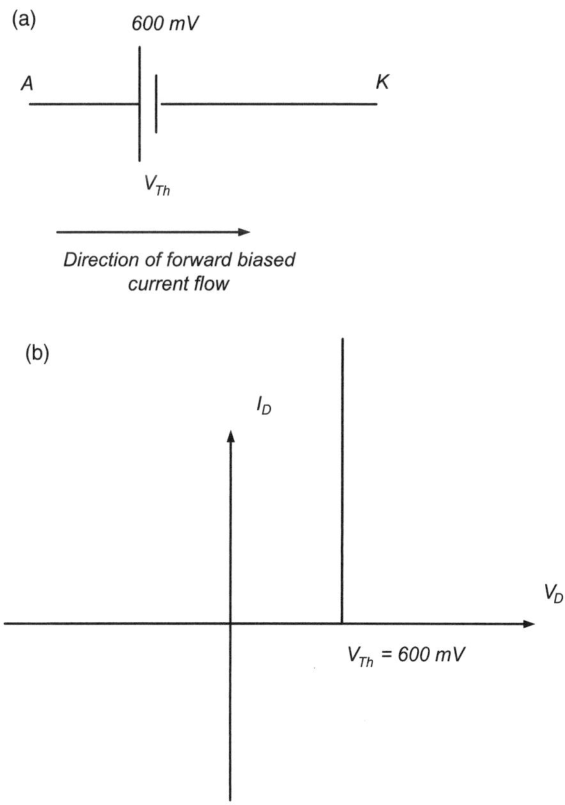

Figure 6.8 (a) A better ideal diode model including the forward voltage drop; (b) *I-V* characteristics of such diode model.

An even better model than the one presented in Figure 6.8 is one that models the forward voltage drop with a line that starts at zero current at 600 mV and has a positive slope that mimics the diode finite inverse of its forward resistance ($1/r_D$).

Figure 6.9a shows the circuit model including the 600 mV forward drop and diode forward resistance. Figure 6.9b depicts the *I-V* characteristics of such model. Such model is called the diode piecewise linear model. For the diode in reversed bias mode, the model is still an open-switch, not shown in Figure 6.9.

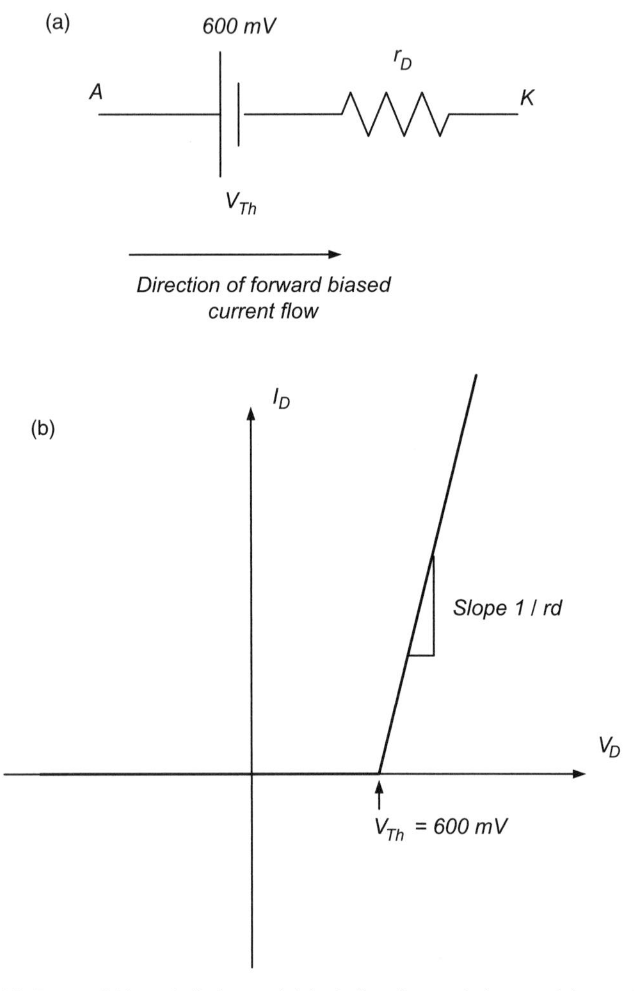

Figure 6.9 (a) Forward biased diode model including forward drop and forward resistance; (b) diode model *I-V* characteristics.

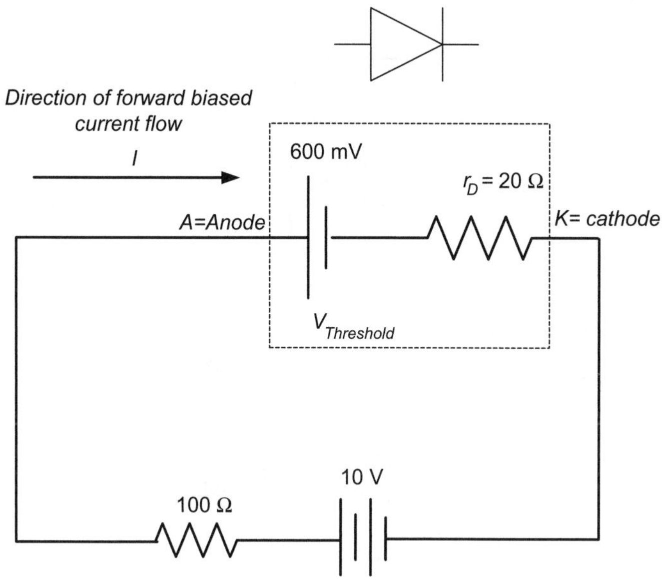

Figure 6.10 Circuit for Example 6.3: DC source in series with diode model.

Example 6.3 Assuming the diode model of Figure 6.9 find the forward current of the diode that has the following characteristics: $r_D = 20\ \Omega$, forward voltage modeling source 600 mV. Use the circuit of Figure 6.10.

From Figure 6.10 we can state the following circuit equation:

$$V_{in} - V_D = \mathrm{I}r_D + IR. \tag{6.6}$$

Using the numerical values from the circuit we obtain:

$$10 - 0.6 = \mathrm{I}(20 + 100). \tag{6.7}$$

From Equation (6.7) we obtain $I = 0.078$ A.

6.2.5 Photodiode

An incident light on a reverse-biased photodiode causes a flow of current through the photodiode. Basically a photon of sufficient energy hitting the photodiode causes an electron to become mobile, causing a photocurrent.

The symbol of a photodiode is presented in Figure 6.11.

6.2.6 Light Emitting Diode (LED)

An LED basically works in the opposite way a photodiode works. A forward current flowing through an LED causes the LED to emit photons or light.

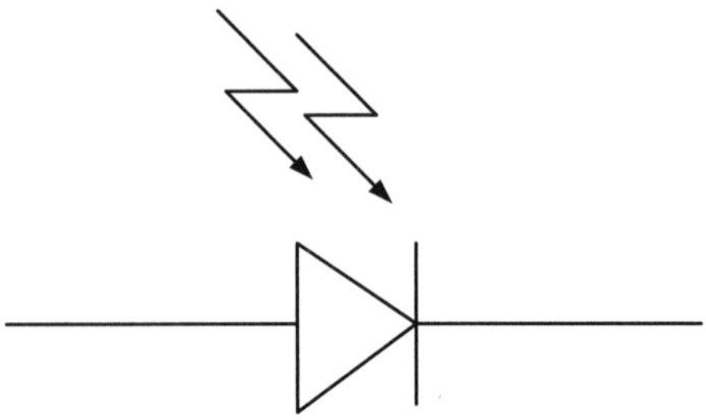

Figure 6.11 Symbol of a photodiode.

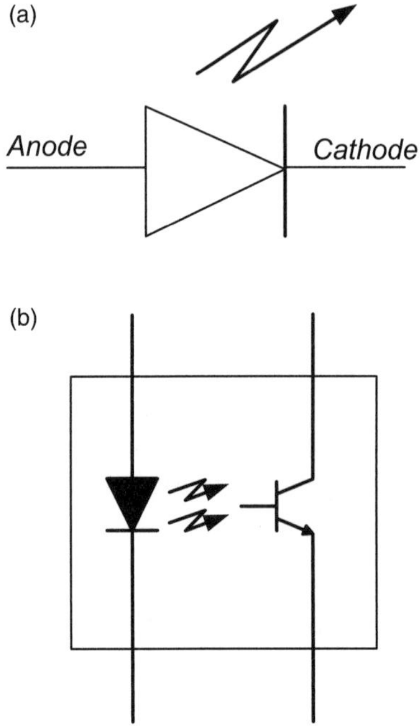

Figure 6.12 (a) LED symbol; (b) optoisolator symbol.

Figure 6.12a depicts the circuit symbol of an LED. A combination of a photodiode and an LED in the same integrated circuit package constitutes what is referred to as an optoisolator. The LED part of the opto converts an electrical signal into light. The photodiode portion receives the light emitted by the LED and produces an electrical signal. Figure 6.12b shows the circuit symbol of a typical optoisolator. This device is suitable to perform electrical-isolation between the input and output of the optoisolating device. This can be done to reduce electronic noise from propagating from one electronic stage to another, or for safety reasons such as in the case of medical instruments. The optoisolator can also be used to reduce the propagation of electromagnetic interference (EMI) in a communication system.

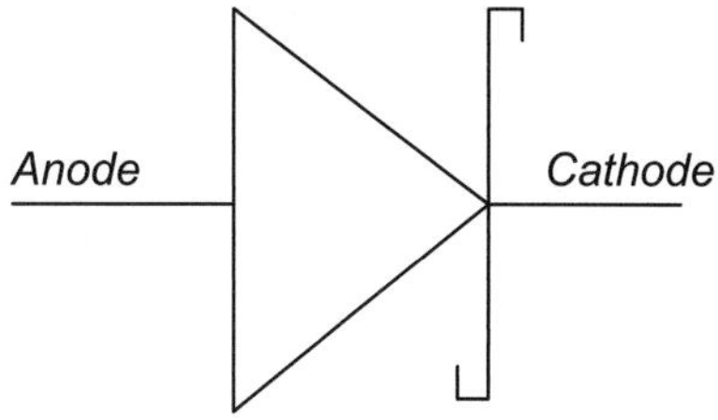

Figure 6.13 Circuit symbol of a Schottky-barrier diode.

6.2.7 Schottky-Barrier Diode

The Schottky-barrier diode is formed by making a metal to *n*-type semiconductor junction. This is unlike a regular junction diode that consists of a semiconductor-to-semiconductor junction formed with *n*-type and *p*-type material. The metal part of the Schottky works as the anode, while the semiconductor part is the cathode. A very distinct characteristic of Schottky-barrier diodes is their much lower forward voltage drop, usually around 200–300 mV. However, there is penalty for using a Schottky diode, their reverse saturation current is two to three orders of magnitude larger than that of their junction diode counterparts. Figure 6.13 shows the circuit symbol of a Schottky-barrier diode.

6.2.8 Another Diode Application: Limiting and Clamping Diodes

Junction and Schottky diodes are good devices to protect integrated circuits inputs. *IC* inputs usually must not be allowed to make voltage levels swings above the *IC* upper power supply rail or below the lower power supply rail. The lower supply rail on cases where the *IC* is only powered by a single positive power supply is the return or ground of that rail.

Example 6.4 Figure 6.14 shows an *IC* powered to a voltage called V_{DD} and V_{SS}. In many cases, V_{DD} can be 3.3 V and V_{SS} 0 V or ground. Diode D_1 protects the input of the *IC* from exceeding a voltage higher than its V_{DD} power supply level. Diode D_2 protects the input from going below the 0 V level. In other words, it prevents the input from going very negative. The purpose of the clamping or limiting diodes is to protect the *IC*. Note that the input line of the protected *IC* may be driven by another *IC* or circuit, the series resistance R in line between the input and the common node between diodes D_1 and D_2 is to prevent a large current surge when the input either is well above V_{DD} or well below V_{SS} (ground). Now let us recall that real diodes have a finite and non-zero forward voltage drop. The diode D_1 protection is actually limited to

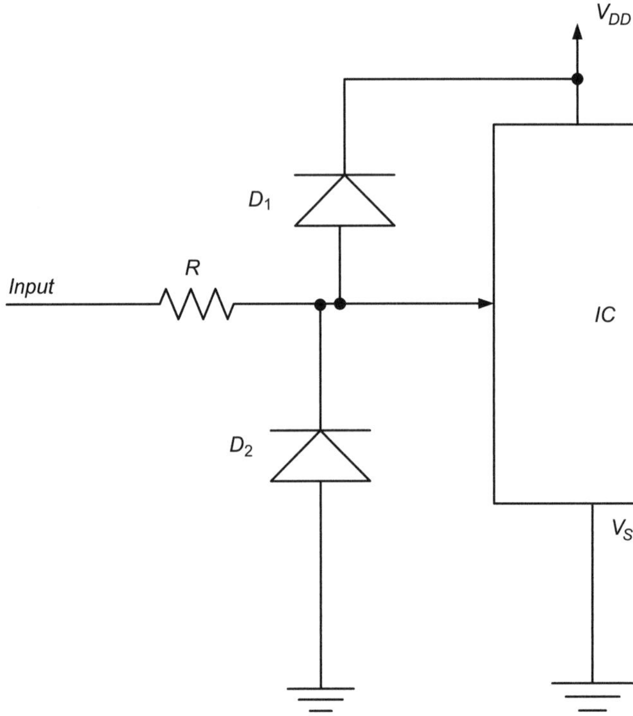

Figure 6.14 Diodes clamping the inputs of an *IC*.

clamping the input to $V_{DD} + V_D$, where V_D is D_1 diode forward voltage drop. Diode D_2 protection is limited to clamping the input to $V_{SS} - V_D$.

Summarizing the upper diode prevents the input signal from going one diode drop above the *IC* V_{DD}. The lower diode prevents the input signals from going more negative than a diode drop below ground.

6.2.9 Diode Selection

The most important parameters of a diode are listed in Table 6.1. They would be advertised by the diode manufacturer as being at one temperature, typically 25°C or valid at a temperature range, for example: 0°C to 70°C.

Upon selecting a diode for an application, none of the specified parameters should ever be exceeded under any conditions. Doing so may not necessarily make the device fail, but it can certainly reduce its useful life. Table 6.1 only lists some of the key parameters; manufacturers also publish curves for some of the parameters to give a better idea to the diode circuit application designer what the limits of the parameters are for slightly different conditions than those published on the table.

Table 6.1 Some diode key parameters

Parameter	Test Condition	Symbol	Min.	Typ.	Max.	Units
Forward voltage	At $I_F = 10\,\text{mA}$	V_F		0.6	1	*V*
Breakdown voltage	At reverse current $I_R = 100\,\mu\text{A}$	V_R	100			V
Peak reverse current	At $V_R = 75\,\text{V}$	I_R			5	μA
Diode capacitance	$V_R = 0$ Freq = 1 MHz	C_D			4	*pF*
Reverse recovery time	$I_F = 10\,\text{mA}$ to $I_R = 1\,\text{mA}$ $V_R = 6\,\text{V}, R_{LOAD} = 100\,\Omega$	t_{RR}			4	*ns*
Power dissipation		P_D			0.5	W

Example 6.5 Let us assume that one wants to design a half-wave rectifier that has to operate under the following conditions: Sinusoidal input waveform: 120 V_{RMS}, frequency 60 Hz, and peak load current 1 A. Assume that the accuracy we want for the rectified output is 120 V ± 1%. Determine the *forward voltage, breakdown voltage, peak reverse voltage, reverse recovery time*, and *power dissipation* of a general-purpose diode that you need to select. Since the RMS value of the input waveform is 120 V, the peak voltage is 120 × 1.41 = 169.2 V peak. Forward voltage: since 1% of 169.2 V is approximately 1.7 V, any general-purpose Silicon diode that has a forward drop of 0.6–0.7 V will meet our requirements.

Since the sinusoidal input has a ± 169.2 V of peak value, we need to ensure that when our diode is reversed biased, it can tolerate without any stress 169.2 V as its reverse voltage. If we picked a diode with about a 15–20% margin, the reverse voltage rating would have to be 200 V. Let us assume that we want more margin than that to obtain excellent reliability of the diode, so we pick a diode with a 400-V repetitive breakdown voltage. This provides more than a generous 100% margin. A diode that handles a 2-A forward peak current will satisfy the requested conditions. Many general-purpose diodes handle more than 2 A, so this is not a hard find. Since the frequency of operation of the diode is 60 Hz, this means that the period of the 60 Hz signal is 16.67 ms. Within the 16.67 ms period, the diode has to be turned on once and turned off once, so the diode shall have timing margin to handle a signals changing every approximately 8 ms. It is easy to find general-purpose diodes for power supply applications that have switching recovery times in the order of *ns*. So speed wise, we can pick any diode that switches at the speed required.

Since the forward peak current shall not be over 1 A, 1 A × 1 V = 1 W power dissipation. Note that 1 V is the maximum forward voltage drop of the diode, so the diode will not dissipate more than 1 W at maximum current.

Summary:

Forward voltage drop $V_D = 0.7$ V to 0.8 V Typ; Max. $V_D \approx 1.2$ V.

Repetitive breakdown voltage $V_R = 400$ V.

Reverse recovery time t_{RR} = anything considerably faster than 8 ms, any value in the *ns* range is trivial to obtain.

Power dissipation $P_D = 2$ W, this allows a nice 100% margin so that the diode not only dissipates less power than it has to but also may not need to be equipped with a heat-sink to remove the heat energy from the diode.

6.3 BIPOLAR JUNCTION TRANSISTORS (BJT)

Transistors and diodes are devices generally fabricated with silicon (Si). Silicon is one of the most abundant elements on the planet. After considerable processing silicon can be obtained from sand. In the earlier years of transistor fabrication, germanium was also used. However, the use of silicon proved to be superior and easier to manufacture transistors with. Silicon is the predominant material used for the fabrication of electronic devices and integrated circuits today.

6.3.1 Basic Concepts on Intrinsic, *n*-type and *p*-type Silicon Materials

We know that silicon is an element of the periodic table of elements. One of its characteristics is that it has an atomic number of 14; that is, it has 14 electrons that spin around the silicon atom nucleus. The electrons are distributed around the nucleus in different energy levels; these levels used to be referred to as electron shells. Elements can have up to seven energy levels. Such levels are 1 through 7, where 1 is the level closest to the nucleus and energy level 7 is the energy level farthest away from the nucleus. The silicon atom has 2 electrons in energy level 1, 8 electrons in energy level 2, and 4 electrons in energy level 3. Energy levels 1 and 2 are complete, but energy level 3 is complete when it contains 8 electrons. Level 3 is silicon's highest energy level and is also referred to as the *valence shell.* Figure 6.15 depicts the seven electron energy levels 1 through 7, each energy level has a maximum number of electrons that can exist in it. Although there are seven levels, only four of those levels are used by the known periodic table elements.

Element silicon with atomic number 14 has the following electron configuration notation:

$$\text{Silicon Electron Configuration}: 1s^2\ 2s^2\ 2p^6\ 3s^2\ 3p^2$$

Silicon atoms have a tendency of uniting with other silicon atoms in order to complete their valence shell with 4 more atoms, so that their *valence shell* becomes complete with 8 electrons. Silicon atoms stick together and each atom has four electrons on its own *valence shell* and four more electrons are shared from neighboring silicon atoms. The sharing of valence electrons between two

(a)

Energy Level	*n*	*Letter*
First	*1*	*K*
Second	*2*	*L*
Third	*3*	*M*
Fourth	*4*	*N*
Fifth	*5*	*O*
Sixth	*6*	*P*
Seventh	*7*	*Q*

(b)

Sublevel	*Electron Possible*	*Orbitals Possible*
s	*2*	*1*
p	*6*	*3*
d	*10*	*5*
f	*14*	*6*

Figure 6.15 Electron energy levels.

or more atoms produces a *covalent bond* between such atoms. *Covalent bonds* hold the atoms together, forming a structure denominated a *crystal.* Because every atom in the crystal structure is bonded to four other atoms, the electrons are not free to move within the crystal. Intrinsic silicon refers in simple terms to pure silicon with no other elements in its crystalline structure. Because of the *covalent bonds* just described, silicon and germanium are not good conductors of electricity. Figure 6.16 depicts a small portion of the periodic table of elements. Silicon and germanium are elements that have four electrons in their valence shell. Also present are what we will refer to as dopants: boron (B) and gallium (Ga) are elements that have three electrons in the valence shells. Phosphor (P) and arsenic (As) are elements that have five electrons in their valence shell. When intrinsic silicon is doped with certain amounts of boron or gallium the doped new structure has a deficit of electrons. It is said that the doped silicon is *p-type* material; the absence of an electron is referred to as a

	Group III	Group IV	Group V	
...	Boron (B)	Carbon (C)		...
...	Aluminum (Al)	Silicon (Si)	Phosphorus (P)	...
...	Gallium (Ga)	Germanium (Ge)	Arsenic (As)	...
...				...

Figure 6.16 Small section of the periodic table of elements.

hole, a hole has a positive charge and its magnitude is equal to the magnitude of the charge of an electron. When intrinsic silicon is doped with certain amounts of phosphor or arsenic, the doped structure has excess of electrons. This silicon structure doped with phosphor or arsenic is referred to as *n-type* material. Again pure silicon is called *intrinsic* silicon. When impurities are added to it, it becomes *extrinsic* silicon. When the impurities added to intrinsic silicon are elements from Group III, like boron, these are called *acceptors*, because they can accept extra electrons into the crystal. When the impurities added to the intrinsic silicon are elements from Group V, like phosphor, these impurities are called *donors*, because they can easily loose or donate an extra electron to the crystalline structure.

There is a vast amount of literature that covers semiconductors physics. In this introductory section we have not even scratched the surface of it. The main intent of this section is to provide the reader with the basic concepts behind intrinsic, n-type and p-type doped silicon. For further details on semiconductors refer to the Further Reading section at the end of this chapter.

6.3.2 The BJT as a Circuit Element

We will study the BJT almost entirely from a circuit behavior point of view. We will be concerned with the voltages across its terminals and its currents, how they change and how they need to be set in order for the transistor to operate as either an amplifier or a linear device, as a switch or a nonlinear device. There are two types of bipolar transistors, the NPN and the PNP. Figure 6.17 shows a simplified diagram of the structure of NPN and PNP transistors.

Bipolar transistors have three terminals, the collector, the base, and the emitter. For an NPN transistor whose schematic symbol is depicted in Figure 6.18, three fundamental voltages and currents are defined: v_{CE}, v_{BE}, and v_{CB} and i_C, i_B, and i_E. Voltage v_{CE} is the collector to emitter voltage drop; this voltage is positive for NPN transistors. v_{BE} is the base to emitter junction voltage, and v_{CB} is the collector to base junction voltage. From KVL it is easy to see that

$$v_{CE} = v_{BE} + v_{CB} \tag{6.8}$$

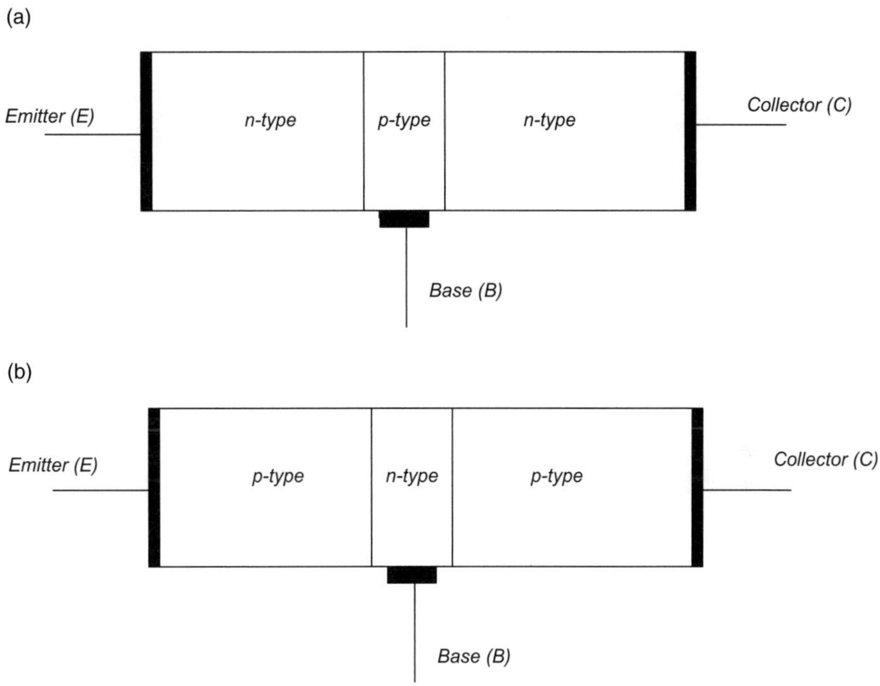

Figure 6.17 Simplified bipolar transistor structures: (a) NPN; (b) PNP.

and from KCL,

$$i_E = i_C + i_B \tag{6.9}$$

where in Equations (6.8) and (6.9) v_{CE} is the collector to emitter voltage drop, v_{BE} is the base to emitter junction voltage drop, v_{CB} is the collector to base junction voltage drop, i_E is the emitter current, i_C is the collector current, and i_B is the base current, respectively.

Note that the differences between voltages and currents in NPN and PNP transistors are that when a voltage is positive in the NPN-type device, the same voltage is negative in its counter part, the PNP device. Similarly, if a current is positive in the NPN, the same current is negative in the PNP.

6.3.3 Bipolar Transistor *I-V* Characteristics

For simplicity we will mostly concentrate describing the NPN transistor first. Later on some material on the PNP will be covered. A family of curves

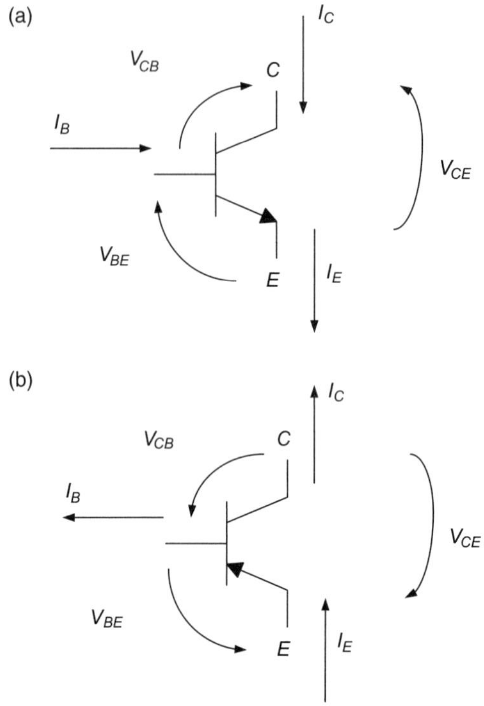

Figure 6.18 (a) Schematics symbols of an NPN transistor, its voltages, and currents; (b) symbol, voltages, and currents for a PNP transistor.

referred to as the current–voltage (*I-V*) characterizes the transistor. Figure 6.19 depicts the *I-V* characteristics of an NPN bipolar transistor.

The transistor *I-V* curves depict collector current versus collector-emitter voltage drop in the horizontal axis. The base current is used as a parameter for collector current versus collector-emitter voltage pairs. If we did not plot the *I-V* curves using the base current as a parameter, we would be forced to use three-dimensional plots, which are much harder to draw and visualize on a flat piece of paper. For the curves depicted in Figure 6.19, $i_{B6} > i_{B5} > \ldots > i_{B1}$.

For example, referring again to Figure 6.19, note that for a base current of 1 mA the collector current starts from zero i_C and zero v_{CE}. i_C ramps up somewhat linearly and then it virtually flattens out with a slight positive slope. We will later see that due to the Early effect, a slight positive slope is observed in the collector current curves. For a base current of 2 mA, i_C versus v_{CE} is plotted somewhat above the curve for 1 mA of base current. The larger is the base current, the higher will be the corresponding $i_C = f(v_{CE})$ curve.

Three key areas of operation are identified in the *I-V* curves of Figure 6.19. The region adjacent to the collector current axis, the vertical axis, is the

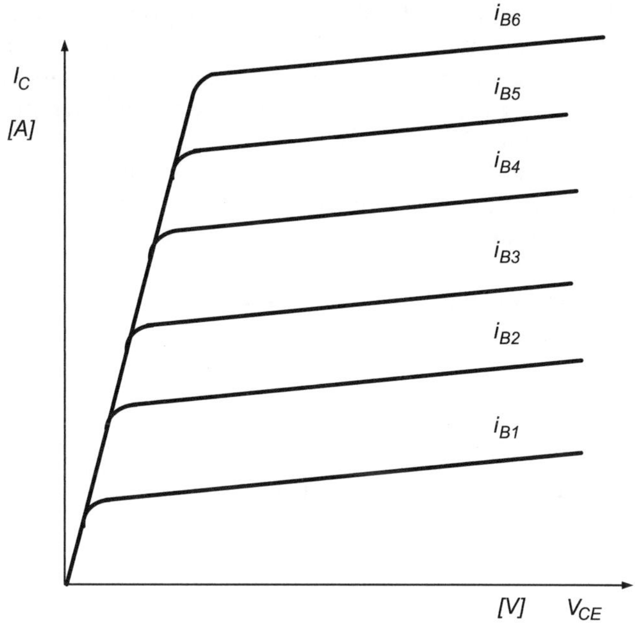

Figure 6.19 *I-V* characteristics curves of an *NPN* transistor.

saturation region; the region adjacent to the horizontal axis or the collector-emitter voltage is the cutoff region. Everything else is basically the so-called active region of the device.

An *NPN* transistor acts as a device that allows collector current, which is almost equal in magnitude to the emitter current; go through the NPN structure. A thin layer of *P* material constitutes the base of the transistor. The voltage that injects current into the base terminal has control over the amount of collector current that goes through the collector to emitter terminals of the device. We have seen that from Equation (6.9), the emitter current is the sum of collector and base current. Practically speaking, the base current is quite small, about two orders of magnitude smaller than the collector and emitter currents. So the approximation:

$$i_E \cong i_C \tag{6.10}$$

is valid for many applications. The reader, however, is cautioned not to use Equation (6.10) liberally. It is a good opportunity to introduce the *DC* current gain factor of the transistor, referred to as its β (Greek letter beta). β relates collector and base currents as follows:

$$i_C = \beta i_B. \tag{6.11}$$

β is generally a number around 100 (i.e., 20–50) for power transistors, but it is usually 100–500 for signal (small) transistors. Note that β has no units, and it is the transistor current gain factor. β is one of the transistor most important figures of merit. β is also referred to as the static or DC current gain of the transistor, and it is sometimes denoted as h_{fe}; the "*h*" stands for the transistor hybrid parameters model, which is beyond the scope of our book. The suffix "$_{fe}$" stands for forward and common emitter configuration.

We will explain later in this chapter what a common emitter configuration is.

Using Equation (6.11) in Equation (6.9) we obtain that

$$\mathrm{i_E} = \mathrm{i_C} + \frac{i_C}{\beta}. \tag{6.12}$$

Doing some algebra on Equation (6.12) we arrive at

$$\mathrm{i_C} = \frac{\beta}{\beta+1} \cdot \mathrm{i_E} \tag{6.13}$$

where the term $\beta/(\beta+1)$ is defined as the transistor's *alpha* (α); note that α always is a number slightly smaller than unity. α also is a dimensionless bipolar transistor parameter.

$$\alpha = \frac{\beta}{\beta+1}. \tag{6.14}$$

Example 6.6 Calculate the value of α for a transistor with a β of 200.

Solution: Using Equation (6.14) we obtain that $\alpha = 200/(200 + 1) = 0.995$.

Unfortunately β is not constant, and it varies for different collector currents and transistor temperature. Figure 6.20 depicts how β changes with collector current and with temperature. Notice the bell-shaped curves, the top curve is for a junction temperature of 125°C, the second one is for 25°C, and the bottom curve is for –55°C. Also note that generally β or h_{fe} grows for larger collector currents, but up to a point, beyond which it decays (Fig. 6.20).

It can be proven from semiconductor physics that the collector current equals

$$I_C = I_S\left(e^{\frac{v_{BE}}{v_T}} - 1\right) \tag{6.15}$$

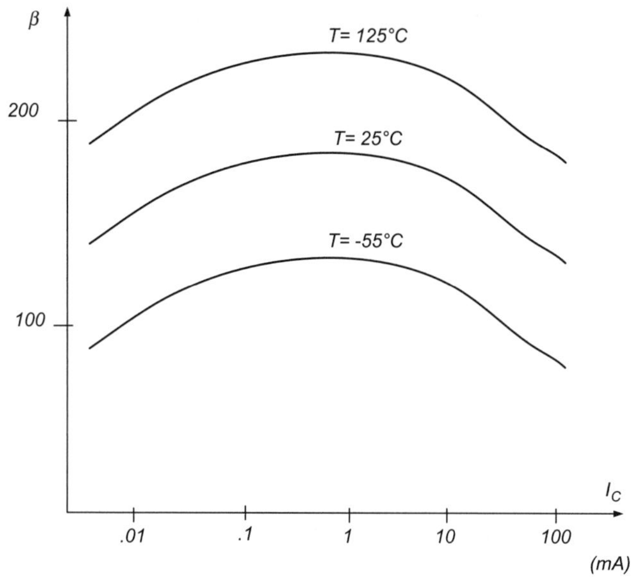

Figure 6.20 β variation with collector current and temperature.

where I_S is the reverse saturation current of the BJT base-emitter junction, v_{BE} is the base-emitter junction forward voltage drop, and v_T is the thermal voltage. v_{BE} typically is 0.6 V or 0.7 V for small signal silicon devices. v_T is the thermal voltage of the junction which at a room temperature of 300 K is approximately 26 mV. Based on the value of exponent $v_{BE}/v_T = 0.6/0.026 \cong 23$ and since $e^{23} >>> 1$ Equation (6.15) is simplified to

$$I_C = I_S e^{\frac{v_{BE}}{v_T}}. \tag{6.16}$$

The junction reverse saturation current is an extremely small current that flows through a reversed biased *pn* junction. For an ideal junction (either a diode junction or a transistor base-emitter junction), the reverse current is zero, for real junctions, it is a finite number typically in the range of 10^{-17} to 10^{-14} A. Using Equation (6.16), knowing the junction temperature, I_S the reverse saturation current of the BJT, and the DC collector current I_C, one can easily find the v_{BE} forward biased voltage drop.

Example 6.7 Assume that the collector current of an NPN transistor is 1 mA, $I_S = 10^{-15}$ A, and the junction temperature is 300 °K, determine the value of the forward biased base-emitter junction voltage drop.

Solution: Using Equation (6.16) we rewrite it as:

$$\frac{I_C}{I_S} = e^{\frac{v_{BE}}{v_T}}. \tag{6.17}$$

Taking the natural logarithm of both sides of the equation, and plugging the values given in the example we obtain that

$$ln\frac{I_C}{I_S} = \frac{v_{BE}}{v_T} \tag{6.18}$$

and rearranging terms,

$$v_{BE} = v_T ln\frac{I_C}{I_S}. \tag{6.19}$$

Plugging $v_T = 0.026$ V, $I_C = 0.001$ A, and $I_S = 10^{-15}$ A into Equation (6.19), it yields

$$v_{BE} = 0.718 \text{ V}. \tag{6.20}$$

6.3.4 Biasing Techniques of Bipolar Transistors

When the bipolar transistor operates as an amplifier, it needs to be biased in the active mode. In the absence of input signals to be amplified, the DC operating point of the transistor must be at point Q shown in Figure 6.21. Q is approximately halfway between the collector-emitter voltage spanned and halfway between the collector current spanned. We will elaborate more on this concept shortly.

For a bipolar transistor to be in the *active* mode or region, the base-emitter junction has to be forward biased, and the collector-base junction must be reverse biased. Referring to Figure 6.21, when both referred junctions are forward biased, the transistor is saturated; under saturation the collector to emitter voltage V_{CEsat} is minimal (typically about 0.2–0.3 V) since the transistor is acting as a closed (turned-on) switch. When both of the junctions are reverse biased, the transistor is said to be cutoff. No collector and no emitter currents flow through it. Finally, when the base-emitter junction is reverse biased and the collector-base junction is forward biased, the transistor is said to be in its reverse active mode of operation. This mode is not shown in Figure 6.21. The reverse active mode of operation is not a preferred mode to operate the transistor because its current gain parameter β is not as large as it is in the active mode. This is due to the fact that neither NPN nor PNP transistors are symmetrical. The emitter is more heavily doped than its collector. The reverse-active mode is seldom used.

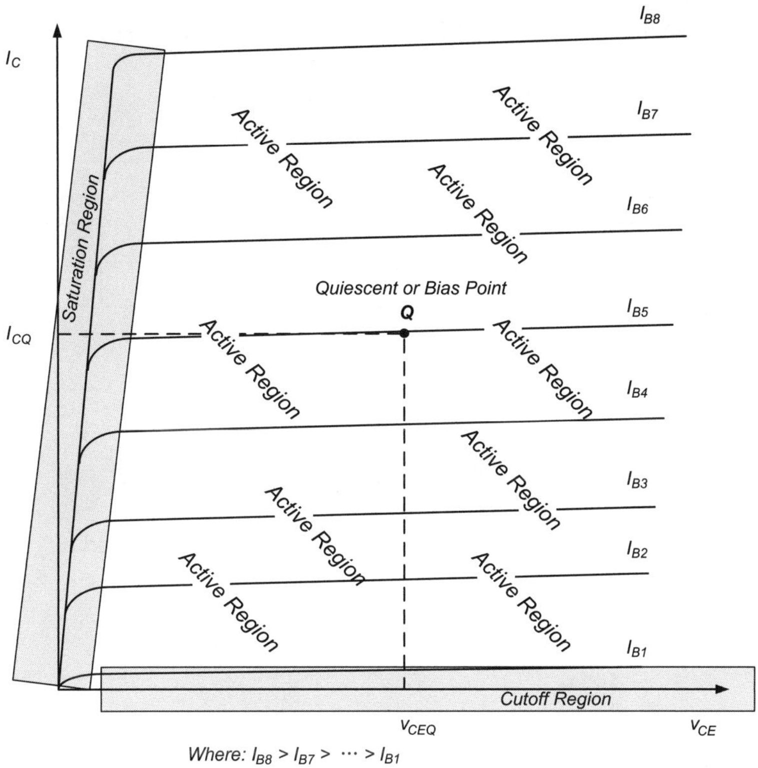

Figure 6.21 NPN transistor operating or quiescent point.

Table 6.2 Bipolar junction transistors mode of operation

Mode of Operation	Emitter-Base Junction	Collector-Base Junction
Active	Forward	Reverse
Reverse-Active[a]	Reverse	Forward
Saturation	Forward	Forward
Cutoff	Reverse	Reverse

[a] Not commonly used.

Table 6.2 shows the four regions of operation of a bipolar transistor. Table 6.2 applies to both NPN and PNP bipolar transistors.

When the BJT is used as a digital element or as a switch, it is mainly used in two modes, either *saturation* or *cutoff*. When the switch or the transistor is *ON*, it is saturated; when the transistor is *OFF*, the transistor is cutoff. When the transistor is used as a switch, it goes from *saturation* to *cutoff* and vice versa. Figure 6.22 depicts an NPN and a PNP BJTs in their active regions. Note

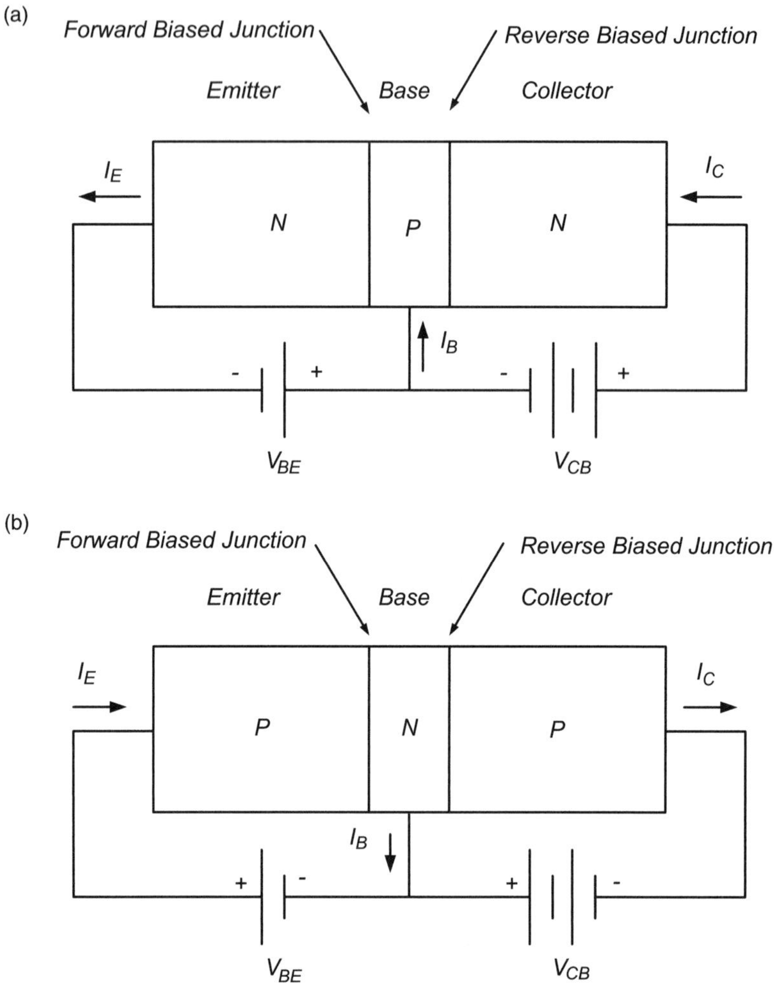

Figure 6.22 (a) NPN BJT in the active region; (b) PNP BJT in the active region.

that the emitter-base junction of both transistors is forward biased, while the collector-base junction of both transistors is reverse biased.

The circuits of Figure 6.22 are just conceptual drawings solely to show the polarities of the transistor junctions; both NPN and PNP transistors comply with the *active* mode described by Table 6.2.

Moreover, circuits of Figure 6.22 are not working biasing circuits.*

The careful observer should note that the NPN BJT of Figure 6.22a has its base-emitter junction forward biased while the collector-base junction is

* Current-limiting resistors should be added, as it will be seen in the following sections on biasing.

reverse biased. Also note that the NPN BJT collector current flows from the collector through the transistor N region, the base current flows into the base P region. Both collector and base currents get added at the base-emitter junction and out of the emitter terminal flows the emitter current. For the PNP case, note that the voltages are reversed in polarity, and the currents flow in the opposite direction to those of the NPN BJT.

6.3.5 Very Simple Biasing

Now let us look at the transistor circuit of Figure 6.23. We will consider the following values: $R_C = 50\ \Omega$, $R_B = 10\ \mathrm{k}\Omega$, and $V_{CC} = 5$ V. Assume the transistor β is 100 and ignore I_S the reverse saturation current of the transistor.

The biasing circuit of Figure 6.23 is the simplest of all the ones that we will cover. Its purpose is to show the reader two methodologies to solve BJT-based biasing circuits. Our first Example 6.8 will show an approximated method. The second Example 6.9 will show a slightly more accurate method of solving the same problem.

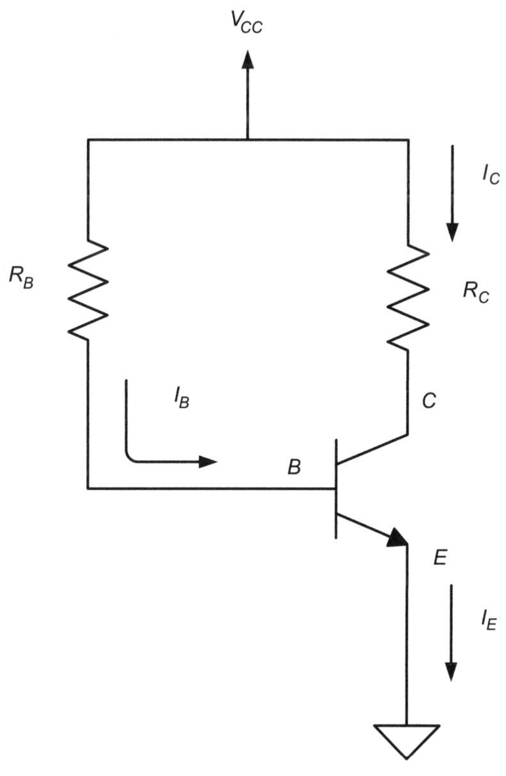

Figure 6.23 Very simple BJT biasing.

Example 6.8 Assume that $V_{CC} = 5$ V, $R_C = 50\ \Omega$, $R_B = 10\ \text{K}\Omega$, and $\beta = 100$. Let us assume that we want to find the collector-to-emitter voltage which also equals the collector-to-ground voltage, since the emitter is grounded. Using the circuit of Figure 6.23 we make an initial guess of the base-emitter junction forward voltage drop. Assuming that $v_{BE} = 0.6$ V (our initial guess), the base current is calculated as follows:

$$I_B = (V_{CC} - v_{BE}) / R_B. \tag{6.21}$$

Plugging the numerical values given into Equation (6.21) we obtain

$$I_B = (5 - 0.6) / 10{,}000 = 440\ \mu\text{A}. \tag{6.22}$$

Since $\beta = 100$, then

$$I_C = 100 \times 440\ \mu\text{A} = 44{,}000\ \mu\text{A} = 0.044\ \text{A}. \tag{6.23}$$

The collector-emitter voltage drop V_{CE} also equal to V_C, because the emitter is grounded, is calculated as follows:

$$V_C = V_{CC} - I_C \times R_C \tag{6.24}$$

Since $R_C = 50\ \Omega$ and $Ic = 0.044$ A from Equation (6.23), then

$$V_C = 5 - 0.044 \times 50 = 5 - 2.2 = 2.8\ \text{V}. \tag{6.25}$$

Example 6.9 Assume the same circuit of Figure 6.23 is used for this example. Assume the same parameters: $V_{CC} = 5$ V, $R_C = 50\ \Omega$, $R_B = 10\ \text{k}\Omega$, and $\beta = 100$. The reverse saturation current is $I_S = 10^{-14}$ A. Check to see if the results obtained for Equations (6.23) and (6.25) become a little more accurate by taking into account I_S.

Using the results given by Equations (6.23) and (6.25) as our initial guess, let us now verify how close to the initial guess of 0.6 V for v_{BE} is. Using Equation (6.19) for v_{BE} we have that

$$v_{BE} = v_T \ln \frac{I_C}{I_S} = 0.026 \ln\left(0.044\ 10^{14}\right) = 0.757\ \text{V}. \tag{6.26}$$

With the new v_{BE} value obtained with Equation (6.26), we reevaluate Equation (6.21) thus:

$$I_B = (V_{CC} - v_{BE}) / R_B = (5 - 0.757) / 10{,}000 = 424.30\ \mu\text{A}. \tag{6.27}$$

Since $\beta = 100$ and $I_C = \beta\, I_B$, then

$$I_C = 42{,}430\ \mu\text{A} = 0.04243\ \text{A}. \tag{6.28}$$

And since $V_C = V_{CC} - I_C R_C$, then

$$V_C = 5 - 0.04243 \times 50 = 5 - 2.1215 = 2.8785 \text{ V}. \quad (6.29)$$

One more time we recalculate the value of v_{BE} using Equation (6.19), thus:

$$v_{BE} = v_T \ln \frac{I_C}{I_S} = 0.026 \ln\left(0.04243\ 10^{14}\right) = 0.756 \text{ V}. \quad (6.30)$$

Since the recalculated value of v_{BE} Equation (6.30) is 155.6 mV off the assumed $v_{BE} = 0.6$ V. If we are satisfied with the result we do not iterate. Otherwise, we iterate the calculations again for a more accurate approximation. The solutions to our example are Equations (6.27) through (6.30).

6.3.6 Resistor Divider Biasing

The biasing scheme seen in the previous section is very sensitive to the variations of β. Since it is common for transistors of the same type and characteristics, to have a wide range of β, which could easily be from 100 to 200, the very simple biasing scheme of Figure 6.23 is not very practical or useful. We will investigate a circuit whose biasing is less sensitive to β variations. Referring to Figure 6.24a we see that the base-emitter voltage v_{BE} is established by the resistor divider formed by R_1 and R_2 if the base current is significantly smaller than the current that flows through the divider top resistor R_1. Thus:

$$v_{BE} = \frac{R_2}{R_1 + R_2} V_{CC}. \quad (6.31)$$

However, if the base current is comparable to the resistor divider current, a more accurate method of calculating the v_{BE} than Equation (6.31) must be used. What we do is apply Thévenin's Theorem to the left of the base node labeled with a "B," Figure 6.24a. The parallel of resistors R_1 and R_2 produces the Thévenin resistance:

$$R_{Thev} = \frac{R_1 R_2}{R_1 + R_2}. \quad (6.32)$$

The Thévenin voltage to the left node B, with the right-hand side of the circuit removed, is calculated with the following expression, and referring to Figure 6.24b,

$$V_{Thev} = \frac{R_2}{R_1 + R_2} V_{CC}. \quad (6.33)$$

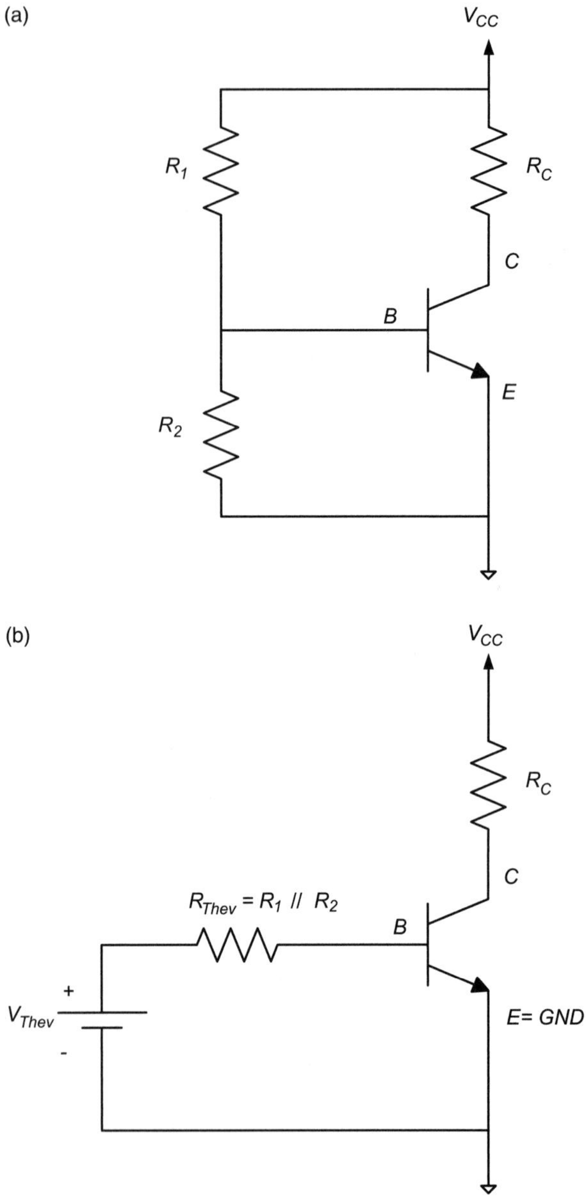

Figure 6.24 Resistor divider biasing method.

Using Equations (6.32) and (6.33) and referring again to Figure 6.24b, we write

$$V_{Thev} = I_B R_{Thev} + v_{BE}. \tag{6.34}$$

Using Equations (6.32) and (6.33) in Equation (6.34) and rearranging terms we obtain

$$v_{BE} = \frac{R_2}{R_1 + R_2} V_{CC} - I_B \frac{R_1 R_2}{R_1 + R_2}. \tag{6.35}$$

Recalling Equation (6.16) which is repeated below for the reader's convenience,

$$I_C = I_S e^{\frac{v_{BE}}{v_T}}. \tag{6.36}$$

Finally, combining Equations (6.35) and (6.36) yields

$$I_C = I_S \exp\left(\frac{V_{Thev} - I_B R_{Thev}}{v_T}\right). \tag{6.37}$$

Rewriting Equation (6.37) to express the base current we obtain

$$I_B = \left(V_{Thev} - v_T \ln \frac{I_C}{I_S}\right) \cdot \left(\frac{1}{R_{Thev}}\right), \tag{6.38}$$

where in Equation (6.38) I_B is the base current, V_{Thev} is given by Equation (6.33), I_C is the collector current, I_S is the reverse saturation current, and R_{Thev} is given by Equation (6.32).

Example 6.10 Using the circuit of Figure 6.24, assume that $R_1 = 65$ kΩ, $R_2 = 15$ kΩ, $R_C = 2$ kΩ, $V_{CC} = 5$ V, $\beta = 100$, and $I_S = 10^{-17}$ A. Calculate I_B, I_C, v_{BE}. Make your first v_{BE} guess equal to 0.8 V, verify that your final v_{BE} is within 40 mV or less than the initial guess. You may have to iterate the process more than once to achieve the result wanted. Once the final value of v_{BE} is computed, find the collector-emitter voltage of the BJT and the voltage drop across resistor R_C.

We will go over this example a little faster since it is similar to the previous one.

We compute V_{Thev} in the usual way:

$$V_{Thev} = \frac{R_2}{R_1 + R_2} V_{CC} = \frac{15}{15 + 65} 5 = 0.9375 \text{ V}. \tag{6.39}$$

$$R_{Thev} = R_1 // R_2 = 12{,}188\ \Omega. \tag{6.40}$$

Using the values found in Equations (6.39) and (6.40) and assuming $v_{BE} = 0.8$ V in Equation (6.34) we get

$$I_B = (V_{Thev} - v_{BE}) / R_{Thev} = (0.9375 - 0.8)/12{,}188 = 11.282\ \mu\text{A}. \quad (6.41)$$

Then, since

$$I_C = \beta I_B = 100 \times 11.282\ \mu\text{A} = 1128.2\ \mu\text{A}. \quad (6.42)$$

And using the value of I_C from Equation (6.42) in Equation (6.19) copied right below for the reader's convenience,

$$v_{BE} = v_T \ln \frac{I_C}{I_S} = 0.026 \ln\left(1128.2\ 10^{-6}\ 10^{17}\right) = 0.8412\ \text{V}. \quad (6.43)$$

The result for v_{BE} is close, but not as close as the example requirements, so we recalculate Equations (6.41), (6.42), and (6.43) using $v_{BE} = 0.8412$ V. Thus,

$$I_B = (V_{Thev} - v_{BE}) / R_{Thev} = (0.9375 - 0.8412)/12{,}188 = 7.9\ \mu\text{A}. \quad (6.44)$$

Then, since

$$I_C = \beta I_B = 100 \times 7.9\ \mu\text{A} = 790\ \mu\text{A}. \quad (6.45)$$

And using the value of I_C from Equation (6.45) in Equation (6.19) copied right below for the reader's convenience,

$$v_{BE} = v_T \ln \frac{I_C}{I_S} = 0.026 \ln\left(790\ 10^{-6}\ 10^{17}\right) = 0.832\ \text{V}. \quad (6.46)$$

Since 0.832 V is within 40 mV of the initial guess for $v_{BE} = 0.8$ V, we are done with the iterations, but not done with the example.

Since $I_C = 790\ \mu$A,

$$V_{R_C} = I_C R_C = 790\ 10^{-6} \times 2000 = 1.58\ \text{V}. \quad (6.47)$$

Since $V_{CC} = 5$ V, then

$$V_{CE} = V_{CC} - V_{R_C} = 5 - 1.58 = 3.42\ \text{V}. \quad (6.48)$$

The resistor divider method is better than the very simple biasing method; however, there is still dependence of the error obtained calculating the collector current I_C if the resistors vary only a small percentage. This can be numerically validated applying Equation (6.36) to a slight change of collector current. The reason should be clear that the exponential Equation (6.36) converts a small deviation of I_C into a large v_{BE} deviation. So this circuit is still of little practical value.

6.3.7 Emitter Degeneration Resistor Biasing

The circuit used for this method is shown in Figure 6.25. We can appreciate that there is just one difference between the circuits of Figures 6.25 and 6.24. The circuit of Figure 6.25 has an emitter resistor, but Figure 6.24 does not. This resistor will make the circuit more independent of β and v_{BE} if the other circuit parameter values are chosen correctly. It is also important to mention that the emitter of the BJT with the emitter degeneration resistor is *not* grounded, like it is in the case of the circuit of Figure 6.24a.

Let us refer to the circuit of Figure 6.25a, without ignoring the base current. Let us partition the circuit to the left of the base of the BJT. Let us find using Thévenin the equivalent of the circuit formed by resistors R_1 and R_2 powered by V_{CC}. Figure 6.25b shows that the voltage at the transistor base with respect to ground is

$$V_B = \frac{R_2}{R_1 + R_2} V_{CC}. \tag{6.49}$$

Assuming that the current through the resistor divider is at least 10 times larger than the transistor base current (Fig. 6.25),

$$V_B = \frac{V_{CC}}{R_1 + R_2} >>> I_B. \tag{6.50}$$

We can also write for the base-emitter loop:

$$V_B = v_{BE} + I_E R_E. \tag{6.51}$$

Since $I_E = I_B + I_C$ and $I_C = \beta\, I_B$, then

$$I_E = (\beta + 1) I_B. \tag{6.52}$$

Plugging Equation (6.52) into Equation (6.51) leads to

$$V_B = v_{BE} + (\beta + 1) I_B R_E. \tag{6.53}$$

Thus: If $V_B >> v_{BE}$ and the voltage drop across R_E is at least a good fraction of v_{BE}, for example 200 mV, R_E "absorbs" the changes in V_B due to variations of v_{BE}. But there is a drawback; we will see later on that this reduces the gain when the circuit operates as an amplifier. So nothing comes for free.

However, at the same time, we have to be careful about keeping the transistor in its active region, that is, forward biased base-emitter junction and reverse biased collector-base junction. So the value of collector voltage with respect

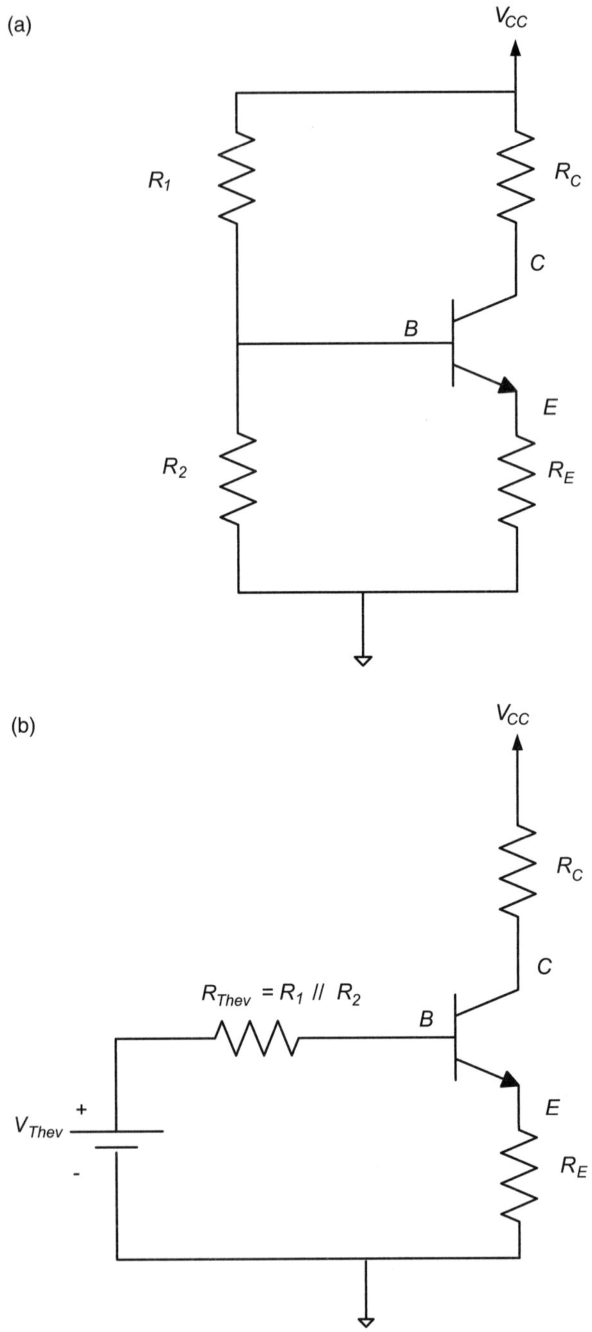

Figure 6.25 (a) Biasing with emitter degeneration resistor; (b) Thévenin equivalent.

to ground shall meet the following requirement for the transistor to be in its active region:

$$V_C = V_{CC} - I_C R_C > V_B. \tag{6.54}$$

Example 6.11 Using the circuit of Figure 6.25a determine the resistor values (R_C and R_E) such that the collector current is 2 mA. Assume that $V_{CC} = 10$ V, $\beta = 70$, and $I_S = 4 \times 10^{-17}$ A all at room temperature: (1) Find suitable values of R_1 and R_2 such that the circuit remains largely insensitive to v_{BE} and β *variations*. (2) Make sure that the transistor is biased and it is in the active region.

We can establish with this example the design criteria for this biasing circuit. In order to have insensitivity to variations of v_{BE} and β, we must design to meet Equations (6.50), (6.52), and (6.54).

First let us calculate v_{BE}:

$$v_{BE} = v_T \ln \frac{I_C}{I_S} = 0.026 \ln\left(2 \times 10^{-3} / 4 \times 10^{-17}\right) = 0.820 \text{ V}.$$

Let us then assume that v_{BE} is 0.8 V.

Since we want to achieve

$$\frac{V_{CC}}{R_1 + R_2} \gg I_B, \tag{6.55}$$

let us make

$$\frac{V_{CC}}{R_1 + R_2} = 10\, I_B. \tag{6.56}$$

Since $I_C = 2$ mA, then $I_B = I_C/\beta = 28.6$ μA and $V_{CC} = 10$ V, plugging these values into Equation (6.56), we obtain that

$$R_1 + R_2 = 34{,}965\ \Omega. \tag{6.57}$$

Since v_{BE} is 0.8 V and imposes a small voltage drop of 0.2 V across R_E, the voltage at the transistor base to ground is thus

$$V_B = \frac{V_{CC} R_2}{R_1 + R_2} = v_{BE} + I_E R_E = \frac{10 R_2}{34{,}965} = 1. \tag{6.58}$$

Thus,

$$R_2 = 3.497\ \text{k}\Omega \quad \text{and} \quad R_1 = 31.469\ \text{k}\Omega.$$

Additionally, using Equation (6.54) repeated below for the reader's convenience,

$$V_C = V_{CC} - I_C R_C > V_B. \tag{6.59}$$

We have to ensure that the transistor collector-base junction is reversed biased.

Let us assume that since $V_B = 1$ V, we want $V_C = V_B + 4$ V $= 5$ V to meet Equation (6.59).

Using Equation (6.59) where $V_C = 5$ V leads to

$$R_C = 2.5 \text{ k}\Omega.$$

For part (b), since the collector is at 5 V and the base is at 1 V, the collector-base junction is reversed biased. Since we assumed a 0.2 V across R_E, the base-emitter junction is forward biased and has a forward v_{BE} drop of 0.8 V. The BJT is in the active region.

Exercise: Use the circuit from the previous example, that is, $R_2 = 1$ kΩ, $R_1 = 9$ kΩ, $R_C = 3$ kΩ, $R_E = 200$ Ω, $V_{CC} = 10$ V. Replace the transistor with another transistor that has a $\beta = 200$ and $I_S = 10^{-17}$ A. Calculate the collector, emitter and base currents, the collector voltage, the emitter voltage, and v_{BE}.

What conclusions do you obtain when you compare these results with the previous example results?

6.3.8 Self-Biased Staged

The circuit of Figure 6.26 depicts a BJT with a self-biasing resistor R_B.

Note that the collector voltage is

$$V_C = V_{CC} - I_C R_C. \tag{6.60}$$

It is also important to notice that V_C always is larger than the base voltage V_B.

By further inspection of the circuit of Figure 6.26, we can also see that

$$V_C = I_B R_B + v_{BE}. \tag{6.61}$$

Merging the right-hand sides of Equations (6.60) and (6.61) and remembering that $I_C = \beta I_B$, we obtain

$$I_C = \frac{V_{CC} - v_{BE}}{R_C + \dfrac{R_B}{\beta}}. \tag{6.62}$$

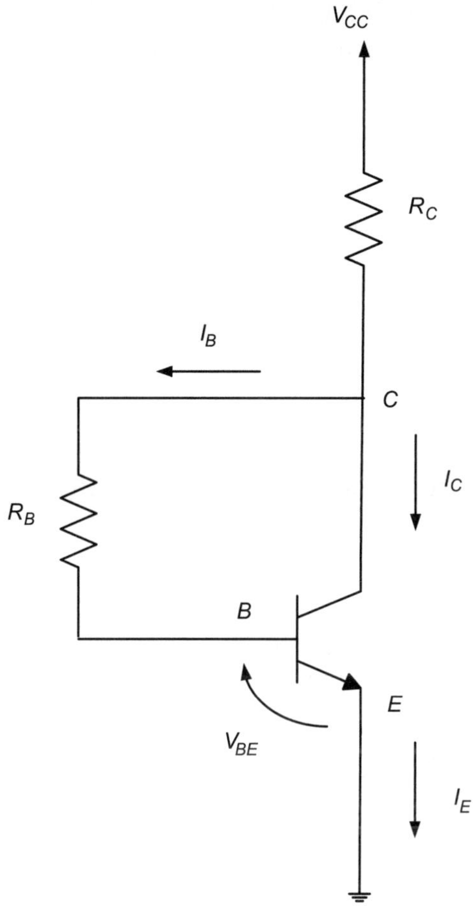

Figure 6.26 Self-biased bipolar transistor.

Example 6.12 Using the circuit of Figure 6.26, assume $V_{CC} = 5$ V, $R_B = 10$ kΩ, $R_C = 1$ kΩ, $\beta = 200$, $V_T = 0.026$ V, and $I_S = 10^{-16}$ A. Determine I_C and v_{BE}.

Using Equation (6.62) and initially assuming that $v_{BE} = 0.8$ V, we obtain

$$I_C = \frac{5-0.8}{1{,}000+\dfrac{10{,}000}{200}} = 4 \text{ mA}. \tag{6.63}$$

Now, using the obtained I_C, let us recalculate the value of v_{BE} using

$$v_{BE} = V_T \ln\frac{I_C}{I_S} = 0.026\ln\frac{4\times10^{-3}}{10^{-16}} = 0.814 \text{ V}. \tag{6.64}$$

Since $v_{BE} = 0.814$ V and the originally guessed value was 0.8 V, we decide not to recalculate the collector current I_C because there is only a 14 mV difference between the initial guess for v_{BE} (0.8 V) and the recalculated v_{BE} (0.814 V).

6.3.9 Biasing Techniques of PNP Bipolar Transistors

Let us take a look at the NPN and PNP transistors voltage and currents. Refer to Figure 6.18a,b, which we repeat here in Figure 6.27 for the reader's convenience.

The NPN transistor, part (a) requires positive v_{BE} for a forward biased *BE*-junction and a positive v_{CB} for a reversed biased *CB*-junction. Currents I_B, I_C, and I_E are positive and flow in the direction shown in Figure 6.27a. On the other hand, *PNP* transistors require a negative v_{BE} for a forward biased *BE*-junction and a negative v_{CB} for a reversed biased *CB*-junction. Currents I_B, I_C, and I_E are negative and flow in the opposite direction as shown in Figure 6.27b.

Taking into account the above considerations, biasing a PNP transistor is not much different from biasing an NPN. The biasing techniques studied for NPNs are valid for PNPs. Due to space reasons, we will only mention the circuit of biasing PNP with emitter degeneration resistor. Figure 6.28 shows such circuit mainly for the purpose of showing voltages and currents. Conceptually, all the concepts that are applicable to NPN transistors biasing also apply to PNPs.

Note that the current flows shown in Figure 6.28 are the actual current directions. Remember that on a PNP transistor all the voltages that were positive for an NPN are negative for a PNP. Similarly, the same concept applies to

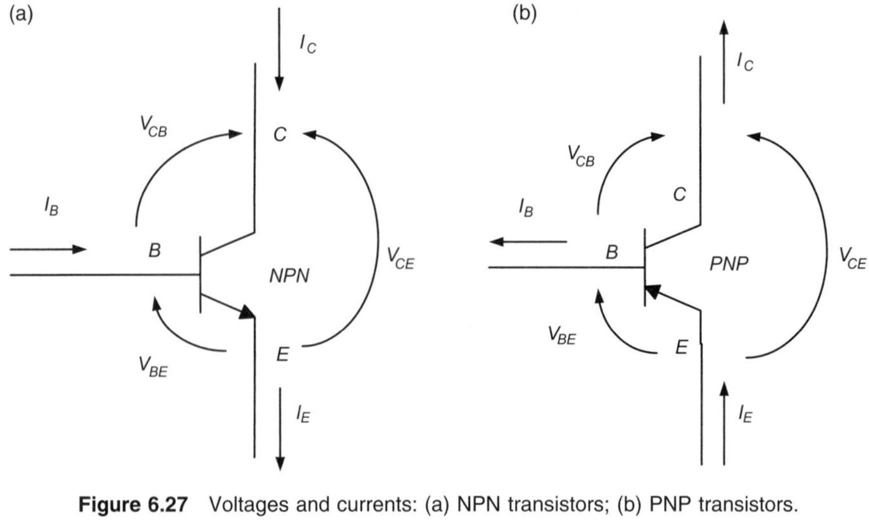

Figure 6.27 Voltages and currents: (a) NPN transistors; (b) PNP transistors.

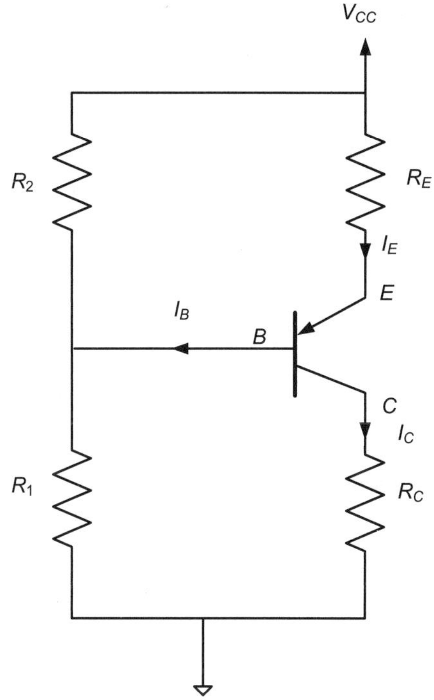

Figure 6.28 PNP transistor biasing with emitter degeneration resistor.

the currents; currents that are positive for the NPN transistors are negative or flow in the opposite direction on the PNP transistors. Carefully compare Figures 6.25a for an NPN and Figure 6.28 for a PNP.

6.3.10 Small Signal Model and Single-Stage Bipolar Amplifier Configurations

In this section we will study the three basic amplifier topologies, the common-emitter, common base, and common collector. In doing so we will use a simple low-to-medium frequency transistor small signal model, called by some authors as the hybrid-π model. Large signal analysis, or amplifier analysis with signals that are comparable in magnitude to the transistor biasing voltages, and multistage amplifiers are beyond the scope of this book.

One of the simplest bipolar transistor models is the one shown in Figure 6.29 even though the model is simple; it is very useful for the understanding of many transistor-based circuits.

The basic model has an input resistance of r_π, between the base and the emitter terminals. The output is modeled with a voltage-controlled current source $g_m V_\pi$, where g_m is the trans-conductance of the transistor and its value

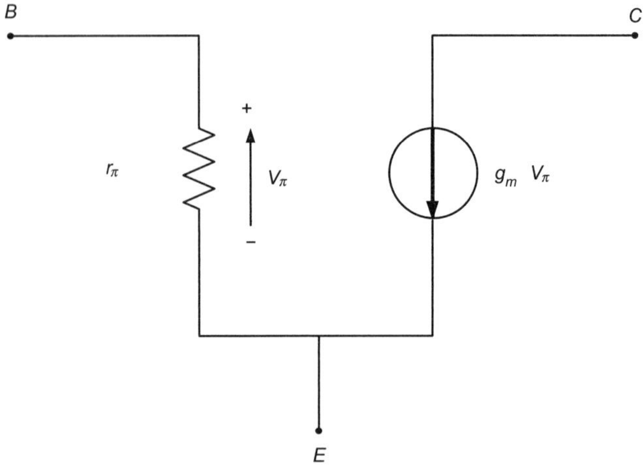

Figure 6.29 Bipolar transistor small signal model.

depends on the bias collector current and the transistor thermal voltage v_T, which is about 26 mV at room temperature of 300 K. Thus,

$$g_m = \frac{I_C}{V_T}. \tag{6.65}$$

So for example, for a 10-mA collector DC current, the transistor transconductance equals

$$g_m = \frac{I_C}{V_T} = \frac{10\text{ mA}}{26\text{ mV}} = 0.385\ \Omega^{-1} = 0.385\text{ S}. \tag{6.66}$$

The voltage V_π, refer one more time to Figure 6.29, is the voltage drop across the input resistance r_π. The input resistance r_π is given by the ratio of β and g_m, that is,

$$r_\pi = \frac{\beta}{g_m}. \tag{6.67}$$

The labels of the model terminals are: *B*, *C*, and *E*, where *B* stands for base, *C* for collector, and *E* for emitter. Finally, it is important to understand that the small signal model is the same whether the transistor is a PNP or an NPN. Table 6.3 summarizes the three key parameters of the BJT small signal model. The output resistance of the model primarily takes into account the fact that the I_C-V_{CE} characteristic curves do not have a zero slope beyond saturation. The characteristic curves of real BJTs have a slightly positive slope even beyond the collector current saturation level. Initially, it is meaningful to

Table 6.3 BJT small signal model: key parameters

Parameter	Parameter Name	Brief Description (only if needed)	Calculation	Units
g_m	Trans-conductance		$g_m = I_C/V_T$	(Siemens)
r_π	Transistor input resistance	Base-to-emitter input resistance	$r_\pi = \beta/g_m$ also: $g_m\, r_\pi = \beta$	(Ω)
r_o	Transistor output resistance	Finite output resistance due to the Early Voltage V_A	$r_o = V_A/I_C$ When the Early voltage is neglected: $r_o \to \infty$	(Ω)

ignore the Early voltage effect in the output resistance r_o, and an infinite r_o may be assumed in parallel with the dependent current source $g_m\, v_\pi$, not shown in Figure 6.26. When greater precision is desired, inclusion of the Early effect finite and non-zero output resistance in parallel with the current source of Figure 6.29 is required.

Small signal models are used to understand the small signal or the AC behavior of the transistor, typically under sinusoidal excitations of small magnitudes within a range of frequencies of interest. How small is a small signal? There are no hard rules, but we can state that a small signal has an amplitude in the order of one-tenth or less of the power supply rail. For example if $V_{CC} = 10$ V, 1-V signals or less are considered small. The purpose of biasing a transistor is to establish a quiescent (Q) or DC operating point. When we analyze an amplifier, under small signal operation, the input signals are applied on top of the DC voltages and currents that bias the transistor. For simplicity and without loss of generality, let us assume that a single sinusoidal signal of one frequency is applied to the input of the amplifier. By the superposition theorem, the signal moves the DC operating point of the amplifier DC voltage and current. Correspondingly, the DC output voltage and current are displaced by some amount that is proportional to the input signal times the gain factor of such amplifier. Let us elaborate more on this in the next section.

6.3.11 Common Emitter (C_E) Configuration

This configuration is the most commonly used stage when designing transistorized amplifiers. The emitter is typically grounded for AC or DC currents. In our example in Figure 6.30, the emitter has an emitter capacitor C_E, whose value is chosen to virtually be a short circuit at the lowest frequencies to be handled by the amplifier. Thus, the emitter is not grounded for DC components because of R_E; however, the emitter is grounded for all *AC* components within

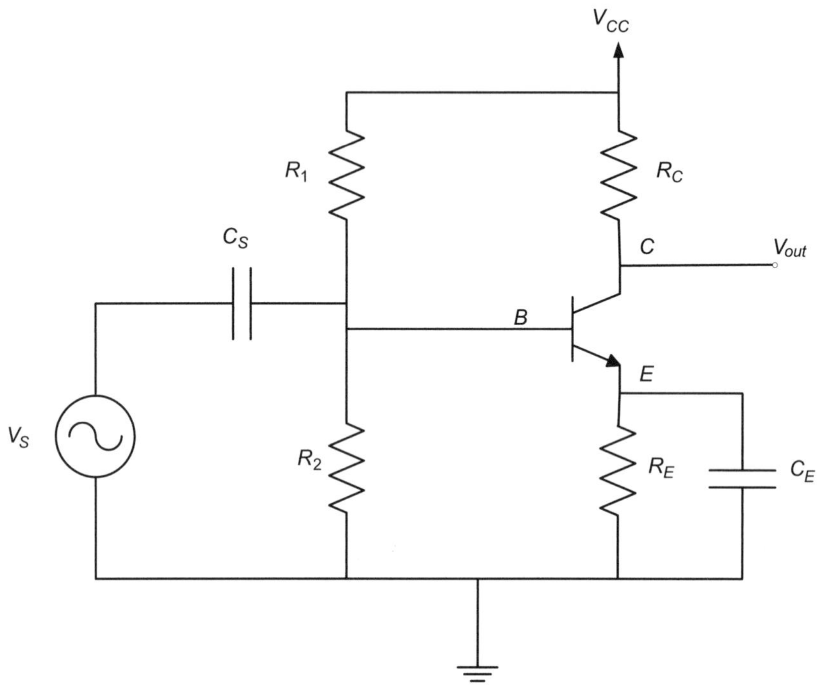

Figure 6.30 Common emitter configuration stage.

the frequency range of interest.* The input signal V_S to be amplified is applied at the base terminal; the amplified output is obtained at the collector V_{out}. Capacitor C_S value is chosen so that the capacitor is a virtual short circuit at the lowest frequency of interest. Figure 6.30 depicts a C_E configuration.

The sinusoidal signal generator is the input signal; the amplified output is obtained at V_{out}. Capacitor C_S AC couples the signal generator into the base of the amplifier, while at the same time the capacitor prevents the DC biasing from being disturbed. Capacitor C_E is a virtual short circuit to AC currents. Thus, the emitter degeneration resistor R_E does not reduce the AC gain of the amplifier. At the same time, R_E provides DC biasing independence from the transistor's β and v_{BE}. Figure 6.31 depicts the AC small signal model of the C_E configuration. The BJT's small signal model has replaced the bipolar transistor in Figure 6.31. The V_{CC} power supply is a short circuit to AC frequencies, thus the collector resistor upper terminal of the model is grounded. Note that capacitor C_S does not appear in the circuit of Figure 6.31 because it behaves as a short circuit to AC frequencies. More interestingly, capacitor C_E and resistor R_E do not show up on the small signal model either, because C_E acts as a short circuit across resistor R_E.

* Note: If a capacitor is a virtual short circuit to some low-frequency signal, it will be an even better short circuit for signals of higher frequency.

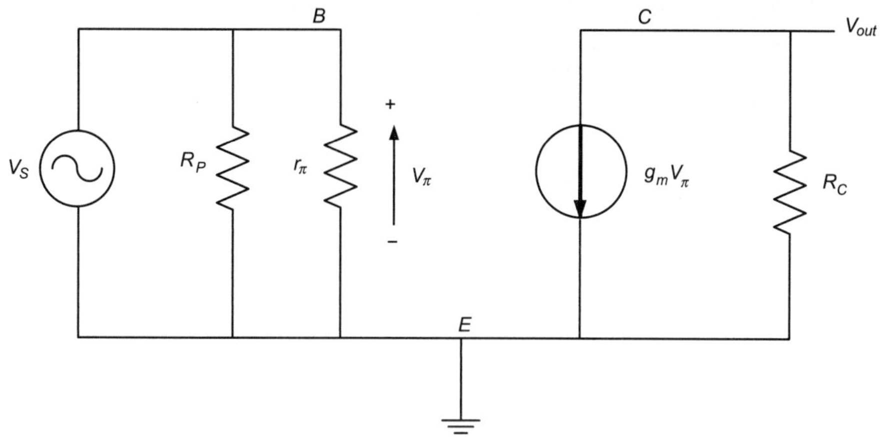

$R_P = R_1 // R_2$

Figure 6.31 Common emitter stage with BJT's small signal model.

Resistors R_1 and R_2 are in effect in parallel with each other and connected between the transistor base and ground because the V_{CC} source is a short to AC frequencies. Now we are ready to start inspecting the circuit of Figure 6.31 to calculate the amplifier gain, its input, and output resistances.

Referring to the small signal common emitter stage of Figure 6.31 we observe that

$$V_{out} = -g_m R_C v_\pi \tag{6.68}$$

Since

$$v_\pi = V_S \tag{6.69}$$

Thus,

$$G_{CE} = \frac{V_{out}}{V_S} = -g_m R_C, \tag{6.70}$$

where in Equation (6.70) G_{CE} is the common emitter stage voltage gain, V_{out} is the stage output or collector voltage, V_S is the input voltage or signal, g_m is the BJT's trans-conductance, and R_C is the collector resistor. Note that the voltage gain has a negative sign; this means that there is a 180° phase shift between the output and the input.

Let us briefly discuss the concepts of input resistance and output resistance of an amplifier. The input resistance of an amplifier is the resistance that a test voltage sees when there is no load at the output of the amplifier. Figure 6.32a depicts the test voltage and the test current applied at the amplifier input with an open-circuited output.

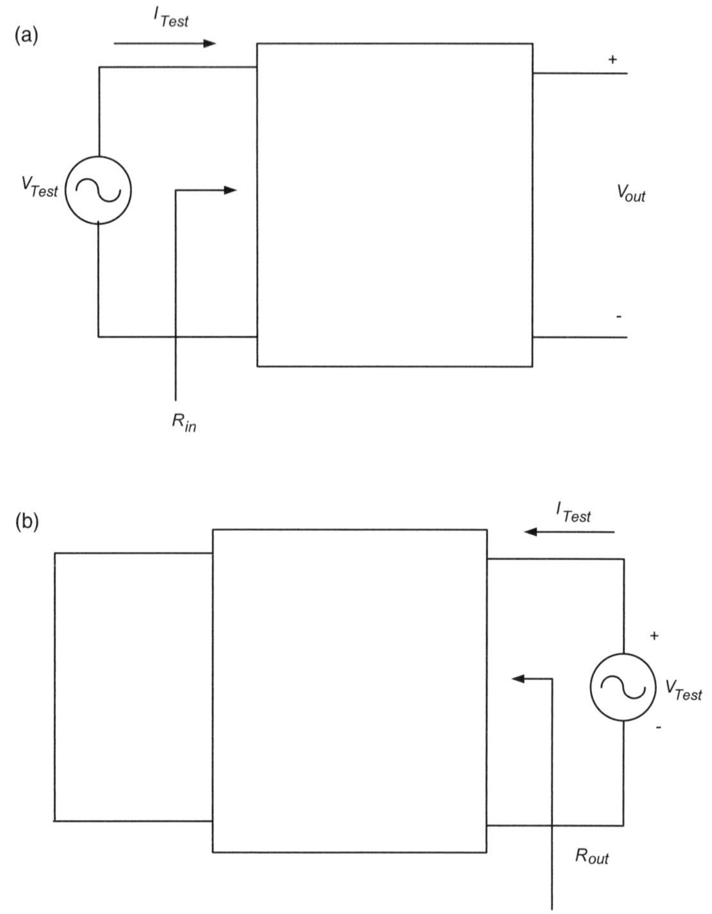

Figure 6.32 (a) Amplifier input resistance; (b) amplifier output resistance.

$$R_{in} = V_{Test} / I_{Test}. \tag{6.71}$$

Using the concept of input resistance from Equation (6.71), we apply it to our circuit of Figure 6.31 and get

$$R_{in} = \frac{r_\pi . R_P}{r_\pi + R_P}, \tag{6.72}$$

where in Equation (6.72) r_π is the BJT's input resistance and R_P is the parallel of resistors R_1 and R_2 in Figures 6.30 and 6.31.

The output resistance of the amplifier is obtained inhibiting sources at the input; since the input source is a voltage source, inhibiting it means to replace it with a short circuit. Then we apply a test voltage at the output of

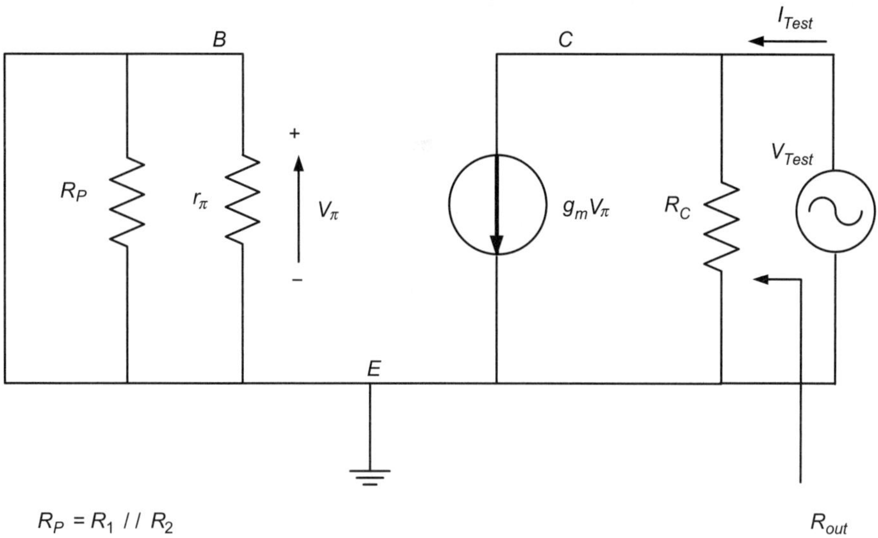

Figure 6.33 Common-emitter output resistance calculation.

the amplifier and the resistance that such test voltage sees is the output resistance of our amplifier. Refer to Figure 6.32b for the conceptualization of output resistance R_{out}.

Now referring one more time to the small signal equivalent circuit of the common emitter configuration of Figure 6.31 we proceed to compute the amplifier output resistance. We first replace the signal generator with a short circuit; we see the results of this in Figure 6.33. Shorting the input source also shorts voltage v_π, then it is easy to see that the current source of the circuit given by $g_m v_\pi$ becomes zero; that is to say, there is no current injected by this voltage-dependent current source. After applying the test voltage and current at the output of this stage, the output resistance seen is just R_C. Thus,

$$\frac{V_{Test}}{I_{Test}} = R_{out} = R_C. \tag{6.73}$$

Inclusion of the Early Effect in the Output Impedance Calculation

In Table 6.3 we presented without proof, that the finite output resistance of a transistor, when better accuracies are desired, is computed as follows:

$$r_o = V_A / I_C,$$

where I_C is the collector bias current and V_A is the Early voltage, a transistor parameter. Equation (6.73) for output resistance of the common emitter amplifier changes to the following when one includes the Early voltage effect:

$$\frac{V_{Test}}{I_{Test}} = R_{out} = R_C \,//\, r_o = \frac{R_C r_o}{R_C + r_o}. \tag{6.74}$$

6.3.12 Common Emitter (C_E) Configuration with Emitter Degeneration

When a resistor is placed between the emitter and ground of the common emitter configuration, the amplifier gain is reduced. This is not necessarily harmful; on the contrary, it benefits the linearity of the amplifier, many times a desirable feature to pay for at the expense of a reduced gain. This common emitter degeneration is applied not only to DC signals but also to AC signals when resistor R_E bypassing capacitor C_E is removed; see Figure 6.31. The gain of the amplifier without common emitter degeneration resistance was addressed by Equation (6.70). Accounting the emitter degeneration resistance in the small signal model (C_E capacitor removed) can be derived from Figure 6.34 recalculating the gain of the amplifier after removing C_E. We just present the result of such gain, which is:

$$G_{CE \text{ with emitter degeneration resistor}} = -\frac{g_m R_C}{1 + g_m R_E} \tag{6.75}$$

which also equals to

$$G_{CE \text{ with emitter degeneration resistor}} = -\frac{R_C}{\dfrac{1}{g_m} + R_E} \tag{6.76}$$

Another important fact in a common emitter configuration with emitter degeneration is that the amplifier input impedance seen between the base and ground becomes

$$R_{in\ CE \text{ with emitter degeneration resistor}} = r_\pi + (\beta + 1) R_E. \tag{6.77}$$

Equation (6.77) is derived from the circuit presented in Figure 6.34.

The importance of Equation (6.77) is that the total input resistance of the amplifier with emitter-degeneration is that the input resistance can be very large because of the $(\beta + 1)$ factor. It also interesting to mention that from an output impedance point of view, the emitter degeneration resistor does not change the output impedance at all, assuming that the Early effect is neglected. Thus,

$$R_{out\ CE \text{ with emitter degeneration resistor}} = R_C.$$

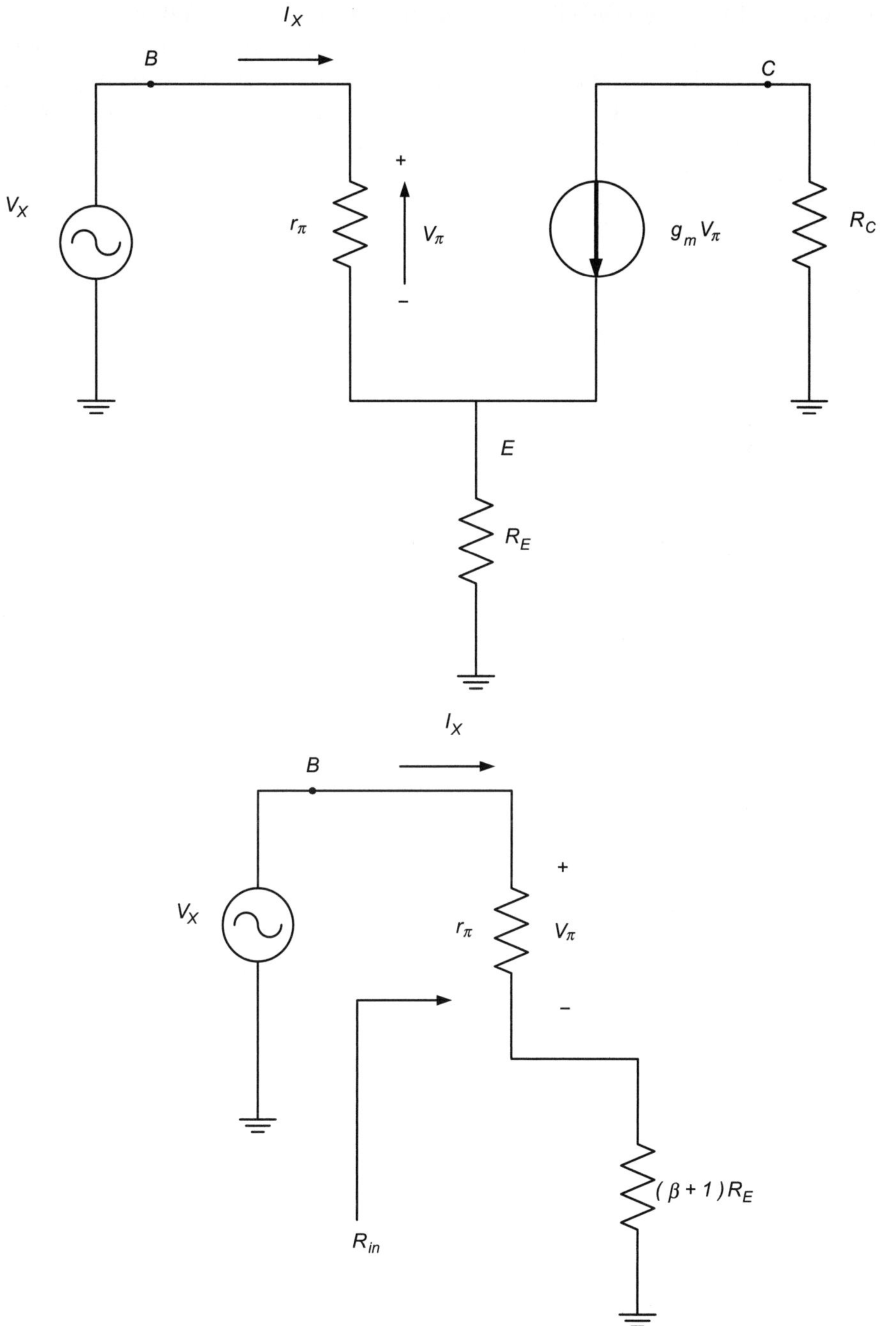

Figure 6.34 Common-emitter amplifier with emitter degeneration resistor.

Example 6.13 Given the circuit of Figure 6.30, assume the following component values: $R_1 = 20\ k\Omega$, $R_2 = 100\ k\Omega$, $R_C = 50\ \Omega$, and $R_E = 100\ \Omega$, the supply is $V_{CC} = 10$ V. (1) Calculate I_C, I_B, V_{CE}, V_{RE}, and β. (2) Also calculate for the found value of β the BJT trans-conductance g_m and r_π of the hybrid-π small signal model. At all times ignore the BJT Early effect; that is, $r_o \to \infty$. Assume the reverse saturation current $I_S = 7.11 \times 10^{-15}$A at room temperature of 300 K and $V_{BE} = 0.763$ V.

Solution to Example 6.13

From Equation (6.16), repeated here for the reader's convenience, we find for $I_S = 7.11 \times 10^{-15}$ A, $v_{BE} = 0.763$ V, and $v_T = 0.026$ V that

$$I_C = I_S e^{\frac{v_{BE}}{v_T}} = 39.5\,\text{mA}. \tag{6.78}$$

Using the above given values in the circuit of Figure 6.30, and applying Thèvenin to the resistor divider on the left of the BJT base node, we obtain

$$R_{Thev} = \frac{R_1 R_2}{R_1 + R_2};\ V_{Thev} = \frac{R_2}{R_1 + R_2} V_{CC}. \tag{6.79}$$

Equations (6.79) were obtained just as Equations (6.32) and (6.33) were obtained, using the *resistor divider method.*

For the Thevenized portion of the circuit we can write KVL equations, to find the base current:

$$I_B = (V_{Thev} - V_{BE} - V_E) / R_{Thev} \tag{6.80}$$

where $V_{Thev} = 8.33$ V and $R_{Thev} = 16{,}667\ \Omega$ are calculated from Equations (6.79) using the given values of R_1, R_2, and V_{CC}. Voltage V_E is the voltage drop across resistor R_E. This voltage is approximately equal to $I_C\ R_E$; a more exact value is $I_E\ R_E$. *Since* $I_C \approx I_E$, the error in the approximation is small, because $I_C + I_B = I_E$, and the value of base current is quite small. Using the value of $I_C = 39.5$ mA, we get that

$$V_E \approx I_C R_E = 0.0395 \times 100 = 3.95\ \text{V}.$$

Plugging the value of $V_E = 3.95\ V$ into Equation (6.80) we obtain: $I_B = 217\ \mu$A.

Since we calculated the collector current I_C and base current I_B, $\beta = I_C/I_B = 182$. Referring one more time to the circuit of Figure 6.30 we can see that the collector-emitter voltage:

$$V_{CE} = V_{CC} - I_C R_C - V_E.$$

Plugging the corresponding values, we obtain that

$$V_{CE} = 10 - (0.0395 \times 50) - 3.95 = 4.08 \text{ V}.$$

For part b of this example, since $g_m = I_C/V_T$ and $r_\pi = \beta/g_m$, we obtain that

$$g_m = 1.52 \text{ S and } r_\pi = 119\ \Omega.$$

The results for part (a) are regrouped and presented here:

$$I_C = 39.5 \text{ mA}$$

$$I_B = 217\ \mu\text{A}$$

$$V_{CE} = 4.08 \text{ V}$$

$$V_{RE} = 3.95 \text{ V and}$$

$$\beta = 182$$

and for part b):

$$g_\text{m} = 1.52 \text{ S}$$

and

$$r_\pi = 119\ \Omega.$$

6.3.13 Common-Base (CB) Configuration

The common-base configuration has a grounded base, the input signal is applied between the emitter and ground and the output of the amplifier is extracted between its collector and ground. Figure 6.35 walks us through the gain calculation of the CB topology. This topology is simplified since biasing is not fully shown. Note that input voltage V_{in} does not have any resistors in series to bias this stage.

Figure 6.35b allows us to see that the stage gain is

$$G_{CB} = V_{out} / V_{in} = g_m R_C. \tag{6.81}$$

It is very important to realize that the CB topology gain is equal to the absolute value of the CE topology gain. The CE gain has, unlike the CB gain, a negative sign; see Equation (6.70).

Next we will inspect Figure 6.36a,b to do a basic calculation of the input resistance of the CB stage. Note that using KCL at node E

$$I_X = -\left(g_m V_\pi + \frac{1}{r_\pi} V_\pi \right) = -V_\pi \left(g_m + \frac{1}{r_\pi} \right). \tag{6.82}$$

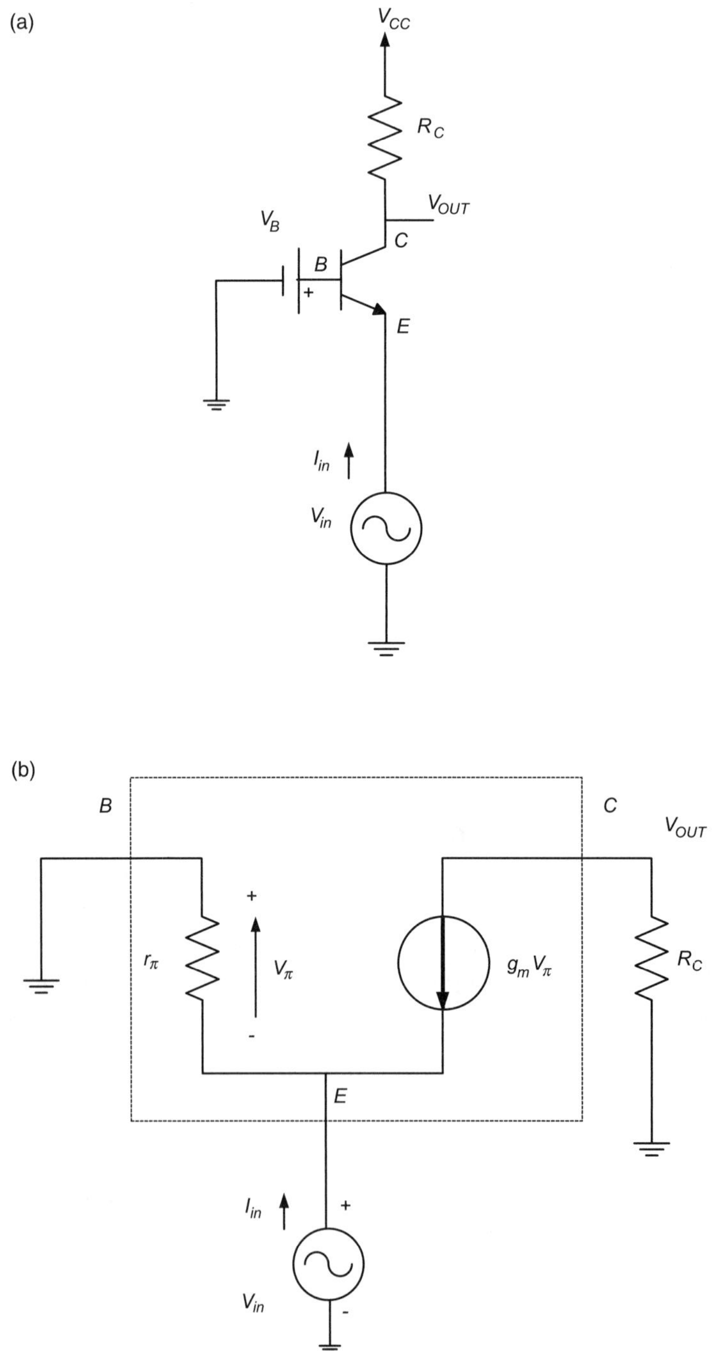

Figure 6.35 (a) CB stage used to calculate the gain; (b) CB stage with the hybrid-π small signal model.

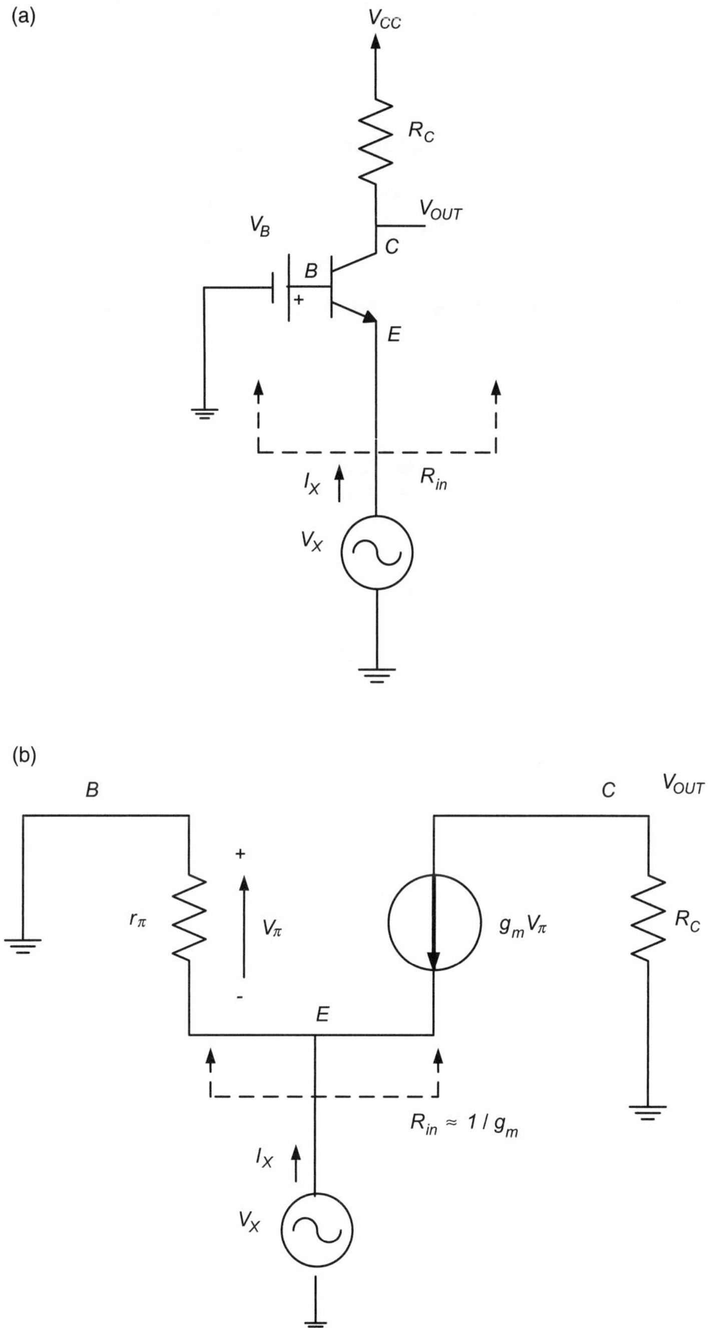

Figure 6.36 (a) Common base stage R_{in} calculation; (b) common base stage with small signal model.

And most importantly and again from Figure 6.36b since:

$$-V_\pi = V_{in}. \tag{6.83}$$

Plugging Equation (6.83) into Equation (6.82) yields

$$R_{in} = V_X / I_X = 1 / \left(g_m + \frac{1}{r_\pi} \right). \tag{6.84}$$

If r_π is large, then $1/r_\pi$ is small and we can approximate Equation (6.84) with

$$R_{in} \approx 1 / g_m. \tag{6.85}$$

Note that it is not unusual for r_π to be about 1 kΩ and for g_m about 50 mA/26 mV ≈ 1.923 S for signal transistors, thus Equation (6.85) is quite accurate. Both Equations (6.83) and (6.84) assume that the Early effect is negligible (i.e., $r_o \to \infty$).

Let us investigate the output resistance of the CB topology. To do that we provide an AC ground to the emitter input, as seen in Figure 6.37a. Then we replace the BJT with its hybrid-π small signal model, apply a test output voltage, which generates a test input current. As usual, R_{out} is given by the ratio of the test voltage and input. It is clear to see from Figure 6.37b that

$$R_{out} = R_C, \tag{6.86}$$

where Equation (6.86) ignores the Early effect, since it assumes that the transistor r_o is infinite. Taking the early effect into account, Equation (6.86) becomes a more accurate expression given by

$$R_{out} = R_C // r_o = \frac{R_C r_o}{R_C + r_o}. \tag{6.87}$$

The previously seen CB topologies covered so far mainly showed their small signal model, but they lacked their biasing circuitry. The next circuit, Figure 6.38a, depicts a CB topology with its biasing circuitry as well as its AC paths, including DC blocking capacitors.

We will speed up the pace a little bit describing this circuit, because of some similarities with the biasing circuits covered for the *CE* topology. Referring to Figure 6.38a we see that resistors R_1 and R_2 provide a biasing voltage to forward bias the base-emitter junction; R_C and R_E provide the means to reverse bias the collector-base junction and to establish the collector and emitter currents. Now from an AC standpoint the input signal generator (V_{gen}) with its internal resistance (R_{gen}) is AC coupled via C_{in} into the emitter input. The purpose of C_{in} is not to disturb the biasing voltage of the stage. At the base terminal capacitor C_B provides an AC short-circuit path to the base. The base

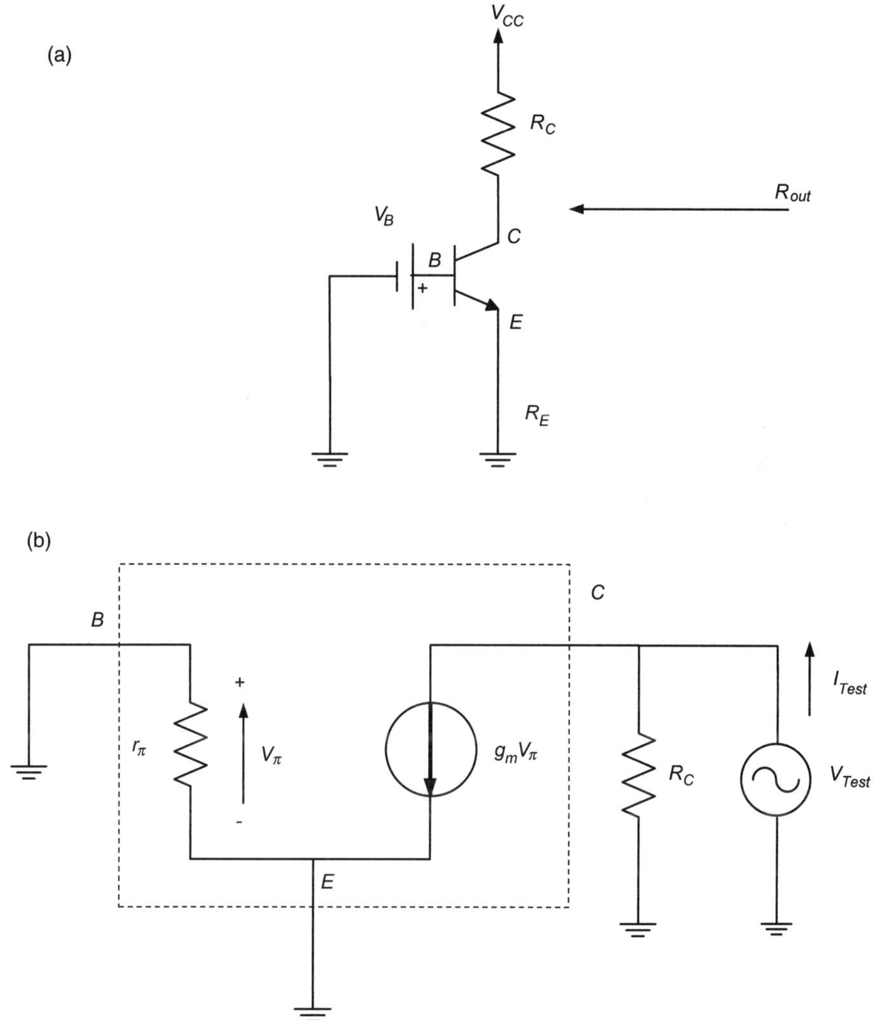

Figure 6.37 (a) CB topology used to calculate the stage output resistance; (b) CB topology after replacing the BJT with its hybrid-π small signal model.

ends up AC grounded as desired. The load resistor R_L that does not participate in the stage biasing is AC coupled via C_L and in effect, it is in parallel with R_C for AC components. It is also important to see that the input resistance seen by the input signal generator is the parallel of the basic CB stage ($1/g_m$) in parallel with the emitter resistor R_E. Thus, in effect, the CB stage R_{in} becomes

$$R_{in} = (1/g_m)//R_E = \frac{R_E}{1+g_m R_E}. \tag{6.88}$$

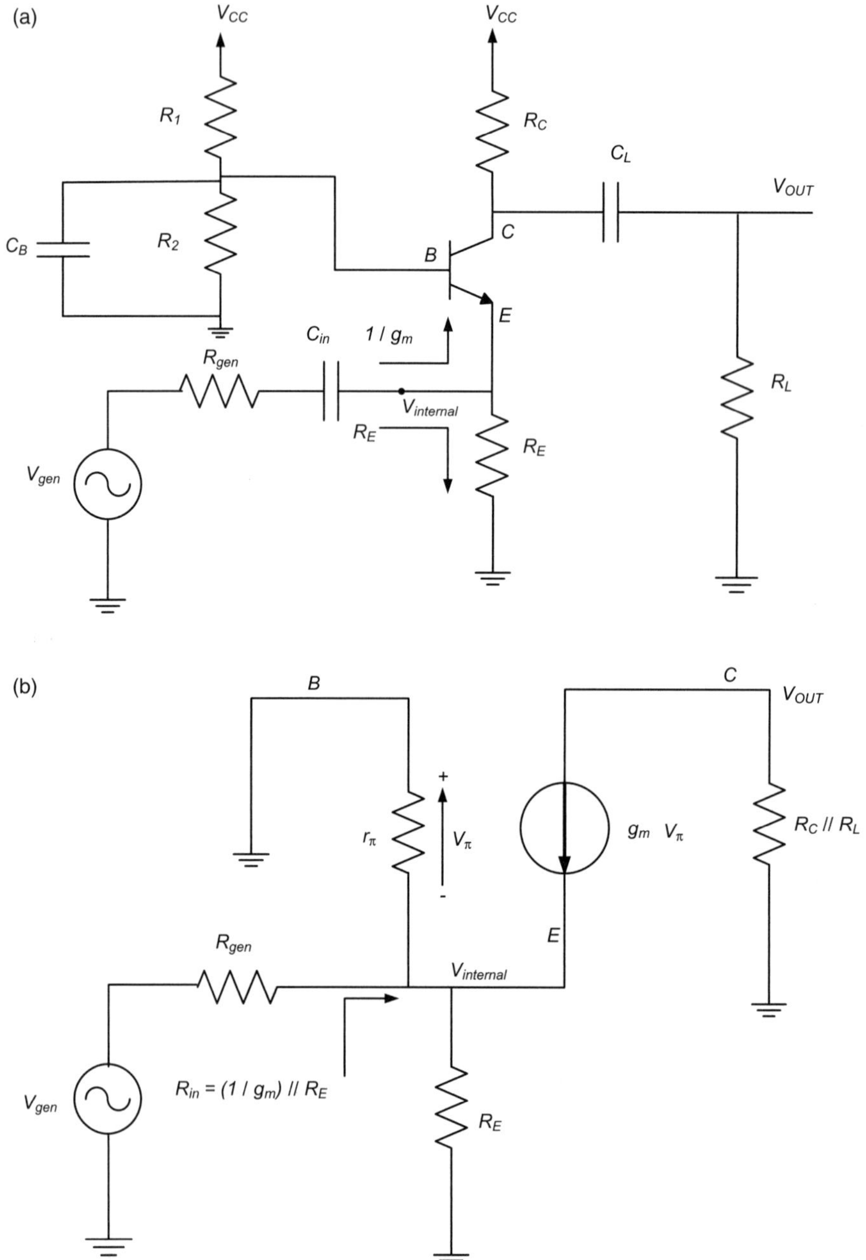

Figure 6.38 (a) CB topology showing biasing circuitry and AC paths; (b) CB topology where the BJT was replaced with its hybrid-π small signal model.

To evaluate the overall voltage gain,

$$G_{CB} = V_{out} / V_{gen} \tag{6.89}$$

We see from Figure 6.38a that voltage $V_{internal}$ equals for AC components to

$$V_{internal} = V_{gen} \frac{R_{in}}{R_{in} + R_{gen}}, \tag{6.90}$$

where R_{in} is the parallel combination of $1/g_m$ and R_E as shown in Figure 6.38b and Equation (6.88).

Since

$$V_{out} / V_{internal} = g_m R_{Par} \tag{6.91}$$

where

$$R_{Par} = \frac{R_C R_L}{R_C + R_L}. \tag{6.92}$$

Combining Equations (6.89) through (6.92) yields

$$G_{CB} = V_{out} / V_{gen} = \frac{1}{1 + (1 + g_m R_E) R_{gen} / R_E} g_m \frac{R_C . R_L}{R_C + R_L}. \tag{6.93}$$

Example 6.14 Let us assume that a 50 Ω coaxial transmission line cable needs to drive the input of amplifier with an input resistance of 10 kΩ. In order to maximize power transfer from one stage to the next, the output impedance of the driving stage, the transmission line in this case, must match the input impedance of the receiving stage, the CB amplifying stage. The output impedance of the CB stage must match the input impedance of the circuit downstream. Figure 6.39a shows this downstream circuit simply as R_{in}. Refer to Figure 6.39a to view the circuit set up. Design a common base amplifier configuration that presents an input impedance of 50 Ω to the coax signals. The amplifier output impedance must be 10 kΩ to match the input impedance R_{in} of the stage that need to be driven.

Solution to Example 6.14

From Equations: (6.86) and (6.88) we know that:

$$R_{out} = R_C.$$

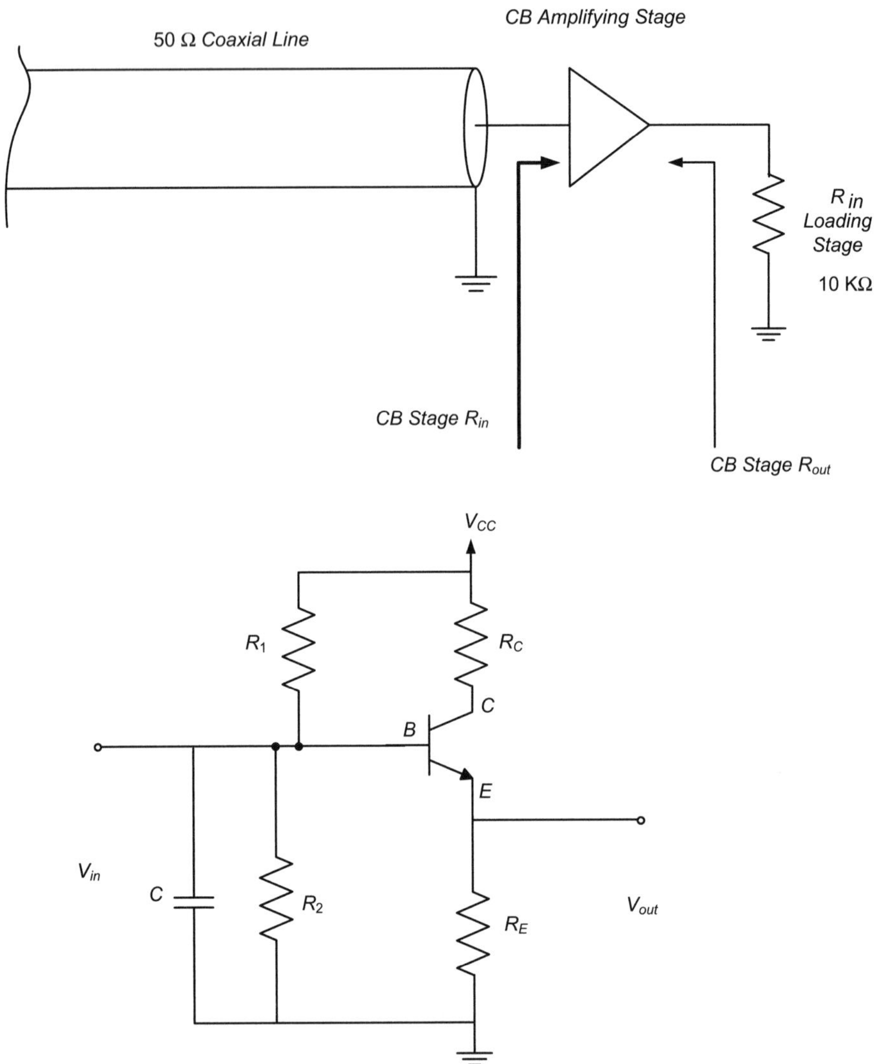

Figure 6.39 Common base (CB) design, Example 6.14, and CB stage.

Ignoring the Early effect of the BJT, and

$$R_{in} = (1/g_m)//R_E = \frac{R_E}{1+g_m R_E}. \tag{6.94}$$

We need the CB stage R_{in} to be 50 Ω, and we need an R_{out} = 10 kΩ. Since $R_{out} = R_C$, if we obtain a BJT with a g_m of 1/53 Ω or 0.0188679 S and R_E = 1 kΩ, it yields

$$R_{in} = (1/g_m)//R_E = 53//1000 \approx 50\ \Omega.$$

And since we are ignoring the Early effect, $r_o \to \infty$, thus R_{out} simple equals R_C, hence:

$$R_{out} = 10\ \text{k}\Omega.$$

Finally, the CB amplifier needs to be biased such that a g_m of 0.0188679 S is obtained. Since $g_m = I_C/V_T$ and $V_T = 0.026$ V at room temperature, then we need a collector biasing current of

$$I_C = g_m V_T = 0.0188679 \times 0.026 = 0.49\ \text{mA}.$$

The selection of resistors R_1 and R_2 are left as an exercise to the reader.

6.3.14 The Common-Collector (CC) Configuration

The CC configuration is also commonly referred to as emitter follower. This configuration has the BJT's collector AC grounded terminal. The input to this stage is applied between the base and ground, the output is sensed between the emitter and ground. The core circuit of the emitter follower is depicted in Figure 6.40. Note that the collector is tied to V_{CC}, thus its AC signals are effectively grounded.

When the input voltage grows, more base current is injected into the BJT, causing the collector and emitter currents to increase. The output voltage V_{out} is never higher than the input voltage V_{in}. Voltage increments in the base cause increments in the voltage V_{out} across R_E than can never keep up with the base voltage because of the base-emitter voltage drop. Figure 6.41 depicts an emitter follower stage BJT small signal model.

Looking at the small signal model of Figure 6.41 we can state KCL equations at node V_{out} and obtain

$$\frac{V_\pi}{r_\pi} + g_m V_\pi = \frac{V_{out}}{R_E}. \qquad (6.95)$$

Doing some algebra on Equation (6.95) and taking into account that: $\beta = g_m r_\pi$ (from Table 6.3),

$$V_\pi = \frac{r_\pi}{\beta + 1} \cdot \frac{V_{out}}{R_E}. \qquad (6.96)$$

From the circuit of Figure 6.41 we can see that

$$V_{in} = V_\pi + V_{out}.$$

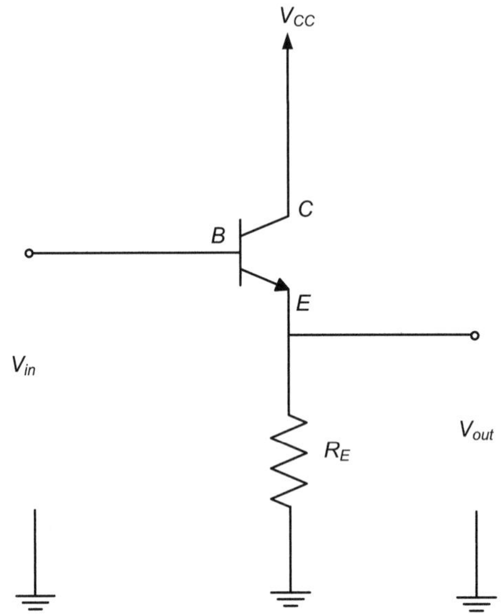

Figure 6.40 Core structure of an emitter follower stage.

Combining V_{in} with Equation (6.96) we obtain

$$G_{CC} = \frac{V_{out}}{V_{in}} = \frac{1}{1 + \frac{r_\pi}{\beta + 1} \cdot \frac{1}{R_E}}, \tag{6.97}$$

and since

$$\frac{r_\pi}{\beta + 1} \approx \frac{r_\pi}{\beta}. \tag{6.98}$$

Using Equation (6.98) in Equation (6.97) yields

$$G_{CC} = \frac{V_{out}}{V_{in}} \approx \frac{R_E}{R_E + \frac{1}{g_m}}. \tag{6.99}$$

From Equation (6.99) it is clear to see that the CC topology voltage gain is always positive and less than *unity*.

Now let us look at the emitter follower stage gain, when fed by an input signal with source resistance (R_S). Figure 6.42a depicts the circuit with the input signal associated with a source resistance. We now proceed to find out

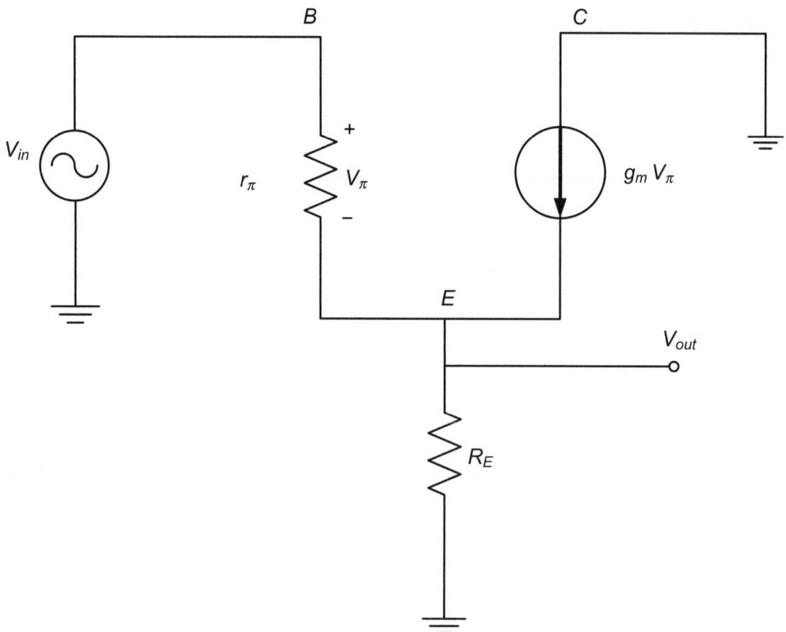

Figure 6.41 Emitter follower stage small signal model.

the gain for this circuit. We apply Thévenin's Theorem to the circuit of Figure 6.42a. We select the emitter resistor to be the element we keep, while we look into replacing the rest of the circuit with its Thévenin's equivalent; as usual we inhibit independent voltage source V_{in}; that is, we replace it with a short circuit. Remember that Thévenin's Theorem does not want you to remove any dependent sources. We calculate the Thévenin resistance injecting a Thévenin voltage source between the emitter and ground nodes. Figure 6.41b depicts the circuit used to calculate the Thévenin's resistance.

Applying KCL at node E to the circuit of Figure 6.42b we obtain

$$\frac{V_\pi}{r_\pi} + g_m.V_\pi = -I_{Thev}. \tag{6.100}$$

By inspection of the circuit we find

$$-V_\pi = \frac{r_\pi}{r_\pi + R_S} V_{Thev}. \tag{6.101}$$

Combining Equations (6.101) and (6.100) it yields

$$Z_{Thev} = V_{Thev} / I_{Thev} = \frac{r_\pi}{\beta+1} + \frac{R_S}{\beta+1}$$

and since $r_\pi/(\beta + 1) \approx r_\pi/\beta = 1/g_m$ (Z_{Thev}) becomes:

$$Z_{Thev} = V_{Thev} / I_{Thev} = \frac{1}{g_m} + \frac{R_S}{\beta+1}$$

The Thévenin voltage is $V_{Thev} = V_{in}$, and using Z_{Thev} from above according to the circuit of Figure 6.42b, which basically is a resistor divider, leads to

$$\frac{V_{out}}{V_{in}} = \frac{R_E}{R_E + \frac{R_S}{\beta+1} + \frac{1}{g_m}}. \tag{6.102}$$

The input impedance of the emitter follower stage is calculated from the transistor and small signal circuit model of Figure 6.42. A voltage V_X is injected into the input of the stage, a current I_X is produced. The input resistance R_{in} is the ratio of V_X and I_X. Applying KCL at node E of the circuit of Figure 6.42b we obtain

$$V_X = V_\pi + (I_X + g_m V_\pi) R_E. \tag{6.103}$$

Since from Figure 6.43b we have that

$$V_\pi = I_X r_\pi. \tag{6.104}$$

Plugging Equation (6.104) in Equation (6.103) we obtain (see also Fig. 6.44b)

$$V_X = I_X r_\pi + (I_X + g_m I_X r_\pi) R_E \tag{6.105}$$

and since from Table 6.3

$$g_m r_\pi = \beta. \tag{6.106}$$

Plugging Equation (6.106) in Equation (6.105) after a little bit of algebra results to

$$R_{in\ emitter\text{-}follower} = \frac{V_X}{I_X} = r_\pi + (\beta+1) R_E. \tag{6.107}$$

It is important to see that the input impedance of an emitter-follower stage is identical to the input impedance of a common-collector stage with emitter degeneration resistance; refer to Equations (6.107) and (6.77).

Now let us calculate the output impedance of an emitter-follower stage. Figure 6.44c,d depicts the output impedance and its components.

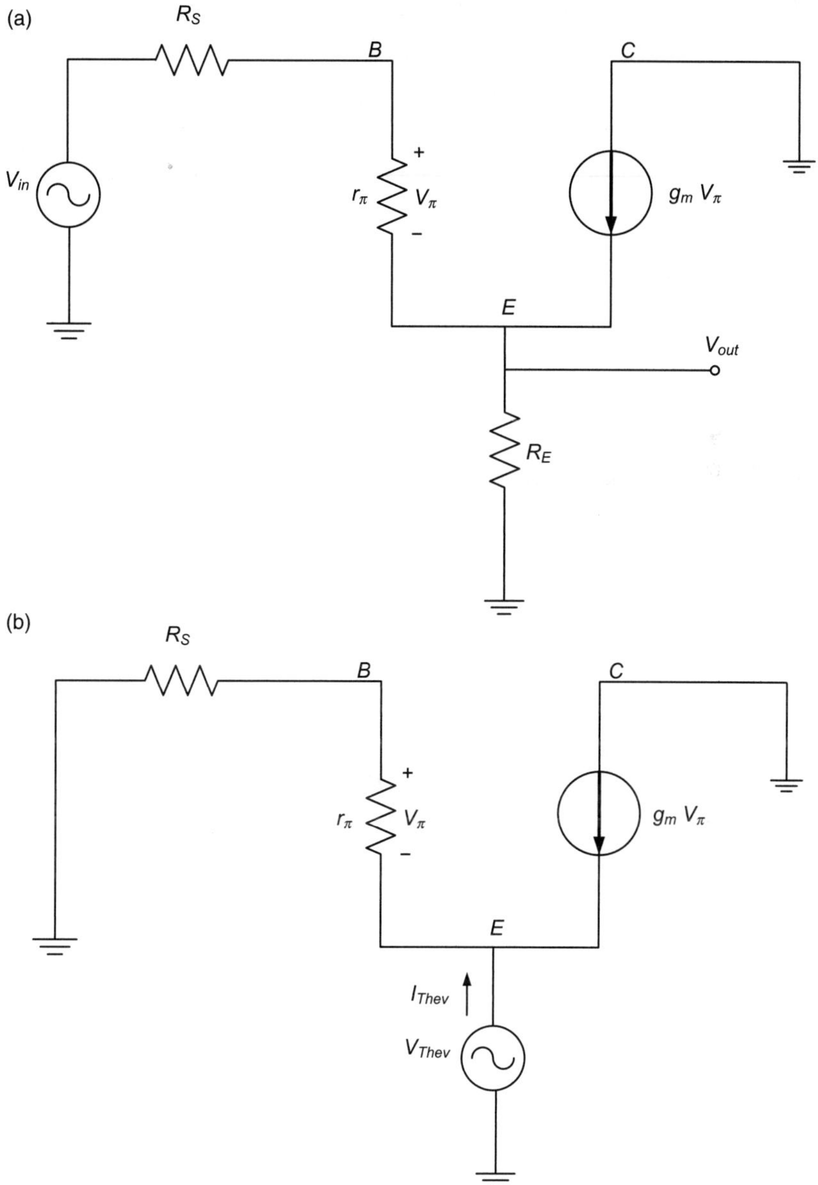

Figure 6.42 (a) Emitter-follower with source resistance small signal circuit model; (b) emitter-follower with source resistance small circuit model used to calculate R_{Thev}.

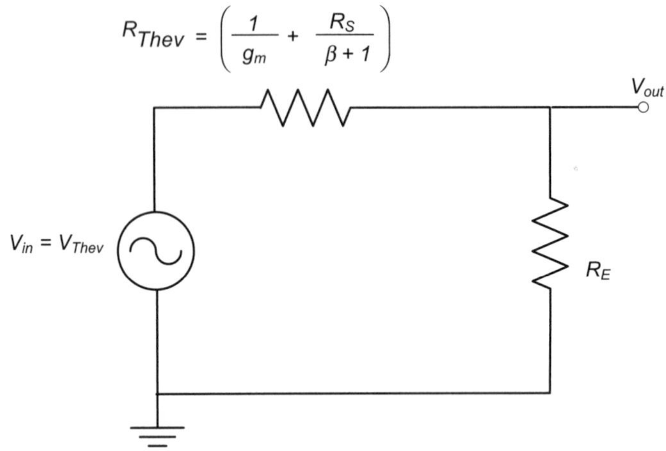

Figure 6.43 Emitter follower gain: input resistance calculation with source resistance using Thévenin's Theorem.

The computation of the emitter-follower output impedance components is performed using Figure 6.44c,d. Referring now to Figure 6.44c,d, the output impedance of the follower stage is simply the parallel of the two output impedance components, hence:

$$R_{out} = \left(\frac{1}{g_m} + \frac{R_S}{\beta+1}\right) // R_E \tag{6.108}$$

Equation (6.108) ignores the Early effect. Taking this effect into account, the output impedance becomes

$$R_{out} = \left(\frac{1}{g_m} + \frac{R_S}{\beta+1}\right) // R_E // r_o$$

where g_m is the transistor's transconductance, R_S is the input signal internal resistance also called the source resistance, β is the transistor current gain parameter, R_E is the emitter resistor, and r_o is the transistor model output resistance due to the Early effect.

6.4 METAL OXIDE FIELD EFFECT TRANSISTOR (MOSFET)

The *n-channel* enhancement mode MOSFET, also called an NMOS transistor is discussed next. Later on we will briefly discuss the *p-channel* enhancement mode MOSFET (or PMOS) and the two depletion type MOSFETs (*n-channel* D-MOSFET and *p-channel* D-MOSFET).

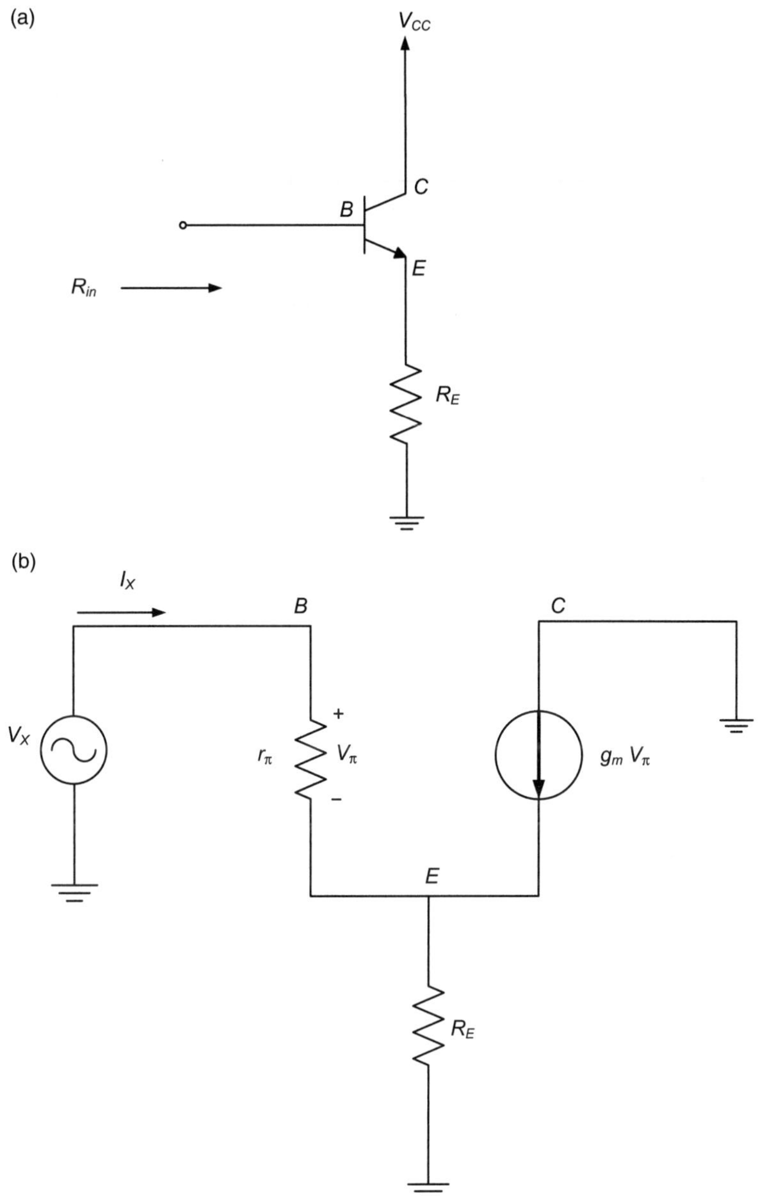

Figure 6.44 (a) Emitter-follower circuit and input impedance calculation; (b) follower small signal model to calculate input resistance; (c) emitter-follower for the calculation of output resistance; (d) emitter-follower with calculated output resistance.

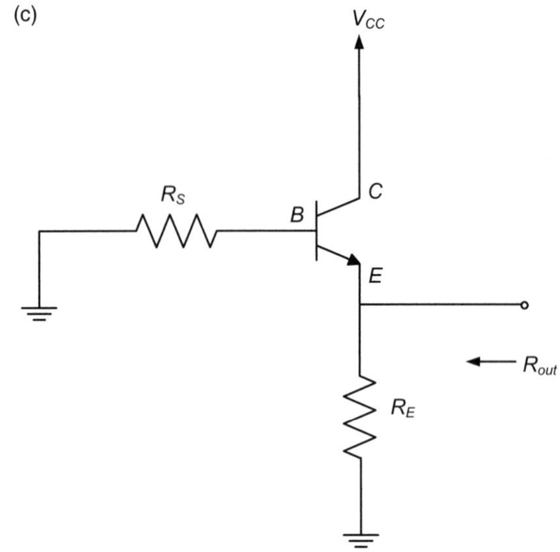

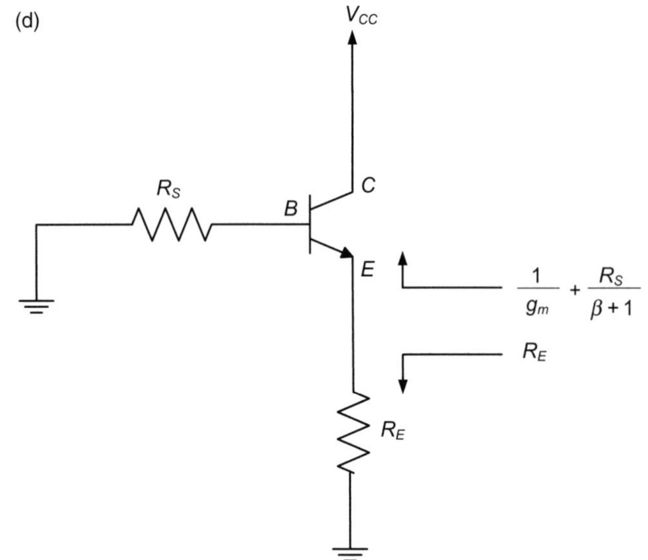

Figure 6.44 (*Continued*)

The enhancement mode *n-channel* MOSFET may conduct an electric current between its *drain* and *source* terminals when a positive voltage is applied to the *gate*. Figure 6.45 depicts the basic structure of an *n-channel* enhancement mode MOSFET. The drain-to-source current is controlled by the magnitude of the gate voltage. For the time being, we are describing the

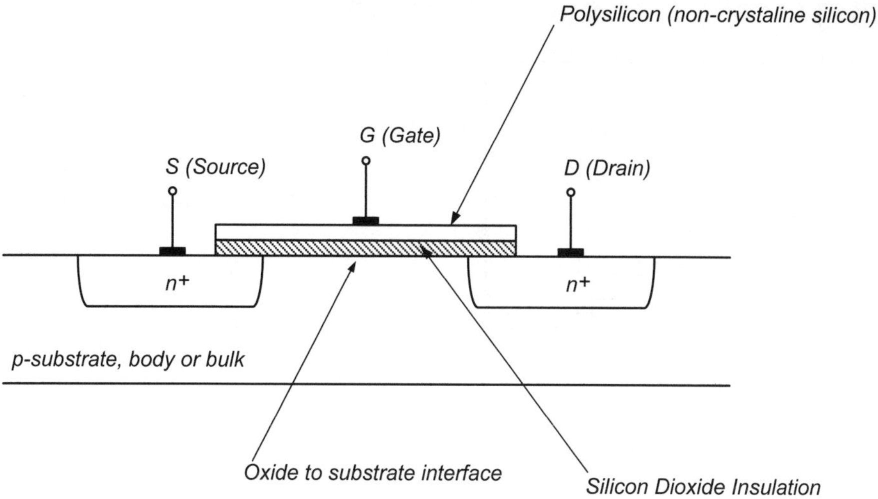

Figure 6.45 Basic MOSFET structure.

Table 6.4 MOSFET types

Enhancement Mode		Depletion Mode	
n-channel Induced channel	*p*-channel Induced channel	*n*-channel Implanted channel	*p*-channel Implanted channel

enhancement mode *n-channel* MOSFET. Table 6.4 lists the four MOSFET types that are available.

MOSFETs have either three or four terminals. *Gate, Drain* and *Source* in all cases and in some cases discrete MOSFETs have a fourth terminal or the body terminal that may be externally connected to the desired voltage. For medium to low frequencies, the MOSFET gate current is zero because the silicon dioxide behaves as an insulator (Fig. 6.45).

Only upon operating the MOSFET at high frequencies there may be a nonzero or significant gate current due to the internal parasitic capacitances within the device structure. Within our coverage of MOSFETs we will not deal with the MOSFET operating at high frequencies. High frequencies are considered those frequencies well above audio frequencies or 20 kHz. For example, in applications where MOSFETs are used at hundreds of kHz, such as in switching power supplies, MOSFET gate currents are not negligible. Unlike the bipolar transistor, which conducts majority as well as minority carriers, MOSFET currents are unipolar and its current consists of majority carriers. *N-channel* MOSFETs conduct currents of electrons while *p-channel* types conduct currents of holes.

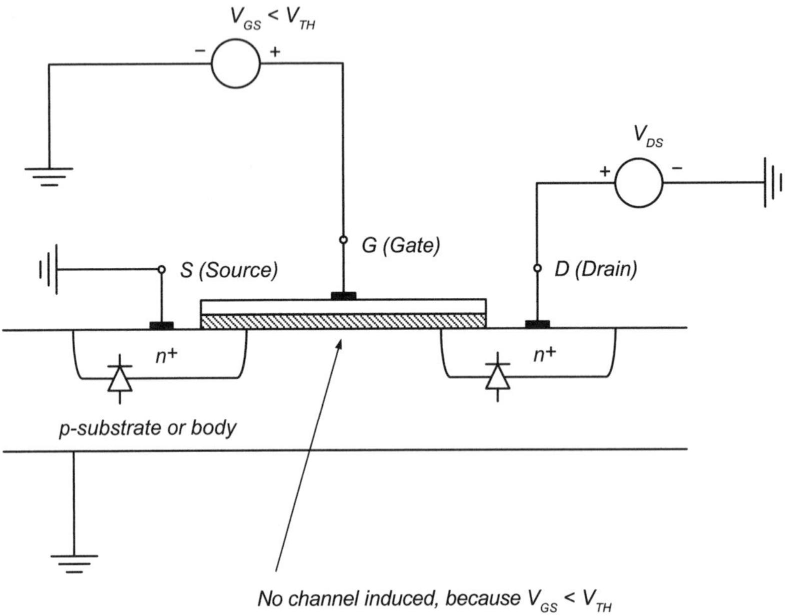

Figure 6.46 MOSFET operating with $V_{GS} < V_{TH}$, $I_D = 0$.

6.4.1 MOSFET *I-V* Characteristics

Let us look into the behavior of the *n-channel* MOSFET from "currents and voltages" point of view. Figure 6.46 depicts an *n-channel* enhancement mode MOSFET with its *source* terminal grounded and a positive voltage V_{DS} applied to its drain. If the *gate* voltage applied is under the so-called MOSFET threshold voltage V_{TH}, no drain-to-source current flows through the MOSFET substrate (Fig. 6.46) and no channel between the *drain* and *source* has been formed yet. More positive charge accumulates on the gate and negative ions form in the substrate with higher positive gate voltages. This is referred to as the depletion region.

The MOSFET is said to be *"off."* As the gate voltage V_G increases beyond the threshold voltage (V_{TH}) free electrons are attracted to the region between the silicon dioxide and the substrate and creates a channel. It is said that the channel is *induced*. When this channel is created, the MOSFET starts conducting a current from *drain* to *source*. The larger V_{GS} becomes, the deeper the channel. Note that the gate never conducts any current (in low and mid-frequencies operation), and the gate acts like a capacitor controlling the channel length and depth. Referring again to Figure 6.46, it is important to note that there are two diodes, one of them between the p-substrate and the drain and a second diode between the p-substrate and the source. It is important not to allow these diodes to conduct any current when the MOSFET is still off, thus the p-substrate is grounded, preventing such diodes from becoming forward biased and current flowing through them.

The current voltage relationship of the NMOS MOSFET is given without proof and it is

$$I_D = \frac{1}{2}\mu_n C_{ox} \frac{W}{L}[2(V_{GS} - V_{TH})V_{DS} - V_{DS}^2]. \tag{6.109}$$

In Equation (6.109) μ_n is the carrier mobility of the electrons in the *n-channel*, C_{ox} is the capacitance formed by the oxide in between gate polysilicon and the p-substrate, this capacitance is greatly controlled by the thickness of such oxide. W is the width of the channel, L is the length of the channel, W/L is referred to as the aspect ratio of the MOS transistor, V_{GS} is the gate to source voltage applied, V_{TH} is the MOSFET threshold voltage, and V_{DS} is the MOSFET drain-to-source voltage. Once a MOSFET technology is chosen, the oxide thickness (t_{ox}) is fixed and cannot be changed, the aspect ratio W/L is under control of the integrated circuit designer.

As the gate voltage is gradually increased, the MOSFET practically acts as a variable resistor. During this region of operation, the drain current (I_D) maintains a linear relationship with respect to the drain-to-source voltage (V_{DS}). The slope of this part of the I-V curve equals $1/R_{DSon}$, where R_{DSon} is the drain-to-source on-resistance of the MOSFET. Equation (6.109) is a nonlinear relationship between I_D and V_{DS}. When:

$$V_{DS} \ll 2\,(V_{GS} - V_{TH}) \tag{6.110}$$

Equation (6.109) reduces to

$$I_D \approx \mu_n C_{ox} \frac{W}{L}(V_{GS} - V_{TH})V_{DS} \tag{6.111}$$

Since

$$R_{DSon} = V_{DS} / I_D \tag{6.112}$$

Plugging Equation (6.111) into Equation (6.112) we obtain:

$$R_{Dson} = \frac{1}{\mu_n C_{ox} \dfrac{W}{L}(V_{GS} - V_{TH})}. \tag{6.113}$$

Equation (6.113) means that for very small V_{DS} values the parabola given by Equation (6.109) is approximated by the linear behavior of Equation (6.112). By inspection of Equation (6.113), it is interesting to see that the transistor R_{Dson} is directly proportional to the channel length (L) and inversely proportional to the channel width (W). This should also be an intuitive conclusion. Equation (6.109) is the expression of the drain current for all values of V_{DS}, and it is the equation of a parabola. If we took the first derivative of Equation (6.109), find its zero and then evaluate its second derivative at the same point, which would yield a negative result. So without doing the mathematical derivation, it can easily be stated that Equation (6.109) has a maximum at

$$V_{DS} = V_{GS} - V_{TH}. \tag{6.114}$$

So evaluating Equation (6.109) using the value of V_{DS} from Equation (6.114) we find

$$I_{D,MAX} \approx \frac{1}{2}\mu_n C_{ox}\frac{W}{L}(V_{GS} - V_{TH})^2 \tag{6.115}$$

The maximum value of I_{DS} shown by Equation (6.115) is the value of current when

$$V_{DS} > V_{GS} - V_{TH}. \tag{6.116}$$

When the MOSFET operates under such conditions, that is, Equation (6.116), it is said to be in its saturation region of operation as depicted by Figure 6.47.

It is important to observe that Equation (6.115) is independent of V_{DS}. This says that the curves for the drain current become constant, and independent of V_{DS} after the pinch-off voltage, that is, $V_{DS} = V_{GS} - V_{TH}$. After the MOS transistor reaches the pinch-off voltage, the drain current does not significantly change with V_{DS}. Such statement is true provided that we do not take into account a second-order effect referred to as *channel length modulation*. Figure 6.47a,b describes the different regions of the MOSFET operation given by Equations (6.109) through (6.116).

The channel length moves with the change of V_{DS}. This effect can be visualized in Figure 6.48: (a) shows the induced channel, (b) shows the pinched-off channel, and (c) shows channel length modulation.

The *channel length modulation effect* is accounted for in Equation (6.115) by multiplying it by the factor: $(1 + \lambda V_{DS})$ that yields

$$I_D = \frac{1}{2}\mu_n C_{ox}\frac{W}{L}(V_{GS} - V_{TH})^2(1 + \lambda V_{DS}). \tag{6.117}$$

In Equation (6.117), λ is called the channel length modulation coefficient.

The effect that channel length modulation produces on the drain current characteristics is a slight positive slope as shown in Figure 6.49. Another second-order effect is the body effect; this takes place when the substrate potential grows above zero, and this causes the threshold voltage to increase. We will not consider the body effect in this book.

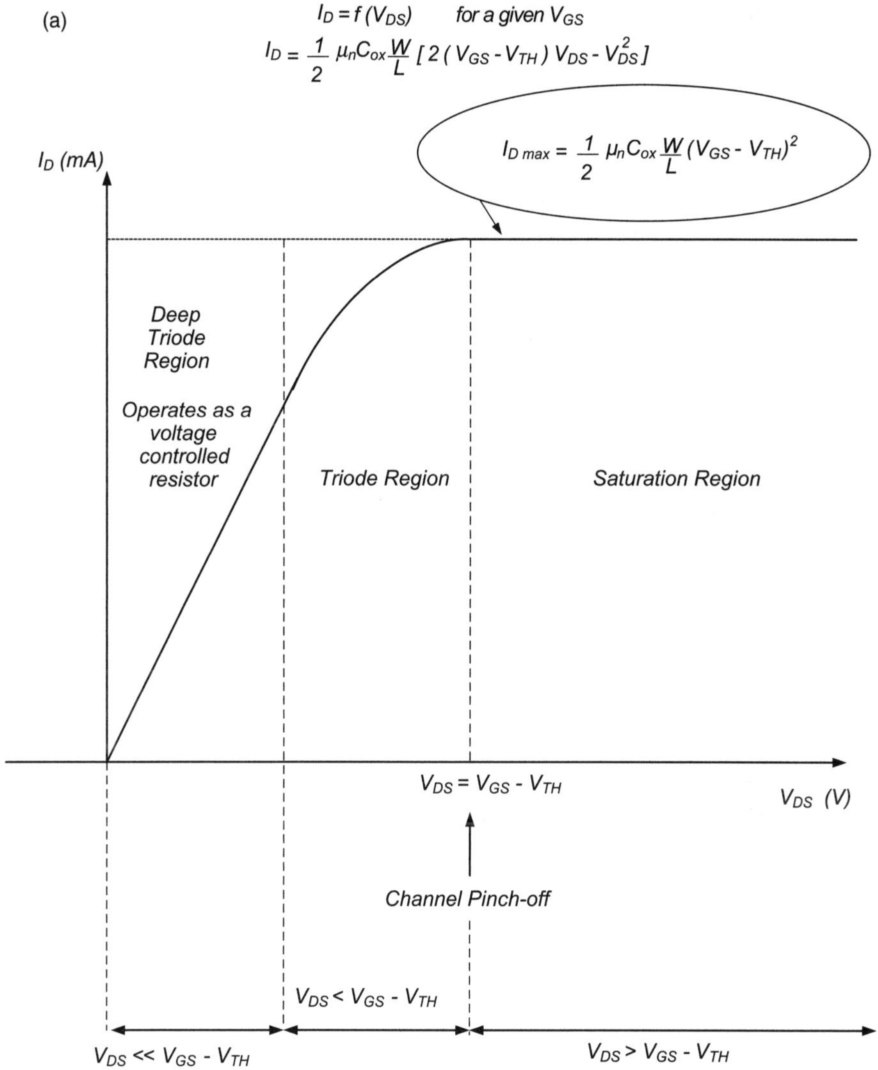

Figure 6.47 (a) MOSFET triode and saturation regions; (b) MOSFET triode region.

6.4.2 MOSFET Small Signal Model

The small signal model of a MOSFET is similar to the small signal model of the bipolar transistor. The basic model consists of a voltage-controlled current source (*VCCS*), its current value is $g_m V_{GS}$, where g_m is the MOSFET *trans-conductance* and V_{GS} is the gate to source controlling voltage. The transistor trans-conductance is given by

$$g_m = \frac{\partial I_D}{\partial V_{GS}}. \tag{6.118}$$

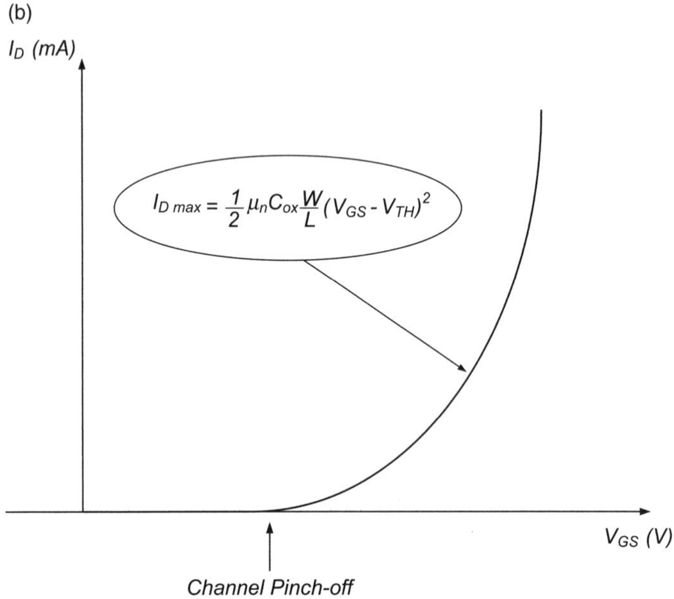

Figure 6.47 (*Continued*)

Using Equation (6.118) for the saturation region without the effect of channel length modulation we obtain:

$$g_m = \mu_n C_{ox} \frac{W}{L}(V_{GS} - V_{TH}). \tag{6.119}$$

The accountability of the channel length modulation, that is, the dependence of the saturation current with V_{DS}, is included with the addition of an output resistor r_o, where

$$r_o = \left(\frac{\partial I_D}{\partial V_{DS}}\right)^{-1} = \frac{1}{\frac{1}{2}\mu_n C_{ox}\frac{W}{L}(V_{GS} - V_{TH})^2 \lambda} \approx \frac{1}{\lambda I_D}. \tag{6.120}$$

Figure 6.51 depicts the small signal, low, and medium frequencies equivalent model of a MOSFET.

6.4.3 MOSFET Biasing Techniques

Let us start by pointing out the fundamental differences and similarities between bipolar and MOS transistors at the terminal voltage and current levels. Table 6.5 basically summarizes key differences and similarities between BJTs and MOSFETs. However, Table 6.5 does not address differences and similarities directly pertaining to semiconductor physics.

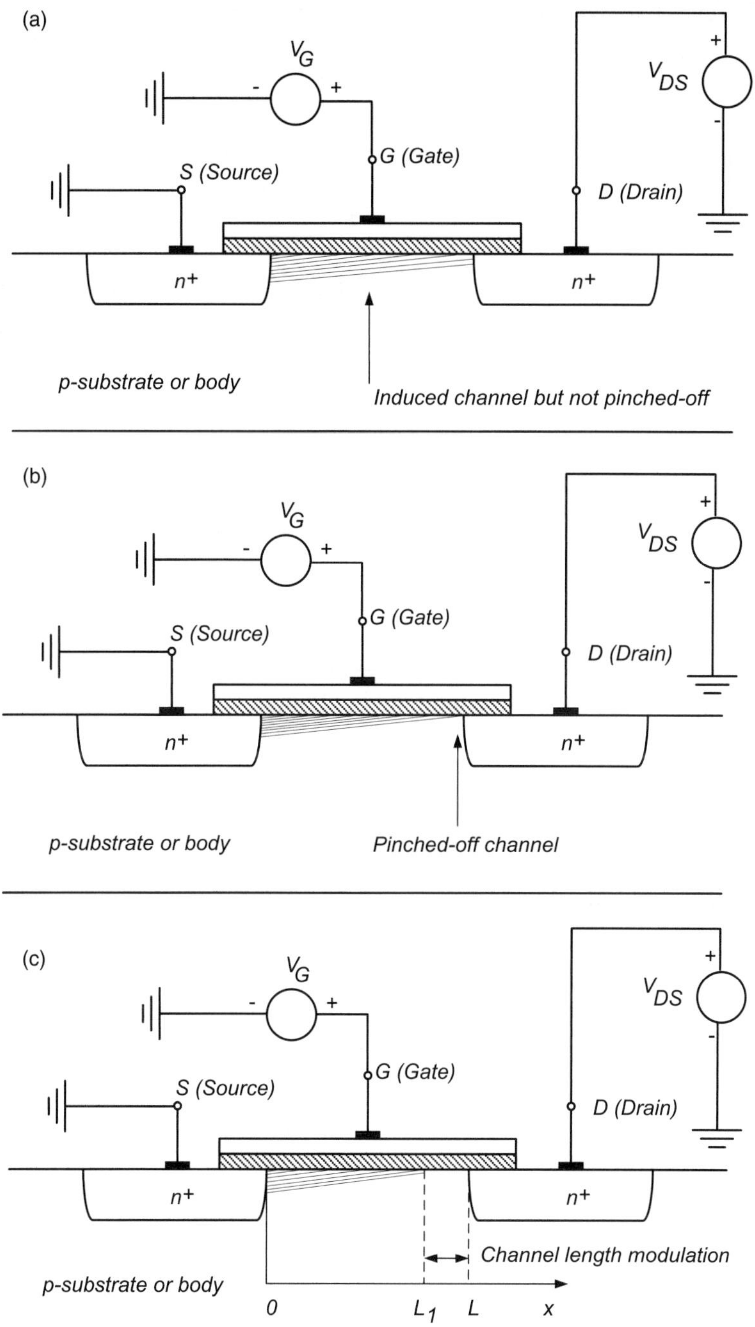

Figure 6.48 (a) Induced channel but not pinched-off yet; (b) pinched-off channel; (c) channel length modulation effect.

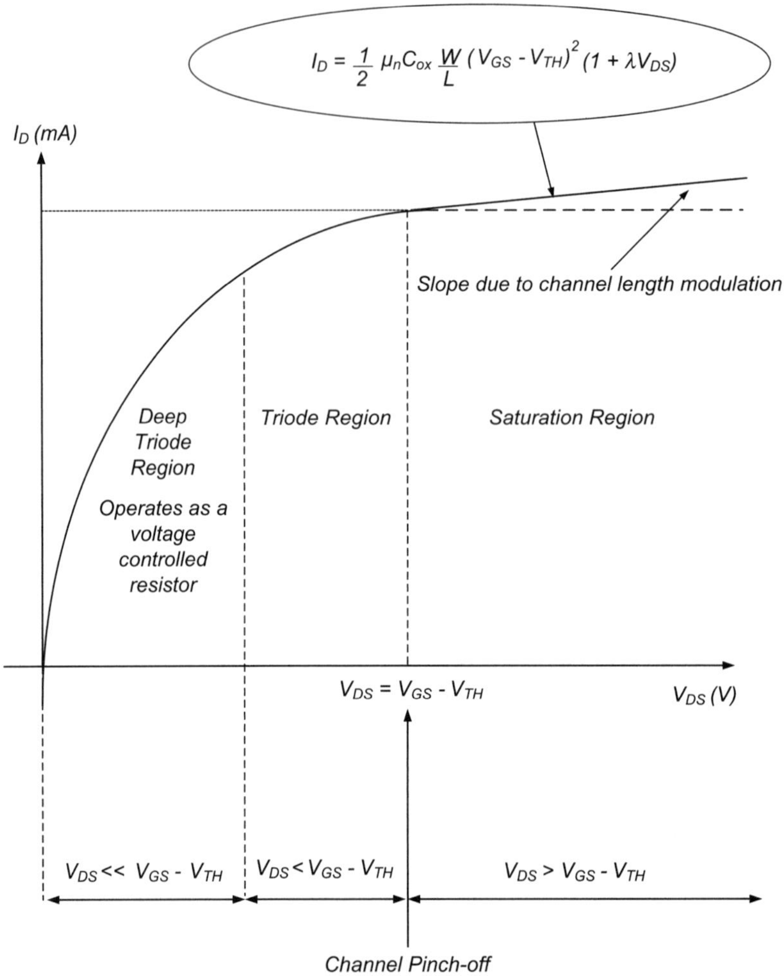

Figure 6.49 N-channel MOSFET: I-V characteristics depicting the effect of channel length modulation.

Taking into account the facts listed in Table 6.5, it is possible to find some similarities as well as differences biasing BJTs and MOSFETs. The remainder of this chapter will leverage on the previously addressed BJT material and is presented in a more speedy fashion.

Let us consider the MOSFET circuit of Figure 6.50; we want to bias the transistor such that it operates in the saturation region. Ignoring channel length modulation we have that

$$V_G = \frac{R_2}{R_1 + R_2} V_{DD}. \tag{6.121}$$

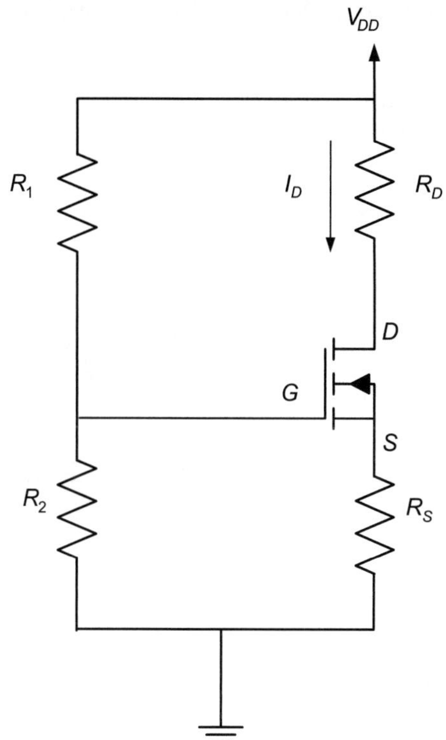

Figure 6.50 N-channel MOSFET biasing.

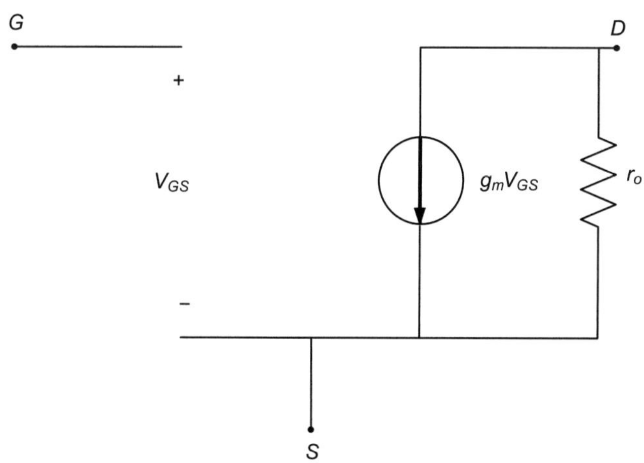

Figure 6.51 MOSFET small signal equivalent model for low and mid frequencies.

Table 6.5 Some fundamental circuit differences and similarities between BJTs and MOSFETs

BJT	MOSFET
Difference: BJT has no body or substrate	Difference: MOSFET has a substrate
Difference: When $V_{BE} = V_{CE}$ BJT is at the edge of the active region	Difference: MOSFET is at the edge of saturation if V_D is below $V_G - V_{TH}$
Difference: Finite (nonzero) base current	Difference: Zero gate current at mid and low frequencies
Difference: Exponential $I_C - V_{BE}$ characteristics	Difference: Square law dependence between I_D and V_{GS}
Difference: Most BJTs have the same I_S reverse saturation current	Difference: MOSFETs have selectable-by-design W/L aspect ratios
Difference: Two BJT types: NPN and PNP	Difference: Four MOSFET types: 2 enhancement mode types: *n*-channel and *p*-channel and 2 depletion mode types: *n*-channel and *p*-channel
Three regions of operation: saturation, active, and cutoff (*)	Three regions of operation: triode (includes deep triode region), saturation, and cutoff (*)
(*) May qualify as a similarity and as a difference	(*) May qualify as a similarity and as a difference
Difference: BJT saturation region is not the same as MOSFET saturation region	Difference: MOSFET saturation region is not the same as BJT saturation region
Similarity: base, collector, and emitter	Similarity: gate, drain, and source
Similarity: Voltage controlled-current source-based small signal model	Similarity: Voltage controlled-current source-based small signal model
Similarity: common emitter amplifier	Similarity: common source amplifier
Similarity: common collector amplifier	Similarity: common drain amplifier
Similarity: common base amplifier	Similarity: common gate amplifier
Similarity: BJT can operate as a switch	Similarity: MOSFET can operate as a switch

It is important to recognize that voltage V_G at node G in Figure 6.50 refers to the gate voltage with respect to ground and not to the gate voltage with respect to the MOSFET source terminal.

From KVL we can see that

$$V_G = V_{GS} + I_D R_S. \tag{6.122}$$

Combining Equations (6.121) and (6.122) we obtain

$$\frac{R_2}{R_1 + R_2} V_{DD} = V_{GS} + I_D R_S. \tag{6.123}$$

Since the saturation current of the MOSFET, from Equation (6.115) is

$$I_D = \frac{1}{2}\mu_n C_{ox}\frac{W}{L}(V_{GS} - V_{TH})^2, \tag{6.124}$$

combining Equation (6.123) with Equation (6.124) it yields

$$\left(\frac{R_2}{R_1 + R_2}V_{DD} - V_{GS}\right)\frac{1}{R_S} = \frac{1}{2}\mu_n C_{ox}\frac{W}{L}(V_{GS} - V_{TH})^2. \tag{6.125}$$

Performing some algebraic operations on Equation (6.125) yields

$$V_{GS} = -(V_a - V_{TH}) + \sqrt{(V_a - V_{TH})^2 - V_{TH}^2 + \frac{2R_2}{R_1 + R_2}V_a V_{DD}} \tag{6.126}$$

$$V_{GS} = -(V_a - V_{TH}) + \sqrt{V_a^2 + 2V_a\left(\frac{R_2 V_{DD}}{R_1 + R_2} - V_{TH}\right)} \tag{6.127}$$

where

$$V_a = \frac{1}{\mu_n C_{ox}\frac{W}{L}R_S}.$$

Finally, we must verify that: $V_{DS} > V_{GS} - V_{TH}$ to satisfy the saturation condition.

When we solve problems it will be more straightforward and faster to use Equation (6.115) to calculate drain current. Making initial guesses of V_{GS} to calculate I_D will take longer to converge.

Example 6.15 Using the circuit of Figure 6.50 assume the following values: $V_{DD} = 10$ V, $R_1 = 40$ kΩ, $R_2 = 100$ kΩ, $R_S = 200$ Ω, and MOSFET parameters: $V_{TH} = 0.5$ V, $\mu_n C_{ox} = 100\ \mu A/V^2$, $W/L = 50$, and $\lambda = 0$. Calculate the maximum allowable value of R_D for the MOSFET to remain on the edge of saturation. Assume $V_{GS} = 3$ V.

Solution to Example 6.15

From inspection of Figure 6.50, we can state that

$$V_{GG} = V_{GS} + I_D R_{DS}. \tag{6.128}$$

In Equation (6.128) V_{GG} is the gate voltage to ground. Since the gate current is negligible

$$V_{GG} = \frac{R_2}{R_1 + R_2}V_{DD} = 7.14\text{ V}. \tag{6.129}$$

And since

$$I_D = \frac{1}{2}\mu_n C_{ox}\frac{W}{L}(V_{GS} - V_{TH})^2, \tag{6.130}$$

using the values provided by the example, Equation (6.130) yields

$$I_D = 15.63\ \text{mA}.$$

The condition for the MOSFET to be on the edge of saturation is

$$V_{DS} = V_{GS} - V_{TH},$$

and since $V_{GS} = 3$ V and $V_{TH} = 0.5$ V, it yields

$$V_{DS} = 2.5\ \text{V}.$$

By inspection of Figure 6.50, we see that $V_{DD} = R_D I_D + V_{DS} + R_S I_D$. Using the given values in the above equation, $10 = R_D\, 0.01563 + 2.5 + 200\, 0.01563$. From the above equation we find the value of R_D to be

$$R_D \cong 280\ \Omega.$$

6.4.4 Common Source (CS) Configuration

The MOSFET CS configuration is very similar to the BJT common emitter configuration. The gain of this circuit turns out to be $-g_m R_D$ which is basically the same expression given for the common emitter. Figure 6.52 depicts a common source amplifier and its small signal model. Figure 6.52b shows that the channel length modulation coefficient $\lambda = 0$, thus $r_o \to \infty$, calculation of the voltage gain leads to

$$G_{CS} = -g_m R_D \tag{6.131}$$

ignoring channel length modulation.

Taking into account channel length modulation, that is, finite and nonzero r_o, the voltage gain becomes

$$G_{CS} = -g_m(\mathrm{R_D} // r_o). \tag{6.132}$$

It is important and interesting to observe that for the CS stage R_{in} approaches infinity,

$$R_{in} \to \infty \tag{6.133}$$

at mid and low frequencies of operation.

The CS stage output impedance is $R_{out} = R_D$ ignoring channel length modulation, or $R_{out} = R_D // r_o$ do taking into account channel length modulation.

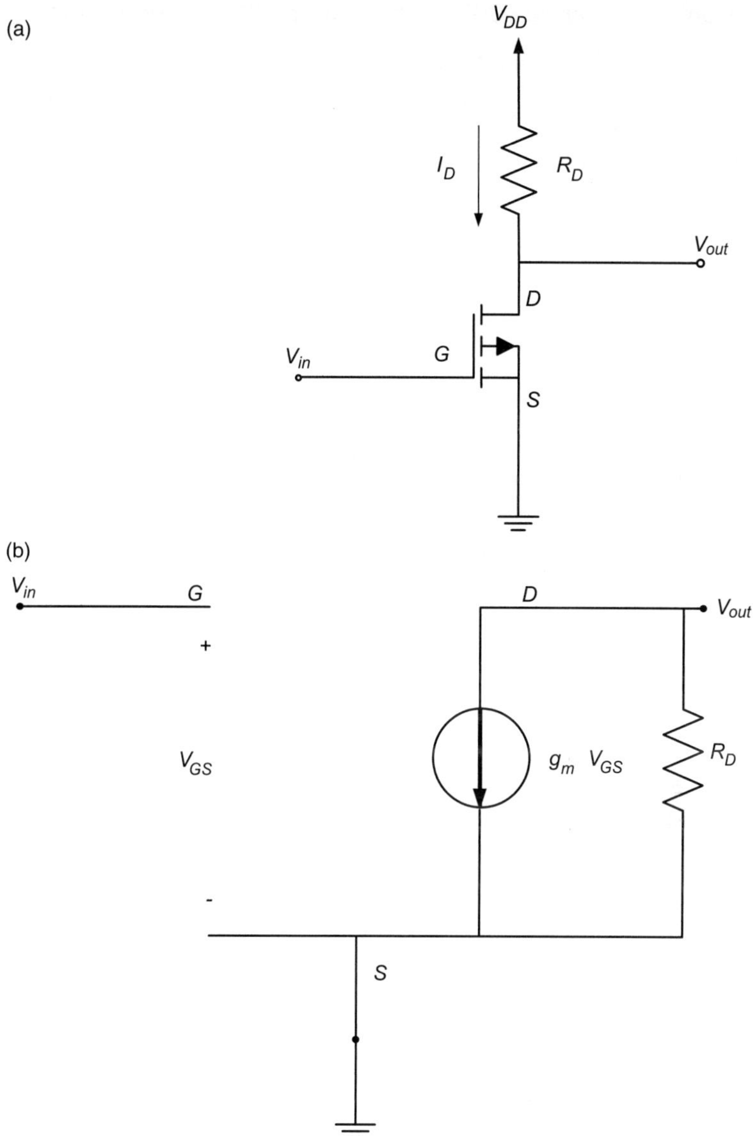

Figure 6.52 MOSFET CS amplifier: (a) MOSFET-based circuit; (b) small signal equivalent model at low and mid frequencies.

6.4.5 Common Source (CS) Configuration with Degeneration

The source terminal degeneration resistor has the same effect in the MOSFET amplifier as it does in the BJT. Figure 6.53 depicts a CS amplifier with emitter degeneration resistor.

From the circuit of Figure 6.53b we have

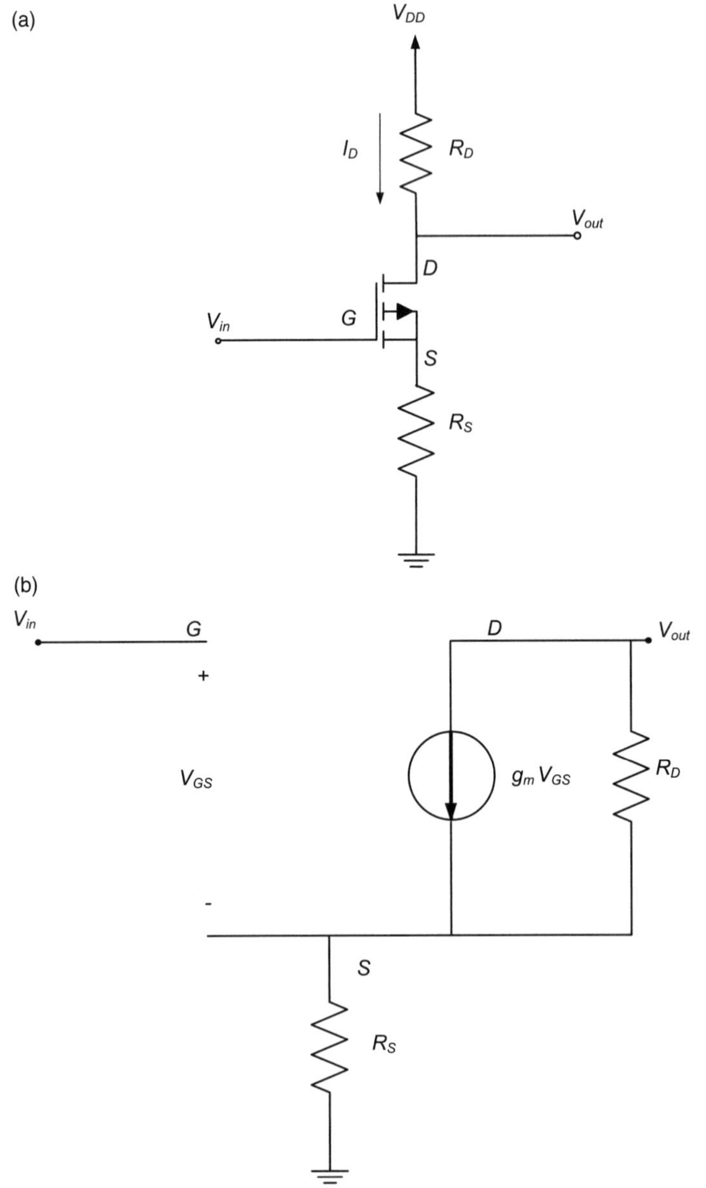

Figure 6.53 CS amplifier with source degeneration: (a) MOSFET-based circuit; (b) small signal equivalent model.

$$V_{in} = V_{GS} + g_m V_{GS} R_S \tag{6.134}$$

Thus,

$$V_{GS} = \frac{V_{in}}{1 + g_m R_S}. \tag{6.135}$$

Since current $g_m V_{GS}$ flows through resistor R_D we have that

$$V_{out} = -g_m V_{GS} R_D \tag{6.136}$$

and

$$G_{CS\ \text{with source degeneration}} = \frac{V_{out}}{V_{in}} = -\frac{g_m R_D}{1 + g_m R_S} = -\frac{R_D}{\frac{1}{g_m} + R_S}. \tag{6.137}$$

The reader is encouraged to compare MOSFET Equation (6.137) with the bipolar transistor expression given by Equation (6.76).

6.4.6 Common Gate (CG) Configuration

The MOSFET CG configuration resembles the BJT CB topology. Looking at the circuit of Figure 6.54, the circuit virtually operates like the BJT-based CB. It can easily be seen that the voltage gain of the CG topology is

$$G_{v\text{-}CG} = g_m R_D \tag{6.138}$$

Note that the gain for this topology does not have an inverting sign.

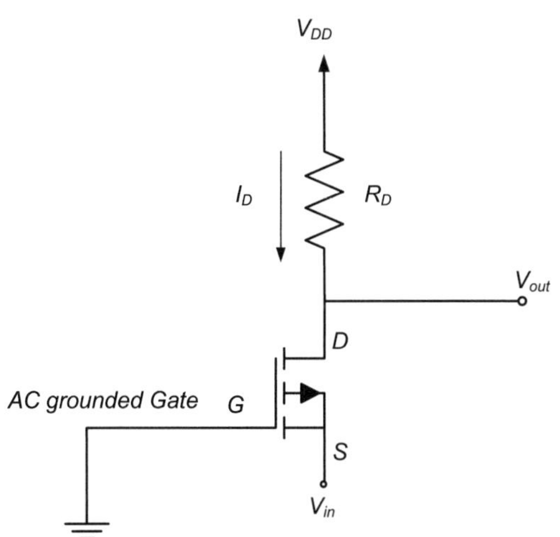

Figure 6.54 MOSFET CG topology.

Let us look at the input and output impedances of the *CG* topology. Neglecting the channel length modulation effect we come up with the small signal equivalent model and apply a voltage V_{Test} at the *source* input, the gate is AC grounded and we find R_{in} of the stage. Similarly we apply V_{Test} at the *drain* output of the stage with both the *gate* and the *source* grounded. This is virtually identical to what we did with the BJT-based CB configuration. Both circuits used to calculate R_{in} and R_{out} are shown in Figure 6.55a,b respectively.

As expected, due to the similarity with the bipolar-based CB circuit, we obtain

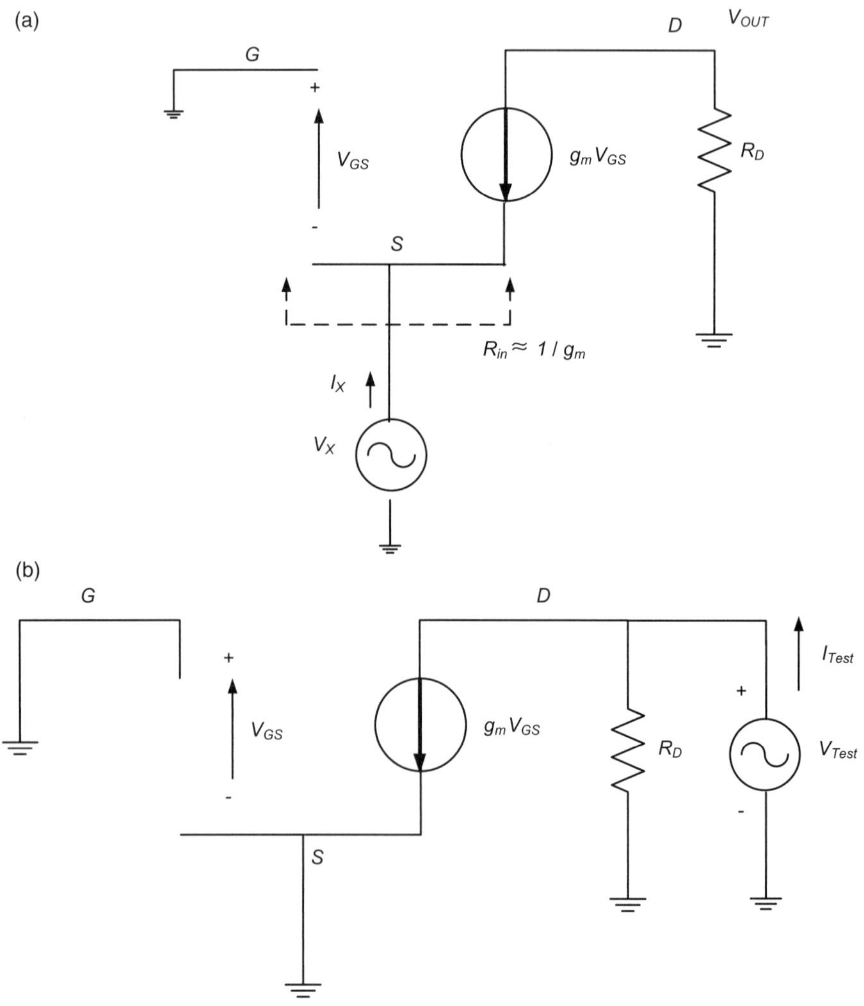

Figure 6.55 (a) CG small signal model to calculate R_{in}; (b) CG small signal model to calculate R_{out}.

$$R_{in} = \frac{1}{g_m} \tag{6.139}$$

and

$$R_{out} = R_D. \tag{6.140}$$

Moreover, when the input voltage applied at the source has a source resistance in series, the gain voltage gain stage is

$$G_{CG\text{ with source resistance}} = \frac{R_D}{\frac{1}{g_m} + R_S}. \tag{6.141}$$

6.4.7 Common Drain (CD) Configuration or Source Follower

The source follower amplifier receives the input signal at the *gate* terminal, and it senses the output signal at the *source* terminal. The *drain* is grounded for AC signals.

The voltage gain of this stage can be derived by inspection of the circuits in Figure 6.56. As expected, the voltage gain of the source follower is by similarity with the BJT follower equal to

$$\frac{V_{out}}{V_{GS}} = g_m(R_S // r_o) \tag{6.142}$$

$$V_{in} = V_{GS} + V_{out} \tag{6.143}$$

$$G_{CD} = \frac{V_{out}}{V_{in}} = \frac{g_m(R_S // r_o)}{1 + g_m(R_S // r_o)} \tag{6.144}$$

6.4.8 Other MOSFETs: Enhancement Mode *p*-Channel and Depletion Mode (*n*-Channel and *p*-Channel)

The enhancement mode *p-channel* MOSFET, also called a PMOS transistor is fabricated on an *n-type* substrate or body. Heavily doped *p*+ regions are created in the substrate to form the *drain* and the *source*. The PMOS device operates just like the NMOS, but some important differences exist. The PMOS transistor operates with negative V_{GS} and V_{DS}. The threshold voltage V_{TH} is negative. The current I_D enters the source terminal and leaves through the drain terminal. A *p-channel* is induced when

$$|V_{DS}| > |V_{GS}| - |V_{TH}| \tag{6.145}$$

The drain saturation current for the PMOS or enhancement mode *p*-channel MOSFET is

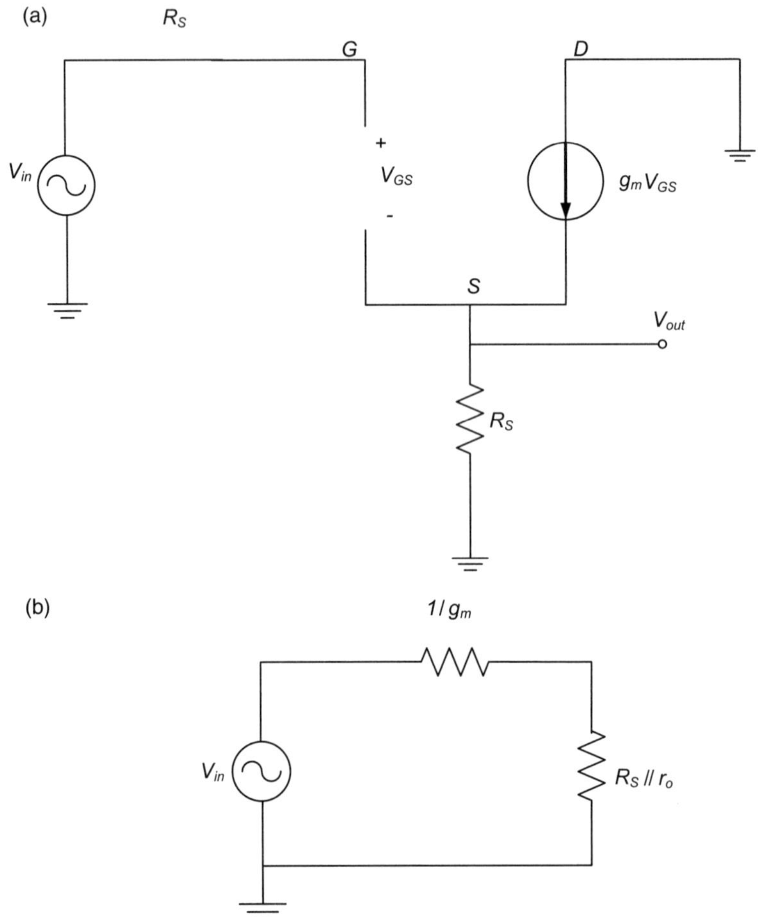

Figure 6.56 Circuits used to calculate the source follower gain and it includes the effects of channel length modulation ($\lambda > 0$).

$$I_{D,Sat} = -\frac{1}{2}\mu_p C_{ox}\frac{W}{L}[2(V_{GS} - V_{TH})V_{DS} - V_{DS}^2] \tag{6.146}$$

μ_p in Equation (6.146) is the majority carriers (holes) mobility. Equation (6.146) also assumes that the channel length modulation factor λ is zero.

Now taking into account a nonzero λ channel length coefficient, the PMOS transistor drain current becomes

$$I_{D,Sat} = -\frac{1}{2}\mu_p C_{ox}\frac{W}{L}[2(V_{GS} - V_{TH})V_{DS} - V_{DS}^2](1 + \lambda V_{DS}) \tag{6.147}$$

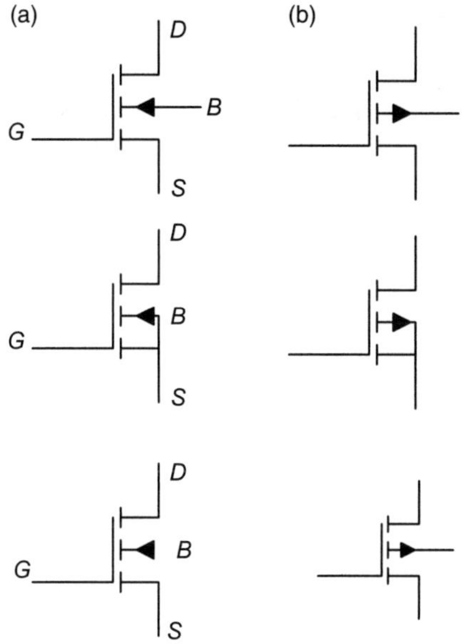

Figure 6.57 Enhancement mode MOSFET symbols: (a) *n-channel*; (b) *p-channel*.

Similarly to the NMOS transistor, the PMOS transistor has a triode region current:

$$I_{D,triode} = -\frac{1}{2}\mu_p C_{ox}\frac{W}{L}[2(V_{GS} - V_{TH})V_{DS} - V_{DS}^2] \tag{6.148}$$

The *p-channel* MOSFET has holes as charge carriers. Remember that the *n-channel* MOSFET has electrons as charge carriers. Symbols for all four MOSFET types listed in Table 6.4 are shown in Figures 6.57 and 6.58. Different authors use slightly different schematic symbols. Figures 6.57 and 6.58 address some of the most common symbols used.

Note in Figure 6.59 the equation for the saturation *drain* current as a function of V_{GS} is also quadratic like the $I_D - V_{GS}$ curve for the NMOS transistor, but it is rotated 180° around the current axis; refer to Figure 6.57b. The reason is that PMOS transistors have negative V_{TH} and negative V_{GS}. Strictly speaking, the current drawn should be negative, but for simplicity of the graphic representation it is not; that is, I_D is drawn as a positive current as it is most commonly done in the MOSFET literature.

All equations for NMOS transistors are applicable to PMOS transistors provided that the *electron mobility* (μ_n) used for *n-channel* devices is replaced with the *hole mobility* (μ_p) for *p-channel* devices. Additionally and hopefully to reduce confusion between positive and negative voltages, one can simply

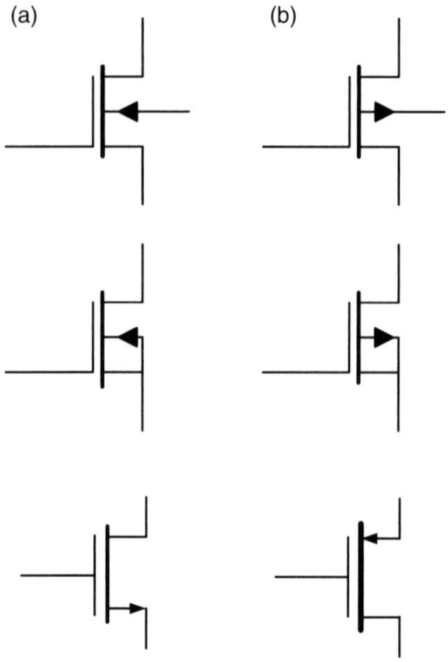

Figure 6.58 Depletion mode MOSFET symbols: (a) *n-channel*; (b) *p-channel*.

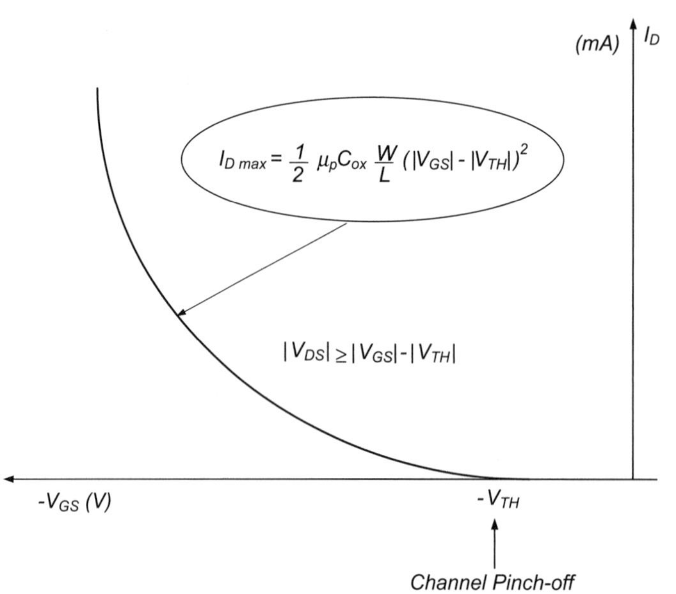

Figure 6.59 Depicts the I_D-V_{GS} transfer curve characteristic of the PMOS transistor.

take the absolute values of voltages V_{DS}, V_{GS}, and V_{TH}, so that the equations basically look the same. The reader is encouraged to rewrite the *n-channel* enhancement mode equations for I_D in both the triode and the saturation regions for the *p-channel* enhancement mode device. In integrated circuits NMOS and PMOS transistors are used, such technology is referred to as Complementary MOS technology (CMOS). CMOS is the most prevalent technology of integrated circuits at the time of this writing. Unfortunately, CMOS technology is well beyond the scope of this book.

Table 6.6 summarizes NMOS and PMOS transistors characteristics.

6.4.8.1 Depletion Type MOSFETs For the sake of completeness we address depletion type MOSFETs; however, they are not as commonly used as enhancement-type devices are. The fundamental difference between the enhancement and the depletion device is that the depletion device does not need a *gate* voltage to induce a channel. Depletion type devices have a *physically implanted channel.* So, for example, when dealing with an *n-channel* depletion device, it just takes a positive V_{DS} voltage to be applied with $V_{GS} = 0$, and the device will conduct current through the implanted channel. Again the channel is not induced like it is for the enhancement-type device. Figure 6.60 depicts the I_D versus V_{GS} transfer characteristic for an *n-channel* depletion device and for an *n-channel* enhancement device. In order not to overlap both curves, the absolute values of the threshold voltages are assumed to be different.

Figure 6.61 depicts the I_D versus V_{GS} transfer characteristic for a *p-channel* depletion device and for a *p-channel* enhancement device.

6.5 SUMMARY

This chapter is quite long and covers key electronic devices from the bottom up. It has been the intent of the author not to cover a great deal of semiconductor physics, but just enough of it to understand circuit-level operation of diodes, bipolar, and MOS transistors. Some of the most important applications with diodes were covered. Biasing and the most common amplifiers configurations were addressed with bipolar and MOS transistors. Because of the similarities of some bipolar and MOSFET-based amplifiers, the MOSFET material heavily relies on having done circuit equations and derivations with the bipolar junction transistor examples. The junction field effect transistor or JFET was not covered because of space reasons. The JFET was a predecessor of the MOSFET. MOSFETs are more heavily used than any other transistor, including bipolars. NMOS and PMOS transistors are used in CMOS technology, the dominating IC technology at this time.

Table 6.6 NMOS and PMOS transistors characteristic

Enhancement Mode *n*-channel (NMOS) characteristics		
1	Drain current for the triode region (no channel length modulation)	$I_{D,\text{triode}} = \frac{1}{2}\mu_n C_{ox}\frac{W}{L}[2(V_{GS}-V_{TH})V_{DS}-V_{DS}^2]$
2	Drain current for the triode region for V_{DS} <<< 2 $(V_{GS} - V_{TH})$ (no channel length modulation)	$I_{D,\text{triode}} \approx \mu_n C_{ox}\frac{W}{L}(V_{GS}-V_{TH})V_{DS}$
3	Drain current at the beginning of the saturation region for: $V_{DS} = V_{GS} - V_{TH}$	$I_{D,\text{Max}} = \frac{1}{2}\mu_n C_{ox}\frac{W}{L}(V_{GS}-V_{TH})^2$
4	Drain current in the saturation region including channel length modulation.	$I_{D,\text{Sat}} = \frac{1}{2}\mu_n C_{ox}\frac{W}{L}(V_{GS}-V_{TH})^2(1+\lambda \text{V}_{\text{DS}})$
5	R_{DSon} For $V_{DS} << 2\ (V_{GS} - V_{TH})$	$R_{DSon} = \frac{1}{\mu_n C_{ox}\frac{W}{L}(V_{GS}-V_{TH})}$ (see [a])
6	Turn-on and turn-off conditions	If: $V_{GS} < V_{TH}$, then $I_D = 0$ regardless of the value of V_{DS} If: $V_{GS} > V_{TH}$, then $I_D > 0$
	Threshold voltage V_{TH} (is a positive quantity for NMOS)	
7	Transconductance g_m	$g_m = \mu_n C_{ox}\frac{W}{L}(V_{GS}-V_{TH}) = \sqrt{2\mu_n C_{ox}\frac{W}{L}I_D}$
Enhancement Mode *p*-channel (PMOS) characteristics		
1	Drain current for the triode region (no channel length modulation)	$\text{I}_{D,\text{triode}} = -\frac{1}{2}\mu_p C_{ox}\frac{W}{L}[2(V_{GS}-V_{TH})V_{DS}-V_{DS}^2]$
2	Drain current for the triode region for V_{DS} <<< 2 ($V_{GS} - V_{TH}$) (no channel length modulation)	$I_{D,\text{triode}} \approx -\mu_p C_{ox}\frac{W}{L}(V_{GS}-V_{TH})V_{DS}$
3	Drain current at the beginning of the saturation region for: $V_{DS} = V_{GS} - V_{TH}$	$I_{D,\text{Max}} = -\frac{1}{2}\mu_p C_{ox}\frac{W}{L}(V_{GS}-V_{TH})^2$
4	Drain current including channel length modulation)	$I_{D,\text{Sat}} = \frac{1}{2}\mu_p C_{ox}\frac{W}{L}(V_{GS}-V_{TH})^2(1-\lambda V_{DS})$
5	r_{DSon} For $V_{DS} << 2\ (V_{GS} - V_{TH})$	$r_{Dson} = \frac{1}{\mu_p C_{ox}\frac{W}{L}(V_{GS}-V_{TH})}$ (see [a])
6	Turn-on and turn-off conditions	If $V_{GS} < V_{TH}$, $I_D = 0$ regardless of the value of V_{DS}
	Threshold voltage V_{TH} (is a negative quantity for PMOS)	Turn on condition: $V_{GS} > V_{TH}$, $I_D < 0$
7	Trans-conductance g_m	$g_m = \mu_p C_{ox}\frac{W}{L}(V_{GS}-V_{TH}) = \sqrt{2\mu_p C_{ox}\frac{W}{L}I_D}$

[a] Note that since electron mobility is larger than hole mobility, for a given oxide thickness C_{ox} and aspect ratio W/L an *n*-channel (NMOS) r_{DSon} is smaller than a *p*-channel (PMOS) r_{DSon}, under the same voltage conditions.

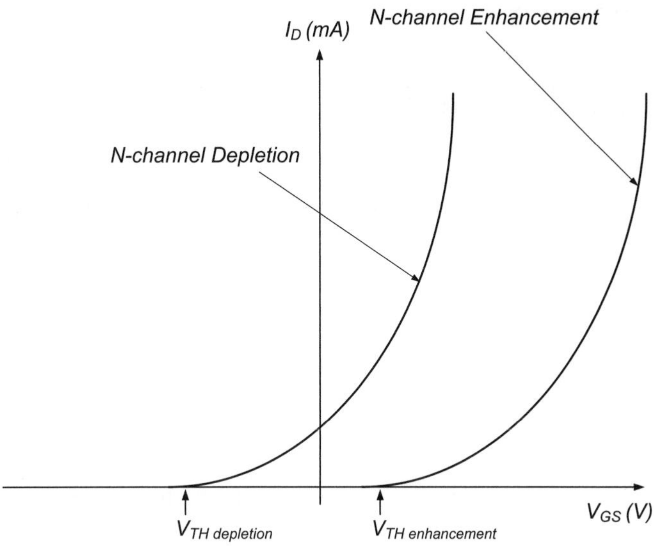

Figure 6.60 *n-Channel* depletion and enhancement MOS transistors operating in saturation.

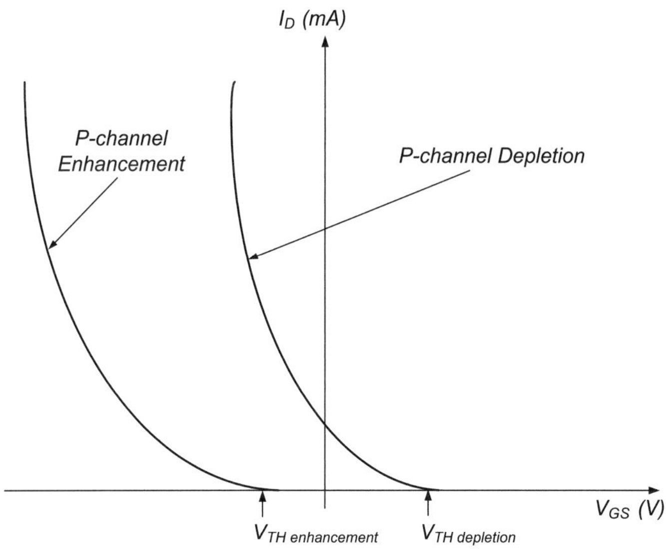

Figure 6.61 *p-Channel* depletion and enhancement MOS transistors operating in saturation.

FURTHER READING

1. Robert F. Pierret, *Semiconductor Fundamentals*, Addison-Wesley Publishing Company, Reading, MA, 1983.
2. T. E. Price, *Introduction to VLSI Technology*, Prentice Hall International (UK) Ltd, Reading, MA, 1994.
3. Behzad Razavi, *Fundamentals of Microelectronics*, John Wiley & Sons, Inc., Hoboken, NJ, 2008.
4. Adel S. Sedra and Kenneth C. Smith, *Microelectronic Circuits*, 5th ed., Oxford University Press, New York, 2007.
5. Jacob Millman and Christos C. *Halkias, Electronic Devices and Circuits*, McGraw Hill, Tokyo, 1967.

PROBLEMS

6.1 The circuit of Figure 6.62 find the following voltages and currents: (a) I_1, (b) I_2, (c) I_{D1}, (d) I_3, (e) I_{D2}, (f) V_{R1}, (g) V_{R2}, (h) V_{R3}. Assume the diode $D1$ forward voltage drop is 0.61 V and $D2$'s forward drop is 0.53 V.

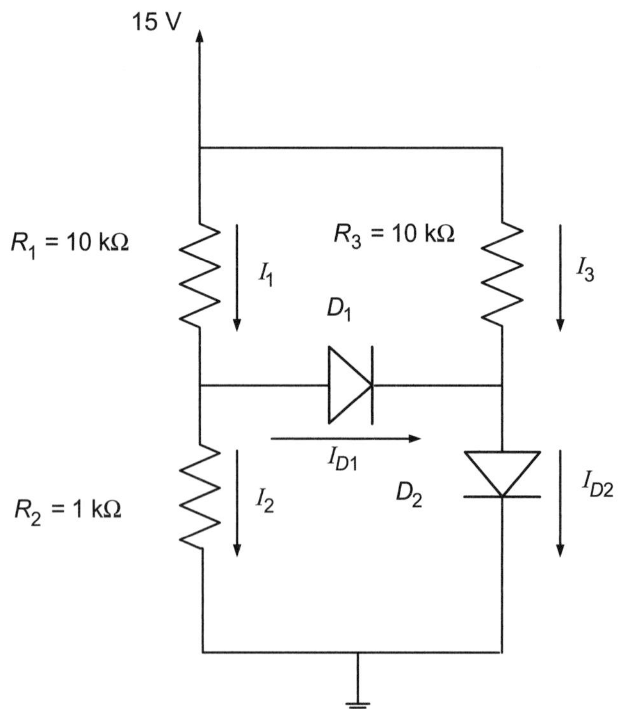

Figure 6.62 Circuit for Problem 6.1.

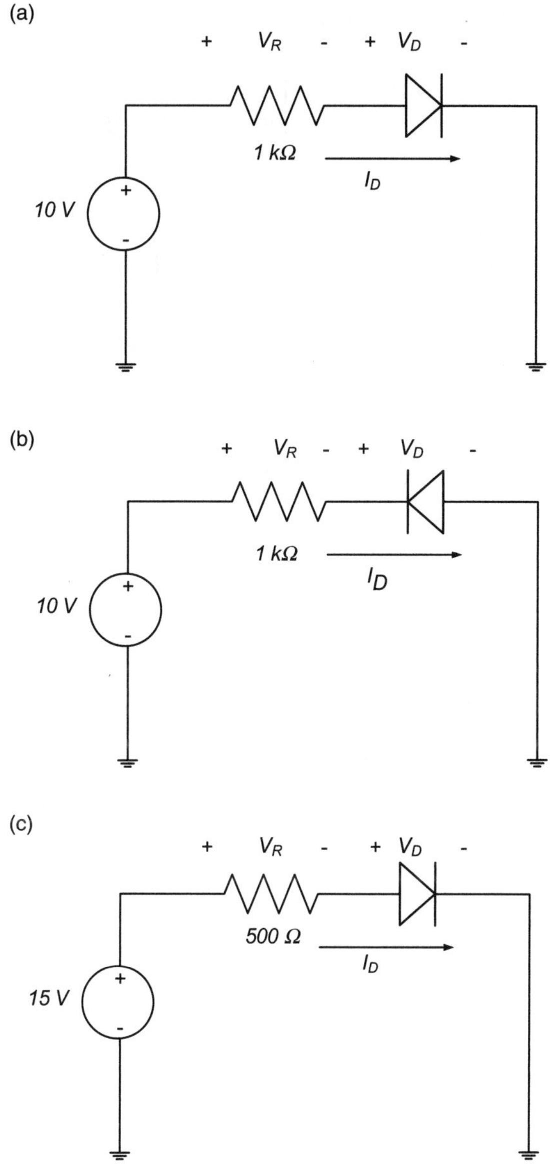

Figure 6.63 Circuits for Problem 6.2.

6.2 Given the circuits of Figure 6.63, assuming all diodes are ideal, determine the voltages V_D, V_R, and currents I_D indicated for circuits (a), (b), and (c).

6.3 For the circuit of Figure 6.64, the square wave V_g is applied to circuit as shown. Assuming that the diodes and op amp are real devices, draw the following waveforms: (a) V_g, (b) I_R, (c) I_{D1}, (d) I_{D2}, (e) V_{in}, and *(f)* V_{out}. Hint: Assume the diode forward drop is 0.6 V and the op amp saturates at ±13 V.

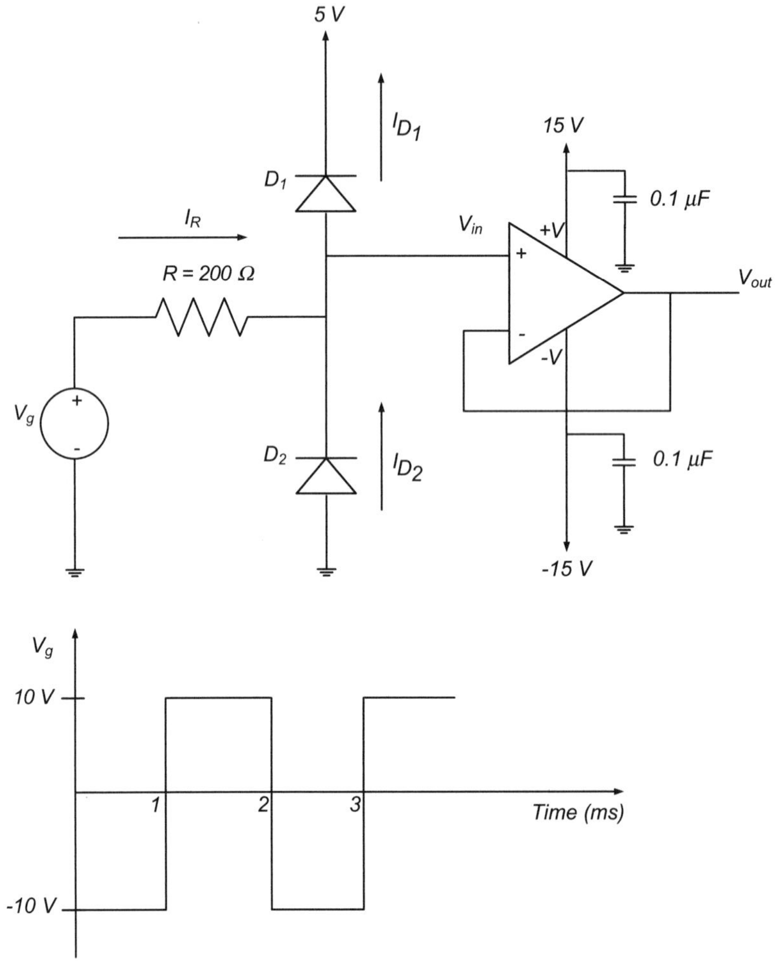

Figure 6.64 Circuits for Problem 6.3.

6.4 For the circuit of Figure 6.65, draw the following waveforms: (a) V_{sig}, (b) I_R, (c) I_{D1}, (d) I_{D2}, (e) V_{in}, and (f) V_{out}. Assume that the op amp is ideal and assume that the diodes forward voltage drop is 600 mV.

6.5 Assume that you have a 2N3904 NPN transistor. Using a biasing circuit topology such as the one presented in Figure 6.26, assume your V_{CC} supply is 10 V. Find the resistor values for R_B and R_C to bias the transistor with a collector current of 10 mA and a V_{CE} of 5 V at a 25°C ambient temperature. Assuming that the resistors have a zero ppm/°C temperature coefficient, and that the 10 V supply does not change due to temperature variations, find: (a) V_{CE} and I_C at –55°C, and (b) V_{CE} and I_C at 123°C. Hint: β, the DC current gain is given as by h_{FE} on the

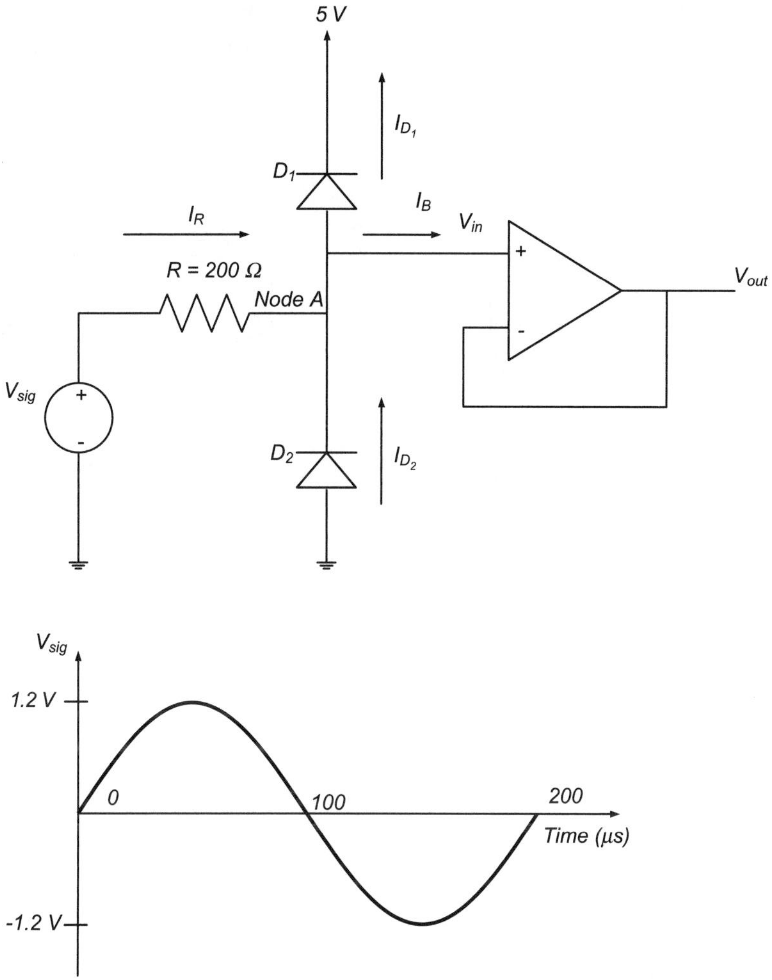

Figure 6.65 Circuits for Problem 6.4.

datasheet. Search for National's, or Fairchild's 2N3904 datasheet to find h_{FE}.

6.6 Assume that you have a 2N3904 NPN transistor. Using a biasing circuit topology such as the one presented in Figure 6.26, assume your V_{CC} supply is 10 V. Find the resistor values for R_B and R_C to bias the transistor with a collector current of 10 mA and a V_{CE} of 5 V at a 25°C ambient temperature. Assuming that the resistors have a 200 ppm/°C temperature coefficient, and that the 10 V supply does not change due to temperature variations, find: (a) V_{CE} and I_C at –55°C, and (b) V_{CE} and I_C at 123°C. Hint: β, the DC current gain is given as h_{FE} on a Fairchild

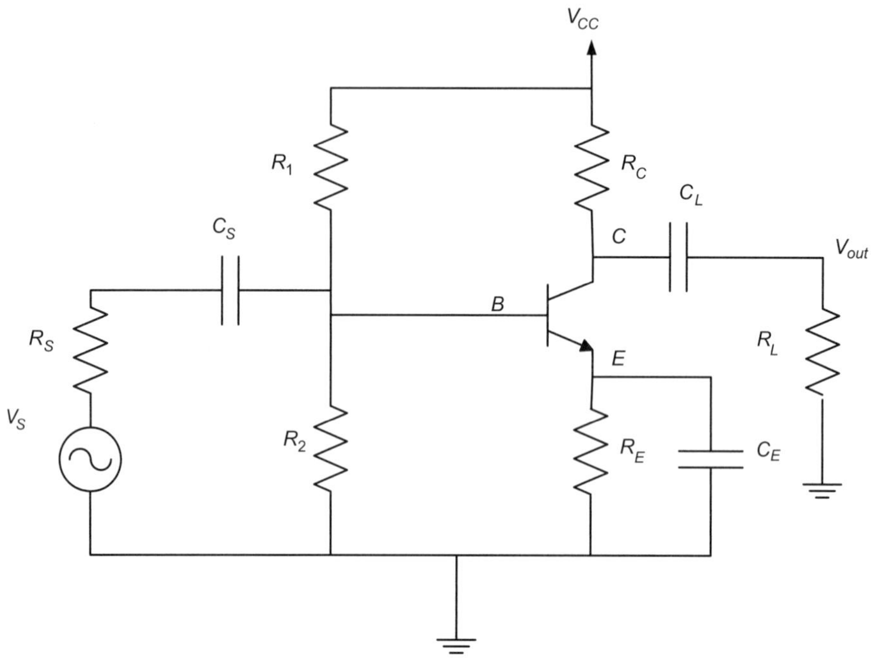

Figure 6.66 Circuit for Problems 6.8, 6.9, 6.10 and 6.11.

Semiconductors bipolar transistor datasheet. Practically speaking β, the DC current gain and h_{FE}, an H-model parameter, are interchangeable.

6.7 Repeat Problem 6.6 using the circuit biasing topology of Figure 6.23.

6.8 Using the circuit topology of Figure 6.66, assume the following values: $V_{CC} = 15$ V, $V_S = 1$ m V peak sine-wave, with 0 DC offset, and 1 kHz frequency, $R_1 = 140$ kΩ, $R_2 = 140$ kΩ, $R_C = 100$ Ω, $R_E = 50$ Ω, $C_S = 10$ μF, $C_L = 10$ μF, and $R_L = 8$ Ω. Assume a transistor $\beta = 300$ and $V_{BE} = 0.74$ V; determine: (a) the transistor operating point Q, that is, V_{CE}, I_B, I_C, and I_E, and (b) the small signal voltage gain of the amplifier circuit.

6.9 Using the circuit of Figure 6.66 calculate the amplifier input impedance. Refer to Figure 6.66.

6.10 Using the circuit of Figure 6.66 calculate the amplifier output impedance.

6.11 Using the circuit of Figure 6.66, establish all the effects that short circuiting the R_E and C_E parallel combination has on: (a) the biasing point of the transistor, (b) the small signal gain of the amplifier. Hint: for part (a) determine the new values of V_{CE}, I_B, I_C, and I_E.

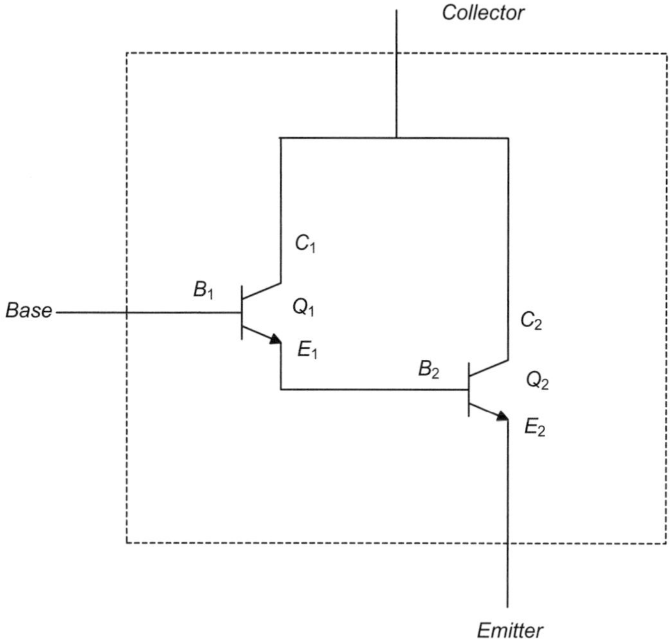

Figure 6.67 Circuit for Problems 6.12 and 6.13.

6.12 The two-transistor circuit depicted by Figure 6.67 is called a Darlington pair configuration. (a) Knowing that the DC current gains for Q_1 and Q_2 are respectively β_1 and β_2 determine the overall DC current gain of the Darlington pair. (b) Determine the Darlington pair base-emitter voltage drop when the pair is in the active region.

6.13 Derive an equivalent Darlington pair configuration using two PNP bipolar transistors. Refer to Figure 6.67.

6.14 The circuit depicted by Figure 6.68 can be used to drive an LED. Let us assume that the base will be driven with a square wave that switches between 0 V and 5 V. Knowing that the LED reaches maximum brightness for 10 mA, determine: (a) the value of resistor $R_{LED,}$ and (b) a reasonable value for resistor R_{BASE}.

Assume that you are asked to use a 2N3904 NPN transistor.

6.15 For the circuit of Figure 6.69, assume $R_1 = R_2 = 550$ kΩ, $R_D = R_S = 4$ kΩ, $V_{DD} = 12$ V. (a) Calculate the drain current I_D and (b) determine in which region the MOSFET operates in. Hint: assume the MOSFET is in saturation and validate this condition, else assume the MOSFET is in its triode region and validate its operation. (c) if the value of resistor R_2 is changed to 0 Ω, determine without using any equations, the operating

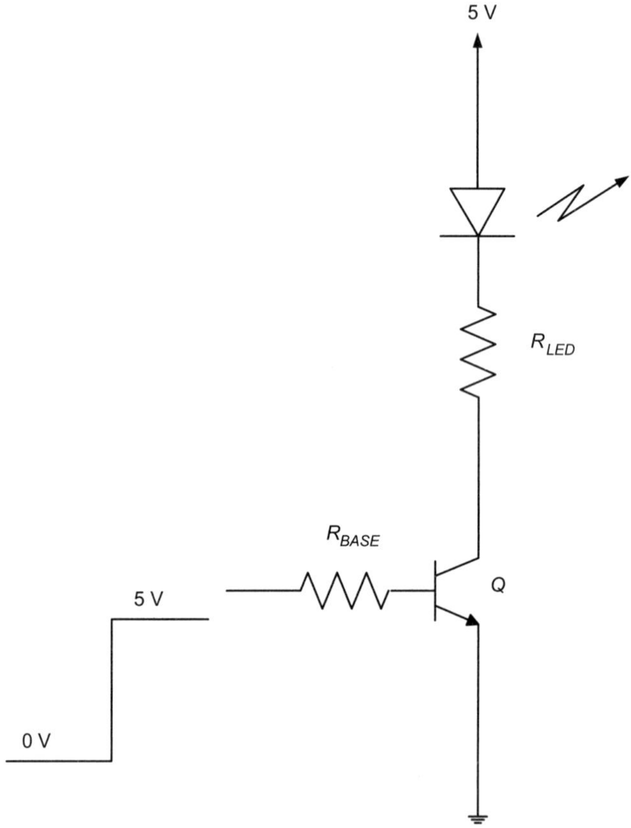

Figure 6.68 Circuit for Problem 6.14.

region of the MOSFET. Assume the MOSFET has $V_{TH} = 2$ V, $\mu_n C_{ox} =$ 100 μA/V^2, $W/L = 50$ and $\lambda = 0$.

6.16 For the circuit of Figure 6.70, calculate the appropriate values for R_D and R_S to keep the transistor *at* $I_D = 1$ mA and $V_D = 2$ V. Assume the following MOSFET parameter:

$$V_{TH} = 2 \text{ V}.$$

6.17 For the circuit of Problem 6.16 determine the DC gate current.

6.18 For the MOSFETs of Figure 6.71 determine the transistor region of operation. Assume that: $V_{TH} = 1.5$ V.

6.19 For the MOSFETs of Figure 6.72 determine the transistor region of operation. Assume that: (a) $\mu_p C_{ox}(W/L) = 100$ μA/V^2, (b) $V_{TH} = -0.4$ V, and (c) $\lambda = 0$.

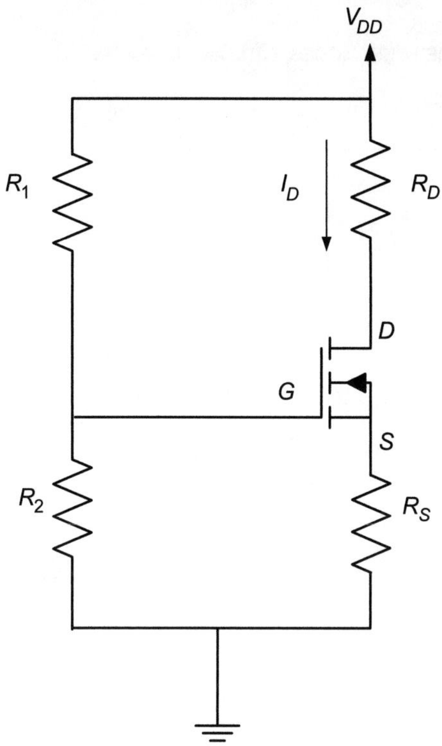

Figure 6.69 Circuit for Problem 6.15.

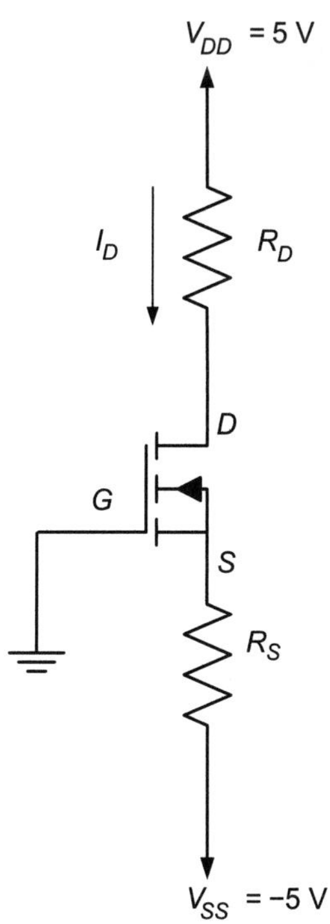

Figure 6.70 Circuit for Problem 6.16.

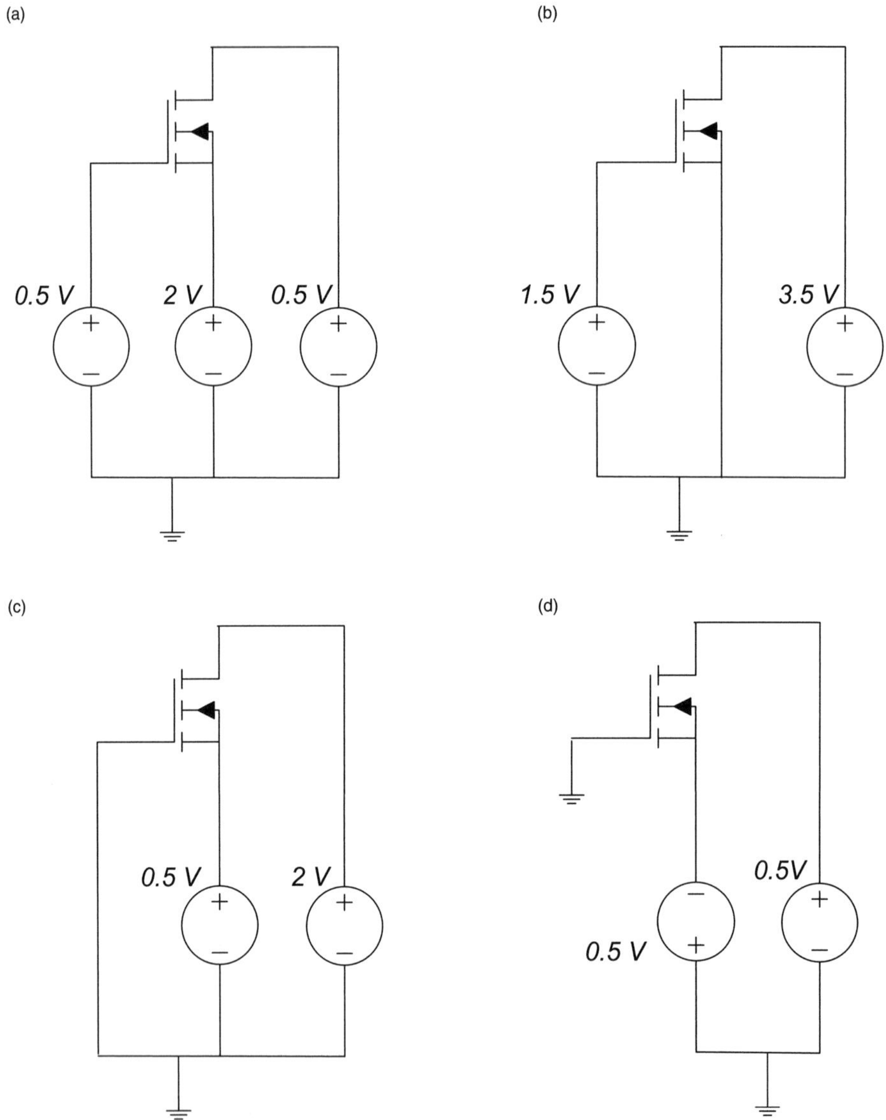

Figure 6.71 Circuits for Problem 6.18.

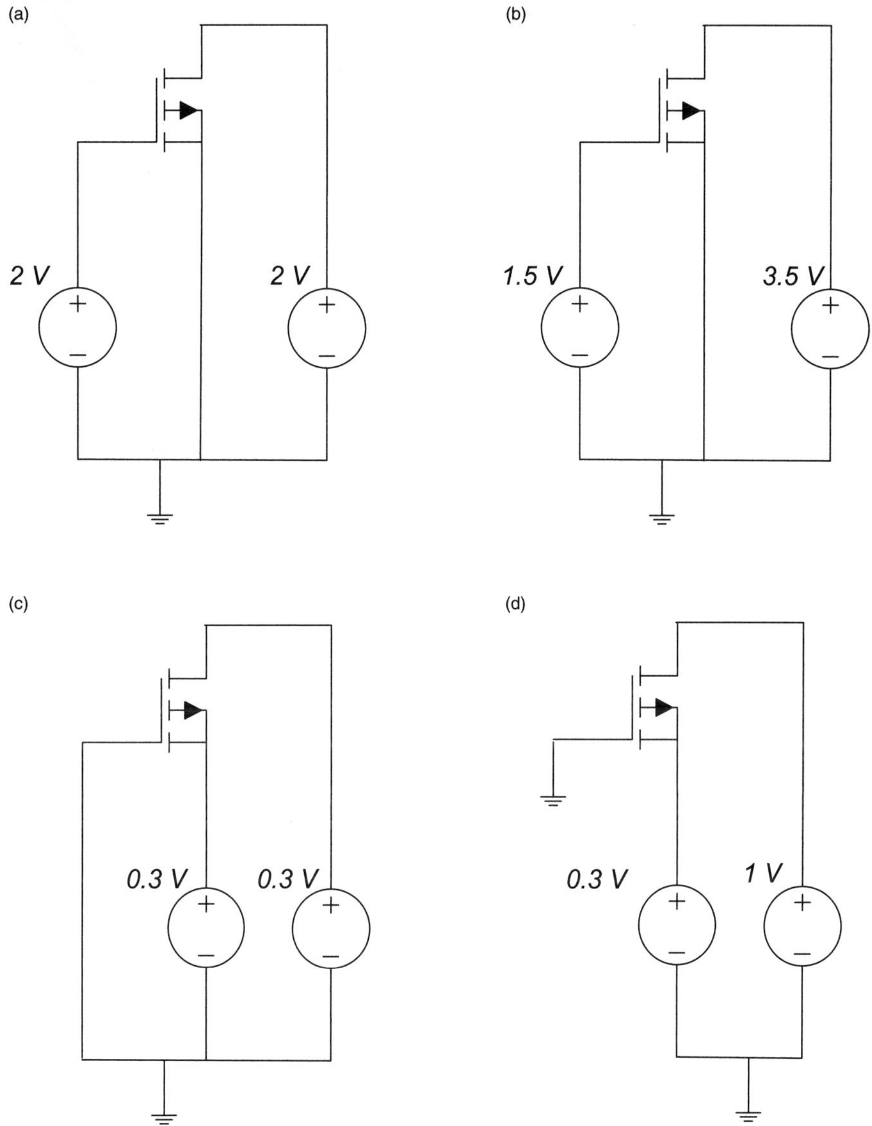

Figure 6.72 Circuits for Problem 6.19.

7

COMBINATIONAL CIRCUITS

7.1 INTRODUCTION TO DIGITAL CIRCUITS

Digital design is concerned with the design of digital electronic circuits. Digital circuits are required to handle just two voltage levels, a *true* level and a *false* level. Because these circuits handle two basic levels as opposed to infinitely many voltage levels as analog circuits do, they are more reliable. They last longer. Also, they are more consistent than analog circuits by repeatedly generating the same results under the same input conditions. The best-known digital system today is the computer; many computer-based products are manufactured today. The low cost, the reliability, the versatility of such circuits allows incorporating computers in virtually all intelligent products at the present time. Two main classes of digital circuits cover the world of digital design. The first is combinational circuits. They are digital circuits that produce outputs when the inputs are presented to them. Such circuits have no memory. The second kind of digital circuit is the sequential circuit, or those digital circuits that have memory capability. Combinational circuits will be the subject of Chapters 7 and 8. Sequential circuits will be addressed in Chapter 9.

7.2 BINARY NUMBERS: A QUICK INTRODUCTION

This chapter assumes that the reader has some knowledge about numbering systems, in particular binary and hexadecimal numbering systems. We will

Electrical, Electronics, and Digital Hardware Essentials for Scientists and Engineers, First Edition. Ed Lipiansky.

Table 7.1 List of all possible three-bit unsigned binary numbers with their decimal equivalents

3-Bit Binary Numbers (Base 2)	Decimal (Base 10)
000	0
001	1
010	2
011	3
100	4
101	5
110	6
111	7

move at a fast pace throughout this subject, hopefully not to bore anyone and at the same time present the needed fundamentals. A binary number is represented with two uniquely defined digits, *ones* and *zeros*. A binary digit is generically referred to as a *bit*, which stands for *binary term*. Any integer number can be represented with the appropriate number of ones and zeros.

Let us consider three-bit binary numbers first. If we exhaustively come up with all the binary combinations of three binary terms it is easy to see that the list in Table 7.1 contains all the possible binary combinations. In this chapter we will only address positive or unsigned binary numbers. In the next chapter we will cover positive as well as signed or negative numbers.

The algorithm to generate base 10 or decimal numbers is simple and we use it all the time, without even giving it a second thought. In base 10 we have 10 uniquely defined digits, 0 through 9. With those 10 digits we can write all possible integer numbers as long as we have the freedom of having enough digits to represent the largest number that we are interested in. For example, if we are asked to write all the possible 3-decimal digit integers; the first decimal number is $(000)_{10}$ while the largest one is $(999)_{10}$. We know that after $(000)_{10}$ comes $(001)_{10}$ then $(002)_{10}$ and so on until $(009)_{10}$. Now we ran out of uniquely defined digits so we reset the least significant decimal digit to zero and set to 1 the digit left to the rightmost decimal digit or the least significant decimal digit. We now compose $(010)_{10}$, then $(011)_{10}$, and so on until we reach $(019)_{10}$. This algorithm is repeated and we clearly know from grade school how to come up with all the 3-digit decimal numbers all the way up to $(999)_{10}$ or any other larger sequence of them.

Now if we want to do the same thing for a different base number, like for integer binary numbers, the algorithm is no different from what we already do with decimal numbers. The possibly "new" thing is that we only have two uniquely defined *bits*, so that we exhaust the use of each bit sooner than we do when dealing with the decimal numbering system. The sequence of all the possible 3-bit integer binary number was listed in Table 7.1. Note that the binary number 01100111 as an 8-bit number consists of eight bits; each bit position is referred to as: $b_7\ b_6\ b_5\ b_4\ b_3\ b_2\ b_1\ b_0$, where bit b_0 is referred to as the *least significant bit* or *LSB* and b_7 is the *most significant bit* or *MSB* of our 8-bit wide number.

So since $b_7\ b_6\ b_5\ b_4\ b_3\ b_2\ b_1\ b_0 = 01100111$, that means that every bit is weighted in the following fashion:

$$0\times2^7+1\times2^6+1\times2^5+0\times2^4+0\times2^3+1\times2^2+1\times2^1+1\times2^0$$
$$=64+32+4+2+1=(103)_{10}.$$

After computing the sum above the equivalent decimal number is $(103)_{10}$

$$=64+32+4+2+1=(103)_{10}.$$

Continuing with 8-bit binary integers, 8 bits span $2^8 = 256$ uniquely defined 8-bit binary combinations. The reader should convince herself that zero in 8-bit binary is: 0000_0000 and 255 is 1111_1111, the largest possible 8-bit unsigned binary integer. The underscores used to separate four-bit groups is simply to enhance the readability of the number. The reader not too familiar with binary sequences is encouraged to write down the complete binary sequence starting at $(0000\_0000)_2$ ending at $(1111\_1111)_2$.

Hexadecimal numbers have 16 uniquely defined symbols:

0, 1, 2,…, 9, A, B, C, D, E, F. Table 7.2 lists the first 16 hexadecimal (or hex) numbers, their binary and decimal equivalents.

Table 7.2 List of 16 uniquely defined hex digits, their binary and decimal equivalents

Hexadecimal (Hex, Base 16)	Binary (Base 2)	Decimal (Base 10)
0	0000	0
1	0001	1
2	0010	2
3	0011	3
4	0100	4
5	0101	5
6	0110	6
7	0111	7
8	1000	8
9	1001	9
A	1010	10
B	1011	11
C	1100	12
D	1101	13
E	1110	14
F	1111	15

Example 7.1 Represent the decimal number $(183)_{10}$ in binary and in hex.

Solution to Example 7.1

Initially we are not sure how many bits the number $(183)_{10}$ requires in binary representation. To sort of figure out the number of bits required, let us list the powers of two:

Note from Table 7.3 that the number $(128)_{10}$ can be represented in binary simply by writing:

$$(1000\_0000)_2 = (128)_{10}.$$

Similarly $128 + 64 = (192)_{10}$ is written as:

$$(1100\_0000)_2 = (192)_{10}.$$

Since we want to write the number $(183)_{10}$ in binary representation we know that *8* bits will suffice. To convert from decimal to binary, algorithmically we proceed as follows:

We divide $(183)_{10}$ by 2 to give an integer quotient of *91* and a remainder of ½. This process is repeated until the integer quotient becomes zero. We record all the operations as shown below:

Integer quotient		Remainder	Bit position	Weight
183/2 = 91	+	½	b0 = 1(LSB)	1
91/2 = 45	+	½	b1 = 1	2
45/2 = 22	+	½	b2 = 1	4
22/2 = 11	+	0	b3 = 0	8
11/2 = 5	+	½	b4 = 1	16
5/2 = 2	+	½	b5 = 1	32
2/2 = 1	+	0	b6 = 0	64
1/2 = 0	+	½	b7 = 1(MSB)	128

From above we conclude that $(183)_{10} = (1011\_0111)_2$.

Table 7.3 Some powers of two and bit position in a binary number

Powers of 2	Binary Bit Position
$2^0 = 1$	b_0
$2^1 = 2$	b_1
$2^2 = 4$	b_2
$2^3 = 8$	b_3
$2^4 = 16$	b_4
$2^5 = 32$	b_5
$2^6 = 64$	b_6
$2^7 = 128$	b_7

Now to convert $(183)_{10}$ to hex we simply translate each group of four bits into their hex equivalent starting with the LSB position according to Table 7.2, thus:

$$(183)_{10} = (1011\_0111)_2 = (B7)_{16}.$$

In Chapter 8 we will address some interesting ways of representing positive and negative binary numbers. This will be useful to design digital arithmetic circuits.

7.3 BOOLEAN ALGEBRA

In 1854, English mathematician and philosopher George Boole developed what is known today as Boolean algebra. Later on in 1938, American engineer and mathematician Claude Elwood Shannon also introduced a two-valued algebra he denominated switching algebra. Boolean algebra, also known as switching algebra, consists of binary variables and the logical operations among them. All logic variables that we will deal with have a binary value; that is, they can only take one out of two possible values, either 1 or 0, which we can associate with a true value and a false value, respectively, or vice-versa. Why do we need to deal with logic that only handles two values or two logic levels? Because it is easier and it is more reliable to develop, build, and use circuits that handle two values rather than circuits that handle infinitely many or many more values than just two. Circuits that handle infinitely many values are commonly referred to as analog circuits. Analog circuits are not as reliable, repeatable, and maintainable as digital circuits are.

The three most important logic operations are:

AND, OR, NOT

If we group these operators as follows: group (1) *AND, NOT* group (2) *OR, NOT* it is interesting to state that all possible binary or combinational functions, regardless of their length or complexity, can be implemented with just the operators of either group (1) or (2). We will come back to this concept once we study the fundamental logic rules and operations.

7.3.1 *AND* Logic Operation

Given a two-variable switching function $f(A, B)$, where A and B are binary-valued variables, function f can be exhaustively represented with the aid of a *truth table.*

Table 7.4 Truth table for the *AND* logic operator

Input A	Input B	Output *f(A,B)* = A B
0	0	0
0	1	0
1	0	0
1	1	1

Binary-Valued Functions: Example 7.2 Let *A* and *B* be two-valued binary independent variables. Let us assume that variable *A* means: *it-is-a-sunny-day* and variable *B* means: *the-soil-is-dry.* Further assume that when *A* is true it takes the value 1, when *B* is true it also takes the value 1. We want to come up with a binary-valued function *f* of variables *A, B* that is true (1) when both *A and B* are both true, else *f* is false (0). Table 7.4 explicitly and fully describes that requirement.

Now let us assume that the meaning of function *f* is: "turn-on-watering-system." So it seems intuitive to think that *f* = *A And'ed with B* is true when both *A* and *B* are true, else *f* is false. The logic symbol for the *AND* operator is a dot or the absence of it. For example:

$$f(\text{A, B}) = \text{A And-ed with B} = \text{A.B} = \text{AB}.$$

We will interchangeably place the *AND* "." (dot) or leave it out trying to make the logic expression more readable.

A true means: *it-is-a-sunny-day* while, *A* false means: NOT *it-is-a-sunny-day,* or grammatically more pleasant, false *A* means "it is not a sunny day."

This can be written in two ways: $\overline{it - is - a - sunny - day}$ = (it-is-a-sunny-day)'

Similarly B true means: "the-soil-is-dry" while B false means: NOT "the-soil-is-dry"

Finally, it is also important to see that function *f* is binary-valued as well; refer to Table 7.4.

Example 7.3 Derive the truth table for of three-input variable binary-valued function *g (A, B, C).* The *And-ing* of three or more variables does not change the significance of the *AND* operator. In general the *And-ing* of any number of binary-valued variables is true when the true value of each independent variable *(A, B, C, . . .)* is true. For all other cases our function is false. This is succinctly listed in Table 7.5.

7.3.2 *OR* Logic Operation (Also Called *Inclusive OR, or XNOR*)

Let us begin with a little more advanced logic function, the 3-variable *OR* or *inclusive OR.* Given the three binary-valued independent variables *A, B, C,* function *h(A,B,C)* = *A* + *B* + *C* is true if and only if any one or more

Table 7.5 Three-variable *AND* truth table

Input C	Input B	Input A	Output g (A, B, C) = A.B.C
0	0	0	0
0	0	1	0
0	1	0	0
0	1	1	0
1	0	0	0
1	0	1	0
1	1	0	0
1	1	1	1

Table 7.6 Three-variable *OR* truth table

Input C	Input B	Input A	Output h (A, B, C) = A + B + C
0	0	0	0
0	0	1	1
0	1	0	1
0	1	1	1
1	0	0	1
1	0	1	1
1	1	0	1
1	1	1	1

independent variables are true, else $h\ (A,B,C)$ is false. The "+" signs are not arithmetic signs, they represent the logic *OR* (or *inclusive OR*) operation. This *OR* is referred to as being inclusive because the output function is true not only whenever each independent variable is true, but also includes the case when one or more than one independent variable is true. We will cover shortly the two-variable exclusive *OR*, which requires that both independent variables have the opposite true value for the *exclusive OR* function to be true.

Table 7.6 depicts a three-variable inclusive OR function.

Exercise: Derive the truth table of a two-variable inclusive *OR* function.

7.3.3 NOT Logic Operation or Inversion—*NAND* and *NOR*

Inversion is the simplest of all logic operations. Given a binary-valued function f of an arbitrary number of independent binary-valued variables and its associated truth table, *NOT* f *or* $\bar{f}$ truth table is generated by changing function f output column *ones* with *zeros* and *zeros* with *ones*.

Table 7.7 Truth table for the *AND* & *NAND* logic operators

Input A	Input B	Output f(A,B) = AB	Output $\overline{f}(A,B)=\overline{AB}$
0	0	0	1
0	1	0	1
1	0	0	1
1	1	1	0

Table 7.8 Truth table for the *OR* & *NOR* logic operators

Input A	Input B	Output g(A,B) = A + B	Output $\overline{g}=\overline{A+B}$
0	0	0	1
0	1	1	0
1	0	1	0
1	1	1	0

Example 7.4 Let us define a function $\overline{f}=\overline{AB}$ which we will refer to as *A NAND'ed* with *B*. The letter "*N*" in the acronym "*NAND*" stands for negation or not.

Referring to our originally studied function $f = AB$, in Table 7.4, function $\overline{f}=\overline{AB}$ is simply annotated in *truth table* of Table 7.7.

In the above table, output *f* column, *f(A,B)* = *AB*, has been complemented bit-by-bit to form the column of our *NAND* function that is $\overline{f}=\overline{AB}$.

Example 7.5 Let us define a function $\overline{g}=\overline{A+B}$ which we will refer to as *A NOR'ed* with *B*. The letter "*N*" in the acronym "*NOR*" stands for negation or not. Table 7.8 presents the truth table of a two-variable *OR* under column g and *NOR*, under column $\overline{g}$.

7.3.4 Exclusive *OR* Logic Operation or *XOR*

The two-variable *XOR* is defined as true whenever an independent variable is true while the other one is false. Additionally when both variables have the same true value the *XOR* is false. Table 7.9 below presents the truth table of a two-variable *exclusive-or* function and the two-variable *equivalence* function.

Another way of defining the two-variable *XOR* is:

$$A\ XOR\ B = A \oplus B = A\overline{B} + \overline{A}B. \tag{7.1}$$

Table 7.9 Exclusive or and equivalence truth tables

Input A	Input B	Output $f(A,B) = A \oplus B =$ A XOR B	Output $\bar{f}(A,B) = \overline{A \oplus B} =$ A EQUIVALENCE B = A XNOR B
0	0	0	1
0	1	1	0
1	0	1	0
1	1	0	1

Equation (7.1) is interpreted from Table 7.9 as follows: A exclusive-or B is true whenever (A is true and B false) or (when A is false and B is true). The parentheses used in the previous sentence emphasize the precedence of the logical operations. Note that $A \oplus B$ is not true when both A and B have the same logic value. This is the reason why the *XOR* is referred to as an *exclusive-OR*, it excludes the cases for which both A and B have the same true value. (Two binary variables have the same true value whenever both are true or both are false.) Remember that $A + B$ (A inclusive-OR B) is true when both A and B are true in addition to being true when either only **A** or **B** is true.

The negation of $A \oplus B$ or $\overline{A \oplus B}$ is referred to as *A equivalence B* or *A XNOR B*. Refer one more time to Table 7.9. Similarly, $\overline{A \oplus B} =$ *A equivalence B* is defined as true whenever ($\bar{A}$ and $\bar{B}$ are true) or (A and B are true), else A equivalence B is false.

7.3.5 DeMorgan's Laws, Rules, and Theorems

DeMorgan's laws are the most powerful Boolean algebra rules. There are two of them. First we will state the two-variable laws, then we will present the generalized multivariable rules. Let A and B be two-valued independent variables, then

$$\text{Rule (1) } \overline{A+B} = \bar{A}.\bar{B} \tag{7.2}$$

$$\text{Rule (2) } \overline{A.B} = \bar{A}+\bar{B}. \tag{7.3}$$

It is appropriate to prove these rules and we will do that using truth tables.

We build Table 7.10 containing independent variables A and B, then we will generate columns corresponding to the following functions: $\bar{A}$, $\bar{B}$, $A + B$, $\overline{A+B}$, $\bar{A}\bar{B}$. Shall the column corresponding to $\overline{A+B}$ equal to column $\bar{A}\bar{B}$ we can affirm that Rule (1) holds.

Adding functions: $\overline{A.B}$, $(\bar{A}+\bar{B})$ under columns *viii* and *ix* we also prove Rule (2).

Looking at the results of Table 7.10 we observe that columns *vi* and *vii* are identical, this proves that: *Rule (1)* $\overline{A+B} = \bar{A}\bar{B}$ is true. Similarly, we observe that columns *viii* and *ix* are identical, thus proving *Rule (2)* $\overline{A.B} = \bar{A}+\bar{B}$.

Table 7.10 Truth table to prove 2-variable DeMorgan's rules

i	ii	iii	iv	v	vi	vii	viii	ix
A	B	$\bar{A}$	$\bar{B}$	A + B	$\overline{A+B}$	$\bar{A}\bar{B}$	$\overline{A.B}$	$(\bar{A}+\bar{B})$
0	0	1	1	0	1	1	1	1
0	1	1	0	1	0	0	1	1
1	0	0	1	1	0	0	1	1
1	1	0	0	1	0	0	0	0

Both Rules (1) and (2) can be generalized for *n* binary valued variables W_0, W_1, W_2, . . . W_{n-1}, where *n* is an integer.

Generalized DeMorgan Rule (1)

$$\overline{W_0 + W_1 + W_2 + \ldots + W_{n-1}} = \overline{W_0}.\overline{W_1}\,\overline{W_2}\ldots.\overline{W_{n-1}}. \qquad (7.4)$$

Generalized DeMorgan Rule (2)

$$\overline{W_0.W_1.W_2\ldots.W_{n-1}} = \overline{W_0} + \overline{W_1} + \overline{W_2} + \cdots + \overline{W_{n-1}}. \qquad (7.5)$$

Exercise: Prove using truth tables that DeMorgan's Rules (1) and (2) hold for four variables. Hint: Four variables will span 16 unique binary combinations.

7.3.6 Other Boolean Algebra Postulates and Theorems

We present in this section some other basics postulates and theorems used in Boolean algebra. Most of them are quite intuitive and a few others are not so intuitive. Postulates need not be proven, theorems generally are. Only some less intuitive theorems will be proven.

Assume that Table 7.11 uses binary-valued variables: *X, Y,* and *Z*.

Example 7.6 Using Boolean postulates and theorems, find the complement the following Boolean expressions. Reduce the expressions as much as possible.

(a) $f(X,Y) = \bar{X}Y + \bar{Y}X$

(b) $f(W, X, Y, Z) = (W\bar{X} + Y)\bar{Z}$

(a) We first apply DeMorgan's rules to the complement of equation (a) and then apply the distributive postulate to the complemented function, we obtain:

$$\bar{f} = XY + \bar{X}\bar{Y}.$$

Table 7.11 Boolean algebra postulates and theorems

	Part (1)	Part (2): Its Dual
Identity postulate	$X + 0 = X$	$X \,.\, 1 = X$
Idem-potent postulate	$X + \bar{X} = 1$	$X \,. X = X$
Neutral element postulate	$X + 1 = 1$	$X \,.\, 0 = 0$
Complements postulate	$X + \bar{X} = 1$	$X \,.\, \bar{X} = 0$
Distributive postulate	$X\,(Y + Z) = XY + XZ$	$X + YZ = (X + Y)(X + Z)$
Involution theorem	(Double negation) $\bar{\bar{X}} = X$	
Commutative theorem	$X + Y = Y + X$	$X \,.\, Y = Y \,. X$
Associative theorem	$X + (Y + Z) = (X + Y) + Z$	$X\,(YZ) = (XY)Z$
Absorption theorem	$X + XY = X$	$X\,(X{+}Y) = X$

Note that all the postulates and theorems only apply to *ANDs, ORs, or NOTs operators.* Some of the above postulates and theorems may or may not apply to exclusive- and/or equivalence operations.

(b) Applying distribution, complementing, and applying De Morgan's rule to Equation (b), we obtain:

$$f = W\bar{X}\bar{Z} + Y\bar{Z}, \text{ then } \bar{f} = \overline{W\bar{X}\bar{Z}}.\overline{Y\bar{Z}} = (\bar{W} + X + Z)(\bar{Y} + Z).$$

Let us use distribution one more time, thus:

$$\bar{f} = (\bar{W} + X + Z)(\bar{Y} + Z) = \bar{W}\bar{Y} + X\bar{Y} + \bar{Y}Z + \bar{W}Z + XZ + Z =$$

Observe the four right-most terms above; let us apply absorption three consecutive times, among terms: $\bar{Y}Z + \bar{W}Z + XZ + Z$.

Hence:

$$\bar{Y}Z + \bar{W}Z + XZ + Z = \bar{Y}Z + \bar{W}Z + Z = \bar{Y}Z + Z = Z.$$

Finally, $\bar{f} = \bar{W}\bar{Y} + X\bar{Y} + Z$.

7.3.7 The Duality Principle

Table 7.11 lists the most important postulates and theorems in Boolean algebra. The previously covered DeMorgan Laws present a good example of duality. Let us look back at Equations (7.2) and (7.3). The duality principle states that one rule (e.g., rule 1) may be obtained for the other one (2) by interchanging operators and identity elements. For our example, using DeMorgan Laws, we interchange *OR* and *AND* operators and replace 1's with 0's and 0's with 1's. Also refer to the complements postulate to see how operators are interchanged and 1's and 0's are swapped.

7.3.8 Venn Diagrams

In order to prove some of the theorems in Table 7.11, instead of using *truth tables*, we will use *Venn diagrams. Venn diagrams* provide a graphical methodology to visually perform logic operations in a very intuitive fashion. Each variable such as A, B, and so on is represented with a circle. All variables must be within a single rectangular area, which is referred to as the *universal* set, which is a frame of reference where all variables reside. The purpose of having a *universal* set is to easily draw A and its complement $\bar{A}$. Figure 7.1 (a) depicts variable A, which is represented by the area within the circle ($\bar{A}$ is the area outside of A but within the universal set), (b) shows A *inclusive-or* B (note the complete area of both variables A, B), (c) depicts the common area to both A and B, thus area AB is cross-hatched, (d) shows A *XOR* B (observe that the cross-hatched areas are also equivalent to $A\bar{B}+\bar{A}B$), and finally (e) shows the complement of A *XOR* B.

Using the concepts just learned about Venn diagrams we will use them to prove graphically some of the theorems listed in Table 7.11.

Figure 7.2 depicts the graphical justification of the absorption theorem, part 1. Part (a) shows X, part (b) shows XY, and part (c) shows the *ORing* of $X + XY = X$.

Figure 7.3 shows the graphical representation of the dual of the first absorption part 2. $X\,(X + Y) = X$ (from Table 7.11).

Exercise: Show using Venn diagram the validity of DeMorgan Rules (1) and (2).

7.4 MINTERMS: STANDARD OR CANONICAL SUM OF PRODUCTS (SOP) FORM

Binary variables may appear in their normal form, sometimes referred to as their true form, and their complemented form. For example, given A a binary-valued variable, we can have A and $\bar{A}$. If we consider two binary variables, such as A and B, since each one of them can take its true and complemented value, both variables together span four unique binary combinations. Table 7.12 depicts two variables and their four combinations and also three variables and their eight unique binary combinations.

Each of those binary combinations of the *ANDed* variables is referred to as a *minterm*. Generalizing the preceding concept given n variables, such n variables can span 2^n unique binary combinations. Each of those combinations is shown in Table 7.12 for 3 and 2 variables.

Note that each minterm is the logic product or the *ANDing* of the variable in question, not complemented when they represent ones and complemented when they represent zeros. It is also important to say that variable A was

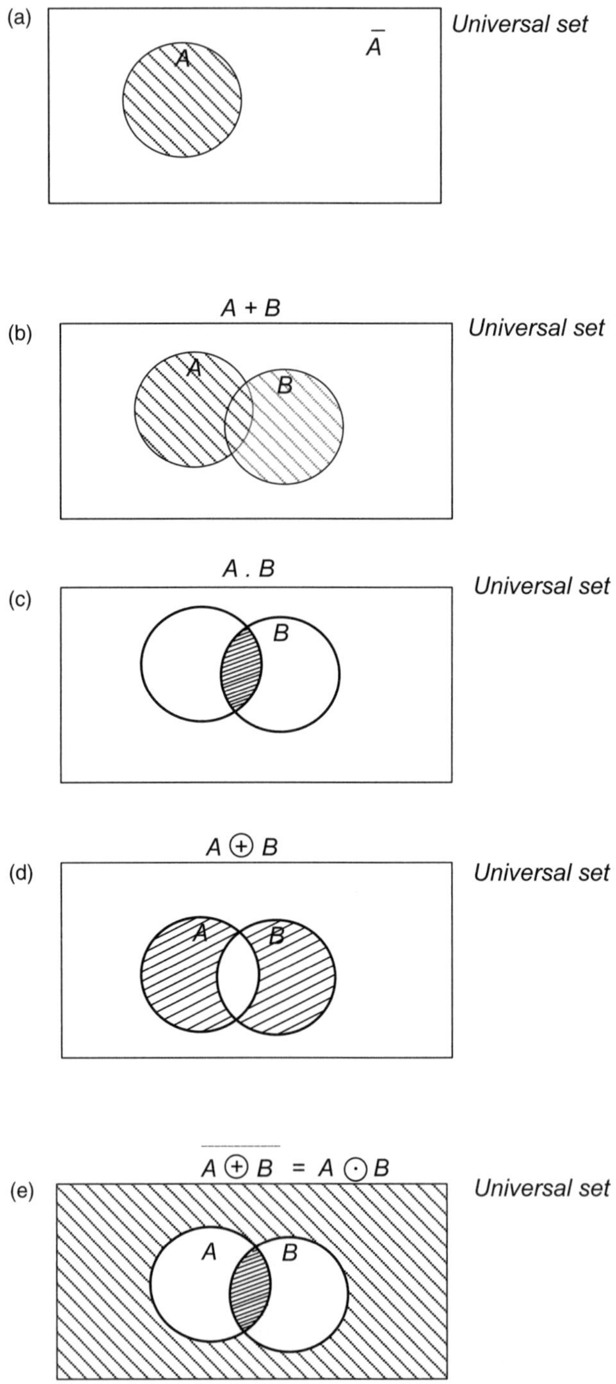

Figure 7.1 Venn diagrams (a) *A* and its complement $\bar{A}$; (b) $A + B = A$ *Inclusive-or B*; (c) $AB = A$ *and B*; (d) $A \oplus B = A$ *Exclusive-or B*; (e) $\overline{A \oplus B} = A$ *Equivalence B*.

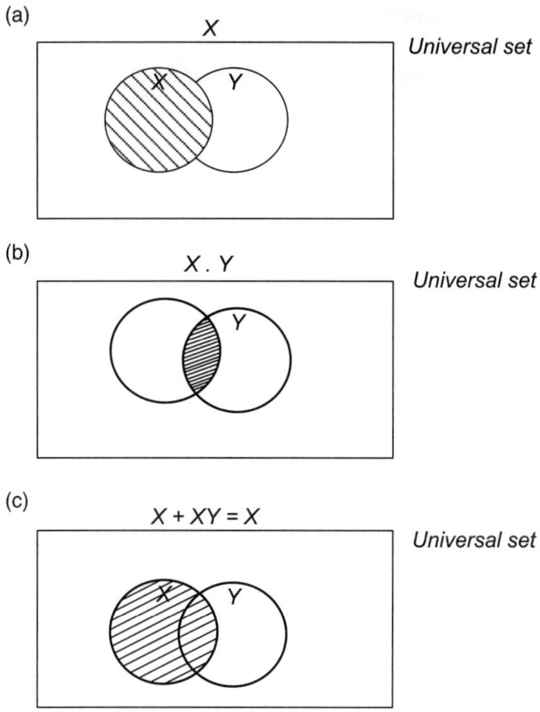

Figure 7.2 Absorption Theorem (a) X; (b) XY; (c) $X + XY = X$.

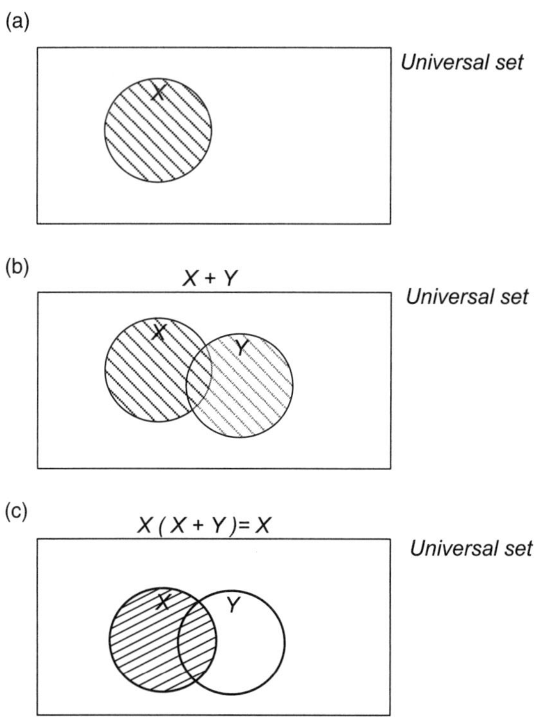

Figure 7.3 Absorption theorem part (2) (a) X; (*b*) $X + Y$; (c) $X\,(X + Y) = X$.

Table 7.12 Three-variable and two-variable *minterms*

3-Variable Minterms					2-Variable Minterms			
C	*B*	*A*	Minterm	Acronym	*B*	*A*	Minterm	Acronym
0	0	0	$\bar{C}\bar{B}\bar{A}$	m_0	0	0	$\bar{B}\bar{A}$	m_0
0	0	1	$\bar{C}\bar{B}A$	m_1	0	1	$\bar{B}A$	m_1
0	1	0	$\bar{C}B\bar{A}$	m_2	1	0	$B\bar{A}$	m_2
0	1	1	$\bar{C}BA$	m_3	1	1	BA	m_3
1	0	0	$C\bar{B}\bar{A}$	m_4				
1	0	1	$C\bar{B}A$	m_5				
1	1	0	$CB\bar{A}$	m_6				
1	1	1	CBA	m_7				

chosen to represent the *LSB* of each minterm. Without loss of generality, any variable name can represent any bit position. The ordering is simply a matter of convenience. As long as one chooses a way of doing things, it is better and less error-prone to stick to a methodical way of defining your minterms.

Example 7.7 Given a three-variable function $f(C, B, A) = CBA + \bar{C}\bar{B}A + C\bar{B}\bar{A}$, write the function as a sum of minterms. Table 7.13 presents a 3-variable function table listing all of its minterms. By inspection of function *f(C, B, A)* we can identify that the function contains three minterms that are *ORed* (or logically summed). The first minterm is m_7, the second one is m_1 and the third and last one is m_4.

So our function $f(C, B, A) = CBA + \bar{C}\bar{B}A + C\bar{B}\bar{A}$ can also be written in a so-called sum-of-products form (*SOP*), and after rearranging its minterms in an ascending order we obtain:

$$f(C, B, A) = \mathrm{m}_1 + \mathrm{m}_4 + \mathrm{m}_7 \tag{7.6}$$

which in a more compact form can be written as:

$$f(C, B, A) = \sum(1, 4, 7). \tag{7.7}$$

Function *f(C, B, A)* is written by Equation (7.7) in a *canonical or standard sum-of-products form (SOP).* Each minterm is generically referred to as a *product*, the logic *AND* is equivalently called a logic product because of its similarity with regular arithmetic. And it is a sum-of-product because each minterm present in the function is *ORed* or logically added. The *OR* operation is also called a *logic sum* or simply a *sum* if the context clearly is that of a logic *OR-ing*.

Table 7.13 Three-variable function of Example 7.7

3-Variable Minterms				
C	B	A	Minterm	$f(C, B, A)$
0	0	0	$\bar{C}\bar{B}\bar{A}$	0
0	0	1	$\bar{C}\bar{B}A$	1
0	1	0	$\bar{C}B\bar{A}$	0
0	1	1	$\bar{C}BA$	0
1	0	0	$C\bar{B}\bar{A}$	1
1	0	1	$C\bar{B}A$	0
1	1	0	$CB\bar{A}$	0
1	1	1	CBA	1

Example 7.8 Given function $f(C, B, A) = CA + BA$, expand it to represent it in its canonical *SOP* form.

Solution to Example 7.8

The given function clearly is not already given as a sum of its minterms.

What we need to do is to create "*logic redundancies*" that do not affect the original logic of the function. For example, *ANDing* terms like $(C\bar{C})$, since $(C\bar{C}) = 1$ and *ANDing* 1 to any logical expression does not alter its original logic, is a way of creating such redundancy. Another type of possible redundancy is *ANDing* terms like $(B + \bar{B})$ to the original function, which will not alter the initial logic of the function because $(B + \bar{B}) = 1$. Proceeding with our function *f*:

$$f(C, B, A) = CA + BA. \tag{7.8}$$

Since (Eq. 7.8) term *CA* is missing the literal *B* we *AND* the term $(B\bar{B})$ with the term *CA* without changing the original logic of function *f*. At the same time we create a redundancy to the term *BA* by *ANDing* the term $(C\bar{C})$ with the term *BA*. Hence:

$$f(C, B, A) = CA(B\bar{B}) + (C\bar{C})BA. \tag{7.9}$$

Applying logic product distribution and making sure that variables are consistently organized from *C* down to *A* (e.g., *CBA*)

$$f(C, B, A) = CBA + C\bar{B}A + CBA + \bar{C}BA. \tag{7.10}$$

Eliminating only the second instance of the term *CBA* because it is redundant yields:

$$f(C, B, A) = C\bar{B}A + CBA + \bar{C}BA. \tag{7.11}$$

Rewriting Equation (7.11) in *SOP* form and rearranging terms:

$$f(C, B, A) = m_5 + m_7 + m_3 = \sum(3, 5, 7). \qquad (7.12)$$

7.5 MAXTERMS: STANDARD OR CANONICAL PRODUCT OF SUMS (POS) FORM

Given any logic function in its standard *SOP* form, taking its complement we obtain a *product-of-sum* form (*POS*). Each *POS* term is referred to as a *maxterm*.

Table 7.14 lists all the *minterms* for a three-variable function and its corresponding complements. Such complements are defined as the function *maxterms*.

Binary-valued functions not only can be expressed in a *standard SOP* form, but also in a *standard product-of-sums* (*POS*) form. Let us explain with the following example.

Example 7.9 Given the following function in *standard SOP* form, convert it to its *standard POS* form.

$$f(C, B, A) = m_6 + m_7 + m_3 = \sum(3, 6, 7). \qquad (7.13)$$

The following steps will lead to the expected results:

It is intuitive to see that if a function *f* is given in *standard SOP* form, such as Equation (7.13), then its complement ($\overline{f}$) also in *standard SOP* form will list all those minterms that are not listed in function *f*. That is to say:

$$\overline{f}(C, B, A) = m_0 + m_1 + m_2 + m_4 + m_5 = \sum(0, 1, 2, 4, 5). \qquad (7.14)$$

Table 7.14 Minterms and maxterms for three-variable functions

C	B	A	Minterm	Minterm Acronym $m_i = \overline{M_i}$	Maxterm	Maxterm Acronym $M_i = \overline{m_i}$
0	0	0	$\overline{C}\overline{B}\overline{A}$	m_0	$C + B + A$	M_0
0	0	1	$\overline{C}\overline{B}A$	m_1	$C + B + \overline{A}$	M_1
0	1	0	$\overline{C}B\overline{A}$	m_2	$C + \overline{B} + A$	M_2
0	1	1	$\overline{C}BA$	m_3	$C + \overline{B} + \overline{A}$	M_3
1	0	0	$C\overline{B}\overline{A}$	m_4	$\overline{C} + B + A$	M_4
1	0	1	$C\overline{B}A$	m_5	$\overline{C} + B + \overline{A}$	M_5
1	1	0	$CB\overline{A}$	m_6	$\overline{C} + \overline{B} + A$	M_6
1	1	1	CBA	m_7	$\overline{C} + \overline{B} + \overline{A}$	M_7

Note that from Table 7.11 of postulates and theorems, *idem potent,* which states that:

$X+\bar{X}=1$, does justify f and $\bar{f}$ expressed by Equations (7.13) and (7.14).

Now let us proceed to take the complement of $\bar{f}$, thus from Equation (7.14):

$$\bar{\bar{f}}(C,B,A)=\overline{m_0+m_1+m_2+m_4+m_5}.$$

Applying Boolean algebra on $\bar{\bar{f}}$, using the *minterm* and *maxterm* definitions from Table 7.14 we obtain:

$$\bar{\bar{f}}=(C+B+A)+(C+B+\bar{A})+(C+\bar{B}+A)+(\bar{C}+B+A)+(\bar{C}+B+\bar{A}). \quad (7.15)$$

Identifying every maxterm from Equation (7.15) with the aid of Table 7.14 we find that:

$$\bar{\bar{f}}=f=\mathrm{M}_0\mathrm{M}_1\mathrm{M}_2\mathrm{M}_4\mathrm{M}_5. \quad (7.16)$$

Equation (7.16) can be written in *POS* compact form using the symbol π to indicate multiplication, hence:

$$f=\prod(0,1,2,4,5). \quad (7.17)$$

Equation (7.17) is a standard *POS* form for the originally given function *f*. It is very important to remember that the π symbol indicates that the numerals within the parentheses are *maxterms* and not *minterms.*

Example 7.10 Let us consider designing a logic block that receives three input binary variables *X*, *Y*, and *Z* and we want it to have a single output which detects whenever two or more of the inputs are ones, else we want to output to be zero. Derive the truth table for such circuit.

Let us draw a truth table with three inputs and one output. Simply follow the example requirements, whenever we see two or more ones in any *Z Y X* row we must write a one at the output, all other cases require a zero output. A digital circuit such as the one just described is called a *majority detector* circuit. Refer to Table 7.15 for a truth table of the majority circuit.

The *standard SOP* form for our majority detector circuit is:

$$F(Z,Y,X)=\sum(3,5,6,7). \quad (7.18)$$

7.6 KARNAUGH MAPS AND DESIGN EXAMPLES

Unless we work with one or two variables, logic equations can become quite complex, particularly when we need to simplify them and express them as less

Table 7.15 Truth table of a majority detector circuit

Input Z	Input Y	Input X	Output F
0	0	0	0
0	0	1	0
0	1	0	0
0	1	1	1
1	0	0	0
1	0	1	1
1	1	0	1
1	1	1	1

complicated expressions. Rather than using the postulates and theorems of Table 7.11, which can easily become cumbersome and lengthy, there is a methodology attributed to *Karnaugh,* referred to as solving or simplifying logic equations using *Karnaugh* maps. This is the topic of this section. Let us begin defining the *Karnaugh* (K) map construction. For a two-variable map, we need to have a map with as many cells as minterms the number of variables spans. A two-variable K. map has $2^2 = 4$ cells. A three-variable K. map has $2^3 = 8$ cells, and so on. Figure 7.4 depicts a *2, 3,* and *4-cell Karnaugh* maps. We will start covering 2-variable maps progressing onto 3 and 4-variables.

7.6.1 Two-Variable Karnaugh Maps

The two-variable K. map is shown in Figure 7.4a. The map clearly shows the relationship between its cells (squares) and the two variables A and B.

Note that the 2-variable map is drawn such that the rightmost vertical column corresponds to variable A (i.e., $A = 1$), in true value or noncomplemented. The leftmost vertical column corresponds to $\bar{A}$, A false or A complemented. Similarly the bottom row corresponds to B ($B = 1$), and the top row corresponds to $\bar{B}$ ($B = 0$). The four-cell map becomes fully defined. Note that the cell at $B = 0$ and $A = 0$, corresponds to minterm *0*. Cell at $B = 0$ and $A = 1$ corresponds to minterm 1. Similarly, cell at $B = 1$ and $A = 0$ corresponds to minterm 2, and cell at $B = 1$ and $A = 1$ corresponds to minterm 3. So let us solve some problems to see the 2-variable *Karnaugh* map at work.

We will solve a handful of 2-variable maps in a somewhat intuitive fashion. Problem solving and K. map simplification will become clearer using 3 and 4-variable maps. In some ways, 2-variable maps are too simple to appreciate the properties of K. map method. We will address simplification in the *Karnaugh map* sense more thoroughly after all the 2-variable examples.

Example 7.11 Refer to Figure 7.5a for this example.

Given function $f(B, A) = m_0 + m_1 + m_2 + m_3$, find a maximally simplified *SOP form* logic expression. As we discussed in earlier sections, canonical or standard *SOP* forms express a logic function as a logic sum (*Or-ing*) of the appropriate

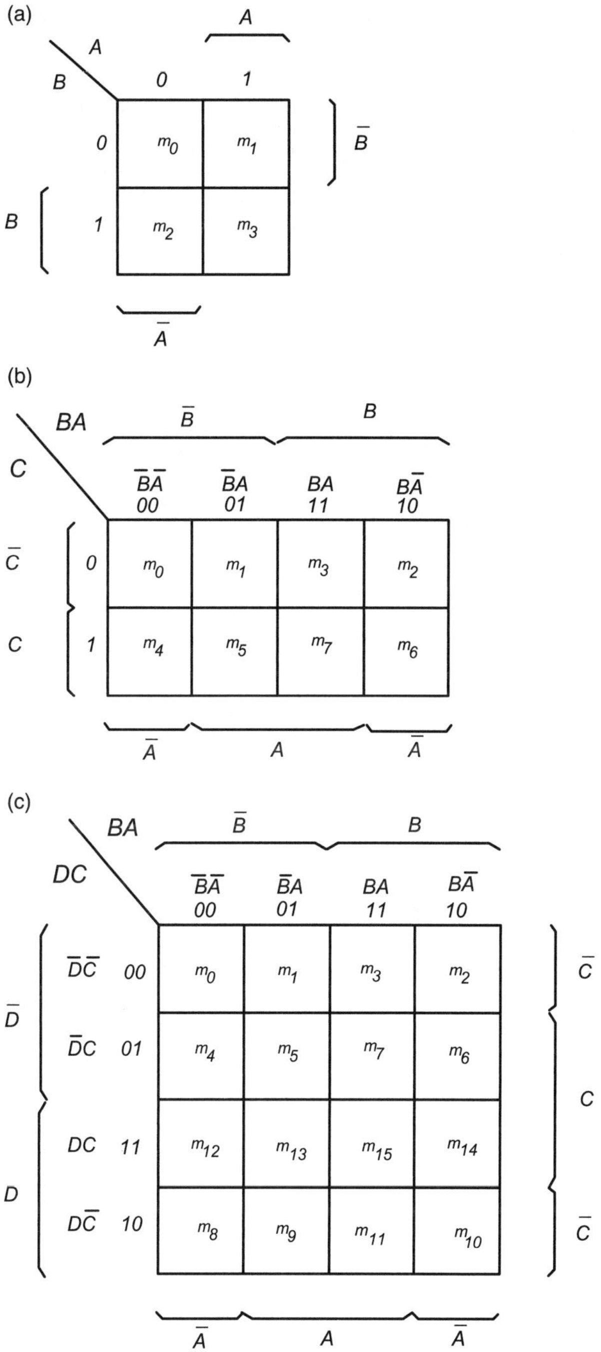

Figure 7.4 Two, three, and four-variable Karnaugh maps definition.

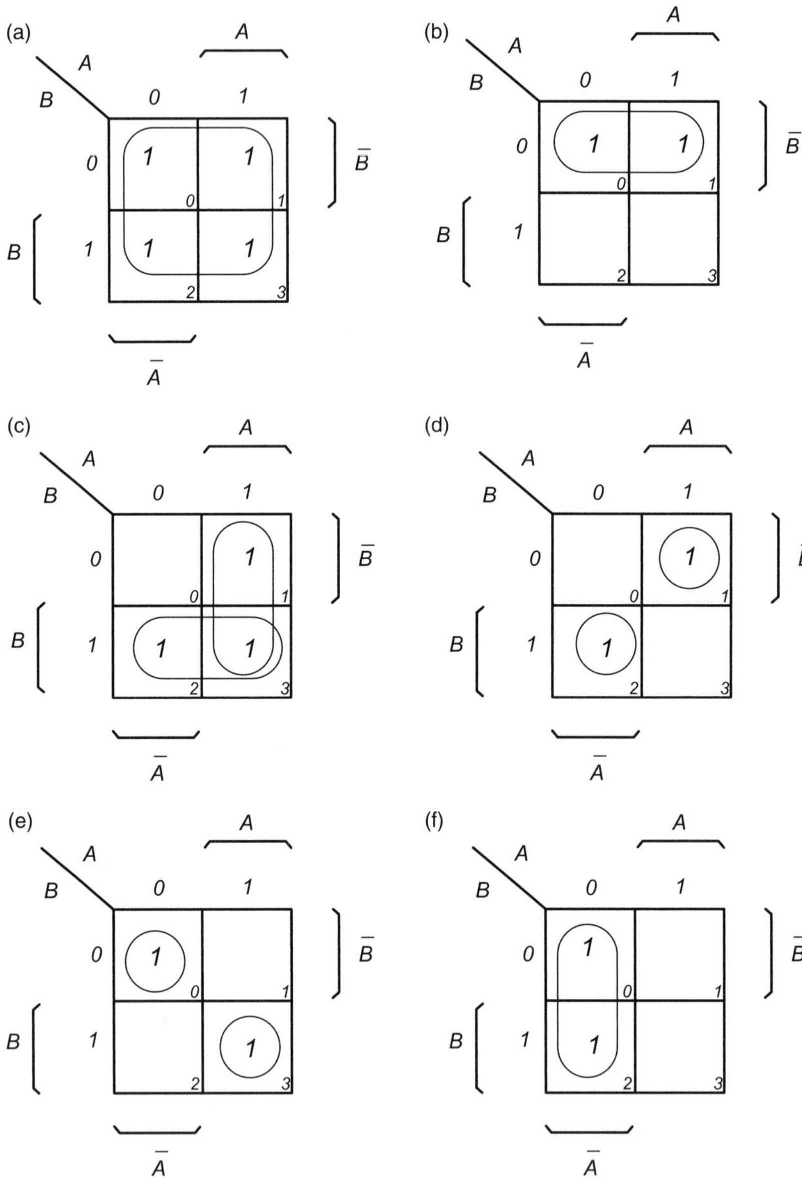

Figure 7.5 Two-variable Karnaugh Maps examples: (a), (b), (c), (d), (e), and (f).

minterms. Standard forms are not maximally simplified from the following criteria. Minterms show all possible variables that the function has. For example, 2-variable functions have two-bit minterms (e.g., such as $B\overline{A}$); three-variable functions have three-bit minterms, and so forth. Some functions can be expressed in a simplified *SOP* form by reducing the number of variables of some or all of its original standard form *SOP* form minterms. Additionally, upon logic simplification, some minterms may disappear from the simplified *SOP* form, this yielding a logic sum of less terms than its corresponding standard *SOP* form. This new form is referred to as a simplified *SOP* form. Let us go over these concepts simplifying our given function $f(B, A)$. By inspection of Figure 7.5a we can see that function f has ones in all of its minterms. It may not be clear right now, but it will become more obvious after we cover a few more examples that function $f(B, A)$ is always true. This means that regardless of the individual values of variables A and B, $f(B, A)$ is always true, that is $f(B, A) = 1$.

This fact is indicated in part (a) of Figure 7.5 by encircling all four minterms.

Summary of what a maximally simplified SOP form is in the K. map sense is:

1. *Not all variables will necessarily show in every term being OR-ed.*
2. *The total number of OR-ed terms will not necessarily be the same number as the number of OR-ed terms in the function's standard SOP form.*
3. *The simplified function will still have an SOP form, this means that it is implemented with just two-levels or logic like a standard SOP form. However, because of points (1) and (2), the number of OR terms will typically (although not all the time) be smaller and not all the variables will be present in each OR-ed term.*

Example 7.12 Given a new function $f(B, A) = m_0 + m_1$, find a maximally simplified *SOP* form. Refer to Figure 7.5b to observe the *K.* map of our function. Note that the *K.* map has four distinct areas: A, $\overline{A}$, B, and $\overline{B}$. Moreover, area A graphically corresponds to the *Or-ing* or *sum* of minterms m_1 and m_3, (i.e., $m_1 + m_3$). Area $\overline{A}$ corresponds to the logic *sum* of m_0 and m_2 (i.e., $m_0 + m_2$). Similarly, area B corresponds to $m_2 + m_3$. Finally, $\overline{B}$ corresponds to $m_0 + m_1$. We can easily identify that our function can simply be represented by area $\overline{B}$. This means that $f(B, A) = m_0 + m_1$ is logically equal to $f(B, A) = \overline{B}$. Other ways of proving that $f = \overline{B}$ is true is by logic simplification or using truth tables or *Venn* diagrams.

For example, let us prove using logic simplification that $f(B, A) = m_0 + m_1 = \overline{B}$. Since:

$$m_0 = \overline{B}\overline{A} \text{ and } m_1 = \overline{B}A,$$

we write:

$$f(B, A) = m_0 + m_1 = \overline{B}\overline{A} + \overline{B}A.$$

Using the distribution property we obtain:

$$f(B, A) = \overline{B}(\overline{A} + A)$$

and since from the idem-potent property from Table 7.11, $\overline{A} + A = 1$, hence:

$$f(B, A) = \overline{B}.$$

Exercise: Prove that $f(B, A) = m_0 + m_1 = \overline{B}$ using *Venn* diagrams.

Example 7.13 Now referring to the *K.* map of Figure 7.5c obtain a maximally simplified *SOP* form for $f(B, A) = m_1 + m_2 + m_3$.

This example is slightly more involved than the previous ones. Now is a good point to start presenting the concept of *adjacent* cells. *Adjacency* in the *K.* map sense are those cells whose binary representation only differs by one bit. Let us inspect any 2-variable *K.* map. Cell *00* (corresponding to minterm m_0) is adjacent to cell *01* because there is only *one bit difference between 00 and 01.* Additionally, cell *00* is adjacent to cell *10* because of the same reason. However, cell *00* is not adjacent to cell *11* (minterm m_3), because two bits differ between binary *00* and *11*. In a generalized and graphical way, adjacent cells are those cells that are above, below, left. and right of a cell. Those cells that are diagonally placed with respect to the cell in question are not adjacent cells. So back to Example 7.13, let us start encircling as many adjacent cells as possible, such that the number of encircled cells is a power of two. When we can no longer encircle more adjacent cells in the first round, we repeat the process again, until we run out of cells to encircle. We must attempt to produce the smallest number of encirclements possible. Every new encirclement of cells may re-encircle previously encircled cells; this process usually ensures that the largest possible number of adjacent cells is obtained. In summary, *Karnaugh* map cells encirclements for the purpose of simplification should follow the following basic steps:

1. *Combine the largest possible number of adjacent cells.*
2. *Such number of cells must be a power of two.*
3. *Minimize the overall number of encircled cells.*
4. *It may be convenient to re-encircle some previously encircled cells to reduce the overall number of variables that a term ends up having.*

It is important to be aware that the methodology described does not necessarily provide a unique solution. It will be up to the design engineer to adopt the most convenient solution for the application. It is also important to know that this technique highly depends on the expertise of the user. The more problems one solves, the better and the easier it will be to obtain a simplified solution. Initially, it is strongly recommended to solve a problem in at least two or more possible ways.

Referring again to Figure 7.5c let us encircle cells *1* and *3*. Clearly, we cannot encircle all three cells because *3* is not a power of *2*. After encircling cells *1* and *3*, only minterm *2* remains uncircled. Let us encircle minterm *2* re-encircling minterm *3* which is adjacent to *2*. We are done encircling all the minterms on the *K.* map that have *ones*. Looking again at Figure 7.5c, we can easily identify that the encirclement of cells *1* and *3* coincides with the area that corresponds to A. The encirclement of cells *2* and *3* corresponds to B. So the simplified function is:

$$f(B, A) = m_1 + m_2 + m_3 = A + B.$$

Example 7.14 Find maximally simplified *SOP* forms for the two functions given by Figure 7.5d,e. Both of these maps are handled at the same time because it is straightforward to see that one is the complement of the other. Addressing first Figure 7.5d, note that cells *01* and *10* are not adjacent in the *K.* map sense, because they differ by more than one bit position. Consequently, the best possible encirclement is to encircle each minterm separately. We can observe that this function:

$$m_1 + m_2 = \bar{B}A + B\bar{A} \qquad (7.19)$$

is $A \oplus B$ from Equation (7.1). In a similar fashion, we can identify that Figure 7.5e function:

$$m_0 + m_3 = \overline{A \oplus B} = A \text{ } equivalence \text{ } B = A \text{ } NXOR \text{ } B. \qquad (7.20)$$

Just as for the previous case, its minterms *00* and *11* are not adjacent.

From Equations (7.19) and (7.20) we conclude that the standard forms for *exclusive or* and *equivalence* cannot be simplified in the *K.* map sense, because for each function, none of its minterms is adjacent to any other minterm within the function.

Example 7.15 For the *K.* map of Figure 7.5f find a maximally simplified *SOP* form.

Now it is easy to see that minterms *00* and *10* are adjacent. We encircle them and notice that they correspond to area $\bar{A}$. Their simplified function is simply, $\bar{A}$.

7.6.2 Three-Variable Karnaugh Maps

The three-variable map was defined in Figure 7.4b. Let us observe the adjacent cells. For example, cell *000* corresponding to minterm m_0 has cells 001, 010, 100; note all these cells differ by no more than one bit position with respect

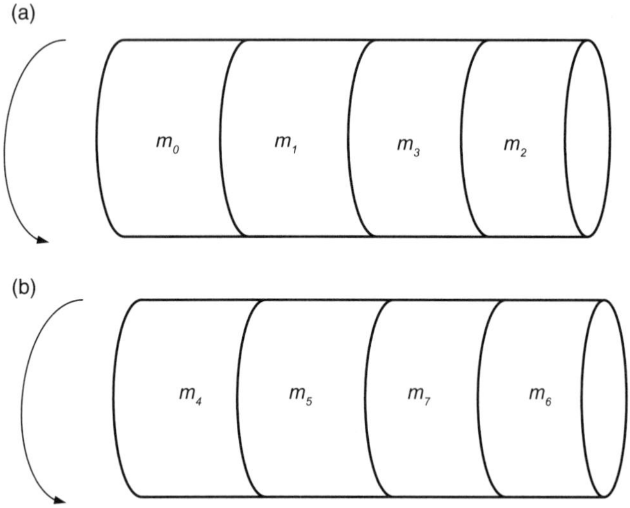

Figure 7.6 Spatial representation of a 3-variable map wrapped around a horizontal axis cylindrical surface. (a) Minterms 0, 1, 3, 2 on the front; (b) minterms 4, 5, 7, 6 on the front.

to cell *000*. Additionally, note that cells *011, 101, 110,* and *111* are not adjacent with respect to cell *000*.

Exercise: Explain why.

It is quite instructive to think about a 3-variable *K*. map as being rolled up on a cylindrical surface with a horizontal axis and also rolled up on a cylindrical surface with a vertical axis. Figure 7.6a depicts a 3-variable rolled up along a horizontal axis cylinder. Minterms m_0, m_1, m_3, *and* m_2 are respectively adjacent to minterms m_4, m_5, m_7, *and* m_6; either looking above or below minterms m_0, m_1, m_3, *and* m_2. Figure 7.6b depicts the same cylindrical surface whose axis is rotated 180°. In Figure 7.6a, minterms m_0, m_1, m_3, *and* m_2 are on the front of the cylinder while minterms m_4, m_5, m_7, *and* m_6 are on the back, thus the latter are not visible from the front. The opposite takes places in Figure 7.6b: minterms m_4, m_5, m_7, *and* m_6 are on the front while minterms m_0, m_1, m_3, *and* m_2 are on the back.

Figure 7.7 depicts a 3-variable map wrapped around a vertical axis cylindrical surface. Since the 3-variable map has *4* cells per row and *2* rows, we can only see two cells of each row in Figure 7.7a.

The front half of the cylinder. This top half contains minterms m_0 *and* m_1 on the top row; and minterms m_4 and m_5 on the bottom row. To aid with the understanding of this spatial description, also refer to Figure 7.4b, which depicts the complete 3-variable map on a flat surface. Back to Figure 7.7a, as we rotate the cylinder 90° in the clockwise direction, while looking at the front of the cylinder, we see minterms m_1 and m_3 on the top front row, and minterms

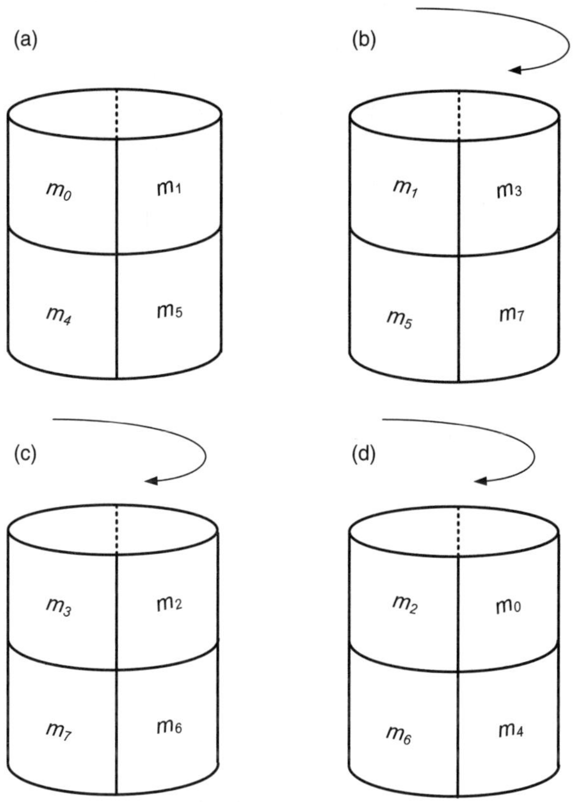

Figure 7.7 Spatial representation of a 3-variable map wrapped around a vertical axis cylindrical surface.

m_5 and m_7 on the second row; see Figure 7.7b. Another 90° rotation in the same direction is depicted in Figure 7.7c, and one last 90° rotation is depicted in Figure 7.7d. What is the purpose of all these cylindrical surfaces rotations for? The ideas that want to be conveyed are an aid to identify adjacent cells without having to figure this out by inspecting every bit of every minterm. Note that adjacent cells in the horizontal direction can be observed using Figures 7.7a through 7.7d, which depict the cylinder along a vertical axis. For example, minterms adjacent to m_1 are m_3 to the right, Figure 7.7b and m_0 to the left, Figure 7.7a. Similarly, this can also be appreciated when looking at the adjacent cells to minterm m_4. Minterm m_5 to the right of m_4 is adjacent to m_4, Figure 7.7a and m_6 to the left of m_4 is also adjacent to m_4, Figure 7.4d. We use the cylindrical picture with a horizontal axis, Figure 7.6a,b when we want to find the adjacent cells to any minterm by looking either above or below the minterm of interest. Obviously, for the 3-variable map case, the cells above and below a minterm of interest are basically the same thing because the

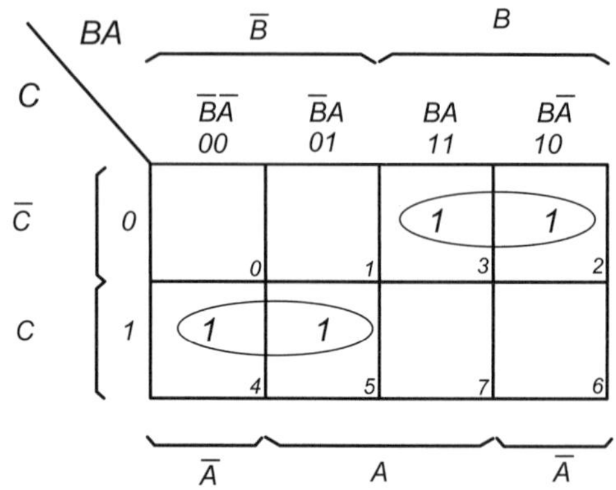

Figure 7.8 Three-variable map for Example 7.16.

3-variable map wraps around the cylinder, and it only has two rows. The 4-variable K. map is really the first *K.* map that we are studying that will exhibit all the features seen in *Karnaugh* maps.

Example 7.16 Given a 3-variable function $g(C, B, A) = \Sigma(2, 3, 4, 5)$, find a maximally simplified *SOP* form using *K.* maps. Figure 7.8 depicts function *g* minterms on a 3-variable *K.* map.

Solution to Example 7.16

Let us begin by encircling the largest possible number of adjacent minterms, such that that number is a power of two. We clearly obtain one grouping with minterms m_2 and m_3 and a second and last grouping produces the grouping of m_4 and m_5. The solution to this problem is particularly straightforward because we do not have other choices of encircling minterms that would produce similar results. Referring to Figure 7.8 we see that the maximally simplified *SOP* form yields:

$$g(C, B, A) = \sum(2, 3, 4, 5) = C\bar{B} + \bar{C}B.$$

Note: The solution of Example 7.16 is the exclusive or of which variables? Answer: *B, C.*

Example 7.17 Given function $w(C, B, A) = \Sigma(3, 4, 6, 7)$, find a maximally simplified *SOP* form using *K.* maps. Representing the given function on the 3-variable *K.* map of Figure 7.9a we encircle m_6 and m_7, then m_3 and m_7 and finally m_4 and m_6. We show this in Figure 7.9a and we obtained the following simplified function:

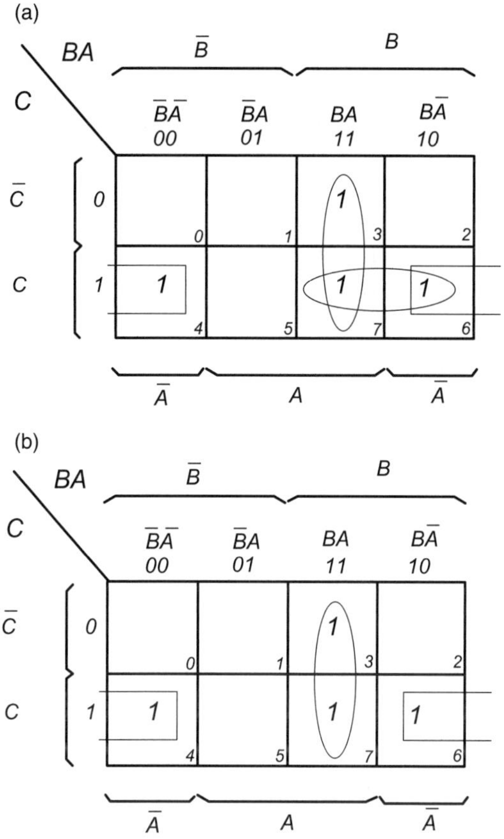

Figure 7.9 Three-variable map for Example 7.16: (a) redundant enclosure m_6 and m_7; (b) redundancy removed.

$$w(C, B, A) = BA + CB + C\bar{A}. \tag{7.21}$$

If we carefully examine Figure 7.9a we can see a redundant enclosure of minterms m_6 and m_7 given by Equation (7.21) term $C\ B$; $C\ B$ does not provide any more logical information. Figure 7.9b shows this term removed, so that our maximally simplified *SOP* form yields:

$$w(C, B, A) = BA + C\bar{A}. \tag{7.22}$$

Exercise: Prove using truth tables that Equations (7.21) and (7.22) are logically equivalent.

7.6.3 Four-Variable Karnaugh Maps

The four-variable map spans 16 cells or minterms. The map is depicted in Figure 7.4c. As usual, the minterms on this map have been encoded with variables:

DCBA, where *A* is the least significant bit variable and *D* is the most significant bit variable. It should be clear that $\bar{D}\bar{C}\bar{B}\bar{A}$ represents m_0, $\bar{D}\bar{C}\bar{B}A$ represents m_1 and so forth. Figure 7.4c also shows the groups of minterms for every one of its four variables. For the purpose of more easily visualizing adjacent cells, we can assume that the map is wrapped around a horizontal axis cylindrical surface as shown in Figure 7.10.

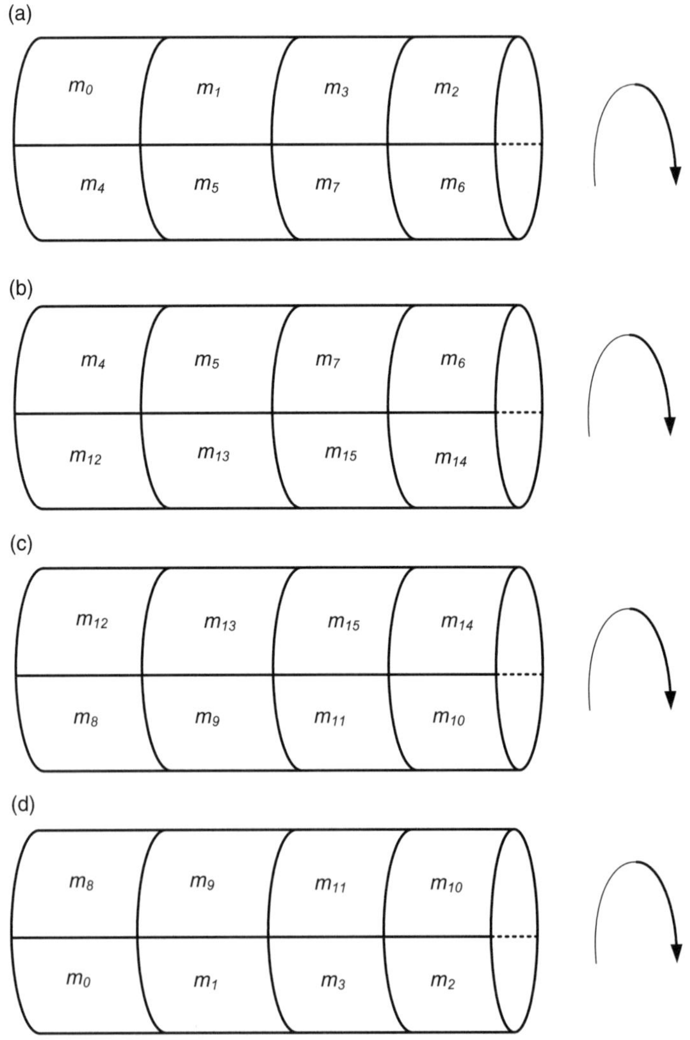

Figure 7.10 Four spatial views of a 4-variable map wrapped around a horizontal axis cylindrical surface.

Figure 7.10a depicts the 4-variable map wrapped around a horizontal cylinder. The two top rows are the only ones visible to the reader. That is, row with minterms m_0, m_1, m_3 & m_2 and row with minterms m_4, m_5, m_7, & m_6. The other two rows are on the back of the cylindrical surface and cannot be seen by the reader. Figure 7.10b depicts the initial cylindrical surface rotated 90 degrees in the direction shown by the arrow. Rows with minterms m_4, m_5, m_7, & m_6 and minterms m_{12}, m_{13}, m_{15}, & m_{14} are visible to the reader; the other two rows of minterms are not visible. Similarly, after rotating another 90 degrees Figure 7.10b we obtain 7.10c which depicts the visible minterms and finally 7.10d depicts the visible minterms after rotating the cylinder one more quarter of a turn.

Figure 7.10 allows us to determine simply by visual inspection adjacent minterm to any cell of interest by looking at the cell above and below the cell in question. For example, for cell m_0, we see that m_8 is adjacent to it because m_0 is right below m_8 according to Figure 7.10d. Minterm m_4 is also adjacent to m_0, since it is located right below m_0 in Figure 7.10a.

To analyze the adjacencies left and right of any cell of interest we develop the spatial view of the 4-variable map wrapped around a vertical axis cylindrical surface; this is depicted in Figure 7.11.

In this case we wrap around a vertical axis cylindrical surface our 4-variable map. Figure 7.11a depicts two of the four columns of minterms that are visible to the reader. These are: m_0, m_4, m_{12}, & m_8 *and minterms* m_1, m_5, m_{13}, & m_9. As before, the other two columns are not visible to the reader, since they are on the back of the cylinder of Figure 7.11a. The next three Figure 7.11b,c,e show the previous view of the cylinder rotated around its vertical axis 90 degrees at a time. Each figure depicts the two front columns that are visible to the reader. Using Figure 7.11, it is easy to visualize adjacent cells to the left and to the right of the cell of interest. Between Figures 7.10 and 7.11, all adjacent cells to any one of the 16-cell, 4-variable map can easily be found.

Example 7.18 Given function $G(D, C, B, A) = \Sigma(1, 4, 6, 7, 8, 9, 10, 11, 15)$, find a maximally simplified *SOP* form. Function *G* is given in Figure 7.12.

Solution to Example 7.18

Using a 4-variable map, we write the unity minterms according to the given function *G*. We encircle groups of adjacent minterms as shown in Figure 7.12. Pay close attention to the choices made grouping adjacent minterms. The groupings shown provide the least number of terms and the least number of variables per term.

The choices made lead to the following *SOP* simplification:

$$G(D, C, B, A) = \bar{C}\bar{B}A + \bar{D}C\bar{A} + CBA + D\bar{C}. \tag{7.23}$$

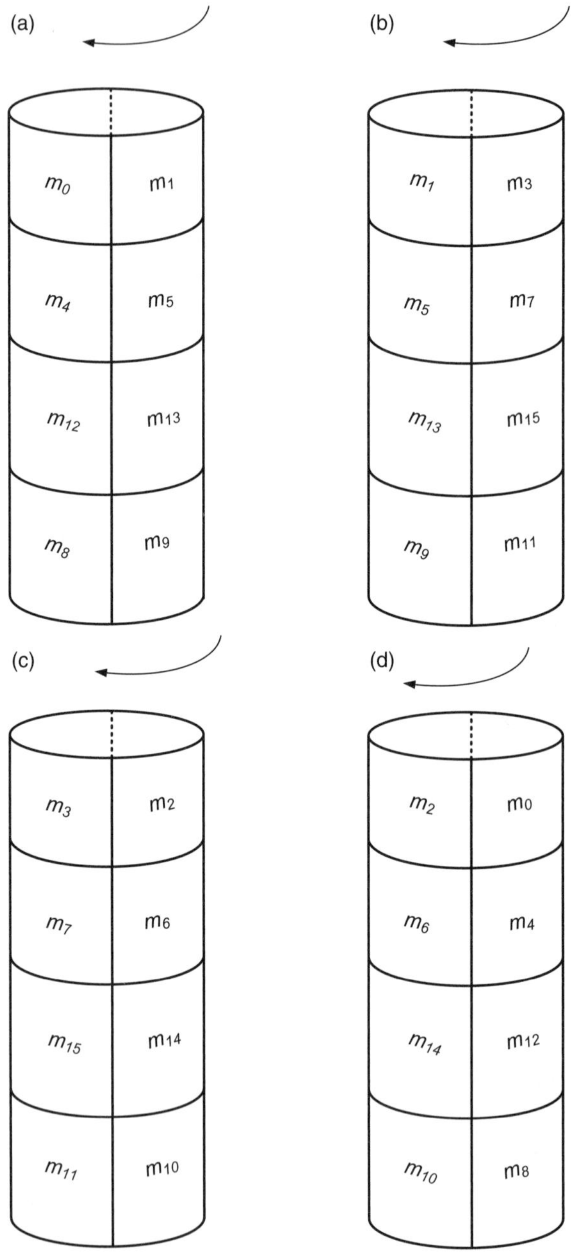

Figure 7.11 Four spatial views of a 4-variable map wrapped around a vertical axis cylindrical surface.

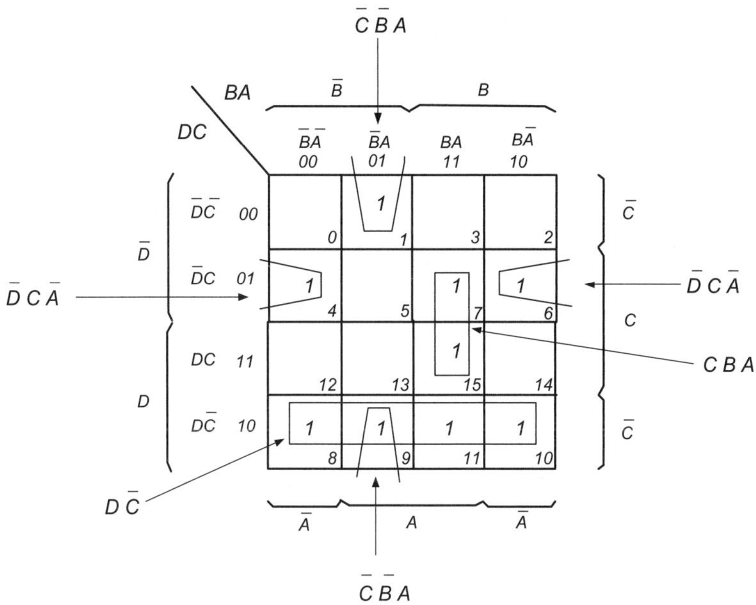

Figure 7.12 Four-variable Karnaugh map for Example 7.18.

7.6.4 Five-Variable Karnaugh Maps

A simple way of handling the 5-variable map, which spans 32 minterms, is to use two 4-variable maps. One of the maps is created for the most significant variable equated to zero (or false), and the second map is created for the most significant variable equated to one (or true). We will use variables *E, D, C, B, A* where *E* is the most significant variable bit and *A* is the least significant variable bit. Figure 7.13 depicts the basic *5*-variable, *32*-minterm, *Karnaugh* map. The upper map is used for $\bar{E}$ and the lower map for *E*. The adjacencies within each map are no different than those adjacencies defined for the *4*-variable map. The additional adjacency criterion of the *5*-variable map is across maps.

A cell on the $\bar{E}$ map that has the same relative position on the *E* map is by definition adjacent because their associated minterm would differ by only the most ignificant bit (variable *E*). Let us look at some examples: cell m_0 and m_{16} are adjacent because $m_0 = \bar{E}.\bar{D}.\bar{C}.\bar{B}.\bar{A} = 0\_0000$ and $m_{16} = E.\bar{D}.\bar{C}.\bar{B}.\bar{A} = 1\_0000$, the only difference is their most significant bit (MSB); thus, since there is a single bit difference they are adjacent. Another example is given by minterms m_{15} and m_{31}. $m_{15} = \bar{E}.D.C.B.A = 0\_1111$ and $m_{31} = E.D.C.B.A = 1\_1111$, again they only differ by one bit.

Exercise: List all other adjacent cells between $\bar{E}$ and the *E* maps.

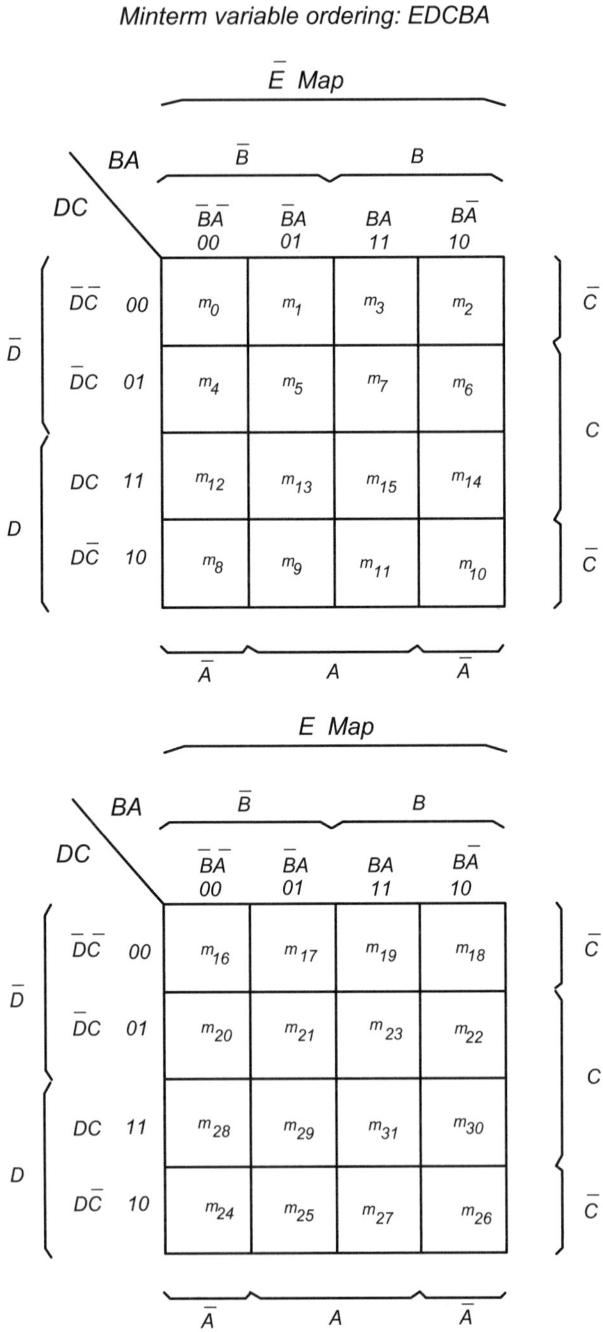

Figure 7.13 Five-variable Karnaugh map definition.

Example 7.19 Given the following function:

$$W(E, D, C, B, A) = \sum (0, 2, 4, 6, 9, 13, 21, 23, 25, 29, 31), \qquad (7.24)$$

find a maximally simplified *SOP* form.

Figure 7.14 shows the minterms of function *W*. Minterms m_0, m_2, m_4, and m_6 encircled on the $\bar{E}$ map yield the simplified term $\bar{E}.\bar{D}.\bar{A}$. Since minterms m_0, m_2, m_4, and m_6 have no adjacent minterms in map *E*, the term $\bar{E}.\bar{D}.\bar{A}$ has

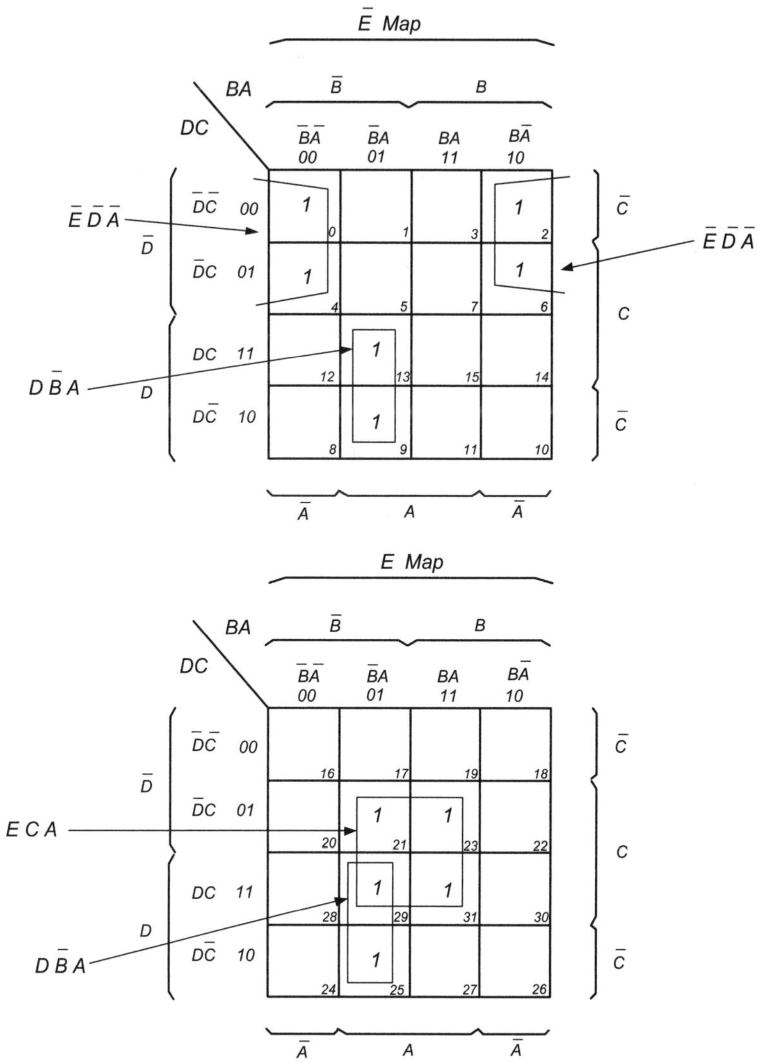

Figure 7.14 Five-variable map for Example 7.19.

a leading $\bar{E}$. Adjacent terms to the $\bar{E}$ map minterms m_0, m_2, m_4, and m_6 are E map minterms m_{16}, m_{18}, m_{20}, and m_{22}, but these last four minterms have zeroes (blank cells) on the E map. The first simplified term of function W is $\bar{E}.\bar{D}.\bar{A}$. The next group of minterms on the $\bar{E}$ map is m_{13} and m_9. The correspondingly adjacent group on the E map is m_{29} and m_{25}. For each of the maps, this group is represented by the term $D.\bar{B}.A$. Since this group is present on both maps, $D.\bar{B}.A$ does not have either a leading E or $\bar{E}$ literal. Alternatively, consider that the $D.\bar{B}.A$ group on the $\bar{E}$ map is annotated as $\bar{E}.D.\bar{B}.A$; also consider the $D.\bar{B}.A$ group on the E map is annotated as $E.D.\bar{B}.A$. Now in the final expression we would have to write:

$$\bar{E}.D.\bar{B}.A + E.D.\bar{B}.A, \tag{7.25a}$$

as part of the overall simplified expression for function W. But from Equation (7.25a), it is easy to observe that:

$$\bar{E}.D.\bar{B}.A + E.D.\bar{B}.A = (\bar{E} + E)(D.\bar{B}.A). \tag{7.25b}$$

Since from Table 7.11 we know that:

$$\bar{E} + E = 1. \tag{7.26}$$

Thus, Equation (7.25b) becomes:

$$D.\bar{B}.A. \tag{7.27}$$

Equation (7.27) is the second term of simplified function W, as shown by (7.28).

Finally, the third and last term of simplified W is obtained from the E map. Grouping minterms m_{21}, m_{23}, m_{29}, and m_{31}. This simplification turns out to be $C.A$ and since it is only present on the E map, the complete term is $E.C.A$, which is the third and last term of the maximally simplified *SOP* form of function W, as shown by (7.28).

The complete maximally simplified function W given by Equation (7.24) is then:

$$W(E, D, C, B, A) = \bar{E}.\bar{D}.\bar{A} + D.\bar{B}.A + E.C.A. \tag{7.28}$$

7.7 PRODUCT OF SUMS SIMPLIFICATIONS

All Karnaugh map simplifications covered so far yielded a *simplified sum of products* form (*SOP*). When we want to produce a *simplified product of sums* form (*POS*) some changes need to be taken into account. Let us address those with the next example.

Example 7.20 Given function:

$$f(D,C,B,A)=\sum(3,4,6,7,11,12,13,14,15) \tag{7.29}$$

The ones marked in Figure 7.15a represent the minterms of function *f.* The cells marked with zeros (actually cells left blank) are all the minterms not included in *f.*

As usual we can obtain a simplified *SOP* form for *f* and this is:

$$f(D,C,B,A)=\bar{C}.A+D.C+B.A. \tag{7.30}$$

Equation (7.30) is obtained with the map of Figure 7.15a.

Now let us consider the map of *f* complement (or simply $\bar{f}$). Refer to Figure 7.15b.

From this figure we obtain a simplified *SOP* form for $\bar{f}$, which is:

$$\bar{f}(D,C,B,A)=\bar{C}.\bar{A}+\bar{C}.\bar{B}+\bar{D}.\bar{B}.A. \tag{7.31}$$

Now let us take the complement of Equation (7.31) and applying DeMorgan rules we obtain:

$$\overline{\overline{f(D,C,B,A)}}=f(D,C,B,A)=(C+A)(C+B)(D+B+\bar{A}). \tag{7.32}$$

Equation (7.32) is a maximally *simplified product of sums* for function *f(D, C, B, A).*

Figure 7.15 depicts the maps for Example 7.20.

7.8 DON'T CARE CONDITIONS

When a logic circuit is designed, we obtain its truth table and we transform the standard sum of products form into a simplified sum of products or product of sums form.

The assumption always has been up until now, that all minterms were defined. This means that minterms were either one's or zero's. There are some applications where not all the possible binary combinations that a number of binary-valued variables spanned are actually used. When this is the case, it is convenient to define the unused binary combination as a third and not previously defined state. We call such state a *don't care.* A *don't care* is typically represented with an *X.* The advantage of defining this *don't care* is convenient because the logic simplification can lead to an easier and more compact simplified *SOP* or *POS* form. Let us address this with an example.

Assume that we have a *4*-bit binary-coded-decimal (*BCD)* number. A single digit *4*-bit *BCD* number ranges from $0000_2=0_{10}$ to $1001_2=9_{10}$. If one wants to

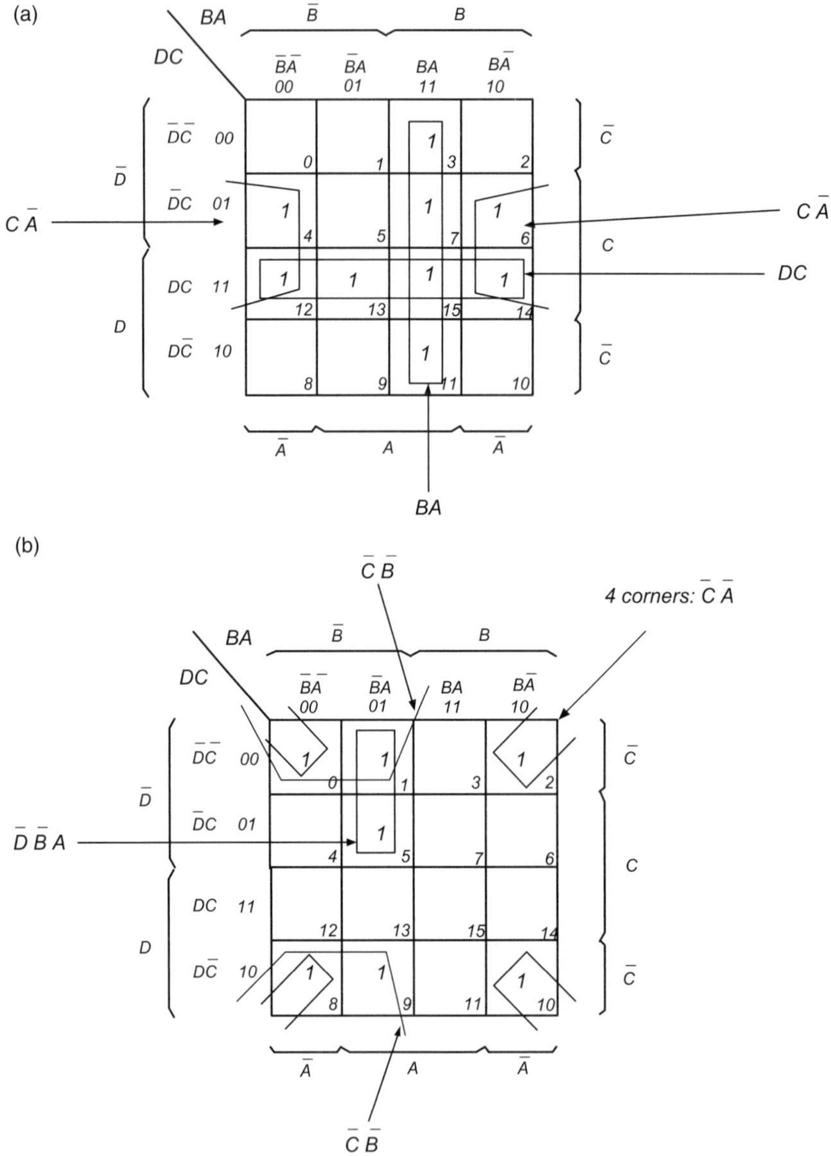

Figure 7.15 Karnaugh maps for Example 7.20: (a) map of f; (b) map of $\bar{f}$.

Table 7.16 Some Binary Coded Decimal Numbers and their decimal and binary representations

BCD	Decimal	Binary
0000_0000	0	0000_0000
0000_0001	1	0000_0001
0000_0010	2	0000_0010
0000_0011	3	0000_0011
0000_0100	4	0000_0100
0000_0101	5	0000_0101
0000_0110	6	0000_0110
0000_0111	7	0000_0111
0000_1000	8	0000_1000
0000_1001	9	0000_1001
0001_0000	10	0000_1010
0001_0001	11	0000_1011
0001_0010	12	0000_1100
⋮	⋮	⋮
0001_1000	18	0001_0010
0001_1001	19	0001_0011
0010_0000	20	0001_0100
0010_0001	21	0001_0101
0010_0010	22	0001_0110

express the number 10_{10} in *BCD*, one needs an extra *BCD* digit (a new *4*-bit set) to represent $10_{10} = 0001\_1001_{BCD}$. Table 7.16 lists some double digit *BCD* numbers, followed by their decimal and binary representations.

Example 7.21 Design a BCD digit range detector: Our problem assumes that we want to design a combinational circuit that receives as input a single-digit *4*-bit *BCD* number. When the BCD digit is either 6, 7, or 8, we want the output of the range detector circuit to be a "*1*," else we want such output to be "*0*." We want to come up with a truth table for the range detector circuit. Additionally provide a maximally simplified *SOP* form for the designed range detector circuit. Based on the requirements the truth table follows below. We define the BCD digit having bits *DCBA,* where *A* is the *LSB* and *D* is the *MSB*. Note that the problem implicitly assumes that the six *4*-bit binary combinations *1010* through *1111* will not be present at the inputs; refer to last six rows of Table 7.17.

From the truth table of Table 7.17 we can easily start filling out a four-variable *Karnaugh* map with the values of output F. Figure 7.16 depicts the four-variable map for our *BDC* range detector. First, carefully observe the six *don't cares* on minterms m_{10} through m_{15}. As expected minterms m_6, m_7, and m_8 are 1's. All other minterms are 0's.

Table 7.17 Truth table for a *BCD* range detector

Input D	Input C	Input B	Input A	Output F
0	0	0	0	0
0	0	0	1	0
0	0	1	0	0
0	0	1	1	0
0	1	0	0	0
0	1	0	1	0
0	1	1	0	1
0	1	1	1	1
1	0	0	0	1
1	0	0	1	0
X	X	X	X	X
X	X	X	X	X
X	X	X	X	X
X	X	X	X	X
X	X	X	X	X
X	X	X	X	X

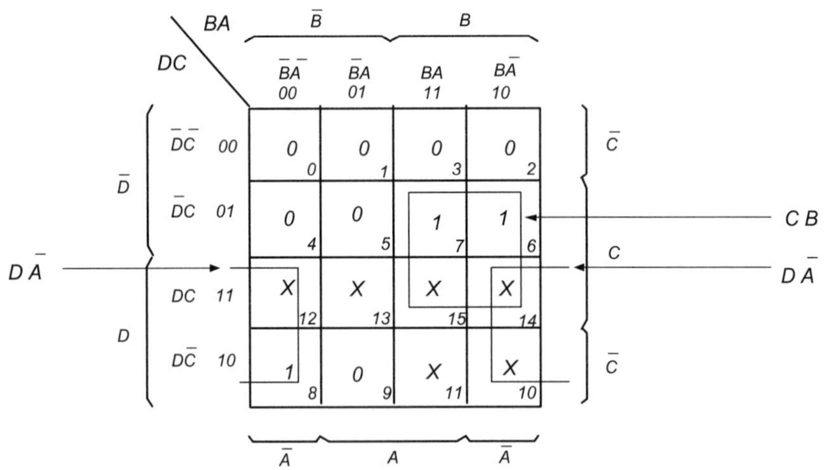

Figure 7.16 Four-variable for map for *BCD* range detector.

Here comes the most important part about simplifying with *don't care* conditions.

Since the *don't care* minterms will never be present at the *DCBA* inputs, it is to the designer's advantage, to most conveniently adopt either a value of 1 or 0 for the *don't cares* in such way that it maximally simplifies the terms to be encircled.

Figure 7.16 shows a possible way of encircling cells. For those *don't cares* that end up within the picked enclosures of minterms we assume that they are valued as 1's. For all other *don't cares* we assume they are 0's. However, in doing that we still do not change the *don't care* notation on the *Karnaugh* map, that is, we leave the *X*.

The maximally *SOP* form for the *BCD* range detector is:

$$F(D, C, B, A) = C.B + D.\bar{A}. \tag{7.33}$$

Exercise: Try other encirclements selecting other don't cares and compare your results against Equation (7.33). What can you tell about your findings?

7.9 LOGIC GATES: ELECTRICAL AND TIMING CHARACTERISTICS

Logic gates are available in integrated circuit packages or as macros or combinational logic building blocks in Application Specific Integrated Circuits (ASICs), Complex Programmable Logic Devices (CPLDs), Field Programmable Gate Arrays (FPGAs), and other devices. Figure 7.17 depicts the most common schematic symbols of the most commonly used logic gates.

All of the above gates are conceptually and sometimes physically available with more than two inputs. There may be three, four, and more inputs in a gate. Using DeMorgan's rules we will justify the logic equivalences given in Figure 7.17. For a positive *AND* gate, *A ANDed* with *B* is *A.B*. From DeMorgan's rule (Eq. 7.3)

$$\overline{A.B} = \bar{A} + \bar{B}. \tag{7.34}$$

Equation (7.34) justifies the logic equivalence between Figure 7.17c,d.

Complementing Equation (7.34) yields:

$$AB = \overline{\overline{A.B}} = \overline{\bar{A} + \bar{B}}. \tag{7.35}$$

Equation (7.35) justifies the logic equivalence of Figure 7.17a,b.

From the other DeMorgan rule (Eq. 7.2) we have that:

$$\overline{A + B} = \bar{A}\bar{B}. \tag{7.36}$$

Equation (7.36) justifies the logic equivalence between Figure 7.17g,h. Now complementing Equation (7.36) yields:

$$\overline{\overline{A + B}} = A + B = \overline{\bar{A}.\bar{B}}. \tag{7.37}$$

Equation (7.37) justifies the logic equivalence between Figure 7.17e,f. Note that neither (i) nor (j) are logically equivalent. The same is true for (k) and (l). Figure 7.17i is the logic complement of Figure 7.17j. So is Figure 7.17k,l.

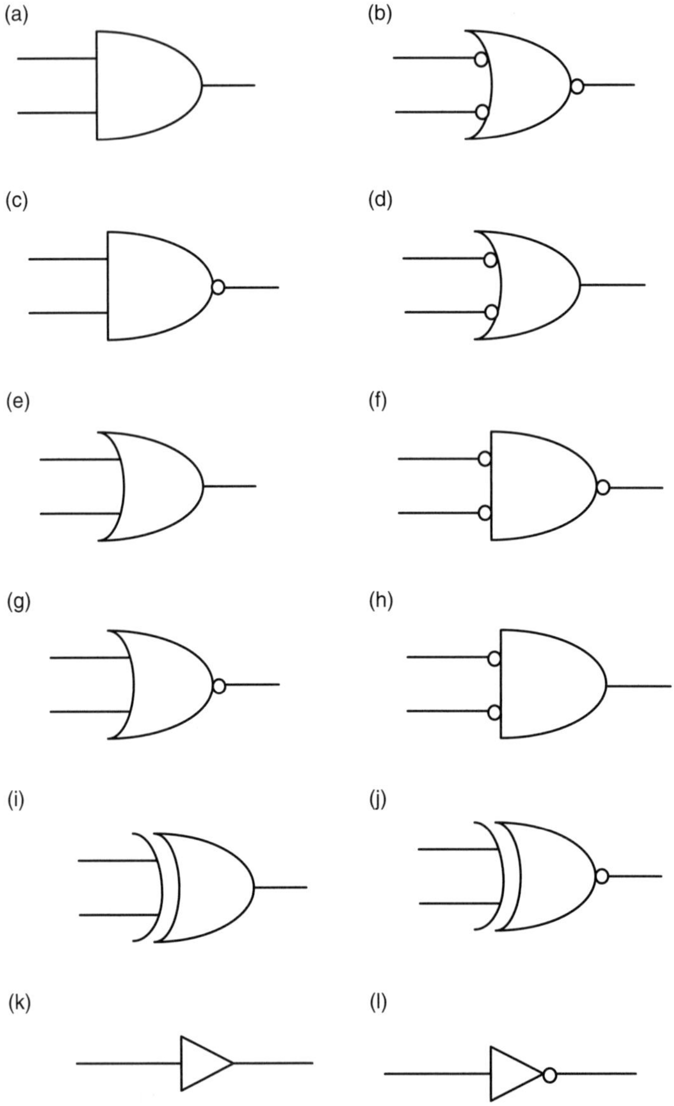

Figure 7.17 (a) Positive *AND* gate; (b) negative *OR* gate; (c) *NAND* gate; (d) DeMorganized *NAND* gate; (e) positive OR gate; (f) negative *AND*; (g) *NOR* gate; (h) DeMorganized *NOR* gate; (i) Exclusive *OR* gate; (j) Exclusive *NOR* gate or Equivalence gate; (k) buffer, no inversion; (i) inverting buffer or inverter.

7.9.1 Gates Key Electrical Characteristics

For the sake of brevity we will only consider gates that operate with 3.3 V TTL logic levels. TTL or *Transistor-Transistor-Logic* is a class of digital circuits built with *bipolar* transistors and resistors. TTL became at one point in time the most widespread logic family used in computers and almost all other electronic equipment. Within our context TTL is used to mean TTL-compatible-logic-levels. The actual logic implementation may not necessarily be TTL, it just means that its input and output logic levels comply with the TTL family of integrated circuits levels. Other families of integrated circuits are *CMOS* and *ECL*. Today one can say that CMOS is the most widespread logic family of integrated circuits.

Table 7.18 defines the voltage logic levels for a *zero (low voltage level)* input and output and for a *one (high voltage level)* input and output. Note that expressing the state of an input or an output with a *voltage level* makes it independent as to whether the application that uses such gate or circuit with high true or low true signals.

Table 7.18 is a simplified real-device data sheet characteristics for the reader's convenience. Figure 7.18 depicts a gate output driving another gate input. Both the high and the low levels are shown. Let us concentrate on the logic

Table 7.18 Some electrical characteristics of low voltage TTL (LVT)

			Limits Temperature Range: –40°C to +85°C			
Symbol	Parameter	Test Conditions	MIN	TYP	MAX	Units
Recommended operating conditions						
V_{CC}	DC supply voltage		2.7		3.3	V
V_{IH}	High-Level input voltage		2.0			V
V_{IL}	Low-level input voltage				0.8	V
I_{OH}	High-level output current				–20	mA
I_{OL}	Low-level output current				32	mA
Electrical characteristics						
V_{OH}	High-level output voltage	$V_{CC} = 2.7$ V, $I_{OH} = -6$ mA	2.4			V
V_{OL}	Low-level output voltage	$V_{CC} = 2.7$ V, $I_{OL} = 32$ mA			0.4	V
I_L	Input leakage current applies to I_{IH} and I_{IL}	$V_{CC} = 3.6$V, $V_I = V_{CC}$ or GND (0 V)			±1	μA

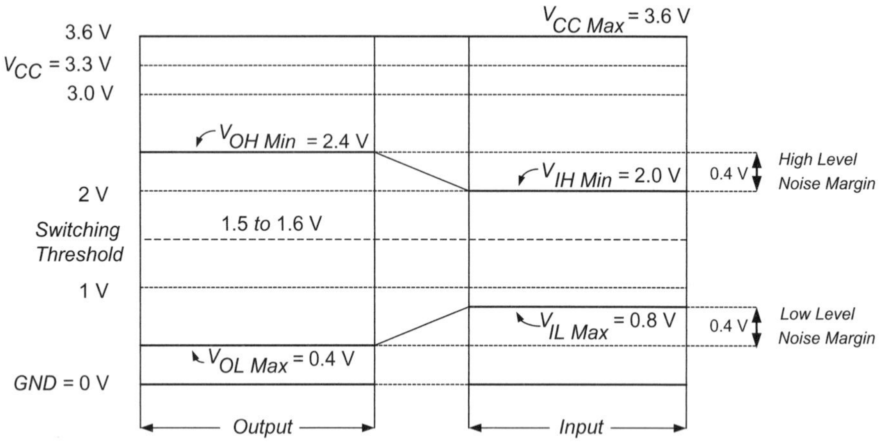

Figure 7.18 TTL output driving a TTL input.

zero or the low level. When a gate output drives a *low level* to the input of another gate, the output must not exceed $V_{OL\ MAX}$ voltage level which is 0.4 V for TTL compatible logic. While at the same time the input gate must be capable of accepting a *low level* that does not exceed a maximum level of $V_{IL\ MAX}$ of 0.8 V. Note that the difference between $V_{IL\ MAX}$ and $V_{OL\ MAX}$ is actually 0.4 V (400 mV), and it is referred to as the low-level noise margin for TTL-compatible logic. Similarly, when the output of a gate drives a *high level* to the input of another gate, the output must not be below $V_{OH\ MIN}$ of 2.4 V. It is also the case that a high output driving an input also has a 400 mV noise margin. The noise margin is a desirable voltage to have to account for system noise, power supply ripple, and other sources of noise that can couple onto the driving and the receiving lines of each gate.

Now what about the current specifications? That is, I_{OH}, I_{OL}, I_{IH}, and I_{IL}? When an output is at a high voltage level, the driving gate *sources* a current to the input gate, the sourced current flows outward from the output. Conventionally, this current is negative (refer to the I_{OH} entry in Table 7.18). When an output is at a low voltage level, the driving gate *sinks* current and sunk current conventionally has a positive sign (refer to the I_{OL} entry in Table 7.18). I_{IH} is the current into an input terminal when a specified high voltage level is applied to it. I_{IL} is the current into an input terminal when a specified low voltage level is applied to it. I_{IH} and I_{IL} are typically found only on devices with bipolar inputs and that significantly have different levels of pull-down current to provide a logic low and pull-up current to provide a logic high. CMOS devices, however, just have an I_L or a leakage current at the input. Such levels of I_L are measured at both low and high bias conditions. Figure 7.18 depicts an output driving an input, indicating all the voltage levels.

For TTL logic levels, the switching threshold is around 1.5–1.6 V.

7.9.2 Gates Key Timing Characteristics

When a digital input signal is applied to a combinational circuit the output will not respond (or change) until the combinational circuit time-propagation delay elapses.

JEDEC, the Joint Electronic Device Engineering Council, is the semiconductor engineering standardization body of the Electronics Industries Alliance, a trade association that represents all areas of industry. JEDEC defines the *propagation delay time* as the time specified between reference points on the input and output voltage waveforms with the output changing from one defined level (either *high* or *low*) to the other defined level. Thus, there will be a t_{PHL} and a t_{PLH}, respectively, a high-to-low propagation delay (t_{PD}), and a low-to-high propagation delay. The maximum value of t_{PD} simply is the worst-case or longest case of t_{PHL} and a t_{PLH}. Figure 7.19 depicts the propagation delay time that exists in a *LVTTL* combinational circuit, or simply just a LVTTL gate between an input and an output. Output *1* depicts both low-to-high and high-to-low t_{PD}. Output *2* depicts the same delays assuming it is the complement of Output *1*. Note that the time references are measured at 1.5 V, or about half of the 3.3 V power supply rail. A 1.5 V is referred to as the LVTTL logic switching threshold. Although the switching threshold of the logic may vary, perhaps as much as ±0.5 V or more, what matters is that all the timing measurements be made consistently with respect to the same 1.5-V reference level.

Examples of t_{PD} of integrated circuit gate delays are anywhere around 10 ns (older TTL technology) to as little as a fraction of a nanosecond (for high speed CMOS technologies and ECL). Reference 5 in the Further Reading section has a discussion on TTL, CMOS, and ECL families of integrated circuits.

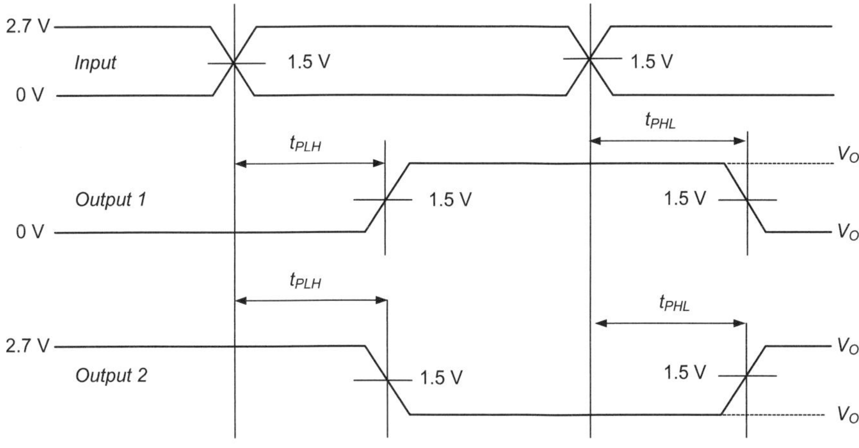

Figure 7.19 Logic gate propagation delay times.

7.10 SUMMARY

This chapter introduced the reader to combinational circuits, which are also referred to as circuits with no memory capability. Binary numbers were presented along with the essential elements of switching or Boolean algebra.

Standard or canonical SOP and POS forms are ways of representing logic functions. For the purpose of logic implementation, where usually we want the number of gates to be reduced as well as the number of inputs per gate to obtain simplified SOP and POS. Methods of simplification were presented by covering *2* through *5*-variable *Karnaugh* maps. *K* maps of *6* or more variables become somewhat impractical to use effectively. We will address other logic design techniques to overcome using huge *K.* maps. Finally, we studied the most basic electrical and timing characteristics of logic gates. The examples were centered around TTL-compatible logic levels gates, not necessarily implemented in TTL technology. Although it is completely true that TTL technology is obsolete, other logic families, like CMOS and BiCMOS, have adopted TTL levels to interface to the many devices, such as line drivers, that continue to use TTL levels.

FURTHER READING

1. M. Morris Mano, *Digital Design*, Prentice Hall, Upper Saddle River, NJ, 1984.
2. John F. Wakerly, *Digital Design: Principles and Practices*, 3rd ed., Prentice Hall, Upper Saddle River, NJ, 2001.
3. Davis Money Harris and Sarah L. Harris, *Digital Design and Computer Architecture*, Morgan Kaufmann Publishers, San Mateo, CA, 2007.
4. Charles H. Roth Jr., *Logic Design*, 2nd ed., West Publishing Company, St. Paul, MN, 1979.
5. Ed Lipiansky, *Embedded Systems Hardware for Software Engineers*, McGraw-Hill, New York, 2011.

PROBLEMS

7.1 Convert the following 16-bit positive binary numbers to decimal:

(a) 1101_1111_1010_0001

(b) 1111_1111_1111_1111

(c) 1000_0000_0000_0000

(d) 1000_1000_1000_1000

(e) 1001_0110_1100_0111

7.2 Convert the following 16-bit 2's complement numbers to decimal:

(a) 1101_1111_1010_0001

(b) 1111_1111_1111_1111

(c) 1000_0000_0000_0000

(d) 1000_1000_1000_1000
(e) 1001_0110_1100_0111

7.3 Convert the following 4-BCD-digit (16-bits) numbers to decimal:
(a) 1101_1111_1010_0001
(b) 1111_1111_1111_1111
(c) 1000_0000_0000_0000
(d) 1000_1000_1000_1000
(e) 1001_0110_1100_0111

7.4 Convert the following decimal numbers into 16-bit 2's complement.
(a) 1537
(b) –10418
(c) 32700
(d) 0
(e) –32700

7.5 Using De Morgan's rules, find simplified logic equivalent Boolean expressions:
(a) $\overline{A+B+C}$
(b) $\bar{A}.\overline{C+B}$
(c) $\bar{A}.\bar{B}+A.\bar{B}+\bar{A}.B+A.B$
(d) $A.\bar{B}.\bar{C}.\bar{D}+A.\bar{B}.\bar{C}.D+A.\bar{B}.C.D$

7.6 Express the following functions in SOP and POS canonical forms:
(a) $F=D(\overline{A+B})+\bar{B}D$
(b) $F=\bar{y}z+wx\bar{y}+wx\bar{z}+\bar{w}\bar{x}z$
(c) $F=(\bar{A}+B)(\bar{B}+C)$
(d) $F=1$
(e) $F=(xy+z)(y+xz)$

7.7 Using Karnaugh maps find simplified sum-of-product forms for the following logic functions:
(a) $f(A,B)=\Sigma(0,1,3)$
(b) $g(A,B,C,D)=\Sigma(0,1,4,58,9)$
(c) $h(A,B,C,D)=\Sigma(0,1,2,3,8,10,11,15)$
(d) $k(A,B,C,D)=\Sigma(1,3,4,7,8,9,10,11,14)$

7.8 Using Karnaugh maps find simplified product-of-sum forms for the following logic functions:
(a) $f(A,B)=\prod(1,2,3)$
(b) $g(A,B,C,D)=\prod(0,1,4,5,8,9)$

(c) $h(A,B,C,D) = \prod(1,2,5,6,9,12,13,14)$

(d) $k(A,B,C,D) = \prod(1,3,4,7,8,9,10,11,14)$

7.9 Graphically depict a 3-variable XOR using Venn Diagrams.

7.10 Graphically depict a 4-variable XNOR using Venn Diagrams.

7.11 Generate the truth table of a 4-variable XOR function.

7.12 Obtain the truth table for the following Boolean function: $F(X,Y,Z) = X.Y + X.\bar{Y} + \bar{Y}.Z$.

7.13 Write the sum-of-products form of a 3-variable XOR.

7.14 Write the product-of-sums form of a 3-variable XOR.

7.15 Obtain the truth table of a 4-variable majority logic circuit.

That is, majority is obtained whenever two or more variables are true, else majority is false.

7.16 Create a 2-level logic implementation of the majority function obtained in Problem 7.15.

7.17 Assume that you have an inverting gate with a 10 ns high-to-low and low-to-high propagation delay. If you connect the output of this gate to its input with a zero-delay wire, sketch the waveform that you would see with an oscilloscope. An oscilloscope is an instrument that allows one to visualize how an electric waveform varies with respect to time.

7.18 Given logic gates which all have a high-to-low and low-to-high 20-ns propagation delay, what is the maximum propagation delay of a function implemented in three levels of logic.

7.19 Given logic gates which all have a high-to-low and low-to-high 20-ns propagation delay, what is the maximum propagation delay of a function implemented in four levels of logic.

7.20 From doing Problems 7.18 and 7.19 what can you generalize when a logic function is implemented with more levels of logic? Clearly the propagation delay increases linearly with each new level of logic.

8

DIGITAL DESIGN BUILDING BLOCKS AND MORE ADVANCED COMBINATIONAL CIRCUITS

8.1 COMBINATIONAL CIRCUITS WITH MORE THAN ONE OUTPUT

Not all combinational circuits have a single output, like it was presented throughout most of Chapter 7. As a matter of fact, many applications have multiple outputs. Let us examine this with an interesting example.

Example 8.1 Design a combinational circuit to decode a four-bit *BCD* number that drives segments of a seven-segment *LED* display. The display must light up showing the corresponding BCD number presented at the input of the decoder. Figure 8.1a depicts a seven-segment LED display. Each segment has been labeled with the letters *a* through *g*. The display has seven inputs, one per segment. Assume that a *high-level* voltage presented at the input of a segment turns such segment *ON*; else when a *low-level* voltage is presented, the segment is *OFF*. When we want the display to show the number *0*, we must ensure to apply high-levels or one's to segments: *a, b, c, d, e,* and *f,* while we need to present a zero to segment *g*. Figure 8.1b shows the wiring of a single segment, and Figure 8.1c shows the schematic representation where the LED is shown with its corresponding symbol. Both (b) and (c) show the current limiting resistor that is placed in-series with the LED so that the appropriate current makes the LED shine when turned ON as the manufacturer specifies.

Electrical, Electronics, and Digital Hardware Essentials for Scientists and Engineers, First Edition. Ed Lipiansky.

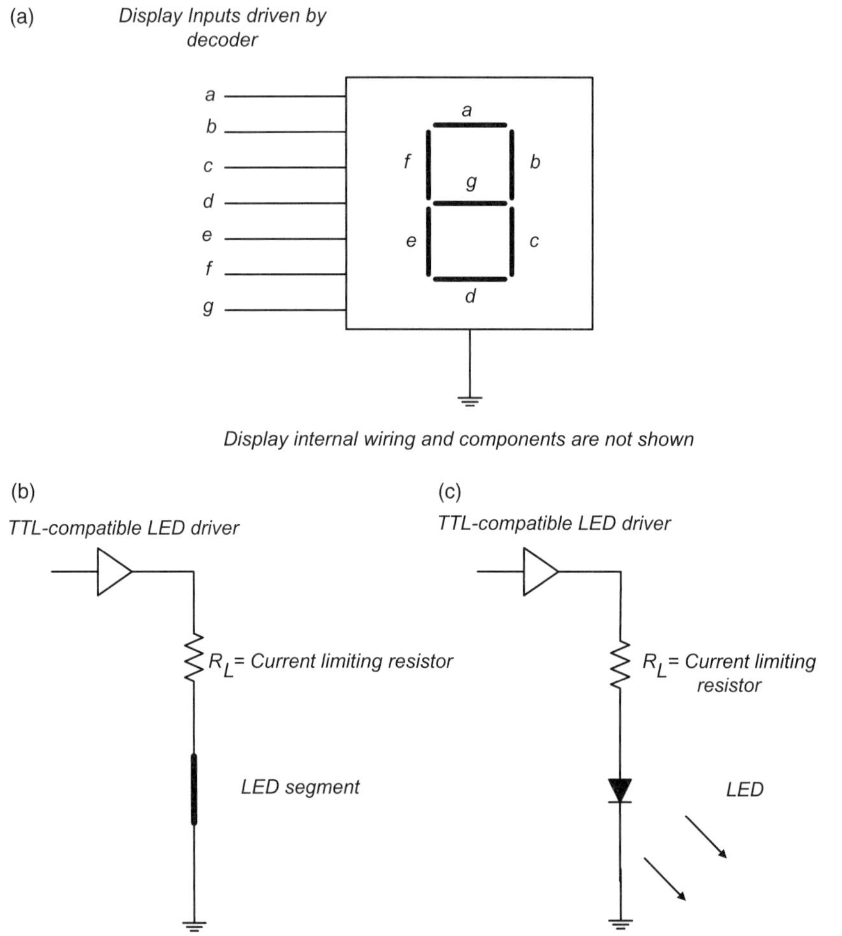

Figure 8.1 (a) Seven-segment display assembly; (b) detailed wiring and connectivity of one segment; (c) detailed wiring as shown in part (b) with the segment replaced with a LED schematic symbol.

The LED driver, LED the current limiting resistor, and the LED segment are all assumed to be part of the display assembly. In addition to the seven LED segments, the assembly contains seven drivers and seven resistors.

Brief Calculation of the Current Limiting Resistor

Assume that the current through the LED for the intended typical luminous intensity required by the manufacturer is 10 mA. The manufacturer also specifies a maximum forward voltage drop. This is $V_{DROPMax} = 2.0$ V. Moreover, assume that our LED driver drives TTL-compatible voltage levels. Since the minimum voltage at the output of the driver is $V_{OHMin} = 2.4$ V sourcing a current of 10 mA, the current limiting resistor value is calculated as follows:

$$R_L = \frac{V_{OHMin} - V_{DROPMax}}{0.010} = \frac{2.4 - 2.0}{0.010} = 40\ \Omega, \tag{8.1}$$

where $V_{OHMin} = 2.4$ V is dictated by the driver TTL compatibility. The driver must be selected so that it can source at least 10 mA. A driver of somewhat higher current source capability may also be selected to do the job. Ultimately, the series resistor will limit the current needed by each segment.

Let us quickly check the amount of power that the resistor will dissipate.

$$P_R = I_{LED}^2 \times R_L. \tag{8.2}$$

Since $I_{LED} = 0.01$ A and R = 40 Ω, thus:

$$P_R = (0.01)^2 \times 40 = 4\ \text{mW}. \tag{8.3}$$

Since some resistors available can handle 1/16 W (62.5 mW) we can use a 1/16 W-rated resistor. The above analysis does not take into consideration variations of LED current and LED voltage forward drop, resistor variability, power supply changes, and temperature changes. The intent of the above calculation is to provide the reader with the basics to calculate the current limiting resistor value.

Figure 8.2 depicts the LED assembly driven by the BCD-to-seven segment decoder that we need to design. The inputs to the decoder are assumed to be BCD *0000, 0001* through *1001*; the other six binary combinations *(1010–1111)* will be assumed not to be present as decoder inputs.

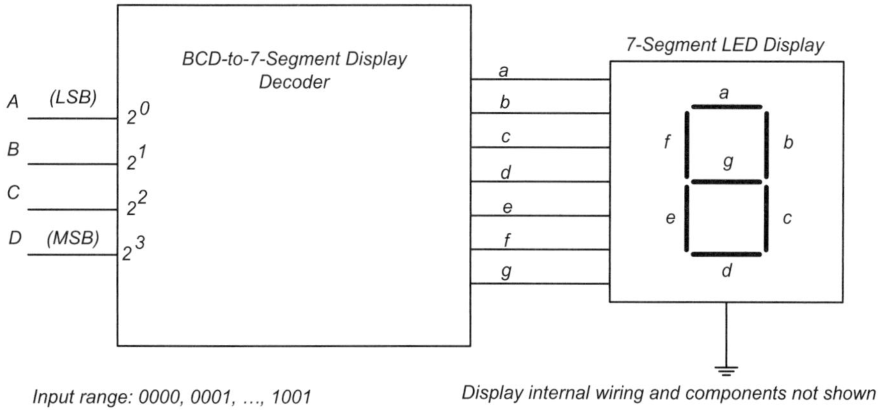

Figure 8.2 Decoder driving a seven-segment LED display.

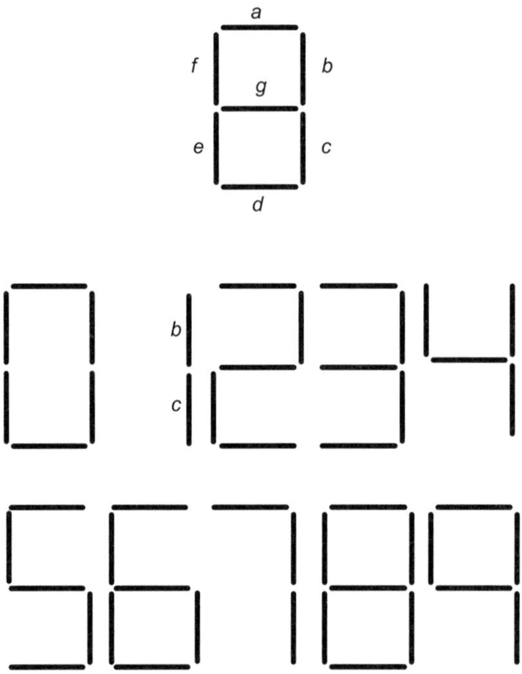

Figure 8.3 Seven-segment LED display segments to numerical mappings.

We will assume that we want the numbers *0* through *9* displayed as depicted by Figure 8.3. Additionally, let us remember that a *high-level voltage* turns a segment ON, while a *low-level* voltage turns it OFF.

At this point we are ready to start working on this example's truth table, which is presented in Table 8.1.

Table 8.1 contains the BCD number bits (*D, C, B, A*) on the four left-hand side columns. Clearly ***A*** is the least significant bit (*LSB*). The columns for each segment are labeled as ***a***, ***b***, and so forth. It is very convenient and important to observe that since the last six binary combinations *1010* through *1111* are not present, because the input number is by definition a BCD number which only spans *0000* through *1001*, it works out to our advantage to place *don't care* conditions (*X's*). So what needs to be done to find a simplified *SOP* forms for each the seven segments? Proceeding we obtain the following seven *K.* maps, depicted by Figure 8.4a through g.

The maximally simplified SOP forms for every segment are given below:

$$\text{Segment a}: a(D, C, B, A) = D + B + C.A + \overline{C}.\overline{A} \tag{8.4}$$

$$\text{Segment b}: b(D, C, B, A) = \overline{C} + D + \overline{B}.\overline{A} + B.A \tag{8.5}$$

$$\text{Segment c}: c(D, C, B, A) = D + C + \overline{B} + A \tag{8.6}$$

Table 8.1 Truth table for Example 8.1, BCD-to-seven-segment decoder

BCD Input bits D: MSB, A: LSB				Outputs to Segments							
D	C	B	A	a	b	c	d	e	f	g	Displays Number
0	0	0	0	1	1	1	1	1	1	0	0
0	0	0	1	0	1	1	0	0	0	0	1
0	0	1	0	1	1	0	1	1	0	1	2
0	0	1	1	1	1	1	1	0	0	1	3
0	1	0	0	0	1	1	0	0	1	1	4
0	1	0	1	1	0	1	1	0	1	1	5
0	1	1	0	1	0	1	1	1	1	1	6
0	1	1	1	1	1	1	0	0	0	0	7
1	0	0	0	1	1	1	1	1	1	1	8
1	0	0	1	1	1	1	0	0	1	1	9
1	0	1	0	X	X	X	X	X	X	X	–
1	0	1	1	X	X	X	X	X	X	X	–
1	1	0	0	X	X	X	X	X	X	X	–
1	1	0	1	X	X	X	X	X	X	X	–
1	1	1	0	X	X	X	X	X	X	X	–
1	1	1	1	X	X	X	X	X	X	X	–

$$\text{Segment d}: d(D, C, B, A) = C.\overline{B}.A + \overline{C}.\overline{A} + \overline{C}.B + B.\overline{A} \tag{8.7}$$

$$\text{Segment e}: e(D, C, B, A) = \overline{C}.\overline{A} + B.\overline{A} \tag{8.8}$$

$$\text{Segment f}: f(D, C, B, A) = C.\overline{A} + C.\overline{B} + \overline{B}.\overline{A} + D \tag{8.9}$$

$$\text{Segment g}: g(D, C, B, A) = \overline{C}.B + B\overline{A} + D + C\overline{B}. \tag{8.10}$$

Each of the seven output functions (*a* through *g*) depends on the same four independent binary variables *A, B, C,* and *D*. Some of the functions have repeated terms, for example, taking a close look at Equations (8.4), (8.7), and (8.8) we see that they have a common term $\overline{C}.\overline{A}$ among them. When we do the logic implementation of functions (a) through (g) we only need to generate the term $\overline{C}.\overline{A}$ once, then feed it into Equations (8.4), (8.7), and (8.8). Before getting into the logic implementation of our seven functions let us identify all other repeated terms. These are: $\overline{B}.\overline{A}$ present in Equations (8.5) and (8.9) and term $\overline{C}.B$ present in Equations (8.7) and (8.10), and term $\boldsymbol{C}.\overline{B}$ present in Equations (8.9) and (8.10).

Following Equations (8.4) through (8.10) these are implemented with logic gates in Figure 8.5a through g. But we are not done yet. Each of the segment functions *a, b,* through *g,* is simplified *SOP* forms in a stand-alone sense. However, since we are implementing all seven functions, which are all functions of input variables *A, B, C,* and *D* there are few other things that we can do in order to reduce the number of logic gates that we use. First by inspection

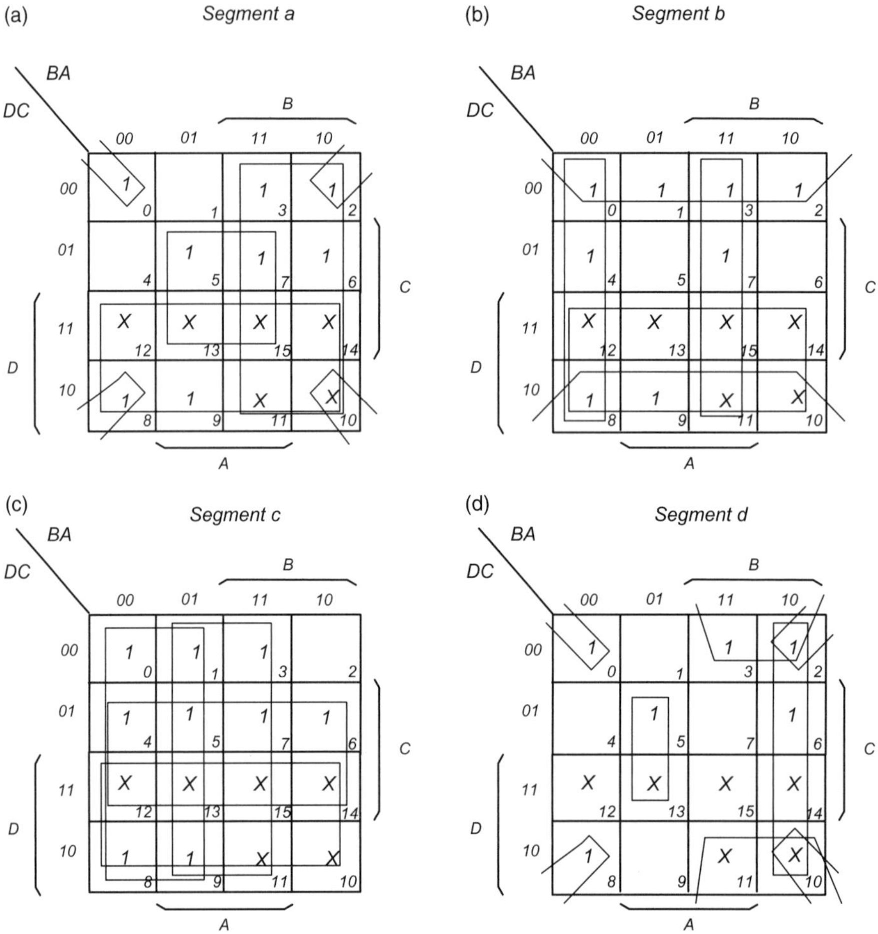

Figure 8.4 Illustration of segments a through g.

of Figures 8.5a through g we can see that each function uses some subset of the variables *A*, *B*, *C*, and *D* and their complements ($\overline{A}$, $\overline{B}$, $\overline{C}$, $\overline{D}$); this means that once we have *A*, *B*, *C*, and *D* and generate their complements once, using four inverters, the variables, and their complements can be connected to each of the functions that require them. For example, referring to Figure 8.5a, we note that $\overline{C}$ is used in the lower *AND* gate. Additionally, $\overline{C}$ is used in Figure 8.5b as an input to the four-input *OR* gate; thus we do not need to use a second inverter to generate $\overline{C}$ again. The same applies to other uses of $\overline{C}$ throughout the rest of the segment functions. Finally. the above is true for all input variables and their complements.

We can still reduce the number of logic gates a little more. Looking further at Figure 8.5a, note that the term $\overline{C}.\overline{A}$ is the fourth input of the *OR* gate for

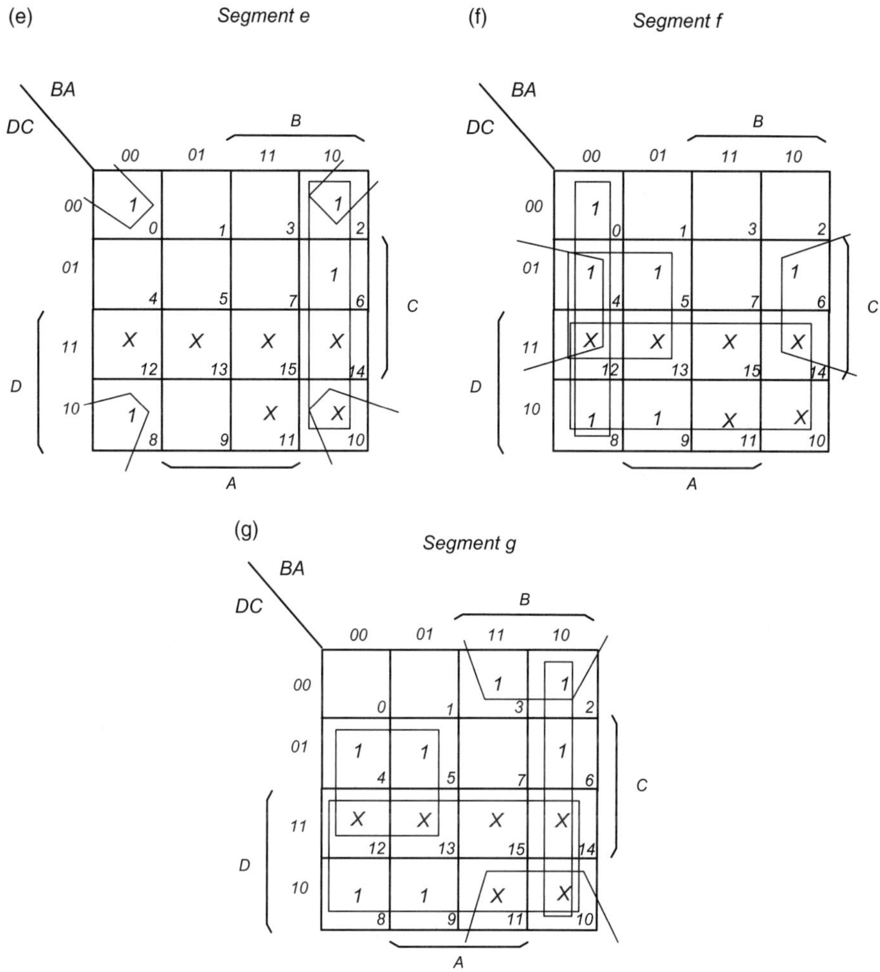

Figure 8.4 (*Continued*)

segment *a*. Term $\overline{C}.\overline{A}$ can also be found as the third input to the *OR* gate of Figure 8.5d and the first *OR* input of the *OR* gate of Figure 8.5e. What does this mean? It means that we do not to repeat the *AND*-ing logic that creates three different $\overline{C}.\overline{A}$ terms in (a), (d), and (e). We actually need just one *AND* gate that produces $\overline{C}.\overline{A}$, and this term is fed to all other users of the $\overline{C}.\overline{A}$ term. This saves us two *AND* gates. Something very similar occurs with terms $\overline{B}.\overline{A}$ and $\overline{C}.B$.

Exercise: Redraw the circuits of Figure 8.5a through g reducing the logic gates by: deleting repetitive logic terms produced by the *AND* gates.

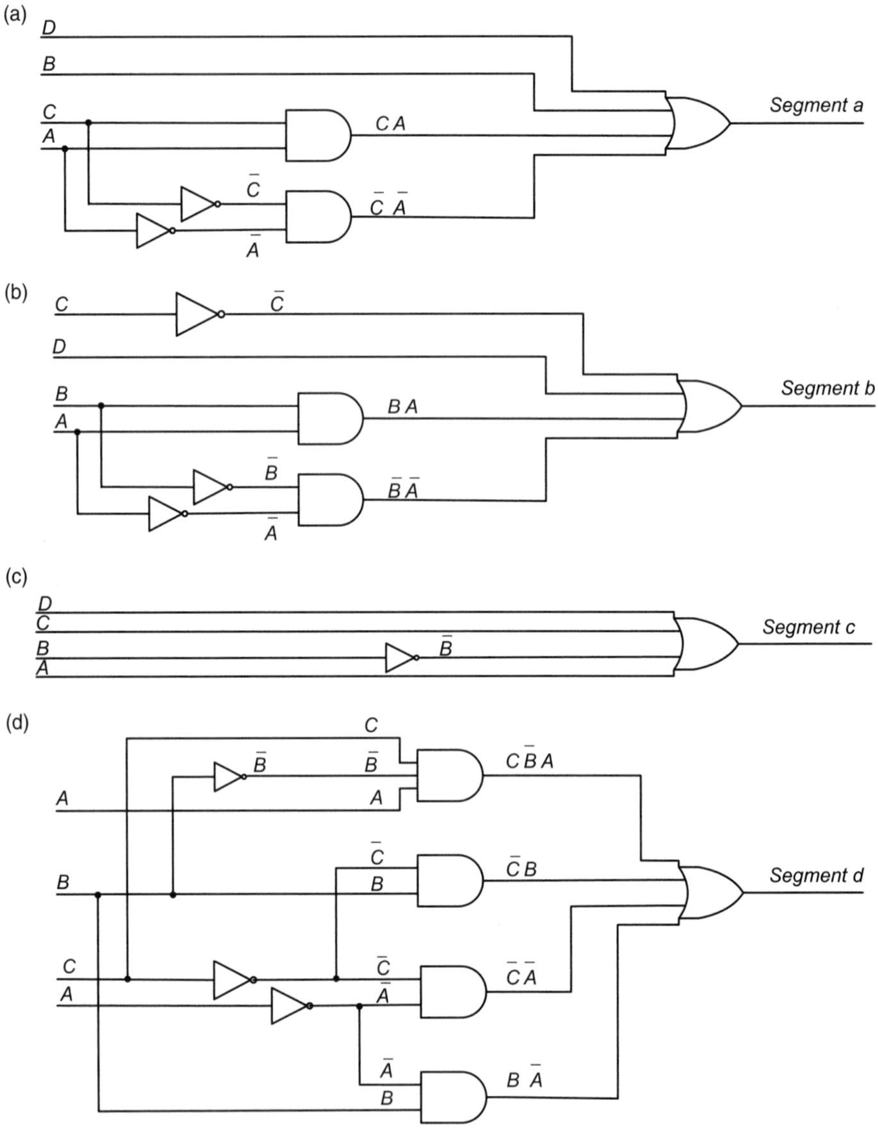

Figure 8.5 Seven-segment decoder logic implementation for segments a through g.

8.2 DECODERS AND ENCODERS

Decoders and encoders are combinational logic circuits. A binary decoder is a digital circuit that has n binary inputs and 2^n outputs. For example, a decoder with three inputs produces eight outputs; this decoder is referred to as a *3-to-8* decoder. Let us assume that the outputs are *active high* or *high-true* signals;

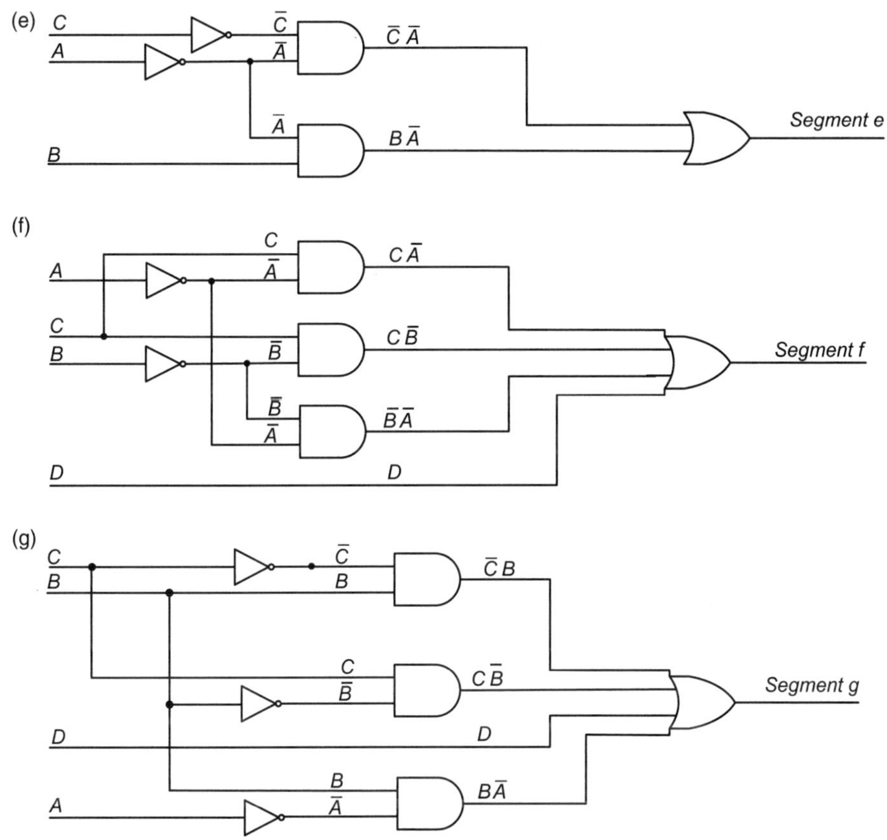

Figure 8.5 (*Continued*)

this means that an asserted signal is interpreted as a high level and this high level is a *one*. Conversely, an inactive or negated output is a low output and such low is a *zero*. Table 8.2 depicts the truth table for such a decoder. The *LSB* input is named A, while the *MSB* is named C. As expected, note that the three inputs span a total of $2^3 = 8$ binary combinations, starting at *000* through *111*. Each of its 8 outputs is associated with each one of the eight binary combinations. In such way that input *000* is associated with Y_0, input *001* is associated with Y_1 and so on.

Each output $Y_0, \ldots, Y_7$ is respectively associated to its output *000, . . . , 111*. Outputs are asserted in a mutually exclusive fashion, that is, one at a time.

By inspection of Table 8.2 we see that if the input code to the decoder is *100*, output Y_4 is *1* while all other outputs are *zero*. The truth table of our decoder has a fourth input that provides a master enable to the component. When the enable is high, the decoder works as we already described. When

Table 8.2 Truth table of a 3-bit decoder with an active high enable

Inputs				Outputs							
C (2^2)	B (2^1)	A (2^0)	E	Y_0	Y_1	Y_2	Y_3	Y_4	Y_5	Y_6	Y_7
0	0	0	1	1	0	0	0	0	0	0	0
0	0	1	1	0	1	0	0	0	0	0	0
0	1	0	1	0	0	1	0	0	0	0	0
0	1	1	1	0	0	0	1	0	0	0	0
1	0	0	1	0	0	0	0	1	0	0	0
1	0	1	1	0	0	0	0	0	1	0	0
1	1	0	1	0	0	0	0	0	0	1	0
1	1	1	1	0	0	0	0	0	0	0	1
X	X	X	0	0	0	0	0	0	0	0	0

the enable is zero or negated, all the decoder output are zero, thus the decoder is disabled. That is, no matter what the values of its three binary input bits are, the outputs remain low as long as the enable is low.

From another point of view, the decoder can be seen as a *minterm* generator. Note that our 3-bit input decoder produces $Y_0 = 1$ upon input combination *000*.

Y_0 is *minterm* m_0 since $\overline{C}$, $\overline{B}$, and $\overline{A}$ are negated. Let us recall from the previous chapter that $m_0 = \overline{C}.\overline{B}.\overline{A}$ when we have a three-bit or three-variable function. The reader should convince herself that that is the case for every one of the eight *minterms*. Based on the decoder truth table, one cannot have more than one output asserted at any given time. Refer to Table 8.2 once more. The enable provides the decoder with a feature to negate all outputs regardless of the input present at inputs *C, B,* and *A*. This enable is useful when we want to make larger decoders with smaller ones. We will see that the enable allows us to interconnect the decoders in the appropriate manner. An example of this will be discussed soon. The decoder logic implementation is straightforward. Initially ignoring the decoder *enable*, we can think of our *3-to-8* decoder having eight three-input *AND* gates into which we present our eight binary combinations *000* through *111*. Let us name each *AND* gate as *AND gate 0, 1, 2,* and so forth. Upon presenting *000* to the inputs of *AND* gate *0* we want *AND* gate *0* output to assert while all other *7 AND* gates we want to see negated. Similarly upon presenting *001* to the inputs of AND gate *1*, we want *AND* gate *1* output to be asserted while all other *AND* gate outputs need to be negated. This procedure is carried for all *8 AND* gates to obtain our *3-to-8* decoder. Now it is time to go back to the decoder's *enable.* Since we want the enable not to interfere with the decoder functionality when *enable* is *1,* we just use four-input *AND* gates instead of the three-input ones used before. So upon *enable* being a *1* or asserted allows decoder operation as usual. When *enable* is negated all outputs are negated because a zero at the input of every one of the 8 4-input *AND* gates negates all outputs. Figure 8.6 depicts a possible logic implementation of a *3-to-8* decoder with an active high master *enable.*

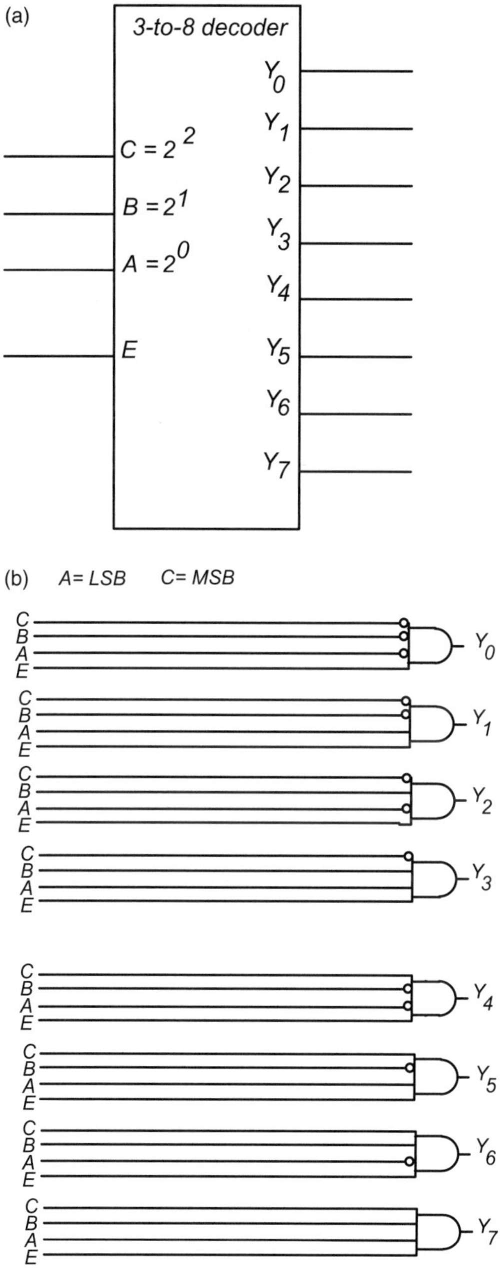

Figure 8.6 (a) Three-to-eight decoder with active high enable symbol; (b) a logic gate implementation of the decoder.

Note that the logic implementation of Figure 8.6b adopted the following notation to offer faster and easier readability of the circuit. The inputs *A, B,* and *C* may or may not have to be inverted depending on which *AND* gate output they need to assert. Instead of drawing explicitly an inverter at the input of every *AND* gate that requires its input to be inverted we draw a *bubble*. A *bubble* represents an inversion in the signal path in which it is drawn. For example, a *NAND* gate has a *bubble* at its output that means that the *NAND* is an *AND* followed by an inverter. Back to our Figure 8.6b explanation, *AND* gate Y_0 has to produce a *1* output upon *enable* = *1* and *C.B.A* = *000,* thus *AND* gate Y_0 has three bubbles to complement all three inputs *A, B,* and *C.* Similarly note that *AND* gate Y_1 has only two *bubbles* to negate inputs *C* and *B*, while input *A* is presented to the *AND* without inversion. A similar reasoning follows for the rest of the *AND* gates. Just remember that decoder input *A* is the least significant bit (2^0), while decoder input *C* is the most significant bit (2^2).

Example 8.2 Using a 3-to-8 decoder implement the following logic function:

$$f(C, B, A) = \sum(2,5,7). \tag{8.11}$$

Since function *f* is a three-variable function, and a 3-to-8 decoder is a 3-bit function minterm generator, the implementation of Equation (8.11) consists simply of *OR-ing* the three minterms m_2, m_5, and m_7. Figure 8.7 depicts this implementation.

This is a good time to talk about the decoder unused outputs. Is there anything wrong with that? From an electrical point of view, there is nothing wrong about leaving combinational circuit outputs *floating* or just simply *not-connected* as Figure 8.7 shows. It is *not* correct though to leave any combina-

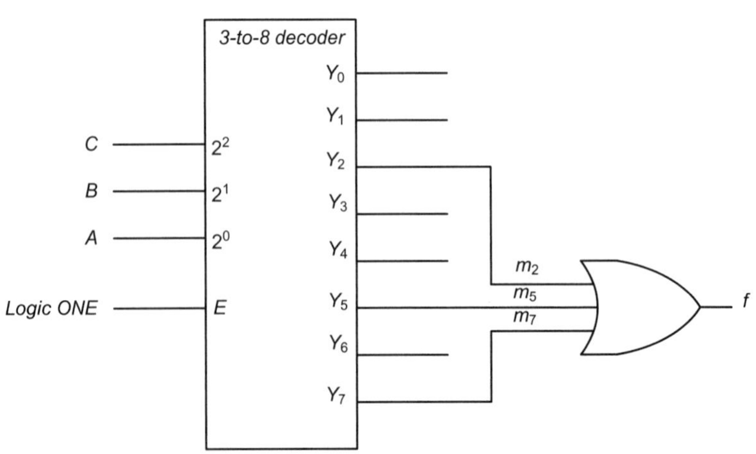

Figure 8.7 Decoder and gates implementation for Example 8.2.

tional circuit inputs *non-connected* or *floating*. Why? Because a floating input has no solid logic or voltage level driving such input. Since logic gates like all electronic circuits are susceptible to electrical and electronic noise, leaving a floating input is an opportunity for random noise to couple into the input and drive the input to the incorrect level. In summary, an unused input must be either tied down to zero ground or tied to a high voltage level or a one, typically the power supply voltage of the logic gate. Of course it is the job of the logic designer making the right choice to what input level to tie the unused input. As a quick example, let us look into a 3-input *AND* gate. Assume the gate is left over logic that we want to use for some other purpose on a board or part of a logic design. However, we only need a 2-input *AND* for this particular application. Can we still use the 3-input *AND* as a 2-input *AND*? The answer is yes, but we should not use the three inputs. Since we have an *AND* gate, tying the unused input to a high voltage level (logic one) in effect removes the third input out of the logic equation. The other two inputs of the *AND* gate behave as a 2-input gate.

Exercise: Prove the above statement with the use of a truth table. Are there any other ways to use a *3*-input *AND* gate, so that it behaves as a *2*-input *AND* gate?

8.2.1 Making Larger Decoders with Smaller Ones

Decoders of larger sizes, such as *5-to-32, 6-to-64* or larger will likely have to be constructed with smaller available decoders. One of the limitations of discrete *IC* decoders is that the larger they are, the larger is their number of pins. It is generally not practical for manufacturers to make huge decoders. Thus, it is usually left to the logic designer to assemble very large decoders using smaller ones or using programmable devices.

Example 8.3 Let us assume that we are given two *2-to-4* decoders with an active high enable input, and somehow we want to build with both of them plus some minimal amount of additional logic a *3-to-8* decoder. Of course for the sake of this example we will assume that we do not have or are not allowed to use a *3-to-8* decoder. Table 8.3 depicts the truth table of a *2-to-4* decoder with active high enable and active high outputs.

What we want to do is somehow connect two *2-to-4* decoders such that both jointly reproduce the truth table of a *3-to-8* decoder such as the one described by Table 8.2 at the beginning of the Decoders and Encoders Section. We will assume that the composite *3-to-8* decoder we are about to build will not necessarily have an enable input. This is not a big imposition; it is just a requirement that we make not to add a few more gates to the logic.

Figure 8.8 depicts the interconnection of two *2-to-4* decoders. Let us understand what such arrangement logically does.

Table 8.3 Truth table of a 2-to-4 decoder with active high enable

Inputs			Outputs			
B (2^1)	A (2^0)	E	Y_0	Y_1	Y_2	Y_3
0	0	1	1	0	0	0
0	1	1	0	1	0	0
1	0	1	0	0	1	0
1	1	1	0	0	0	1
X	X	0	0	0	0	0

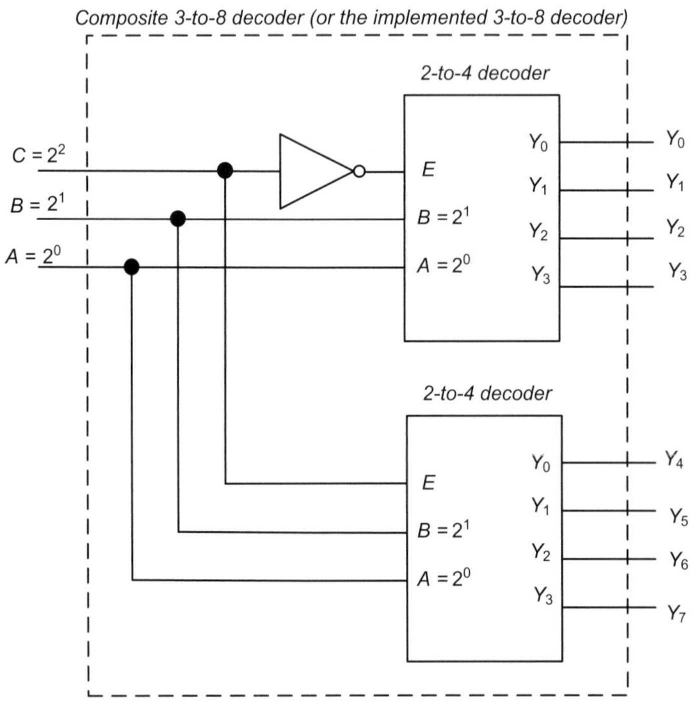

Figure 8.8 Three-to-eight decoder implementation with *2-to-4* decoders and one inverter.

The upper *2-to-4* decoder is wired such that its four outputs Y_0 through Y_3, will behave as the Y_0 through Y_3 outputs of the composite *3-to-8* decoder that we are trying to build. The lower *2-to-4* decoder is wired such that its four outputs Y_0 through Y_3, will behave as the Y_4 through Y_7 outputs of the composite *3-to-8* decoder. Furthermore, notice that both *A* and *B* inputs of each *2-to-4* are tied together and in turn they will also become the composite *3-to-8* decoder *A* and *B* inputs, where *A* is the *LSB*. Finally, the most interesting part of the design of Figure 8.8, is the way in which both enables are handled. The upper *2-to-4* decoder *E* enable input ties through an inverter to input *C*, the *MSB* of the composite decoder. Why? Note that upon *C*, *B*, and *A* binary combinations *000* through *011* being presented to the composite decoder, since *C* the *MSB* is inverted by the external inverter, the upper decoder behaves just like the composite *3-to-8* but just for the first four binary combinations of inputs (0 through 3). On the other hand, since the *E* enable of the lower *2-to-4* decoder is directly connected to input *C* of the composite, the lower decoder operates as the *3-to-8* composite one for *CBA* binary combinations, four through seven.

Exercise: Carefully trace the behavior of the composite decoder of Figure 8.8 and convince yourself that indeed it operates as a *3-to-8* decoder.

8.2.2 Encoders

An encoder is a combinational logic block that performs the inverse operation of a decoder. For example, for a *2-to-4* decoder, the associated encoder is a logic block with 4 inputs and 2 binary encoded outputs. An important combinational block used in embedded systems is the *priority encoder.* This encoder is important because it expands the number of interrupts that a micro controller is capable of handling using a single micro controller interrupt input line.

Example 8.4 Assume that a single interrupt line, an input to a micro controller, needs to have some logic in front of it to allow four interrupts to be *funneled* into the micro controller single interrupt line. Additionally, we want our priority encoding logic to supply the interrupt priority level of the interrupt with the highest priority on the P_1 and P_0 binary encoded outputs. Refer to the priority encoder schematic symbol in Figure 8.9.

Assume that interrupt priority *3* (I_3) is the highest while priority *0* is the lowest. If two interrupts assert at the same time, say *3* and *1*; since *3* has higher priority than *1,* we want the priority encoder to produce an encoded binary *3* at its output P_1 and P_0. In addition we want our priority encoder to assert a

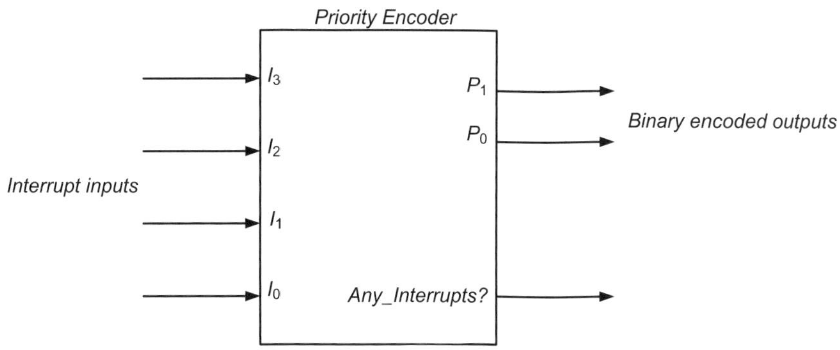

Figure 8.9 Priority encoder schematic symbol.

Table 8.4 Priority encoder truth table

Interrupt Inputs				Outputs		
I_3	I_2	I_1	I_0	P_1	P_0	Any_Interrupts?
0	0	0	0	X	X	0
0	0	0	1	0	0	1
0	0	1	X	0	1	1
0	1	X	X	1	0	1
1	X	X	X	1	1	1

Table 8.5 Depiction of the expansion of Table 8.4

Interrupt Inputs				Outputs		
I_3	I_2	I_1	I_0	P_1	P_0	Any_Interrupts?
0	0	0	0	0	0	0
0	0	0	1	0	0	1
0	0	1	0	0	1	1
0	0	1	1	0	1	1
0	1	0	0	1	0	1
0	1	0	1	1	0	1
0	1	1	0	1	0	1
0	1	1	1	1	0	1
1	0	0	0	1	1	1
1	0	0	1	1	1	1
1	0	1	0	1	1	1
1	0	1	1	1	1	1
1	1	0	0	1	1	1
1	1	0	1	1	1	1
1	1	1	0	1	1	1
1	1	1	1	1	1	1

third output to indicate that no interrupts are asserted. Table 8.4 presents a complete description of how we want our priority encoder logic to work. Note that Table 8.4 has several don't-care conditions. For the first line, when no interrupts are asserted the *Any Interrupts?* output is negated meaning, there are no interrupts, thus the priority code bits P_1 and P_0 are don't cares. For the next line of the truth table when interrupt I_0 asserts, while I_1, I_2, and I_3 are zero, priority code bits P_1 P_0 must become *00*. For the last line of Table 8.4, if I_3 asserts regardless the state of interrupt bits I_0, I_1, and I_2, priority code bits P_1 P_0 must become *11* and the *Any-Interrupts?* output must assert.

Let us now consider the same priority encoder, explicitly assigning its $2^4 = 16$ values, to the four interrupt input lines, priority encoded outputs P_1 and P_0 and *Any_Interrupts?*. As usual output *Any_Interrupts?* indicates the presence of an asserted interrupt at the input of the encoder. Thus, we obtain Table 8.5 for the same logic presented by Table 8.4, without using *don't cares* in an explicit form.

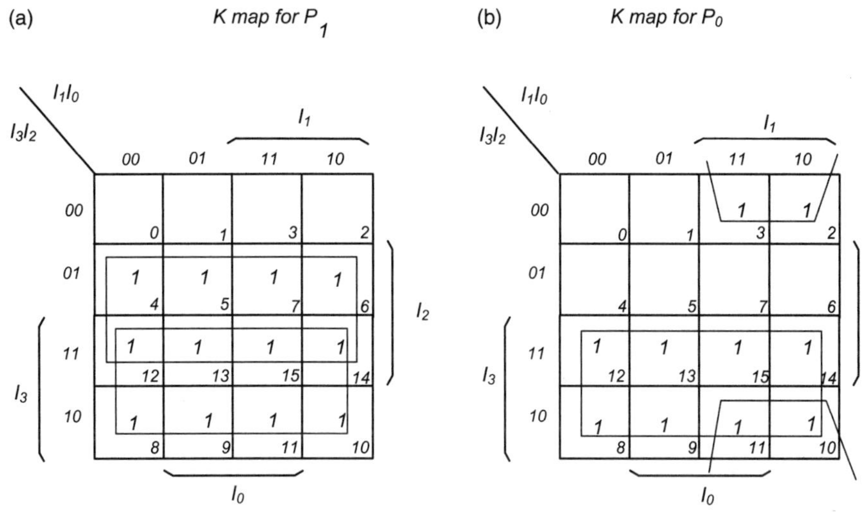

Figure 8.10 (a) Karnaugh map for P_1; (b) Karnaugh map for P_0.

From Table 8.5 we can do the three Karnaugh map to find out the combinational logic of outputs: P_1, P_0, and *Any_Interrupts?* But let us look at the logic of output *Any_Interrupts?*

By carefully inspecting the truth table, it is easy to see that the logic for *Any_Interrupts?* is:

$$\text{Any\_Interrupts?} = I_3 + I_2 + I_1 + I_0. \tag{8.12}$$

This is clear because output *Any_Interrupts?* is zero only when all inputs are zero (Tables 8.4 and 8.5).

For outputs P_1 and P_0 we produce the *K*. maps of Figures 8.10a,b.

By inspection of Figure 8.10a,b we obtain the following:

$$P_1 = I_3 + I_2 \tag{8.13}$$

$$P_0 = I_3 + \overline{I_2}.I_1. \tag{8.14}$$

Drawing the logic gates of Equations (8.12) through (8.13) we obtain Figure 8.11.

8.3 MULTIPLEXERS AND DEMULTIPLEXERS (MUXES AND DEMUXES)

Many years ahead of digital multiplexers and demultiplexers, mechanical versions of them were available. These devices were initially called distributors

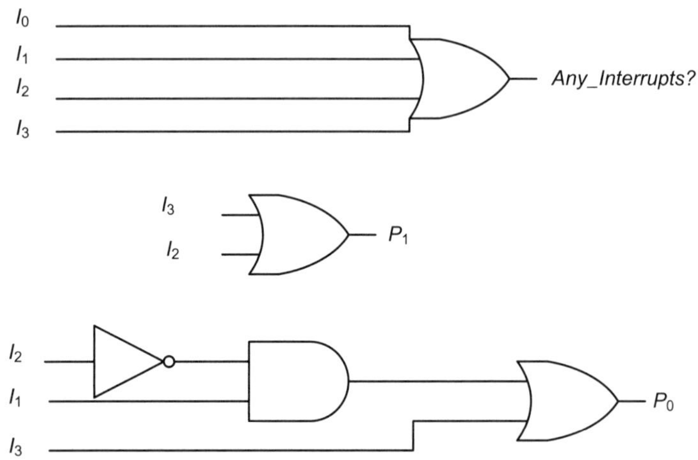

Figure 8.11 Logic implementation of the priority encoder for Example 8.4.

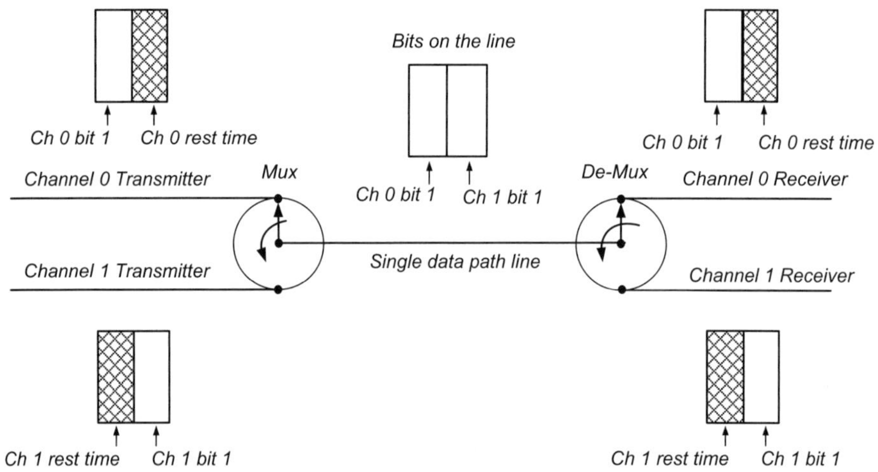

Figure 8.12 Multiplexer (mux) and demultiplexer (de-mux) transmission/reception scheme.

and used mostly as part of telegraph equipment by the end of the nineteenth century. The purpose of these devices is to allow more than one transmitting data source to use a single serial line, connected between the *n* sources and *n* destinations. The serial line between source and destination is time-shared. Let us look at how this works looking at the scheme depicted in Figure 8.12. For simplicity, assume that there are just two transmitting sources, *channel 0* and *channel 1*. Let us assume that each source transmits a bit (either a 0 or a 1) for 1 ms and does not transmit anything for another millisecond. It is conceivable to synchronize the two transmitting sources such that when channel *0* transmits it data bit, channel *1* rests; the next bit time channel *0* rests while channel *1* transmits its data bit. For the sake of simplicity, let us not be concerned with exact timing details of a real implementation.

When *channel 0* transmitter has a bit to send over the line, *channel 0* and its mux rotor must be engaged and *channel 0* receiver and its de-mux rotor must be engaged. This connection has to persist for 1 ms, during which time a bit is transmitted from *channel 0* transmitter (on the left) to *channel 0* receiver (on the right). We are ignoring finite propagation delays over the serial line, rotor rotation times, and several other factors that should not matter at this point. After 1 ms, *channel 1* transmitter has a bit to send over the line, *channel 1* and its mux rotor must be engaged, and *channel 1* receiver and its de-mux rotor must be engaged. During this time bit *1* of *channel 1* gets transmitted. This process repeats indefinitely or until no more transmissions are desired.

Today muxes and de-muxes can be designed to transport analog or digital signals. This chapter focuses on digital devices only.

8.3.1 Multiplexers

Digital multiplexers are devices that allow a number of data sources to route one out of the total data sources to its output. Let us assume that we have a four-input mux, at any given time one input is allowed to pass straight through the mux onto the output. At such time none of the other inputs can go through the mux. This scheme clearly works fine when the data path at the output of the mux can be time-shared by the various inputs to the mux. Multiplexers are referred to as being *1-of-*2^n, where n is the number of input channels. Conceptually we can have *2, 4, 8, 16, . . . ,* 2^n input multiplexers. Table 8.6 depicts the truth table of a 1-of-4 mux in a compact fashion using *don't care* conditions. The same truth table is somewhat expanded in Table 8.7 by explicitly stating the values of each input channel data input. Note that in order to *fully expand* the truth table of Table 8.7, since there are seven inputs, the fully expanded truth table would have $2^7 = 128$ entries! Clearly this is not practical, and it is not too clear to understand either.

Truth Tables 8.6 and 8.7 are easy to understand. They should be read in the same manner as the mux operates. For example when input *0* is selected (select

Table 8.6 Compressed 1-of-4 multiplexer truth table

Enable	Input Channel	Data Select Line		Output
E	I_x	S_1	S_0	Y
1	I_0	0	0	I_0
1	I_1	0	1	I_1
1	I_2	1	0	I_2
1	I_3	1	1	I_3
0	X	X	X	0

Table 8.7 Somewhat expanded or more explicit 1-of-4 multiplexer truth table

Enable	Data Inputs				Data Control Line		Output
E	I_0	I_1	I_2	I_3	S_1	S_0	Y
1	0	X	X	X	0	0	0
1	1	X	X	X	0	0	1
1	X	0	X	X	0	1	0
1	X	1	X	X	0	1	1
1	X	X	0	X	1	0	0
1	X	X	1	X	1	0	1
1	X	X	X	0	1	1	0
1	X	X	X	1	1	1	1
0	X	X	X	X	X	X	0

lines set to select such input), regardless of what all other mux inputs input levels are (*don't cares*), the selected input *0* passes straight through the mux to its output *Y*. The same is true for when input *1* or *2* or *3* is selected. The operation of the enable *E* is such that the mux does its thing (route input data to output), upon *E* being high. However, when *E* is low, the mux output is zero. The *E* input is useful when we want to build larger multiplexers using smaller ones.

Figure 8.13a depicts the schematic symbol of a *1-of-4* mux, and b of the same figure depicts a possible logic implementation of such mux.

It is not too hard to figure out the truth table of virtually any size mux just by similarity with the *1-of-4* mux just covered. For example, a *1-of-8* mux will have eight data inputs, three select lines to choose one out of eight inputs to go through the *Mux*, a master enable *E*, that allows us to concatenate the mux with others to build even larger multiplexers, and one output.

Exercise: Derive the truth table and a logic implementation of a *1-of-8* mux. Assume that one has logic gates of the required number of inputs, to facilitate the task. This last assumption does not preclude generality to the exercise. If gates of the required number of inputs are not available or we are not allowed to use them, we can always build gates with larger number of inputs using multiple gates with a smaller number of inputs. For example an *8*-input *OR* gate can be built in a number of ways according. Figure 8.14 shows two possible implementations of an *8*-input *OR* gate (a) using *4*-input and *2*-input gates (b) using all *2*-input gates.

8.3.2 Building Larger Multiplexers

How do we construct a *1-of-8* mux using just *1-of-2* muxes? Assume that all of our *1-of-2* muxes have a master enable input pin. We can think of a mux having a funneling effect on its input data signals from left to the output on

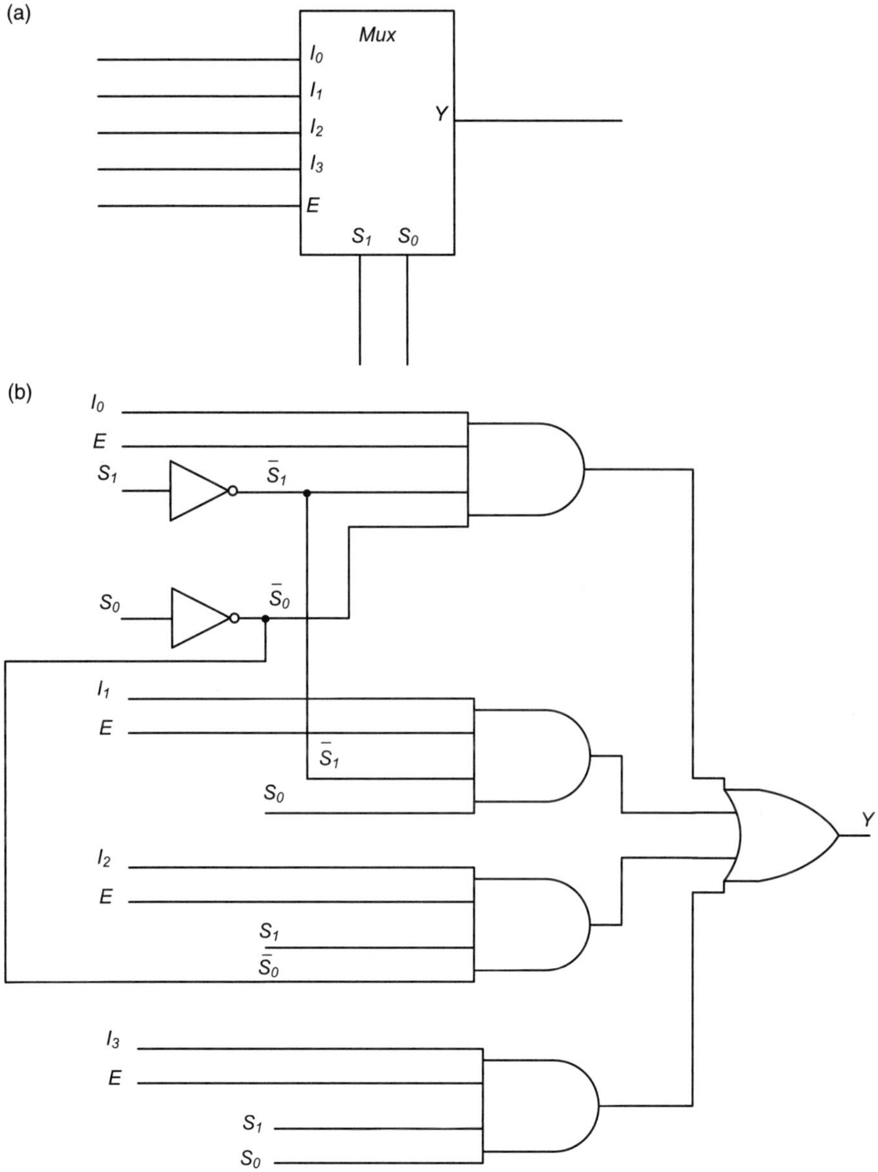

Figure 8.13 (a) *1-of-4* mux schematic symbol; (b) *1-of-4* mux logic implementation.

the right. Let us refer to the graph depicted in Figure 8.15, if we start with a *1-of-2* mux (*Mux 1*) we can feed with two other *1-of-2* muxes (*Muxes 2 and 3*) a total of *4* signals into *Mux 1.* We repeat this process one more time and we can feed *8* signals into *Mux1*, using in addition to *Muxes 2* and *3*, *Muxes 4, 5, 6,* and *7*.

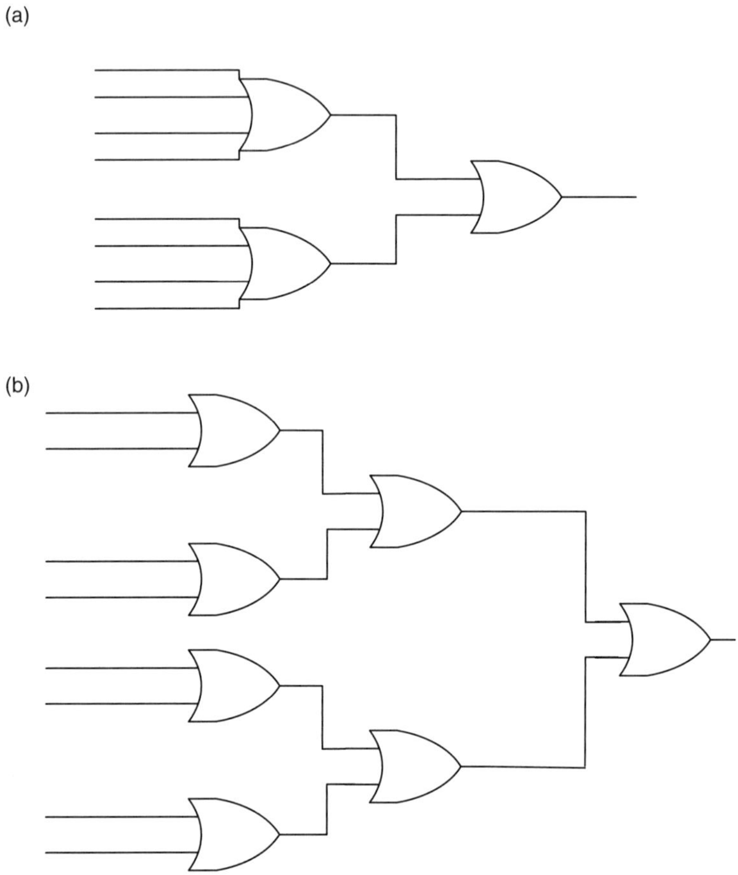

Figure 8.14 Eight-input OR gate implementation (a) using *4*-input and *2*-input gates; (b) using all *2*-input gates.

But we are not done yet; we still need to identify data inputs I_0 through I_7 of the overall composite *1-of-8* mux. So one more time referring to our picture of Figure 8.15, assume that the data select line of *Mux 1* is assigned to be the *MSB* of the select lines of our composite *1-of-8* mux. The select lines of muxes 2 and 3 are tied together and assigned to be the middle bit of the *3*-bit select line group of our *1-of-8* mux. Finally, we assign the select line of muxes *4, 5, 6,* and *7* tied together to the *LSB* of the 3-bit select line group of our composite *1-of-8* mux. Following what was just described can be seen depicted in Figure 8.15. In Figure 8.15 the data paths are highlighted with heavy lines. The select lines are shown with a medium weight line. Finally, the master enable lines are all drawn with a lightweight line. Lines that cross and do not have a heavy dot at their intersection are *not connected.*

The techniques depicted in Figure 8.15 can be generalized to build virtually any mux of any desired number of inputs with other combination of smaller

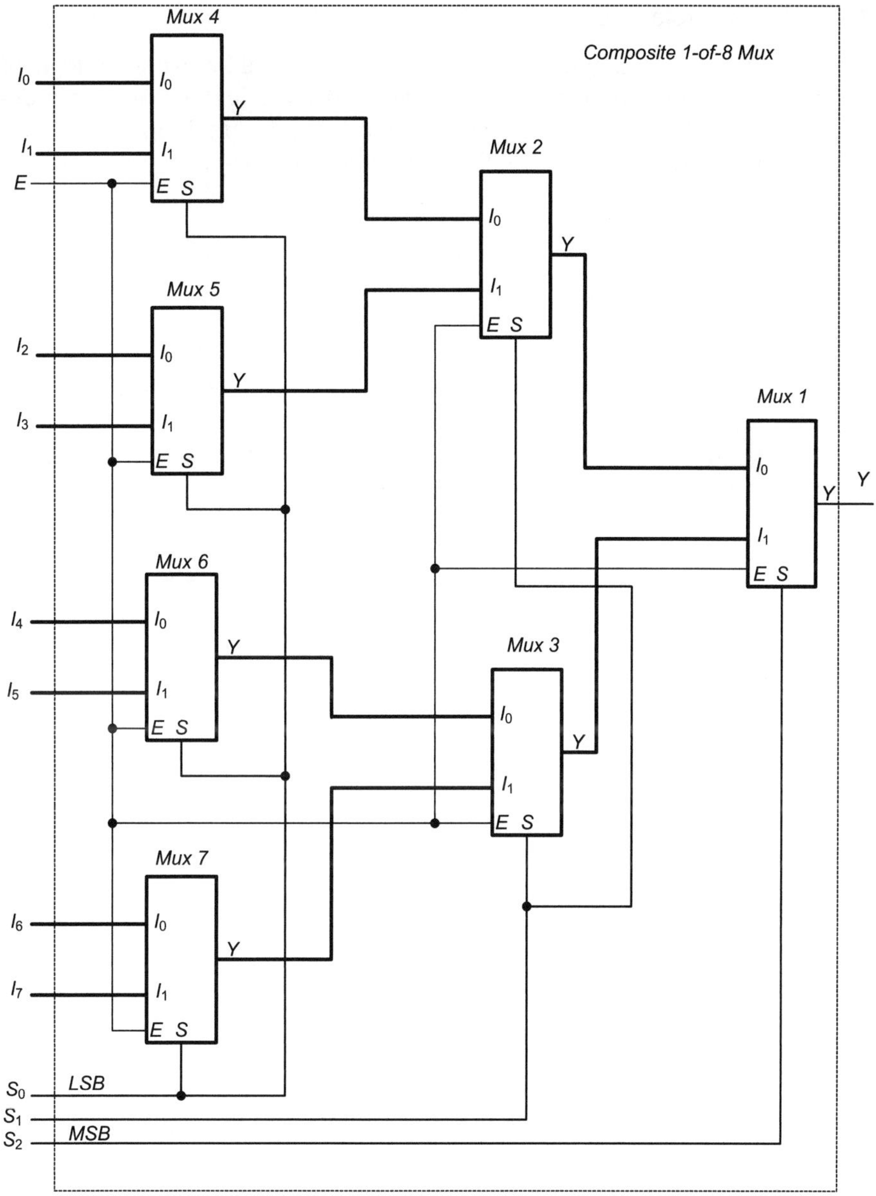

Figure 8.15 A *1-of-8* mux implementation using *1-of-2* muxes.

muxes. Naturally the number of mux inputs is always 2^S, where exponent S is the number of the mux select lines.

Exercise: (1) Using the techniques used for the *1-of-8* mux, build a composite *1-of-32* mux. Hint: Use four *1-of-8* muxes and one *1-of-4* mux. (2) Try a different implementation with another mix of muxes.

8.3.3 De-Multiplexers

From our previous example depicted in Figure 8.12 and knowing the logic of a decoder, we can appreciate that in a way a decoder can be used as a de-multiplexer, in the sense that it reverses what multiplexers do to data. Figure 8.16a shows wedged-shaped symbols for *mux* and *de-mux*. Such wedged

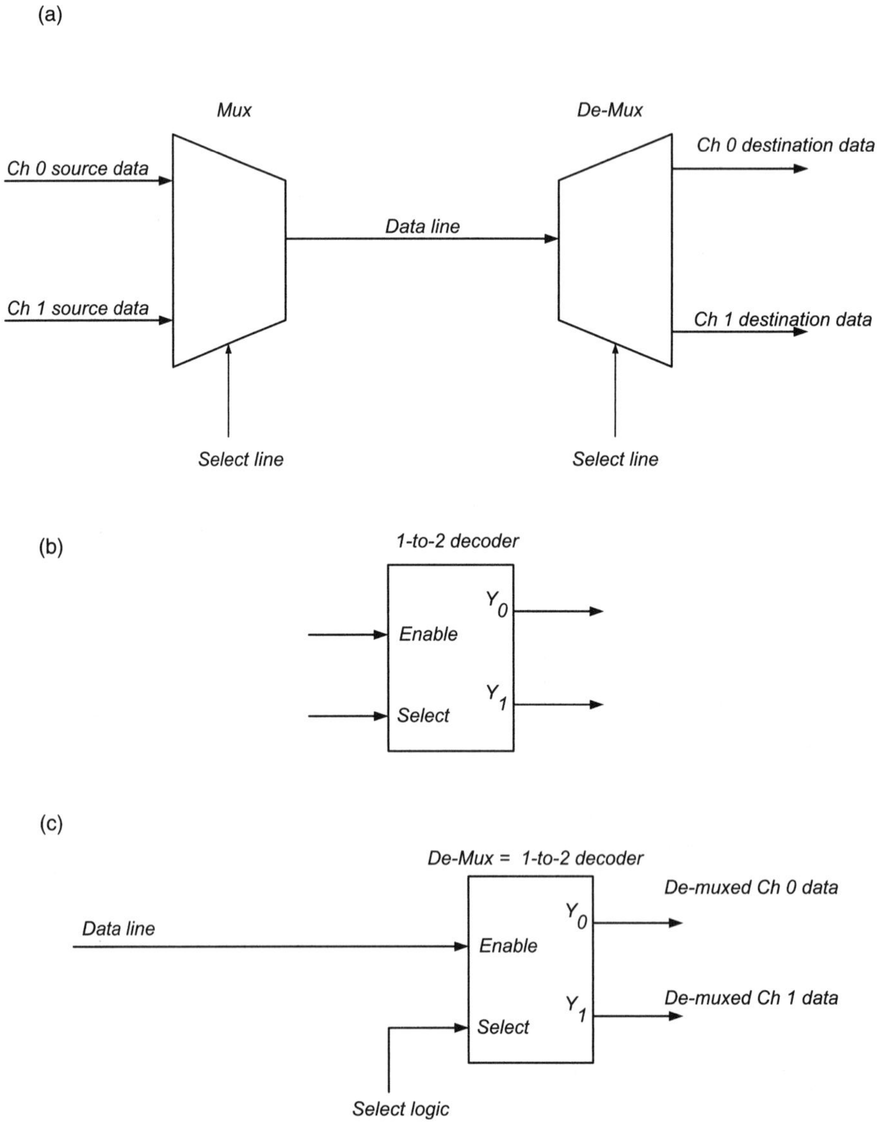

Figure 8.16 (a) System level view of a mux/de-mux application; (b) *1-to-2* decoder with enable; (c) *1-to-2* decoder wired as a de-multiplexing device.

symbols are preferred in computer architecture and systems illustrations. We will use a wedge for a *mux* in later chapters of the book that deal with computer architecture. Figure 8.16b shows the schematic symbol of the simplest decoder one can have a 1-to-2 decoder, with a single select line, two outputs and its master enable line. Finally, Figure 8.16c depicts the use of a decoder as a de-multiplexer in our application of Figure 8.12.

8.4 SIGNED AND UNSIGNED BINARY NUMBERS

The binary numbers that we described on the previous chapter did not have any sign; they were just positive or unsigned binary numbers. If we have n bits to represent a positive number there are 2^n binary combinations of such numbers. Now if we intend to represent positive as well as negative numbers, but continue to use binary-valued terms or *bits,* we must give up some of the positive number binary combinations and allocate them to the negative range. Why? Because we cannot use a negative sign to depict a negative number; this implies the need of three different symbols to represent numbers, the *1*, the *0*, and the "-" sign. We are supposed to represent positive and negative numbers with just *ones* and *zeros*. This will become clearer when we go over some examples.

8.4.1 One's Complement Representation of Binary Numbers: Addition

Let us assume that we are working with *3*-bit binary numbers. The *1's complement* of a binary number is defined as the *bit-to-bit* complementation of every one of its bits. For example, given the *3*-bit binary number *010*, its *1's* complement is *101*. Similarly, given the *3*-bit number *101,* its *1's* complement is *010.* It is easy to see that *1's* complementing a number twice in a row leads to the original number we started with. This is similar to the involution rule covered in the previous chapter (refer to Table 7.11 in Chapter 7).

Now what follows is the most important consideration about *1's* complement numbers, given n bits to represent a *1's* complement number, the most significant bit *(MSB)* is allocated to represent the number sign. A leading *0* means the number is positive, a leading *1* means that the number is negative. The rest or the $(n - 1)$ remaining bits are assigned to represent the number's magnitude. Figure 8.17a depicts the bit assignments of an *n-bit 1's* complement number, Figure 8.17b depicts the bit assignments for a three-bit $(n = 3)$ *1's* complement number.

Now we know how to obtain the *1's* complement of a number and we know how the bits are assigned. Let us look into how we obtain the negative number of *3*-bit positive *001.*

Simply take the *1's* complement of *1*, which leads to *110.*

$$\text{1's C}(001) = 110. \tag{8.15}$$

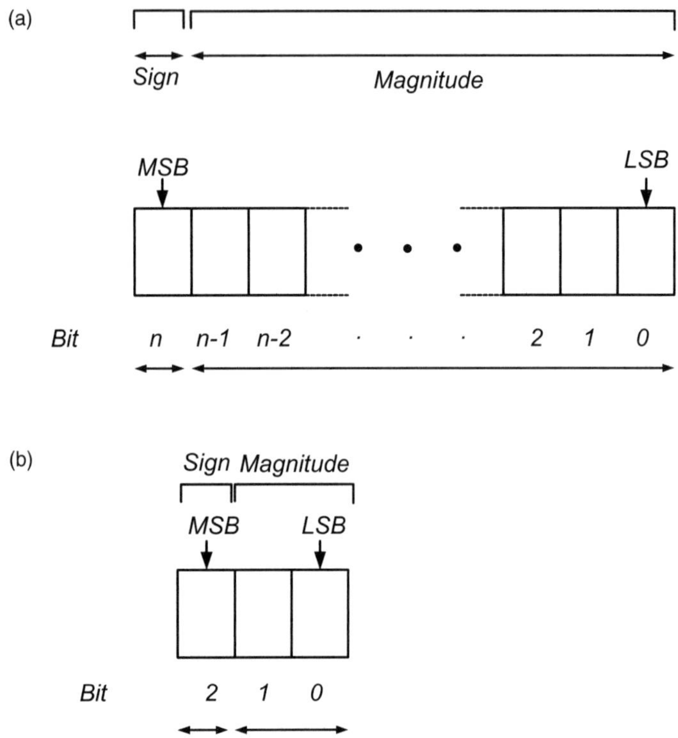

Figure 8.17 One's complement bit assignments (a) for an *n*-bit number; (b) for a *3*-bit number.

Equation (8.15) thus is the representation of decimal number *–1* in three-bit 1's complement form. Conversely, given a *3*-bit *1's c*omplement number such as *110,* by inspection of the number's *MSB* we know that we are dealing with a *3*-bit negative number, check the MSB. One more time in order to find the magnitude of such negative number; again we take the *1's c*omplement of 110.

$$\text{1's C}(110) = 001. \tag{8.16}$$

Hence the given negative number (i.e., *110*) magnitude is *1*. Table 8.8 depicts the *1's c*omplement of all 3-bit positive numbers.

By inspection of Table 8.8 we can tell that if we want to use the 1's complement representation for positive as well as negative numbers, the first four numbers under the *1's Complement* column have to be negative numbers, because they have a *one MSB*; whereas the last four binary combinations of the same column have to represent four positive binary numbers, because they have a leading *zero*.

Note from Table 8.8 that the number zero has two 1's complement representations, that is *000* and *111*; that is *positive zero* and *negative zero*. We will

Table 8.8 3-bit binary numbers and their associated 1's complement representation

Positive 3-Bit Binary Number	Positive 3-Bit Binary Decimal Equivalent	1's Complement	1's Complement Decimal Equivalent
000	0	111	–0
001	1	110	–1
010	2	101	–2
011	3	100	–3
100	4	011	3
101	5	010	2
110	6	001	1
111	7	000	0

Table 8.9 Basic rules for unsigned or positive binary addition

Augend	Addend	Sum	Carry Out
0	0	0	0
0	1	1	0
1	0	1	0
1	1	0	1

see that *2's* complement is a better negative number representation system, which will be the topic of our subsequent section.

Table 8.9 depicts the basic addition of two single bit unsigned numbers, *augend* and *addend*, the results is the sum and the right most column is the *carry out*.

So after all of the above why is *1's* complement good or what is it for? We can add numbers in *1's* complement representation using the fundamental rules of unsigned binary addition given by Table 8.9.

8.4.1.1 Four-Bit 1's Complement Representation

4-bit *1's* complement numbers range from *0000* (decimal *+0*) up to *0111* (decimal *+7*). Negative *4*-bit *1's* complement numbers range from *1000 (decimal –7)* up to *1111 (decimal –0)*.

The algorithm to obtain the 1's complement of an n-bit binary number is simply flipping its *zeros* to *ones* and its *ones* to *zeros*. Given that X is our n-bit binary number:

$$\mathit{1'sComplement}(X) = \mathit{1'sComplement}\,(x_{n-1}, x_{n-2}, \ldots, x_1, x_0) = \overline{x_{n-1}}, \overline{x_{n-2}}, \ldots, \overline{x_1}, \overline{x_0}$$

Table 8.10 lists all *4*-bit numbers in *1's* complement representation.

From Table 8.10 again we see that *4*-bit *1's* complement numbers exhibit plus and minus zero or double representation for the number zero. As a matter of fact, all n-bit *1's* complement numbers will always produce double representation of the number *zero*.

Table 8.10 Four-bit 1's complement numbers

Representation in 4-bit 1's Complement	Assigned Decimal Number
0000	+0
0001	+1
0010	+2
0011	+3
0100	+4
0101	+5
0110	+6
0111	+7
1000	–7
1001	–6
1010	–5
1011	–4
1100	–3
1101	–2
1110	–1
1111	–0

Example 8.5 Given the following *4*-bit *1's* complement numbers, perform the additions indicated below and double check your results using their decimal equivalent.

(a) *0100 + 0011*,
(b) *0101 – 0001*,
(c) *1011 + 0010*,
(d) *1011 + 0110*.

Solutions

With the aid of Tables 8.9 and 8.10 we perform the operation as follows:

(a) *0100 + 0011 = 0111* and in decimal: 4 + 3 = 7
(b) *0101 – 0001 = 0101 + (–1 in 1's C) = 0101 + 1110 = 0011 and a carry of 1* in order to achieve the correct decimal result of *+4* (since we are subtracting *1* from *5*) the carry must be wrapped around and added back to the previous sum. Thus:

$$\begin{array}{lr} & 0101 \\ & +1110 \\ \hline \text{Carry} = 1 & 0011 \end{array}$$

Finally add the carry and let us refer to it as the *End-Around-Carry (EAC)* of *1* so that:

$$\begin{array}{r} 0011 \\ EAC + 0001 \\ \hline 0100 \end{array}$$

in decimal we have that *5 – 1 = +4,* which is the final answer. Note that the first addition and the addition of the *EAC,* the second addition, effectively take two *addition times*, to perform the complete sum.

(c) *1011 + 0010 = 1101* and in decimal, note that *1011* is *–4* in 1's Complement, and *1101* is *–2. Thus: –4 + 2 = –2*

(d) 1011 + 0110 = 0001 and a Carry = 1, treating the Carry as an EAC we obtain:
0001 + 0001 = 0010. In decimal we have that *1011 + 0110 = 0010,* which is *–4* + 6 = +2. Note that when dealing with the *1's* complement addition not wrapping around the carry and adding to the previously obtained addition will not lead to the correct answer.

Note that what we need to do when we want to subtract *B* from *A*, is to add *A* and minus *B,* plus any end-around carry *(EAC)* that comes out of the operation. Because of the subsequent addition of the *EAC* the *1's* complement subtraction method is twice as slow as the *2's* complement subtraction. Because of this fundamental reason *1's* complement subtraction is hardly used.

8.4.2 Two's Complement Representation of Binary Numbers: Addition

Having learned *1's* complement representation well it is reasonably straightforward to understand *2's* complementation. The basic formula to obtain the *2's* complement representation of *X* an *n*-bit number is:

$$2's\,C(X) = 1's\,C(X) + 1 \text{ ignoring the Carry out bit.} \qquad (8.17)$$

The sign and magnitude format for *2's* complement numbers, i.e. *MSB* is the sign bit, rest of the bits are its magnitude, is identical to the bit assignment for *1's* complement numbers (see Fig. 8.17).

Why *2's* complement numbers, we might ask ourselves. There are two reasons for them; first we will see after applying Equation (8.17) that there is a single representation for the number zero. Secondly, *2's* complementation addition never has to add a carry as an *EAC* (like *1's* complement does), the

Table 8.11 Four-bit 2's complement numbers

Representation in 4-bit 2's Complement	Assigned Decimal Number
0000	+0
0001	+1
0010	+2
0011	+3
0100	+4
0101	+5
0110	+6
0111	+7
1000	–8
1001	–7
1010	–6
1011	–5
1100	–4
1101	–3
1110	–2
1111	–1

carry in a *2's* complement addition must be set to zero if there is no C_{in} from a less significant bit position. When carry out is ignored, the *2's* complement representation of the addition is obtained.

Table 8.11 depicts all four-bit 2's complement numbers.

Example 8.6 Shows how to perform binary additions in *2's* complement representation. The decimal equivalents of the same operations are shown. Perform the following four-bit *2's* complement additions:

(a) *0100 + 0011*
(b) *0111 + 1110*
(c) *0101 + 1100*
(d) *1011 + 1110*

Solutions to Example 8.6

Use the algorithm presented by Equation (8.17) to perform *2's* complementation.

(a) *0100 + 0011 = 0111*, in decimal: *+4 + 3 = +7*
(b) *0111 + 1110 = 0101 and a carry of 1*, since carry out has to be ignored the result is: *0101*, which in decimal is: +7 – 2 = +5
(c) *0101 + 1100 = 0001* and a carry out of 1, since the carry has to be ignored the result is *0001*, which in decimal is: +5 – *4 = +1*
(d) *1011 + 1110 = 1001* and a carry out of *1*, since the carry has to be ignored the result is *1001*, which in decimal is: *–5 – 2 = –7*

Table 8.12 Some decimal, octal, and hexadecimal numbers

Decimal	Octal	Hex	Decimal	Octal	Hex
0	0	0	9	11	9
1	1	1	10	12	A
2	2	2	11	13	B
3	3	3	12	14	C
4	4	4	13	15	D
5	5	5	14	16	E
6	6	6	15	17	F
7	7	7	16	20	10
8	10	8	17	21	11

8.4.3 Other Numbering Systems

Interestingly, infinitely many numbering systems exist. In computer software the most interesting and usual numbering systems, which we have not discussed yet, are the octal and the hexadecimal systems. We will very briefly touch on this subject since in the author's experience most scientists and engineers already know those numbers. The octal numbering system is simply based on eight uniquely defined digits, which are: 0, 1, . . . , 7. This numbering system is referred to as base *8*. The hexadecimal numbering system or base 16, has 16 uniquely defined digits, which are: *0* through *9* and *A* through *F*. Table 8.12 depicts the first *18* decimal, octal, and hexadecimal numbers.

The arithmetic rules for adding octal-to-octal and hex-to-hex numbers are pretty similar to those of decimal arithmetic. Care must be exercised knowing the uniquely defined digits for each numbering representation.

8.5 ARITHMETIC CIRCUITS: HALF-ADDERS (HA) AND FULL-ADDERS (FA)

Arithmetic circuits can be designed using the same concepts that we use when designing any other logic circuits. Basically truth tables and simplification methods are used to design them. Let us assume that we want to design the logic implementation of an adding cell. That is, a circuit that reads an augend bit (A), an addend bit (B), and produces the sum bit (S) and its carry out (C_{out}). Such circuit is referred to as a half-adder *(HA)* because it does not handle the carry in bit as full-adders do. The full-adder *(FA)* receives three input bits: augend (A), addend (B), and carry in (C_{in}), and it produces the sum bit of all three input bits and a carry out (C_{out}) bit. Table 8.13 depicts the truth table for a half-adder.

We obtain a maximally *SOP* form for output bits C_{out} and S of our half-adder.

Table 8.13 Half-adder truth table

Augend A	Addend B	(Carry out) C_{out}	Sum S
0	0	0	0
0	1	0	1
1	0	0	1
1	1	1	0

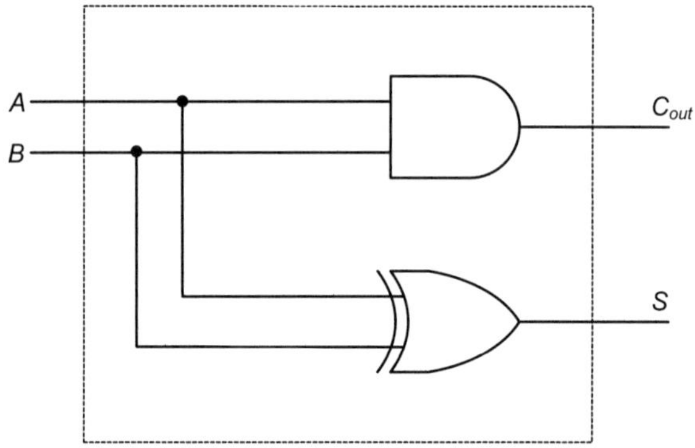

Figure 8.18 Half-adder logic implementation.

Table 8.14 Full-adder truth table

(Carry in) C_{in}	Augend A	Addend B	(Carry out) C_{out}	Sum S
0	0	0	0	0
0	0	1	0	1
0	1	0	0	1
0	1	1	1	0
1	0	0	0	1
1	0	1	1	0
1	1	0	1	0
1	1	1	1	1

Without doing an explicit *2*-variable *K. map* it can be seen that:

$$C_{out} = AB \tag{8.18}$$

and

$$S = A \oplus B \tag{8.19}$$

The logic implementation for the *HA* is given by Figure 8.18.

Table 8.14 depicts the truth table of a full-adder.

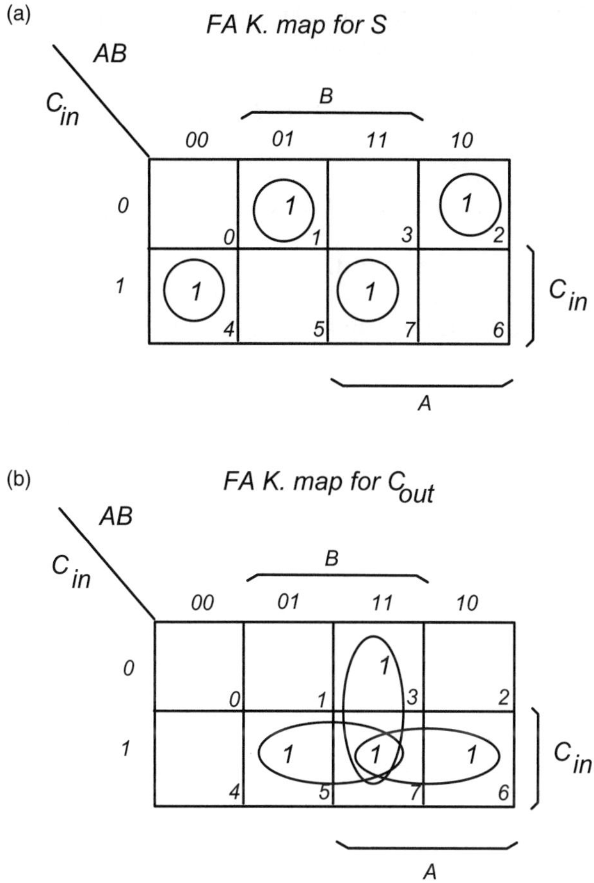

Figure 8.19 (a) Full-adder: *K.* map for C_{out}; (b) full-adder: *K.* map for *S*.

Using a *3*-variable *K.* map we find simplified logic equations to express the sum bit *S* and C_{out} of the *FA*. Figure 8.19 depicts the *K.* maps to obtain the maximally simplified *SOP* form for output bits C_{out} and *S*. From the truth table (Table 8.14) we fill in the *K.* maps for both output bits, C_{out} and *S*, these are depicted in Figure 8.19.

Referring to Figure 8.19a it is evident that none of the minterms m_1, m_2, m_4, and m_7 has any adjacent minterms. So the simplified *SOP* and the canonical *SOP* forms are identical. Moreover, from the canonical equation:

$$S = \sum(1,2,4,7) = \overline{C}_{in}.\overline{A}.B + \overline{C}_{in}.A.\overline{B} + C_{in}.\overline{A}.\overline{B} + C_{in}.A.B \qquad (8.20)$$

and since a two-variable *XOR* is:

$$A \oplus B = A.\overline{B} + \overline{A}.B, \qquad (8.21)$$

Equation (8.21) is found to be logically equivalent to Equation (8.23) after some Boolean algebra manipulations; that is:

$$S=\sum(1,2,4,7)=\overline{C}_{in}.\overline{A}.B+\overline{C}_{in}.A.\overline{B}+C_{in}.\overline{A}.\overline{B}+C_{in}.A.B=A\oplus B\oplus C_{in}. \tag{8.22}$$

The simplified *SOP* form for output bit C_{out} is:

$$C_{out}=AB+(A+B)C_{in}. \tag{8.23}$$

Writing the canonical form of Equation (8.23) by inspection of Figure 8.19b we obtain:

$$C_{out}=\overline{C}_{in}.A.B+C_{in}.\overline{A}.B+C_{in}.A.\overline{B}+C_{in}.A.B. \tag{8.24}$$

Grouping terms:

$$C_{out}=C_{in}(\overline{A}.B+A.\overline{B})+\overline{C}_{in}.A.B+C_{in}.A.B. \tag{8.25}$$

Applying Equation (8.25) to Boolean algebra rules yields:

$$C_{out}=AB+(A\oplus B)C_{in}. \tag{8.26}$$

The C_{out} of the *FA* has two alternate logic Equations (8.23) and (8.26).

The logic implementations of our full-adder S and C_{out} output bits are depicted in Figure 8.20.

Figure 8.21 depicts the schematic symbol diagram of a full-adder.

Note that this is the first time in this text that the schematic symbol of a combinational circuit is drawn in a somewhat nonconventional form. Conventionally circuits are drawn with inputs on the left-hand side and outputs on the right-hand side. Full-adders violate those conventions for exceptionally good reasons. Note that the C_{in} input to the *FA* is drawn on the right hand side, while its C_{out} output is drawn on the left hand side of the symbol. Inputs A and B are drawn on top and output S at the bottom. Inputs at the top and outputs at the bottom of schematic symbols are within the conventional drawing criteria. The reason why C_{in} is on the right and C_{out} is on the left is primarily due to the arithmetic done by an *FA*; similar to hand addition operations carries move from right-hand side digits to more significant or left-hand side digits. In the next section we will see that an interconnection of full-adders allows us to build multi-bit adders.

8.5.1 Building Larger Adders with Full-Adders

When we perform the addition of two numbers, an augend and an addend, regardless of whether these numbers are decimal or binary, the addition algorithm is always the same.

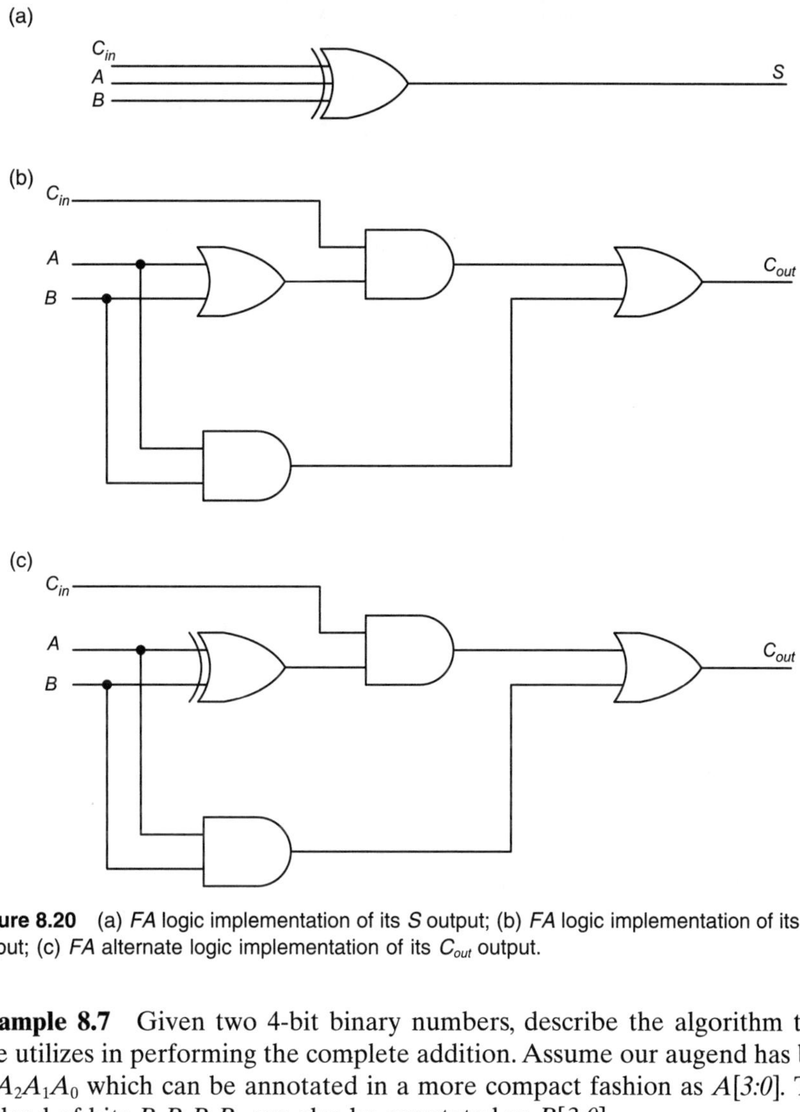

Figure 8.20 (a) *FA* logic implementation of its *S* output; (b) *FA* logic implementation of its C_{out} output; (c) *FA* alternate logic implementation of its C_{out} output.

Example 8.7 Given two 4-bit binary numbers, describe the algorithm that one utilizes in performing the complete addition. Assume our augend has bits $A_3A_2A_1A_0$ which can be annotated in a more compact fashion as $A[3{:}0]$. The addend of bits $B_3B_2B_1B_0$ can also be annotated as $B[3{:}0]$.

To perform the addition of $A[3{:}0]$ and $B[3{:}0]$ we write both numbers as follows:

$$\begin{array}{r} A_3A_2A_1A_0 \\ +\ B_3B_2B_1B_0\,. \\ \hline S_3S_2S_1S_0 \end{array} \tag{8.27}$$

We will refer to the above layout of numbers, Equation (8.27) being formed by four slices, slice *0* the least significant slice, contains A_0, B_0, *and* S_0, then slice

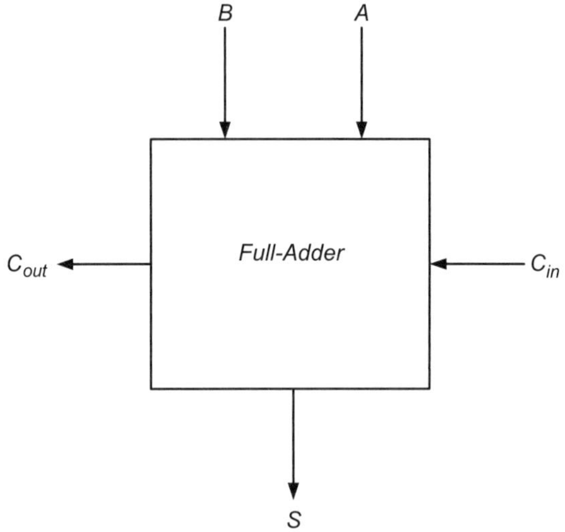

Figure 8.21 *FA* schematic symbol.

1 contains A_1, B_1, and S_1, and similarly for slices *2* and *3*. Slice *3* is the most significant slice of our numbers.

The number arrangement depicted by Equation (8.27) is exactly what we do when we perform an addition with paper and pencil. For now let us assume that there is no C_{in} into slice *0*. We begin the addition from the least significant slice by adding A_0 and B_0 to obtain S_0. This sum may produce a C_{out}* from slice *0* that has to propagate to slice *1*. To obtain the sum for slice *1*, or S_1 we must add C_{out} from slice *0*, which we will name as C_1, to A_1 and B_1. The process continues in the same fashion for all the slices. The last slice, slice *3* of our example, produces C_4, which is the overall C_{out} of the *4*-bit addition.

Example 8.7 actually is the justification for drawing the *FA* inputs and output the way that they are shown in Figure 8.21. Having gone over the algorithm of Example 8.7 we can easily interconnect four FA's to build a 4-bit binary adder. This time however since all *FA's* have the same logic, there will an overall C_{in} to slice *0* and we will refer to it a C_0. Figure 8.22a depicts an interconnection of full-adders that constitute a *4*-bit binary adder. Figure 8.22b shows a more compact manner of showing a 4-bit binary adder. Figure 8.22b does not imply in any way how the adder is internally designed. It can be built with *FA's* or other type of logic. Finally Figure 8.22c depicts the most compact form of all three of representing a *4*-bit binary adder. These types of symbols are very convenient to use when we deal computer architecture issues and micro controllers in general.

* C_{out} is always produced by a preceding slice regarding of its value (0 or 1); unless fast carry logic is used.

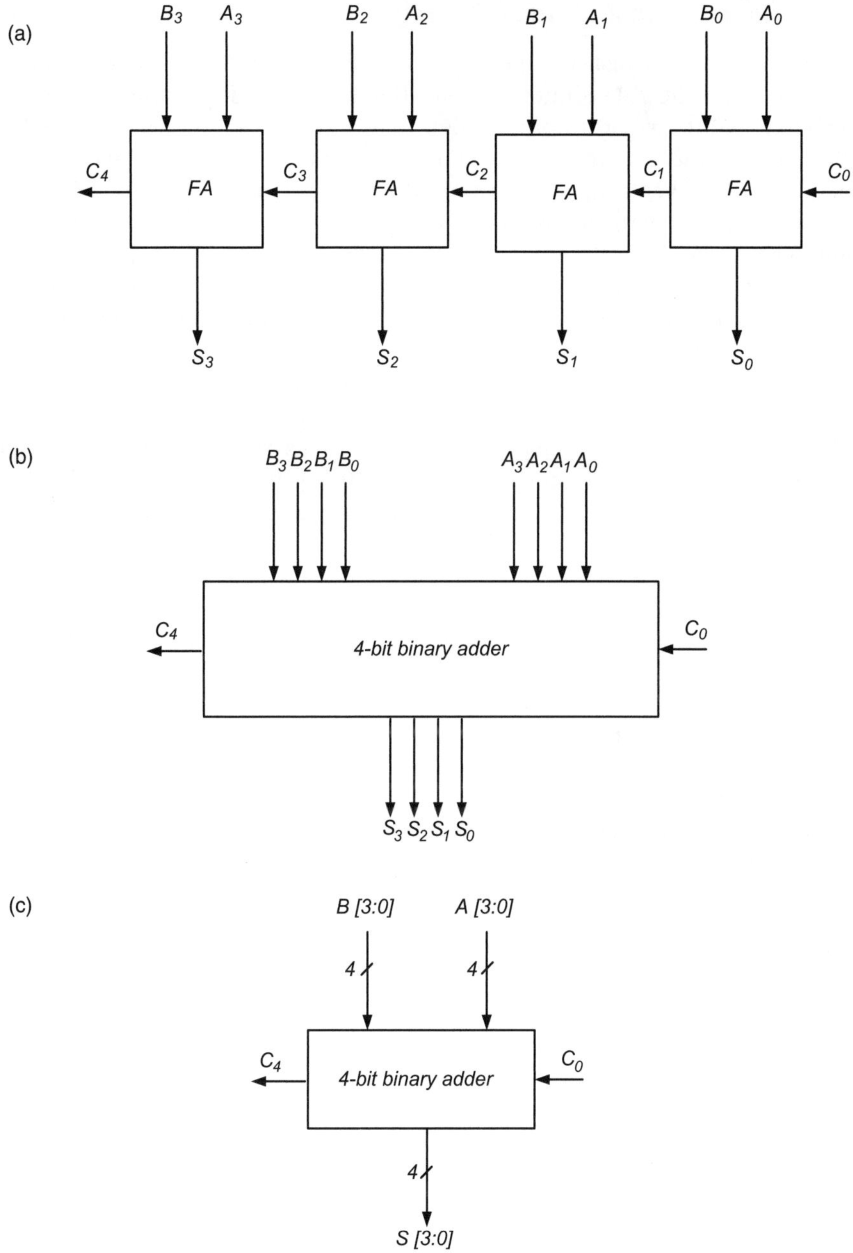

Figure 8.22 (a) *4*-bit binary adders built with *FA*; (b) compact form of a *4*-bit binary adder; (c) an even more compact form of a *4*-bit binary adder.

8.5.2 Notes about Full-Adder Timing

Let us assume a *FA* just like the one depicted in Figure 8.21. When all inputs are applied to the *FA's* simultaneously, there are two delays; one of them is the delay that it takes for output S before to settle down to a valid value. The other delay is the time that it takes the C_{out} output to settle down and become valid. Since it takes some longer time for C_{out} to settle to a valid value, C_{out} practically becomes the gating factor or the slow timing path, for the *complete* sum to be ready. The *complete* sum refers to the availability of valid values for S and for C_{out}. Since S is valid a little earlier that C_{out}, we then say that C_{out} is the long path in the sum. Now let us call this longer delay the full-adder delay, which at the moment we do not care about its absolute value in nano-seconds. When we build a *4*-bit adder like the one shown in Figure 8.22a note that we now call the complete sum the availability of all outputs of the 4-bit adder, i.e. valid S_3, S_2, S_1, S_0, and C_4, which is also the overall carry out of the adder. From the *4*-bit adder point of view we do not care (to a point) about the availability of valid internal carries that propagate through the adder. We do care about them from the perspective that the longer it takes for those carries to propagate through the internal logic the longer it will take to obtain the complete sum. For the *4*-bit adder of Figure 8.22a, implemented with full-adders, the overall adding time is four full-adder delays. So from the time the last input becomes valid at the adder input, we need to wait for the result sum and carry out for four full-adder delays. Should we take the sum reading before such time there is no guarantee about its correctness.

8.5.3 Subtracting with a *4*-bit Adder Using *1's* Complement Representation

Let us continue our example with *4*-bit wide *1's c*omplement numbers. Let us also recall from an earlier section of this chapter that if we need to subtract B from A, where A is the minuend and B is the subtrahend, this can be accomplish using 1's complement arithmetic by adding the *1's* complement of B, the subtrahend to A, the minuend, await for the overall output carry of the 4-bit adder to become valid and add it back into the input carry of the adder. This last step is referred to as adding the End-Around-Carry *(EAC)*. Figure 8.23 shows how that implementation is done.

Referring to Figure 8.23a observe that the carry out of the *4*-bit adder is tied back into the carry in. The subtraction is performed as the sum of A with the *1's* complement of B plus any end-around-carry (*EAC*):

$$A - B = A + (-B) = A + \mathit{1's\ Complement}\,\{B\} + EAC. \tag{8.28}$$

Note that if we deal with 4-bit wide numbers then A = *A[3:0]* and B = *B[3:0]*.

Because of the need to add the *EAC*, we have to wait to obtain the complete subtraction, which is two complete *4*-bit adder delays. Figure 8.23b implement

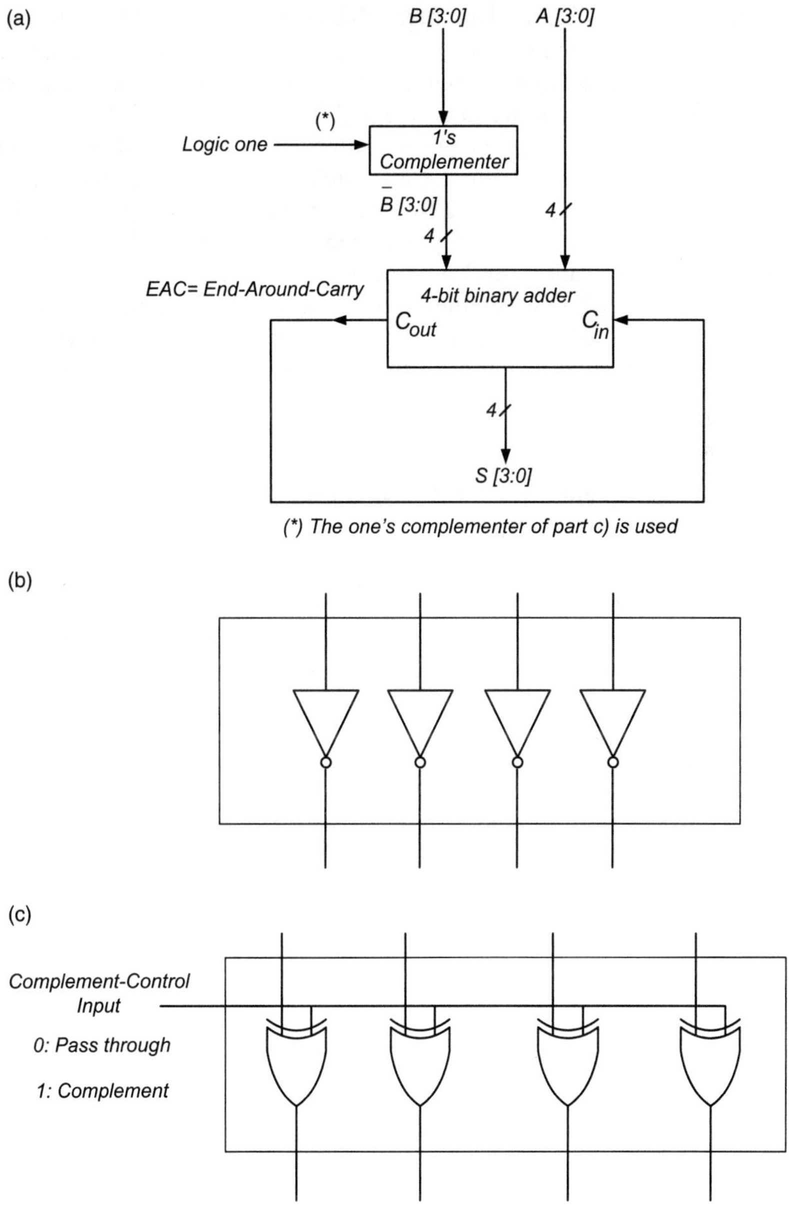

Figure 8.23 (a) 4-bit binary adder configured as a subtractor using *1's* complement arithmetic, (b) hardwired logic for a *1's* complementer, (c) programmable logic for a *1's* complementer. (*) When the complement-control input signal (c) is high, the 4-bit input number at the *1's* Complementer logic will become *1's* Complemented at the *1's* Complementer 4-bit output. When the complement-control input signal (c) is low, the 4-bit input number at the *1's* Complementer logic will pass-through the *1's* Complementer logic to its 4-bit output unchanged.

the simplest possible *1's* complementer, which is just a bit-to-bit inverter. Such one's complementer is said to be *hardwired*. Figure 8.23c shows an implementation using *XOR* gates, why? This implementation allows one to use the same logic as pass-through logic when its control input is zero and it converts the logic into a *1's* complementer when the control input is a one. That means that the logic block depicted by Figure 8.23a could also be used as an adder and not just as a subtractor. Of course for this scheme to be complete we should multiplex or gate the end around carry, opening the *EAC* path when we configure the logic as an adder and provide a path for the output carry to feed into the input carry when we configure it as a subtractor. For the sake of practicality, we simply now move on to the *2's* complement adder/ subtractor which is the most effective way of implementing and adder and a subtractor using a *4*-bit adder.

8.5.4 Subtracting with a *4*-bit Adder Using *2's* Complement Representation

The *1's* complement adder/subtractor is interesting but it is not fast enough. We can do better if we do not have to wait for the carry out to travel its way to the carry in (*EAC*). So the *2's* complement version of the adder/ subtractor is presented in Figure 8.24. Note the *EAC* path, seen for the *1's* complement implementation is now not connected for the 2's complement implementation.

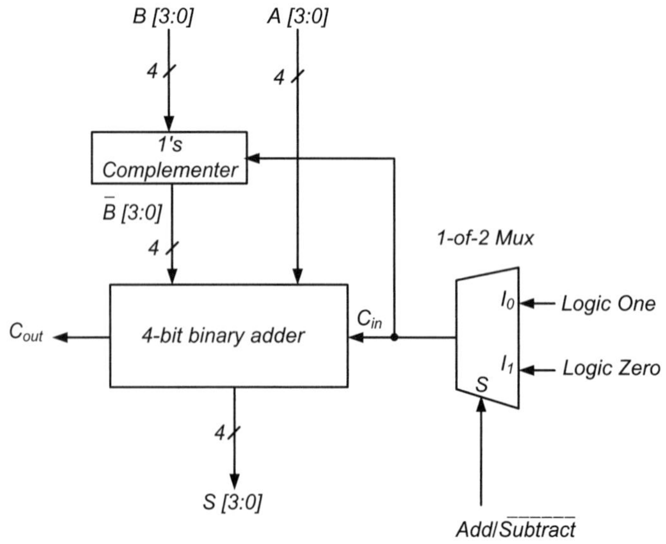

Figure 8.24 2's Complement *4*-bit binary adder/subtractor.

It is important to remember that when using *2's c*omplement arithmetic the carry out has to be ignored.

Example 8.8 Draw a block diagram of a *4*-bit *2*'s complement adder/subtractor that upon its control input being *0* it adds, but if the control input is *1* it subtracts using 2's complement arithmetic. Equation (8.17) repeated below for the reader's convenience shows the algorithm used to obtain the 2's complement of an *n*-bit binary number

$$2\text{'s } c(X) = 1\text{'s } complement\,(X) + 1 \text{ ignoring the Carry out bit.} \qquad (8.29)$$

Figure 8.24 below depicts the solution to Example 8.8.

Referring to Figure 8.24 let us see how the adder part works. Upon setting to zero the control input to the select line of the *1-of-2* mux, the overall adder C_{in} is zero and the *1's* complementer logic box is in pass-through mode, i.e. does not invert its inputs. The logic of Figure 8.24 simply adds with a *high-level* control input. When the control input is zero two things happen, the C_{in} is set to *one* and the one's complementer logic box control input is also set to *one*. The *1's* complementer is set to complement mode and since the C_{in} to the adder/subtractor is set to *1*, the logic is basically executing:

$$A - B = A + (-B) = A + 1\text{'s } C\{B\} + 1 = A + 2\text{'s } C\{B\}. \qquad (8.30)$$

Equation (8.30) in effect performs the subtraction of *B* from *A* in *2's* complement form.

8.6 CARRY LOOK AHEAD (CLA) OR FAST CARRY GENERATION

Let us now go back to our full-adder basic building block with its two logic equations, repeated below for the reader's convenience:

$$S = A \oplus B \oplus C_{in} \qquad (8.31)$$

$$C_{out} = AB + (A \oplus B)C_{in}. \qquad (8.32)$$

Since a basic multi-bit adder can be thought as a concatenation of full-adders, let us generalize Equations (8.31) and (8.32) for a *FA slice*. So let us re-write Equations (8.31) and (8.32) as if they were the equations of the *i*th *slice*.

$$S_{i+1} = A_i \oplus B_i \oplus C_i \qquad (8.33)$$

$$C_{i+1} = A_i B_i + (A_i \oplus B_i)C_i \qquad (8.34)$$

where in Equation (8.33) we can appreciate that C_i is the C_{in} to the *i*th slice in question produced by its immediately less significant and adjacent slice, or slice $(i - 1)_{th}$. Similarly with C_i on Equation (8.34); and C_{i+1} is the C_{out} of the i_{th} slice.

We will describe a 4-bit adder, which has four slices, slice *0* is the least significant slice and *3* is the most significant slice. Personalizing Equations (8.33) and (8.34) for each of our four slices we obtain the following logic expressions, where we are assuming that *slice 0* is the least significant slice and *slice 3* is the most significant slice of our *4*-bit adder.

For all sum bits:

$$Slice\ 0: S_0 = A_0 \oplus B_0 \oplus C_0 \tag{8.35}$$

$$Slice\ 1: S_1 = A_1 \oplus B_1 \oplus C_1 \tag{8.36}$$

$$Slice\ 2: S_2 = A_2 \oplus B_2 \oplus C_2 \tag{8.37}$$

$$Slice\ 3: S_3 = A_3 \oplus B_3 \oplus C_3 \tag{8.38}$$

and for all the carryouts we obtain:

$$Slice\ 0: C_1 = A_0B_0 + (A_0 \oplus B_0)C_0 \tag{8.39}$$

$$Slice\ 1: C_2 = A_1B_1 + (A_1 \oplus B_1)C_1 \tag{8.40}$$

$$Slice\ 2: C_3 = A_2B_2 + (A_2 \oplus B_2)C_2 \tag{8.41}$$

$$Slice\ 3: C_4 = A_3B_3 + (A_3 \oplus B_3)C_3. \tag{8.42}$$

By close inspection of Equations (8.39) through (8.42) we see that since real logic gates have non-zero gate delays, C_1 has to be generated by the C_{out} logic of *slice 0* before the addition can proceed to *slice 1*. Similarly C_2 has to be generated by *slice 1* C_{out} logic before the addition can proceed to *slice 2*. Exactly the same is true for C_3 and for C_4.

Let us now make a couple of definitions, let the $A_i\ B_i$ be *Generate*$_i$ or G_i terms for i ranging from *0* to *3*. We will also define the term $(A_i \oplus B_i)$ as the *Propagate*$_i$ term or P_i for i ranging from *0* to *3*. With those new definitions we rewrite Equations (8.39) through (8.42) and we obtain:

$$Slice\ 0: C_1 = G_0 + P_0C_0 \tag{8.43}$$

$$Slice\ 1: C_2 = G_1 + P_1C_1 \tag{8.44}$$

$$Slice\ 2: C_3 = G_2 + P_2C_2 \tag{8.45}$$

$$Slice\ 3: C_4 = G_3 + P_3C_3. \tag{8.46}$$

Plugging C_1 from Equation (8.43) into Equation (8.44) yields:

$$Slice\ 1: C_2 = G_1 + P_1G_0 + P_1P_0C_0. \tag{8.47}$$

Plugging Equation (8.47) into Equation (8.45) yields:

$$\textit{Slice 2}: C_3 = G_2 + P_2G_1 + P_2P_1G_0 + P_2P_1P_0C_0. \tag{8.48}$$

Plugging Equation (8.48) into Equation (8.46) yields:

$$\textit{Slice 3}: C_4 = G_3 + P_3G_2 + P_3P_2G_1 + P_3P_2P_1G_0 + P_3P_2P_1P_0C_0. \tag{8.49}$$

Now let us carefully look at Equations (8.43), (8.47), (8.48), and (8.49). Note that each one of those equations, regardless of the number of terms and the number of inputs per *AND* gate, are all in *SOP* form, which means that they can all be implemented in just two levels of logic. Refer to Figure 8.25 and look at the logic implementations that produce carryouts: C_1, C_2, C_3, *and* C_4.

Using the defined generate G_i and propagate P_i terms in Equations (8.43), (8.47), (8.48), and (8.49) these can be rewritten as:

$$\textit{Slice 0}: S_0 = G_0 \oplus C_0 \tag{8.50}$$

$$\textit{Slice 1}: S_1 = G_1 \oplus C_1 \tag{8.51}$$

$$\textit{Slice 2}: S_2 = G_2 \oplus C_2 \tag{8.52}$$

$$\textit{Slice 3}: S_3 = G_3 \oplus C_3. \tag{8.53}$$

So based on Equations (8.43), (8.47) through (8.49), and Equations (8.50) through (8.53), the picture of a *4*-bit adder with *carry look ahead* or *fast carry generation* is depicted in Figure 8.25.

Note that the adder has four clearly marked areas with horizontal dotted lines. Each area is a slice of the adder. Now for the sake of simplicity of the timing analysis that we will get into, assume that all logic gates on the diagram have the same propagation delay. This is not true, but the assumption simplifies the analysis without us getting lost in the details. Note that this adder (Fig. 8.25) unlike the adder implemented with full-adders, (Fig. 8.22a), does not have a carry that propagates from the least to the most significant slice.

The adder with fast carry look ahead produces each slice's carryout with three levels of logic gates. Carefully observe that the first level is the one that generates the P_i and G_i terms, the *AND* gates are the second level and the collecting *OR* gate is the third level. Finally is easy to observe that the sum bit requires one additional logic gate delay *(exclusive OR)* to produce the S_i bit. The above statement is true for the entire adder, slices 0 through 3. It is also important to mention that the same adder with carry look ahead can be used to implement a subtractor by adding the *1's* complementing logic to the subtrahend. We could continue to discuss fast adders but because of space reasons, we refer the reader to the references at the end of this chapter.

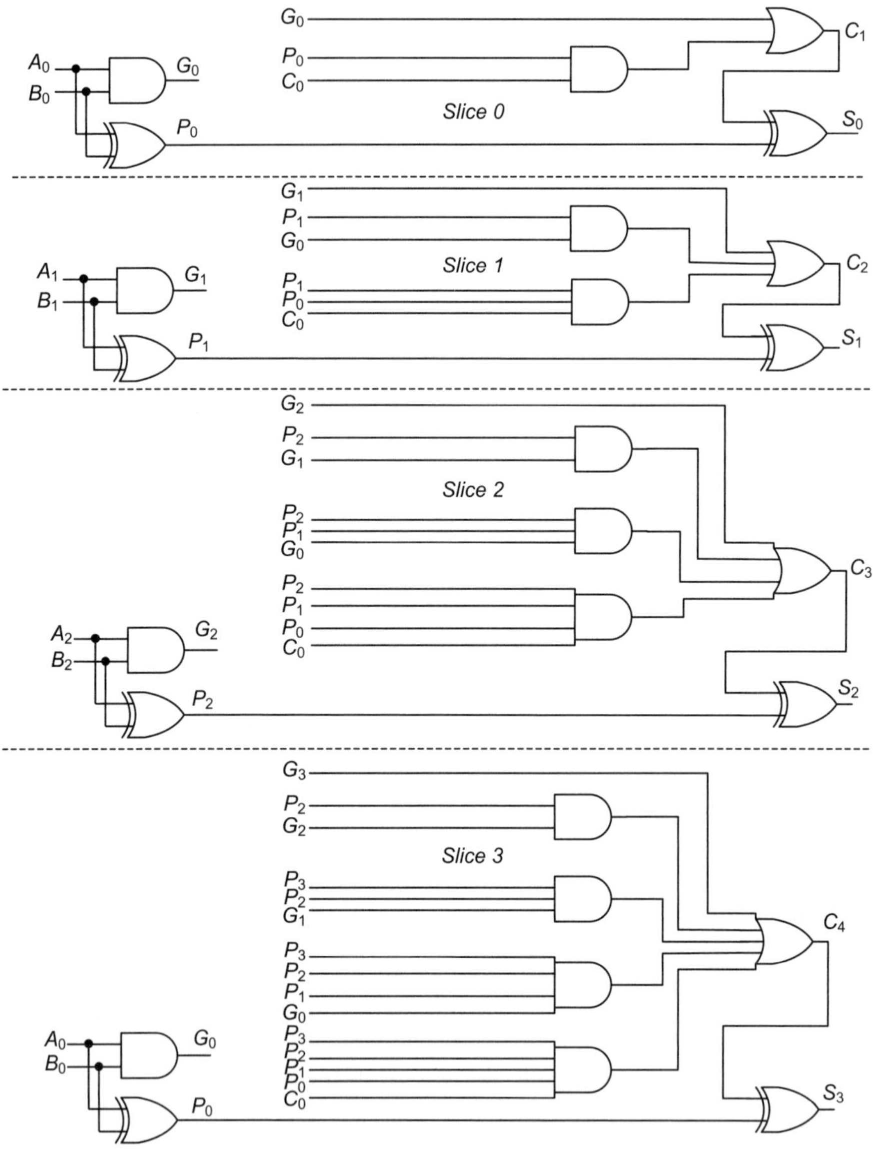

Figure 8.25 Logic implementation of a 4-bit adder with fast carry generation.

8.7 SOME SHORT-HAND NOTATION FOR LARGE LOGIC BLOCKS

We covered several logic blocks like decoders, multiplexers and adders; in most cases we drew every bit of the logic block in an explicit fashion. For example look back at Figures 8.13a and 8.15. In computer architecture literature is common to see a very compact way of drawing multi-bit devices.

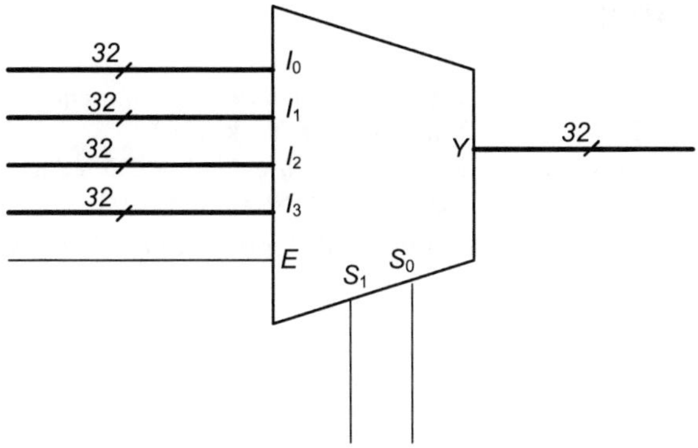

Figure 8.26 Compact graphic representation of a *32*-bit wide *1-of-4* multiplexer.

Assume we need a *1-of-4* multiplexer where each of its input is *32* bits wide. If we drew such device bit by bit it would look cumbersome and hard to read. Instead we draw a single line for each one of the *1-of-4* mux inputs and indicate with a short crossed line the number of bits the mux *leg* has. Note that output *Y* is a 32-bit wide output.

Figure 8.26 is a representation of the *32*-bit wide *1-of-4* multiplexers in compact notation.

Note that such 32-bit wide mux can be constructed with thirty-two regular *1-of-4* multiplexers.

Exercise: Get a large piece of paper and draw a detailed explicit drawing for the 32-bit 1-of-4- mux implemented with 32 individual 1-of-4 muxes.

We draw in a similar fashion very wide adders and subtractors. In the subsequent chapter we will use this compact notation more liberally which we will see that it also applies to registers, counters, and state machines. These are all sequential circuits (circuit with memory capabilities) not covered yet.

8.8 SUMMARY

This chapter covers a good number of combinational circuits: decoders, multiplexers, and arithmetic combinational circuits such as adders and subtractors.

The purpose of this chapter is to make the reader feel at ease designing virtually any simple or complicated combinational circuit by learning how to establish their truth table, how to do the logic simplification, and in some cases, how to partition big circuits into smaller ones to simplify the methodology or even clarify their behavior.

For in-depth coverage of arithmetic circuits the reader is referred to Reference [1].

FURTHER READING

1. Shlomo Waser and Michael J. Flynn, *Introduction to Arithmetic for Digital System Designers*, Holt, Rinehart and Winston, New York, 1982.
2. Robert L. Morris and John R. Miller, *Designing with TTL Integrated Circuits, Texas Instruments Electronic Series*, McGraw-Hill, New York, 1971.
3. Morris Mano, *Digital Design*, Prentice Hall, Upper Saddle River, NJ, 1984.

PROBLEMS

8.1 (a) Define the truth table for a logic circuit that given a three-bit positive binary number input, produces a four-bit output that equals the initial 3-bit number plus 7. This circuit has three inputs and four outputs; (b) Obtain the simplified SOP form of the four outputs; and (c) draw the circuit.

8.2 (a) Define the truth table for a logic circuit that given a four-bit binary number, produces a four-bit output that equals the initial 4-bit number 1's Complemented. This circuit has four inputs and four outputs; (b) Obtain the simplified SOP form of the four outputs; and (c) draw the circuit.

8.3 Implement the logic found for Problem 8.1 with the smallest possible multiplexer and any inverter gates as needed.

8.4 Implement each piece of logic found for Problem 8.2 with the smallest possible multiplexer and any inverter gates as needed.

8.5 Implement a 3-variable XOR function with the smallest possible multiplexer and inverter gates as needed.

8.6 Implement with the smallest possible decoder and minimal number of OR gates a 4-variable XOR function.

8.7 Without simplifying the logic, implement the following function entirely with NAND gates:

$$f(A,B,C,D) = A.B.C + \overline{A.B.\overline{D}} + A.\overline{B.\overline{D}}$$

8.8 Without simplifying the logic, implement the following function entirely with NOR gates:

$$f(A,B,C,D) = A.B.C + \overline{A.B.\overline{D}} + A.\overline{B.\overline{D}}$$

8.9 Write the truth table for a 4-bit positive binary adder with no carry in and no carry out. Implement and draw the circuit of the 4-bit input and

4-bit output positive binary adder with the smallest number of multiplexers of the smallest size possible and inverter gates as needed.

8.10 Implement and draw the circuit of the adder of Problem 8.9 with the smallest size and smallest possible number of decoders and OR gates as needed.

8.11 Draw the circuit of a two 4-bit input binary adder with carry in and carry out with fast carry-look-ahead logic. This circuit has 2 4-bit inputs, one overall carry-in input, one 4-bit output and one overall carry-out output. Use 4 single bit full-adders in addition to the carry-look-ahead logic.

8.12 Design and draw the circuit of a 1-of-8 multiplexer using only 1-of-2 multiplexers. Write the truth table of the circuit to-be-designed.

8.13 Design and draw the circuit of a 1-of-16 multiplexer using some 1-of-2 multiplexers and some 1-of-4 multiplexers. Write the truth table of the circuit to-be-designed.

8.14 Given a 4-bit 2's Complement binary number design the logic that produces the absolute value of the 2's Complement input. This circuit shall have a 4-bit input and 3-bit output. Write the truth table of the circuit to-be-designed.

8.15 Design a combinational circuit that given a 4-bit input produces the bit-to-bit OR of each one of the 16 binary input combinations. This circuit shall have 4 inputs and one output and must be implemented with the smallest possible decoder and OR gates if needed. Write the truth table of the circuit to-be-designed.

8.16 Implement a 3-bit XOR logic block entirely with NAND gates.

8.17 Implement a 3-bit XOR logic block entirely with NOR gates.

9

SEQUENTIAL LOGIC AND STATE MACHINES

9.1 INTRODUCTION

Logic is classified in two main types: combinational and sequential. We covered examples of combinational circuits in the previous chapter. This chapter mainly deals with sequential logic circuits. Let us recall that combinational logic circuits are those whose outputs depend on the current inputs. Such outputs are considered *good* or stable after the gate propagation delays have settled down. Combinational circuits are logic circuits without memory capabilities. Sequential logic circuits' outputs depend not only on the current inputs but also on their past history. This means that somehow sequential circuits must have some sort of memory. Such information in the sequential circuit memory is referred to as a state. Having added the memory concept to the sequential circuit, the outputs of a sequential circuit may depend on the *current inputs* and the *current state* or just on the current state. The terms sequential logic or state machine are often interchangeably used. Now within sequential state machines there are two categories of them: *synchronous* and *asynchronous* state machines. The majority of digital designs are done with synchronous logic. Synchronous designs are very well behaved and controlled by typically a fixed frequency clock, the clock supplies pulses at well-defined intervals of time. Asynchronous designs are not clocked and designers try to stay away from them because of their complexity and debug difficulties. In synchronous designs states can only change upon an active edge of the clock. Asynchronous

Electrical, Electronics, and Digital Hardware Essentials for Scientists and Engineers, First Edition. Ed Lipiansky.

designs are useful when input signals to the circuit may change at any time. Asynchronous circuits must obtain a stable state before an input can change again. Simultaneous changes of more than one input at a time are usually prohibited in asynchronous circuits. When two different micro-controllers communicate to each other, since each one has its own synchronous clock domain, the interfacing between the two is done as if the circuits were asynchronous with respect to each other. Examples of ways of allowing communication between independently synchronous machines are serial interfaces, such as *RS-232*, I^2C, and so on. Another example of asynchronous signals that need to be interfaced to a synchronous machine are the external devices interrupts that need to be routed to a micro-controller interrupt line. To accomplish that, asynchronous circuits, referred to as synchronizers and priority encoders are employed. Throughout this chapter the emphasis is given on synchronous state machines.

It is important to visualize that almost anything built in electronics is or contains one or more state machines. A garden-watering control system with a soil humidity sensor embedded in the soil is a good example of a state machine. The watering system can be programmed to water for 5 minutes every day provided that the humidity sensor detects more soil moisture is needed. However, if the humidity sensor detects enough soil moisture, the watering period for that day can be skipped. Other examples of embedded state machines that we see on a daily basis are traffic lights, washing machines, alarm clocks, computerized controls in automobiles, like anti-lock braking systems (ABS), cash registers in stores, global positioning systems (GPS), all kinds of telephones and many more gadgets. Table 9.1 below summarizes the two types of sequential state machines that exist and some of their fundamental characteristics.

Table 9.1 Types of sequential logic circuits

Sequential Circuits (Have Memory)		
Synchronous		Asynchronous
Clocked		Non clocked
Moore	Mealy	Outputs changes occur on response to a change on an input. Changing more than one input at any given time is avoided.
Outputs depend on present state only	Outputs depend on present state and current inputs	
Simplest to understand		Complicated to understand
Robust design. Preferred design practice. Very reliable behavior.		Hard to debug. Designers avoid them as much as practically possible. Whenever used they are usually interfaced with clocked state machines. Require synchronizing circuits.

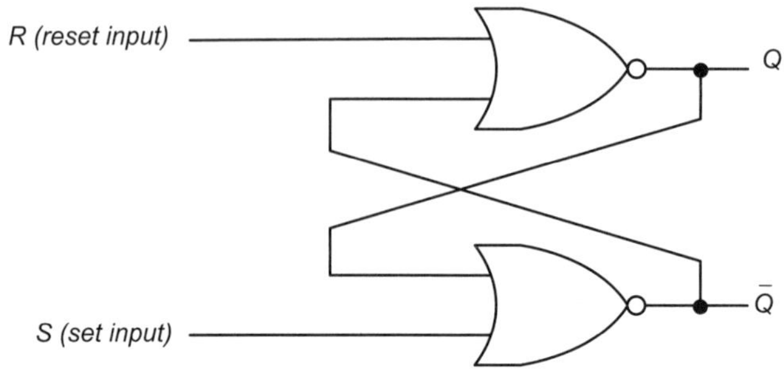

Figure 9.1 *SR-latch* with *NOR* gates.

9.2 LATCHES AND FLIP-FLOPS (FF)

The fundamental memory element is the latch. A latch can memorize a binary state indefinitely as long as there is power to the circuit and no failures occur. The basic latch is built with two *NOR* gates or two *NAND* gates. More elaborate combinations of latches and features can be obtained and are referred to as flip-flops. Figure 9.1 depicts an *SR-latch* implementation that with two *NOR* gates.

From Figure 9.1 we observe that the output of each *NOR* is fed back into one of the inputs of the adjacent *NOR* gate. That scheme is called a cross-coupled *NOR* gate configuration. The *S* (*Set*) and the *R* (*Reset*) inputs are the controlling signals to the latch. The *Q* output or simply the noninverted output, and $\bar{Q}$ or the inverted output indicate the state the latch is in. Assuming one correctly uses the latch, it can only be in one of two possible states at any given time. The *Set* state is when the latched *Q* output holds a *one* ($\bar{Q}$ holds a zero). The *Reset* state is when the latched *Q* output holds a *zero* ($\bar{Q}$ holds a one). The *Set* state is also called the *Preset* state, while the *Reset* state is also called the *Clear* state. For consistency we will continue to talk just about *Set* and *Reset* states. The latch is said to be in a state (*Set or Reset*) after the transients and gate propagation delays effects are over. A very simple example of the use of a latch is to detect if a signal made a change from one state to the other. For example, we leave the house and would like to know if our telephone will ring at least once during our absence. Assume that the latch is initially in a *Reset* state and its *Q* output driving an *LED*, the *LED* is off and we leave the house. An off *LED* means the phone never rang. Assume that during our absence the phone rings, the *LED* will light up. When we come home we see the turned on *LED*. What happens if the phone ringed more than once, nothing would happen, the *LED* continues to be turned on. Note that a latch can only store one bit of information. We need more latches if we want to detect multiple rings. We will get there. The fact that the latch has two outputs it does not mean that it can store two bits; because after transients elapse the outputs are always complements of each other (provided that the latch was used correctly). We will analyze several cases to understand the *SR-latch* operation.

Case 1 The latch is initially Reset and then the SR inputs set it.

Let us remember that the output of a *NOR* gate is *one* only if both of its inputs are *zero*, and the output is *one* when either one of its inputs is *one*.

The following analysis can be followed with the aid of Figure 9.2. Observe that the S and R inputs of the latch are negated or zero. We need to check if the state of its outputs $Q = 0$ and $\bar{Q} = 1$ is consistent with inputs $S = 0$ and $R = 0$. Since $\bar{Q} = 1$ and $R = 0$, the top *NOR* gate produces a 0 at the Q output. Since $Q = 0$ and $S = 0$ the bottom *NOR* gate sustains a 1. Interchangeably, if we analyze the same conditions starting with the bottom gate we have

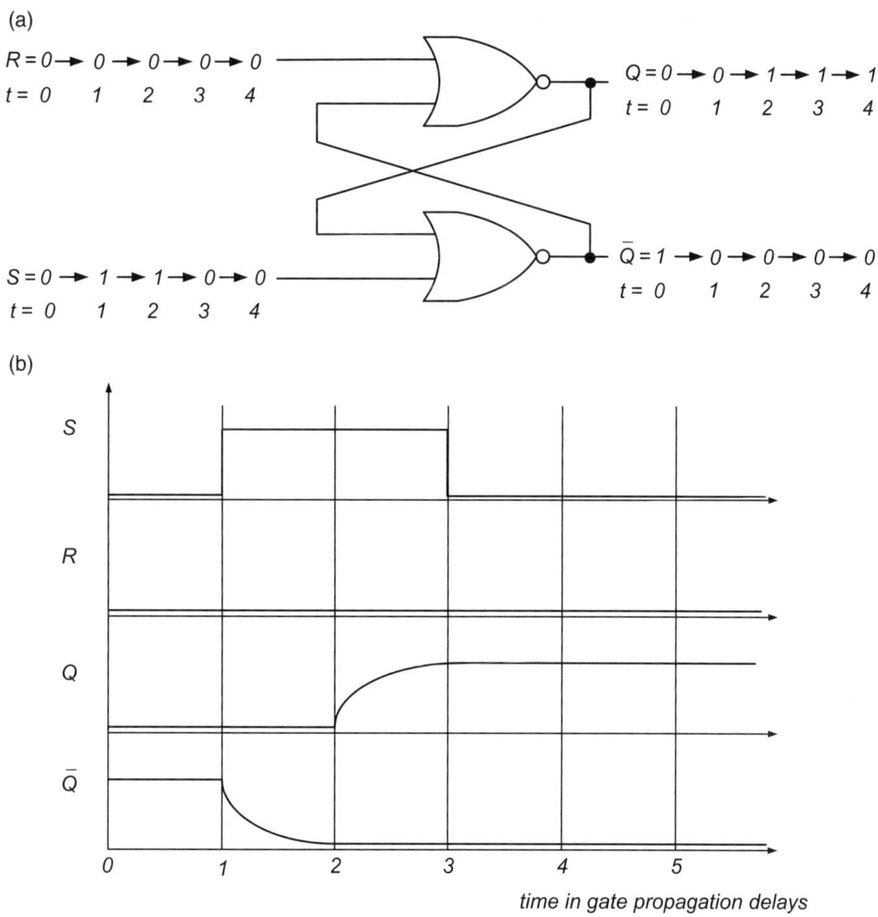

Figure 9.2 Case 1: From a *Reset* latch to a *Set* latch: (a) latch circuit with sequence of inputs and outputs; (b) timing diagram.

that since $S = 0$ and $Q = 0$ the gate output $\bar{Q}$ is 1. Now since the top gate has $\bar{Q} = 1$ and $R = 0$ at its inputs the Q output sustains a 0. This analysis allows us to state that if the latch is already *Reset* and its inputs R and S are *zeros* that *Reset* state is held. We just concluded analyzing the initial state of the latch at time $t = 0$. Note that the timing diagram of Figure 9.2 is in units of gate delays. That is, generally we will assume that both *NOR* gates have the same gate propagation delay. When we will make different assumptions they will be noted. Continuing with Figure 9.2, at time $t = 1$ the S input goes high with a zero rise time (very very quickly), and it stays high for two gate delays. The first change that takes place in the circuit is on the lower *NOR* gate, because one of its inputs is S. $S = 1$ along with the zero from the Q output causes the lower gate output $\bar{Q}$ to change from 1 to 0. This change takes no more that one gate delay, from $t = 1$ to $t = 2$. The new value of $\bar{Q} = 0$ is fed into the upper *NOR* gate along with $R = 0$, this produces after one gate propagation delay a 1 at the upper gate output Q, so at time $t = 3$ the latch is in its new state, the *Set* state. Note that if the S input did not stay high for two gate delays the latch would not function correctly because the signals would not have time to fully propagate through both *NOR* gates. Input S drops back to 0 at time $t = 3$. So S had been high for the latch minimum required length of time. The latch is now Set ($Q = 1$) and remains Set as long as $R = 0$ and $S = 0$ as the timing diagram shows. The reader should validate that *Set* state is stable analyzing the latch like we did at the beginning of Case 1.

Figure 9.2a depicts the timing transitions at times 0, 1, 2, . . .

Case 2 Figure 9.3 shows the operation of latch initially *Set*, then being *Reset* by $R = 1$, $S = 0$. The reader is strongly encouraged to do a similar analysis to that made for Case 1.

Case 3 Figure 9.4 depicts a case with an initially *Reset* latch, is later *Set* once, then its inputs are held low ($R = 0$, $S = 0$) and *Set* again a second time.

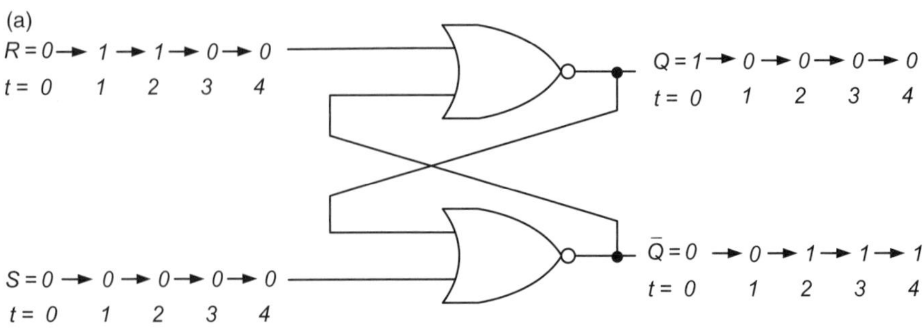

Figure 9.3 Case 2: From a *Set* latch to a *Reset* latch: (a) latch circuit with sequence of inputs and outputs; (b) timing diagram.

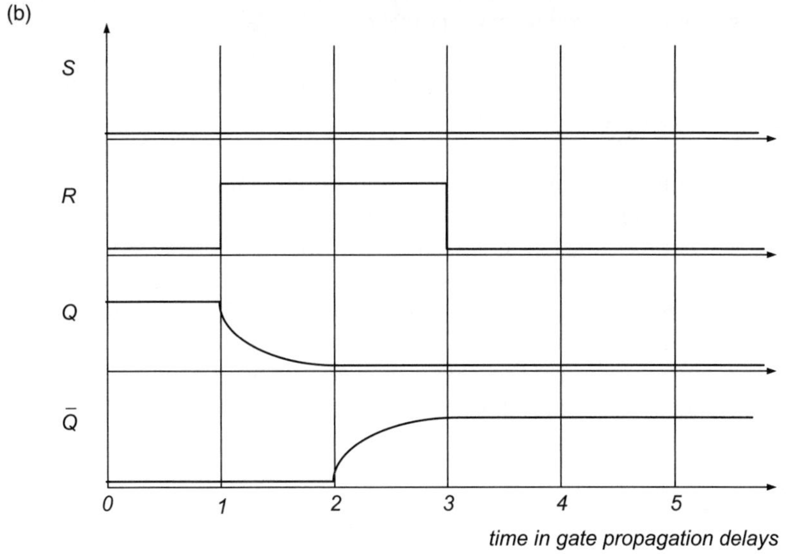

Figure 9.3 (*Continued*)

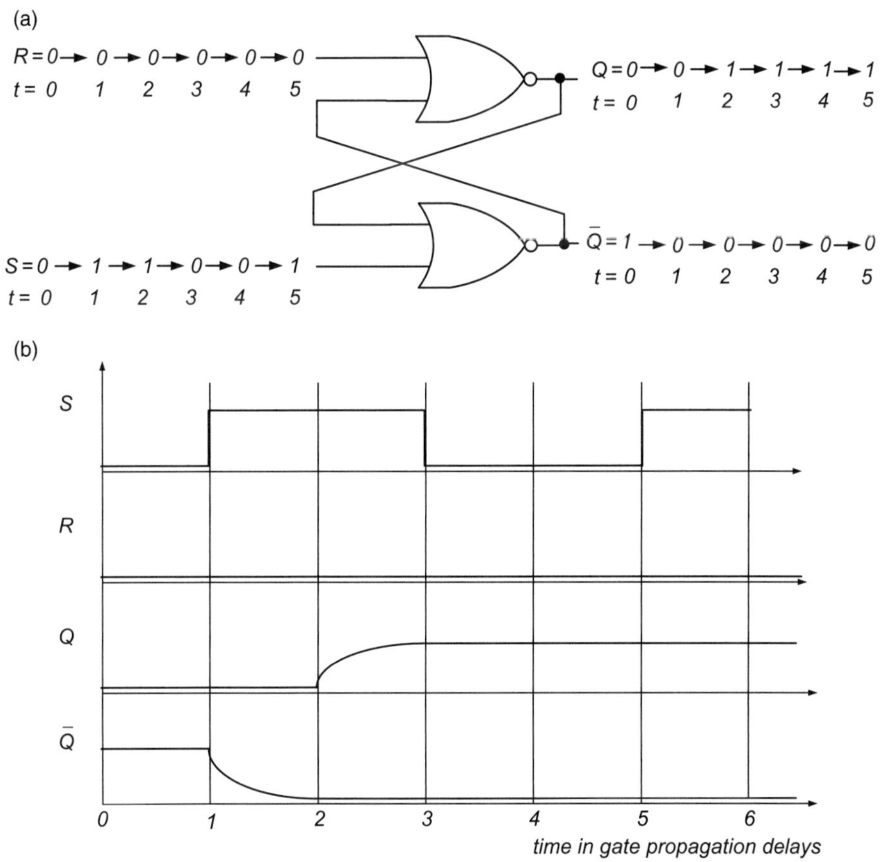

Figure 9.4 Case 3: From a *Reset* latch to a *Set* latch followed by one more *Setting*: (a) latch circuit with sequence of inputs and outputs; (b) timing diagram.

The importance of this case is to show that an already *Set* latch when *Set* again, remains *Set*, no changes.

Case 4 Shows a misuse of the latch. When both inputs are *one* ($S = 1$, $R = 1$) the latch no longer has complementary outputs. But this is not as bad as what follows. Upon both inputs dropping back to 0 a race condition takes place. Figure 9.5b timing diagram expresses that with question marks along the horizontal axis starting at time $t = 3$. If both gates delays are identical it is not possible to determine the state of outputs Q and $\bar{Q}$. Their state is undetermined. However, if one gate is faster than the other one, the faster gate will race and dominate the end state of the latch. Both of these cases are depicted by Figure 9.6, which assumes that the top gate (gate 1) is faster than gate 2, and Figure 9.7 which assumes that the bottom gate

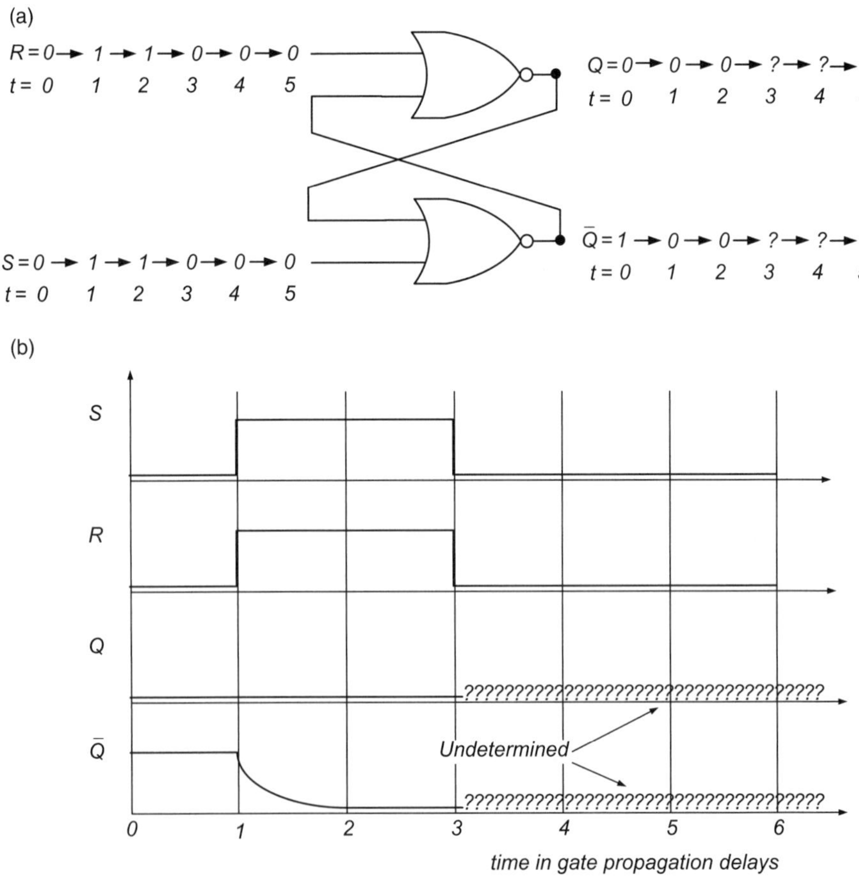

Figure 9.5 Case 4: Nonallowed usage of the latch; both *NOR* gates have identical gate delays so that a final state is undetermined: (a) latch circuit; (b) timing diagram.

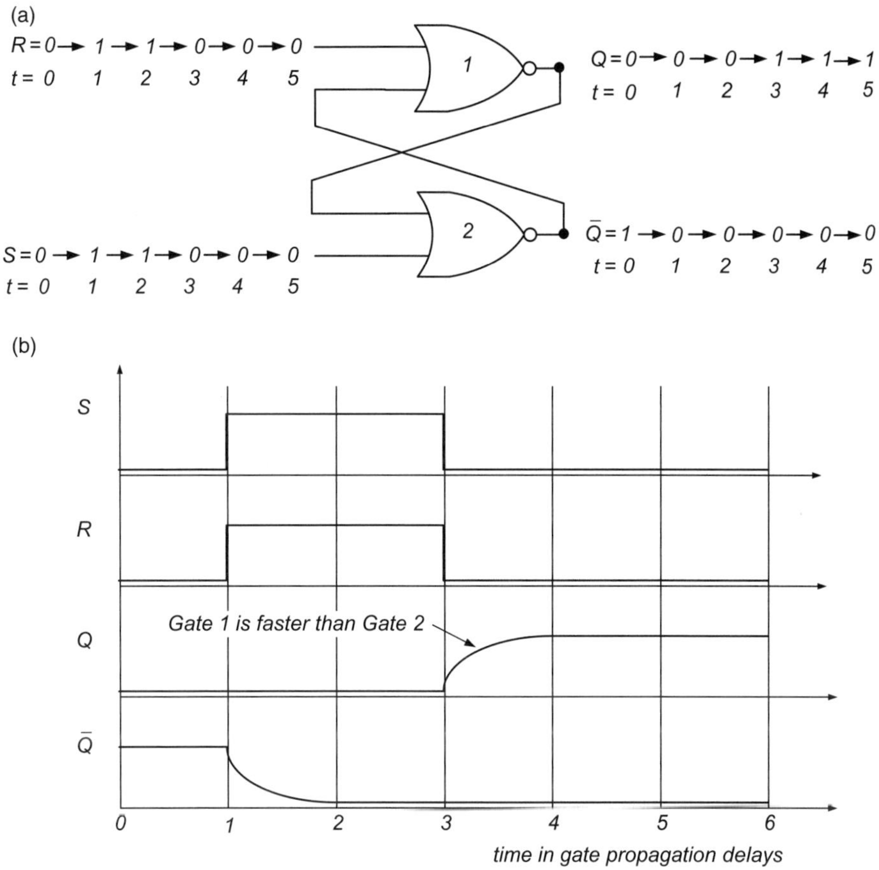

Figure 9.6 Case 4: Nonallowed usage of the latch; top *NOR* gate is faster than bottom *NOR* gate: (a) latch circuit; (b) timing diagram.

(gate 2) is faster than gate 1. In summary, Figures 9.5, 9.6 and 9.7 are all representations of Case 4, the nonallowed case.

Having worked on all of the above examples we are now ready to write the characteristic table of the *SR-latch*. Table 9.2 summarizes the latch behavior.

9.2.1 *NAND*-Implemented $\bar{R}$ / $\bar{S}$ Latch

An *SR-latch* implemented with *NAND* gates turns out to be an *SR-latch* with low-true or active low inputs. $\bar{S}$ becomes active when driven low, else $\bar{S}$ is inactive. The same is true for $\bar{R}$. $\bar{R}$ becomes active when driven low, and inactive when driven high. It is important and also interesting to notice that the noninverted Q output is the output of the *NAND* gate whose input is $\bar{S}$. Unlike

(a)

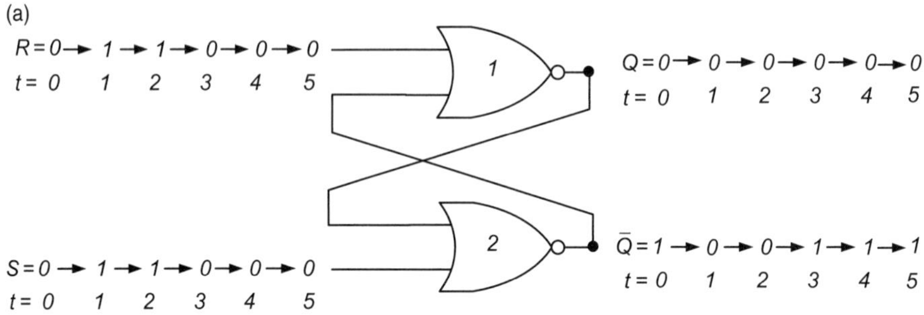

(b)

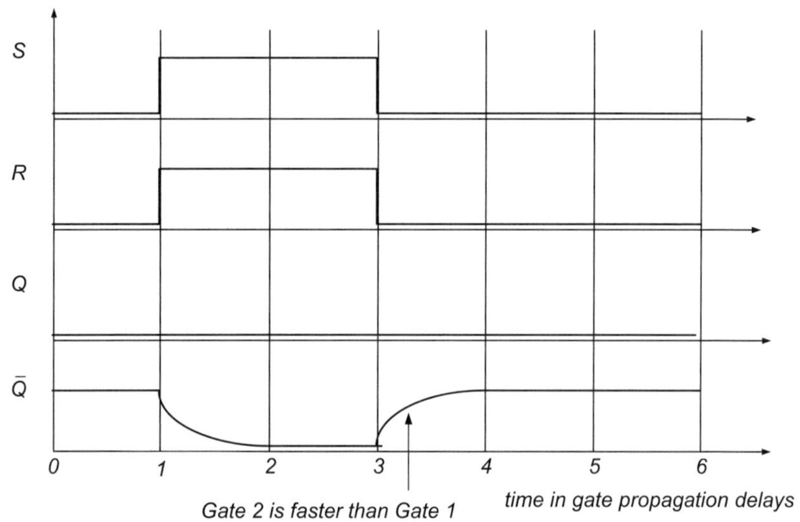

Figure 9.7 Case 4: Nonallowed usage of the latch; bottom *NOR* gate is faster than top *NOR* gate: (a) latch circuit; (b) timing diagram.

Table 9.2 ***NOR*-based *SR-latch* characteristic table**

S	R	Q	$\bar{Q}$	Description
0	0	Retains previously latched state of Q	Retains previously latched state of inverted Q	Holds the previously latched state
0	1	0	1	Reset
1	0	1	0	Set
1	1	0	0	Nonallowed condition. Outputs are no longer complements of each other. A race condition occurs upon both inputs being negated.

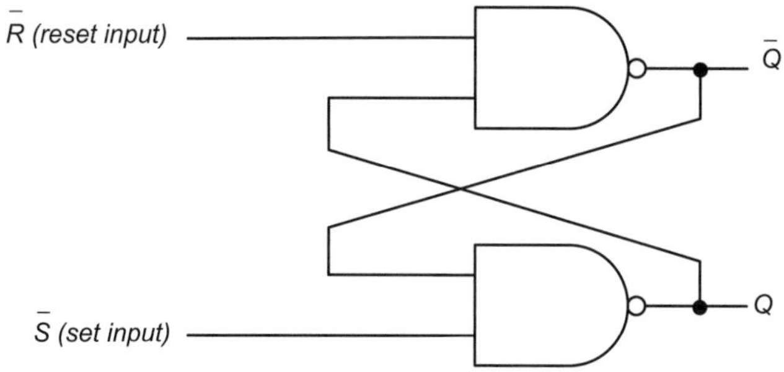

Figure 9.8 *NAND*-based $\bar{R}/\bar{S}$ latch.

Table 9.3 Characteristic cable for the $\bar{R}/\bar{S}$ latch

$\bar{S}$	$\bar{R}$	Q	$\bar{Q}$	Description
0	0	1	1	Nonallowed condition. Outputs are no longer complements of each other. A race condition occurs upon both inputs being negated.
0	1	1	0	Set
1	0	0	1	Reset
1	1	Retains previously latched state of Q	Retains previously latched state of inverted Q	Holds the previously latched state

the *NOR*-based *SR-latch* that has it Q output associated with the active high R input. Figure 9.8 shows the implementation of a *NAND*-based $\bar{R}/\bar{S}$ latch. Table 9.3 presents the characteristic table for a *NAND*-based latch.

Other than the differences observed on Table 9.3, the *NAND*-based latch is not different from the *NOR*-based latch. They both hold the previously latched state when both inputs are negated, they both can be in one of two possible states at any given time; the Q and $\bar{Q}$ loose their complementary nature upon both inputs being asserted at the same time. Finally, both latches exhibit a race condition when the inputs are negated immediately after being both asserted.

9.2.2 *SR-Latch* with Enable

We place an *AND* gate in front of every latch input and a clock pulse gates the flow of the S and R inputs to the latch. When the clock pulse is high the latch is a regular *SR-latch*, but when the clock pulse is low, the latch holds the previously latched state. Figure 9.9 depicts an *SR-latch* with enable.

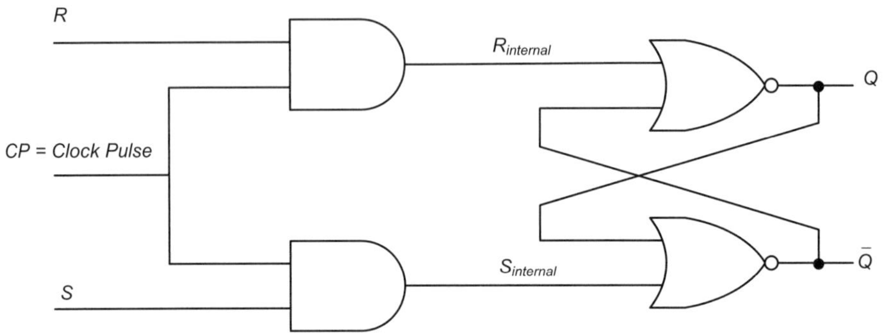

Figure 9.9 *NOR*-based *SR-latch* with clock pulse or enable.

Table 9.4 *NOR*-based *SR-latch* with clock pulse or enable, characteristic table

R	S	Clock Pulse	Q	$\bar{Q}$
0	0	1	Previously latched Q	Previously latched inverted Q
0	1	1	1	0
1	0	1	0	1
1	1	1	Not-allowed	Not-allowed
X	X	0	Previously latched Q	Previously latched inverted Q

Table 9.4 shows the *SR-latch* with enable characteristic table. Note that we called the enable (E) is also called *clock pulse* (CP). The emphasis is on the fact that the enable is a level sensitive control line, unlike *flip-flops*, which we will cover in other sections of this chapter, are clock-edge sensitive devices.

The *NAND*-based version of the latest follows for reference. Figure 9.10 depicts the *NAND*-based latch. Table 9.5 contains the *NAND*-based latch characteristic table. Note that regardless of the clock pulse or enable control line, both latches implementations, either with *NOR* or *NAND* gates still have the nonallowed state that would lead to a *race condition* upon their inputs negating at the same time. One more time let us remember that the *NOR*-based latch has active high inputs (R, S) while the *NAND*-based latch has active low inputs ($\bar{R}, \bar{S}$). However, when we add the two NAND gates to gate the $\bar{R}, \bar{S}$ inputs into the NAND-based SR-latch, refer again to Figure 9.10, it is worth mentioning that the composite latch, which includes the gating *NAND* gates, acts as if it was an active high input device, whereas the internal *NAND*-based *SR-latch* is still an active low input device. Refer to the annotations for the internal $\bar{R}, \bar{S}$ inputs (active low) and the external R,S inputs (active high) in Figure 9.10.

Whenever we want to refer to a latch with active high or active low inputs, with enable or without it, there are four new schematic symbols for them. Refer to Figure 9.11, which depicts the three types of latches that we have been discussing. It is important to note that regardless of the internal implementation of

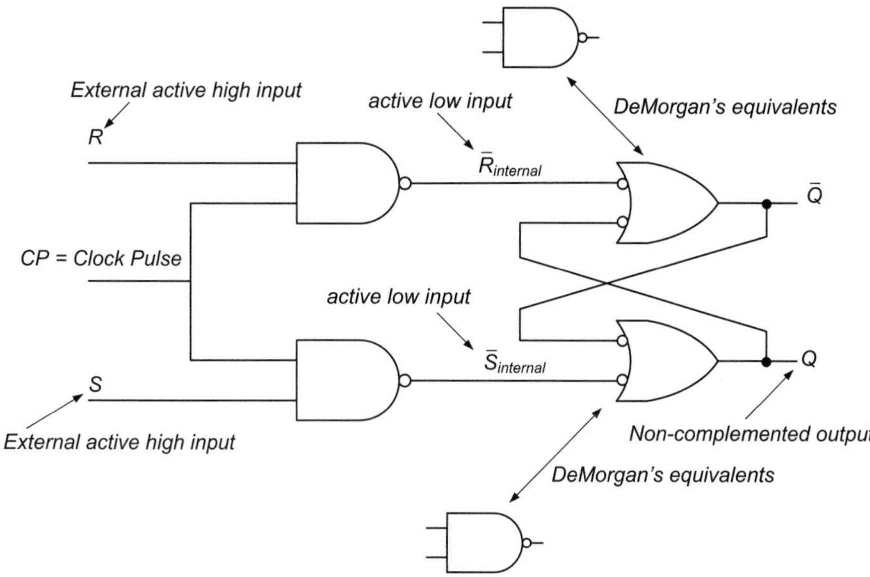

Figure 9.10 *NAND*-based *SR-latch* with clock pulse or enable.

Table 9.5 ***NOR*-based *SR-latch* with clock pulse or enable characteristic table**

R	S	Clock Pulse	$Q(t+1)$	$\bar{Q}(t+1)$
0	0	1	Previously latched Q(t)	Previously latched $\bar{Q}(t)$
0	1	1	1 (Set)	0
1	0	1	0 (Reset)	1
1	1	1	1 (Not allowed)	1 (Not allowed)
X	X	0	Previously latched Q(t)	Previously latched $\bar{Q}$(t)

circuit presented by a schematic symbol, its inputs and its characteristic table govern its behavior.

9.2.3 Master/Slave *SR*-Flip-Flop

*SR-latch*es are useful in control applications. Latches with and without clock pulse enable are still not very precise because their Q outputs will not settle to their stable state as long as the enable is active or as long as the inputs do not settle. What we would like to have are devices that respond to either a low-to-high or a high-to-low going edge of the clock pulse or enable. Such devices, which are clock-edge sensitive or master slave devices, change state at the active edge of their clock. The clock is no longer called enable, it is just the clock input. An enable has the connotation of level sensitivity, whereas clocks have the connotation of edge sensitivity.

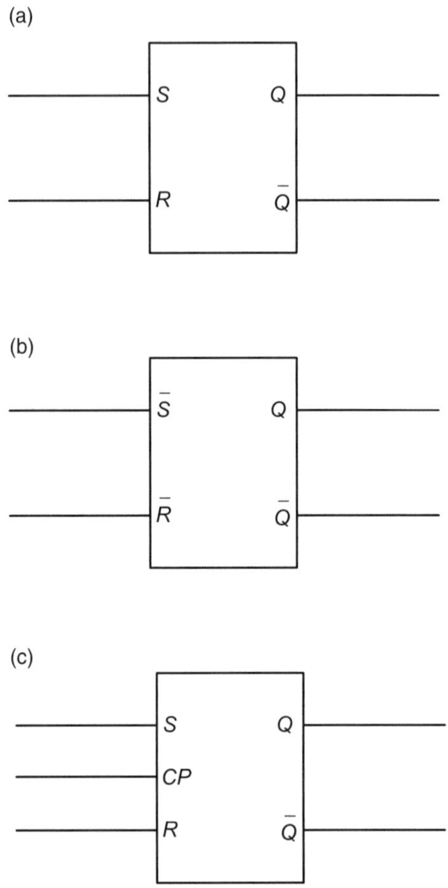

Figure 9.11 Latches schematics symbols: (a) *SR-latch* with active high inputs and no enable; (b) *SR-latch* with active low inputs and no enable; (c) *SR-latch* with active high inputs and an active high enable.

One master/slave configuration can be implemented with two cascaded SR-latches. Figure 9.12 shows the interconnection of both latches. Negative and a positive edge triggered flip-flops with their respective schematic symbols are shown.

Note a few differences between the symbols for latches and for flip-flops. All positive edge-sensitive devices show their clock with a small triangular symbol adjacent to the clock input line inside the device symbol. Negative edge-sensitive devices show their clock with a small triangular symbol adjacent to the clock input line and within the device symbol. Additionally, a bubble (inverting circle) is drawn at the base of the triangular symbol, just outside the symbol perimeter. Since clocks are inputs to flip-flop, clocks are drawn on the left-hand side of the schematic symbol. Figure 9.12 c and d depict, respectively,

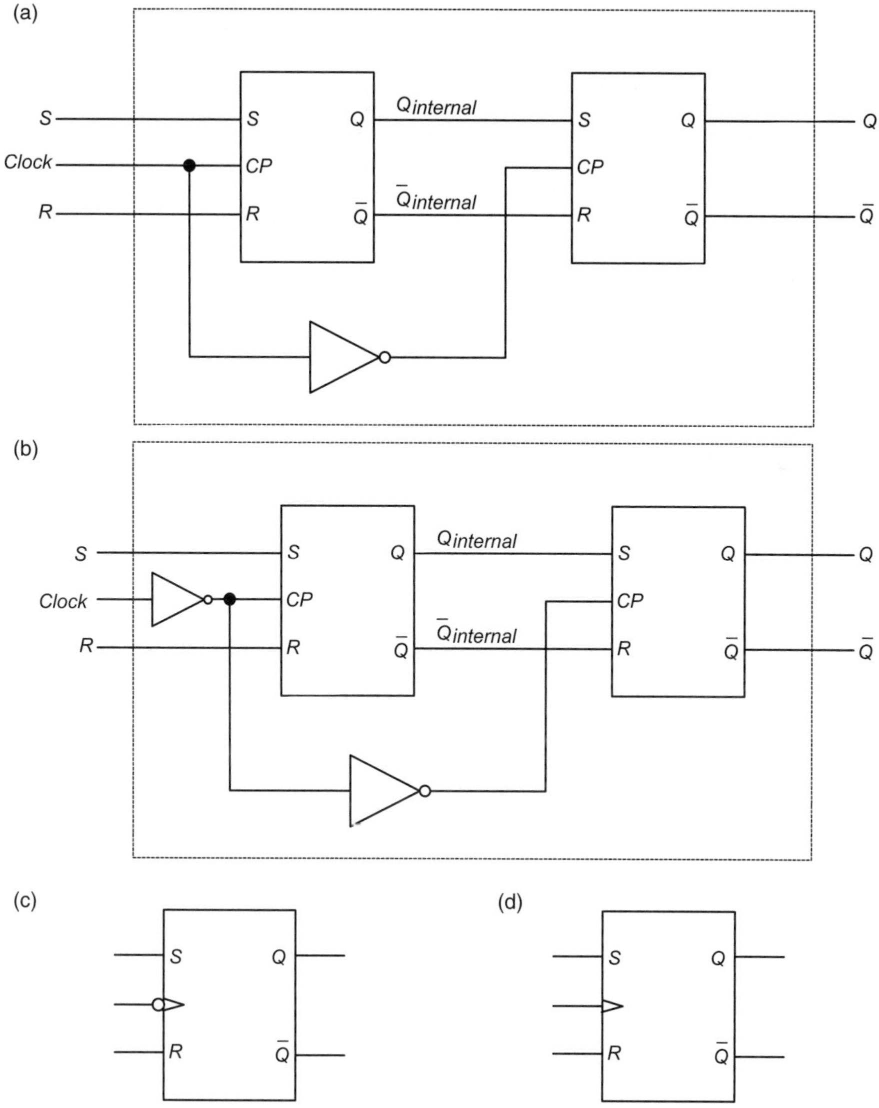

Figure 9.12 *SR-latch*es forming an *SR* master/slave flip-flop: (a) negative edge triggered *SR* flip-flop; (b) positive edge triggered *SR* flip-flop; (c) negative edge triggered flip-flop symbol; (d) positive edge triggered flip-flop symbol.

symbols of a negative edge triggered *SR* flip-flop and a positive edge triggered *SR* flip-flop.

Let us analyze how the *SR* flip-flop of Figure 9.12a operates. The first *SR-latch* is referred to as the master latch, while the right-hand side latch is the slave device. Note that both latches are simply active high inputs latches with

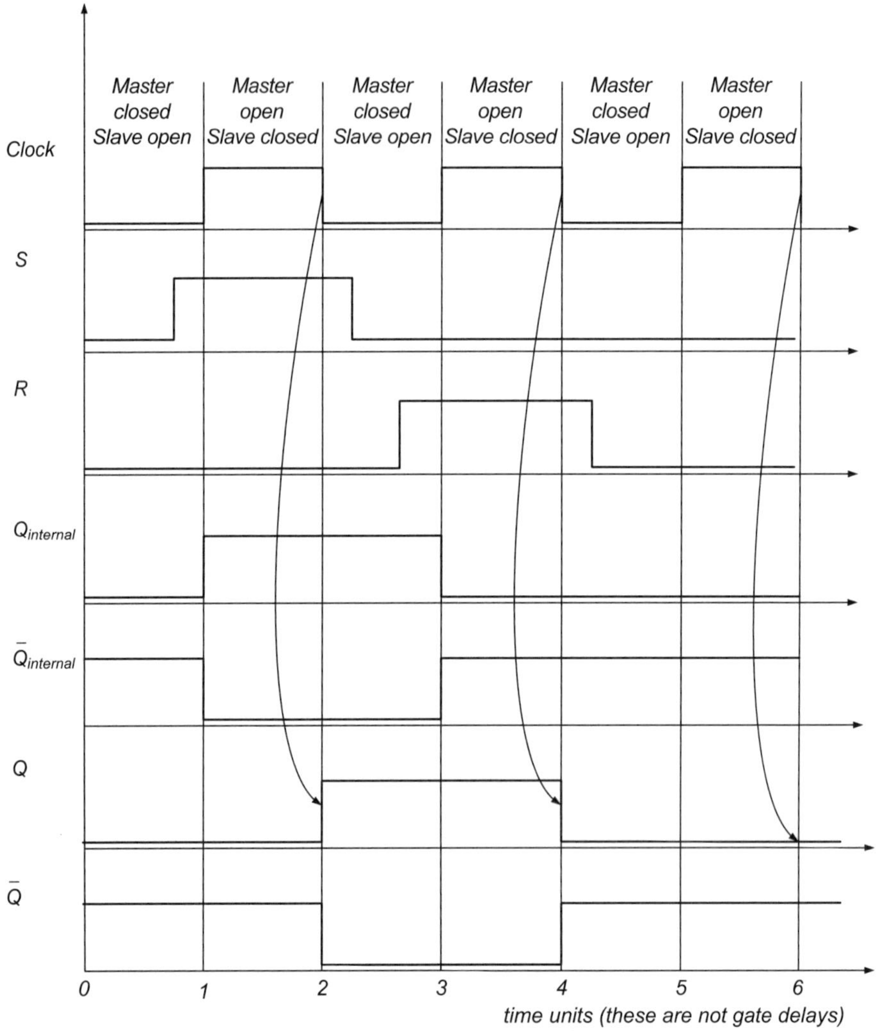

Figure 9.13 Positive and negative edge triggered *SR* flip-flop timing.

active high enable input. The inverter in the clock line of the latches causes them to become enabled in a mutually exclusive fashion. When the left-hand side latch is enabled or open, the right-hand side one is disabled or latched, and vice versa. A latched SR-latch refers to the latch holding or preserving its output value due to its negated enable.* So let us present any of the three valid input combinations (i.e., $R = S = 0$, *or* $R = 1$, $S = 0$, *or* $R = 0$, $S = 1$) to the left-hand side latch. The rightmost latch will preserve or hold its outputs at

* A latch is said to be enabled when its outputs may change due to changes of its inputs. A latch is said to be disabled when its outputs are latched and will not change due to changes of its inputs.

whatever state was previously latched, but the leftmost latch will act according to the setting of its inputs since its enable is active. The master latch transitions to the commanded new state;* the enable goes low, thus placing the master latch in *hold* mode. Now the slave latch sees a high clock because of the inverter in the clock line. The master holds it previously latched contents now and the slave gets commanded by the Q outputs of the master latch to change to the state commanded. Now let us look at this entire process as if the complete master/slave configuration was a whole device to the external user, such user is the one that observes only the master inputs and the slave Q outputs. The change seen on the master-slave Q outputs occur as if the slave state outputs were changing on the negative transition of the clock. Figure 9.13 depicts a timing diagram of a master/slave *SR* flip-flop. Although the R & S inputs, Q and $\bar{Q}$ internal output signals and the Q and $\bar{Q}$ outputs of the slave device are shown, it is important to look at the R & S inputs of the master device and the outputs of the slave device to appreciate the effect of the outputs changing at the negative edge of the clock. The $Q_{internal}$ and $\bar{Q}_{internal}$ are important for the correct operation of the flip-flop, but at the flip-flop high level view, the most important signals to observe are *R and S* inputs and the flip-flop Q and $\bar{Q}$ outputs.

It is also important to see that the master/slave scheme did not in any way suppress the *nonallowed conditions* of both latches. That means that if both inputs R & S became asserted, the master would loose complementary outputs, both $Q_{internal}$ and $\bar{Q}_{internal}$ will go to *zero* at time unit 2. Upon closing the master and opening the slave, both negated outputs of the master stage will propagate to the slave device, time unit 3. Between time units 2 and 3 both inputs R & S drop to zero. This will cause indeterminate output of the master stage upon its enable going high at time unit 3. On the next and low clock level the instability propagates to the slave stage at time unit 4. Figure 9.14 shows the operation of the *SR-latch*-based flip-flop under the *nonallowed conditions*, that is: $R = S = 1$ and then both negating simultaneously ($R = S = 0$).

9.2.4 Master/Slave *JK* Flip-Flop

It is meaningful to ask ourselves why don't we suppress or fix the *nonallowed condition* of the *SR* latches and flip-flops. This exactly is what a *JK* flip-flop does. The *JK* (in short) is sort of the Cadillac of the flip-flops, as we will see very soon. The *JK* flip-flop not only has a master and a slave stages but also has some additional logic that blocks the *nonallowed condition* being present at its inputs. Figure 9.15 depicts the implementation of a *JK* using *SR-latch*es, and inverter for the clock line and two *AND* gates to block the disallowed condition that would otherwise cause a race.

* Setting the R & S inputs at the desired levels and applying the clock to the flip-flop causes the issuing of a command. Such command can make the flip-flop transition to another state.

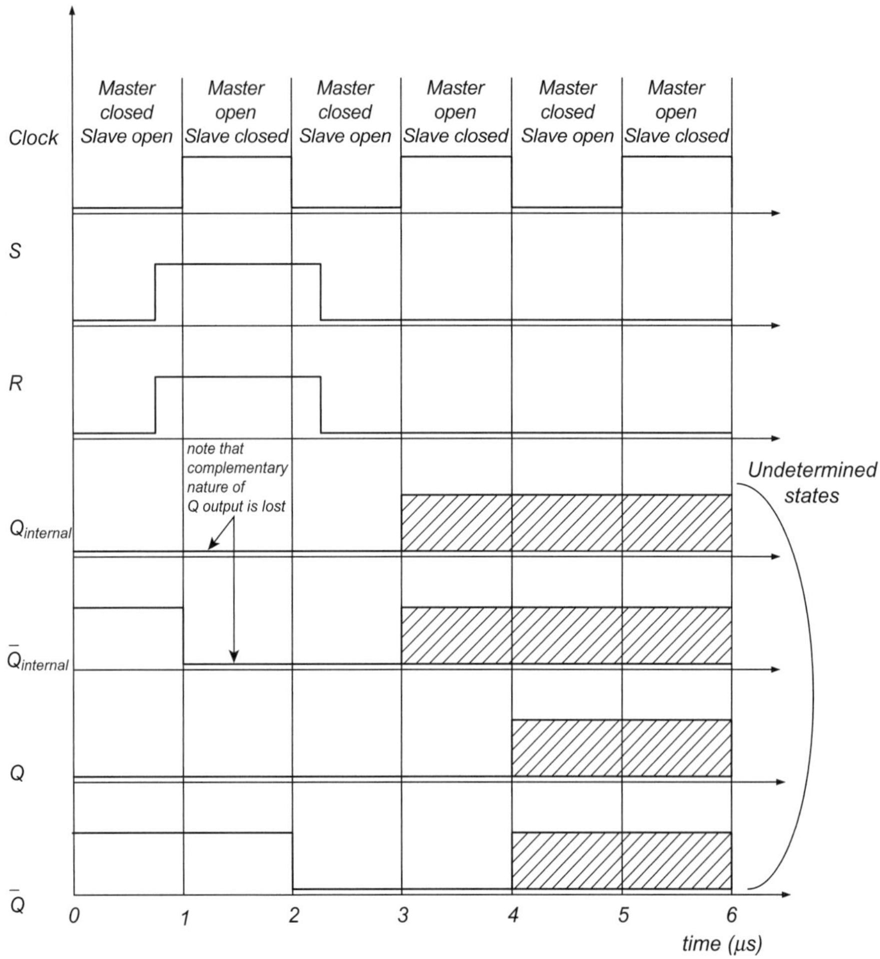

Figure 9.14 *SR* flip-flop operating under nonallowed conditions.

Let us inspect the *JK* and find out if it operates like the *SR* flip-flop at least under some input conditions. Assuming the *JK* is initially in the *Set* state and that its inputs *J* and *K* are *zero*, we know that the master latch should also be holding a *Set* state. Remember that a *Set* state means that $Q = 1$ and $\bar{Q} = 0$. Notice the two *AND* gates, one fed by the *Q* output of the slave latch and the other *AND* fed by the $\bar{Q}$ output of the slave latch. Since we are assuming that both *J* & *K* are equal to 0 (this is our initial condition), the outputs of both *AND* gates produce zeros. These *zeros* feed the *R* and *S* inputs of the master stage latch. Thus, upon clocking this device as long as we want, while the *JK* inputs are negated, the flip-flop will preserve or *hold* the previously captured state. Such previous state in our example was the *Set* state. If we start the analysis all over again with just a minor change that the *JK* initial state is a

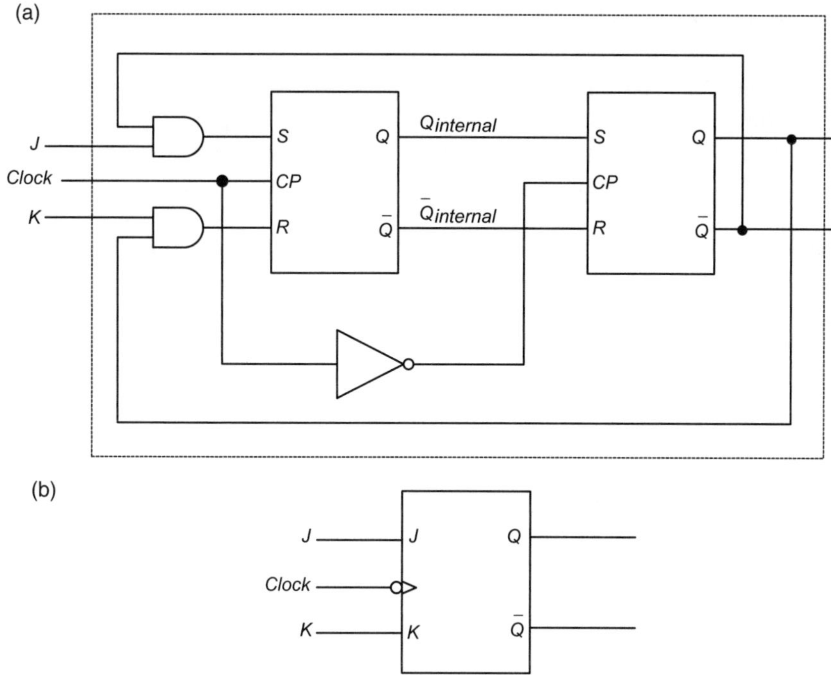

Figure 9.15 Negative edge-triggered master/slave *JK* flip-flop: (a) flip-flop logic circuit; (b) schematics symbol.

Reset we will arrive at the same conclusion. This means that our newly defined *JK* flip-flop holds the previous state for negated inputs *J* and *K*. This is so far no different than an *SR* flip-flop.

Assuming the *JK* flip-flop is again *Set,* if we bring the *K* input high and keep *J* low, note that the *JK* lower *AND* gate, is fed by a one from the *Set* state (or the *Q* output) and a 1 from the fact that *K* is high. Thus, the master stage *SR* latch sees the condition $R = 1$, $S = 0$, which after the clock allows to propagate the output of the master stage to the slave stage the *JK* flip-flop ends up in a *Reset* state. Similarly, if the *JK* is already *Reset*, clocking the condition $R = 1$, $S = 0$ will continue to *Reset* the *JK*, thus it stays Reset for as long as we keep clocking the flip-flop.

Now if we assume that out original *JK* flip-flop is either *Set* or *Reset*, input conditions are: $R = 0$, $S = 1$,the flip-flop will end up in a *Set* state for as long as we keep clocking the *JK*. So far the *JK* flip-flop behaves just like an *SR* flip-flop for the given conditions.

Now the *JK* becomes more interesting, assume an initially *Set JK* flip-flop and inputs *J* and *K* are both high, one more time following the logic of Figure 9.15a the *one* at the *Q* output of the slave stage along with *K* input that is *one*, produces a one at the output of the lower *AND* with inputs *Q* and *K*. Remember that *J* is one and with its associated *AND* gate that receives a zero from the

$\bar{Q}$ output of the slave stage, thus it produces a zero at the S input of the master stage SR latch. In summary, the master latch sees $R = 1$ and $S = 0$. These conditions produce the master latch to *Reset*. This *Reset* state propagates to the slave stage so that after a complete clock, the JK flip-flop Q output goes from 1 to 0.

Now what happens if we clock the *JK FF* one more time, while both inputs J and K are still 1? Now since the Q output of the slave stage is 0, with $K = 1$ into the bottom *AND* gate produces a zero into the R input of the slave stage. However, since $\bar{Q}$ is 1 and J is 1, the top *AND* generates a 1 into the J input of the JK flip-flop. Upon such state of the inputs propagating through the master and slave stages, after one complete clock cycle, the Q output of the JK sets again. If we allow the clock to run indefinitely, the JK flip-flop Q output will toggle from 1 to 0 and from 0 to 1. Similarly, $\bar{Q}$ toggles from 0 to 1 and from 1 to 0. The above function of the JK just described does not preclude both inputs to the JK of being *one*.

Next we summarize the complete behavior of our negative edge triggered master/slave JK flip-flop. Table 9.6 summarizes the *JK FF* characteristic table.

Table 9.6 applies to negative clock edges. The same table applies to positive edge triggered JK flip-flop if under the *Clock* column we indicate an up-going arrow. The logic for a positive edge triggered JK and its schematics symbols are depicted in Figure 9.16.

The *SR FF* is very similar to the *JK FF;* the difference is that the SR does not support the simultaneous assertion of its R and S inputs. By inspection of the *JK FF* characteristic table, it is easy to see that only the last row of the JK would be equivalent to a nonallowed condition for the SR. We present the SR FF characteristic table in Table 9.7. Since generally we will be using positive edge triggered devices, the SR characteristic table is presented for rising clock edges.

9.2.5 Master/Slave *T* and *D* Type Flip-Flops

We will now study two more flip-flops and we will address them as particular cases of a JK. If we tie both inputs of a JK flip-flop together as shown in Figure 9.17a, the JK is renamed T for Toggle flip-flop and its schematic symbol can be found in Figure 9.17b. Table 9.8 lists the characteristic table of a T flip-flop.

Table 9.6 Negative edge triggered master/slave *JK* flip-flop characteristic table

Clock	J	K	$Q(t+1)$
↓	0	0	Hold (no change)
↓	0	1	0 (Reset)
↓	1	0	1 (Set)
↓	1	1	Complement (toggle)
0	X	X	Hold (no change)

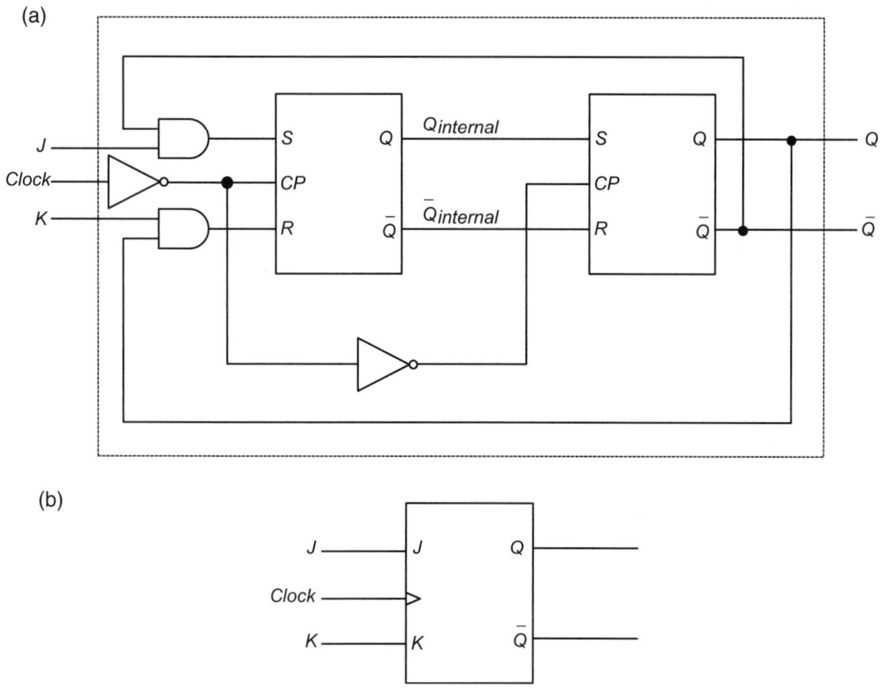

Figure 9.16 Positive edge-triggered master/slave *JK*: (a) Flip-flop logic circuit; (b) schematics symbol.

Table 9.7 Positive edge triggered master/slave *SR* flip-flop characteristic table

Clock	S	R	$Q(t+1)$	Comments
↑	0	0	$Q(t)$	Hold (no change)
↑	0	1	0	Reset
↑	1	0	1	Set
↑	1	1	Not-allowed	Unpredictable
0	X	X	$Q(t)$	Hold (no change)

The next and last flip-flop that we will cover is the *D-type* flip-flop or the *Data* flip-flop. The *D*-type flip-flop can be constructed by tying the *JK* inputs as depicted by Figure 9.17c; its schematic symbol is presented in Figure 9.17d.

The *D*-type flip-flop characteristic table is presented in Table 9.9.

The *D* flip-flop is commonly used to store a bit of information. It is in essence a 1-bit register. Remember that the *SR* latch studied earlier also stores one bit of information, but the latch is not a clocked device like the *D* flip-flop, the latch is referred to as an asynchronous device. Multi-bit registers are made with *D* flip-flops; a flip-flop per bit of storage is required. With the current

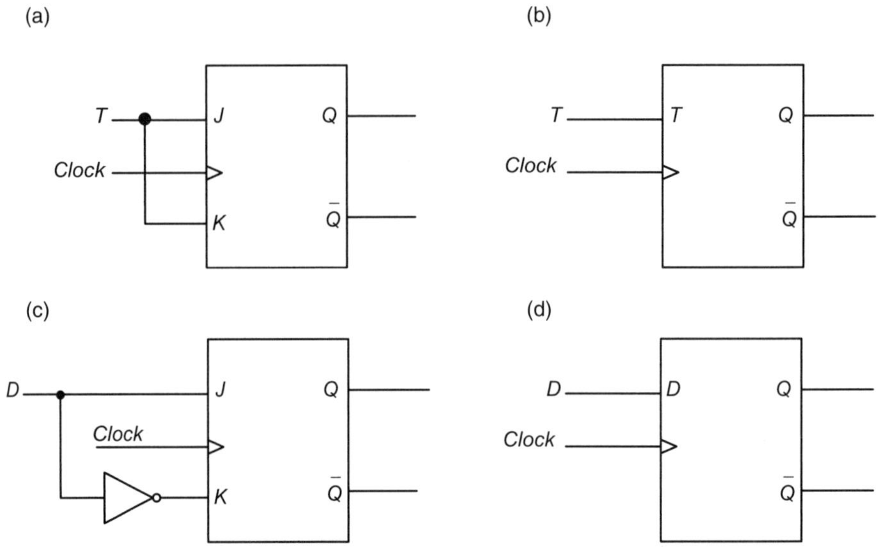

Figure 9.17 (a) Positive edge triggered master slave *T* flip-flop circuit diagram; (b) schematic symbol; (c) positive edge triggered master slave *D* flip-flop circuit diagram; (d) schematic symbol.

Table 9.8 ***T*-flip-flop characteristic table**

Clock	T	$Q(t+1)$	Comment
↑	0	$Q(t)$	Hold (no change)
↑	1	$\bar{Q}(t)$	Complement (toggle)
0	X	$Q(t)$	Hold (no change)

Table 9.9 ***D*-flip-flop characteristic table**

Clock	D	$Q(t+1)$	Comment
↑	0	0	Reset
↑	1	1	Set
0	X	$Q(t)$	Hold (no change)

state-of-the-art technology, the *D-FF* is the most widely used sequential device. Field programmable gate arrays (FPGAs), application specific integrated circuits (ASICs), and programmable logic devices (PLDs) make use of the *D-FF* extensively. The *D* flip-flop is the most commonly used device. The other flip-flops (*SR, T, and JK*) were more heavily used when medium scale integration (MSI) circuits use was more prevalent. Those *MSI ICs* were the very popular *7400 series*, manufactured by many different IC manufacturers for more than two decades.

Those were the days when there were practically no FPGAs, ASICs, and PLDs. When these types of devices became available, they were not as widely used and their cost was high. That situation is practically reversed today.

9.3 TIMING CHARACTERISTICS OF SEQUENTIAL ELEMENTS

This section deals with the fundamental timing parameters that clocked sequential devices must meet in order to operate correctly. Given a flip-flop like an *SR*, *JK*, *T*, or a *D*-type, before an active edge of the clock, it is required that the inputs be *stable before and after* the active edge of the clock makes its active transition. A positive edge-triggered device active transition of its clock is a low-to-high clock transition. A negative edge-triggered device active transition of its clock is a high-to-low clock transition. The time required by the device inputs to be stable prior to the active transition of the clock is called the *set-up* time (t_{SU}).

Additionally, the data inputs to the flip-flops are required to stay at the level that they were set up for a period of time immediately after the active edge of the clock. This time is referred to as the input data *hold-time* (t_H). For example, if one wants to clock a *high level* into a *D* flip-flop, the *high level* must be stable before, during, and after the active edge of the clock by a total time given by $t_{SU} + t_H$. These two timing parameters insure that the delays and timing requirements of the master and slave stages within the flip-flops are met. When we studied *SR-latch*es we learned that the minimum required delay for a single latch to produce a stable output is at least two-gate delays. To obtain some positive margin, the latch inputs should be stable some time longer than that minimum requirement. Luckily integrated circuit flip-flop manufacturers and ASIC manufacturers dictate the timing parameters required by their internal flip-flops. Another important timing parameter of a flip-flop is the *clock-to-output* propagation delay also called the *clock-to-Q* delay. Generally speaking the *clock-to-Q* (t_{CQ}) and the *clock-to-*$\bar{Q}$ ($t_{C\bar{Q}}$) need not and they are usually not the same. Engineers calculating timing requirements usually pick the longest of both *clock-to-output* times. Figure 9.18 shows the set-up; hold time and clock-to-output parameters of a clocked device and how they are related to the active edge of the clock.

9.3.1 Timing of Flip-Flops with Additional Set and Reset Control Inputs

The four flip-flops that we studied (i.e., *SR, JK, D, and T*) may be available with two additional control inputs, *Set* and *Reset*. What is the purpose of these control inputs? Since flip-flops are used to design state machines, many times it is convenient to start or restart a flip-flop to a known state. Such state may be a *Reset* or a *Set*, depending on the application. Flip-flops available in integrated circuits are usually made with asynchronous reset and set control

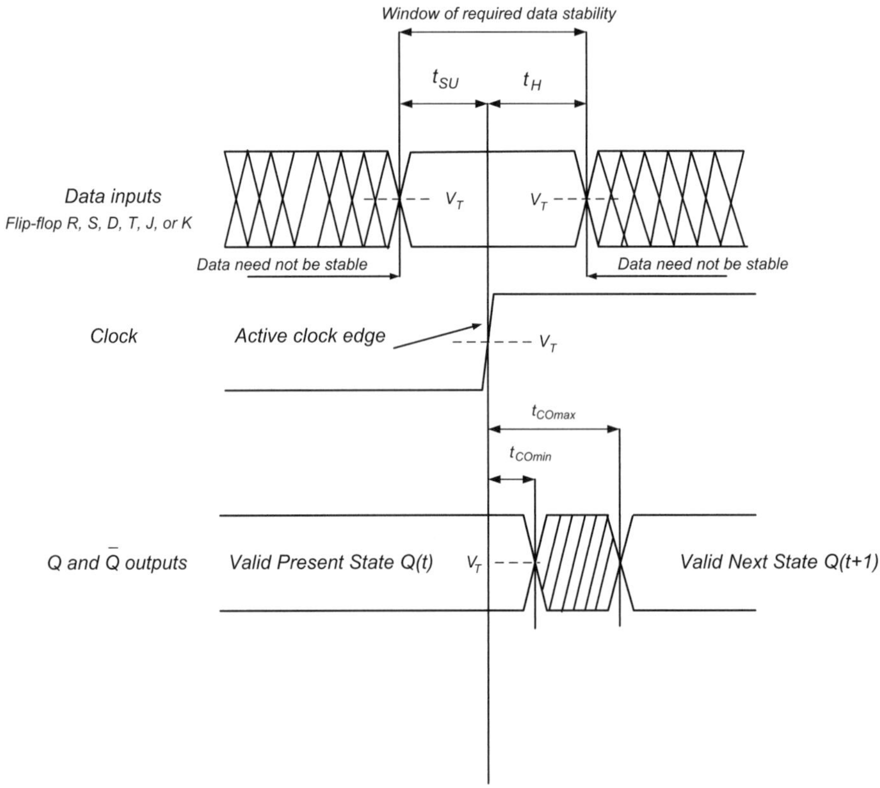

Figure 9.18 Flip-flop timing parameters.

inputs. Other flip-flops are available with synchronous inputs. Asynchronous reset and set control inputs act on the state output of the flip-flop (the Q outputs) completely asynchronously with respect to the clock. We learned that flip-flops will only change state on or immediately after the active edge of the clock; and asynchronous control input will make the flip-flop change state virtually at any point between active clock edges. Synchronous set and reset inputs are like any other inputs of the flip-flop, like J or K inputs, they must meet the set-up and hold time requirements of the flip-flop and the flip-flop makes a transition to either the set or reset state on the next active clock edge. Figure 9.19 depicts a timing diagram of a flip-flop with asynchronous reset and one with synchronous reset. A flip-flop asynchronous reset or set input is required not to assert during an active clock transition, else possible timing malfunction may occur.

Note that in Figure 9.19a upon the assertion of the active high asynchronous reset, the Q output goes low virtually immediately or shortly after the Q output delay without waiting for the next active edge of the clock. This means

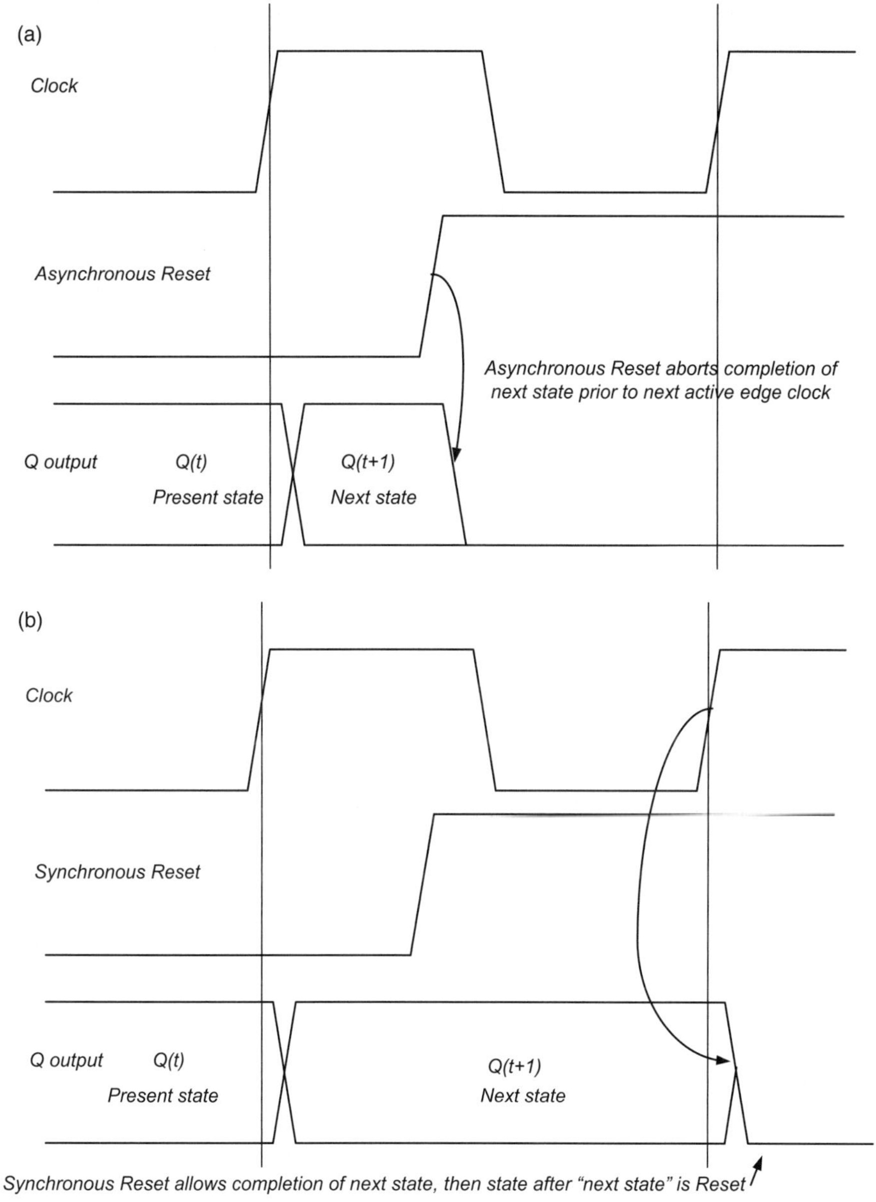

Figure 9.19 Timing diagram showing (a) asynchronous reset and (b) synchronous reset.

that the present state is abruptly interrupted and the reset is applied at the Q output. In Figure 9.19b upon the assertion of the active high synchronous reset, the *Q* output waits until the active edge of the clock arrives, and then the *Q* output gets reset then. Naturally there is also a clock-to-output delay before the *Q* output changes to the *zero* state. It is important to see that in the synchronous case the currently being executed state reaches completion and the reset state synchronously takes place on the subsequent active clock edge.

9.4 SIMPLE STATE MACHINES

Instead of defining what state machines are, we will present a simple example of a synchronous state machine; understand what it does, how it is described, and how it is designed. Later on we will make general statements as to what state machines are, but not without having gone over one simple but complete example.

Example 9.1 Define a 2-bit synchronous up binary counter with an asynchronous reset: (1) Write its state table and (2) state diagram. (3) Perform a logic implementation of the counter using positive edge-triggered *JK* flip-flops.

Solution to Example 9.1

A *2*-bit synchronous up binary counter is a *2*-bit state machine. Upon clocking the state machine the counter will go through states *00, 01, 10, 11*, and it will repeat that sequence of four states indefinitely, as long as it continues to be clocked. Figure 9.20 shows the state machine state diagram, while Table 9.10 depicts the state table of the *2*-bit counter. The purpose of the asynchronous reset is for external logic to reset the state machine upon power-up. If there was no reset the initial state of the state machine after power-up is unpredictable; in other words with no reset initializing the flip-flop, one cannot predict which will be the starting state.

Note that the state machine Q_1 state bit is the *MSB* and Q_0 is the *LSB*. The state machine has an asynchronous input, its *Reset* which we can easily implement using the asynchronous reset of the flip-flops to be used. Note that the state machine upon *Reset* being negated and receiving clocks it will walk through states *00, 01, 10, 11, 00, . . .* indefinitely. If at any time *Reset* goes high the state machine will abruptly go to state *00*. Notice that all state transitions are conditioned by the *Reset* input, the present state and upon the reception of a clock the machine will move to its next state. Unfortunately, the state diagram does not show in a *clean way* the fact that *Reset* is asynchronous. If the design requirements would have been to do the same design with a synchronous *Reset*, the state diagram would not change. Usually in state machine design *Reset* is one of the few or sometimes the only asynchronous control signal in the system. It is customary to synchronize asynchronous signals into the clock domain of the state machine that one is dealing with.

State machines in a general sense have two main parts, its sequential logic, that is the flip-flops that memorize the state $Q(t)$. They also have their

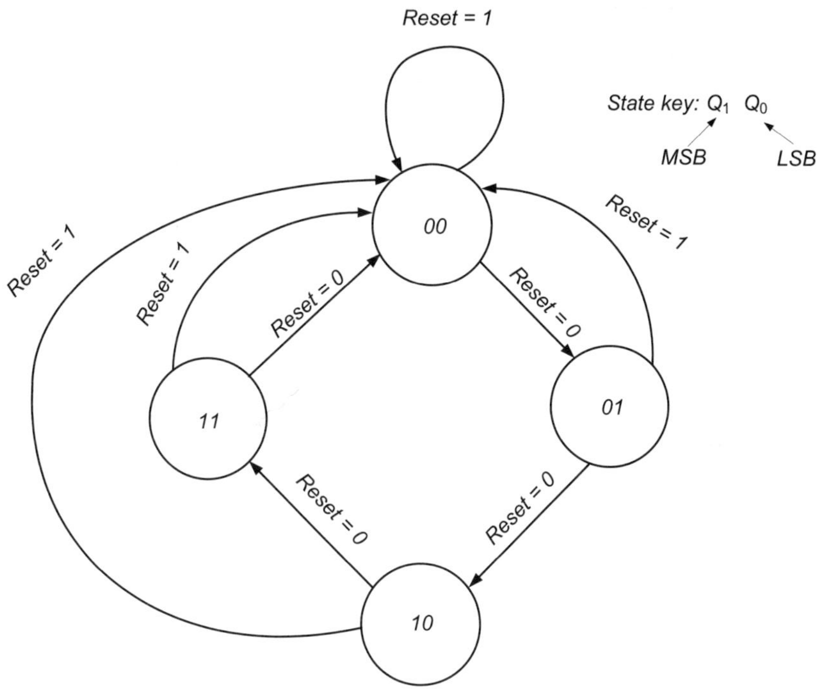

Figure 9.20 2-bit synchronous up binary counter state diagram.

Table 9.10 State table for the *2*-bit counter of Example 9.1

	Present State	Next State	Async. Reset Input
Clock	$Q_1(t)\ Q_0(t)$	$Q_1(t+1)\ Q_0(t+1)$	(Active high) Reset
↑	0 0	0 1	0
↑	0 1	1 0	0
↑	1 0	1 1	0
↑	1 1	0 0	0
0	X X	0 0	1

combinational logic, which is the logic used by the sequential portion of the machine to determine the inputs to the flip-flops that will generate the next state. Figure 9.21 depicts a high-level circuit diagram showing the state machine pieces. The block with the shape of a cloud represents combinational logic, or simply circuitry without memory. Note: having said that reset is needed to initialize the state of a machine, other methodologies are used to initialize state machines. One such method is designing *scannable* machines. All the state machine registers upon power on can be configured like a giant shift register and a known state is clocked in every flip-flop. Once all flip-flops are initialized

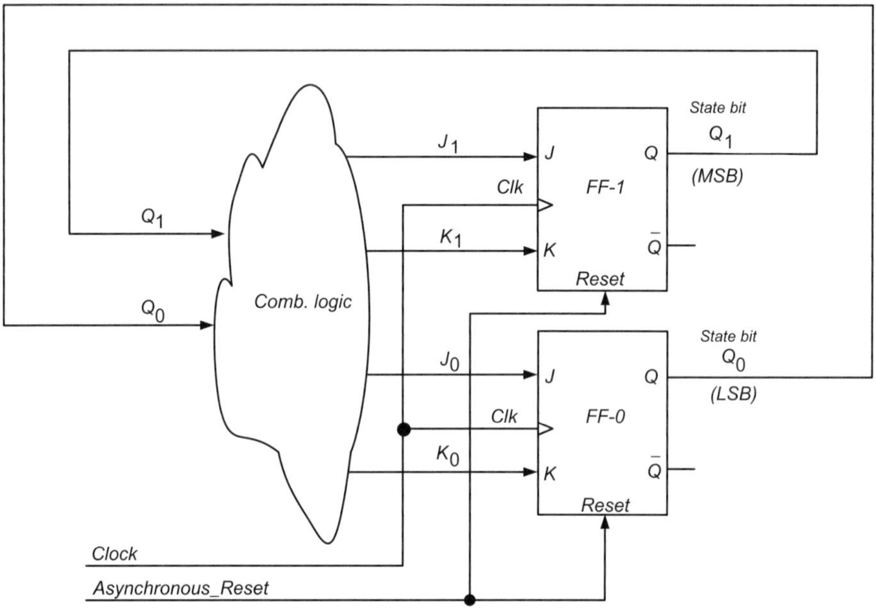

Figure 9.21 2-bit up synchronous counter logic diagram.

the state machine is placed back in its normal operating mode and begins to run. For a good reference in *scannable* systems refer to [1].

By inspection of Figure 9.21 we can observe that the present state through the combinational logic produces the outputs for flip-flops 1 and 2 inputs. *Reset* as stated earlier is directly applied to the asynchronous reset input that we assume the *JK* flip-flops already have. So we need to come up with four logic equations, which are the equations of the flip-flop inputs as a function of the present state. The equations follow:

$$J_1 = \mathrm{f}(Q_1, Q_0) \tag{9.1}$$

$$K_1 = \mathrm{g}(Q_1, Q_0) \tag{9.2}$$

$$J_0 = \mathrm{h}(Q_1, Q_0) \tag{9.3}$$

$$K_0 = \mathrm{i}(Q_1, Q_0). \tag{9.4}$$

With the present state information and the combinational logic the inputs to each flip-flop is presented so that upon the next active edge of the clock the state machine goes to its next state. So before doing the design we need a different form of the *JK* flip-flop characteristic table that facilitates the design process. Such new table is the *JK* flip-flop excitation table. Actually, the excitation table presents the same information provided by the characteristic table but in the following form:

"What do the FF inputs need to be to make a transition from a determined present state to a desired next state when the active clock edge is present?" Review Table 9.6 with the JK FF characteristic table. Using Table 9.6 we compose the excitation table for the JK FF. Table 9.11 depicts such excitation table.

It is important to emphasize that since the excitation table has a present state and a next state column without showing the clock explicitly, the same table applies to positive as well as to negative edge-triggered devices. Let us recall from the previous chapter that X refers to a *don't care* condition, either a 1 or a 0.

State Machine Design Process Our 2-bit counter has four states and we are using two flip-flops to implement it. A circuit designed with n flip-flops can support a maximum number of 2^n states; this is the case in our example: 2-bit machine and four states. At times some sequential circuit designs have less than 2^n states. So we need to be careful about what we do if the state machine accidentally lands in one of those *unused states*. For the design process we will merge the state table of the desired state machine (Table 9.10) with the excitation table (Table 9.11) of the flip-flops to be used. We construct a new table that has present state, next state information, and the outputs of the combinational circuit of Figure 9.21; such outputs are the flip-flop inputs. The new table is the excitation table for the complete design and Table 9.12 shows it.

Table 9.11 ***JK FF* excitation table**

Present State	Next State	JK FF inputs	
$Q(t)$	$Q(t+1)$	J	K
0	0	0	X
0	1	1	X
1	0	X	1
1	1	X	0

Table 9.12 Excitation table for the *2*-bit counter design of Ex. 9.1 using *JK FF*

Inputs of Combinational Circuit				Outputs of Combinational Circuit			
Present State		Next State		Flip-Flop Inputs			
$Q_1(t)$	$Q_0(t)$	$Q_1(t+1)$	$Q_0(t+1)$	J_1	K_1	J_0	K_0
0	0	0	1	0	X	1	X
0	1	1	0	1	X	X	1
1	0	1	1	X	0	1	X
1	1	0	0	X	1	X	1

Notice that the asynchronous *Reset* input was left out of the excitation table because it is directly hardwired to the asynchronous reset of the flops (Fig. 9.21).

The present and next state columns of Table 9.12 are identical to those of Table 9.10.

To fill in the columns for J_1, K_1, J_0, and K_0 of Table 9.12 we will proceed to work with *FF*-1 inputs first and *FF*-0 later. Let us ask ourselves what do inputs J_1 and K_1 need to be to go from a present state $Q_1(t) = 0$ to next state $Q_1(t + 1) = 0$. The answer to this is on Table 9.11 third line from the top, that is, $J_1 = 0$ and $K_0 = X$. So we proceed to fill in the first line under the J_1 and K_1 columns with 0 and X, respectively. For the next entry: what do inputs J_1 and K_1 need to be to go from a present state $Q_1(t) = 0$ to next state $Q_1(t + 1) = 1$. Again from Table 9.11 the answer is $J_1 = 1$ and $K_1 = X$. This is the next entry under columns for J_1 and K_1. We continue doing the same process until we are done with columns J_1 and K_1. We do the same for flip-flop 0 columns J_0 and K_0. Having completed Table 9.12 we look at it from a different perspective when we need to design the combinational logic that the state machine requires, refer one more time to Figure 9.21. Imagine that we remove from Table 9.12 the two columns that correspond to the *next state* bits $Q_1(t + 1)$ and $Q_0(t + 1)$. What is left of Table 9.12 should be seen as a combinational logic truth table. This table has four outputs: J_1, K_1, J_0, and K_0. These four outputs are Equations (9.1) through (9.4), which are functions of the present state bits $Q_1(t)$ and $Q_0(t)$. To solve the four logic equations we can do four *K. maps,* one per output. Figure 9.22 shows the *K.* maps.

The *SOP* simplifications from each output follow:

$$J_1 = Q_0 \tag{9.5}$$

$$K_1 = Q_0 \tag{9.6}$$

$$J_0 = 1 \tag{9.7}$$

$$K_0 = 1. \tag{9.8}$$

Note that the *SOP* functions for J_1 and K_1 became independent of Q_1. J_0 and K_0 end up being constants (Equations (9.7) and (9.8)). The actual circuit diagram for the complete state machine of Example 9.1 is redrawn with the actual logic found with Equations (9.5) through (9.8). Figure 9.23 shows the circuit.

9.4.1 *SR* Flip-Flop Excitation Table

To derive the excitation tables of the *SR* flip-flop we refer back to characteristic table provided by Table 9.7. This table is repeated below for the reader's convenience.

Starting with the *SR* flip-flop, this device basically works like the *JK* flip-flop with the exception of the $S = 1$, $R = 1$ which is not allowed for the *SR*. Tables 9.13 and 9.14 show the *SR FF* characteristic and excitation tables, respectively.

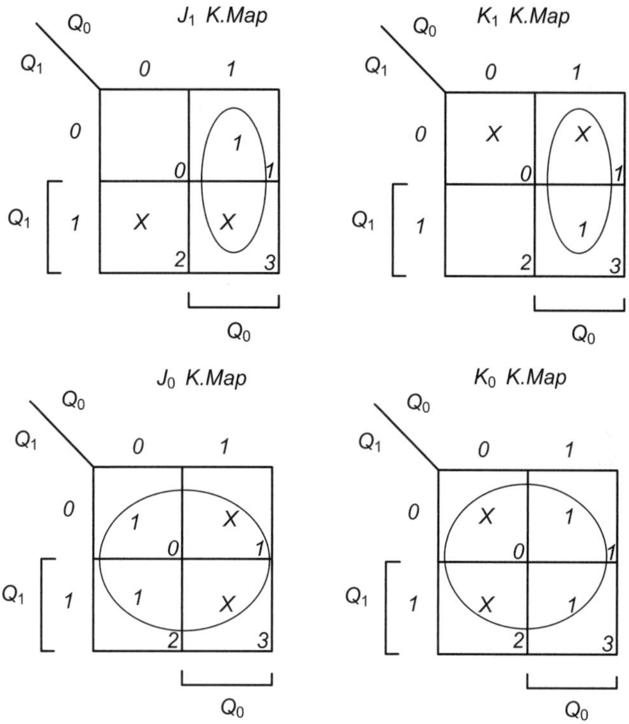

Figure 9.22 Karnaugh maps for outputs J_1, K_1, J_0, and K_0.

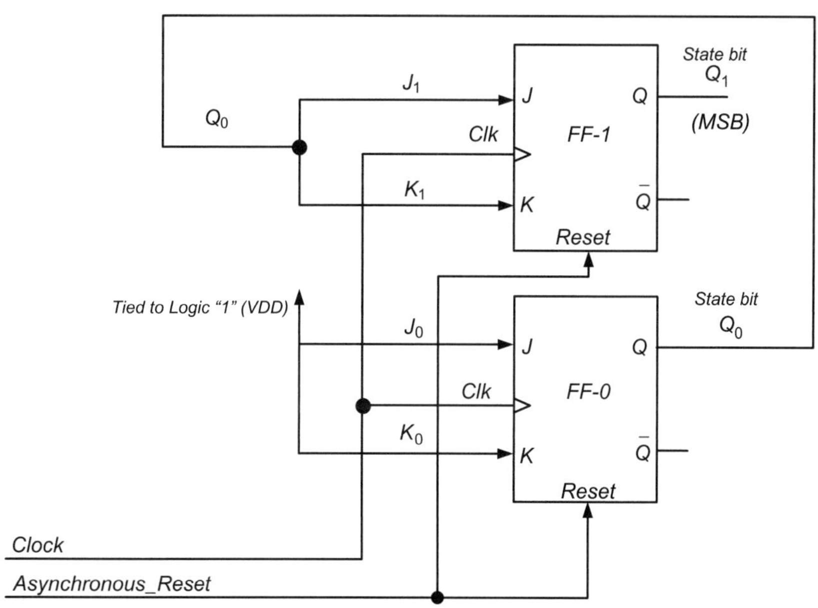

Figure 9.23 State machine for Example 9.1 logic implementation.

Table 9.13 ***SR FF*** **characteristic table**

Row	Clock	S	R	$Q(t+1)$
1	↑	0	0	Hold last $Q(t)$
2	↑	0	1	0 (Reset)
3	↑	1	0	1(Set)
4	↑	1	1	Not-allowed
5	0	X	X	Hold last Q(t)

Table 9.14 ***SR FF*** **excitation table**

	Present State	Next State	SR FF inputs	
Row	$Q(t)$	$Q(t+1)$	S	R
1	0	0	0	X
2	0	1	1	0
3	1	0	0	1
4	1	1	X	0

Recall that to construct the excitation table of an *SR FF* we need to ask ourselves what do inputs S and R need to be prior to the active clock edge, to cause a transition from a given the present state to a desired next state? From Table 9.13, Row 1, states that $S = 0$ and $R = 0$ will do the job, because it holds the previous state. Also $S = 0$ and $R = 1$ (Table 9.13, Row 2) will do the same thing because this condition *Resets* the *FF*. Thus, $S = 0$ and $R = X$ produces a transition from present state zero to next state zero; refer to Table 9.14, Row 1 for $S = 0$, $R = X$. It is easy to see that $S = 1$ and $R = 0$ (Table 9.13, Row 3) *Sets* the *FF* and that $S = 0$ and $R = 1$ (Table 9.13, Row 2) *Resets* the *FF*. Finally, having a present state of one to transition to a next state of one we either Set the *FF* ($S = 1$ and $R = 0$) or hold the previous state ($S = 0$ and $R = 0$); thus, $S = X$ and $R = 0$; refer to Table 9.14, Row 4.

9.4.2 *T* Flip-Flop Excitation Table

In a similar fashion, to derive the T FF excitation table we start with the *T FF* characteristic table (Table 9.15).

Again we ask ourselves the question: "what does input T need to be prior to the active clock edge, to cause a transition from a given the present state to a desired next state?" We start constructing the *T FF* excitation Table 9.16. So using the *T FF* characteristic Table 9.15, we see that to obtain a *zero* present state to *zero* next state transition, the T input needs to be a *zero* (hold, no state

Table 9.15 *T*-flip-flop characteristic table

Row	Clock	T	$Q(t+1)$	Comment
1	↑	0	Q(t)	Hold (no change)
2	↑	1	$\bar{Q}(t)$	Complement (toggle)
3	0	X	Q(t)	Hold (no change)

Table 9.16 *T FF* excitation table

	Present State	Next State	T FF input
Row	$Q(t)$	$Q(t+1)$	T
1	0	0	0
2	0	1	1
3	1	0	1
4	1	1	0

change). This gets written as a zero entry under Row 1 for *T FF* input column of Table 9.16. To fill in the *T* input for Rows 2 and 3 of Table 9.16, inspecting Table 9.15 we see that in both cases input *T* must be a *one*. Finally, for Table 9.15, Row 1, for the *T FF* not to change from present state 1 to next state 1, then *T* input needs to be a *zero*. This is shown on Table 9.16, Row 4 and under the *T FF* input column.

9.4.3 *D* Flip-Flop Excitation Table

Similar to the way we did with the *SR* and *T* excitation tables we create the *D FF* excitation table. Note that the *D FF*, which is the simplest one to understand, simply copies whatever the *D* input is, to the *Q* output upon the active edge of the clock.

Since we want an excitation table to cause the usual four-row present state to next state transitions, the *D* input needs to be whatever we want the next state to become. Tables 9.17 and 9.18 depict the characteristic and the excitation tables for a *D FF*.

Table 9.17 *D* Flip-flop characteristic table

Clock	D	$Q(t+1)$	Comment
↑	0	0	Reset
↑	1	1	Set
0	X	Q(t)	Hold (no change)

Table 9.18 ***D FF* excitation table**

	Present State	Next State	*D* FF input
Row	$Q(t)$	$Q(t+1)$	D
1	0	0	0
2	0	1	1
3	1	0	0
4	1	1	1

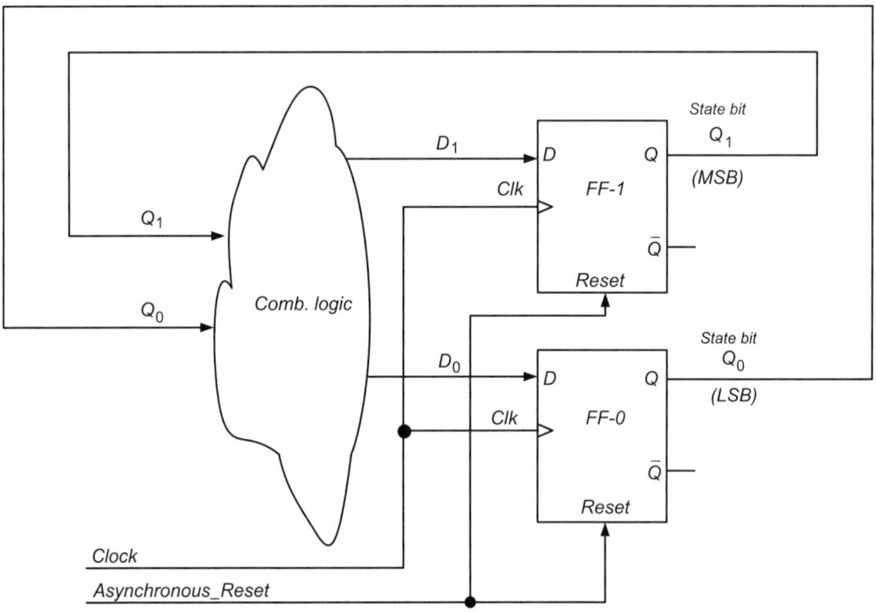

Figure 9.24 2-bit binary up counter for Example 9.2.

Example 9.2 Let us now use the *D* flip-flop to implement a 2-bit up binary counter just like the one that we covered in Example 9.1. The counter will have an asynchronous reset input that upon being asserted it will force the counter to go to state $Q_1Q_0 = 00$. Implement the counter only with *D* flip-flops and the required next state combinational logic.

Clearly the state diagram for this counter is identical to the state diagram depicted in Figure 9.20. The logic implementation though is structurally similar to that of Figure 9.21, but the *JK FF's* are replaced with *D FF's*. Figure 9.24 depicts the circuit diagram of our counter to be implemented with *D FF's*. The design of the state machine control logic is implemented generating the excitation table of the complete circuit using the *D FF* excitation tables and the state machine state diagram. Table 9.19 depicts Example 9.2 excitation table.

Proceeding to do the *K*. maps for D_0 and D_1 both as functions of present state bits: $Q_1(t)$ and $Q_0(t)$ we obtain the K. maps shown in Figure 9.25

Table 9.19 Excitation table for the *2*-bit up counter design of Example 9.2 using *D FF*

Inputs of Combinational Circuit				Outputs of Combinational Logic	
Present State		Next State		Flip-flop inputs	
$Q_1(t)$	$Q_0(t)$	$Q_1(t+1)$	$Q_0(t+1)$	D_1	D_0
0	0	0	1	0	1
0	1	1	0	1	0
1	0	1	1	1	1
1	1	0	0	0	0

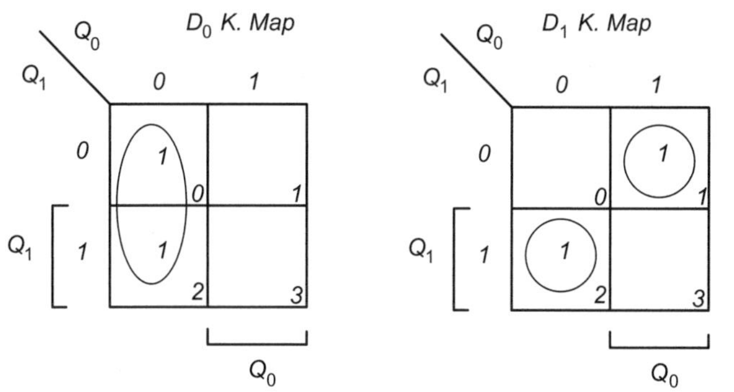

Figure 9.25 *K.* maps for *D FF* inputs D_0 and D_1 as a function of the present state bits.

$$D_0 = f[Q_1(t), Q_0(t)] = \bar{Q}_0 \tag{9.9}$$

$$D_1 = g[Q_1(t), Q_0(t)] = Q_0 \oplus Q_1 \tag{9.10}$$

Equations (9.9) and (9.10) lead to the logic implementation presented in Figure 9.26.

The previous examples dealt with a synchronous state machine that made state transitions in an unconditional fashion. Except for the asynchronous *Reset* input to the flip-flops, (and of course the clock) there were no other inputs to the state machine. Let us now consider slightly more complex state machines. Assume that we want to design a machine that makes state transitions upon being on a certain state and upon an input being *one* or *zero*. Such state transitions are referred to as *conditional state transitions*, because they depend on the state of an external state machine input in addition to the present state of the machine. *Unconditional state transitions* occur when the present to next state transition takes places irrespective of the level of the state machine external input. In a general sense, state machines can have a mix of *conditional* and *unconditional* state transitions or only one of the two kinds.

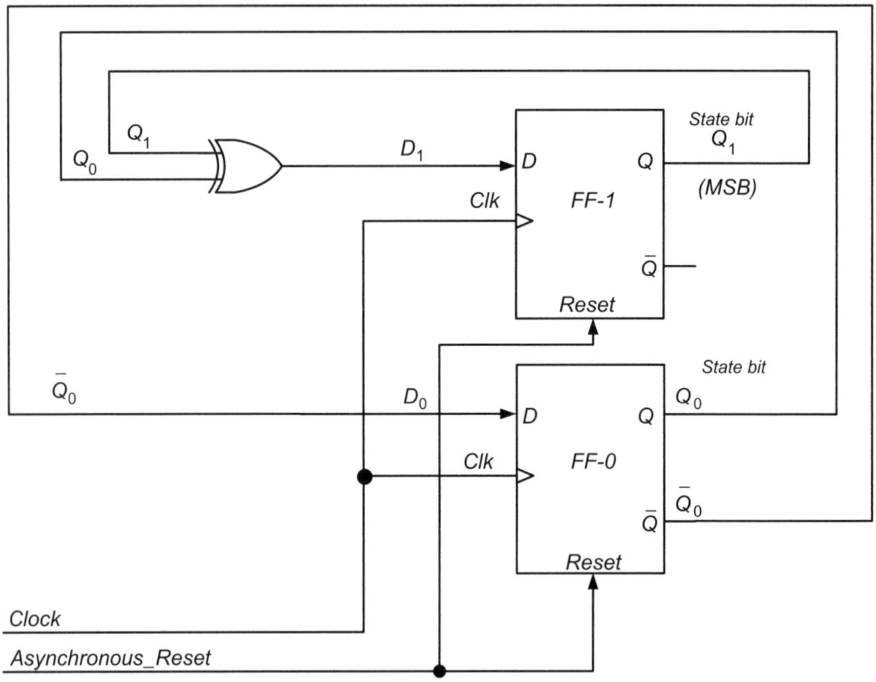

Figure 9.26 Logic implementation of 2-bit up counter with *D FFs*.

Example 9.3 Given the state diagram of Figure 9.27, (a) find the state table of the circuit and (b) perform a logic implementation of the state machine using *D FF*. Assume that the external state machine input *A* is synchronous to the state machine clock. This concept has to do with state machine timing requirements, which will be discussed in another section within this chapter. So do not worry a whole lot of the previous sentence right away, we will get to it in greater detail.

Figure 9.27 state diagram exhibits two different types of state transitions: *conditional* as well as *unconditional* transitions. The state machine has two bits of state and a single external and clock-synchronized input *A*. State transitions from state *00* to *11* or *01* are *conditional transitions*. The state diagram should be read as follows for those transitions: If the present state is *00* and input *A* is *high* the next state is *01*, else if present state is *00* and input *A* is *low* the next state is *11*. All other transitions, that is, from state *01* to *00*, *11* to *00* and *10* to *00*, are *unconditional state transitions*. Let us make clear before the circuit is drawn that the state key is defined as $Q_1\ Q_0$, where Q_0 is the *LSB* of state. From the state diagram in Figure 9.27 we will construct the excitation table for the circuit to be designed. Note that the only conditional state transitions that exist are Rows 1 and 2 of Table 9.20; state transition from *00* to *11* occurs only upon input *A* being *low*. On Row 2 the state transition from *00* to *01* occurs on *A* being *high*. Rows 3 through 8 all show *unconditional* state transitions. For example, for the state transition *01* to *00* we can see in Rows 3 and 4 that take place regardless of whether *A* is *low* or *high*.

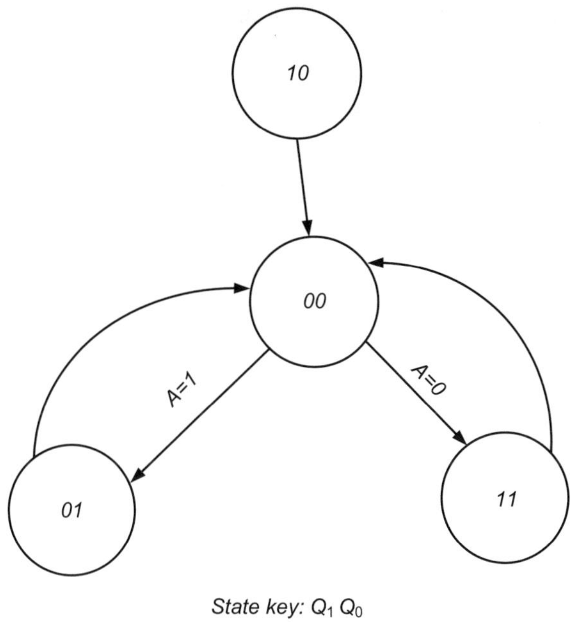

Figure 9.27 State diagram for Example 9.3.

Carefully inspecting excitation Table 9.20 we have to obtain combinational equations for the *D FF* inputs D_1 and D_0 as functions of the present state and the external state machine input *A*. So imagining that the next states columns are gone from Table 9.20 we use such table as a truth table to find out the simplified products of sum logic equations for the *D FF* inputs D_0 and D_1.

$$D_1 = f[Q_1(t), Q_0(t), A] \tag{9.11}$$

$$D_0 = g[Q_1(t), Q_0(t), A]. \tag{9.12}$$

We proceed creating the *K.* maps for each *FF* input, but note that for this example the *K.* maps are 3-variable maps which can be observed from Equations (9.11) and (9.12).

From the *K.* maps of Figure 9.28 we obtain:

$$D_1 = \bar{Q}_1\bar{Q}_0\bar{A} \tag{9.13}$$

and

$$D_0 = \bar{Q}_1\bar{Q}_0. \tag{9.14}$$

The logic implementation of our state machine using logic Equations (9.13) and (9.14) is depicted in Figure 9.29.

Table 9.20 Excitation table for the state machine design of Ex. 9.3 using *D FF*

	Combinational Circuit Inputs					Combinational Circuit Outputs	
	Present State		External Input	Next State		Flip-Flop Inputs	
Row	$Q_1(t)$	$Q_0(t)$	A	$Q_1(t+1)$	$Q_0(t+1)$	D_1	D_0
1	0	0	0	1	1	1	1
2	0	0	1	0	1	0	1
3	0	1	0	0	0	0	0
4	0	1	1	0	0	0	0
5	1	0	0	0	0	0	0
6	1	0	1	0	0	0	0
7	1	1	0	0	0	0	0
8	1	1	1	0	0	0	0

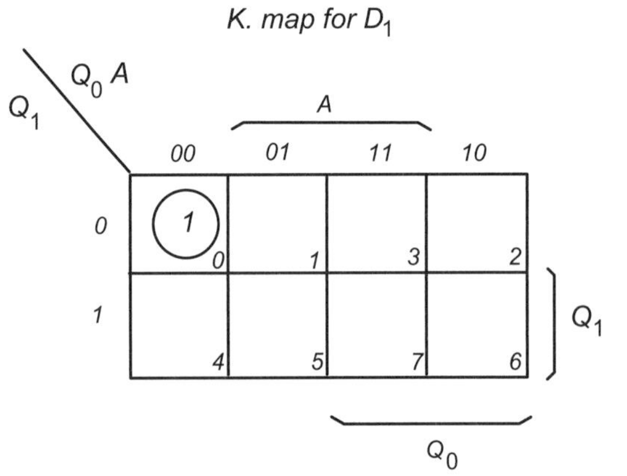

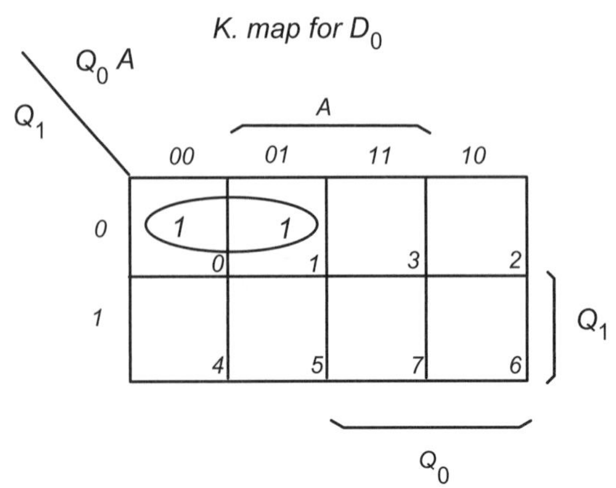

Figure 9.28 Karnaugh maps for Example 9.3.

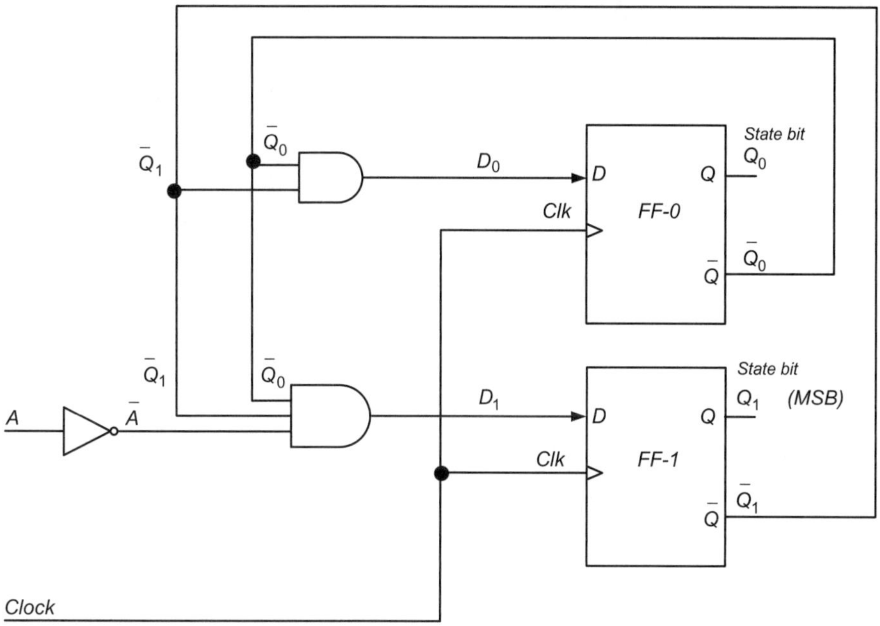

Figure 9.29 Logic implementation of state machine for Example 9.3.

Example 9.4 Design a *3*-bit down binary counter. Design an output called *MIN-COUNT* that detects when state *0 (000)* is reached. The output must be a function of the state bits. Use *D FFs.* (1) generate the state diagram of the counter, (2) generate the circuit excitation table, (3) find the logic for output *MIN-COUNT* and (4) find the complete logic for the counter state machine. We will move a little faster throughout this example, since most of the concepts have already been explored in previous examples. Refer to Figure 9.30 for the state machine state diagram.

The excitation table (Table 9.21) follows from the state diagram. The state diagram of our down counter has *unconditional* state transitions. This means that no external input to the counter exists to prevent or allow any of the state transitions. The output logic has to detect state *000*; this output must remain asserted during the time the counter is at state *000.* Once the counter is in any other state, the output has to be negated. Thus, the logic for the *MIN-COUNT* output is a three input *AND* gate that receives inverted state bits: $Q_2(t)$, $Q_1(t)$, *and* $Q_0(t)$. By DeMorgan's rule such gate is a three-input *NOR* gate.

Now to design the state machine we use excitation Table 9.21 come up with the excitation functions or the input equations of each state flip-flop: D_2, D_1, and D_0 as a function of the three state bits $Q_2(t)$, $Q_1(t)$, *and* $Q_0(t)$. Proceeding as we did before, we need to produce three *3*-variable *Karnaugh* maps to obtain a simplified *SOP* expression for each *FF* input.

From Figure 9.31 we obtain a simplified *SOP* forms for D_2, D_1, and D_0:

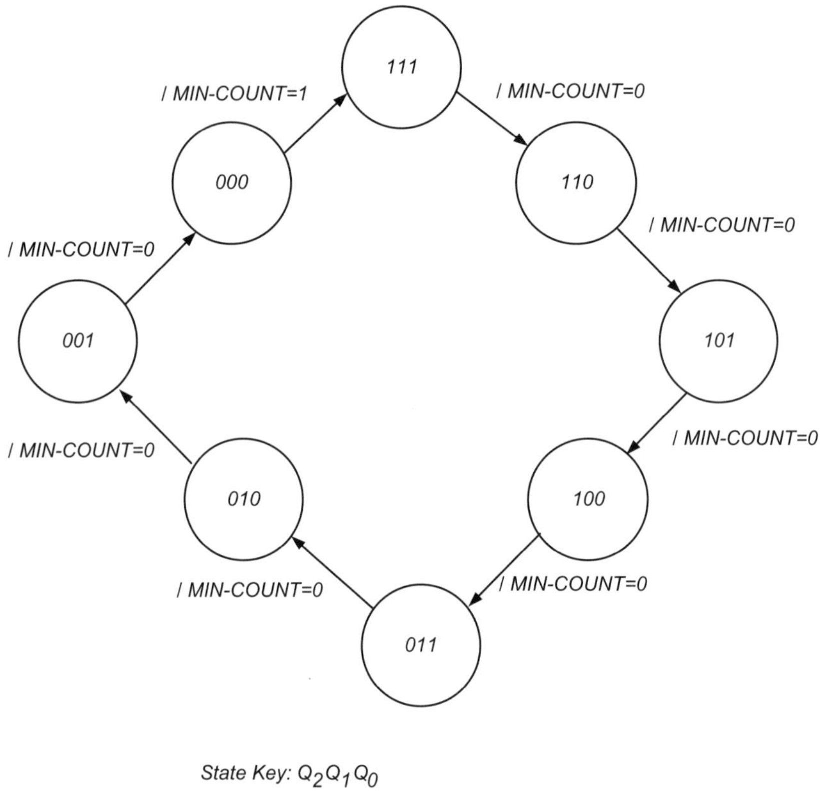

Figure 9.30 State diagram of 3-bit down counter for Example 9.4.

Table 9.21 Excitation table for the down counter of Example 9.4

Present State			Output	Next State			*D* Flip-Flop Inputs		
$Q_2(t)$	$Q_1(t)$	$Q_0(t)$	MIN-COUNT	$Q_2(t+1)$	$Q_0(t+1)$	$Q_0(t+1)$	D_2	D_1	D_0
0	0	0	1	1	1	1	1	1	1
0	0	1	0	0	0	0	0	0	0
0	1	0	0	0	0	1	0	0	1
0	1	1	0	0	1	0	0	1	0
1	0	0	0	0	1	1	0	1	1
1	0	1	0	1	0	0	1	0	0
1	1	0	0	1	0	1	1	0	1
1	1	1	0	1	1	0	1	1	0

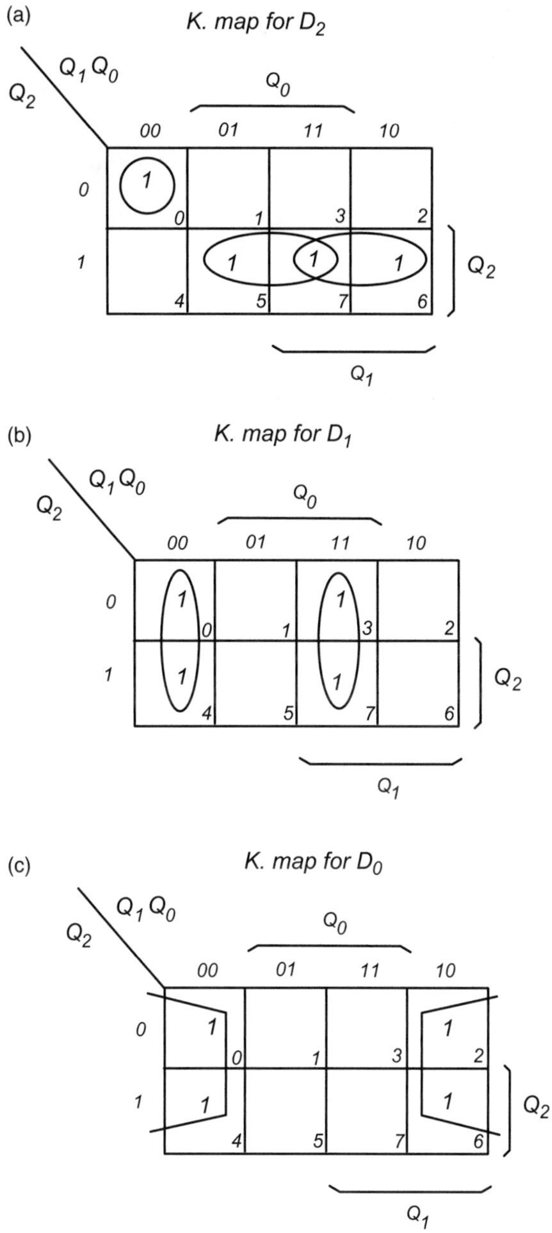

Figure 9.31 Karnaugh maps for *FF* next state equations for Example 9.4: (a) D_2 map; (b) D_1 map; (c) D_0 map.

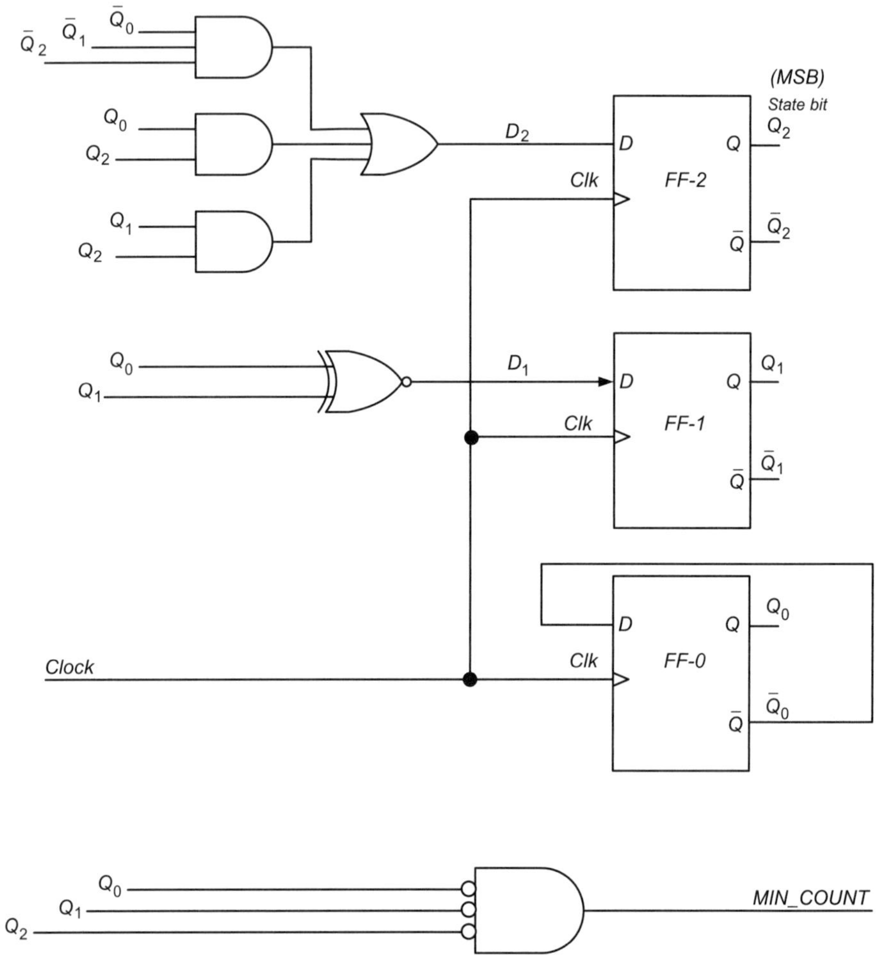

Figure 9.32 Implementation of state machine for Example 9.4.

$$D_2 = \bar{Q}_2\bar{Q}_1\bar{Q}_0 + Q_2Q_1 + Q_2Q_0 \tag{9.15}$$

$$D_1 = \bar{Q}_1\bar{Q}_0 + Q_1Q_0 = \overline{Q_1 \oplus Q_0} \tag{9.16}$$

$$D_0 = \bar{Q}_0. \tag{9.17}$$

Figure 9.32 depicts the implemented state machine of Example 9.4 with next state Equations (9.15), (9.16), and (9.17).

Example 9.5 A shift register: An *n-bit* shift register is an *n-bit* register that makes provision of shifting its stored data by one bit position at every active clock edge. It is possible to design shift registers that shift bits to the right or that can shift bits to the left. More elaborate shift registers can be designed to shift left or right upon asserting a shift direction control input. In this example we will address a 4-*bit* shift register that shifts data from its *LSB* one bit position to the left to a more significant bit. The *LSB* will get loaded with a zero. For example, given the 4-bit binary number *1010*, shifting it left by one bit position leads to *0100*; another shift left will produce *1000*; another shift left produces *0000*. Figure 9.33 depicts the left shifting of the example just described.

Shift registers are used in arithmetic and control type operations. Multiplication and division algorithms use shift registers. Control applications can use

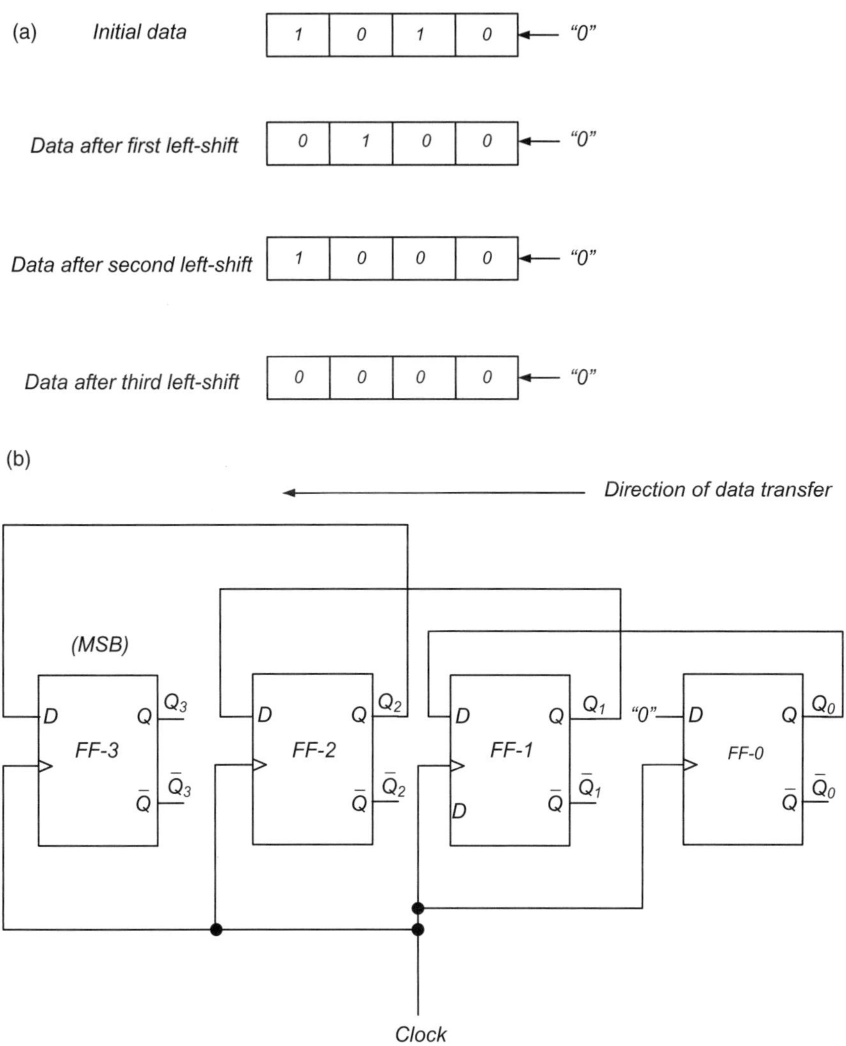

Figure 9.33 Left shifting a four-bit number, loading the *LSB* with a *zero*: (a) left-shift operation; (b) four-bit shift register with *D FF*.

shift registers. Other uses of shift registers include serial-to-parallel and parallel-to-serial data conversion for data communication applications. Advanced central processing units (CPUs) use shift registers in conjunction with multiplexers to implement *barrel shifters. Barrel shifters* allow a CPU to perform shifts either left or right by any amount of desired bits, such that the amount of bits is within the width of the CPU registers, in a single clock cycle.

9.5 SYNCHRONOUS STATE MACHINES GENERAL CONSIDERATIONS

State machines consist of three fundamental pieces of logic. The flip-flops, which serve as the memory elements, store state information. The *next state logic* is the logic that receives state information and external state machine inputs if any. The third and last piece is the logic that generates the machine outputs. The next state logic produces the correct signals to the flip-flops, so that the transition to the desired next state occurs. Figure 9.34a,b depict the two most important state machine architectures, the *Moore* and the *Mealy* state machines. Referring to Figure 9.34a we see a *Moore* machine block diagram. This machine has next state logic and flip-flops, which does not differ much from a *Mealy* (Fig. 9.34b) state machine. The fundamental difference between the *Moore* and *Mealy* is the way in which both state machines produce the control outputs. *Moore* machines produce their outputs with combinational logic that is only function of the present state bits. *Mealy* state machines produce their control outputs with combinational logic that is a function not only of the present state bits but also the external inputs to the state machine.

This implies that the control outputs not only may change when the state is updated but also when the any of the state machine external inputs change. Note that it is customary in computer architecture literature to indicate clocked elements with memory (i.e., flip-flops) with a heavy trace where their Q outputs are. Sometimes when the heavy trace is shown the clock does not need to be explicitly drawn, as we did in Figure 9.34.

9.5.1 Synchronous State Machine Design Guidelines

The single most important step of designing a state machine is the complete, accurate and concise description of the problem that needs to be solved. This is in essence producing a specification. Now we need to translate such description into a state diagram. Determine the external state machines inputs and the control outputs. We also need to determine the number of states needed, try to minimize them if possible. Choose the flip-flops type to be used. Do the state assignments. There are three basic ways of making state assignments. The simplest one is to do binary state assignments, usually takes the smallest number of flip-flops but with more next state combinational logic. If we know that some state machine outputs need to be glitch free during their transition

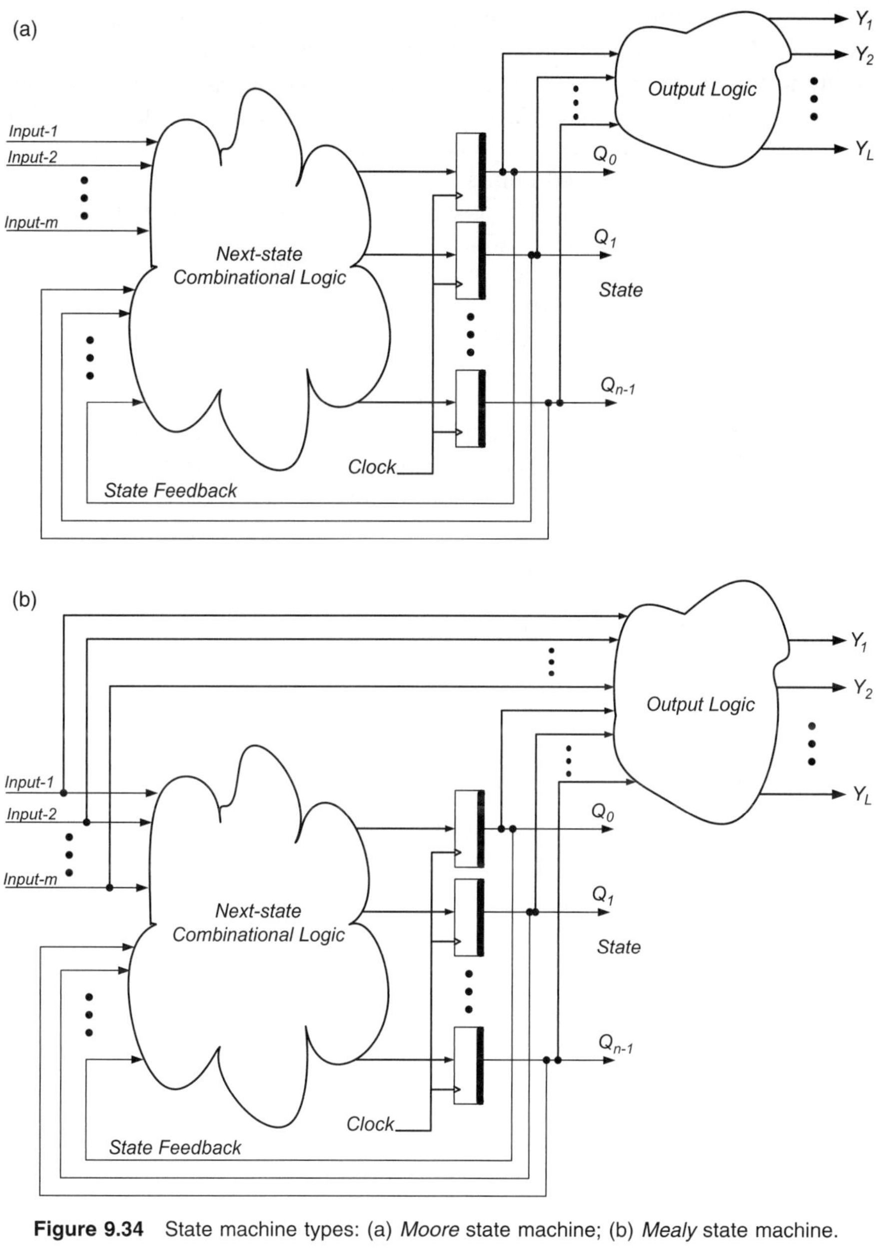

Figure 9.34 State machine types: (a) *Moore* state machine; (b) *Mealy* state machine.

to their active level, it is beneficial to Gray code encode the states. Gray code encoding is a binary code that only varies by one and only one bit between adjacent codes. Note that the following 3-bit Gray code sequence *000, 001, 011, 010,110, 111, 101, 100* differs between adjacent terms by no more than one bit position. So if in addition to Gray code encoding the states, a state bit is chosen to be a needed state machine output, then such output will be glitch-free. For example, if one of our state machine outputs is a *WRITE* signal, we definitely want this signal not to glitch, ever. State one-hot encoding is another technique that uses a flip-flop per state. Although this scheme uses more flip-flops that it would require if we binary encoded the states, it has the advantage that it is very simple to design and debug. Usually when designing state machines FPGAs using a few more flip-flops is not a problem when compared to the benefits that it provides. Care must be exercised upon resetting the state machine to a desired initial state. We create the excitation table of the state machine. That means to construct a table that shows the required excitation (flip-flop inputs) to obtain the desired next state for state/ input combination. This leads to the design of the next state logic. Finally, we decide how we want the outputs to be generated. Do we want a *Moore* machine, where the outputs are strictly functions of the state bits? Alternatively, do we need a *Mealy* machine, where the outputs depend not only on the state bits but also on the state machine inputs? Finally, we draw a complete schematics or logic diagram of the design. The above steps assume a fairly manual procedure to design state machine. These days *CAD* tools such as hardware description languages (HDL), Verilog and VHDL being the two most popular ones, allow a designer to design and simulate the behavior of the state machine before it is actually implemented with logic gates. HDLs are beyond the scope of this book. References to HDLs are cited in the Further Reading section of this chapter.

In summary, we can list the basic steps required to design synchronous state machines.

(a) Produce a complete and succinct state machine specification.
(b) Determine number of states needed, state machine inputs and outputs.
(c) Produce a state diagram and state assignments.
(d) Optionally minimize the number of states.
(e) If any unused states are present in the state machine designed, ensure that if for some undesired reason the state machine got into one or more of such states, that it finds a graceful way to continue operating, to recover or to stop.
(f) Choose the flip-flop type to be used.
(g) Produce the circuit excitation table.
(h) Design the next state logic using the excitation table.
(i) Design the output logic.
(j) All or some of the steps above will have to be repeated and refined until you reach at a satisfactory solution that meets the requirements.

It is important to keep in mind that there are three major factors that are present in any practical design that is done with the purpose of becoming a large volume product. From an engineering point of view the natural factor is *quality*. Quality is associated with the idea that the machine works, reliably and consistently. The other two factors are that the project has to meet a given *schedule* and meet its *cost* targets. In essence *quality*, *cost,* and *schedule* are practically inseparable factors that at one point or another throughout the design cycle, will force engineering to make tradeoffs to meet the overall goals.

9.5.2 Timing Considerations: Long and Short Path Analyses

Synchronous state machines need to operate at some intended clock frequency. But this is not all; most importantly, every clocked device has its own set-up and hold time requirements that need to be met at all times. Given a simple synchronous state machine such as the one in Figure 9.35, the timing path between to consecutive clock edges is:

$$t_{COmax} + t_{PDmax} + t_{SUmin} \leq T_{CLKmin} \tag{9.18}$$

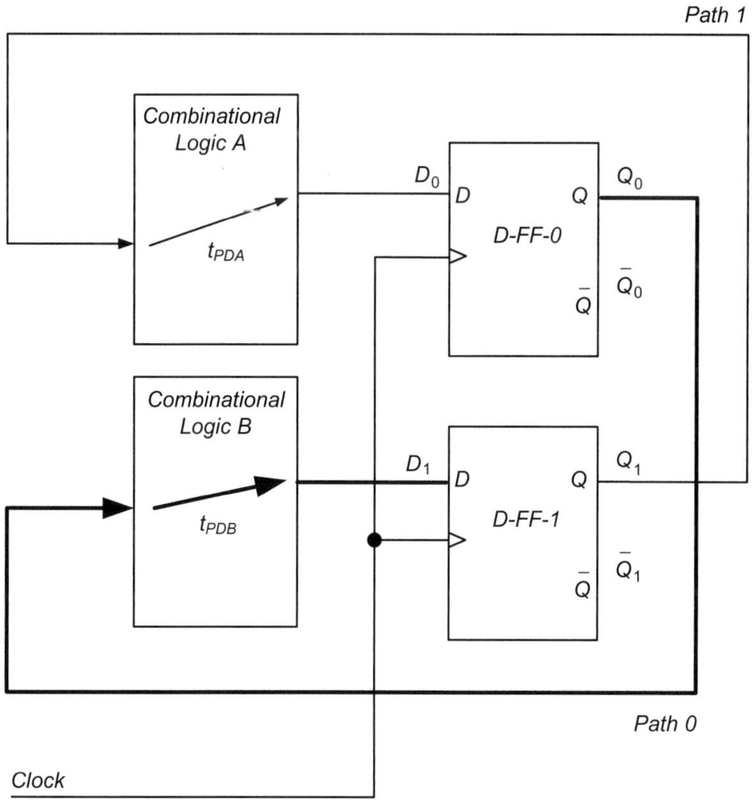

Figure 9.35 Long path analysis of a simple state machine.

where t_{COmax} is the flip-flop maximum clock-to-output delay, t_{PDmax} is the combinational logic maximum propagation delay, t_{SUmin} is the minimum required set-up time that the manufacturers specifies for its flip-flop, and T_{CLKmin} is the minimum required clock period to allow the required minimum set-up time to the flip-flop. Equation (9.18) has to be met individually for every one of the timing paths that exist in the circuit designed.

In particular and referring to the two-path state machine of Figure 9.35, Equation (9.18) can be written for each one of the long paths. One path, Path 0, shown with very heavy lines, begins at the Q_0 output of *DFF*-0, continues to the input of *combinational logic B,* the output of logic B ends at the D_1 input of *DFF-1*.The second long path, Path 1, is drawn with standard weight lines, begins at the Q_1 output of *DFF-1*, continues through combinational logic A, and ends at the D_0 input of *DFF*-0. Rewriting Equation (9.18) for path 0 which starts at the Q_0 output of *DFF*-0 and ends at the D_1 input of *DFF-1*, we obtain.

$$t_{CO0max} + t_{PDBmax} + t_{SU1min} \leq T_{CLKmin}. \tag{9.19}$$

For the other path, beginning at the Q_1 output of *DFF-1* and ending at the D_0 input of *DFF*-0, we have:

$$t_{CO1max} + t_{PDAmax} + t_{SU0min} \leq T_{CLKmin}. \tag{9.20}$$

In Equation (9.19) t_{CO0max} is the maximum or longest clock-to-output delay of *FF*0, t_{PDBmax} is the maximum propagation delay of combinational logic *B* and t_{SU1min} is the minimum required set-up time needed by *FF*1, which is the destination flip-flop of this path. Similarly, in Equation (9.20) t_{CO1max} is the maximum or longest clock-to-output delay of *FF*1, t_{PDAmax} is the maximum propagation delay of combinational logic *A* and t_{SU0min} is the minimum required set-up time needed by *FF*0, which is the destination flip-flop of this path. Note that in order to meet set-up time requirements both long paths must be strictly less than T_{CLKmin}. The state machine uses the same clock (at least from a logical standpoint) ; whichever path is the longest is the one that determines the maximum frequency of operation. This example ignores wire delays, transmission lines effects, clock skews, and signal integrity issues.

When a state machine path, such as the one given by Equations (9.19) or (9.20) is not met, we refer to this as being a *long path* (*also referred to as the critical path of the state machine*). Excessive *long paths* cause set-up time violations. Set-up and/or hold time violations of a flip-flop cause the device to temporarily go into an undefined or metastable state. Looking at Equation (9.18) it is clear to see that given a T_{CLKmin} is equivalent to specifying a maximum operating frequency. So once T_{CLKmin} is fixed, the three left-hand side terms in Equation (9.18) need to be such that their sum is strictly less than T_{CLKmin}. When t_{COmax} + t_{PDmax} + t_{SUmin} equals T_{CLKmin}, the set-up time of the flip-flop is just marginally met. In practice we want to have a small but positive margin to account for other factors such as noise, ground bounce and other electrical effects not accounted for by Equation (9.18). Long path analysis or set-up time calculations are an edge-to-edge phenomenon. In practical designs usually the term that can be significantly reduced in order to meet set-up time is the combinational logic propagation delay (t_{Dmax}). The reason is that once a family or type of flip-flop is selected for the design, the state machine designer has little or no control over t_{CO} and t_{SU} since these are flip-flop timing parameters. On the other hand, when every effort was made to meet Equation (9.18) either by reducing t_{Pdmax} or by changing flip-flops for faster ones (with smaller t_{COmax} and t_{SUmin}), there is still a possibility of rearranging the circuit logic and topology. Shall this last attempt fail to meet Equation (9.18), there is no more option other than reducing the operating frequency of the state machine. Stretching out the clock cycle of the state machine is the last and least desirable mean to implement, when the machine violates set-up time.

We have not mentioned anything about hold time requirements yet. Hold time analysis is also referred to as short path analysis. Let us consider the simple circuit of Figure 9.36.

The circuit of Figure 9.36 is a synchronous divide-by-two counter. Let us look into the hold time requirements by its flip-flop. Since the *complemented Q* output is tied back into the *D* input at the time an active clock edge occurs the data present at the D input has to be held constant or stable not just before the clock edge but also immediately after the clock edge. So we then write this requirement as:

$$t_{COmin} \geq t_{Hmin}. \tag{9.21}$$

At first Equation (9.21) seems a bit awkward to understand. One of the reasons is that neither the clock period nor the set-up time show up in Equation (9.21). Since the data into the flip-flop is provided by its complemented *Q* output, fed directly with a zero delay wire, it is under the control of how fast is the *clock-to-inverted-Q-output* delay. This also determines how long the current data at the *D* input stays around (hold time requirement) immediately after the clock edge. So that is the reason why it is important that the

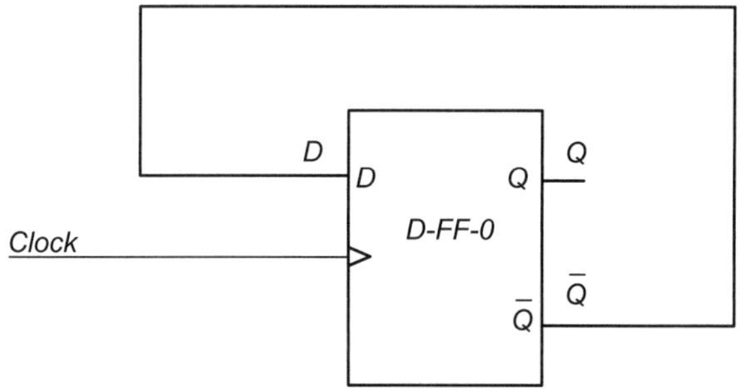

Figure 9.36 Short path analysis of a simple state machine.

clock-to-output minimum delay be larger than the hold time required by the flip-flop. Shall the flip-flop be too fast, that is, $t_{COmin} < t_H$, a hold time violation takes place. Analyzing hold time violations or the lack of them is referred to as short path analysis. Just like in the set-up case, for hold time requirements Equation (9.21) wants the inequality to be "greater than or equal". When t_{COmin} *equals* t_H the flip-flop just barely meets hold time requirements. It is practically desirable for Equation (9.21) to exceed the hold time required by the flip-flop. In other words:

$$t_{COmin} + \text{positive-margin} = t_H. \tag{9.22}$$

Example 9.6 Given the simple state machine of Figure 9.36 assume the following flip-flop parameters are: $t_{COmin} = 1$ ns $t_{COmax} = 5$ ns, $t_{SUmin} = 3$ ns and $t_{Hmin} = 1.5$ ns.

(a) Determine doing long path analysis, the highest frequency at which the flip-flop can be clocked reliably. Assume that *zero* timing margin on set-up is acceptable.
(b) Doing short path analysis, determine if there are any hold time violations. Assume that *zero* margin on hold time is acceptable.

Solution to Example 9.6

(a) For long path analysis for the circuit of Figure 9.36 we have that:

$$t_{COmax} + t_{PDmax} + t_{SUmin} \leq T_{CLKmin} \tag{9.23}$$

where $t_{COmax} = 5\ ns$, $t_{PDmax} = 0$ ns and we assume that the wire has zero delay and $t_{SUmin} = 3\ ns$. We rewrite Equation (9.23) using the numerical values and it becomes:

$$5 + 0 + 3 \leq t_{CLKmin}. \tag{9.24}$$

Thus $t_{CLKmin} = 8\ ns$, which corresponds to a frequency of 125 MHz.

(b) Doing short path analysis and not to have a hold time violation the following is required:

$$\mathrm{t_{COmin}} \geq \mathrm{t_{Hmin}} \tag{9.25}$$

where $\mathrm{t_{COmin}} = 1$ ns and $\mathrm{t_{Hmin}} = 1.5$ ns.

Clearly Equation (9.25) cannot be met with the values given so the circuit has a hold time violation. Note that regardless of the clock frequency you cannot fix the hold time violation. In ASIC design what engineers do is place a buffer between the D input and the $\bar{Q}$ output with a propagation delay of at least 0.5 ns, for this particular example just to comply with the hold time required by the flip-flop. However, care must be exercised because helping the hardware meet hold time, degrades or takes away timing margin out of the set-up time. Now Equation (9.23) no longer has a 0 ns t_{PDmax} term and it becomes 0.5 ns, and that reduces the maximum frequency of operation from 125 MHz down to 117.6 MHz (t_{CLK} = 8.5 ns).

9.6 SUMMARY

This chapter covers sequential devices or devices with memory. Latches and flip-flops are the fundamental memory elements with which to build state machines. It is important to distinguish the *latch* from a *flip-flop*. Generally, a *latch* is basically an asynchronous device, it typically does not have a clock and is not sensitive to a clock edge. A *flip-flop* typically is referred to as a master/slave clock edge sensitive device. When in doubt, the reader should read the context carefully to determine which type of device the author is referring to. State machines are logic circuits that have combinational logic, that is, gates without feedback, plus sequential logic, or devices built using combinational logic with feedback. Remember that the basic latch was built with *2- NOR* gates (alternatively with *2 NAND* gates) with feedback in a cross-coupled configuration. Synchronous state machine are circuits whose flip-flops get clocked at the same time. State machines have state bits, and their next state upon applying the subsequent clock edge depends not only on the present state but also on the machine external inputs. We generally described two basic synchronous state machines, the *Moore* and the *Mealy* types. The basic difference between them is that the *Mealy* state machine produces outputs that are function of the state and external inputs. The *Moore* state machine produces outputs, which are only functions of the state. We designed state machine deriving excitation tables for the circuits, using any flip-flop type we really prefer to use. The *D flip-flop* is the predominant type used in programmable logic and ASICs. A general section on how to design state machines was provided. Such design emphasizes the state assignment an encoding types available: binary, Gray code and one-hot. Finally designers try to stay away from total asynchronous designs for reasons of design difficulty and not being easy to test. Generally designs are quasi-synchronous. Each state machine has its own clock domain; however, both of them need to interface to each other. Timing analysis can be broken down in two major pieces: long-path and short-path timing analyses. It is a requirement that every path in a circuit has to comply with both to have a timing error-free (or metastability-free) design.

FURTHER READING

1. Parag K. Lala, *Fault Tolerant and Fault Testable Hardware Design*, Prentice Hall International, London, 1985.
2. Frederick J. Hill and Gerald R. Peterson, *Digital Systems Hardware Organization and Design*, 3rd ed., John Wiley and Sons, New York, 1981.
3. David J. Comer, *Digital Logic and State Machine Design*, 3rd ed., Oxford University Press, New York, 1995.
4. Donald E. Thomas and Philip R. Moorby, *The Verilog Description Language*, 4th ed., Kluwer Academic Publishers, Norwell, MA, 1998.

PROBLEMS

9.1 Draw the circuit of a master-slave JK flip-flop entirely with NAND gates.

9.2 Implement the state machine whose state table is shown in Table 9.22 with JK flip-flops and combinational gates. Draw the circuit schematics.

9.3 Implement the state machine whose state table is shown in Table 9.22 with two D-type flip-flops and logic gates. Draw the circuit schematics.

9.4 Implement the state machine whose state table is shown in Table 9.22 with a ROM of the smallest possible size; and using D-type flip-flops. Draw the circuit schematics.

9.5 Implement the state machine whose state table is shown in Table 9.22 with 1-of-8 muxes and D-type flip-flops. Draw the circuit schematics.

9.6 For the state machine of Figure 9.37 derive a complete state table. Note that the state machine has 2 bits of state and two external inputs: A and B. Draw the circuit schematics.

9.7 Design the state machine, whose state diagram is shown by Figure 9.37 using 2 JK flip-flops and some minimal number of logic gates. Draw the circuit schematics.

Table 9.22 State table for Problems 9.2 through 9.5

Clock	Input A	Present State	Next State
↑	1	00	10
↑	0	00	01
↑	X	01	10
↑	0	10	11
↑	1	10	00
↑	1	11	10
↑	0	11	01

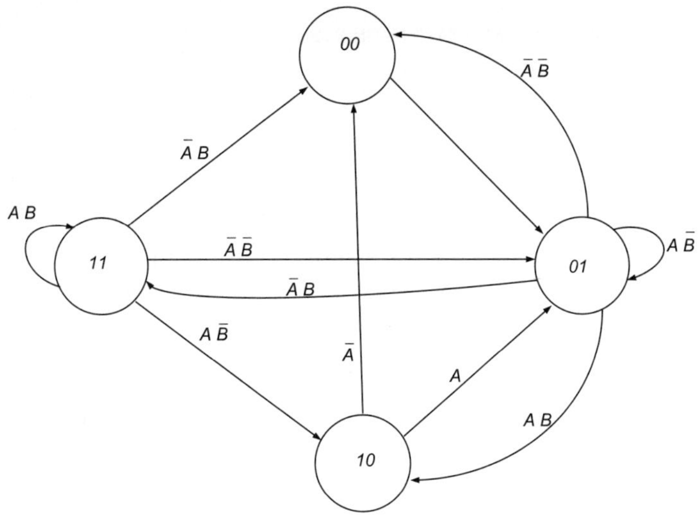

Figure 9.37 State diagram for Problems 9.6 through 9.10.

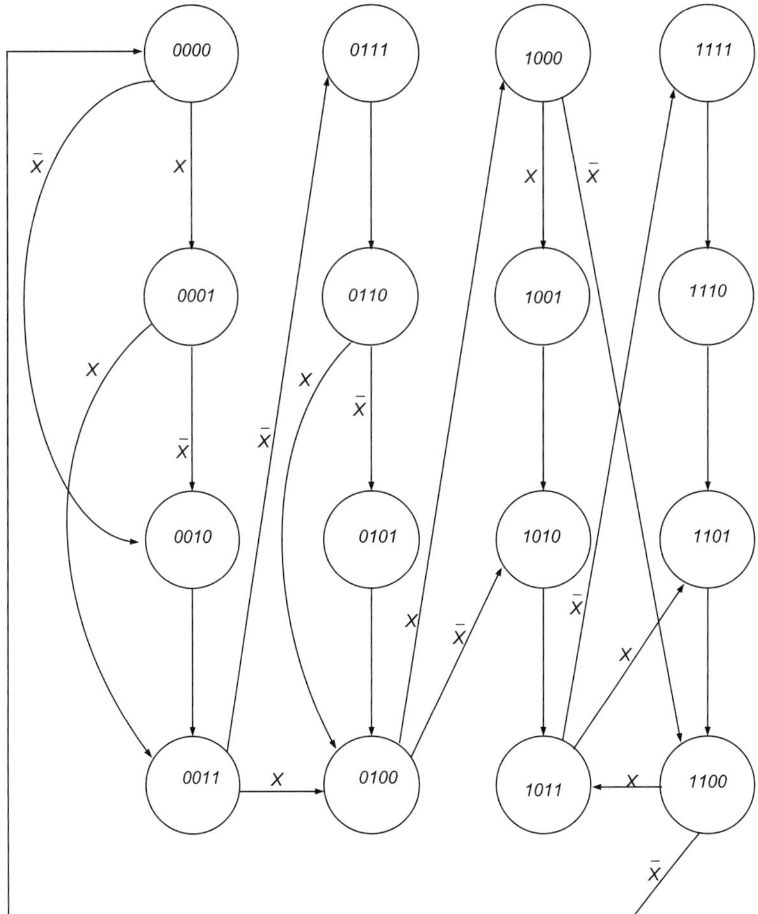

Figure 9.38 State diagram for Problem 9.12.

9.8 Design the state machine of Figure 9.37 using 2 T-type flip-flops and some minimal number of logic gates. Draw the circuit schematics.

9.9 Design the state machine of Figure 9.37 using 2 D-type flip-flops and some minimal number of logic gates. Draw the circuit schematics.

9.10 Design the state machine of Figure 9.37 using the smallest size ROM and two D-type flip-flops. Draw the circuit schematics.

9.11 Design a 3-bit decrementing binary counter. Write the state table of the to-be-designed counter. Draw the circuit schematics.

9.12 Design a state machine that whose state diagram is given by Figure 9.38. (a) Do an implementation using the smallest possible ROM and the smallest possible number of D-type flip-flops. (b) Write a table with the micro-code for the ROM. Assume that input X is already synchronized to the state machine clock. Draw the circuit schematics.

10

A SIMPLE CPU DESIGN

The design of a simplified central processing unit (CPU) is covered in this chapter. This design exemplifies a somewhat more involved and practical design than the examples studied in the previous chapter. This entire chapter basically is a huge example that shows the most important considerations when designing a simple CPU. We start defining the CPU instruction set and the machine instruction word. What the instructions do. The registers, memory, and combinational logic blocks are the components that the CPU requires to be able to execute the defined instruction set. We will also cover the design of the sequencer or control section of the machine with the details of its state diagram and circuit implementation. Finally, a system section covers some of the most important aspects, and sometimes overlooked issues of embedded system design: clocks, resets, power decoupling, and timing. The goal of this chapter is not only to cover a simple CPU but at the same time a complex enough design example that is more comprehensive than previously covered design examples. The basic approach taken is mostly bottom up.

10.1 OUR SIMPLE CPU INSTRUCTION SET

This section introduces the reader to our small CPU instruction set. The instruction set is carefully picked such that various types of the most popular machine language instructions are represented. We will not categorize this

Electrical, Electronics, and Digital Hardware Essentials for Scientists and Engineers, First Edition. Ed Lipiansky.

design neither as a *CISC* or *RISC* example. *CISC* stands for *Complex Instruction Set Computer* and *RISC* stands for *Reduced Instruction Set Computer*. From a computer architecture point of view our design is closer to a *von Neumann* machine. This is an architecture that consists of a stored-program digital machine that has a central processing unit and a single separate memory unit that holds program instructions as well as data. An example of a *RISC* and *CISC* is covered in References 3, 4, and 6 in the Further Reading section.

The instruction set architecture (ISA) that our simple CPU supports consists of a few but very significant instructions that all real-world machines support. The purpose of studying a very simple CPU is to prove basic architecture concepts to the reader, which later on we will use to add real world factors that embedded systems face. Such factors over and above the computer architecture are timing analysis of the CPU, how to clock the machine, the reset logic, and integrated circuit power decoupling.

The basic instructions that our simple CPU supports are: *LOAD, STORE, AND, ADD*, (unconditional jump) *JMP*, (conditional branch) *BRNA*, and (complement) *CMPA*.

Our CPU has a single accumulator register or simply register *A*. Register *A* is a *16-bit* wide register. The computer memory has *4096 16-bit* wide words, that is, *4K* words. Since the memory has *4K* memory locations, the address width required to address each word uniquely, is *12* since 2^{12} is *4096*. The program counter register or the PC is *12* bits wide and it is used to store the address of the instruction to be executed immediately after the currently being executed instruction. Summarizing our CPU has *16*-bit wide data paths between register *A* and its memory. All memory accesses are done with a *12-bit* wide address. The PC stores the address of the to-be executed instruction. Our simple CPU has a single *16-bit* word instruction word. The lower *12* bits are used as an address to memory for those instructions that require such address, while the upper four bits are allocated as operational code bits *(opcode bits)* format. Opcode bits are unique binary codes defined for every unique instruction that the CPU supports. Table 10.1 below depicts the organization of our CPU instruction word format.

$$\text{Opcode} = \text{IWF}\,[15{:}12] \tag{10.1}$$

$$\text{Address X} = \text{IWF}\,[11{:}0]. \tag{10.2}$$

Now we are ready to explain what each instruction does from a programmer's model point of view.

Table 10.1 Instruction word format (IWF) bit assignments

MSB															LSB
15	14	13	12	11	10	9	8	7	6	5	4	3	2	1	0
Operational code				Memory address X											

10.2 INSTRUCTION SET DETAILS: REGISTER TRANSFER LANGUAGE (RTL)

Micro-operations are the most basic actions that digital computers make. Examples of micro-operations are: register to register transfers, register to memory location transfers, memory location to register transfers, perform a logic or arithmetic operation between a register and the contents of a memory location, storing the result in the register or in the memory location. Data transfers are indicated as: $X \leftarrow Y$, where X and Y are registers. The contents of register Y, the source register, get transferred (in actuality is copied) to register X, the destination register. The original contents of Y are preserved. In addition to the data transfer itself; there may be conditions under which a transfer takes place. For example: If "a bit of some register is set" (i.e., the condition) transfer the contents of *Memory* location whose address is in the *memory address register MAR* into register X. This is indicated as: $X \leftarrow M\ [MAR]$, that is, specific sequences of micro-operations constitute macro-instructions or machine language instructions. Such instructions, in binary form, are loaded into the main memory of our simple CPU and the CPU fetches, decodes, and executes them. Assembly language is a symbolic language that allows programmers to more easily write low-level machine language. An assembler typically translates the assembly language instructions into machine language before execution. Let us get started with our very simple CPU instruction set.

LOAD Instruction: Syntax *LDA A, (X)*. This *opcode* is defined as 0000_2. This instruction reads the contents of memory location whose address is X and copies such contents into register A. It is important to recall that X is a *12-bit* address and that the memory contents at any memory address and the contents of register A are all *16* bits wide. The *16*-bit data transfer that the *LOAD* instruction produces is indicated as:

$$\mathrm{A} \leftarrow (\mathrm{X}). \tag{10.3}$$

The *LDA* instruction does not affect the state of the *S* (Sign) bit. The S bit is usually part of a condition codes register (CCR) or Processor Status Word (PSW). This register holds bits that are set or reset based on some arithmetic or logic operation outcomes. Our simple CPU will one have a *Sign bit* in its CCR. The sign bit is the *MSB* of the results produced by the Arithmetic and Logic Unit (ALU), to be discussed shortly.

STORE Instruction: Syntax *STA (X), A:* Its opcode is 0001_2. This instruction reads the contents of register A and copies them into memory location whose address is X. The *16*-bit data transfer that the *STORE* instruction produces is indicated as:

$$(\mathrm{X}) \leftarrow \mathrm{A}. \tag{10.4}$$

The *STA* instruction does not affect the state of the *S* (Sign) bit.

ADD Instruction: Syntax *ADD A, (X):* Its opcode is 0010_2. The *ADD* instruction reads the contents of a memory location whose address is *X*, adds them to the value contained in register *A*, prior to the execution of the *ADD*, and produces the sum of these two *16*-bit quantities, storing the result in register *A*. It is the programmer's responsibility to have some desired value in register *A* prior to the execution of the *ADD* instruction. After the *ADD* instruction is executed the original contents of *A* are overwritten. The *16*-bit addition and data transfer that the *ADD* instruction produces is indicated as:

$$\mathrm{A} \leftarrow \mathrm{A} + (\mathrm{X}). \tag{10.5}$$

Execution of this instruction sets the sign *(S)* flag. The *S* flag is a registered copy of the accumulator *MSB*. When the accumulator is zero or positive the *S* bit is zero, when the accumulator is negative the *S* bit is one. The *S* bit has the same meaning as the *MSB* in a 2's Complement number.

AND Instruction: Syntax *AND A, (X):* Its opcode is 0011_2. The *AND* instruction reads the contents of a memory location whose address is *X*, performs a bit-to-bit logical *AND* of the read memory contents and register *A* bits, and stores the *ANDing* of these two *16*-bit quantities in register *A*. It is the programmer's responsibility to have some desired value in register *A* prior to the execution of the *AND* instruction. After the *AND* instruction is executed the original contents of *A* are overwritten. The *16*-bit data transfer that the *AND* instruction produces is indicated as:

$$\mathrm{A} \leftarrow \mathrm{A}.(\mathrm{X}). \tag{10.6}$$

Execution of the *AND* instruction sets the sign flag accordingly.

JMP Instruction: Syntax *JPM X:* Its opcode is 0100_2. Upon execution of this unconditional instruction the PC gets loaded with address *X*, which is *bits 11:0* from the fetched instruction.

$$\mathrm{PC} \leftarrow \mathrm{X}. \tag{10.7}$$

The *JMP* instruction does not affect the *S* flag.

BRNA Instruction: Syntax *BRNA X*: Its opcode is 0101_2. This instruction is called branch on negative accumulator. This instruction looks at the *Sign* (*S*) bit, if the *S* bit is *1* (i.e., a negative *2's* Complement number is in the accumulator) the PC gets loaded with address *X*, which are bits *11:0* from the fetched instruction. When the *S* bit is zero (accumulator is zero or positive) no change to the PC takes place. That is the PC remains incremented by one from the fetch cycle. This may sound a little confusing but it will be understood better when we will study the data path architecture of our simple CPU:

$$\text{If } (\mathrm{S} = 1) \text{ then } \mathrm{PC} \leftarrow \mathrm{X}. \tag{10.8}$$

Table 10.2 Simple CPU instruction set

Instruction Syntax	Opcode (Binary) IWF [15:12]	Address X IWF [11:0]	Description of What Gets Executed	Affects Sign Flag?
LD A, (X)	0000	A valid address	A ← (X)	No
STA (X), A	0001	A valid address	(X) ← A	No
ADD A, (X)	0010	A valid address	A ← A + (X)	Yes
AND A, (X)	0011	A valid address	A ← A . (X)	Yes
BRNA X	0101	A valid address	If S = 1 then PC ← X	No
JPM X	0100	A valid address	PC ← X	No
CMP A	0110	Bits [11:0] are ignored	A ← $\bar{A}$	Yes

The *BRNA* instruction does not affect the *S* flag; however it uses the *S* setting made by some prior instruction to the *BRNA* to make a decision.

CMP Instruction: Syntax *CMP A:* Its opcode is 0110_2. Upon execution of this instruction the contents of accumulator register *A* become *1's* complemented.

$$\bar{A} \leftarrow A. \tag{10.9}$$

Execution of this instruction sets the *S* bit accordingly.

Table 10.2 summarizes the instruction set of our simple CPU.

10.3 BUILDING A SIMPLE CPU: A BOTTOM-UP APPROACH

Our CPU is required to have the registers and memory access that support the above-described instruction set. The CPU registers are part the programmer's model of the CPU. However, there will invariably be other registers, mechanisms and devices that are totally transparent (or not visible) to the programmer. So before looking at the big view of the data path we will study bits and pieces of the fundamental elements that constitute such data path architecture.

10.3.1 The Registers

Our CPU needs registers. Registers are used to hold memory addresses, memory data read or memory data to be written. Registers also hold the operand of the *ADD* and *AND* instructions. Registers are typically built with *D* type flip-flops; however they are not just a free running group of flip-flops. Why not? If the registers were free running they would get loaded on every

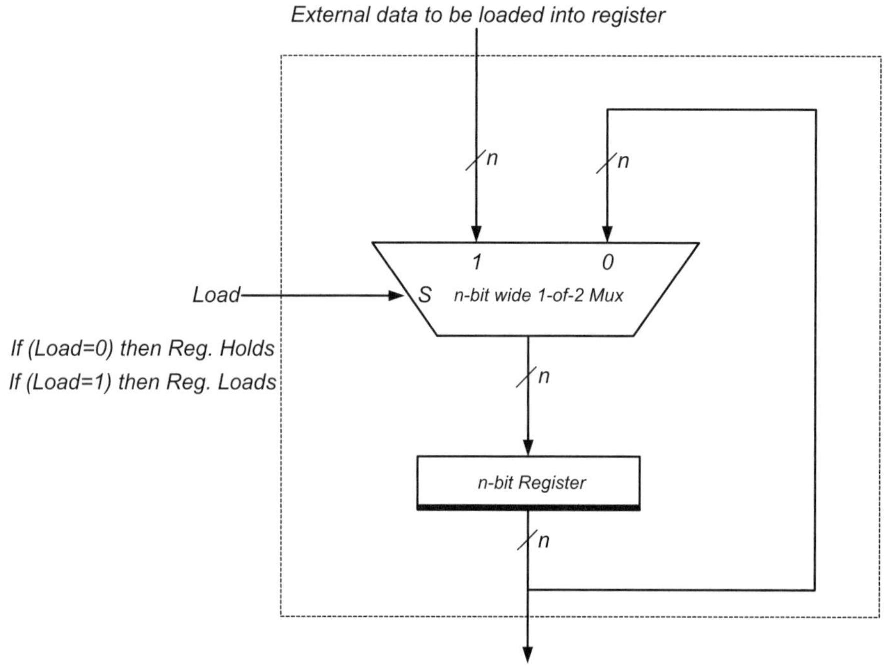

Figure 10.1 Implementation of a register with synchronous load control.

single active edge of the clock. To achieve a selective load of a register we place a *1-of-2 Mux* as shown in Figure 10.1. When the CPU control circuit drives a *zero* onto the *mux* select line, the register holds it data indefinitely because it keeps reloading itself with its own outputs for as long as clocks keep coming into the register. When a control circuit drives a *one* onto the *mux* select line, the external data placed on *channel 1* of the mux gets loaded upon the active edge of the clock clocking the register.

For example if we have a data path like the one shown in Figure 10.2, which has four 8-bit registers *A, B, C, and D*. Assume that somehow all four registers have their own initial values. Let us assume that on the next clock edge we would like the contents of register *A* to get transferred to register *B*, overwriting the current value of *B* and preserving the current values of *A*, *C*, and *D*. The control required to do that has to assert the load input for register *B* and keep the load inputs of registers *A, C*, and *D* negated. The mux select lines have to select input channel 0, which feeds the contents of register *A* onto the inputs of all four registers. Upon the active clock edge only register *B* will get written with the contents of *A* because register *B* has its load control input asserted while the other three registers do not. So upon the active edge of the clock clocking all four registers synchronously perform the following data transfers:

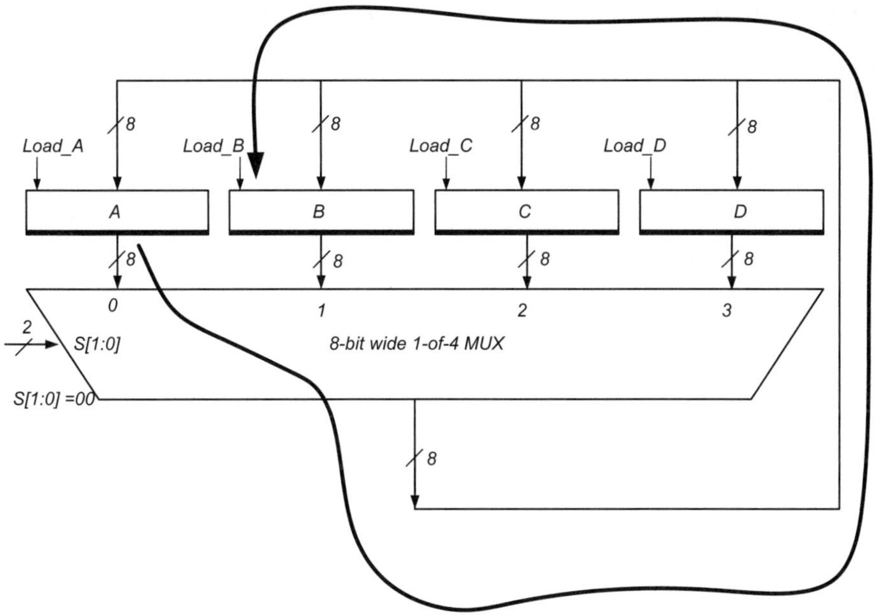

Figure 10.2 Simple data path architecture to show synchronous data transfers.

$$B \leftarrow A \tag{10.10}$$

$$A \leftarrow A \tag{10.11}$$

$$C \leftarrow C \tag{10.12}$$

$$D \leftarrow D. \tag{10.13}$$

Note from Equations (10.10) through (10.13) only B gets loaded with the contents of register A. Data transfers (Eqs. 10.11 through 10.13) show that A, C, and D preserve or hold their original contents, because their load control inputs are negated upon the assertion of the clock edge. It is very important to observe that although data transfers (Eqs. 10.10 through 10.13) are written in a sequence, actually all four of them take place concurrently. This is what a synchronous state machine does. For correct data transfers to take place, set-up and hold times of all flip-flops need to be met. We will address timing when we get to the control section of our simple CPU. Note that based on the simple data path depicted by Figure 10.2 we can transfer the contents of any one register to any two, or any three or all four registers. This is accomplished by asserting the load lines of all the registers we want the new data to get loaded into and by selecting the mux channel of the register that we want to source or provide the data.

10.3.2 The Memory Access Path or Memory Interface

Our CPU main memory or simply its memory is an array of 4096 16-bit words or 4Kx16. Twelve address lines are required to access 4096 locations since $2^{12} = 4096$. Each word is 16 bits wide so the data path to memory has 16 data lines. Memory is designed in such way that one 16-bit data word out of 4096 words can be accessed at any given time. Our memory has two control input lines: a READ and a WRITE. READ and WRITE can be both negated (non-asserted), but only one control input can be asserted at any given time. This means that we can only read a word from memory or write a word into memory at a time.

Practically two registers are needed to interface the CPU with its memory. The Memory Address Register or MAR register holds a memory address. The Memory Buffer Register or MBR receives the data read from memory on memory READS; or holds the data to-be-written into memory on memory WRITES. The interfacing protocol between the CPU and its memory is performed via the MAR and MBR registers.

Using the *MAR* and the *MBR* the data path transfers below are required to read memory:

1. *MAR* ← *Address;* places desired address to read memory from, in the *MAR*.
2. *MBR* ← *Memory [MAR];* retrieves the contents of memory location whose address is in the *MAR*.

For a memory write:

1. *MAR* ← *Address;* places desired memory address to write to in the *MAR*.
2. *MBR* ← Data; places desired data to be written into memory in the *MBR*.
3. *Memory [MAR]* ← *MBR;* performs the write. Transfers data from the *MBR* into the memory location whose address is in the *MAR*.

Figure 10.3 depicts the data path and interfacing registers with our memory array. Note that the *MAR* is *12* bits wide, because it has to hold a *12*-bit address. The *MBR* is *16* bits wide because the memory data is *16* bits wide. The memory access path not only shows the path for the memory data but also the memory address path. Data path pictures usually do not include control signals, like the *READ* and *WRITE* controls for the memory. Register *Q* outputs are represented with a heavy line, refer to *MAR* and *MBR* in Figure 10.3. The register clock is implicit. So when we see a rectangle with one of its sides being very thick, it means that we have a register. Usually, the width of the register is indicated with a little forward slash followed by the width of the input and output buses in bits.

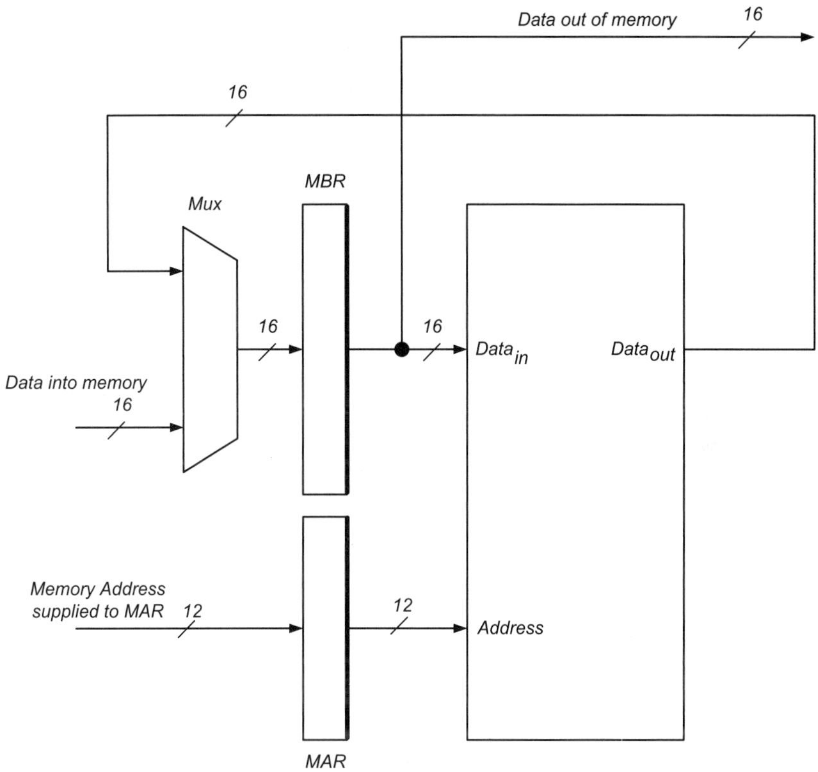

Figure 10.3 Memory data path and interfacing registers.

10.3.3 The Arithmetic and Logic Unit (ALU)

Instructions *ADD* and *AND* respectively require arithmetic and logic to be performed on its operands. We will see within this chapter that two registers will hold the operands that the ALU receives on its two input legs *P* and *Q*. Both the *ADD* and the *AND* operations are performed with combinational logic. Such logic constitutes the ALU. Figure 10.4 shows a high-level block diagram of our ALU. Essentially our ALU has two *16*-bit wide input legs, *P* and *Q* and a *16*-bit output leg *Z*. The ALU is designed to perform the operations listed in Table 10.3. Although it may not be clear as to why we need all those operations now, please hang on and everything will come together when we stitch together all the components of our simple CPU.

10.3.4 The Program Counter (PC)

We mentioned earlier that the *12*-bit wide PC holds the address of the to-be-fetched instruction when an instruction is currently being executed. The PC

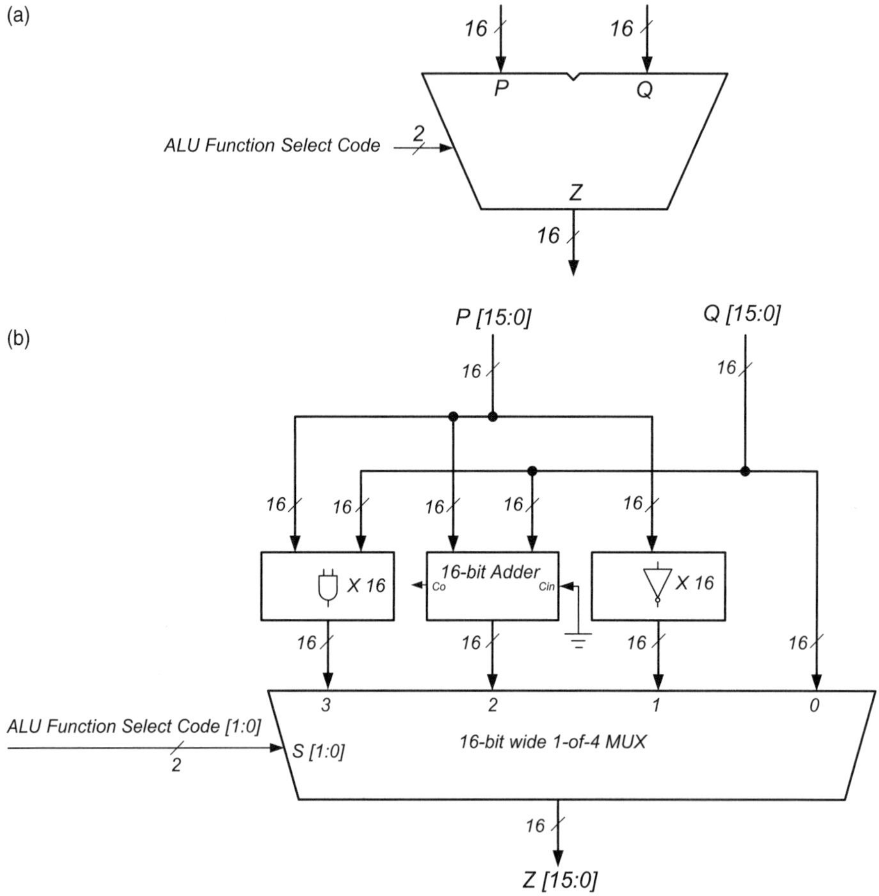

Figure 10.4 (a) ALU high-level block diagram; (b) ALU block diagram.

Table 10.3 ALU Basic operations

ALU Operation	High Level Description	ALU Function Select Code
Connect input leg Q to output leg Z	PASS Q to Z	00
Complement input leg P and connect it to output leg Z	$\overline{P}$ to Z	01
Arithmetic ADD of input legs P and Q send result to output leg Z	(P + Q) to Z	10
Logical bit-to-bit AND of input legs P and Q send result to output leg Z	(P . Q) to Z	11

keeps track of which instruction within a program the CPU is at. A program: "a sequence of instructions with a defined purpose" is an oversimplified but correct view of what a program is. During the execution of most of the CPU instructions the PC gets incremented by one, since all of our instructions are one *16*-bit word in length. So unless our program encounters an unconditional jump or a conditional branch instruction, the PC is always incremented by one. Upon a jump or branch instruction the PC wants to be loaded with a new destination or jump-to address. Such address, called address *X*, is fetched from memory along with the four-bit opcode. Once the PC gets loaded with this destination address *X* the jump will always fetch the next instruction from this new destination address. In the case of the *BRNA (branch on negative accumulator)* the branching will occur if the condition of a negative accumulator (*S* flag set) was met before the BRNA instruction. If the branch condition is met the next instruction fetch takes place from the destination address *X*. If the branch condition is false the branch does not occur and the PC remains loaded with the address of the previous instruction incremented by one. From all of the above we need to have the PC perform at least three distinct functions: hold its contents, auto-increment by one or get loaded with a new destination address. Additionally the PC will have an asynchronous *Reset* input line to ensure that its contents are cleared upon power-up reset. Because of all of that functionality the PC is a little more involved that the accumulator *A*, *MAR* or *MBR* registers. One way of implementing the PC is with a counter, to obtain the auto-increment feature. The PC could use the CPU ALU to increment its contents by one, but this is not desirable because the PC would be using the ALU, which is a valuable resource of the CPU. Figure 10.5 depicts an implementation of the PC. Table 10.4 describes all the operations that the PC performs. Note that the PC implementation in itself is a simple synchronous state machine. Its basic functions are: *(1) hold, (2) increment by one, (3) external data synchronous load*, and *(4) synchronous reset (or clear)*. Note that the combinational logic between the PC input control lines *(Incr, Load & Clear)* and the *12*-bit wide *1-of-4 mux* has two select lines S_1 and S_0, is designed according to Table 10.4, look under columns PC *Control Inputs* and *Mux Select Inputs*. Note that Rows *1* through *4* of Table 10.4 defines the main functions of the PC combinational logic. In general control lines *Incr, Load & Clear* must be asserted in a mutually exclusive fashion, when all control inputs are negated the PC holds its previously clocked state. Assertion of two or three PC control inputs is not meaningful and to avoid this illegal condition whenever they are asserted the PC register will simply hold its previously clocked state. Refer to Table 10.4 Rows *5* through *8*. Finally Row *9* indicates that when the clock into the PC is not active the PC register holds the previously clocked state. The asynchronous *Reset* line is not shown in Table 10.4 since this table is busy enough as it is. Figure 10.5 depicts a functional block diagram of the PC register architecture.

Table 10.4 describes the operation or characteristic table of the PC register. Carefully read Table 10.4 while and also inspect Figure 10.5.

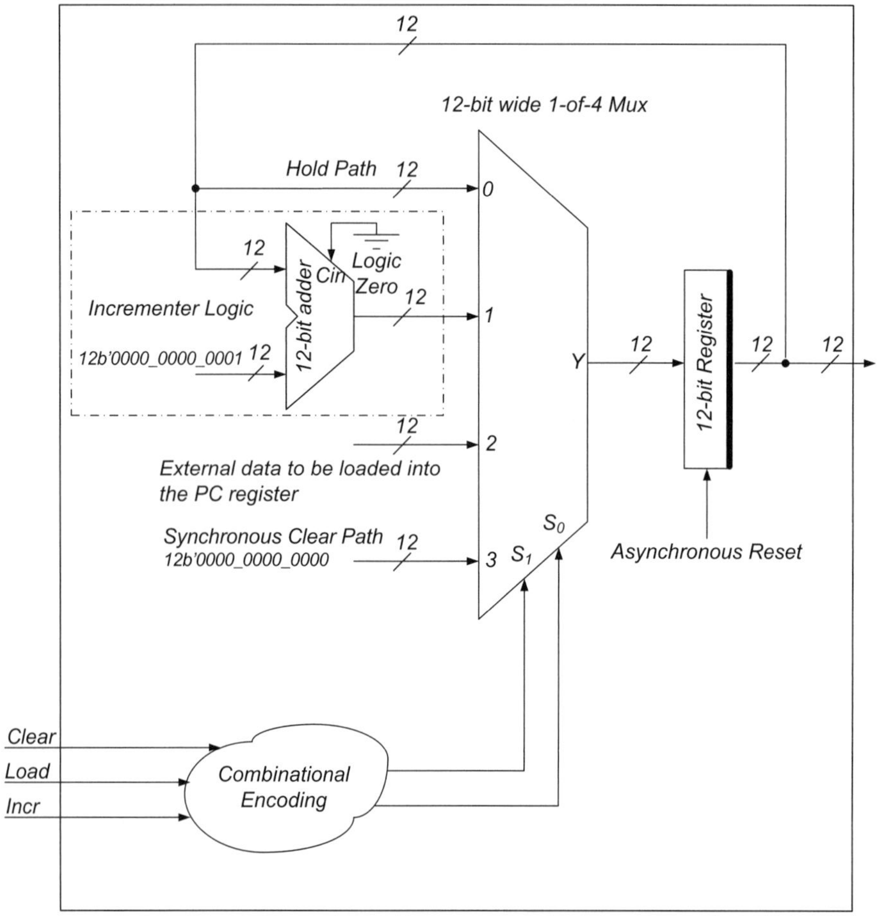

NOTE:Register clock signal not explicitly shown.

Figure 10.5 The program counter register (PC).

Table 10.4 PC register characteristic table

Row	Clock	High-true PC Control Inputs			Mux Select Inputs		PC Next State Output Becomes	
#	Active Edge	Clear	Load	Incr	S_1	S_0	$Q_{n+1}[11:0]$	PC Function
1	↑	0	0	0	0	0	$Q_{n+1} \leftarrow Q_n[11:0]$	Hold
2	↑	0	0	1	0	1	$Q_{n+1} \leftarrow Q_n+1$	Increment
3	↑	0	1	0	1	0	$Q_{n+1} \leftarrow$ Ext-Data$_n[11:0]$	Load
4	↑	1	0	0	1	1	$Q_{n+1} \leftarrow 0$	Synchronous Clear
5	↑	0	1	1	0	0	$Q_{n+1} \leftarrow Q_n[11:0]$	Hold
6	↑	1	1	0	0	0	$Q_{n+1} \leftarrow Q_n[11:0]$	Hold
7	↑	1	0	1	0	0	$Q_{n+1} \leftarrow Q_n[11:0]$	Hold
8	↑	1	1	1	0	0	$Q_{n+1} \leftarrow Q_n[11:0]$	Hold
9	Inactive	X	X	X	X	X	$Q_{n+1} \leftarrow Q_n[11:0]$	Hold

10.4 DATA PATH ARCHITECTURE: PUTTING THE LOGIC BLOCKS TOGETHER

Do not read this section until you have a good understanding of everything in this chapter that precedes this section. We will be heavily referring to previous sections, figures and tables. Once you are are ready we will begin to discuss how the individual logic blocks from Section 10.3 fit together. When our CPU has to execute an instruction it does it in three basic stages: *(1) instruction fetch, (2) instruction decode (3) instruction execution*. From a digital design standpoint each of the stages mentioned may be have one or more states.

10.4.1 Data Path: *LDA* Instruction Fetch, Decode and Execution RTL

When an instruction has to be fetched from memory the *Program Counter Register* (PC), which should have the address of the to-be-executed instruction has to transfer its contents to the *MAR*. To fetch an instruction means that the instruction has to be read from memory and be placed in some register within the CPU. Remember that our CPU instructions are only one *16*-bit word long and it is not a multi-word instruction like in some advanced machines. The instruction upper four bits are the opcode and the lower *12* bits are address *X*. Refer to the *ISW* in Table 10.1. Upon being fetched, the instruction needs to be decoded; this tells the CPU what instruction was just fetched from memory and what else it needs to do. Upon the CPU figuring out which instruction it fetched, and assuming that in our example it was a *LDA A, (X)*; the CPU knows that it requires bringing a word of data from a memory location whose address is *X*. Such data are copied from memory into the memory *MBR*. Lastly the CPU transfers such word, now in the *MBR*, to accumulator register *A*. This last data path transfer finalizes the execution phase of the instruction. That is, $A \leftarrow (X)$, refer to Table 10.2.

Let us look at the data path architecture diagram of Figure 10.6. Our complete simple CPU data path and its main memory interface consist of four registers and the ALU. The PC register holds the address of the to-be executed instruction for the currently being executed instruction. The *MAR* register holds the address of a memory location the CPU wants to access. The *MBR* register is used to read data from and write data to memory. Remember that the PC register shown in Figure 10.6 is actually all the logic of Figure 10.5. We will explain the need for the *MAR* and the *MBR* multiplexers in the data path as we explain the operation of key instructions that use such muxes. Going back one more time to our *LDA* instruction, the fetch cycles consist of:

$$1.\ \text{MAR} \leftarrow \text{PC}. \tag{10.14}$$

For this transfer to happen the selection of channel *1* of the *MAR_MUX* enables the PC to the *MAR* path. The transfer of the PC contents into the *MAR* is setting up the address from where to do the *LDA* instruction fetch

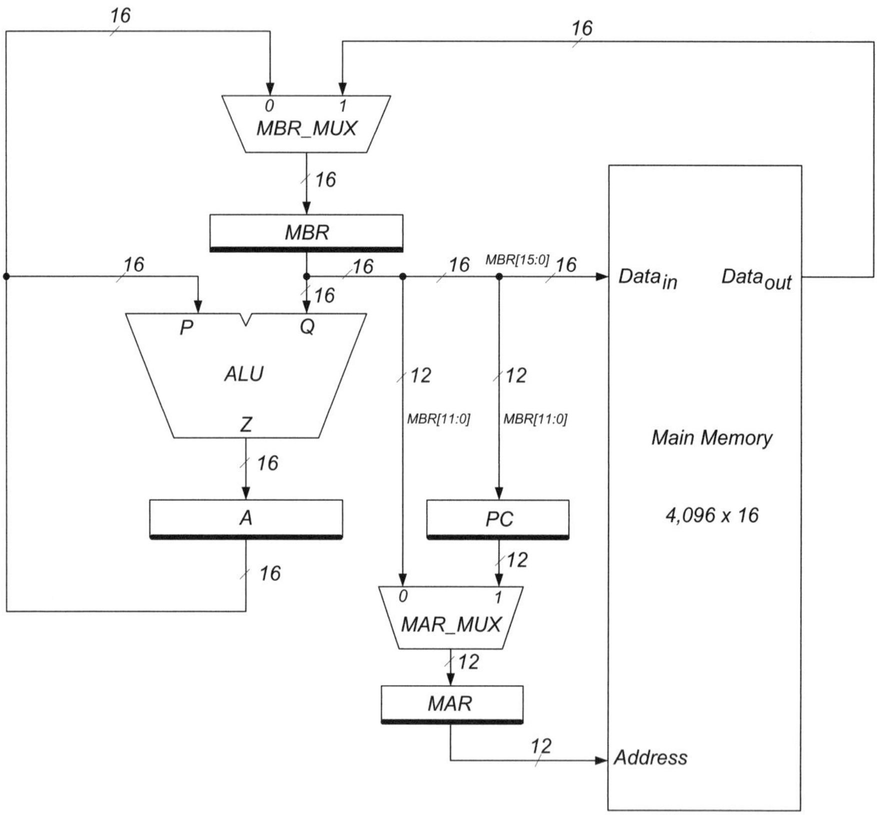

Figure 10.6 Data path architecture of our simple CPU.

from memory. Remember our instructions are all *16* bits wide and one word long. If Step (1) above happens to be the very first cycle that the CPU has to perform upon power-up reset or cold-start, the PC is previously cleared via its asynchronous clear line by the power-up reset circuitry (not shown), refer to Figure 10.5 to see the asynchronous clear line into the PC. The second data transfer that the *LDA* instruction requires is a memory read (or actual instruction fetch) that is:

$$2.\ \mathrm{MBR} \leftarrow \mathrm{M[MAR]};\ \mathrm{PC} \leftarrow \mathrm{PC} + 1. \quad (10.15)$$

Also at this time the contents of the PC are incremented by one. When the instruction completes with all its micro-operations the PC will already be pointing to the next instruction in memory. There is no reason to delay incrementing the PC or yet worse, to do it with a whole separate microinstruction. We will see shortly that if the PC needs to get loaded with a different value (instead of its incremented value) when *JMP* and *BRNA* instructions are executed, the PC will get overwritten with the address that these instructions jump or branch to. After micro-operation (2) the *MBR* has fetched complete instruction, *4*-bit opcode and *12*-bit address *X*. Notice that at the completion of step (2) both the *MBR* gets loaded with the contents of memory pointed to by the address held in the *MAR* and the PC is incremented by one. The is no resource conflict for those two operations to be performed on the same state (or clock); that is because we designed the increment PC function such that it does not use the CPU ALU to do this. Refer to Figure 10.5 note that the PC has its own increment control line.

The next step for the CPU is to decode the opcode bits, which are in bits *MBR [15:12]*. Hence:

$$3.\ \text{Decode MBR }[15{:}12];\ \text{MAR} \leftarrow \text{MBR }[11{:}0]. \tag{10.16}$$

Step 3 is a good time to transfer *MBR [11:0]* (address *X*) to the *MAR*, since address *X* will be needed to get the operand from memory. The operand refers to the data in memory that needs to be copied into the *MBR*. The two micro-operations in step three occur concurrently. Refer to Equation (10.16).

Now the CPU knows that the fetched instruction was an *LDA* and that data from memory location whose address is *X* has to be brought into the CPU *MBR*. Thus:

$$4.\ \text{MBR} \leftarrow \text{M [MAR]}. \tag{10.17}$$

Remember that the address in the *MAR* is already *X* from Step 3.

The heart of the CPU data path is its ALU and accumulator register *A*. Note that the ALU can *PASS* the contents of the *MBR* connected to the ALU *Q* input leg straight into its accumulator register *A*. This portion of the data transfer is required for the final execution path of the *LDA* instruction. So the ALU is placed in *PASS Q* mode by the control logic and the data from memory, now in the *MBR* gets transferred to register *A*, in one clock cycle. Hence:

$$5.\ \text{A} \leftarrow \text{MBR}. \tag{10.18}$$

Summarizing, the complete sequence of micro-operations to fully fetch, decode and execute our *LDA* instruction follows:

$$1.\ \text{MAR} \leftarrow \text{PC} \tag{10.19}$$

$$2.\ \text{MBR} \leftarrow \text{M [MAR]};\ \text{PC} \leftarrow \text{PC} + 1 \tag{10.20}$$

$$3.\ \text{Decode MBR }[15{:}12];\ \text{MAR} \leftarrow \text{MBR }[11{:}0] \tag{10.21}$$

$$4.\ \text{MBR} \leftarrow \text{M [MAR]} \tag{10.22}$$

$$5.\ \text{A} \leftarrow \text{MBR}. \tag{10.23}$$

We will see later that the five micro-instructions given by Equations (10.19) through (10.23) occur in five clocks.

10.4.2 All Other Instructions: Fetch, Decode and Execution: RTL

Having gone through the *LDA* instruction fetch, decode and execution in detail we will go over the rest of the instructions a little faster. We will emphasize the differences that each instruction presents with respect to a previously described instruction. To start with, it is important for the reader to know that all seven instructions of our simple CPU have the same identical fetch and decode steps. We will see in the control section that what we are calling steps are actually states of the controller state machine that steers the data transfers throughout the data path of the machine. In summary all instructions perform the same three states to do the instruction fetch (first two states) and instruction decode (third state). For the reader's convenience these three states are repeated here:

$$1.\ \mathrm{MAR} \leftarrow \mathrm{PC} \tag{10.24}$$

$$2.\ \mathrm{MBR} \leftarrow \mathrm{M\,[MAR]};\ \mathrm{PC} \leftarrow \mathrm{PC}+1 \tag{10.25}$$

$$3.\ \mathrm{Decode\ MBR\ [15{:}12]};\ \mathrm{MAR} \leftarrow \mathrm{MBR\ [11{:}0]}. \tag{10.26}$$

10.4.2.1 Store Instruction Having said that, let us look into the *STA* instruction and its difference with respect to *LDA*. Referring to Table 10.2 *STA* stores data from register *A* into a memory location whose address is *X*. That is:

$$(\mathrm{X}) \leftarrow \mathrm{A}. \tag{10.27}$$

The *STA (X), A* instruction stores or writes the contents of register *A* into memory location whose address is *X*. *LDA* on the other hand reads memory from address *X*, refer to Equation (10.22). Continuing with our *STA* instruction, in its fourth state we need to transfer register *A* data into the *MBR* and in the fifth state we write *A* to memory. Note that the *MAR* is already loaded with address *X* from state (3), refer to Equation (10.26). Summarizing the five states of the *STA (X), A:*

$$1.\ \mathrm{MAR} \leftarrow \mathrm{PC} \tag{10.28}$$

$$2.\ \mathrm{MBR} \leftarrow \mathrm{M\,[MAR]};\ \mathrm{PC} \leftarrow \mathrm{PC}+1 \tag{10.29}$$

$$3.\ \mathrm{Decode\ MBR\ [15{:}12]};\ \mathrm{MAR} \leftarrow \mathrm{MBR\ [11{:}0]} \tag{10.30}$$

$$4.\ \mathrm{MBR} \leftarrow \mathrm{A} \tag{10.31}$$

$$5.\ \mathrm{M\,[MAR]} \leftarrow \mathrm{MBR}. \tag{10.32}$$

Basically states (4) and (5) (Eqs. 10.31 and 10.32) respectively, accomplish Equation (10.27), that is,

$$(X) \leftarrow A.$$

10.4.2.2 Add Instruction Referring one more time to Table 10.2 the *ADD A, (X)* performs:

$$A \leftarrow A + (X). \tag{10.33}$$

ADD needs to read contents of memory location whose address is *X* and then add them to the existing contents of register *A* and store the results in *A*. So state four is identical to state four of the *LDA* instruction, which reads memory from location *X* into the *MBR*. The fifth state adds the read data now in the *MBR* to *A* and places the result in *A*. This last step overwrites the previous contents of *A* and sets the *S* bit accordingly. Thus *ADD* is:

$$1.\ \text{MAR} \leftarrow \text{PC} \tag{10.34}$$

$$2.\ \text{MBR} \leftarrow \text{M}[\text{MAR}];\ \text{PC} \leftarrow \text{PC}+1 \tag{10.35}$$

$$3.\ \text{Decode MBR}[15{:}12];\ \text{MAR} \leftarrow \text{MBR}[11{:}0] \tag{10.36}$$

$$4.\ \text{MBR} \leftarrow \text{M}[\text{MAR}] \tag{10.37}$$

$$5.\ \text{A} \leftarrow \text{A}+\text{MBR}. \tag{10.38}$$

The produce Equation (10.38) the CPU controller has to select the ALU *ADD* function select lines. Refer to Table 10.3.

10.4.2.3 And Instruction From a data path or register transfer language (*RTL*) viewpoint *AND* is virtually identical to *ADD*. The sole difference is that the CPU controller selects the *AND* function of the ALU instead of the *ADD*. So for the AND instruction we have that:

$$1.\ \text{MAR} \leftarrow \text{PC} \tag{10.39}$$

$$2.\ \text{MBR} \leftarrow \text{M}[\text{MAR}];\ \text{PC} \leftarrow \text{PC}+1 \tag{10.40}$$

$$3.\ \text{Decode MBR}[15{:}12];\ \text{MAR} \leftarrow \text{MBR}[11{:}0] \tag{10.41}$$

$$4.\ \text{MBR} \leftarrow \text{M}[\text{MAR}] \tag{10.42}$$

$$5.\ \text{A} \leftarrow \text{A}.\text{MBR}. \tag{10.43}$$

10.4.2.4 Conditional Branch Instruction Branch on negative accumulator needs to check the state of the *S* bit, upon this bit being one it loads the PC with address *X*, which as usual is already in the *MAR* from state 3. If the

S bit is zero then the PC remains with its previous contents, which are PC + 1 from state 2. Hence:

$$1.\ \mathrm{MAR} \leftarrow \mathrm{PC} \tag{10.44}$$

$$2.\ \mathrm{MBR} \leftarrow \mathrm{M}\,[\mathrm{MAR}];\ \mathrm{PC} \leftarrow \mathrm{PC}+1 \tag{10.45}$$

$$3.\ \text{Decode MBR}\,[15{:}12];\ \mathrm{MAR} \leftarrow \mathrm{MBR}\,[11{:}0] \tag{10.46}$$

$$4.\ \text{If S bit} = 1? \tag{10.47}$$

$$5.\ \text{Then: } \mathrm{PC} \leftarrow \mathrm{MAR}. \tag{10.48}$$

10.4.2.5 Unconditional Jump Instruction Simply requires loading the PC with the MAR that already has address X from state 3. This instruction only has four states, which are:

$$1.\ \mathrm{MAR} \leftarrow \mathrm{PC} \tag{10.49}$$

$$2.\ \mathrm{MBR} \leftarrow \mathrm{M}\,[\mathrm{MAR}];\ \mathrm{PC} \leftarrow \mathrm{PC}+1 \tag{10.50}$$

$$3.\ \text{Decode MBR}\,[15{:}12];\ \mathrm{MAR} \leftarrow \mathrm{MBR}\,[11{:}0] \tag{10.51}$$

$$4.\ \mathrm{PC} \leftarrow \mathrm{MAR}. \tag{10.52}$$

10.4.2.6 Complement Accumulator Instruction This is an instruction that already has its operand, that is, register A, in the CPU itself, it does not need to go to memory to read a location to get the operand like the *AND* and *ADD* instructions do. To generate $A \leftarrow \bar{A}$, the controller simply needs to select the ALU *PASS* $\bar{P}$ *to Z mode* to complement the accumulator and store it back into itself. Thus the *CMP A* instruction looks like:

$$1.\ \mathrm{MAR} \leftarrow \mathrm{PC} \tag{10.53}$$

$$2.\ \mathrm{MBR} \leftarrow \mathrm{M}\,[\mathrm{MAR}];\ \mathrm{PC} \leftarrow \mathrm{PC}+1 \tag{10.54}$$

$$3.\ \text{Decode MBR}\,[15{:}12];\ \mathrm{MAR} \leftarrow \mathrm{MBR}\,[11{:}0] \tag{10.55}$$

$$4.\ \mathrm{A} \leftarrow \bar{A}. \tag{10.56}$$

Figure 10.7 depicts a complete state diagram of all seven instructions. The state assignment of each state is not done in Figure 10.7 yet. This will be addressed when we design the CPU controller state machine. Notice that for all seven instructions we used the numbers (1), (2), etc. just to indicate the sequence of states in time. Those numbers should not be construed as the state number assignment.

10.5 THE SIMPLE CPU CONTROLLER

The controller is the state machine that orchestrates the functioning of the data path data transfers, registers loading, PC incrementing, synchronous

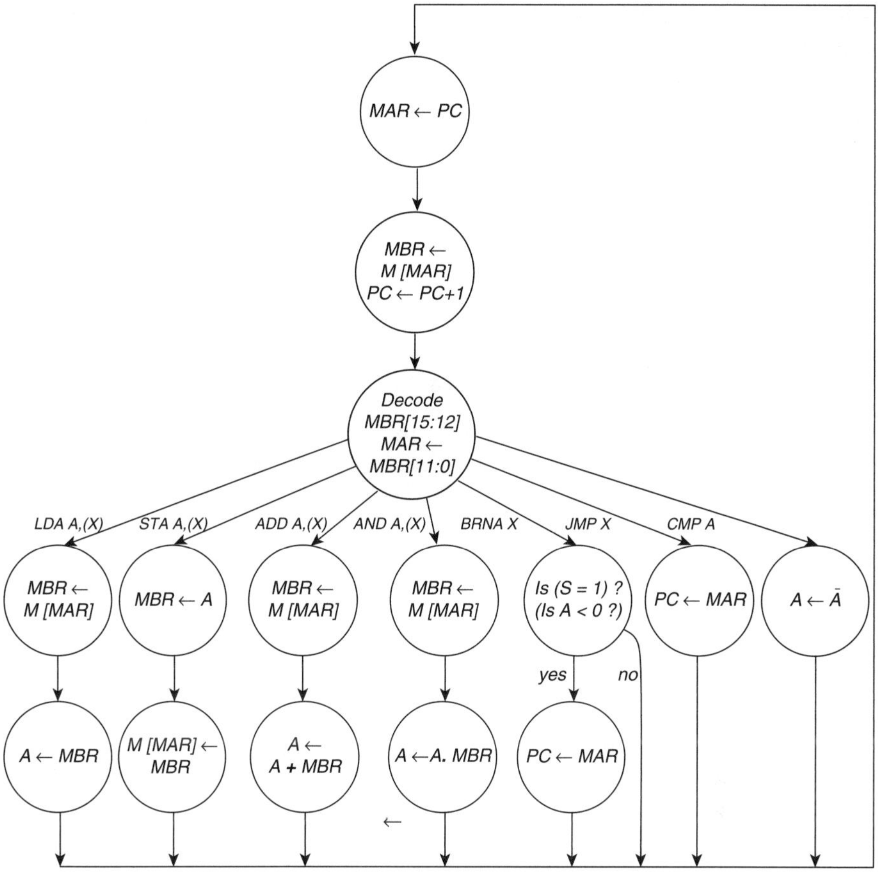

Figure 10.7 Simple CPU state diagram.

clearing, ALU function selection, *MAR* and *MBR* multiplexers steering, memory reads and writes, so that the instruction set is executed as described by its state diagram (Fig. 10.7). We need to first identify all the inputs and the outputs that our controller needs. This process has to be done by mainly careful inspections of Figures 10.6 and 10.7 and it is greatly a comprehensive process. Input signals to our controller are: (1) *MBR [15:12]* the opcode of each instruction. (2) The accumulator *MSB* stored in the *S* bit flip-flop (not explicitly shown on the data path diagram). (3) A system level asynchronous reset to clear the PC upon power-up. Outputs of the controller are listed by functional block and are the following: (1) For register *A*: *LOAD_A*, (2) for the *MAR: LOAD_MAR*, (3) for the *MBR: LOAD_MBR*, (4) for the PC: *INCR_PC*, *LOAD_PC* and *CLEAR_PC* all three signal being synchronous controls. (5) The asynchronous clear control for the PC: *ASYNC_CLEAR_PC* will be generated by the reset logic (not covered yet) and not by the controller. (6)

The *MAR mux* needs a *MAR_MUX_SEL* line, (7) the *MBR mux* needs a *MBR_MUX_SEL* line, (8) the ALU needs two bits of *ALU_FUNCTION_SELECT* bits. (9) We need a *READ* and a *WRITE* control signals for the memory array. Finally four bits of state are needed for our controller because since our state diagram has *16* states or less (actually *15*). In summary we have:

Five input bits for the controller and one input bit for the asynchronous *CLEAR* for the PC from the reset logic. Twelve control outputs and four state bits. State bits are outputs too. It is interesting to mention that a relatively large number of control signals are needed even though we are dealing with a very simple CPU.

10.5.1 State Assignments and Controller Implementation

Carefully reviewing *RTL* micro-operations (Eqs. 10.14 through 10.56) and the state diagram of Figure 10.7 we will make the following state assignments and justify their selection later. Starting with the first state at the top of Figure 10.7 we assign to it the value of *0*, then state *1* and *2* for the fetching and decoding states. For the *LDA* instruction, we assign states *3* and *4*. States *5* and *6* for *STA*; states *7* and *8* for *ADD*; states *9* and *10* for *AND*; states *11* and *12* for the *BRNA*, state *13* for *JMP*, and state *14* for *CMP A*. There is only one state that remains unassigned, state *15*. Since state *15* is an unused state we can design our controller state machine assigning unused state 15 to unconditionally go to the machine instruction fetch, state *0*. Another option is to make state *15* an isolated state (Fig. 10.8). Alternatively, if the CPU reaches state *15*, for example upon power up reset or due to some failure mechanism, we may choose to force the user to re-start or power cycle the CPU, since the CPU would freeze. Although this may seem a little unreasonable, it may be a better choice than letting the computer start fetching instructions from perhaps not the correct memory address. Or perhaps with some other register contents corrupted.

Let's talk about the state transitions that exist with the current state assignment. All transitions from state *0* to *1*, from state *1* to *2*, from state *3* to *4*, from state *5* to *6*, from state *7* to *8*, from state *9* to *10* and from state *11* to *12*, are achieved by incrementing the previous state by one. For example, you reach state 10 by incrementing state 9 by one. All state transitions from state *4 to 0, 6 to 0, 8 to 0, 10 to 0, 12 to 0, 13 to 0*, and *14 to 0* are reached by clearing a state register of the to-be-designed state machine controller. Finally all transitions from state *2* to state *3, 5, 7, 9, 11, 13*, and *14* are attained by loading the corresponding state into the state register. This has an important implication when we need to design the controller for our simple CPU. The reason is that the hardware implementation of the controller is greatly simplified because there are only three types of state transitions: that is, *Increment, Load or Clear*. This is the same register architecture as the one used for the PC register, refer one more time to Figure 10.5. We will implement the controller with a *4*-bit state register and three *1-of-16* multiplexers. Figure 10.9 depicts the hardwired

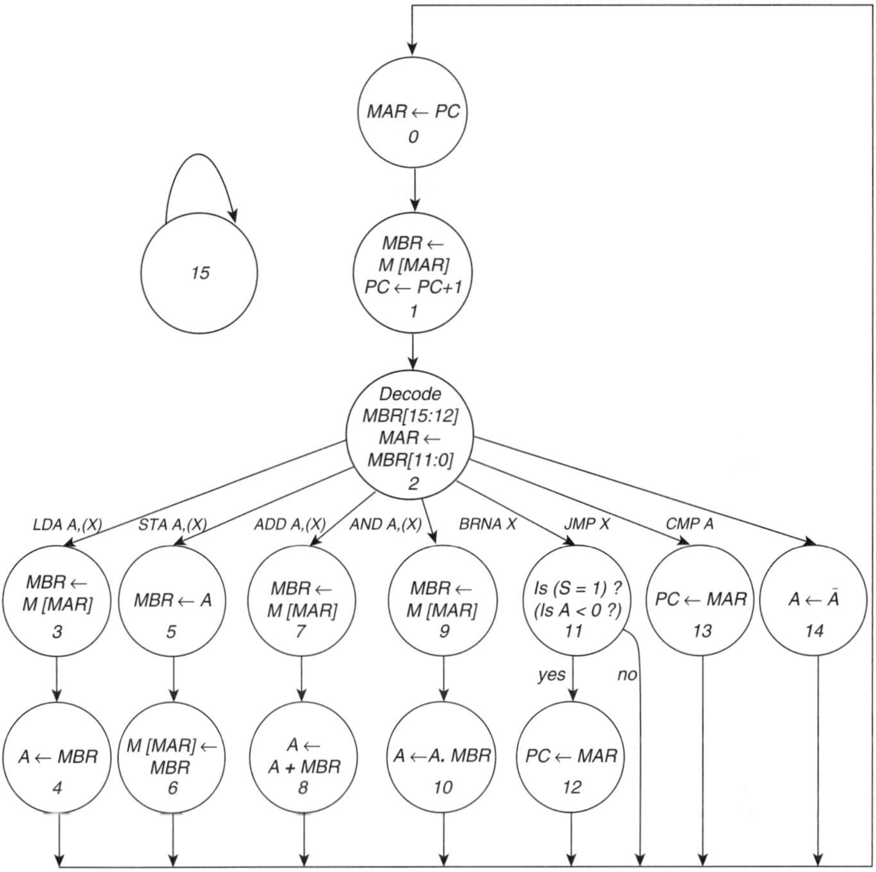

Figure 10.8 State assignments of our simple CPU state diagram.

implementation of the controller. It is called a hardwired implementation because it is not programmable or as easily changeable as it would be if we used a programmable memory device, such device is referred to as the micro-sequencer or controller micro-store. Because of space reasons this book does not deal with micro-store based controllers. However, the reader can find material for further study under the Further Reading section at the end of this chapter.

The hardwired controller of Figure 10.9 is a clean and simple implementation. The logic has not been minimized in any way. The multiplexers are there to emphasize the function we want the state counter to take; these functions are increment, load or clear, all of them synchronous functions. The initial clearing of the state counter or register is accomplished with it asynchronous reset supplied by the reset or power-on circuit, not shown. Careful analysis of

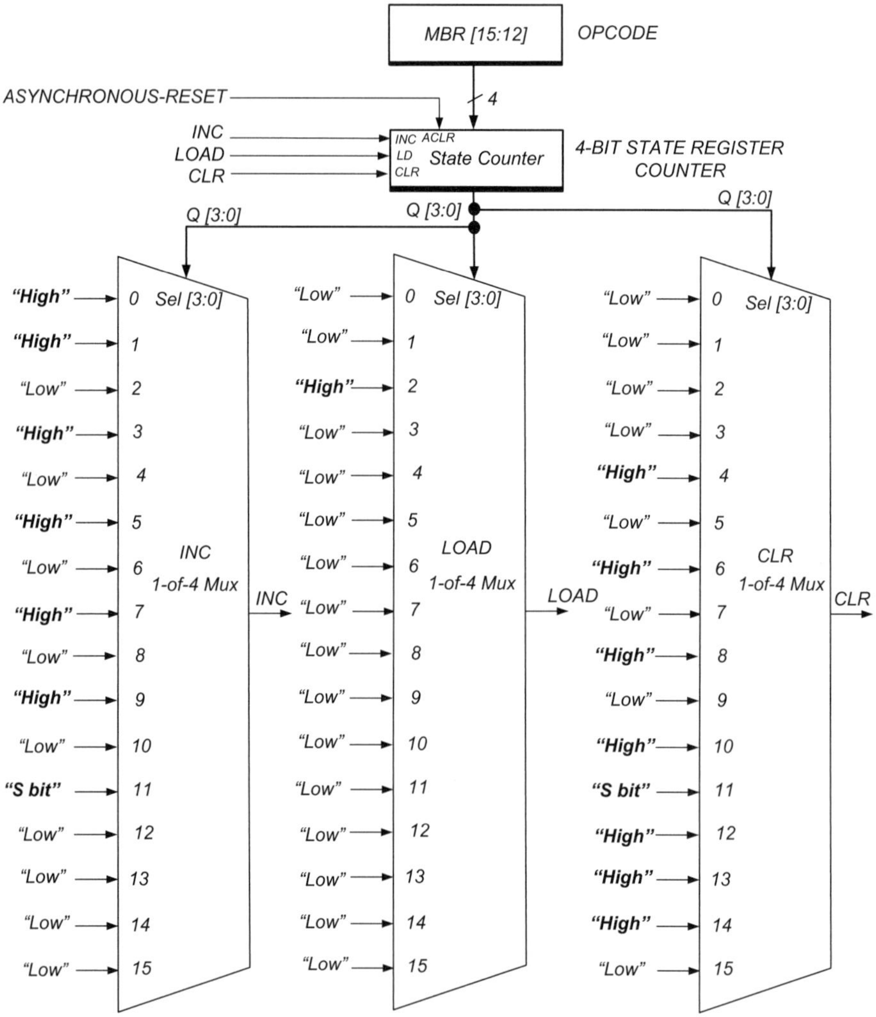

Figure 10.9 Hardwired implementation of our simple CPU controller or sequencer.

Figure 10.9 in conjunction with 10.8 allows the reader to understand the way the controller *walks* through each and every one of the assigned states, depending on the instruction that is presented for external loading to the state counter. Table 10.5 shows the mapping that needs to be produced from each state to the appropriate assertion of the output control signals. The state bits to output signal mapping can be implemented with a *Read Only Memory (ROM)* or combinational logic gates to do the decoding to assert the appropriate control outputs. The reader is asked, as an exercise, to design the combinational logic described by Table 10.5. Hint: The inputs of the logic should be *Q[3:0]* the state bits of the controller state counter.

Table 10.5 Combinational logic or ROM table to drive the CPU controller control outputs

State	Control Outputs									
Q[3:0]	LD_A	LD_MAR	LD_MBR	LD_PC	INC_PC	CLR_PC	MRB_ MUX_SEL	MAR_ MUX_SEL	ALU_FUNC_ SEL[1:0]	M_RD	M_WR
0	0	1	0	0	0	0	X	1	XX	0	0
1	0	0	1	0	1	0	1	X	XX	1	0
2	0	1	0	0	0	0	X	0	XX	0	0
3	0	0	1	0	0	0	1	X	XX	1	0
4	1	0	0	0	0	0	X	X	00	0	0
5	0	0	1	0	0	0	0	X	XX	0	0
6	0	0	0	0	0	0	X	X	XX	0	1
7	0	0	1	0	0	0	1	X	XX	1	0
8	1	0	0	0	0	0	X	X	10	0	0
9	0	0	1	0	0	0	1	X	XX	1	0
10	1	0	0	0	0	0	X	X	11	0	0
11	0	0	0	0	0	0	X	X	XX	0	0
12	0	1	0	0	0	0	X	1	XX	0	0
13	0	1	0	0	0	0	X	1	XX	0	0
14	1	0	0	0	0	0	X	X	01	0	0
15	0	0	0	0	0	0	X	X	XX	0	0

In Table 10.5 several control output names have been abbreviated: for example *LOAD_A* is renamed: *LD_A*, to have more room on the table. Let us refer to the *LDA* instruction *RTL* given by Equations (10.19) through (10.23). Referring to Table 10.5, Figure 10.6 and the *LDA RTL* repeated for the reader's convenience, we verify how Table 10.5 was filled in.

$$\text{State 0: MAR} \leftarrow \text{PC} \tag{10.57}$$

$$\text{State 1: MBR} \leftarrow \text{M[MAR]; PC} \leftarrow \text{PC}+1 \tag{10.58}$$

$$\text{State 2: Decode MBR [15:12]; MAR} \leftarrow \text{MBR [11:0]} \tag{10.59}$$

$$\text{State 3: MBR} \leftarrow \text{M[MAR]} \tag{10.60}$$

$$\text{State 4: A} \leftarrow \text{MBR.} \tag{10.61}$$

In state *0:* we need to assert *LD_MAR* = *1* and *MAR_MUX_SEL* = *1* to transfer data from the PC into the *MAR*, and all other control signals need to be negated upon receiving the active edge of the clock. Once in state *1: LD_MBR* = *1, MBR_MUX_SEL* = *1* and *M_RD* = *1* reads memory location pointed to by the *MAR* and saves the read data in the *MBR*. Simultaneously the *INC_PC* = *1* to increment the PC by one. Once on state *2*: The *MBR[15:12]* get looked at (decoded) by the controller and *MBR[11:0]* transferred to the *MAR*, so that *LD_MAR* = *1, MAR_MUX_SEL* = *0*. This micro-operation just copied address *X* into the *MAR*. The controller asserts its *LOAD Mux* output to load a jump to state *3* to go to the *LDA* instruction execution sequence. On state *3 M_RD* = *1, MBR_MU_SEL* = *1* and *LD_MBR* = *1* reads memory data from address *X*. Finally once in state *4 ALU_FUNC_SEL* = *00 (PASS Q to Z)* and *LD_A* = *1* transfers the contents in the *MBR* to register *A*.

For the rest of the instructions the reader should refer to the state diagram with state assignments of Figure 10.8 and with the aid of Figure 10.6 (data path architecture) start verifying the correctness of the contents of the rest of Table 10.5. Please allocate, as long a time as you need, the first time going through the complete table may take more than a few hours and more than one sitting. I have been there; please do not feel frustrated, this will only make sense once you go over the material exhaustively.

10.6 CPU TIMING REQUIREMENTS

This section will be concerned with identifying the timing paths in the machine data path, memory interface and controller. To study these paths we will make a reasonable assumption, which is that all clocks arrive at their registers clock inputs at the same time. This means that we assume that there is zero clock-skew. Starting with the data path architecture of Figure 10.6 timing paths are simply identified starting from the *Q* outputs of a register working your way though a combinational data path or simply through a bus wire into the *D* inputs of either the same or another register.

The following are the long paths in Figure 10.6:

1. *PC Q* outputs through *MAR_MUX* into *MAR D* inputs.
2. *MRB Q* outputs through *MAR_MUX* into *MAR D* inputs.
3. *MBR Q* outputs through *ALU leg Q* into *A* register *D* inputs.
4. Register *A Q* outputs through *ALU leg P* into register *A D* inputs.
5. Register *A Q* outputs through *MBR_MUX* into *MBR D* inputs.
6. *MBR Q o*utputs into *PC D* inputs.
7. *MAR Q* outputs through memory address to data out through *MBR_MUX* into *MBR D* inputs.
8. *MBR Q* outputs though memory data in and data out, through *MBR_MUX* into *MBR D* inputs.

For the logic of the CPU controller of Figure 10.9 we can identify the following paths:

9. *MBR Q* outputs to state counter *D* inputs.
10. State counter *Q* outputs to *INC MUX* select lines to state counter *INC* input.
11. State counter *Q* outputs to *LOAD MUX* select lines to state counter *LOAD* input.
12. State counter *Q* outputs to *CLR MUX* select lines to state counter *CLR* input.
13. *S*-bit flip-flop *Q* output through *INC Mux INC* output into state counter *INC* input.
14. *S*-bit flip-flop *Q* output through *CLR Mux CLR* output into state counter *CLR* input.

When we want to study for each timing path their corresponding long path, to calculate if the set-up time is met we proceed as follows with path (1): We use the maximum clock-to-output time of the sourcing register, the maximum propagation delay of the combinational logic and the minimum set-up time required by the *D* input of the receiving register.

So when we want to study path (1) as a long path it becomes:

$$1.\ \mathrm{PC}\ t_{\text{clock-to-output max delay}} + t_{\text{max pd MAR\_MUX}} + \mathrm{MAR}\ t_{SU}.$$

Finally to determine if such path meets the set-up time requirement of the MAR register we need to check if:

$$\mathrm{PC}\ t_{\text{clock-to-output max delay}} + t_{\text{max pd MAR\_MUX}} + \mathrm{MAR}\ t_{SU} \le T_{\text{clkmin}}.$$

When we want to study the effect of the same path (1) as a short path, we consider the fastest or shortest delay at which the *Q* outputs of the PC travel plus the shortest delay through the *MAR_MUX* combinational logic. Since we want to analyze the corresponding short path we want the check if the following inequality is met:

$$\text{PC } t_{\text{clock-to-output min delay}} + t_{\text{min pd MAR\_MUX}} \geq \text{MAR } t_{\text{HOLD}}.$$

The reader is strongly encouraged to establish all 14 long path and 14 short path equations for the complete CPU. In general we find that the longest path usually is the path through the ALU because it has the largest amount of combinational logic. The usual short paths are those that connect registers *Q* outputs through short wires straight into registers *D* inputs. One such example in our CPU design is the *MBR[15:12]* to state counter inputs.

10.7 OTHER SYSTEM PIECES: CLOCK, RESET AND POWER DECOUPLING

10.7.1 Clock

The clock of a synchronous state machine is typically generated with a crystal-based oscillator with good stability. Clocks are buffered and distributed to its loads in a point-to-point fashion. With the current sub-nanosecond rise/ fall times and clock frequencies of today, virtually all signals are interconnected point-to-point for optimal signal integrity and timing behavior. Clock buffers used should be preferably packaged within the same *IC* package and the wire lengths should be matched. When the clock frequencies are high, like several hundred megahertz crystal oscillators are not available so *Phase Lock Loop* circuits *(PLL)* are used to generate gigahertz range frequencies. PLLs are not within the scope of this book and an excellent reference [7] to this topic is given in the Further Reading section of this chapter.

10.7.2 Reset

Upon good power being applied to a CPU-based system, reset is the first hardware signal that the system requires to power-up correctly. In our simple CPU, reset clears the PC and the CPU controller state counter register. In real world systems reset initializes or clears all the appropriate registers on board or within programmable devices such as *CLPDs* and *FPGAs*, it also allows the clock generating circuit to start running. It is extremely important for the reset or *Power-On Reset (POR)* signal not to glitch, because that may cause registers to come out of reset at different times and the system to behave unpredictably. It is desirable to assert reset asynchronously to ensure that all resettable devices are cleared regardless of the state of their clock. However,

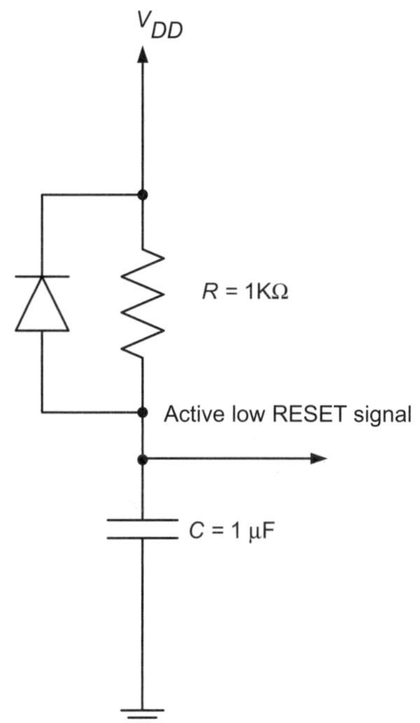

Figure 10.10 A poor example of a reset circuit.

it is more advisable to negate reset synchronously. Reset should be released (negated) synchronously, because releasing reset in an uncontrolled environment (i.e., asynchronously) may cause its flip-flops or registers to go metastable. Two problems may occur upon an asynchronous reset release: (a) the reset recovery time may be violated and (b) reset negation may occur at different clock cycles for different clocked elements. The reset recovery time is the time between when reset is negated and the time that the clock edge signal goes active again. Figure 10.10 depicts a very bad example of a reset circuit. It has multiple problems. Upon power-on, assuming a discharged capacitor, the low-true RESET signal will start at zero and has an exponentially increasing waveform. This waveform may not be suitable for the ICs that receive the reset, because it may stay at the IC logic threshold too long. It may not apply reset for the required time because the exponential ramp up time is not very precise. Chips generally have a normal operating voltage plus and minus 5 or 10 percentage points of such voltage. The assertion of reset with the circuit of Figure 10.10 is not very well defined. The diode that the circuit has across the 10 kΩ resistor is to provide a quick discharge path of the capacitor if a user turns off the system and turns it back on quickly. A better scheme would use a Schottky diode since it has a lower forward voltage drop.

A somewhat improved reset is depicted in Figure 10.11.

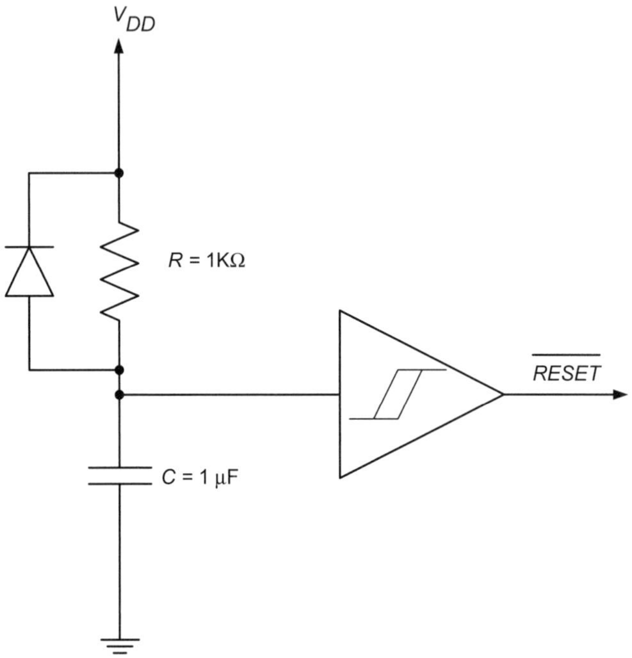

Figure 10.11 Improved reset circuit.

The advantage of this circuit is that a Schmitt trigger logic gate is used to shape the reset pulse. A Schmitt trigger gate has built-in hysteresis properties, thus filtering out short-lived glitches and reset pulse variations.

An even better approach can be implemented with two D flip-flops synchronously clocked, but having asynchronous reset inputs. The second flip-flop reduces metastability. The operation of this reset circuit is clean. Reset out is active low. It produces an asynchronous active high reset, even before the clock runs. But upon active low Reset_Out being released a low level (i.e., no reset) is synchronously clocked into active low reset_out. The two synchronizing flip-flops reduce the probability of metastable behavior to practically negligible levels. Figure 10.12 depicts such reset circuit. This circuit is commonly used at board level as well as in programmable devices level designs.

At the board level reset chips are available from several IC manufacturers, such as Analog Devices (ADI), Intersil, Maxim, ST Microelectronics, Texas Instruments (TI) and several others. Such ICs are referred to as *supply voltage supervisors or reset chips*. Supervisor circuits monitor system voltages from a range that may vary from about 0.5V to some upper voltage limit like 5 V. When the voltage dips below a preset threshold or when a manual reset (typically a pushbutton) drops to a logic low (active low manual reset) the active low open drain reset output asserts. It usually remains asserted low for a user-programmed time delay. An external resistor and capacitor time

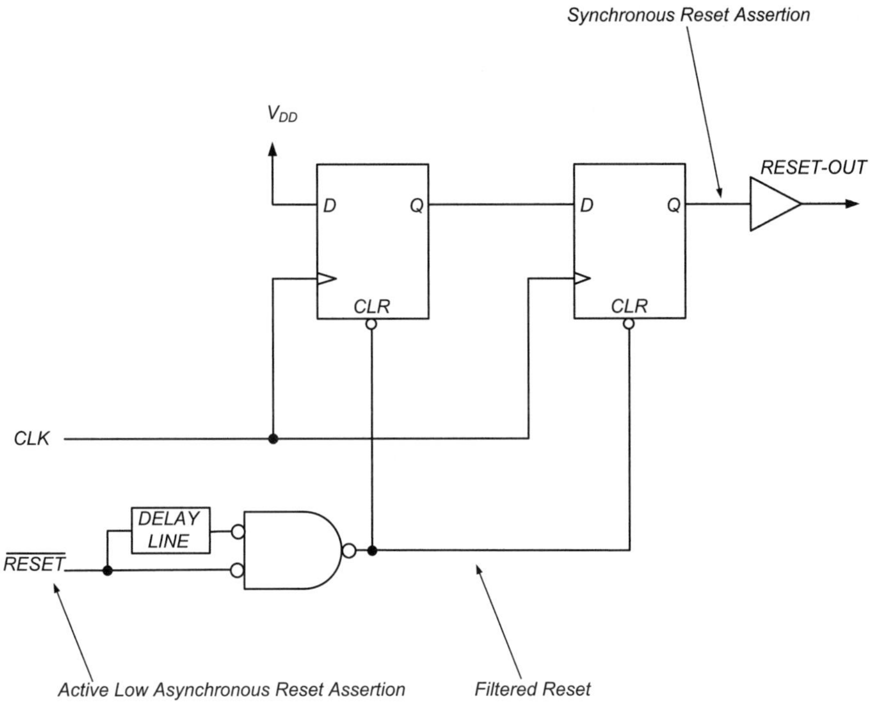

Figure 10.12 Reset circuit with asynchronous assertion and synchronous negation.

constant usually control such time delay. These supervisor chips use a precision reference voltage to achieve good threshold accuracy (typically 1% or better). When the *DC* voltage to an embedded system dips below the required minimum voltage for proper operation, the supervisor circuit asserts the reset signal; this initiates a system shut down. Current flow stops and the voltage to the supervisor may increase due to decreased IR drop. This produces a false reset negation from the supervisor circuit, that is, the supervisor incorrectly turns the system back on. To mitigate that problem sensing voltage hysteresis is provided to the supervisory circuit. For more details on supervisory circuits refer to the websites of the IC manufacturers mentioned above. Examples of some power-on reset or supervisory ICs are: Intersil ISL6131, Maxim MAX691A, MAX700 and MAX800 series, Analog Devices ADM63xx series, TI TPS3808 series, and many others.

10.7.3 Power Decoupling

Just like any *IC* on a system good decoupling with low equivalent series resistance (*ESR)* and low lead inductance capacitors must be supplied to every *IC* on the board. What is decoupling for? Power supplies provide voltage and current to integrated circuits over the time they need to be operational. IC's make sudden transitions, usually in the nanosecond range, of one or many of their inputs and output pins. Such fast signals transitions produce high current demands as the power across the IC droops a few millivolts. The power across the IC power pins droops, because there is no power supply that can have a time response to such a fast power demand. The power supply and its power distribution scheme, cables, traces and wires that route the power to the point of consumption, both have a finite and definitely non-zero response to current transients. For example *300 kHz* switching power supplies, may have a bandwidth or capacity to respond to current transients of about *10* microseconds. During these *10* microseconds the decoupling capacitors, placed in extremely close proximity to the IC being decoupled, provide the amount of current for the amount of time that the power supply requires to react and start to provide current to the IC. This indicates that the decoupling capacitor in a very first pass approximation has to be able to provide enough current for a certain minimum time allowing its voltage to droop no more than a predetermined limit. Such calculation is based on the basic equation than links current voltage and time in a capacitor. From Chapter 1 we know that such relationship is:

$$i_C(t) = C\frac{dv_C(t)}{dt} \tag{10.62}$$

Let us consider the following numerical example to illustrate how to calculate the value of decoupling capacitance needed. Assume that our power supply has a bandwidth of *100 kHz*, which means that it will be able to respond to current transients after *10 microseconds* from the beginning of the current transient event. Assume that we want to have a maximum voltage droop across the IC power pins of no more than 300 mV. Finally assume that the current transient demanded by the IC is *1 A*. From Equation (10.62) we calculate C as:

$$C = \frac{i_C d(t)}{dv_C(t).} \tag{10.63}$$

Replacing differentials with finite time and voltage increments or decrements in Equation (10.63) we obtain:

$$C = \frac{1\times 10\times 10^{-6}}{0.300} = 33.333\ \mu\text{F}.$$

When decoupling calculations need to done more accurately equivalent series resistance *(ESR)* and lead inductance *(ESL)* of the capacitor should be taken into account. A very comprehensive treatment of this topic can be found in [1,2].

10.8 SUMMARY

This chapter defined the instruction set of a very simple CPU. A data path architecture was presented to support the defined instruction set. Using such data path we covered the microinstructions of every machine language instruction or simply called a macroinstruction. We further developed the state diagram of the complete instruction set, and the sequencer design.. The longest instructions take five states or clocks to fully execute and the shortest ones take four clocks. We further look into some system level issues: timing, clocks, and their distribution, resets and power decoupling.

FURTHER READING

1. Larry Smith, *Decoupling Capacitor Calculations for CMOS Circuits*, Electrical Performance of Electronic Packaging Conference, 1994.
2. Larry Smith, et al., *Power Distribution System Design Methodology and Capacitor Selection for Modern CMOS Technology*, 1999.
3. David Money Harris and Sarah L. Harris, *Digital Design and Computer Architecture*, Morgan Kaufmann Publishers, San Francisco, CA, 2007.
4. David A. Patterson and John L. Hennessy, *Computer Organization and Design the Hardware/Software Interface*, Morgan Kaufmann Publishers, San Mateo, CA, 1994.
5. John Mick and Jim Brick, *Bit-Slice Microprocessor Design*, McGraw-Hill, New York, 1980.
6. Carl Hamacher, Zvonko Vranesic, and Safwat Zaky, *Computer Organization*, 5th ed., McGraw-Hill, New York, 1978.
7. Roland Best, *Phase Lock Loops*, 6th ed., McGraw-Hill, New York, 2007.

PROBLEMS

10.1 Design the two new instructions shown at the bottom of Table 10.6
Note:

1. The new instruction set supports two more instructions: *LD B, (X)* and *STA (X), B.*

2. The CPU has one new 16- bit register: that is, register B.

(a) Make any needed modifications to the data path architecture of Figure 10.6.

(b) Draw a complete state diagram only of the two new instructions shown in Table 10.6.

10.2 In reference to Figure 10.9 the CPU controller:

(a) Redesign the control logic of Figure 10.9 using a minimally sized ROM and D-type registers for the new instruction set of Table 10.6.

Table 10.6 New simple CPU instruction set

Instruction Syntax	Opcode (Binary) IWF [15:12]	Address X IWF [11:0]	Description of What Gets Executed	Affects Sign Flag?
LD A, (X)	0000	A valid address	A ← (X)	No
STA (X), A	0001	A valid address	(X) ← A	No
ADD A, (X)	0010	A valid address	A ← A + (X)	Yes
AND A, (X)	0011	A valid address	A ← A. (X)	Yes
BRNA X	0101	A valid address	If S = 1 then PC → X	No
JPM X	0100	A valid address	PC ← X	No
CMP A	0110	Bits [11:0] are ignored	A ← $\bar{A}$	Yes
LD B, (X)	0111	A valid address	B ← (X)	No
STA (X), B	1000	A valid address	(X) ← B	No

(b) Draw the circuit schematic of the new simple CPU controller logic.

(c) Write the micro-code that the ROM needs to perform the complete instruction set given by Table 10.6.

10.3 Assume we want to add another new instruction to Table 10.6 New Instruction Set. This new instruction is:

$$\text{OR A, (X); A} \leftarrow \text{A} + \text{(X)}$$

The *OR* instruction does a bit-to-bit *OR* operation between the contents of memory location whose address is *X* and the contents of register *A*, it finally stores the result in register A, overwriting its original contents. Remember that the new instruction has to increment the contents of the PC just as any other instruction does.

(a) Enumerate and describe all the required changes to the data path architecture, and (b) the state diagram.

10.4 Assume we want to add another new instruction to Table 10.6 New Instruction Set. This new instruction is:

NOP; 5 clock cycles doing nothing

The NOP should have a normal instruction fetch and decode cycles. The actual execute cycle should be long enough to use 5 cycles of the clock, including the fetch and decode phases.

Remember the new instruction has to increment the contents of the PC just as any other instruction does. However, the NOP must *not* change the contents of registers A and B and must not affect the ALU flag bit S (sign bit).

(a) Enumerate and describe all the changes to the data path architecture, and the state diagram that apply.

10.5 Create an algorithm that allows one to add two 4-bit unsigned binary numbers and produce the resulting sum in BCD.

For example: 1001 + 0001 = 1010, binary nine plus binary one equals binary ten. Your algorithm should report the sum as: 0001_0000, which stands for ten in binary coded decimal (BCD). Similarly 0110 + 0101 = 1011 should be reported as 0001_0001 (eleven in BCD). And a last example, 0011 + 0010 = 0101, binary 3 plus binary 2 equals equals five in BCD.

10.6 Design the logic hardware to implement the algorithm found in Problem 10.5. Draw the circuit schematics of the logic.

10.7 Let us assume that we want to design a MOVE instruction that copies the contents of a memory location whose address is X, into another memory location whose address is Y. Such instruction must not affect ALU flags, it must increment the PC just as any other instruction does. The original (and likely unknown) contents of memory location Y are overwritten with the data copied from address X. Original contents of memory location X remain unchanged. The syntax for such new instruction is:

$$\text{MOVE (Y), (X); (Y)} \leftarrow \text{(X)}$$

(a) Enumerate and describe any (and all) required changes to the data path architecture, (b) instruction word format (IWF), and (c) the state diagram.

For part (a) draw and show needed changes to the data path architecture of Figure 10.6. For part (b) generate the complete state diagram for the new instruction only.

10.8 For the circuit schematics of Figure 10.13 assume the following timing parameters:

D-type Flip-flop:

$$t_{\text{CLK-to-QMAX}} = 2\text{ ns}$$

$$t_{\text{CLK-to-Qmin}} = 1\text{ ns}$$

$$t_{SU} = 1.5\text{ ns}$$

$$t_{H} = 0.5\text{ ns}$$

The circuit represents a hardwired shift register without external reset line or controls.

Assume that the wires are ideal and have no delays. Assume that the clock arrives at all flip-flop clock inputs at the same time, no clock skew.

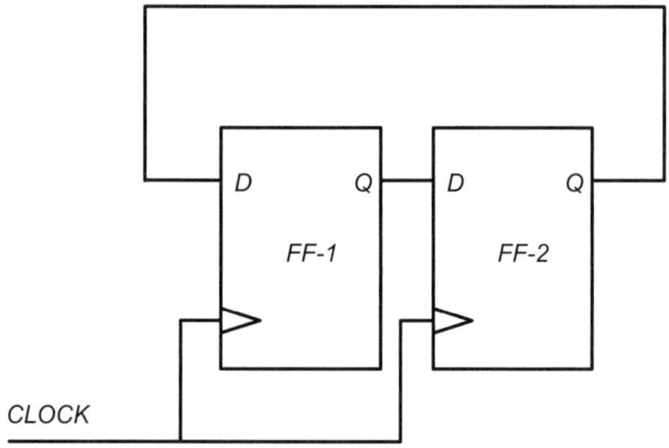

Figure 10.13 Circuit for Problem 10.8.

(a) Determine the maximum frequency at which the shift register can be reliably clocked.

(b) Determine the available hold time (short path) available to each flip-flop.

(c) Assume that flip-flop 1 clock has a rising edge at time 0 ns and flip-flop 2 receives the same logically rising edge 1 ns later. Determine the available hold time (short path) available to each flip-flop. (This point and the next one no longer assume zero clock-skew.)

(d) Draw the following waveforms: clock 1, clock 2, FF1 input data, FF1 Q output, FF2 input data, and FF2 Q output, showing clock skews.

INDEX

acceptors, 376
adders, *see* full-adders; half-adders
alternating current (AC) circuits, 98
amplifier, 287, 292–294
  bipolar, 397–420
  linear, 155–165, 290–294
  linear applications of op amps, 294, 299, 306, 314, 316, 321, 322, 326
  non-linear applications of op amps, 331, 332, 335, 336
  small signal model, 397
apparent, *see under* power
arithmetic, 533
arithmetic and logic unit (ALU), 611
atoms, 1, 2

battery, 5, 6, 30–33
BiCMOS, 500
binary numbers, 456, 527, 531
Biot-Savart law, 49
bipolar junction transistor (BJT), 374
  biasing, 378–382
  biasing PNP transistors, 396
  biasing resistor divider, 387
  as a circuit element, 377
  common base configuration, 407
  common collector configuration, 415
  common emitter configuration, 399
  electronic devices, 354
  emitter degeneration resistor biasing, 391
  follower configuration, *see* common collector configuration
  self-biased, 394
  small signal model, 397
BJT, *see* bipolar junction transistor
Bode plots, 238, 241, 244, 254, 257, 275, 278
Boolean algebra, 460
Boolean algebra postulates and theorems, 465

canonical forms, *see* standard canonical forms
capacitive reactance, 131–133, 152, 153
capacitor, 111, 114, 123, 124, 126, 127, 131–133, 136–139, 142, 152
  decoupling, 42–44, 304–306, 603, 628, 631–633
  dielectric constant, 35, 36
  electric component, 12
  parallel-plate, 33, 35, 36, 94
  power drawn by, 137, 139
  pure, 131
  time domain equations, 115
  under sinusoidal steady-state, 114
  voltage and current phasors, 134
carry look ahead, 543
central processing unit (CPU)
  design, 603
  instruction set, 603
characteristic tables, *see* flip-flops

Electrical, Electronics, and Digital Hardware Essentials for Scientists and Engineers, First Edition. Ed Lipiansky.

charge, 1–3, 5, 6
circuit, 156
  analysis, 155
  first order, 233, 236, 241, 246, 279
  linear, *see* amplifiers
  second order, 233, 265, 266, 275, 278, 279
  theorems, 155
clock, 550, 551, 559–567, 571–575, 584, 587, 590, 591–593, 595–599
  cpu, 603, 604, 608, 609, 614, 617, 618, 626–628
CMOS, 443, 497, 498–500
coil, *see* inductor
complex number, 115–122, 126, 128, 130, 131, 133–135
controller, 618–620
counter, program, 604, 611, 614, 615
current
  electric, 3–11
  electronic, 6
  source, 3, 5, 7–11
current controlled current source (CCCS), 145–149, 154
current controlled voltage source (CCVS), 145, 146, 147

D flip-flop, 568, 581
data path architecture, 615
De Morgan laws, rules and theorems, 464
decoder, 510–515
decoupling, 628, 631
de-multiplexers, 526
dependent sources
  current controlled current source (CCCS), 148, 149
  current controlled voltage source (CCVS), 145–147
  finding the Norton equivalent circuit, 175
  establishing mesh equations when there are, 196
  establishing nodal equations when there are, 207
  Norton's theorem highlights, 211
  Thevenin, 167
  Thevenin theorem highlights, 211
  voltage controlled current source (VCCS), 145, 146, 147
  voltage controlled voltage source (VCVS), 145
depletion mode, *see* metal oxide semiconductor field effect transistor
dielectric constant, 35, 36
difference amplifier, *see* operational amplifiers
differentiator, *see* filters
digital circuits, 456
diode, 354–357, 363, 367, 369, 371, 372
direct current (DC) circuits, 5, 87
donors, 376
don't care, 491
dot rule, 68
duality principle, 466

effective value, 99, 105, 149; *see also* root mean square (RMS)
electric charge, 1, 3, 34
electric current, 3, 6, 12, 37, 45, 53, 88, 422
electric voltage, 3, 27
electromotive force (EMF), 51, 53, 54, 59
electron, 1–3, 5, 6
electronic devices, 354
EMF, *see* electromotive force
encoder, 510, 517
enhancement mode, *see* metal oxide semiconductor field effect transistor
Euler's form, 119–123
excitation tables, *see* flip-flops

Faraday's Induction Law, *see* induction
fast carry generation, 543
filters
  differentiator, 263, 265, 279
  first order, 233, 235
  high-pass (HPF), 252–265, 279, 294
  integrator, 251, 252, 279
  low-pass (LPF), 235, 236, 238, 240–245, 246, 248–252, 254, 257, 260, 279, 280, 285
  second order, 266, 275
flip-flops
  characteristic tables, 568–570, 576
  D-type, 568, 581

excitation tables, 576–578, 580–583, 586, 588
JK, 565
master/slave SR, 561
timing, 572
toggle or T-type, 568
frequency domain, 111, 123, 235, 253
frequency response plots, *see* Bode plots
full adders, 533, 536
full-wave rectifier, 360

gates, 495
logic, electrical and timing characteristics, 495, 497
MOSFET, 424, 427, 429, 432

half-adders, 533
half-wave rectifier, 357
hysteresis, 70, 336, 338–341, 630, 631

impedance, 123, 124, 126, 127, 128, 129, 130–136, 140, 144, 151–153
induction, 49–51, 53
electromagnetic, 53, 54, 62
Faraday law of, 53, 62, 64, 66
inductive reactance, 130, 151, 152
inductor
electric component, 12
mutual inductance, 62–70
non-linearity, 70–73
parallel, 60–62
power drawn by, 136, 139, 140
pure, 127, 128, 129, 130, 133
series, 58, 59
time domain equations, 115
under sinusoidal steady-state, 113
voltage and current phasors, 134
integrator, *see* filters

JK flip-flop, 565

Karnaugh maps, 473, 474, 479, 484, 487
design examples, 473, 474, 479, 484, 487
Kirchoff's current law (KCL), 73, 81, 173, 199
Kirchoff's voltage law (KVL), 73, 79, 173

latch, 552–559
characteristic table, 558, 560
latch SR, 559
linear amplifiers, *see* amplifiers
linear function, 156–160
load line, 108, 109
logic gates, 495, 497, 499
logic operation, *see* Boolean algebra
loop, 77, 78
Lorentz law, 49
LVTTL, 499

magnetic characteristic, 71
magnetic field, 45–51, 53, 54, 58, 59–61, 70
magnitude Bode plot, *see* Bode plot
master/slave, 561, 565, 568
maxterms, 472
memory, 610
mesh method, 179–181, 189, 196, 199, 210, 212, 213, 223, 225
with dependent sources, 196
metal oxide semiconductor field effect transistor (MOSFET)
biasing techniques, 428
common drain or source follower, 439
common gate configuration, 437
common source configuration, 434
common source configuration with emitter degeneration, 436
electronic device, 354, 420
enhancement mode p- channel, and depletion mode n-channel and p-channel, 439
I-V characteristics, 424
small signal model, 427
minterms, 467
MOSFET, *see* metal oxide semiconductor field effect transistor
multiplexers, 519, 521
mutual inductance, *see* inductor

nodal method, 199–207, 210
non-linear applications, *see* amplifiers
Norton's method, 172–175, 210

Ohm's law, 25, 26, 29, 101
one's complement, *see* binary numbers

op amp, *see* operational amplifier
operational amplifier, 287, 288, 294, 331, 341, 342
  buffer, 309
  data sheet, 344
  decoupling, *see* capacitor
  difference, 316
  inverting, 294
  inverting differentiator, 322
  inverting integrator, 321
  non-idealities, 341
  non-inverting, 316
  power dissipation, 344, 373, 374
  selection, 341
  unity gain, 309

permeability, 45, 52, 53, 71, 95
phase Bode plot, *see* Bode plots
phasor, 115, 123–135, 138–140, 149, 153
photodiode, 369
polar form, 117–121, 123, 129, 132, 134, 135
power, 98
  AC instantaneous, 137, 138
  active, 136, 137, 140, 141, 142, 143
  apparent, 136, 137, 140, 141, 142, 143, 144, 149
  average, 103, 106, 136, 137, 138, 140, 141
  decoupling, *see* capacitor
  instantaneous, 103, 104, 112, 113, 114, 136, 137, 138, 139, 140
  reactive, 136, 137, 139, 140
  triangle, 137, 140, 141, 142
primary, *see* transformer
program counter, 611

RC circuit, *see* filters, high-pass and low-pass
rectangular form, 115–122, 129, 130, 132, 135
register, 547, 551, 569, 575, 605, 607
reset, 322, 457, 552, 553–555, 558, 559, 561, 567–576, 578, 580, 581, 594, 603–605, 628–630
    asynchronous, 571,572, 582, 583, 584, 613, 614, 621, 622
    synchronous, 572–574, 613, 631
resistance, *see* resistor
resistor, 98, 101, 102, 103, 104, 105, 106, 107, 111, 115, 123, 124, 126, 128, 134, 136, 137, 144, 146, 152, 153, 154
    electric component, 12
    internal, 9–11
    parallel, 19
    polarities, 13
    powered by a DC source, 4
    powered by an AC voltage, 105
    power dissipation, 18, 105, 106
    series, 19, 42, 355
    temperature coefficient, 14–17
    tolerances, 15
    under sinusoidal steady-state, 112
RLC circuit, 266, 271–275, 277, 286
root mean square (RMS), 98, 99, 101, 103, 105, 106, 137, 138, 140, 143, 144, 149

Schottky diode, 371, 629
secondary, *see* transformer
shift register, 575, 590–592
simplification, *see* Karnaugh maps
sinusoidal, 98, 99, 101, 102, 103, 106, 111, 113, 114, 124, 126
  current, 106, 111, 127, 131, 137, 152
  excitation, 112, 113, 114, 115
  steady-state, 111, 112, 113, 115, 123, 130, 133, 134, 141
  voltage, 99, 101, 103, 105, 111, 113, 123, 126, 127, 128, 131, 151, 152, 153
  waveform, 98, 99, 101, 105, 111, 124, 125
source
  AC voltage, 108, 111
  current, 3
  DC voltage, 108
  dependant voltage, 145, 147
  dependant current, 145, 146
  follower, BJT, *see* common collector configuration
  follower, MOSFET, *see* common drain configuration
  ideal voltage, 98, 102, 103
  real voltage, 100, 109
  sinusoidal voltage, 103, 111

transformation method, 174, 175, 177, 178, 210, 212, 213, 221, 222, 225, 229
voltage, 3, 5, 7, 30, 73, 79, 82, 85, 88, 94, 98, 99, 100, 101, 103, 108
source transformations, 174, 175, 177, 178, 210, 212, 213
SR flip-flop, 561
SR latch, 559
standard canonical forms
product of sums (POS), 472
sum of products (SOP), 467
state machines, 574
subtracting, 540, 542
superposition method, 156, 169, 171, 210, 213, 226
synchronous state machines, 592

T flip-flop, 568, 580
Thevenin method, 165, 168, 172, 210
time domain, 123, 244, 245, 257, 258
timing, 571, 595, 626
transformer, 67, 68, 70
transistor
BJT, 354, 374, 420
MOSFET, 354
small signal model, 398, 399, 427
TTL, 497, 498, 499
two's complement, *see* binary numbers

Venn diagrams, 467
voltage, 1, 3, 29
electric, 3–11
polarities, 13
source, 25, 27
voltage controlled current source (VCCS), 147, 148, 153
voltage controlled voltage source (VCVS), 145

IEEE PRESS SERIES ON MICROELECTRONIC SYSTEMS

The focus of the series is on all aspects of solid-state circuits and systems including the design, testing, and application of circuits and subsystems, as well as closely related topics in device technology and circuit theory. The series also focuses on scientific, technical and industrial applications, in addition to other activities that contribute to the moving the area of microelectronics forward.

R. Jacob Baker, *Series Editor*

---

1. *Nonvolatile Semiconductor Memory Technology: A Comprehensive Guide to Understanding and Using NVSM Devices*
   William D. Brown, Joe Brewer
2. *High-Performance System Design: Circuits and Logic*
   Vojin G Oklobdzija
3. *Low-Voltage/Low-Power Integrated Circuits and Systems: Low-Voltage Mixed-Signal Circuits*
   Sanchez-Sinencio
4. *Advanced Electronic Packaging, Second Edition*
   Richard K. Ulrich and William D. Brown
5. *DRAM Circuit Design: Fundamental and High-Speed Topics*
   Brent Keith, R. Jacob Baker, Brian Johnson, Feng Lin
6. CMOS: Mixed-Signal Circuit Design, Second Edition
   R. Jacob Baker
7. *Nonvolatile Memory Technologies with Emphasis on Flash:A Comprehensive Guide to Understanding and Using NVM Devices*
   Joseph E. Brewer, Manzur Gill
8. *Reliability Wearout Mechanisms in Advanced CMOS Technologies*
   Alvin W. Strong, Ernest Y. Wu, Rolf-Peter Vollertsen, Jordi Sune, Giuseppe La Rosa, Timothy D. Sullivan, Stewart Rauch
9. *CMOS: Circuit Design, Layout, and Simulation, Third Edition*
   R. Jacob Baker
10. *Quantum Mechanics for Electrical Engineers*
    Dennis M. Sullivan
11. *Nanometer Frequency Synthesis Beyond Phase-Locked Loop*
    Liming Xiu
12. *Electrical, Electronics, and Digital Hardware Essentials for Scientists and Engineers*
    Ed Lipiansky